Safety at Work

Safety at Work

Sixth edition

Edited by

John Ridley and John Channing

AMSTERDAM BOSTON HEIDELBERG LONDON NEW YORK OXFORD
PARIS SAN DIEGO SAN FRANCISCO SINGAPORE SYDNEY TOKYO

Butterworth-Heinemann
An imprint of Elsevier
Linacre House, Jordan Hill, Oxford OX2 8DP
200 Wheeler Road, Burlington, MA 01803

First published 1983
Second edition 1985
Reprinted 1987
Third edition 1990
Reprinted 1991, 1992, 1993
Fourth edition 1994
Reprinted 1996, 1998
Fifth edition 1999
Reprinted 2000
Sixth edition 2003

British Library Cataloguing in Publication Data
A catalogue record for this book is available from the British Library

Library of Congress Cataloguing in Publication Data
A catalogue record for this book is available from the Library of Congress

ISBN 0 7506 5493 7

Composition by Genesis Typesetting Limited, Laser Quay, Rochester, Kent
Printed and bound in Great Britain BY

Contents

Foreword

Bill Callaghan, *Chairman – Health & Safety Commission*

In 1972 Lord Robens wrote

> [We are] in no doubt that the most important single reason for accidents at work is apathy. There is a curious paradox here. Society as a whole reacts keenly to major disasters. There is also some ephemeral reaction to the annual statistics of industrial death and injury....Even rarer is personal awareness of diseases which manifest themselves long after periods of exposure in an unhealthy working environment.

We have made much progress since the Robens Report and the Health & Safety at Work Act. Our record is a good one internationally.

However, we need to improve performance to meet the challenging targets of revitalising Health & Safety. The benefits of good health and safety provision are not universally appreciated; poor risk management often only becomes apparent once it is too late. Some employers hope that incidents of occupational ill health or injury will not happen to them. They overlook the pure financial cost of such events – in terms of lawyers' fees, insurance, legal claims, downtime, recruiting and training replacements, and even their own reputation.

Not enough employers appreciate the value of a committed, interested workforce that knows it is cared for and involved. Obvious concern for health and safety is one way employers can show they value their workforce. We need to invest in a culture that involves employers and employees working together in partnership. A committed workforce brings continuity, drives down costs of recruitment and training, and reduces absence. A good reputation for health and safety suggests reliability and encourages investors, shareholders and customers.

Since Lord Robens wrote his report, the world has moved on. The demographics of the country and the workplace have changed; education levels have risen; the balance of work has shifted from manufacturing to service industries. With developing technology and research, new risks and hazards have evolved. But so too has our understanding and ability to combat them.

We need to make work a better place to be. This book is a support for those of you who want to take forward this work. Together we can improve the competitiveness of British industry. Together we can make good health and safety the cornerstone of a civilised society.

Preface to sixth edition

Since publication of the 5th edition there have been significant developments in health and safety. First, the importance of the roles of directors and managers in health and safety has been highlighted by a number of major incidents, in some of which lives were lost. Subsequent investigation established serious shortcomings in the commitment and support for health and safety at very senior levels in the organisations concerned. Safety is like water, it flows downhill and if the spring at the top is dry it is unlikely that the seeds planted below will be nurtured and flourish. There is a growing conviction that the attitudes at the most senior levels in an organisation are crucial to its safety performance. We have taken this on board and added two new chapters on the management of risk, one of which emphasises the need for a sound organisational structure to achieve effective risk management. Other existing chapters have been significantly enlarged. They reflect developments in the behavioural approach to risk management and current thinking on effective management systems including the development of a international standard on safety management that will inevitably emerge.

Second, what has been a fringe subject but is rapidly becoming part of the safety manager's responsibilities is the quality of the environment – both in the workplace and in the wider context of global pollution. Interest in this subject is stimulated by the growing concerns for what we are doing to the global environment through man's thoughtless and barely controlled emission of 'greenhouse' and ozone-depleting gases and fumes. But this has in turn focused attention on more local issues such as the quality of the working environment and how we dispose of our waste – solid, liquid and gases. To reflect the growing importance of this subject we have added a completely new part (Part 5) dealing solely with environmental issues.

Health, safety and the environment tend to be linked together and frequently fall to the safety department to administer, whereas quality management remains firmly in the hands of the design and production departments. Each management system demands the same attitudes and approaches to attaining high standards without increasing costs or burdens beyond an acceptable limit and we see the development of the integration of the three management systems as an inherent and essential part of an enterprise's operational procedures.

The official recognition of IOSH as a professional body through the granting to them of a Royal Charter has made membership an even greater valued goal. This, in turn, will demand high standards of academic attainment in the qualifying examinations. While the universities have established their own standards of excellence that they demand of their graduates – standards that are moderated by IOSH – other teaching and training establishments have to rely on nationally available standards. Following discussions, led by the National Training Organisation for Employment (ENTO) in which a range of interested parties participated, a set of occupational health and safety standards was developed. These have been integrated with the Scottish and National Vocational Qualifications (S/NVQ) to form the basic academic background leading to membership of the Institution of Occupational Safety and Health (IOSH). Examinations against these new standards will be administered by the National Examination Board in Occupational Safety and Health (NEBOSH) and certified through the awarding of a Diploma. Much of the content of the book is pertinent to those studying for this Diploma and to guide students to the relevant parts, a suggested reading list has been included as an appendix.

Major changes have and still are taking place in the procedures and conduct of cases in the courts following the recommendations of the Woolf report. The aim is to make the administration of justice more dynamic through the more flexible use of resources and representations. Those changes in court procedure that have already been made have been incorporated into chapter 1 *Explaining the law*. It is to be hoped that we, and in particular claimants for compensation following injuries, will benefit from these changes through the quicker settlement of litigations and other matters of legal contention put before the courts.

Although we have included much that is new in the way that health, safety and environmental responsibilities are viewed and accepted, we have not ignored the solid foundation on which the past high levels of health and safety in the UK have been based. The text includes those basic concepts, techniques and practices that have served so well the cause of health and safety of workpeople for so many years. Not least among these are the relationships that have developed between employers and employees that are so essential for good performance in, not only health and safety, but all the aspects of occupational activities.

Risk assessments have been a growing and central feature of health and safety activities through their application to general employment and chemical hazards. The gay abandon with which the phrase *risk assessment* is used gives rise to the risk that it is becoming a 'buzz' word and that the essential nature of the technique is being brought into disrepute by misunderstanding of its role and purpose in the overall pattern of health and safety activities. Our approach has been to put the technique into perspective and view it in the overall context of all the other actions that are taken to reduce the risks faced by workpeople and to improve the quality of their working life.

Many injuries and much of the ill health from work activities can be laid at the door of inappropriate physical demands being made on the operators. Not only through strains and sprains from overloaded limbs but subjection to unsuitable working environment, difficult manual operations, etc. A new chapter on *Applied ergonomics* has been included that approaches the subject from the point of view of what the body and its limbs can reasonably do. This will allow the manager and practitioner to design processes and operations in the confidence that the work to be carried out will not overstretch the operator beyond the limits of what the human body is physically capable of doing.

While there appears to have been a reduction in the rate of propagation of new EU directives – a respite resulting from the updating of many of the existing directives to bring them in line with technological developments since they were first adopted – there has been no relief from the flow of European and international standards that put the meat on the bones of legislative requirements. A number of these standards are based on UK originals that have been modified (some might say diminished) to relate to the employment and industrial climates of other EU member states. Once a standard is adopted as harmonised (European) or international, it takes precedence over the equivalent national standard. A notable casualty in this process has been BS 5304 *Safety of Machinery*. However, because it is so well known and provides such a breadth of guidance on good safety practices, BSI have re-issued it as a 'Published Document' PD 5304 which is advisory and cannot be used as evidence of conformity with legislative requirements and as such it is referred to in the text.

An objective of the book has always been, and still is, to provide information and guidance in understanding health and safety legislative requirements and standards. Also to assist all those involved in this field in attaining the highest standards, whether as a practising manager, safety practitioner or student. We hope that this new edition carries this objective forward and in so doing points the way for future developments.

John Ridley
John Channing
September 2002

but subjection to unsuitable working environment, difficult manual operations, etc. A new chapter on *Applied ergonomics* has been included that introduces the subject from the point of view of what the body and its limbs can reasonably do. This will allow the designer and practitioner to assess processes and operations in the confidence that the work to be carried out will not overstretch the operatives (within the limits of what the human body is physically capable of doing).

While these appear to have been [illegible]

[illegible] Another of these standards [illegible] that have been modified [illegible] to [illegible] the improvement and industrial climates of other EU [illegible] adopted these standards as adopted and harmonised [illegible] over the equivalent national standard. [illegible] in this process has been BS 5304 *Safety of Machinery* [illegible] well known and provides such a breadth of [illegible] safety practice. This has now been [illegible] but has been retained [illegible] advice in [illegible] used as a reference [illegible]

[illegible] dangers that the individual [illegible] this edition [illegible]

Preface to first edition

Since the first welfare Act was put on the Statute Book in 1802 there has been a steady development in safety and health legislation aimed at improving the lot of those who work in mills, factories, and even in offices. In the past two decades official concern has increased, culminating in 1974 in the Health and Safety at Work etc. Act. 'Safety at Work' has now taken on a new and more pertinent meaning for both employer and employee.

Developments in the field of safety extend throughout much of the world, indicating an increasing concern for the quality of working life. In Europe the number of directives promulgated by the European Economic Community are evidence of this growing official awareness of the dangers that the individual faces in his work.

Health and safety laws in the UK are the most complex and comprehensive of all employment laws. Consequently employers are looking to a new breed of specialists, the occupational safety advisers, for expert advice and guidance on the best means for complying with, and achieving the spirit of, the law. These specialists must have the necessary knowledge of a wide range of disciplines extending from safety and related laws to occupational health and hygiene, human behaviour, management and safety techniques, and of course, the hazards inherent in particular industries or pursuits. With this demand for expert advice has grown a need for a nationally recognised qualification in this new industrial discipline. With this in mind the Institution of Occupational Safety and Health (IOSH) published in 1978 a syllabus of subjects for study by those seeking to become professionally qualified in this field. This syllabus now forms the foundation upon which the National Examination Board in Occupational Safety and Health (NEBOSH) sets its examinations.

Prepared in association with IOSH, this book covers the complete syllabus. It is divided into five parts to reflect the spectrum of the five major areas of recommended study. Each part has a number of chapters, which deal with specific aspects of health or safety. To enable readers to extend their study of a particular subject, suitable references are given together with recommendations for additional reading. Further information and details of many of the techniques mentioned can also be obtained through discussions with tutors. A table is given in Appendix II to guide students in their selection of the particular chapters to study for the appropriate levels and parts of the examination.

A major objective of this book has been to provide an authoritative, up-to-date guide in all areas of health and safety. The contributing authors are recognised specialists in their fields and each has drawn on his or her personal knowledge and experience in compiling the text, emphasising those facets most relevant to the safety advisers' needs. In this they have drawn material from many sources and the views they have expressed are their own and must not be construed as representing the opinions or policies of their employers nor of any of the organisations which have so willingly provided material.

It has been common practice to refer to the safety specialist as the 'safety officer', but this implies a degree of executive authority which does not truly indicate the rôle he plays. Essentially that rôle is one of monitoring the conditions and methods of work in an organisation to ensure the maintenance of a safe working environment and compliance with safety legislation and standards. Where performance is found wanting his function is to advise the manager responsible on the corrective action necessary. Reflecting this rôle, the safety specialist is throughout the book referred to as the 'safety adviser', a title that more closely reflects his true function.

The text has been written primarily for the student. However, a great deal of the content is directly relevant to the day-to-day work of practising managers. It will enable them to understand their safety obligations, both legal and moral, and to appreciate some of the techniques by which a high standard of safe working can be achieved. It will also provide an extensive source of reference for established safety advisers.

The text of any book is enhanced by the inclusion of tables, diagrams and figures and I am grateful to the many companies who have kindly provided illustrations. I would also like to acknowledge the help I have received from a number of organisations who have provided information. Particularly I would like to thank the journal *Engineering*, the Fire Prevention Association, the Health and Safety Executive, the British Standards Institution and the International Labour Office.

I also owe grateful thanks to many people for the help and encouragement they have given me during the preparation of this book, in particular Mr J. Barrell, Secretary of IOSH, Mr D.G. Baynes of Napier College of Science and Technology, Mr N. Sanders, at the time a senior safety training adviser to the Road Transport Industry Training Board, Dr Ian Glendon of the Department of Occupational Health and Safety at the University of Aston in Birmingham, Mr R.F. Roberts, Chief Fire and Security Officer at Reed International's Aylesford site and David Miskin, a solicitor, for the time each gave to check through manuscripts and for the helpful comments they offered.

I am also indebted to Reed International P.L.C. for the help they have given me during the editing of this book, a task which would have been that much more onerous without their support.

John Ridley

Contributors

John Adamson, MRSC FBIOH, Dip.Occ.Hyg., ROH, MIOSH, RSP
Manager Health, Safety, Hygiene and Fire, Kodak Chemical Manufacturing

L. Bamber, BSc, DIS, FIRM, FIOSH, RSP
Consultant, Risk Solutions International

G. N. Batts BSc, M.Phil., PhD, DIC
Environmental Adviser, Kodak European Region

Dr A. J. Boyle, BSc, MSc, PhD, CPsychol, AFBPhS, FIOSH, RSP

Chris Buck, BSc(Eng), MIEE, FIOSH, RSP
Consultant

J. E. Caddick, BSc(Hons), CEng, MIMechE
Zurich Risk Services

Ray Chalklen, MIFireE
Fire consultant

John Channing, MSc(Safety), MSc(Chemistry) FIOSH, RSP
Formerly Manager Health, Safety and Environment, Kodak Manufacturing
Director, Pharos Consultancy Services Limited

Dr A. R. L. Clark, MSc, MB, BS, MFOM, DIH, DHMSA

Dr T. Coates, MB, BS, FFOM, DIH, DMHSA

Nick Cook, MRSC, CChem, MIOSH
Occupational Hygienist, Kodak Ltd

Jonathan David, BSc
The Chartered Institution of Building Services Engineers

Frank S. Gill, BSc, MSc, Ceng, MIMinE, FIOSH, FFOM(Hon), Dip.Occ.Hyg.
Consultant ventilation engineer and occupational hygienist

Professor Andrew Hale, PhD, CPsychol., MErgS, FIOSH
Safety Science Group, Delft University of Technology, Netherlands

Dr Chris Hartley, PhD, MSc, MIBiol
Visiting Fellow to Aston University and a consultant

R. W. Hodgin, LL.M
School of Law, University of Birmingham

Edwin Hooper

Roland Hudson, FIOSH, RSP, FRSH, ASSE
Construction safety consultant

Amanda Jones, Solicitor
Pattinson and Brewer

Dr R. G. Lawson, LL.M, PhD
Consultant in marketing and advertising law

Dr M. Maslanyj
National Radiological Protection Board

John McMullen, BSc, CEng, MIMechE
Zurich Risk Services

R. D. Miskin
A solicitor

Samantha Moss, BSc(Hons), MIOSH, AMIEMA
HSE Manager, Business Operations

John Ridley, BSc, CEng, MIMechE, FIOSH, DMS

Peter Shaw
National Radiological Protection Board

Stan Simpson, CEng, MIMechE, FIOSH

Eric J. Skellett
Solicitor

Ron W. Smith, BSc(Eng), MSc(Noise and vibration)
Hodgson & Hodgson Group Ltd

Brenda Watts, MA, BA
Barrister, Senior Lecturer, Southampton College of Higher Education

Ashton West, BA(Hons), ACII
Claims Director, Rubicon Corporation

Dr A. D. Wrixon, DPhil, BSc(Hons)
International Atomic Energy Agency

Part I

Law

Laws are necessary for the government and regulation of the affairs and behaviour of individuals and communities for the benefit of all. As societies and communities grow and become more complex, so do the laws and the organisation necessary for the enforcement and administration of them.

The industrial society in which we live has brought particular problems relating to the work situation and concerning the protection of the worker's health and safety, his employment and his right to take 'industrial action'.

This section looks at how laws are administered in the UK and the procedures to be followed in pursuing criminal actions and common law remedies through the courts. It considers various Acts and Statutes that are aimed at safe working in the workplace and also some of the influences that determine the content of new laws. Further, the processes are reviewed by which liabilities for damages due to either injury or faulty product are established and settled.

Chapter 1.1

Explaining the law

Brenda Watts

1.1.1 Introduction

To explain the law an imaginary incident at work is used which exemplifies aspects of the operation of our legal system. These issues will be identified and explained with differences of Scottish and Irish law being indicated where they occur.

1.1.2 The incident

Bertha Duncan, an employee of Hazards Ltd, while at work trips over some wire in a badly lit passageway, used by visitors as well as by employees. The employer notifies the accident in accordance with his statutory obligations. The investigating factory inspector, Instepp, is dissatisfied with some of the conditions at Hazards, so he issues an improvement notice in accordance with the Health and Safety at Work etc. Act 1974 (HSW), requiring adequate lighting in specified work areas.

1.1.3 Some possible actions arising from the incident

The *inspector*, in his official capacity, may consider a prosecution in the criminal courts where he would have to show a breach of a relevant provision of the safety legislation. The likely result of a successful safety prosecution is a fine, which is intended to be penal. It is not redress for Bertha.

The *employee*, Bertha, has been injured. She will seek money compensation to try to make up for her loss. No doubt she will receive State industrial injury benefit, but this is intended as support against misfortune rather than as full compensation for lost wages, reduced future prospects or pain and suffering. Bertha will therefore look to her employer for compensation. She may have to consider bringing a civil action, and will then seek legal advice (from a solicitor if she has no union to turn to) about claiming compensation (called damages). To succeed,

Bertha must prove that her injury resulted from breach of a legal duty owed to her by Hazards.

For the *employer*, Hazards Ltd, if they wish to dispute the improvement notice, the most immediate legal process will be before an employment tribunal. The company should, however, be investigating the accident to ensure that they comply with statutory requirements; and also in their own interests, to try to prevent future mishaps and to clarify the facts for their insurance company and for any defence to the factory inspector and/or to Bertha. The company would benefit from reviewing its safety responsibilities to non-employees (third parties) who may come on site. As a company, Hazards Ltd has legal personality; but it is run by people and if the inadequate lighting and slack housekeeping were attributable to the personal neglect of a senior officer (s. 37 HSW), as well as the company being prosecuted, so too might the senior officer.

1.1.4 Legal issues of the incident

The preceding paragraphs show that it is necessary to consider:

criminal and civil law,
the organisation of the courts and court procedure,
procedure in employment tribunals, and
the legal authorities for safety law: legislation and court decisions.

1.1.5 Criminal and civil law

A *crime* is an offence against the State. Accordingly, in England prosecutions are the responsibility of the Crown Prosecution Service; or, where statute allows, an official such as a factory inspector (ss. 38, 39 HSW). Very rarely may a private person prosecute. In Scotland the police do not prosecute since that responsibility lies with the procurators-fiscal, and ultimately with the Lord Advocate. In Northern Ireland the Director of Public Prosecutions (DPP) initiates prosecutions for more serious offences, and the police for minor cases. The DPP may also conduct prosecutions on behalf of Government Departments in magistrates' courts when requested to do so. The procurators-fiscal, and in England and Northern Ireland the Attorney General acting on behalf of the Crown, may discontinue proceedings; an individual cannot. The Justice (Northern Ireland) Bill[1] provides for a Prosecution Service to undertake all prosecutions and for the DPP to discontinue proceedings, not the Attorney General (see also section 1.1.14).

Criminal cases in England are heard in the magistrates' courts and in the Crown Court; in Scotland mostly in the Sheriff Court, and in the High Court of Justiciary. In Northern Ireland criminal cases are tried in magistrates' courts and in the Crown Court. In all three countries the more serious criminal cases are heard before a jury, except in Northern Ireland for scheduled offences under the Northern Ireland (Emergency Provisions) Act of 1996.

The burden of proving a criminal charge is on the prosecution; and it must be proved beyond reasonable doubt. However, if, after the incident at Hazards, Instepp prosecutes, alleging breach of, say s. 2 of HSW, then Hazards must show that it was not reasonably practicable for the company to do more than it did to comply (s. 40 HSW). This section puts the burden on the accused to prove, on the balance of probabilities, that he had complied with a practicable or reasonably practicable statutory duty under HSW.

The rules of evidence are stricter in criminal cases, to protect the accused. Only exceptionally is hearsay evidence admissible. In Scotland the requirement of corroboration is stricter than in English law.

The main sanctions of a criminal court are imprisonment and fines. The sanctions are intended as a punishment, to deter and to reform, but not to *compensate* an injured party. A magistrates' court may order compensation to an individual to cover personal injury and damage to property. Such a compensation order is not possible for dependants of the deceased in consequence of his death. At present the upper limit for compensation in the magistrates' court is £5000[2].

A *civil action* is between individuals. One individual initiates proceedings against another and can later decide to settle out of court. Over 90% of accident claims are so settled.

English courts hearing civil actions are the county courts and the High Court; in Scotland the Sheriff Court and the Court of Session. In Northern Ireland the County Court and the High Court deal with civil accident claims. Civil cases rarely have a jury; in personal injury cases only in the most exceptional circumstances.

A civil case must be proved on the balance of probabilities, which has been described as 'a reasonable degree of probability . . . more probable than not'. This is a lower standard than the criminal one of beyond reasonable doubt, which a judge may explain to a jury as 'satisfied so that you are sure' of the guilt of the accused.

In civil actions the claimant, formerly the plaintiff, (the pursuer in Scotland) sues the defendant (the defender) for remedies beneficial to him. Often the remedy sought will be damages – that is, financial compensation. Another remedy is an injunction, for example, to prevent a factory committing a noise or pollutant nuisance.

1.1.6 Branches of law

As English law developed it followed a number of different routes or branches. The diagram in *Figure 1.1.1* illustrates the main legal sources of English law and some of the branches of English law.

Criminal law is one part of public law. Other branches of public law are constitutional and administrative law, which include the organisation and jurisdiction of the courts and tribunals, and the process of legislation.

Civil law has a number of branches. Most relevant to this book are contract, tort (delict in Scotland) and labour law. A contract is an agreement between parties which is enforceable at law. Most commercial

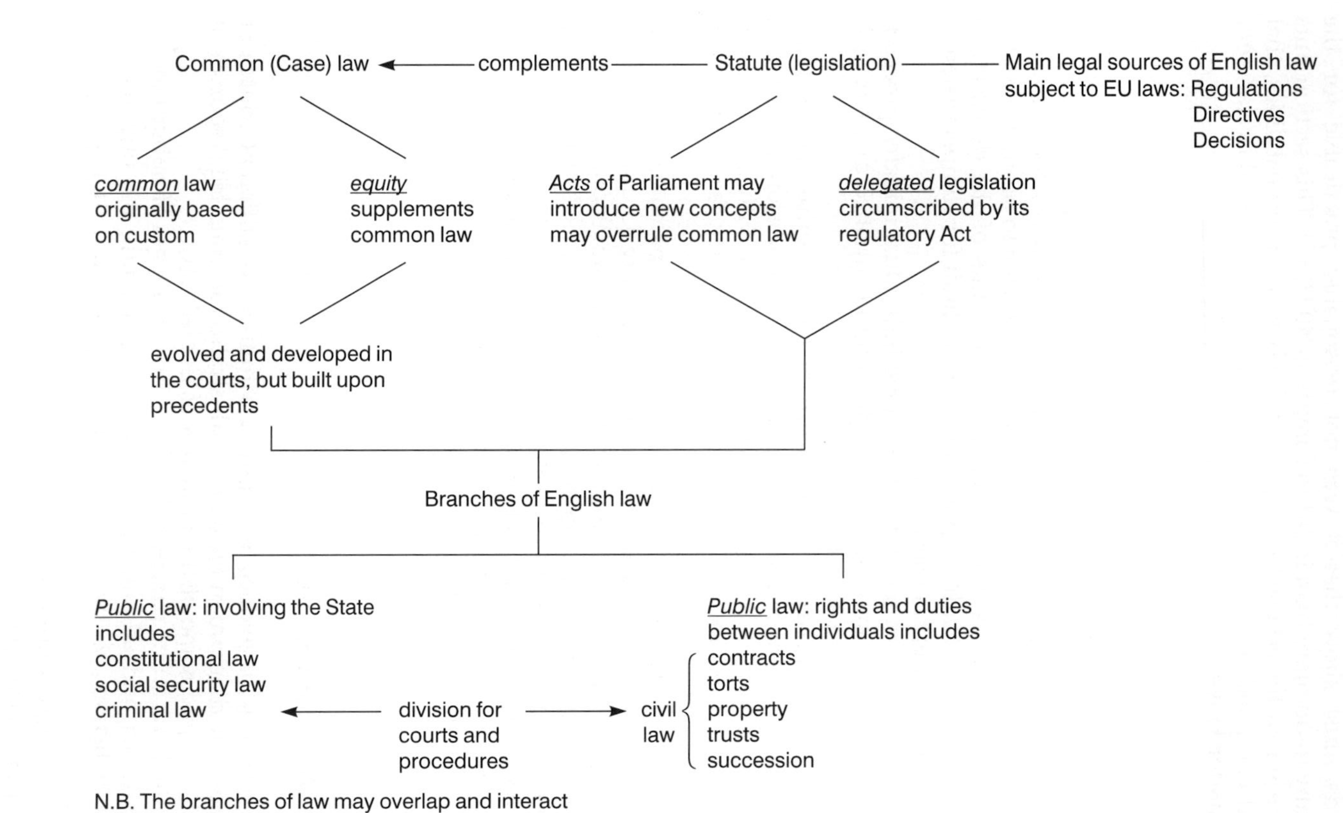
Common (Case) law
complements
Statute (legislation)
Main legal sources of English law
subject to EU laws: Regulations
Directives
Decisions
common law
originally based
on custom
equity
supplements
common law
Acts of Parliament may
introduce new concepts
may overrule common law
delegated legislation
circumscribed by its
regulatory Act
evolved and developed in
the courts, but built upon
precedents
Branches of English law
Public law: involving the State
includes
constitutional law
social security law
criminal law
division for
courts and
procedures
civil
law
Public law: rights and duties
between individuals includes
contracts
torts
property
trusts
succession
N.B. The branches of law may overlap and interact

Figure 1.1.1 Sources and branches of English law

law (for example, insurance) has a basis in contract. A tort is a breach of duty imposed by law and is often called a civil wrong. The two most frequently heard of torts are nuisance and trespass. However, the two most relevant to safety law are the torts of negligence and of breach of statutory duty.

The various branches of law may overlap and interact. At Hazards, Bertha has a *contract* of employment with her employer, as has every employee and employer. An important implied term of such contracts is that an employer should take reasonable care for the safety of employees. If Bertha proves that Hazards were in breach of that duty, and that in consequence she suffered injury, Hazards will be liable in the *tort* of negligence. There could be potential *criminal* liability under HSW. Again, Hazards might discipline a foreman, or Bertha's workmates might refuse to work in the conditions, taking the situation into the field of *industrial relations* law.

1.1.7 Law and fact

It is sometimes necessary to distinguish between questions of law and questions of fact.

A jury will decide only questions of fact. *Questions of fact* are about events or the state of affairs and may be proved by evidence. *Questions of law* seek to discover what the law is, and are determined by legal argument. However, the distinction is not always clear-cut. There are more opportunities to appeal on a question of law than on a question of fact.

Regulation 12 of the Workplace (Health, Safety and Welfare) Regulations 1992 (WHSW)[3] requires an employer (and others, to the extent of their control) to keep, so far as reasonably practicable, every floor in the workplace free from obstructions and from any article which may cause a person to slip, trip or fall. In the Hazards incident Bertha's tripping, her injury, the wire being there, the routine of Hazards, are questions of fact. However, the meaning of 'obstruction', of 'floor', of 'reasonably practicable' are questions of law.

1.1.8 The courts

1.1.8.1 First instance: appellate

A court may have *first instance* jurisdiction, which means that it hears cases for the first time; it may have *appellate* jurisdiction which means that a case is heard on appeal; or a court may have both.

1.1.8.2 Inferior: superior

Inferior courts are limited in their powers: to local jurisdiction, in the seriousness of the cases tried, in the sanctions they may order, and, in England, in the ability to punish for contempt.

In England the superior courts are the House of Lords, the judicial Committee of the Privy Council, and the Supreme Court of Judicature. Magistrates' and county courts are inferior courts.

For Scotland the Sheriff Court is an inferior court while the superior courts are the House of Lords, the Court of Session and the High Court of Justiciary.

In Northern Ireland the superior courts are the House of Lords and the Supreme Court of Judicature of Northern Ireland. The inferior courts are the magistrates' courts and the county courts.

1.1.8.3 Criminal proceedings – trial on indictment; summary trial

The indictment is the formal document containing the charge(s), and the trial is before a judge and a jury (of 12 in England and N. Ireland, of 15 in Scotland). A summary trial is one without a jury.

The most serious crimes, such as murder, or robbery, must be tried on indictment (or solemn procedure in Scotland). Some offences are triable only summarily (for example, most road traffic offences), others (for example, theft) are triable either way according to their seriousness. Most offences under HSW are triable either way, but in practice are heard summarily.

1.1.8.4 Representation

A practising lawyer will be a solicitor or a barrister (advocate in Scotland). Traditionally, barristers concentrate on advocacy and provide specialist advice. A qualification for senior judicial appointment is sufficient experience as an advocate. A barrister who has considerable experience and thinks he has attained some distinction may apply to the Lord Chancellor to 'take silk'. A solicitor is likely to be a general legal adviser. Until the Courts and Legal Services Act 1990 (as amended by the Access to Justice Act 1999), a solicitor's right to represent in court was limited to the lower courts. That Act provides for the ending of the barrister's monopoly appearances in the higher courts. Solicitors are able to appear in the High Court and before juries; and be appointed judges in the High Court. Qualified Fellows of the Institute of Legal Executives now have certain rights of audience, particularly in county courts and tribunals. A party may always defend himself, but there are restrictions on an individual personally conducting a private prosecution in the Crown Court or above (*R.* v. *George Maxwell Ltd*[4]). There is no general right of private prosecution in Scotland.

1.1.8.5 An outline of court hierarchy in England

There is a system of courts for hearing civil actions and a system for criminal actions. These are shown diagrammatically in *Figures 1.1.2* and *1.1.3*. However, some courts have both civil and criminal jurisdiction[19].

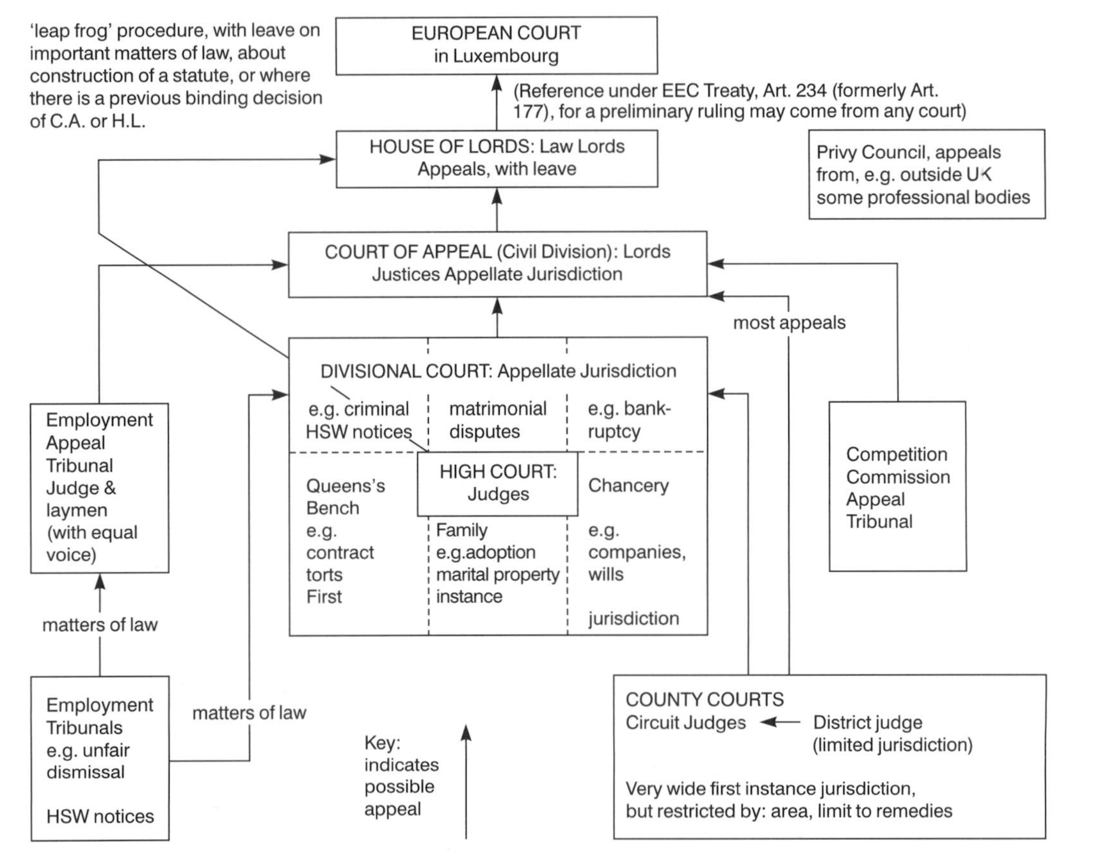

Figure 1.1.2 The main civil courts in England

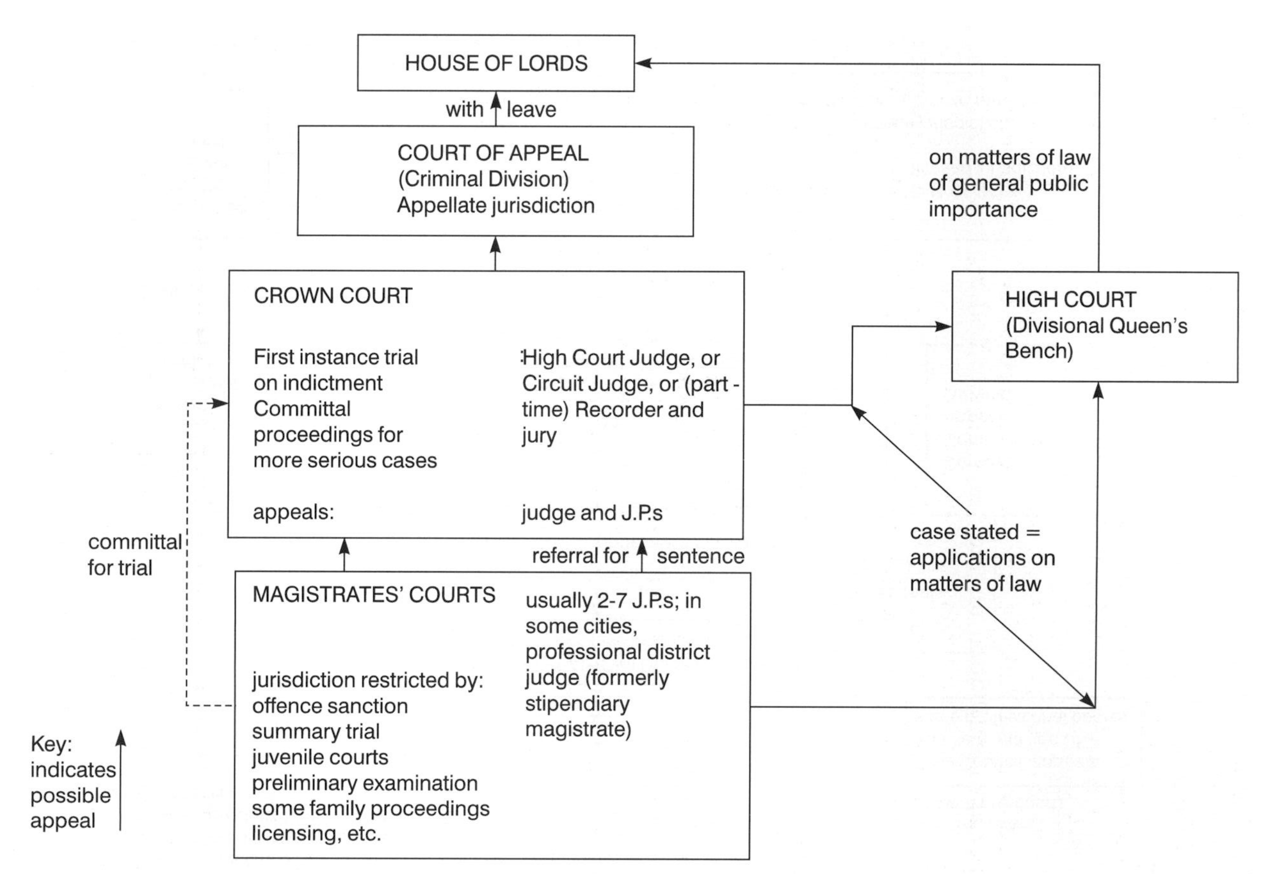

Figure 1.1.3 The main criminal courts in England

The lowest English courts are the magistrates' courts, which deal mainly with criminal matters; and the county courts, which deal only with civil matters.

Magistrates determine and sentence for many of the less serious offences. They also hold preliminary examinations into other offences to see if the prosecution can show a prima facie case on which the accused may be committed for trial. Serious criminal charges (and some not so serious where the accused has the right to jury trial) are heard on indictment in the Crown Court. The Crown Court also hears appeals from magistrates.

There are extensive proposals for the reform of criminal court procedure in the Auld Report[5], acknowledged by the Government as an important contribution towards establishing modern, efficient criminal courts. Among the recommendations is unification of the magistrates' courts and Crown Court in a single criminal structure with a new third intermediate District Division. Cases would be tried, according to their seriousness and the likely maximum sentence, either by magistrates, by judge and jury or in a District Division by a judge and two lay magistrates. The court, not the defendant, would decide which tribunal should hear the case. Jury trial should be limited to allegations of more serious crimes. The report also proposes increasing magistrate's powers to impose prison sentences up to 12 months and the formalising of a system of sentence discounts to encourage defendants to plead guilty, as an attempt to reduce the number of jury trials. For civil cases, the Courts and Legal Services Act increases the jurisdiction of the county courts. All personal injury claims for less than £50 000 will start in the county court; there is no upper limit but county court jurisdiction depends on the complexity of the case. District judges attached to the small claims courts may deal with personal injury cases for less than £5000. More important civil matters, because of the sums involved or legal complexity, will start in the High Court of Justice. The High Court has three divisions:

Queen's Bench (for contract and torts),
Chancery (for matters relating to, for instance,
  land, wills, partnerships and companies),
Family.

In addition the Queen's Bench Division hears appeals on matters of law:

1 from the magistrates' courts and from the Crown Court on a procedure called 'case stated', and
2 from some tribunals, for example the finding of an employment tribunal on an enforcement notice under HSW.

It also has some supervisory functions over lower courts and tribunals if they exceed their powers or fail to carry out their functions properly, or at all.

The High Court, the Crown Court and the Court of Appeal are known as the Supreme Court of Judicature.

The Court of Appeal has two divisions: the Civil Division which hears appeals from the county courts and the High Court; and the

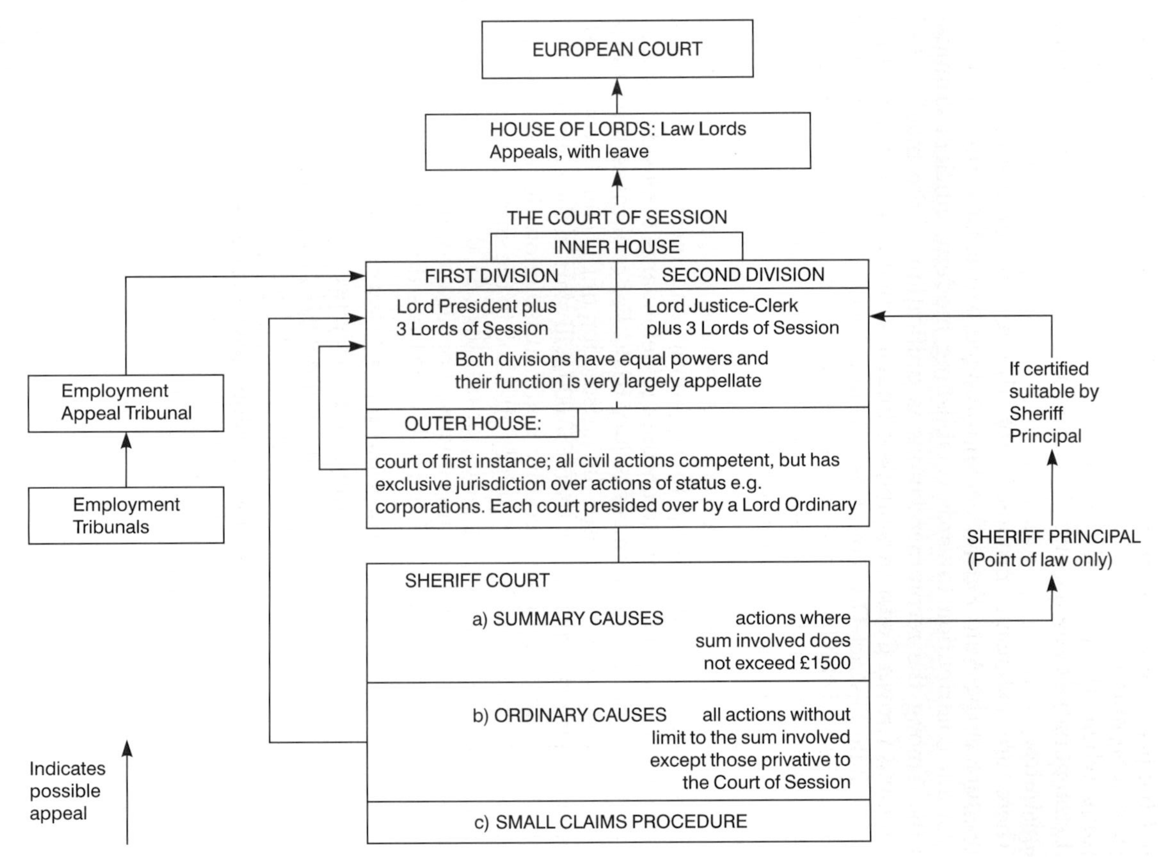

Figure 1.1.4 The main civil courts in Scotland

Criminal Division which hears appeals from the Crown Court. Further appeal, in practice on important matters of law only, lies to the House of Lords from the Court of Appeal and in restricted circumstances from the High Court. The Judicial Committee of the Privy Council is not part of the mainstream judicial system, but hears appeals, from, for instance, the Channel Islands, some Commonwealth countries and some disciplinary bodies.

Since our entry into the European Community, our courts must follow the rulings of the European Court of Justice. On an application from a member country, the European Court will determine the effect of European directives on domestic law. Potentially, the involvement is far-reaching in industrial obligations, including safety.

1.1.8.6 Court hierarchy in Scotland

Scotland also has separate but parallel frameworks for the organisation of its civil and criminal courts. These are shown diagrammatically in *Figures 1.1.4* and *1.1.5* and are discussed below.

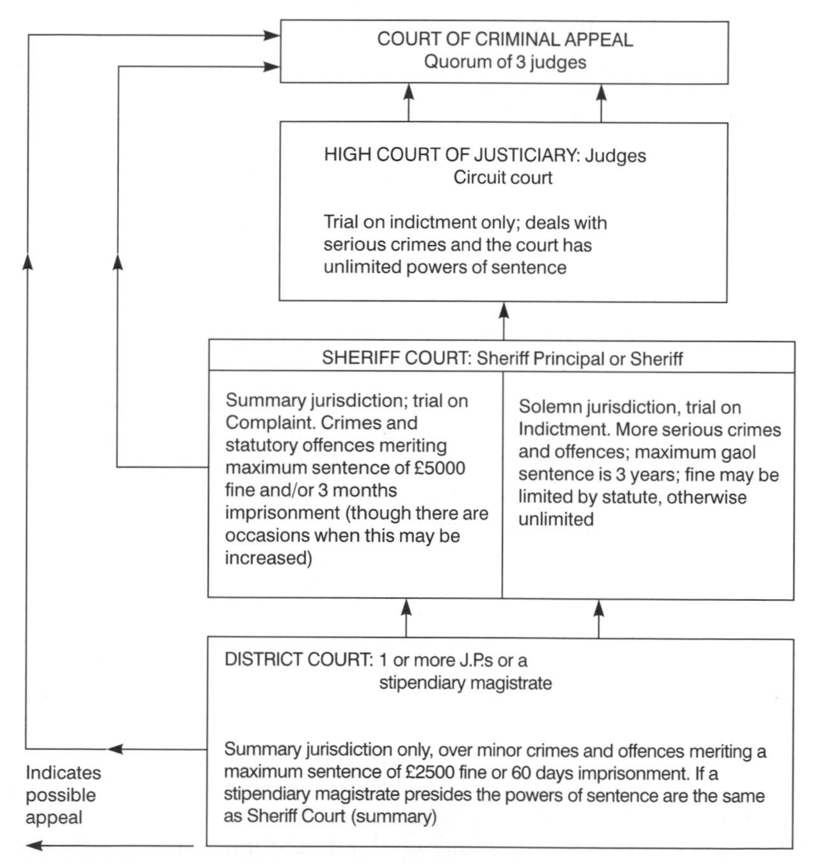

Figure 1.1.5 The main criminal courts in Scotland

The court most used is the local Sheriff Court which has wide civil and criminal jurisdiction. Civilly it may sit as a court of first instance or as a court of appeal (to the Sheriff Principal from a sheriff's decision). For criminal cases the sheriff sits with a jury for trials on indictment, and alone to deal with less serious offences prosecuted on complaints, when its jurisdiction encompasses that of the restricted district court.

The Court of Session is the superior civil court. The Outer House, sometimes sitting with a jury, has original jurisdiction; the Inner House hears appeals from the Sheriff Court and from the Outer House. Matters of law may be referred to the Inner House for interpretation, and it also hears appeals on matters of law from some committees and tribunals, such as decisions on HSW enforcement notices. Appeal from the Inner House is to the House of Lords. For criminal cases the final court of appeal is the High Court of Justiciary, with three or more judges. When sitting with one judge and a jury it is a court of first instance, having exclusive jurisdiction in the most serious criminal matters and unrestricted powers of sentencing. The High Court of Justiciary hears appeals from the first instance courts but only on matters of law in cases tried summarily in the Sheriff Court and the district courts. The judges of the High Court are the same persons as the judges of the Court of Session. They have different titles and wear different robes.

1.1.8.7 Court hierarchy in Northern Ireland

The hierarchy of courts in Northern Ireland is different from that for the English courts and is shown in *Figures 1.1.6* and *1.1.7*.

Most criminal charges are heard in the magistrates' courts. Magistrates try summary accusations or indictable offences being dealt with summarily. They also undertake a preliminary examination of a case to be heard in the Crown Court on indictment (committal proceedings).

Following trial in a magistrates' court, the defendant may appeal to the county court; or, on matters of law only, by way of 'case stated' to the Court of Appeal. The prosecution may appeal only to the Court of Appeal and only on a matter of law by way of 'case stated'. Trial on indictment, for more serious offences, is in the Crown Court, before a judge and jury (except for scheduled offences under the emergency legislation when cases are heard before a judge alone).

Appeal from the Crown Court is to the Court of Appeal. The defendant needs leave unless he is appealing only on a matter of law. The prosecution may refer a matter of law to the Court of Appeal, but this will not affect an acquittal. Final appeal by either side is to the House of Lords, but only with leave and only on matters of law of general public importance. Some civil proceedings take place in a magistrates' court before a resident magistrate[6] (RM). County courts have a wider and almost exclusive civil first instance jurisdiction. The procedure is less formal than in English county courts. Appeal from a County Court is to the High Court for a rehearing, or to the Court of Appeal on a matter of law only.

The High Court has unlimited civil jurisdiction. Appeal by way of rehearing is to the Court of Appeal; or in exceptional circumstances on

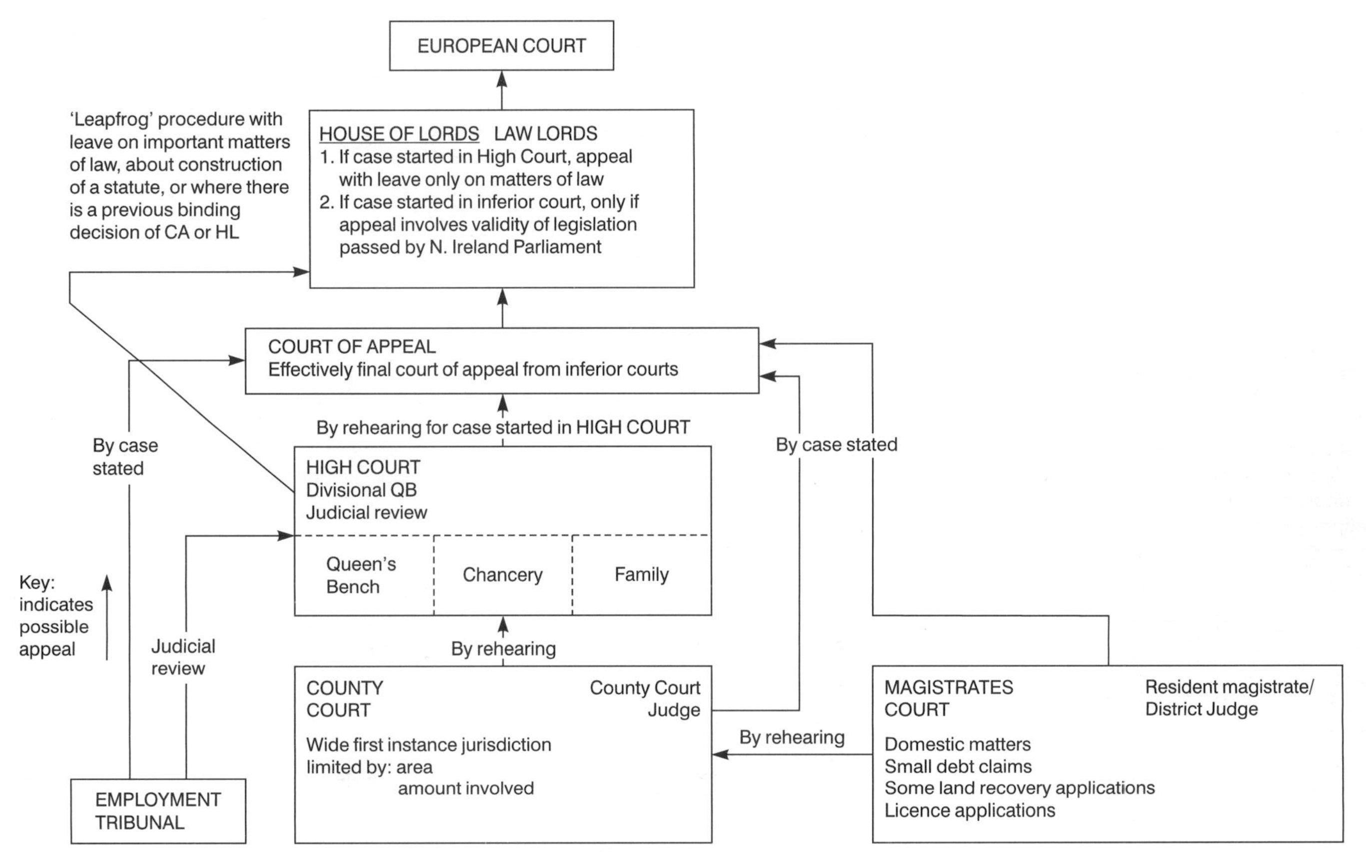
EUROPEAN COURT
'Leapfrog' procedure with leave on important matters of law, about construction of a statute, or where there is a previous binding decision of CA or HL
HOUSE OF LORDS LAW LORDS
1. If case started in High Court, appeal with leave only on matters of law
2. If case started in inferior court, only if appeal involves validity of legislation passed by N. Ireland Parliament
COURT OF APPEAL
Effectively final court of appeal from inferior courts
By case stated
By rehearing for case started in HIGH COURT
By case stated
HIGH COURT
Divisional QB
Judicial review
Queen's Bench
Chancery
Family
Key: indicates possible appeal
Judicial review
By rehearing
COUNTY COURT
County Court Judge
Wide first instance jurisdiction limited by: area
amount involved
By rehearing
MAGISTRATES COURT
Resident magistrate/ District Judge
Domestic matters
Small debt claims
Some land recovery applications
Licence applications
EMPLOYMENT TRIBUNAL

Figure 1.1.6 The civil courts in Northern Ireland

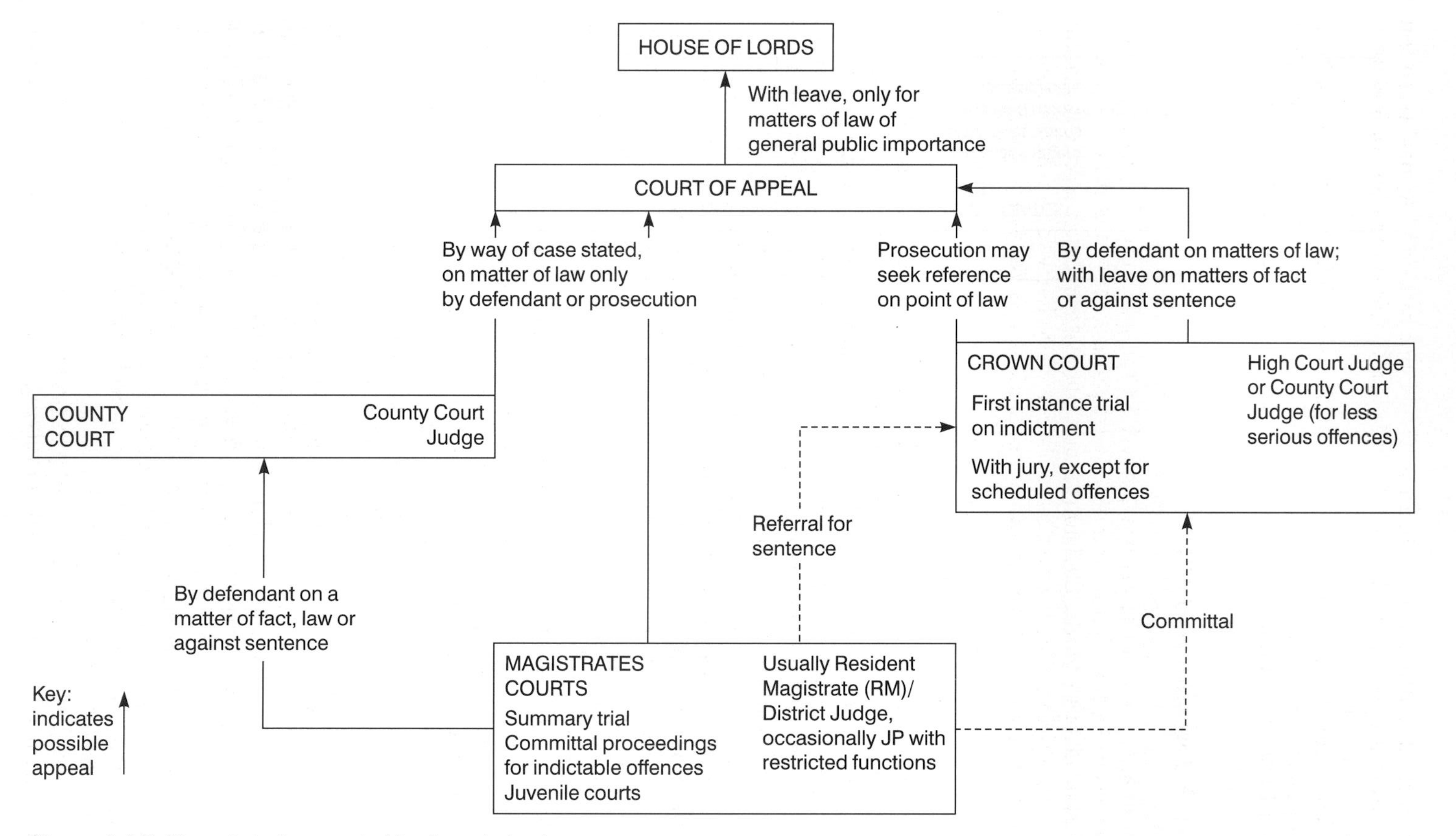
HOUSE OF LORDS
With leave, only for matters of law of general public importance
COURT OF APPEAL
By way of case stated, on matter of law only by defendant or prosecution
Prosecution may seek reference on point of law
By defendant on matters of law; with leave on matters of fact or against sentence
COUNTY COURT
County Court Judge
CROWN COURT
First instance trial on indictment
With jury, except for scheduled offences
High Court Judge or County Court Judge (for less serious offences)
Referral for sentence
Committal
By defendant on a matter of fact, law or against sentence
MAGISTRATES COURTS
Summary trial
Committal proceedings for indictable offences
Juvenile courts
Usually Resident Magistrate (RM)/ District Judge, occasionally JP with restricted functions
Key: indicates possible appeal

Figure 1.1.7 The criminal courts in Northern Ireland

important matters of law, direct to the House of Lords. Appeal from the Court of Appeal to the House of Lords is possible on matters of law only and with leave.

The Divisional Court hears application for judicial review and habeas corpus in contrast to the wider jurisdiction on 'case stated' of the English court and the English Divisional Courts for Chancery and Family.

1.1.9 Judicial precedent

Previous court decisions are looked to for guidance. English law has developed a strong doctrine of judicial precedent (sometimes referred to as *stare decisis* – let the decision stand). Some decisions (precedents) **must** be followed in a subsequent case. Other precedents are only persuasive. To operate the doctrine of judicial precedent it is necessary to know:

1 the legal principle of a judgement, and
2 when a decision is binding and when persuasive.

Higher courts bind lower courts, and superior courts usually follow their own previous decisions unless there is good reason to depart from them. Only since 1966 has departure been possible for the House of Lords, and the Civil Division of the Court of Appeal is not expected to depart from its own properly made previous decisions. The Criminal Division has more latitude because the liberty of the accused may be affected.

Decisions of the superior courts which are not binding are *persuasive*, judicial decisions of other common law countries or from the Judicial Committee of the Privy Council (see below: 1.1.14, para. 2) are also persuasive. The judgements of inferior courts are mostly on questions of fact and are not strict precedents. Decisions of the Court of Justice of the European Communities bind English courts on the interpretation of EC legislation.

The legal principle of a judgement, the actual findings on the particular facts, is called the *ratio decidendi*. Any other comments, such as what the likely outcome would have been had the facts been different, or reference to law not directly relevant, are persuasive but not binding. They are called *obiter dicta* – 'comments by the way'. The *obiter dicta* can be so persuasive that they are incorporated into later judgements and become part of the *ratio decidendi*. This happened to the *dicta* in the famous negligence case of *Hedley Byrne* v. *Heller & Partners*[7] (see below: 1.1.18, para. 4). Also, *obiter* is any dissenting judgement.

A precedent can bind only on similar facts. A court may *distinguish* the facts in a present case from those in an earlier case so that a precedent may not apply. A previous decision which has been distinguished may still be *persuasive*. An appeal court may *approve* or *disapprove* a precedent. A higher court may *overrule* a precedent, i.e. overturn a principle (though not the actual decision) of a lower court in a different earlier case. If a decision of a lower court is taken to appeal, the higher court will confirm or *reverse* the specific original decision.

The English doctrine of judicial precedent has evolved to give certainty and impartiality to a legal system relying upon case law decisions. Other advantages of the doctrine are the range of cases available and the practical information therein is said to provide flexibility for application to new circumstances and at the same time detailed guidance. Criticisms of the doctrine are that it is not always easy to discover the *ratio decidendi* of a judgement. One way in which a court may avoid a previous decision is to hold that it is *dicta* and not *ratio*. Other criticisms are that the doctrine leads to rigid compliance in a later case unless the previous decision can be distinguished; and that trying to avoid or distinguish a precedent can lead to legal deviousness. The doctrine of binding judicial precedent applies similarly in N. Ireland. In Scotland precedent is important, but there is also emphasis on principle. The European Court of Justice regards precedents but is not bound by them.

For the doctrine of precedent to operate there must be reliable law reporting. Important judgements are published in the Weekly Law Reports (WLR), some of which are selected for the Law Reports. Another important series is the All England Reports (All ER). Important Scottish cases are reported in Sessions Cases (SC) and Scots Law Times (SLT). In N. Ireland the two main series of law reports are the Northern Ireland Law Reports (NI) and the Northern Ireland Judgements Bulletin (NIJB), sometimes called the Bluebook. There are various specialist law reports, to which reference may be made when considering safety cases. A list of their abbreviations is published in *Current Law*[8] which also summarises current developments and current accident awards.

To assist international electronic searching and citation, High Court and Court of Appeal judgements are to be cited in a neutral form, identified by a unique number, for example [2001] EWHC number (QB); [2001]EWCA Civ number. The neutral citation may be followed by report citations, for example Smith v Jones [2001] EWCA Civ10 at 30, [2001] QB 124, [2001] All ER 364, etc.

Legal terminology in Law Reports includes abbreviations such as LJ (Lord Justice), MR (Master of the Rolls), per Mr Justice Smith (meaning 'statement by'); *per curiam* means statement by all the court; *per incuriam* means failure to apply a relevant point of law.

A decision of a higher court is a precedent, even though it is not reported in a law report. As well as written law reports, there are computerised data bases. An important example is Lexis[10], which includes unreported judgements of the Civil Division of the Court of Appeal. This very useful development may also accentuate a practical problem of the doctrine of judicial precedent. The volume of cases which may be cited may unnecessarily complicate a submission and lengthen legal hearings. This danger has been recognised in the House of Lords[11].

1.1.10 Court procedure

1.1.10.1 Stages in the proceedings

English, Irish and Scottish law follow an 'adversary' system, in which each side develops its cases and answers the contentions of the other.

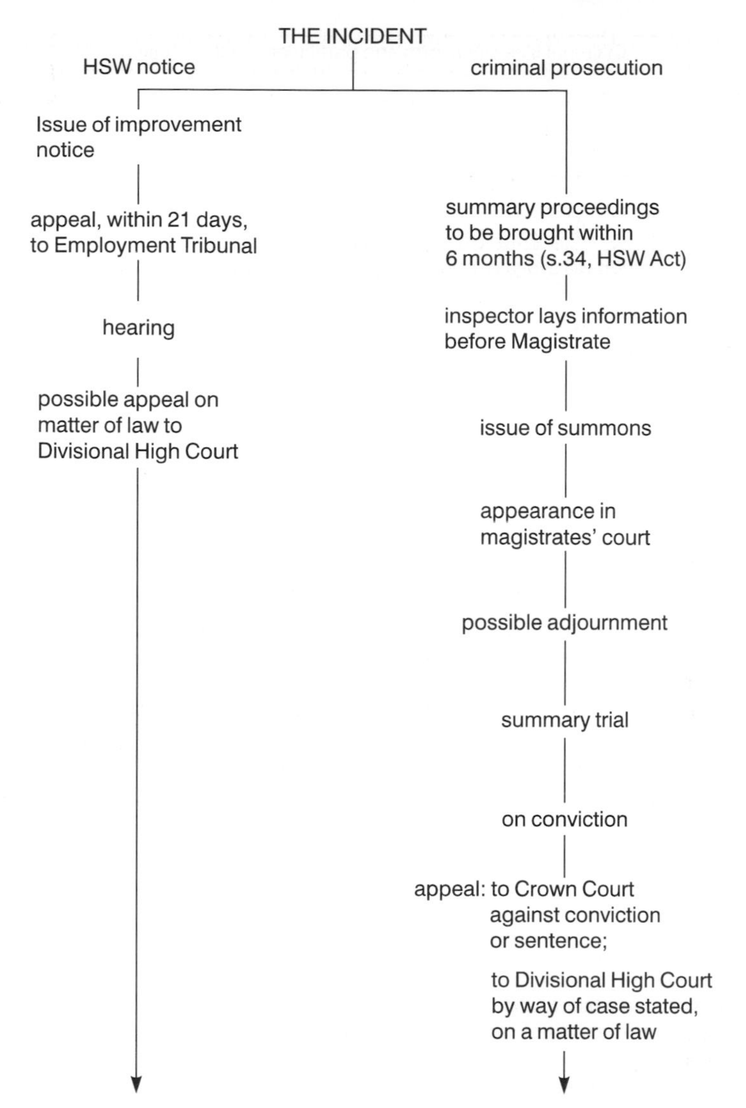

Figure 1.1.8(a) Possible enforcement legal proceedings following an accident at work

The judge's functions are to ensure that the correct procedures are followed, to clarify ambiguities, and to decide the issue. He may question, but he should not 'come down into the arena' and enter into argument.

An indication of the possible proceedings that could arise following an accident to an employee at work are shown in *Figure 1.1.8(a)* and *(b)* and considered below.

Important changes to the Civil Justice System are being implemented following the Woolf Report *Access to Justice* (1996)[12]. The objectives include a speedier and simpler procedure with tighter case management,

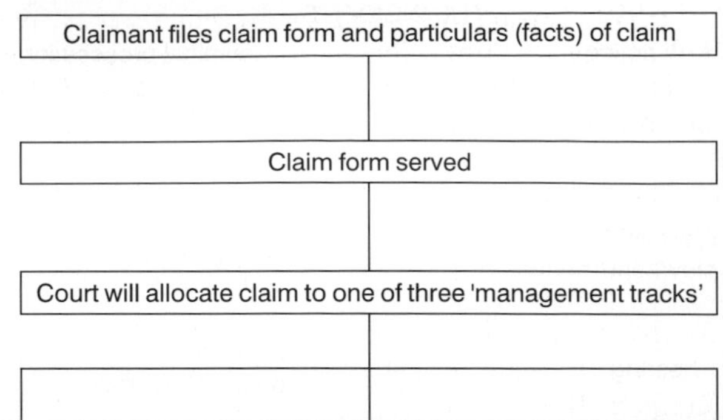

'Small Claims'	Fast track	Multi track
For not more than £5000	Defended claims for not more than £15 000; trial likely to last a day or less	Defended claims if amount or complexity requires this procedure
County Court	High Court if personal injury claim over £5000	High Court
Court gives directions, fixes dates for preliminary hearing. May suggest dealing with claim without a hearing.	Court gives directions for case management and sets timetable. Possible listing questionnaire form to ensure that all issues are resolved and parties ready for trial.	Court gives directions for case management and may set case management conference. Listing questionnaire form to ensure that all issues are resolved and parties ready for trial.
Final hearing information, without strict rules of evidence.		

In each track:

- Pre-action protocol directed towards co-operation between litigants before proceedings.
- Claimants' and defendants' statements (called 'statement of case') must be verified by a 'statement of truth'.
- Possible settlement before trial.
- Possible appeal from County Court or from High Court to Court of Appeal.

Figure 1.1.8(b) Possible civil legal proceedings following an accident at work

clarification of terminology and discouragement of the use of Latin expressions, encouragement of alternative dispute resolution and avoidance of litigation wherever possible. Litigation is to be less adversarial and more co-operative, and single experts, appointed by both parties, should be used whenever practicable. There are proposals for the effective use of information technology. Extensive proposals for reform of criminal court procedure are the subject of the Auld Report (see section 1.1.8.5).

Referring to the incident, should criminal proceedings be instituted against Hazards, in England and Wales any information stating the salient facts is laid before a magistrate.

Section 38 of HSW requires this to be by an inspector or by or with the consent of the Director of Public Prosecutions. The magistrate will issue a summons to bring the defendant before the court, and this would be served on Hazards at their registered office. Since a company has no physical existence, and therefore cannot represent itself, it would act through a solicitor or barrister. In a company, documents may be verified by a person holding a senior position.

In Scotland offences are reported to the local procurator-fiscal who decides whether to prosecute (and in what form when offences are triable either way). With serious cases he would consult with the Crown Office. If there is to be a summary trial a complaint is served on the accused stating the details of the charge.

Most HSW prosecutions are heard summarily, and then trial may commence when the accused is before the magistrates (in England and N. Ireland) or the sheriff (in Scotland). In England and N. Ireland, if the trial is to be on indictment, the magistrates will sit as examining justices to see if there is a case to answer before committing the accused for trial at the Crown Court. A magistrate may issue a witness summons and a procurator fiscal a citation if it appears that a witness will not attend voluntarily.

In a civil claim in the High Court or Court of Session Bertha Duncan, the claimant (plaintiff) (pursuer), starts her action by obtaining a statement of claim (writ of summons) and then serving this on Hazards Ltd. Hazards would consult their solicitors who would acknowledge service and indicate whether they intend to contest proceedings (if they do not, there may be judgement in default).

The claimant details the grounds of her claim and the damages she is claiming; and the defendant replies to the specific allegations.

Before trial each side must disclose to the other the existence of documents relevant to its case. The other side is allowed to inspect documents which are not privileged. An important ground of privilege is the protection of communication between a party and his legal advisers. In 1979 the House of Lords in *Waugh* v. *British Railways Board*[13] held that legal advice must be the dominant purpose of a document for it to be privileged. In this case disclosure was ordered of the report of a works accident, incorporating witnesses' statements, which while intended to establish the cause of the accident was intended also for the Board's solicitors.

An order (subpoena) requiring the attendance of a witness may be obtained. In N. Ireland witnesses may remain in court during the hearing of evidence, unlike England.

Proceedings in the inferior courts are similar to those in the High Court and Court of Session, but quicker, cheaper and more under the direction of the court administrators.

Usually a criminal case is decided before a related civil hearing comes on. The Civil Evidence Act 1968 (1971 for N. Ireland) allows a conviction to be used in subsequent civil proceedings. The conviction and the intention to rely on it must be set out in the formal civil claim. If this happened with Hazards then it would be for the company to file a defence and to prove (on the balance of probabilities) that the conviction is irrelevant or was erroneous. Dispute resolution is encouraged. In civil personal injury claims, settlement rather than court trial is a likely outcome, under the guidance of insurers.

If Bertha Duncan (*née* Smith) is suing in Scotland her case is referred to as *Smith (or Duncan)* v. *Hazards Ltd*, though for brevity it may be quoted as *Duncan* v. *Hazards Ltd*. The latter is also the English and N. Ireland practice (in speech the case is referred to as Duncan *and* Hazards Ltd).

On appeal, the party appealing, who may have been the defendant in the earlier trial, may be called the appellant and the other party the respondent.

1.1.10.2 The burden of proof

The phrase 'burden of proof' may be used in two senses. The underlying burden is on the prosecution or claimant to prove liability, sometimes called the 'legal' or 'final' burden of proof. However, during the trial the defendant may, for example, dispute evidence or argue a defence. The 'evidentiary' burden of proof then shifts to the defendant, but will shift back to the prosecution if it wishes to dispute that evidence of the defendants. The defendant's evidentiary burden of proof is on the balance of probabilities, even in a criminal trial.

1.1.10.3 The accused

With a criminal prosecution, normally the accused must attend court to answer the allegation(s) put to him. However, with offences triable only summarily (before magistrates) carrying a maximum penalty of three months or less[14], the accused may plead guilty in writing. The accused must answer every count (offence) alleged. Any acknowledgement of guilt must be unmistakeable and made freely without undue pressure from counsel or the court. If a guilty plea is made in error, it may be withdrawn at any time before sentence. A plea of not guilty may be changed during the trial with the judge's leave. It is possible for plea arrangements to be made between prosecuting and defence counsel where a plea of guilty to a lesser charge is accepted in return for the prosecution not proceeding with a more serious charge; or for a guilty plea to allow consideration of a sentence concession.

The accused has a right to silence, but since the Criminal Justice and Public Order Act 1994 there are greater risks in maintaining that position[15]. However, there can be no conviction on silence alone. There are statutory restrictions on questioning the accused about any criminal past and bad character[16]; and there are strict rules as to the admission as evidence of confessions of guilt[17].

1.1.10.4 Witnesses

The function of a witness is to inform the judge or the jury of facts, not opinions, unless the witness is called as an expert witness. Most people can be compelled to be witnesses[18]. Failure to comply with a witness order is contempt of court. A witness will be questioned by counsel who called him/her and may then be cross-examined by counsel for the other side. Counsel who called the witness may re-examine but may not raise new issues. 'Leading' questions (a question suggesting an answer) may not be asked. A witness cannot be compelled to answer a question which may incriminate him/her. A witness's evidence is usually given orally in open court, but in certain circumstances (e.g. illness) evidence is allowed by witness statement (affidavit, a sworn written statement).

Expert evidence is opinion evidence on a technical point(s). Opinion evidence is admissible from an expert but not from an ordinary witness. There should normally be pre-trial disclosure of expert evidence, in order to save court expense. A party will not normally be allowed to call expert evidence at trial if there has not been disclosure, unless the other side agrees.

In a criminal trial, the prosecution must inform the defence of the name and address of any person who has made a statement related to the prosecution but is not being called as a witness; of the existence of any previous witness statements which are inconsistent with those that person made at the trial; and of any known previous convictions of prosecution witnesses.

1.1.10.5 Reform

Litigation, whether civil or criminal, is time consuming and expensive. There is ongoing critical discussion about the need for reform in various contexts. The following are among the proposals bringing about or likely to bring about change for health and safety cases:

- The Woolf Report on Access to Justice (see section 1.1.10.1), advocates greater judicial (rather than lawyer) control before trial including over-calling expert evidence.
- The Auld Review of the Criminal Courts of England and Wales[19] (October 2001).
- The encouragement of alternative dispute resolution (negotiation and arbitration) with civil disputes.
- Proposals to restrict and target legal aid.
- The statutory power[20] for conditional fee agreements linked to a successful outcome, including for personal injury cases.
- Proposals[21] for the introduction of a special offence of corporate killing where a company's management failure in causing a death fell far below what could be reasonably expected.
- Proposals[22,23] for punitive damages to be allowed where employers show 'a blatant disregard of the health and safety of their workforce'.

1.1.11 Identity of court personnel

1.1.11.1 The English system

Court personnel include the bench, that is judges or magistrates; counsel for either side (see section 1.1.8.4); and the court usher appointed to keep silence and order in court, and to attend upon the judge. All judges are appointed by the Crown, and the appointment is salaried and pensionable.

In the Magistrates' Court there are 2–7 Justices of the Peace; or, in London and some large cities, possibly a District Judge (formerly stipendiary magistrate). Justices of the Peace are lay persons appointed by the Lord Chancellor on behalf of the Queen. The office dates back to the thirteenth century, but is now mainly regulated by the Justices of the Peace Act 1997. Justices sit part-time. They are not paid, but are reimbursed for financial expenses incurred from the office. A stipendiary magistrate is appointed by the Lord Chancellor, and is a qualified solicitor or barrister of at least seven years' standing. The office is salaried and full-time.

A Clerk to the Justices advises justices on questions of law, procedure and evidence; but should not be involved in the magistrates' function of trying the case. Legislation specifies the qualifications for justices' clerks.

Officiating in the county court is a Circuit judge; or a District judge for small claims and interlocutory (pre-trial) matters. A Circuit judge may also sit in the Crown Court. As a result of the Courts and Legal Services Act 1990, eligibility for appointment to the bench is based on having sufficient years of right of audience (qualification) in the courts. A Circuit judge must have 10 years' county court or Crown Court qualification, or be a Recorder, or have held other specified appointments. A District judge requires a 7 year general qualification (i.e. right of audience in any court).

First instance cases in the Crown Court are tried before a judge (to decide on matters of law); and a lay jury (for matters of fact). The Crown Court has three kinds of judge according to the gravity of the offence: a High Court judge, a Circuit judge or a Recorder. A High Court judge (necessary for a serious case) will be a Circuit judge with at least two years' experience, or have a 10 year High Court qualification. A Recorder is part-time, with a 10 year county court or High Court qualification. For appeals to the Crown Court, there will be no jury, but possibly the judge will sit with 2–4 justices.

For the Court of Appeal, normally three judges sit. They are called Lord Justices of Appeal. Appointments are normally made from High Court judges. An alternative prerequisite is 10 years' High Court qualification. High Court judges may also be asked to assist in the Court of Appeal. The Master of the Rolls is president of the Civil Division of the Court of Appeal. The Lord Chief Justice presides in the Criminal Division.

The Appellate Committee of the House of Lords as the final court of appeal sits with at least three 'Law Lords'. The Law Lords include the Lord Chancellor, the Lords of Appeal in Ordinary (who must have held high judicial office for two years or have 15 years' Supreme Court (see section 1.1.8.5) qualification, and Peers of Parliament who hold or have held high judicial office.

The head of the judiciary and president of the House of Lords is the Lord Chancellor. He is also a government minister, and the Speaker of the House of Lords. He is exceptional in combining judicial, executive and legislative functions.

The Attorney General is the principal law officer of the Crown. He is usually an MP and answers questions on legal matters in the House of Commons. He may appear in court in cases of exceptional public interest. His consent is required to bring certain criminal actions, for example in respect of offences against public order. The Solicitor General is immediately subordinate to the Attorney General.

The Director of Public Prosecutions must have a 10 year general qualification. He undertakes duties in accordance with the directions of the Attorney General. He will prosecute cases of murder and crimes amounting to an interference with justice.

1.1.11.2 Legal personnel in Scotland

In Scotland the Lord Advocate is the chief law officer of the Crown and has ultimate responsibility for prosecutions. He and the Secretary of State for Scotland undertake the duties which in England and Wales are the responsibility of the Home Secretary, the Lord Chancellor and the Attorney General. The Lord Advocate is assisted by the Solicitor General.

Judicial appointment, to the Supreme Court and the Sheriff Court, is by royal authority on the recommendation of the Secretary of State. Judges in the District Courts are lay justices of the peace, apart from some stipendiary magistrates in Glasgow.

The two branches of the legal profession are solicitors and advocates. As in England, advocates no longer have exclusive rights of audience in the higher courts. Traditionally a Scottish solicitor is more a manager of his client's affairs than in England.

1.1.11.3 Legal personnel in Northern Ireland[24]

The Lord Chancellor, and the English Attorney General and Solicitor General act also for Northern Ireland. The Director of Public Prosecutions is appointed by the Attorney General, and has 10 years' legal practice in Northern Ireland. His chief function is responsibility for prosecutions in serious cases (compare the Crown Prosecution Service in England, and the Lord Advocate and procurators-fiscal in Scotland).

As in England, appointment to the bench and advocacy in the superior courts is at present restricted to barristers. A major difference between the legal system of Northern Ireland and England is the appointment of resident magistrates (RM). They are full-time and legally qualified, with responsibility for minor criminal offences, committal proceedings, and some civil matters. The powers of lay Justices of the Peace in Northern Ireland are limited in comparison with JPs in England and Wales.

1.1.12 Employment Tribunals

Industrial Tribunals, now called Employment Tribunals, were set up in 1964 to deal with matters arising under the Industrial Training Act of that year. Now they have statutory jurisdiction in a range of employment matters, such as unfair dismissal, redundancy payments, equal pay, sex and race discrimination and claims for breach of contract of employment. Such jurisdiction does not include a claim in respect of personal injuries[25]. In the context of HSW they hear appeals against prohibition and improvement notices, and applications by statutory safety representatives about payment for time off for training.

The burden of proof is on the inspector to satisfy the Tribunal that the requirement for a notice is fulfilled: *Readmans Ltd* v. *Leeds City Council*[17] (a prohibition notice under s. 3). The High Court held that the notice alleged a breach of a criminal duty and it was for the council who had issued the notice to establish the existence of the risk of serious personal injury not for the appellant to have the burden of proving that there was no such risk. The burden of proof is then on an appellant who wishes to show that it was not, for example, practicable or reasonably practicable (according to legislation) to carry out certain measures. This must be proved on the balance of probabilities.

Tribunals sit locally and consist of a legally qualified chairman plus a representative from each side of industry. Proceedings begin with an originating notice of application in which the applicant sets out the name and address of both parties and the facts of the claim. The application must be made within the prescribed time limit. This varies. It is 21 days with enforcement notice; three months for unfair dismissal and paid time off for union duties; six months for redundancy applications.

Proceedings are on oath, but they are more informal than in the courts and the strict rules of evidence are not followed. Legal aid is not available for representation. A friend or union official may represent (which is not possible in the courts). Costs are rarely awarded. Like the courts, Tribunal proceedings are open to the public, and visits are the best way to understand their working.

An appeal is possible from an Employment Tribunal decision, but only on a matter of law. In respect of enforcement notices it is to the High Court in England; and to the Court of Session in Scotland. In respect of other matters it is to the Employment Appeal Tribunal except in N. Ireland.

The Employment Appeal Tribunal is a superior court associated with the High Court. It sits with a judge and 2–4 lay members, and all have equal voice. Parties may be represented by any person they wish, and legal aid is available. Further appeal is to the Court of Appeal (in Scotland to the Inner House of the Court of Session). In N. Ireland there is no Employment Appeal Tribunal but a Tribunal's decision may be challenged by review by the Tribunal itself, by judicial review by the High Court, or by way of case stated to the Court of Appeal.

1.1.13 European Community Courts (ECJ)

1.1.13.1 The Court of Justice of the European communities

The European Court is the supreme authority on Community law. Its function is to 'ensure that the law is observed in the interpretation and application of the European Community Treaty 1957' (art. 220, formerly art. 164). The EC Treaty[28] and the Single European Act 1986, are concerned with matters such as freedom of competition between Member States; and aspects of social law, including health and safety at work.

The Treaty of European Union 1991 (the Maastricht Treaty) re-emphasised these Community aims and added further goals of economic and monetary union, and these are developed in the Treaty of Amsterdam 1997 (ToA). The Treaty of Amsterdam consolidates the administration of the Communities and institutions of the Union and provides for the coherent renumbering of Treaty Articles. The ToA is also concerned with developing the concept of European citizenship, common strategies for employment and the co-ordination of national policies; consolidation of environmental policy, provisions for high standards of public health and clarification of consumer protection policy, for example. The Treaty of Nice[29] is in preparation for the enlargement of the European Union to include countries of central and eastern Europe, the Mediterranean and the Baltic. It will enter into force once it has been ratified by all Member States of the Union. Operation of the European Union is also considered in section 1.1.16.4.

In *R* v. *Secretary of State for Transport* v. *Factortame Ltd*[30], the ECJ directed the House of Lords that any provision of a national legal system which might impair the effectiveness of EU law is incompatible with the requirements of EU law. UK regulations made under the Merchant Shipping Act 1988, to prevent 'quota hopping' by Spanish fishermen, were struck down as being contrary to EU law. In *Factortame No. 5*[31], the Court of Appeal held that the breaches of Community Law were sufficiently serious to give rise to liability for damages to individuals[32].

The European Court has two types of jurisdiction, direct actions, and reference for preliminary rulings.

Direct actions may be:

- against a Member State for failing to fulfil its obligations under Community law and be brought by the Commission or by another Member State;
- against a Community institution, for annulment of some action, or for failure to act (judicial review);
- against the Community for damages for injury by its institutions or servants;
- against a Community institution brought by one of its staff.

References for preliminary rulings are requests by national courts for interpretation of a Community provision. Article 234 (formerly 177) provides that any court or tribunal may ask the European Court for a ruling; but only the final court of appeal (the House of Lords in the UK) must ask for a ruling if a party requests it. In the English case of *Bulmer* v. *Bollinger*[33] the Court of Appeal held that the High Court and the Court of Appeal may interpret Community law.

The European Court is based at Luxembourg. There are 15 judges (to include one from each Member State), assisted by eight Advocates General. The function of an Advocate General is to assist the Court by presenting submissions, in which he analyses the relevant issues and makes relevant recommendations for the use of the Court. The judgement itself is a single decision, thus an odd number of judges is required. With the increase in workload, there is a facility for the Court to sit in subdivisions called Chambers. Cases brought by a Member State or by a Community institution must still be heard by the full court. Although the Court seeks to have consistency in its findings, precedents are persuasive rather than binding on itself. Decisions are binding on the particular Member State.

Referrals to the Court of Justice are requests to it to rule on the interpretation or applicability of particular parts of Community law. Where the Court of Justice makes a decision, it not only settles the particular matter at issue but also spells out the construction to be placed on disputed passages of Community legislation, thereby giving clarification and guidance as to its implementation.

It keeps under review the legality of acts adopted by the Council and the Commission and also can be invited to give its opinion on an agreement which the Community proposes to undertake with a third country, such opinions become binding on the Community.

Through its judgments and interpretations, the Court of Justice is helping to create a body of Community law that applies to all Community institutions, Member States, national governments and private citizens. Judgements of the European Court of Justice take primacy over those of national courts on the matters referred to it.

Although appointed by the Member States, the Court of Justice is not answerable to any Member State or to any other EC institution. The independence of the judges is guaranteed.

Under the Single European Act 1986, the Council of Ministers has the power to create a new Court of First Instance. This Court was established by Council decision in 1988 and became effective in September 1989. It has 15 members, appointed by common accord of the Member States. Members may also be asked to perform the task of an Advocate General. It may sit with three or five judges.

The Court of First Instance is responsible for hearing all direct actions against the Community, such as seeking annulment because of illegality, or damages because of legal liability or actions by Community staff. It is subject to the legal supervision of the Court of Justice.

There is also a Court of Auditors, which supervises the implementation of the budget.

At the beginning of this section, the *Factortame* litigation illustrated the interaction of ECJ decisions with UK courts. Another illustration of the effect of an ECJ decision on national law comes from the UK challenge to the *Working Time Directive*. The ECJ rejected the UK argument that the legal basis of the directive was defective, and also considered that the directive did not breach the principle of *subsidiarity* (the aims could not be achieved by Member States alone), nor *proportionality* (the requirements were not excessive). The directive is now being implemented in the UK by the Working Time Regulations 1998, which have been extended by the Working Time (Amendment) Regulations 2001 as a result of the EU finding incomplete fulfilment of the Directive by the UK.

1.1.14 Human Rights Courts

1.1.14.1 The European Court of Human Rights

This Court should not be confused with the Court of Justice of the European Communities. The Court of Human Rights sits at Strasbourg. Its function is to interpret the European Convention for the Protection of Human Rights and Fundamental Freedoms, drawn up by the Council of Europe in 1950. The Council of Europe comprises 44 European states (replacing 23 Western European states). It is active on social and cultural fronts rather than economic. The United Kingdom ratified the Convention in 1951, so that it is binding on the UK internationally. UK legislation incorporated the Convention by the Human Rights Act 1998[34] which was entirely in force by October 2000[35]. The articles of the Convention provide for matters such as the right not to be subjected to inhuman or degrading treatment, the right to freedom of peaceful assembly, the right to respect for family life, home and correspondence.

An example of a decision directed to the UK was the '*Sunday Times* thalidomide case' in 1981. A drug prescribed for pregnant women caused severe abnormalities in the children. The manufacturers sought an injunction to prevent the *Sunday Times* publishing an article about the drug. The Court of Human Rights ruled that the House of Lords' confirmation of an injunction was a violation of the right of freedom of expression[36]. *R. v Francois Pierre Marcellin Thoron*[37] is an example of an abortive attempt to use the Human Rights Convention in the context of health and safety.

1.1.14.2 International Criminal Court

The establishment of an International Criminal Court (ICC) was confirmed in April 2002, following ratification of the Rome Statute of the International Criminal Court by a required 60 countries. The UK ratified the Rome Statute in October 2001[38], becoming the 42nd state to do so. The ICC came into existence in July 2002 and enables prosecution of human rights abusers worldwide.

1.1.15 Sources of English law

The two main sources of UK law are legislation, and legal principles developed by court decisions (common or case law).

English common law, based on custom and evolving since the eleventh century, developed indigenous concepts, and unlike most European countries was little influenced by Roman law. In Scotland Roman law was an important influence from the sixteenth to the eighteenth century, particularly on the law of obligations, which includes contract and delict. In Ireland, before the seventeenth century, Brehan law (of early Irish jurists) or English common law predominated according to political control at the time. Since the seventeenth century the law in Ireland and England developed along similar lines in general, with some exceptions such as marriage and divorce. English common law concepts were applied in former British territories. Today most of the United States, Canada (other than Quebec), Australia, New Zealand, India and some African countries remain and are called common law countries.

England, Scotland and N. Ireland do not have codified legal systems. Nearly all of our law of contract and much of the law of tort or delict is case law. This will gradually change with the production and implementation of Law Commission reports.

As with most subjects, law has specific terminology. The historic development of our law is illustrated by the Latin, old French and old English phrases which are sometimes used. This chapter contains some Latin words, for example, *obiter dicta* and *ratio decidendi* (section 1.1.9); and some coming from the French, such as tort and plaintiff (sections 1.1.5, 1.1.6). The Woolf Report[39] reforms include replacing Latin phrases where possible, clarifying and simplifying terminology so that, for example, the *plaintiff* becomes the *claimant*. The most straightforward rule for legal Latin or French is to pronounce words as though they were English. Other words and phrases met with have a particular legal meaning, such as damages, contract of employment, relevant statutory provision; and abbreviations such as JP or *v.* (as in Donoghue *v.* Stevenson). There are a number of law dictionaries to explain or to translate words and these are listed at the end of this chapter.

1.1.16 Legislation

1.1.16.1 Acts of Parliament and delegated legislation

Since the eighteenth century increasing use has been made of legislation. Legislation comprises Acts of Parliament and delegated legislation made by subordinate bodies given authority by Act of Parliament. Examples of delegated legislation are ministerial orders and regulations (Statutory Instruments), local authority byelaws and court rules of procedure. All legislation is printed and published by The Stationery Office Ltd. Often, but not always, delegated legislation requires the approval of Parliament, for example by negative resolution (that is by not receiving a negative vote of either House); or, more rarely, by affirmative resolution (that is by requiring a positive vote of 'yes').

HSW and its associated regulations is an example of how extensive subordinate legislation may be. HSW is an enabling Act. Section 15, schedule 3 and s. 80 give very wide powers to the Secretary of State to make regulations. The regulations are subject to negative resolution (s. 82). They may be made to give effect to proposals of the Health and Safety Commission (in N. Ireland the Health and Safety Agency); or independently of such proposals, but following consultation with the Commission and such other bodies as appear appropriate (s. 50). The Commission may also issue Approved Codes of Practice (s. 16 HSW) for practical guidance. Such codes are not legislation and s. 17 confirms that failure to observe such codes cannot of itself ground legal proceedings. However, failure to comply is admissible evidence and will be proof of failure to comply with a legislative provision to which the code relates unless the court is satisfied that there is compliance in some other way.

Delegated legislation is suitable for detailed technical matters. By avoiding the formality required for an Act of Parliament the legislation can be adapted, and speedily (for example, the maximum unfair dismissal payment may be increased quickly by an Order).

Long drawn out consultation may slow down any legislation. In 1955 the decision in a famous case of *John Summers & Sons Ltd* v. *Frost*[40] virtually meant that an abrasive wheel was used illegally unless every part of that dangerous machinery was fenced. Regulations were required to allow its legal use. There were drafts and consultations, but it was 1971 before the Abrasive Wheels Regulations came into operation[41].

During its passage through Parliament and before it receives the Royal Assent an intended Act is called a *bill*. Most government bills start in the House of Commons, but non-controversial ones may start in the House of Lords. Ordinary public bills such as that for HSW go through the following process. The bill is introduced and has a formal first reading. At the second reading there is discussion on the general principles and purpose of the bill. It then goes to committee. After detailed consideration the committee reports the bill to the House, which considers any amendments. The House may make further amendments and return the bill to committee for further consideration. After the report stage the bill is read for the third time. At third reading in the Commons only verbal alterations may be made.

The bill now goes through similar stages in the other House. If the second House amends the bill it is returned to the first House for consideration. If the Lords reject a bill for two sessions it may receive the Royal Assent without the Lords' agreement. Practically, the Lords can delay a bill for a maximum of one year.

After being passed by both Houses the bill receives the Royal Assent, which conventionally is always granted, and thus becomes an Act. A statute will normally provide at the end whether it is to apply in Scotland and N. Ireland as well as in England and Wales. Subsequent legislation may apply provisions to Scotland or N. Ireland, for example the Health and Safety at Work (NI) Order 1978. Increasingly, various sections of Acts are effected by later Statutory Instrument(s), which can cause uncertainty.

Parliament has supreme authority. It may enact any measure, other than binding future Parliaments. It is not answerable to the judiciary.

The United Kingdom is now part of the European Community (EU) and subject to the Community's regulations and directives (see sections 1.1.13.1 and 1.1.16.4). These require Member States to implement agreed standards on, among other concerns, safety and health at work and the environment.

The ultimate sovereignty of the UK Parliament is theoretically retained in that Parliament could repudiate agreement to EU membership[42]. Also, since the Single European Act there has been increased emphasis on *subsidiarity*. This is the principle that decisions should be taken at the most suitable level down the hierarchy of power, that is at national rather than EC level where appropriate.

1.1.16.2 Statutory interpretation

Inevitably some legislation has to be interpreted by the courts, to clarify uncertainties, for example, and substantial case law may attach to a statute. Judicial consideration of the effect of legislation for the fencing of dangerous machinery is an example of this (see sections 1.1.17.3 and 1.1.19.1).

Statutes normally contain an interpretation section. There is also the Interpretation Act 1978 which provides, for example, that unless the contrary is stated, then male includes female, the singular includes the plural, writing includes printing, photography and other modes of representing or reproducing words in visible form. In modern legislation, the detail is often relegated to Schedules at the end of the Act.

Parliamentary discussions are reported verbatim in *Hansard*. In 1992 the House of Lords decided that if there is an ambiguity, a minister's clear explanation to Parliament, as published in *Hansard*, may be used to interpret a statute[43].

As a result of Article 10 (formerly 5) of the EEC Treaty 1957, which requires Member States to 'take all appropriate measures to ensure fulfilment of the obligations arising out of the treaty', UK courts give a *purposive* interpretation where the purpose of UK legislation is to give meaning to a directive. An example is *Pickstone* v. *Freeman plc*[44]. The House of Lords interpreted regulations amending the Equal Pay Act 1990 against their literal meaning to allow a female warehouse operative to use as a comparison a man doing a different job of equal value.

1.1.16.3 White Papers and Green Papers

Proposed legislation may be preceded by documents presented by the government to Parliament for consideration. A Green Paper is a discussion document. A White Paper contains policy statements and explanations for proposed legislation. Such papers are published as Command Papers.

On a narrower basis the Government also consults with outside interests when drafting legislation, bodies such as the CBI and TUC on industrial and economic matters. Legislation may require such consultation, for example s. 50 HSW.

1.1.16.4 European Union (EU) legislation

Originally known as the European Economic Community (EEC) and then as the European Community (EC) it is now usual to refer to the Community as the European Union (EU). The primary legislation is the Treaty of Rome 1957 which established the Community and was incorporated into UK law by the European Communities Act 1972; the Single European Act 1985 was incorporated into UK law in 1986 and the Treaty of European Union 1991 (TEU) (the Maastricht Treaty), incorporated into UK law by the European Communities Act 1992. The TEU and the Treaty of Amsterdam 1997[46], strengthen the role of the European Parliament (EP) by extending the scope of the *co-decision* procedure (outlined below). The Treaty of Nice 2001[47] is directed to enlargement of the European Union.

The supreme body of the EU is the Council of the European Union – the Council of Ministers – with one member from each state but with weighted voting rights according to size. Decisions are prepared within Working Parties and within the permanent representatives committee (COREPER), made up of the Permanent Representatives (Ambassadors) of the Member States of the European Union.

The administration of the EU is in the hands of the European Commission which has 20 members, one from each Member State but with the larger Member States having two. The Council and the Commission are assisted by an Economic and Social Committee (EcoSoC) whose members represent various categories of economic and social activity, and by a Committee of the Regions (COR) formed from representatives of regional and local bodies. The Commission's functions include initiating proposals for legislation and managing and executing EU policies, such as communicating policies on health and safety at work.

Legislative power in the Community is exercised either by the Council of the European Union or jointly by EP and the Council. Legislation is normally initiated by the Commission[48], and requires statutory consultation with the Council and EP. Adoption of legislation is by the Council with EP having a considerable say in what is adopted. Also, Article 192 (formerly 138b) provides that the EP may, acting on a majority vote of its members, request the Commission to prepare appropriate proposals on any matter on which it considers that Community legislation is required for the purposes of implementing the Treaty.

Secondary Community legislation takes three forms: Regulations which are binding on Member States; Directives which require national implementation (see section 1.1.16.1 and *Figure 1.1.9*) and Decisions of the Council or Commission. Such a Decision is specific rather than general. Its main use is if a State asks permission to depart from the EC Treaty, for example in respect of competition policy.

The legislative process for secondary legislation is complex. There are six different procedures, principally distinguished by the degree of power which the European Parliament has in each process. The procedure that now applies to most legislative proposals is the co-decision procedure[49], which shares decision making power equally between the EP and the

Community adopts Directive
(member States must implement within time limit)

1. HSE prepare proposals
2. Limited consultations with e.g. CBI, TUC, professional bodies
3. Draft regulations to HSC
4. HSC issues consultative document for public comment (consultation period of some 4 months)
5. Comments co-ordinated by HSE
6. Finalised proposals, taking account of public comment
7. Final proposals submitted to HSC
8. HSC submit proposals to Secretary of State
9. Proposals placed before parliament; negative resolution
10. Put on Statute Book and becomes UK law
11. Effective on date announced

Figure 1.1.9 Stages of internal UK procedure for implementing a directive

European Council of Ministers. Article 137 (formerly 118) is important for health and safety measures and requires the co-decision procedure. The article refers, in particular, to improvements of the working environment to protect workers' health and safety, working conditions and the information and consultation of workers.

When the Council adopts a proposal it places obligations on Member States to incorporate its requirements into national laws within a stated time scale. Adopted legislation is published in the *Official Journal of the European Communities* (OJ). *Figure 1.1.9* shows the internal UK procedure for incorporating a directive into UK law.

1.1.16.5 The co-decision procedure

The co-decision procedure involves up to three readings in EP and in the Council and, should there be disagreement between the two institutions, requires a Conciliation Committee (of Council members and a like number of Parliamentary representatives) to resolve it. The Commission takes the necessary initiatives to administer the procedure.

When the Commission submits a proposal to the EP and the Council, the Council, acting by a qualified majority[50], may adopt the proposal either if EP has no amendments or if the Council approves EP's amendments to the proposal. Otherwise the Council will adopt a common position and inform EP giving full reasons for its decision.

If within three months of such communication, EP has approved the common position or has not taken a decision, the proposal is deemed to have been adopted in accordance with the common position. However, within three months EP can either, by an absolute majority, reject the Council's common position when the proposal fails, or propose amendments to the common position. The texts of the amendments are referred to the Council and the Commission for an opinion on them.

Within three months of such referral, the Council, acting by a qualified majority, may approve all the amendments proposed by EP when the proposal is deemed to be adopted as amended. The Council must act unanimously on any amendments on which the Commission has delivered a negative opinion. If the Council does not approve all the amendments, the President of the Council in agreement with the President of EP must, within six weeks, convene a meeting of the Conciliation Committee to try to develop a joint text that is agreed by a qualified majority of Council and a majority of representatives of EP. Any such agreement being based on Council's common position as amended by EP. Failure by the Conciliation Committee to reach agreement results in the proposal failing. The Commission acts to facilitate the conciliation proceedings.

If, within six weeks of its being convened, the Conciliation Committee approves a joint text, EP acting by absolute majority and Council acting by qualified majority each have six weeks in which to adopt the joint text when the proposal as amended is deemed to be adopted. If either of the two bodies fails to approve the joint text, the proposal fails. *Figure 1.1.10* illustrates the co-decision procedure.

The extension, by the Single European Act, of qualified majority voting to proposals concerning the health and safety of workers was the stimulus for a great increase in EU health and safety directives from 1989 onwards.

1.1.16.6 The European Agency for Safety and Health at Work

The European Agency for Safety and Health at Work[51] was officially inaugurated in 1997. It is located in Bilbao (Spain) and managed by a board with Government, employer and worker representatives from all EU Member States as well as representatives from the European Commission. The Agency's functions include assessment of the impact of health and safety legislation on small and medium enterprises and the establishment of a network to share health and safety information within the EU and more widely.

1.1.16.7 Application of EU legislation to an individual

The Treaty and Community legislation must be recognised in the Member States, but an individual can only enforce it, if at all, in the national courts; and only if it has 'direct effect' for that individual. Community legislation takes two main forms, regulations and directives (see also section 1.1.16.4). A regulation is a law in the Member States to which it is directed; it is said to be 'directly applicable' to that State. According to its content a Community regulation may impose obligations and confer rights on individuals enforceable in the national courts; it is then said to have 'direct effect'. A directive must be enacted by the Member State, and then, according to how it is enacted, may give enforcement rights to individuals in the national courts. Sometimes a directive, even before implementation by the Member State, may have 'direct effect' for an individual to rely on it against the State. This could be so if the date of

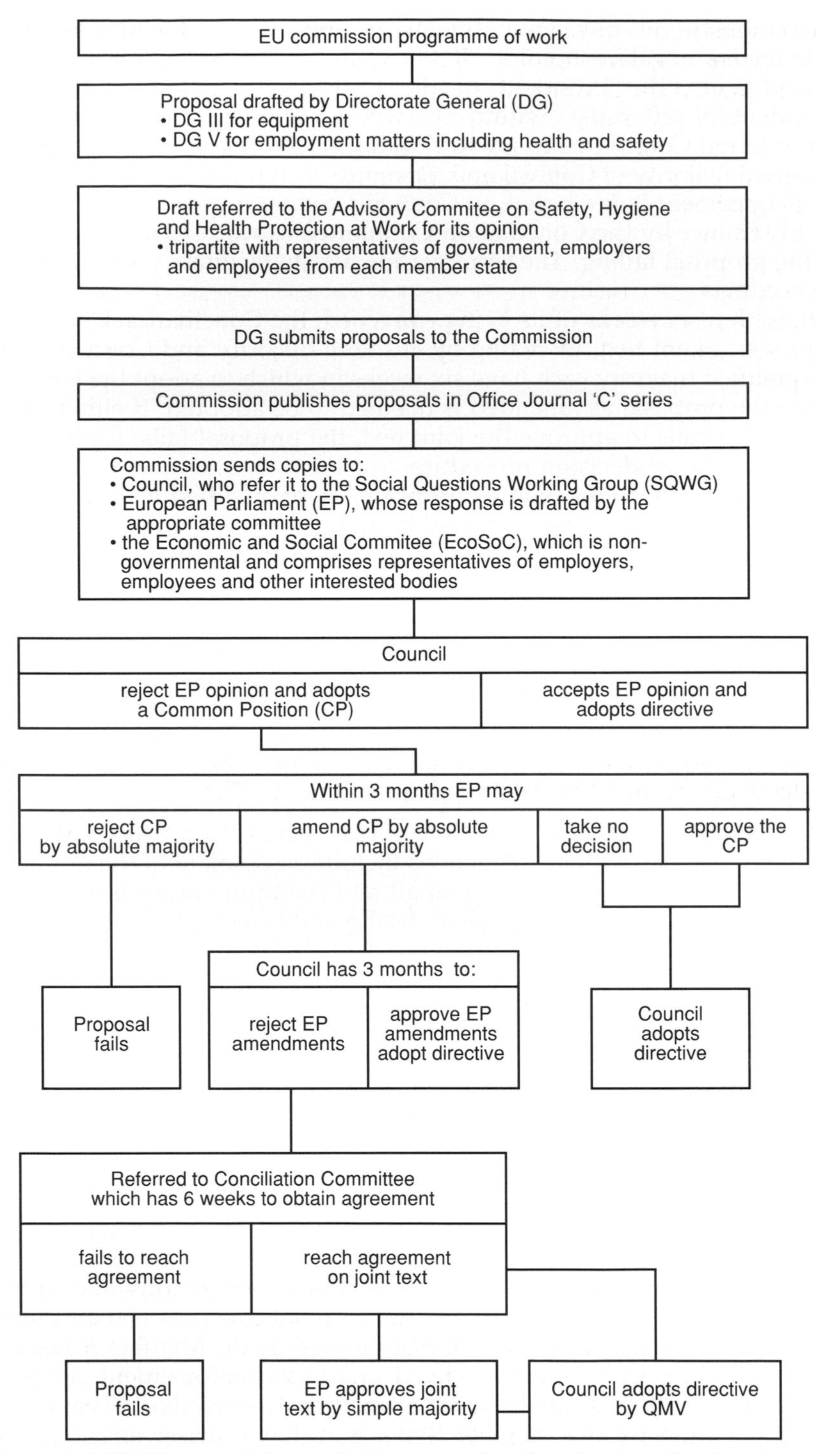

Figure 1.1.10 Diagram of the co-decision procedure for adoption of a directive

implementation had passed and the existing law of the Member State contravenes the directive[52]. The directive must be sufficiently clear, precise and unconditional.

Any such direct effect of a directive does not give rise to obligations between individuals. However, in *Marshall* v. *Southampton and South West Hampshire Area Health Authority (Teaching)*[53], Mrs Marshall successfully challenged the health authority's compulsory retiring age of 65 for men and 60 for women as being discriminatory. An individual may not enforce such a decision against a private employer but can against a government body[54]. See also the repercussions of the *Factortame* case outlined in section 1.13.1. However, the European Court of Justice has required national courts to *interpret* national legislation to be consistent with directives.

1.1.17 Safety legislation before the Health and Safety at Work etc. Act

1.1.17.1 Factories

Early factory legislation, in the nineteenth century, concerned the textile and allied industry. It was directed towards the protection of young persons and women and was motivated by concern for moral welfare and sanitation as much as for safety. Between 1875 and 1937 there were attempts to unify the increasing but fragmented legislation, but subsequent inadequacies resulted in patchwork amendments. The Factories Act 1937 was intended as a coordinating measure. It brought together health, safety and welfare in all factories: and introduced some new requirements such as those for floors, passages and stairs, and for safe access.

But regulations made under previous legislation continued in force as though made under the 1937 Act. This practice was repeated by the Factories Act 1961 so that some of the provisions and standards were outdated. The HSW and consequent regulations, including those implementing EC directives, have replaced much of the Factories Act and associated legislation.

Similarly, HSW regulations have superseded or augmented other workplace-specific provisions, such as for offices, agriculture, mines and quarries.

1.1.17.2 Offices

In 1949 the Gower Committee report made recommendations about the health, welfare and safety of employed persons outside the protections of existing legislation. In 1960 an Offices Act was passed. Before it became operative, however, it was repealed and replaced by the Offices, Shops and Railway Premises Act 1963. This adopted much of the structural content of the Factories Act 1961 but not the regulations, which apply only to factories.

1.1.17.3 Mines, quarries etc.

The law relating to safety and management in mines and quarries was examined in the 1950s and the principal Act is now the Mines and Quarries Act 1954. HSW regulations are more likely to augment and update rather than absorb rules for this very particular work environment. There is wide power to make regulations. Other Acts refer to work practices in agriculture, aviation and shipping.

1.1.18 Safety legislation today

1.1.18.1 Health and Safety at Work etc. Act 1974

In 1970 the Robens Committee was set up to review the provision made for the safety and health of persons in the course of their employment. At that time safety requirements were contained in a variety of enactments (as the list of relevant statutory provisions in schedule 1 of HSW indicates). An estimated five million employees had no statutory protection. Protection was uneven. Administration was diverse and enforcement powers were considered inadequate. The wording and intent of the legislation were not directed towards personal involvement of the worker; and in parts it was obsolete.

HSW corrects many of these defects. General principles are enacted, to be supplemented by regulations. The provisions apply to employments generally to protect persons at work and those at risk from work activities.

The Act was intended to be wide to facilitate changing circumstances. Examples of development are the sanctions for non-compliance; and the use of the extensive powers to make regulations under s. 15 and Schedule 3.

Magistrates may now impose a fine up to £20 000 for breach of ss. 2–6 HSW or for a breach of an improvement or prohibition notice or a court remedy order. In addition, magistrates may imprison individuals for up to six months for breach of an improvement or prohibition notice or court remedy order[55].

Sections 2–6 were selected because they contain the main health and safety duties of those responsible for workplace safety. It was considered that a company charged with breach of one of these sections is probably responsible for a systematic failure to meet these general duties and is putting its employees and possibly others at risk. Failure to comply with a notice indicates a deliberate flouting of health and safety law.

The maximum magistrates' fine for other offences is £5000[56].

In the Crown Court there is no maximum fine. Up to two years imprisonment may be imposed for breach of a prohibition or improvement notice or remedy order or contravening a licence requirement. HSC Enforcement Policy[57] includes a statement that 'wherever appropriate' enforcing authorities should seek disqualification of directors under the Company Directors Disqualification Act 1986. Disqualification is possible on conviction for an indictable offence in connection with the *management* of a company[58]. In a health and safety context disqualification could follow conviction under s. 37 HSW because a company's offence was committed with a director's/manager's consent, connivance or attributable to his neglect (or under s. 36 if it is a manager whose default caused the offence of another).

After over 25 years and in a new century, there is critical analysis of the relevance and effectiveness of the 1974 Act to today's changing workplace[59]. The Government considers that the basic framework set up by the 1974 Act has stood the test of time, but that it is necessary to give a new impetus to health and safety at work. *Revitalising Health and Safety Strategy* (HSC 346) was published in 2000 with a 10-point strategy and a 44-point action plan, to set the direction for health and safety over the next 10 years with subsequent 'Revitalising Progress' (RHS) reports.

1.1.18.2 EU influence

The Single European Act 1986, with the objective of a single market by 1 January 1993, has had a dynamic effect on the introduction of health and safety legislation. The implementation of effective common health and safety standards is considered conducive to attaining a 'level playing field' for employers across the Community; and to the participation of the workforce in the intended resulting economic benefits.

Article 138 (formerly 118A) (introduced by the 1986 Act) provided that Member States shall 'pay particular attention to *encouraging improvements, especially in the working environment, as regards the health and safety of workers,* and shall set as their objective the harmonisation of conditions in this area, while maintaining the improvements made'.

A change in EU approach has been the use of Framework and related 'Daughter' Directives. The Framework Directive on the introduction of measures to encourage improvements in the safety and health of workers at work, with five daughter directives is an example[61,62]. The directive has been implemented in the UK as the Management of Health and Safety at Work Regulations 1992 (MHSW) now 1999. The core of these regulations is the duty to assess the risks to health and safety to employees and anyone who may be affected by the work activity, and to follow through with appropriate measures of planning, care and information.

Implementation has been possible under HSW. Section 1(2) provides for the progressive replacement of existing legislation by a system of regulations and approved codes of practice 'designed to **maintain or improve** the standards of health, safety or welfare established by or under those enactments'.

There are a number of further directives and draft directives relevant to health and safety. National consultation on EU proposals and draft directives concerned with health and safety will usually be co-ordinated by the Department of Transport, Local Government and the Regions, or, where appropriate, by other lead Departments such as the Department of Trade and Industry or the Home Office. The HSE and HSC co-operate with the Commission Directorates of the EU and their advisory committees and working groups and the European Agency for Health and Safety at Work and with other involved organisations. The HSE provides the HSC with policy, technological and professional advice, using expert information from the various HSE advisory committees which include a balance of employer and employee representatives from, for example, the CBI and TUC. Local Authorities are consulted through the HSE/Local Authority Enforcement Liaison Committee (HELA).

1.1.18.3 Standards of duty

In criminal and in civil actions the person alleging a breach has the burden of proof, i.e. must prove the wrongdoing. This burden is more easily discharged if an offence is 'absolute' which means that proof of the commission of the act is enough for liability. In criminal law the prosecution must normally prove guilty intent (*mens rea*) in addition to the guilty act (*actus reus*). If, exceptionally, guilty intent need not be proved, the crime is described as absolute. In that sense, the Health and Safety at Work Act (HSW) imposes absolute duties. This was emphasised in *R.* v. *British Steel plc*[63] where the Court of Appeal held that it was not necessary to find a company's 'directing mind' (its senior management) at fault in order to prove the company's liability.

Although corporate liability is absolute in the above sense, most of the general duties of HSW (and some of the duties of the regulations[64]) are qualified by the defence that steps must be 'reasonably practicable'. This has been interpreted to mean that the risk should be balanced against the 'cost' of the measures necessary to avert the risk (whether in money, time or trouble) to see if there is gross disproportion[65].

Other duties are qualified by 'practicable'. This is a stricter duty than reasonably practicable and has been interpreted to mean not as arduous as physically possible. A measure is practicable if it is possible in the light of current knowledge and invention[66].

In *Stark v Post Office, 2000*[67] the Court of Appeal ruled that regulation 6(1) of the *Provision and Use of Work Equipment Regulations 1992* (now r. 5(1) of PUWER 1998) imposes on an employer *absolute* liability for defective equipment.

The description 'strict' liability is sometimes used in the same sense as 'absolute' liability (to apply to criminal offences where there is no requirement of *mens rea*). However, 'absolute' and 'strict' are sometimes differentiated so that absolute is used in a narrow sense to mean that there is no defence if the act is proved, although there may be a defence in strict liability. Section 9 HSW, the duty not to charge an employee for things provided because of a specific statutory requirement, has been suggested as a rare example of 'absolute' in the narrow sense. In contrast, an employer's duty to undertake a suitable and sufficient risk assessment of his/her undertaking for employees and others is strict. However, the approved code of practice[68] suggests risk 'reflects both the likelihood that harm will occur and its severity'. That will affect whether the assessment is suitable and sufficient. In *Drage v Grassroots Ltd, 2000*[69] it was held that under regulations 12(1) and 12(2) of the *Workplace (Health, Safety and Welfare) Regulations 1992* every floor and traffic surface in a workplace . . . shall be . . . suitable for the purpose for which it is used, and this imposed **strict liability** on the employer. In contrast, Regulation 12(3) imposes a qualified duty. However, where there is water or oil or some other slippery substance on the floor, the employer only has to exercise such care as is reasonably practicable[70].

In civil law involving personal accidents (the law of tort) strict liability is unusual. A plaintiff must normally prove fault, in the form of negligent conduct of the defendant, which is assessed objectively.

Some apparently strict duties of EU health and safety directives have been transposed into UK legislation as being reasonably practicable. The HSE has explained that this is to avoid conflict of two absolute duties. For example Article 3 of the EU manual handling of loads directive requires the employer to use appropriate means to avoid manual handling and to take steps to control manual handling that does take place. European law is accustomed to deal with such conflicts with the doctrine of *proportionality,* that is balancing consequences to see whether an absolute ban is disproportionate to a goal which could be achieved by less restrictive means. The HSE issue guidance on the interpretation, in context, of reasonably practicable[71].

1.1.19 Principles developed by the courts

1.1.19.1 Case law interpretation

Case law interpretation has had an adverse effect on some safety legislation. A notorious example is the fencing requirements for dangerous machinery (then s. 14 FA), as illustrated by, for example, *Close* v. *Steel Company of Wales*[72]. With reluctance judges interpreted the statute so that s. 14 could not be used where parts of the machine or of the material being worked on have been ejected at a workman. This interpretation has now been remedied by reg. 12(3) of PUWER 1998 (repeating reg. 12(3) of 1992).

Such interpretations affect the scope of legislation, and of civil action for breach of statutory duty. Breach of statutory duty and the tort of negligence are the two most frequent grounds for civil claims following accidents at work. As identified in section 1.6, an employee's contract of employment is important for the duties owed by the employer.

1.1.19.2 Tort of negligence

Negligence is a relatively modern tort, but today it is probably the most important in number of cases and for the amount of damages which may be awarded for serious injury.

The tort consists of a breach by the defendant of a legal duty to take care not to damage the plaintiff or his property and consequent damage from that breach. From early times the common law has placed on the employer duties towards his employees. In 1932, Lord Atkin, in the leading case of *McAlister (or Donoghue) v. Stevenson*[73] suggested a general test for when a duty is owed. It is owed to persons whom one ought reasonably to have in mind as being affected by the particular behaviour. In 1963 the persuasive precedent of *Hedley Byrne* v. *Heller & Partners*[7] extended the duty to include financial loss resulting from some careless statements.

Since 1988[74] the potentially wide scope of the duty of care has been narrowed so that there are now four indicators: foresight of damage, proximity of the defendant to the plaintiff, policy and whether it is just and reasonable to impose a duty. A court will not necessarily refer to them all in the same case, but will look at the particular relationship. An important one is that of employer and employee. The duty of care owed to an employee is an implied term of the contract of employment (see section 1.1.19.4). In respect of premises, the common law duty of care owed by the occupier is now statutory (see section 1.1.19.5).

Examples of health concerns, developed in the civil tort of negligence and which are receiving increasing attention in the courts and by the HSE, are workplace stress[75]; repetitive strain injury (RSI)[76] and (WRULD); and vibration white finger (VWF)[77].

In addition to grounding a civil action, the statutory requirements in various regulations for employers to assess and to have a policy to deal with risks could now be relevant to other situations.

1.1.19.3 Tort of breach of statutory duty

When a statutory duty is broken there is liability for any penalty stipulated in the statute. In addition a person suffering damage from the breach may sometimes bring a civil action in tort to obtain compensation. Sometimes the Act specifies this (for example, the Consumer Protection Act 1987). Sometimes the Act is silent but the courts allow the action, as happened with FA and related regulations; or the Act is silent but the courts deny a civil action. This happened with the Food and Drugs Act 1955 (which has now been consolidated with other enactments relating to food into the Food Safety Act 1990) when it was decided that the statute was not intended to add to a buyer's civil remedies for breach of contract or of negligence.

Section 47 of HSW provides that breach of the Act will not give rise to a civil action, but breach of any regulation made under the Act is actionable, unless the regulations say otherwise. So far the only regulations to provide otherwise are the MHSW[78] and the *Fire Precautions (Workplace) Regulations 1997*[79] (FP(W)), but this restriction is likely to end. In December 2001 the HSC published proposals[80] to amend the MHSW 1999 and the FP(W) Regulations to allow employees to claim from their employer in a civil action where they suffer injury as a result of the employer breaching the legislation. The HSC explains that this proposal is consistent both with the commitment the UK has given the EC to provide employees with the rights of civil action against their employers, and with the scope of the EC Framework Directive, which is concerned with employers' responsibilities towards their employees.

Negligence and breach of statutory duty are two different torts, but both may be relevant following an incident. Bertha, injured at work because of an obstruction of the factory floor, might allege negligence plus breach of reg. 12 of the Workplace (Health, Safety and Welfare) Regulations 1992 (WHSW), and possibly succeed in both torts. She would not recover double damages because the remedy is compensation for the actual loss suffered.

1.1.19.4 The contract of employment

Implied terms of the contract of employment include the common law requirements that *employers* take reasonable care of the safety of employees and do not undermine the trust and confidence of the employee. The former duty has three connected requirements – the provision of competent fellow workers, safe premises, plant and equipment and a safe system of work. An employer cannot delegate this duty to another[81].

This implied contractual duty is the basis of the legal duty of care to an employee in the tort of negligence. The concept has extensive implications. For example, the Court of Appeal has said that a contract requiring long hours of work from a junior doctor is subject to the implied duty of care not to harm an employee[82]. In a successful constructive dismissal claim based on passive smoking[83], the Employment Appeal Tribunal (finding guidance from s. 2(2)(e) HSW) suggested that the implied contractual duty in any employment contract encompassed an implied term that the employer will provide and maintain, so far as is reasonably practicable, a working environment that is reasonably suitable for the performance of an employee's duties.

1.1.19.5 Duty to third parties on site

Third parties may be on premises with the occupier's *express* consent. Examples include customers, independent contractors and their employees, business associates or non-executive directors. Others such as an inspector or the postman may be on the premises with the occupier's *implied* consent. There may also be trespassers on the premises without express or implied permission, this category including those exceeding their invited purpose, such as customers entering the stock room, for instance.

The common law duty of care owed to visitors by an occupier in respect of premises is now statutory and was clarified in the *Occupier's Liability Act 1957* which ended the previous (often subtle) distinction between persons invited to enter (called *invitees*) and those allowed to enter (*licensees*), a distinction which previously affected the standard of duty. Under the 1957 Act, both categories are visitors to whom an occupier owes the 'common duty of care' once the relationship of occupier and visitor is established. The duty is to take such care as in all the circumstances of the case is reasonable to see that the visitor will be reasonably safe in using the premises for the purposes for which he is invited or permitted to be there. An example is *Cunningham* v. *Reading Football Club*[84]. Due to the football club's failure to maintain its terraces, football hooligans were able to use lumps of masonry as missiles. A policeman on duty at the club was injured and successfully sued that club.

The 1957 Act makes specific reference to visitors present in the exercise of their calling who may be expected to appreciate and guard against any special risks incidental to that calling, and to child visitors.

The *Occupier's Liability Act 1984* now applies to 'persons other than visitors'. As well as trespassers, this category also includes persons merely exercising a right of way across premises. The 1984 Act provides that there is a duty owed to uninvited entrants if the occupier has reasonable grounds to believe a danger exists on his premises and the consequent risk is one against which, in all the circumstances, he/she may reasonably be expected to provide some protection.

Aside from the duties as occupier, the tort of negligence continues to apply for whoever creates a source of danger. In the criminal context, HSW[85] and the relevant regulations also apply when a contractor is employed, for example.

1.1.19.6 Defences

There are two general defences to a civil action for the torts of negligence and breach of statutory duty. The defence that the negligent behaviour of the plaintiff contributed to the result allows the court to reduce a damage award proportionately. The defence of consent to the risk (*volenti non fit injuria*) negates liability. Consent is more than knowledge and this defence rarely succeeds against an employee, because employees may feel constrained in how they undertake tasks. Additionally, there may be specific defences to allegations of breach of statutory duty such as the defence of reasonable practicability.

Statute limits the time within which an action may be brought. For personal injuries the time limit is three years from the date of the breach or from the date of knowledge (if later) of the person injured. The plaintiff must prove every element of an allegation, including that the injury (physical or financial) was consequent on the breach. Thus 'no causation' may be a defence[86].

In criminal prosecutions, the absence of any element of an offence will provide a specific defence to a criminal charge. The time limit for a prosecution of a summary offence in a magistrates' court is six months from the date of the offence. (There is no time limit for Crown Court prosecutions). Statute may provide specific defences, for example HSW generally allows 'not reasonably practicable' as a defence. Some of the health and safety regulations (though not HSW) have a 'due diligence' defence, for example the Control of Substances Hazardous to Health Regulations 2002 (COSHH) provide that 'it shall be a defence for any person to prove that he took all reasonable precautions and exercised all due diligence to avoid committing an offence'.

The fact that an accident has occurred and resulted in legal action being taken is unsatisfactory. An award cannot repair an injury; the outcome of an action is uncertain; and the considerable cost and ingenuity expended in the investigation, developing the pleadings and the trial itself, could have been used more positively in trying to avoid such accidents. Such avoidance is an objective of HSW; and of the EC Directives, which are having increasing importance.

Because of the constraints of space, this chapter can be an outline only. Students are recommended to complement the chapter with further reading (see below) and visits to courts and tribunals.

References and endnotes

1. Introduced in the House of Commons on 18 December 2001 (Bill 75). The purpose of the Bill is to implement the recommendations of the Criminal Justice Review Group, set up in June 1998 under the Belfast Agreement. The Government's target is to devolve policing and justice after the Assembly elections scheduled for May 2003
2. Powers of the Criminal Courts Act 1991, The Stationery Office, London (1991), but see proposals for Auld Report (section 1.1.11)
3. Updated Workplace (Health, Safety and Welfare) Regulations are proposed for late 2002 (HSC paper HSC/01/93), but not to affect the application in the text
4. R. *v.* George Maxwell Ltd (1980) 2 All ER 99
5. 'Review of the Criminal Courts of England and Wales' (October 2001) by the Right Honorable Lord Justice Robin Auld
6. To be renamed as District Judge (Magistrate's Court) under proposals of Justice (Northern Ireland) Bill
7. Hedley Byrne & Co. Ltd *v.* Heller & Partners Ltd (1964) AC 463
8. *Current Law*: a monthly publication from Sweet and Maxwell
9. Practice Direction (Superior Court: Judgements: Form and Citation) 1 WLR 194
10. Operated by Butterworth.com
11. For example: Roberts Petroleum Ltd *v.* Bernard Kenny Ltd (1983) 1 All ER 564 HL
12. Lord Woolf is Lord Chief Justice
13. Waugh *v.* British Railways Board (1979) 2 All ER 1169
14. See s. 33(2) HSW for such offences
15. Criminal Justice and Public Order Act 1994. The court may 'draw such inferences as appear proper from a failure to mention facts relied on in his defence (s. 34), and if the accused does not give evidence or answer questions without good cause (s. 35)'
16. Criminal Evidence Act 1898 section 1.f
17. The Police and Criminal Evidence Act 1984
18. There are special rules about children, the accused and the accused's spouse
19. The Courts and Legal Services Act 1990
20. Access to Justice Act 1999; Conditional Fee Agreement Regulations 2000; Collective Conditional Fee Agreement Regulations 2000. The Stationery Office, London
21. Law Commission, Consultation Paper No. 135, *Manslaughter* (1994) and No. 237, *Legislating the Criminal Code; Involuntary Manslaughter* (1996)
22. Law Commission No. 247, *Aggravated, exemplary and restitutionary damages*, HC 346
23. Health and Safety Commission, *Revitalising Health and Safety Strategy*, HSE Books, Sudbury (2000)
24. Note proposed changes of draft Justice (Northern Ireland) Bill. op.cit. sections 1.1.5 and 1.1.14
25. Employment Tribunals Act 1996, The Stationery Office, London (1996)
26. Employmrnt Tribunals Extension of Jurisdiction (England and Wales) Order SI No. 1994/1623; for Scotland, SI No. 1994/1624. The Stationery Office, London 1994

27. Readmans Ltd *v.* Leeds City Council (1992) COD 419
28. The European Economic Community Treaty 1957 was renamed the European Community Treaty in 1991 by the Treaty of European Union (the Maastricht Treaty)
29. Ratified in the UK by the European Communities (Amendment) Act 2002
30. R. *v.* Secretary of State for Transport *v.* Factortame Ltd C 221/89; (1991) 1 AC 603; (1992) QB 680
31. Factortame Ltd No. 5, Times Law Reports, 28 April 1998
32. Liability for any damage caused to trawler owners and managers refused registration
33. Bulmer *v.* Bollinger (1974) 4 All ER 1226
34. 1998 Chapter 42
35. Human Rights Act 1998 (Commencement No. 2) Order 2000, The Stationery Office, London (2000)
36. AG *v.* Times Newspapers Ltd (1979) 2 EHRR 245, European Court of Human Rights
37. In R. v Francois Pierre Marcellin Thoron, CA (Criminal Division) 30 July 2001, as a ground of appeal, the appellant's counsel had suggested that sections 2 and 40 of the Health and Safety at Work Act 1974 created an offence of strict liability subject to a defence of non-practicability in respect of which the onus of proof runs on the defence. So interpreted, it was argued, the provisions are incompatible with article 6(1) and (2) of the European Convention on Human Rights by imposing a reversed burden of proof in what is otherwise an offence of strict liability without the need of any proof of any intention to commit the offence. However this ground of appeal was withdrawn before trial
38. The International Criminal Court Act 2001 and the International Criminal Court (Scotland) Act 2001 received the Royal Assent on 11 May and 28 September 2001 respectively
39. Op.cit. section 1.1.10.1
40. John Summers & Sons Ltd *v.* Frost (1955) AC 740
41. The Abrasive Wheels Regulations 1970 were revoked by the Provision and Use of Work Equipment Regulations 1998 and previously partially revoked by the Provision and Use of Work Equipment Regulations 1992 and the Workplace (Health, Safety and Welfare) Regulations 1992
42. McCarthys Ltd *v.* Smith (1979) 3 All ER 325
43. Pepper *v.* Hart (1992) NLJ Vol. 143 p. 17
44. Pickstone *v.* Freeman plc (1989) 1 AC 66
45. The European Community – the single market – is one of the three pillars of the European Union. The other two pillars are common foreign and security policy and Justice and home affairs
46. op.cit. section 1.1.13.1
47. op.cit. section 1.1.13.1
48. In exceptional cases, the initiative is shared with Member States or the Council can act on its own initiative
49. The co-decision process applies whenever the Treaty refers to Article 251 EC (formerly 189b) for implementation procedure; an example is Article 137 of the Treaty
50. Qualified majority decision requires 62 votes out of 87 (71%). Member State's votes are weighted on the basis of their population and corrected in favour of the less-populated countries
51. European Agency for Safety and Health at Work authorised by European Council Regulation No. 1643/95
52. Van Duyn *v.* Home Office (Case 41/74) (1975) 3 All ER 190
53. Marshall *v.* Southampton and South West Hampshire Area Health Authority (Teaching) (1986) case 152/84 1 CMLR 688; (1986) QB 401
54. Rolls Royce plc *v.* Doughty (1992) ICR 538
55. by s. 4 of the Offshore Safety Act 1992
56. Effective since October 1992 by the Criminal Justices Act 1991
57. HSC Enforcement Policy Statement, January 2001
58. The period for disqualification is 2 years minimum and 5 years maximum by a lower court and 15 years maximum by a higher court
59. Including suggestions that it is time for a substantial review and possible replacement of the present occupational health and safety legislative framework, as, for example, in *Regulating Health and Safety at Work: The Way Forward*, edited by Phil James and Davis Walters, published by the Institute of Employment Rights, December 1999

60. See now Article 137 (formerly 118)
61. EC Directive No. 89/391/EEC, adopted 12.6.89 with five daughter directives
62. Consolidating amendments of 1994 and 1997.
63. R. *v.* British Steel plc (1995) ICR 587. This was a prosecution under s. 3 HSW following the death of two employees of a subcontractor employed by British Steel to reposition a steel platform. The contractor's procedure was inherently dangerous but the contract provided for the supervision of the work by a British Steel employee
64. For example, the Manual Handling Operations Regulations 1992
65. Edwards *v.* National Coal Board (1949) 1 All ER 743
66. Adsett *v.* K & L Steelfounders and Engineers Ltd (1953) 1 All ER 97; 2 All ER 320
67. Stark *v.* The Post Office [2000] ICR 1013. A successful damage claim by a postman who was seriously injured when the front wheel of the bicycle locked and he was propelled over the handlebars. Applied in Green *v.* Yorkshire Traction Co Ltd [2002] EWCA Civ 1925
68. Health and Safety Executive, Legal series publication no: L 21, *Management of health and safety at work. Management of Health and Safety at Work Regulations 1999. Approved Code of Practice*, HSE Books, Sudbury (2000)
69. Drage *v.* Grassroots Ltd, Watford County Court (2000). Current Law Yearbook, 2967
70. See also section 1.1.7
71. To help duty holders reduce risk to as low as is reasonably practicable (ALARP) and to ensure health and safety as far as is reasonably practicable (SFAIRP)
72. Close *v.* Steel Company of Wales (1962) AC 367
73. Donoghue *v.* Stevenson (1932) AC 562
74. Murphy *v.* Brentwood District Council (1991) AC 398
75. **Stress**. Sutherland *v.* Hatton; Baker Refactories Ltd *v.* Bishop; Sandwell MBC *v.* Jones; Somerset CC *v.* Barber (Court of Appeal) [2002] EWCA Civ 76; Walker *v.* Northumberland CC [1995[1 All ER 737
76. **RSI**. Alexander and others *v.* Midland Bank plc (Court of Appeal) [2000] ICR 464; Pickford *v.* ICI plc (House of Lords) (1998) 1 WLR 1189
77. **Vibration white finger**. Smith *v.* Wright & Beyer Ltd (Court of Appeal) [2001] EWCA Civ 1069; Allen *v.* British Rail Engineering Ltd (Court of Appeal) [2001] PIQR Q10
78. Regulation 22 of the Management of Health and Safety at Work Regulations 1999
79. Regulation 17 of the Fire Precautions (Workplace) Regulations 1997 as it amended section 27A of the Fire Precautions Act 1971
80. CD177 C40 12(01)
81. Wilsons & Clyde Coal Co. Ltd *v.* English (1938) AC 57, HL
82. Johnstone *v.* Bloomsbury Health Authority (1992) QB 333
83. Waltons & Morse *v.* Dorrington (1997) IRLR 488
84. Cunningham *v.* Reading Football Club (1991) *The Independent*, 20 March 1991
85. ss. 3 and 4 HSW, for example
86. Corn *v.* Wier's Glass Ltd (1960) 2 All ER 300

Further reading

Atiyah, P.S. and Cane, P., *Accidents, Compensation and the Law*, 6th edn, Butterworths, London (1999)

Barrett, B. and Howells, R., *Occupational Health and Safety Law Cases and Materials*, 2nd edn. Cavendish (2000)

Clinch, P., *Using a law library*, 2nd edn. Blackstone (2001)

Dickson, B., *The Legal System of Northern Ireland*, SLS Legal Publications (NI), Belfast (2001)

Encyclopaedia of Health and Safety at Work, Sweet and Maxwell, London (loose-leaf)

Hutchins, E.L. and Harrison, A., *History of Factory Legislation*, F. Cass, London (1996)

Keenan, D., *Smith and Keenan's English Law*, 13th edn. Pearson Education (2001)

Marshall, E., *General Principles of Scots Law*, 7th edn. W. Green (1999)

Selwyn, N., *Law of Health and Safety at Work*, Croner (2000)

Selwyn, N., *The Law of Employment*, 11th edn. Butterworths, London (2000)

Smith, Bailey and Gunn on *The Modern English Legal System*, Sweet and Maxwell, (2002)

Stranks, J., *Manager's Guide to Health and Safety at Work*, 6th edn. Kogan Page (2001)

Tolley's Health and Safety at Work Handbook, Tolley (2002)

Ward, R., *Walker and Walker's English Legal System*, 8th edn. Butterworth (1998)
Walker, R.J., *The Scottish Legal System*, 8th edn, W. Green (2001)

Law dictionaries

Curzon, *A Dictionary of Law* 5th edn, Pitman, London (1998)
Mozley and Whiteley's Law Dictionary, 12th edn, Butterworths, London (2001)
A Dictionary of Law, edited by E. Martin, 5th edn., Oxford University Press

Chapter 1.2

Principal health and safety Acts

S. Simpson

UK health and safety legislation consists of a number of main or principal Statutes or Acts which are supported by a great deal of subordinate legislation in the form of Regulations and Orders. This chapter deals with the more commonly applied main Acts that are concerned with protecting the health and safety of the working population and those who may be put at risk from the manner in which the work is carried out.

1.2.1 The Health and Safety at Work etc. Act 1974

1.2.1.1 Pre-1974 legislation

For more than a century health and safety legislation for persons at work in the UK had developed a piece at a time, each piece covering a particular class of person and not in a consistent manner each time. Separate legislation with variations in details and in the methods of enforcement would apply to a process or requirement when undertaken in a factory, as opposed to an office, a mine or a quarry. For example, an air receiver situated in a factory would be required to be examined for safety reasons by a competent person at least once every 26 months, but the same receiver moved to a shop would not require examination nor would the same receiver need to be inspected in the factory if, instead of air, another gas at the same or even higher working pressure was substituted.

In the main, the principal Act affecting the particular groups of persons, usually on the basis of the kind of premises in which they worked, was supplemented by regulations. The Act and its regulations would be enforced by a particular inspectorate (e.g. by factory inspectors for factories and notional factories such as construction sites, mines inspectors for mines and quarries and local authority inspectors for offices and shops). Any breach of the appropriate legislation could lead to a prosecution by an inspector which in turn could lead to a fine usually imposed on the company or other organisation rather than an individual.

The major responsibility for observing the requirements of the legislation was that of the employer with some responsibilities falling on the occupier, if he was not the employer, and on the employees. Only in mining legislation was there also a criminal liability placed on managers and other officials. On the whole, legislation tended to look to the protection of plant and equipment as a way of preventing injuries to workers. Visitors, contractors, neighbours and other third parties were mainly ignored in the drafting of these earlier Acts and regulations, as were many employees who did not work on premises (e.g. roadsweepers) or worked in premises not covered (e.g. schools, research establishments, hospitals, etc.).

By 1970 many organisations, especially the trade unions, were questioning whether the existing legislation was either sufficient or effective in providing proper protection for work people.

The effect that workers' organisations could have on workshop safety was limited and large sections of the working population were not covered.

A Private Member's Bill aimed at providing for the compulsory involvement of workers in accident prevention was withdrawn when in 1970 a committee was set up under the chairmanship of Lord Robens to look at safety and health at work. After studying the whole problem in depth the committee reported in 1972[1] making many recommendations of a wide ranging nature.

The essence of the 'Robens Report' recommendations was to:

1 Replace the mass of existing safety legislation with one Act applying generally to all persons at work.
2 Replace the mass of detail with a few simple and easily assimilated precepts of general application.
3 Change methods of enforcement so that prosecution is not always the first resort.
4 Ensure that occupational safety should also protect visitors and the public.
5 Place more emphasis on safe systems of work rather than technical standards.
6 Actively involve the workers in the procedures for accident prevention at their place of work.

In spite of changes of Governments, the main recommendations of the Robens Committee were accepted by Parliament and were incorporated in the Health and Safety at Work etc. Act 1974 (HSW).

1.2.1.2 The Health and Safety at Work etc. Act 1974

Drafted as an enabling Act, it permitted the Secretary of State or other Ministers to make regulations with a view to replacing the existing piecemeal legislation, typified by those Acts listed in schedule 1 of HSW, by regulations and codes of practice requiring improved standards of safety, health and welfare. It established a co-ordinating enforcement authority, the Health and Safety Commission (HSC), giving its inspectors greater powers than hitherto. It also extended legislative protection for health and safety to everyone who was employed, whether paid or not (except domestic servants), and imposed more general but wider duties on both employer and employee.

The Act makes provision for protecting others against risks to health and safety from the way in which work activities are carried out. It also seeks to control certain emissions into the atmosphere, as did the Control of Pollution Act 1974, and to control the storage and use of dangerous substances. In addition, the Act ensures the continuation of the Employment Medical Advisory Service.

Although mostly superseded there is still a need to comply with the requirements of parts of the pre-1974 legislation which remain in effect but which apply only to those work activities covered previously.

1.2.1.3 General duties on employers and others

These duties are outlined in ss. 2–5 where the obligations are qualified by the phrases 'so far as is reasonably practicable' and 'best practicable means'. Interpretations of these phrases have been made[2] which indicate that 'reasonably practicable' implies a balance of the degree of risk against the inconvenience and cost of overcoming it, whereas 'best practicable means' ignores the cost element but recognises possible limitations of current technical knowledge.

In common law, employers have had, and still have, duties of care with regard to the health and safety of their employees, duties which are now incorporated into statute law as part of s. 2 of this Act.

The first part of s. 2 contains a general statement of the duties of employers to their employees while at work and is qualified in subsection (2) which instances particular obligations to:

1 Provide and maintain plant and systems of work that are safe and without risks to health. Plant covers any machinery, equipment or appliances including portable power tools and hand tools.
2 Ensure that the use, handling, storage and transport of articles and substances is safe and without risk.
3 Provide such information, instruction, training and supervision to ensure that employees can carry out their jobs safely.
4 Ensure that any workshop under his control is safe and healthy and that proper means of access and egress are maintained, particularly in respect of high standards of housekeeping, cleanliness, disposal of rubbish and the stacking of goods in the proper place.
5 Keep the workplace environment safe and healthy so that the atmosphere is such as not to give rise to poisoning, gassing or the encouragement of the development of diseases. Adequate welfare facilities should be provided.

In this section 'work' means any activities undertaken as part of employment and includes extra voluntary jobs for which payment is received or which are accepted as part of the particular job, i.e. part-time firemen, collecting wages etc.

Further duties are placed on the employer by:

s. 2(3) To prepare and keep up to date a written safety policy supported by information on 'the organisation and arrangements for carrying out the policy. The safety policy has to be brought to the notice of employees. Where there are five or less employees this section does not apply.

s. 2(6) To consult with any safety representatives appointed by recognised trade unions to enlist their co-operation in establishing and maintaining high standards of safety.

s. 2(7) To establish a safety committee if requested by two or more safety representatives.

The general duties of employers and self-employed persons include, in s. 3, a requirement to conduct their undertakings in such a way that persons other than their employees are not exposed to risks to their health and safety. In certain cases information may have to be given as to what these risks are.

Landlords or owners are required by s. 4 to ensure that means of access or egress are safe for those using their premises and these are defined in s. 53 as any place and, in particular, any vehicle, vessel, aircraft or hovercraft, any installation on land, any offshore installation and any tent or movable structure. However, safety in workplaces, on vehicles etc. and on offshore installations are being overtaken by EU directives.

Those in charge of premises are required by s. 5 to use the best practicable means for preventing noxious or offensive fumes or dusts from being exhausted into the atmosphere, or that such exhausts are harmless. Offensive is not defined and may depend upon an individual's opinion.

Duties are placed by s. 6 on everyone in the supply chain, from the designer to the final installer, of articles of plant or equipment for use at work or any article of fairground equipment to:

1 ensure that the article will be safe and without risks to health at all times when it is being set, used, cleaned or maintained,
2 carry out any necessary testing and examination to ensure that it will be safe, and
3 provide adequate information about its safe setting, use, cleaning, maintenance, dismantling and disposal.

These duties are further extended in detail for machinery[3], electrical and electronic apparatus[4], gas appliances[5], lifts[6] and other equipment that is required to carry the CE mark before it can be put on the EU market.

There is obligation on designers or manufacturers to do any research necessary to prove safety in use. Erectors or installers have special responsibilities to make sure that when handed over the plant or equipment is safe to use. Obligations on designers are reinforced in regulations covering construction[7], offshore installations[8] etc.

Similar duties are placed on manufacturers and suppliers of substances for use at work to ensure that the substance is safe when properly used, handled, processed, stored or transported, to provide adequate information and do any necessary research, testing or examining. There are regulations detailing how substances and preparations should be classified, packaged and labelled with, in addition, the need for safety data sheets to be provided[9].

Where articles or substances are imported, the suppliers' obligations outlined above attach to the importer, whether a separate importing business or the user himself.

Often items are obtained through hire-purchase, leasing[10] or other financing arrangements with the ownership of the item being vested with the financing organisation. Where the financing organisation's only function is to provide the money to pay for the goods, the suppliers' obligations do not attach to them.

The employees' duties are laid down in s. 7 which state that, whilst at work, every employee must take care for the health and safety of himself and of other persons who may be affected by his acts or omissions. Also employees should co-operate with the employer to meet legal obligations. Section 8 requires that no one, whether employee or not, shall either intentionally or recklessly, interfere with or misuse anything, whether plant equipment or methods of work, provided by the employer to meet obligations under this or any other related Act.

The employer is not allowed by s. 9 to charge any employee for anything done or provided to meet statutory requirements.

1.2.1.4 Administration of the Act

The Act through s. 10 caused the establishment of two bodies to direct and enforce legislative matters concerned with health and safety. The Health and Safety Commission (HSC), appointed by the Secretary of State, consists of a chairman and six to nine members. Three of the members are appointed after consultation with the employers' organisations, three after consultation with employees' organisations and two after consulting local authorities.

It is the duty of the Commission (s. 11) to:

1 assist and encourage persons in furthering safety,
2 arrange for the carrying out of research and to encourage research and the provision of training and information by others,
3 provide an information and advisory service,
4 submit proposals for regulations, and
5 report to and act on directions given to it by the Secretary of State.

It also liaises with local authority and fire authority organisations to whom it has delegated[11,12] (s. 18) some of its duties.

Whereas the Commission has the function of formulating policies, the Health and Safety Executive (HSE) is responsible for their implementation. The Executive which is appointed by the Commission and consists of three persons, one of whom is the director, has a duty to exercise on behalf of the Commission such functions as the Commission directs. If so requested by a Minister, the Executive shall provide him with information of the activities of the Executive on any matter in which he is concerned and to provide him with advice.

The Commission may direct the Executive or authorise any other person to investigate or make a special report on any accident, occurrence, situation or other matter for a general purpose or with a view to making regulations.

The duties of the Commission and the Executive are contained in ss. 11–14 of HSW.

1.2.1.5 Regulations and Codes of Practice

The enabling powers of this Act are exercised through s. 15 whereby the appropriate Secretary of State or Minister may without referring the matter to Parliament require regulations to be drawn up by the Executive and submitted through the Commission to him. Such regulations may need to be submitted to Parliament for ratification. Although there is a general requirement for the Commission and Executive to keep interested parties 'informed of and adequately advised on, such matters' (s. 11(2)c) there is no obligation to consult. However, in drafting regulations that affect workplace safety, extensive consultation does occur.

The regulations may repeal or modify any of the existing regulations and matters related to ss. 2–9 of the Act. They can also approve or refer to specified documents, such as British Standard Specifications. A list of 22 subject matters that can be covered by regulations is given in schedule 3 of the Act.

The need to provide guidance on the regulations is recognised in s. 16 which gives the Commission power to prepare and approve Codes of Practice on matters contained not only in the regulations but also in ss. 2–7 of the Act.

To implement the EU framework and its daughter directives, a 'six-pack' of regulations was introduced in 1992, some of which have since been updated and replaced, covering management[13], work equipment[14], display screens[15], manual handling[16], personal protective equipment[17] and health, safety and welfare[18]. The 'management' regulations extend HSW by requiring employers to:

- carry out risk assessments
- (apply the principles of prevention
- have arrangements for the planning and control of protective and preventive measures
- appoint competent persons to give health and safety assistance
- have procedures to cope with serious and imminent danger
- give information to employees
- co-operate and co-ordinate with other employers sharing the same premises
- take into account the employee's capabilities and training when entrusting tasks

- protect both young workers and pregnant workers
- give special consideration to workers who have recently given birth
- provide information to temporary workers.

Through the Fire Precautions (Workplace) Regulations 1997[19] these regulations encompass fire safety.

Before approving a code, the Executive acting for the Commission must consult with any interested body. The Commission have powers to approve codes prepared by bodies other than themselves, and some British and harmonised Standards have been approved.

An Approved Code is a quasi-legal document and although non-compliance with it does not constitute a breach, if the contravention of the Act or a regulation is alleged, the fact that the code was not followed will be accepted in court as evidence of failure to do all that was reasonably practicable. A defence would be to prove that something equally as good or better had been done (s. 17(2)). To supplement the Approved Codes of Practice, the Executive issue guidance notes which are purely advisory and have no standing in law.

1.2.1.6 Enforcement

1.2.1.6.1 General

The enforcement of the Act (s. 18), with some exceptions in respect of noxious and offensive emissions[20] (s. 5), is the responsibility of the HSE through its constituent inspectorates with certain premises delegated to local authorities[11] and for certain fire matters to the Fire Authority[12].

Actual enforcement is carried out by inspectors (s. 19) who should have suitable qualifications and be authorised by a written warrant outlining the powers they may exercise. An inspector must produce his warrant on request; without it he has no powers of enforcement.

1.2.1.6.2 Powers of inspectors

By virtue of his warrant an inspector has the powers outlined in s. 20 which relate only to the field of the inspectorate authorising him and include:

1. The right to enter premises and if resisted to enlist the support of a police officer.
2. To inspect the premises.
3. To require, following an incident, that plant is not disturbed.
4. Taking measurements and photographs although in the latter case it is usual to obtain permission first.
5. Taking samples of suspect substances.
6. Require tests to be carried out on suspect plant or substances.
7. Requiring the dismantling of plant.
8. Require those with possible knowledge relevant to his investigation to give it either verbally or in a written statement. The inspector has discretion to allow another to be present during questioning and the taking of a written statement.

9 The right to inspect and take copies of books or documents required to be kept by safety or other legislation if it is necessary for him to see them as part of his investigation but he has no right to examine documents for which legal privilege is claimed.
10 Requiring assistance within a person's limits of responsibilities.

Where an inspector takes samples of substances he must leave a similar identified sample with a responsible person or leave a conspicuous notice stating that he has taken a sample.

Information contained in an answer to an inspector cannot be used in criminal proceedings against the giver.

A customs officer may seize any imported article or imported substance and detain it for not more than two working days on behalf of an inspector (s. 25A).

Where an employer suffers damage to property or business, as a result of actions of an inspector, the inspector can be sued personally for recompense against which he may be indemnified by the enforcing authority.

After an inspector has completed his investigation he has a duty to inform representatives of the workpeople of actual matters he has found (s. 28(8)) and must give the employer similar information.

1.2.1.6.3 Notices

If an inspector is of the opinion that a breach has, or is likely to, occur he may serve an Improvement Notice (s. 21) on the employer or workman. The notice must state which statutory provision the inspector believes has been contravened and the reason for his belief. It should also state a time limit in which the matter should be put right.

However, if the activity involves immediate risk of serious personal injury, the inspector may serve a Prohibition Notice (s. 22) requiring immediate cessation of the activity. This notice must state what, in the inspector's opinion, is the cause of the risk and any possible contravention. If the risk is great but not immediate a deferred Prohibition Notice may be served stating a date after which the activity must cease unless the matter has been put right. Where corrective work cannot be completed in time, the inspector may extend the period of the notice. There is no procedure for certifying that a notice has been complied with.

Appeals against a notice may be made to an Industrial Tribunal[21]. On entering an appeal an Improvement Notice is suspended until the appeal is disposed of or withdrawn, whereas a Prohibition Notice continues in effect unless the Tribunal directs otherwise.

1.2.1.7 Offences

Offences listed in s.33 include:

1 failing to discharge a duty imposed by ss. 2–7,
2 contravening ss. 8 and 9, any regulation or notice,
3 making false entries in a register,

4 obstructing or pretending to be an inspector, and
5 making false statements etc.

If an inspector decides to institute legal proceedings, he must do so within six months of learning of the alleged contravention (s. 34(3)). Cases can be heard either summarily which attracts a fine not exceeding level 5 on the standard scale on conviction, or on indictment where the penalty can be imprisonment and/or an unlimited fine. Offences concerned with interfering with the powers or work of an inspector (s. 33(1)d,f,h and n) are to be dealt with summarily but for all the other offences listed in s. 33(1) plus in certain circumstances contravention of a requirement imposed by an inspector in the exercise of his powers (s. 33(1)e) the case can be tried either summarily or, if the offence is serious enough and the parties agree, on indictment, when the penalty on conviction can be an unlimited fine.

Responsibility for an offence usually attaches to the employer but may attach to an employee (ss. 7–8). However, where the contravention was caused with the consent or knowledge or be due to the neglect of a director, manager, company secretary or other officer (s. 37) then he too can be prosecuted.

In proceedings alleging a failure to use reasonably practicable or best practicable means the prosecution only has to state the suspicion and it is up to the accused to prove that what was done was as good as, if not better than, the duty required (s. 40).

Penalties were increased by the Offshore Safety Act 1992 so that failing to discharge a duty under ss. 2–6 attracts a liability on summary conviction to a fine not exceeding £20 000 and on conviction on indictment to an unlimited fine. For specified offences, a person (such as a director, manager etc.) found guilty of the offence shall be liable on summary conviction to imprisonment, for a term not exceeding six months or a fine not exceeding £20 000 but for conviction on indictment, to imprisonment for a term not exceeding two years or a fine or both. Fines for other offences are set at level 5 (at present, through the Criminal Justices Act 1991, this is a sum not exceeding £5000).

1.2.1.8 Extensions

Part 1 of the Act has been extended to include:

1 the protection of the public from danger associated with the transmission and distribution of gas through pipelines,
2 securing the health, safety and welfare of persons on offshore installations engaged in pipeline works,
3 securing the safety of such installations and preventing accidents on or near them,
4 securing the proper construction and safe operation of pipelines and preventing damage to them,
5 securing the safe dismantling, removal and disposal of offshore installations or pipelines, and
6 the police.

1.2.1.9 Parts II to IV and Schedules

Part II of the Act allows for the continuation of the Employment Medical Advisory Service, defines the purpose and responsibilities of the service, allows for fees to be charged, for payments to be made and for the keeping of accounts.

Part III, except for s. 75, has been repealed by the Building Act 1984.

Part IV is a miscellaneous and general part amending the Radiological Protection Act 1970, Fire Precautions Act 1971, Companies Act 1967 and stating such matters as the extent and application of the HSW Act.

The following schedules of the Act cover:

1 Relevant existing enactments.
2 The constitution etc. of the Commission and Executive.
3 Subject matter of health and safety regulations.
4–7 Repealed.
8 Transitional provisions with respect to Fire Certificates.
9 Repealed.
10 List of repealed Acts.

1.2.1.10 Definitions

Sections 52 and 53 contain a number of definitions aimed at clarifying part I of the Act:

> 'Work' means an activity a person is engaged in whether as an employee or as a self-employed person. An employee is considered to be at work all the time he is following his employment whether paid or not and a self-employed person is at work throughout such time as he devotes to work as a self-employed person. Regulations can extend the meaning of 'work' and 'at work' to other situations such as to the storage and use of genetically modified organisms and biological agents and to YTS training[22].

Other definitions include:

> 'Article for use at work' includes any plant designed for use at work and any article designed for use as a component in such plant.
> 'Code of practice' includes a standard, a specification or any other documentary form of practical guidance.
> 'Domestic premises' means premises occupied as a private dwelling (including gardens, yards, garages etc.).
> 'Employee' means an individual who works under a contract of employment.
> 'Personal injury' includes any disease or any impairment of a person's physical or mental condition.
> 'Plant' includes any machinery, equipment or appliance used at work.

'Premises' include any place, vehicle, vessel, aircraft, hovercraft, installation on land, offshore installation, installation resting on the sea bed or other land covered by water and any tent or movable structure within territorial waters. This definition has been extended by the Health and Safety at Work etc. Act 1974 (Application outside Great Britain) Order 1995 to include offshore installations, wells and pipelines, mines under the sea etc.
'Self-employed person' is an individual who works for gain or reward otherwise than under a contract of employment, whether or not he employs others.
'Substance' means any natural or artificial substance whether solid, liquid, gas or a vapour and includes micro-organisms.

1.2.2 The Factories Act 1961

The Factories Act 1961 was in the main a consolidating Act, bringing together earlier Factories Acts. Very few of the major provisions with regard to health, safety and welfare continue in force.

However, those sections that do remain in effect refer to particular safety requirements but apply only to factories as defined in the Act.

1.2.3 The Fire Precautions Act 1971

The Act furthers the provisions for the protection of persons from fire risks. If any premises are put to use and are designated, a certificate is required from the fire authority. Although classes of use cover the provisions of sleeping accommodation; use as an institution; use for the purposes of entertainment, recreation, instruction, teaching, training or research; use involving access to the premises by members of the public and use as a place of work, so far only the provision of sleeping accommodation and use as a place of work have been designated.

Houses occupied as single private dwellings are exempt, but the fire authority have powers to make it compulsory for some dwellings to be covered by a fire certificate.

Applications for fire certificates must be made on the prescribed form and the fire authority must be satisfied that the means of escape in case of fire, means of fire fighting and means of giving persons in the premises warning in case of fire are all adequate. Every fire certificate issued shall specify particular use or uses of the premises, its means of escape, details of the means of fire fighting, and of fire warning and, in the case of factories, particulars of any explosive or highly flammable materials which may be stored or used on the premises. The certificate may impose such restrictions as the fire authority considers appropriate and may cover the instruction or training of persons in what to do in case of fire or it may limit the number of persons who may be in the premises at any one time. In certain circumstances the fire authority may grant exemption from the requirements to have a fire certificate, otherwise a copy of the fire certificate is sent to the occupier and it must be kept on the premises. The owner of the building is also sent a copy of the certificate.

It is an offence not to have or to have applied for a fire certificate for any designated premises. Contravention of any requirement imposed in a fire certificate is also an offence. A person guilty of an offence (with some exceptions) shall be liable on summary conviction to a fine not exceeding level 5 on the standard scale and on conviction on indictment a fine or imprisonment or both.

So long as a certificate is in force, the fire authority may inspect the premises to ascertain whether there has been a change in conditions. Any proposed structural alterations or extensions to the premises, major changes in the layout of furniture or equipment or, in factories, to begin to use or store or increase the extent of explosive or flammable material shall, before the proposals are begun, be notified to the fire authority.

It is also necessary while the certificate is in force, or an exemption has been granted under s. 5A, for the occupier to give notice of any proposed material extension or alterations to the premises or its internal arrangement and, in the case of factories, to store or use or to materially increase the amount of explosive or highly flammable materials. Within two months of receiving notice, the fire authority must, if they regard the requirements of the relevant fire certificate as becoming inadequate, inform the occupier, or owner, and give such directions as they consider appropriate. If the directions are duly taken the fire authority will amend the certificate or issue a new one. Not giving suitable notice or contravening a direction are offences that on conviction could lead to a fine or imprisonment, or both. The rights of appeal are detailed in s. 9.

The coming into effect of the Fire Safety and Safety of Places of Sport Act 1987 amended but did not replace the FPA and gave the Fire Authority much wider powers. These include the power to charge a reasonable fee for the initial issue, or the amendment or the issue of a new fire certificate (s. 8B). Even though premises may be exempt from the requirements for a fire certificate there are duties to provide both means of escape and means of fighting fire (s. 9A). In order to assist occupiers to meet these duties the Secretary of State may issue Approved Codes of Practice and the fire authority may serve Improvement Notices if they think a code is not being met (ss. 9A–9F).

Should the fire authority be of the opinion that, in the event of fire, the use of premises involves or will involve so serious a risk to persons on the premises that continuing use ought to be prohibited or restricted, the authority may serve a Prohibition Notice on the occupier. There are rights of appeal against these notices (ss. 10–10B).

The Secretary of State has powers under the Act to make regulations about fire precautions in designated premises other than those in which manufacturing processes are carried on (s. 12). Requirements have been further extended by the Fire Precautions (Workplace) Regulations 1997 which apply particularly to premises for which a fire certificate is not required. Amendments to the 1997 Regulations require the carrying out of risk assessments of the fire hazards.

The Act deals with matters pertaining to building regulations (s. 13), the duties of consultation between local authorities (ss. 15 et seq.), fire authorities and other authorities such as the HSE, the enforcement of the Act (s. 18), the powers of inspectors to enter premises (s. 19), offences, penalties and legal proceedings (ss. 22–27) and the amendment of other Acts (ss. 29 et seq.).

Schedule 1 has the effect of making special provisions for factory, office, railway or shop premises, that do not form part of a mine, in relation to leasing, part ownership, the issue of licences under the Explosives Act 1875 and the Petroleum (Consolidation) Act 1928. It also has an effect on the proposed or actual storage or use of explosives or highly flammable material in factory premises.

1.2.4 The Mines and Quarries Acts 1954–71

The main Acts laying down the general safety duties of mines and quarries personnel (i.e. owners, managers, undermanagers, surveyors and officials) were the Mines and Quarries Act 1954, the Mines and Quarries (Tips) Act 1969 and the Mines Management Act 1971. The latter Act in particular and the mines sections of the 1954 Act were revoked by the Management and Administration of Safety and Health in Mines Regulations 1993.

Parts of the 1954 Act dealing with quarries have also been replaced by the Quarries Regulations 1999[23].

1.2.5 The Environmental Protection Act 1990

To prevent the pollution from emissions to air, land or water from scheduled processes the concept of Integrated Pollution Control has been introduced. Authorisation to operate the relevant processes must be obtained from the enforcing authority which, for the more heavily polluting industries, is the Environment Agency. Control for pollution to air from the less heavily polluting processes is through the local authority.

Regulations also place a 'duty of care' on all those involved in the management of waste, be it collecting, disposing of or treating *Controlled Waste* which is subject to licensing. Sections of the 1990 Act have been superseded by the Pollution Prevention and Control Act 1999. More detailed requirements of the 1999 Act are contained in the subsequent *Pollution Prevention and Control Regulations* for England and Wales and for Scotland.

In addition to extending the Clean Air Acts by including new measures to control nuisances, the Regulations introduce litter control; amend the Radioactive Substances Act 1960; regulate genetically modified organisms; regulate the import and export of waste; regulate the supply, storage and use of polluting substances and allow the setting up of contaminated land registers by the local authority. In 1991 the Water Act 1989 which controlled the pollution and supply of water was replaced by five separate Acts (see section 5.4.3).

The Environment Agency was set up by the Environment Act 1995 which also makes provision for contaminated lands, abandoned mines, control of pollution and the conservation of natural resources and the environment.

1.2.6 The road traffic Acts 1972–91

The road traffic Acts, including the Road Traffic Regulation Act 1984, together with the Motor Vehicle (Construction and Use) Regulations 1986, Road Vehicle Lighting Regulations 1981, Goods Vehicles (Plating and Testing) Regulations 1988, the Motor Vehicle (Tests) Regulations 1982 and numerous other regulations, form comprehensive safety legislation not only of the occupants of the vehicles but also for members of the general public who may be affected by the driving and parking of vehicles.

In the construction of vehicles, safety features include the provision of suitable braking systems; burst-proof door latches and hinges; material for fuel tanks; types of lamps and reflectors; the fitting of audible warnings, mirrors, safety glass windscreens, seat belts; acceptable tyres, the driver's view of the road and the lighting of vehicles. Noise and smoke emissions are also topics related to safety and covered by the legislation.

When loading a vehicle care must be taken to ensure that the load is evenly distributed to conform to the vehicle's individual axle weight and where necessary the driver must make suitable corrections on multi-delivery work to ensure that no axle becomes overloaded due to transfer of weight. Since it is an offence to have an insecure load, all loads must be securely fixed and roped and, if necessary, sheeted. Restrictions are placed on projecting loads, extra long or extra wide loads and abnormal indivisible loads. The carriage of dangerous goods, be they toxic, flammable, radioactive or corrosive, is covered by regulations made under other Acts (e.g. Petroleum (Consolidation) Act, HSW).

The road traffic Acts also deal with offences connected with the driving of motor vehicles and of traffic generally, accidents, road safety, licensing of drivers, driving instruction, restrictions on the use of motor vehicles, periodic testing of vehicles to ensure that they are roadworthy etc.

1.2.7 The Public Health Act 1936

This is another consolidating Act and in Part III statutory nuisances and offensive trades are dealt with.

Statutory nuisances are any premises in such a state as to be prejudicial to health or a nuisance, likewise the keeping of any animal, allowing any accumulation or deposit and causing any trade, business or process dust or effluvia to affect inhabitants of the neighbourhood. Not ventilating, not keeping clean and not keeping free from noxious effluvia or overcrowding any workplace are also statutory nuisances.

Where a statutory nuisance exists, the local authority can serve an abatement notice on the appropriate person, owner or occupier. If the abatement notice is disregarded, the court has powers to make a nuisance order.

Consent of the local authority is required before specified trades or business can be carried on. These offensive trades include blood boiling and drying, bone boiling, fat extracting and melting, fell mongering, glue making, soap boiling, tripe boiling and dealing in rags and bones.

The Act gives local authorities power to make bye-laws with regard to offensive trades and of fish-frying.

An allied piece of legislation is the Food Safety (General Food Hygiene) Regulations 1995 made under the Food Safety Act 1990. These regulations apply in England and Wales only but similar food hygiene regulations also exist for Scotland and for Northern Ireland.

The principal requirements of the Regulations relate to:

(a) the cleanliness of premises and the equipment used for the purpose of a food business;
(b) the hygienic handling of food;
(c) the cleanliness of persons engaged in the handling of food;
(d) the construction of premises used for the purposes of a food business and their repair and maintenance;
(e) the provision of water supply and washing facilities;
(f) the proper disposal of waste;

1.2.8 Petroleum (Consolidation) Act 1928

Little remains of this Act which is now restricted to licensing for the keeping of petroleum spirit, the making of byelaws for filling stations and canals and to testing petroleum. Extant regulations cover compressed gas cylinders, the keeping of petroleum spirit and extending the provisions of the Act to other substances such as carbide of calcium and liquid methane.

1.2.9 Activity Centres (Young Persons Safety) Act 1995

With the growth of centres providing facilities where children and young persons can engage in adventure activities, and as a result of tragedies due to poor management of such centres, the need to control these centres became apparent. The Act allows for the making of regulations such as those prescribing the type of person who should hold a licence to provide and run the centre, the duties of the licensing authorities and enforcement of the Act[24,25].

1.2.10 Crown premises

Although s. 48 of the HSW makes provision for binding the Crown to the provisions of part of the Act it excepts ss. 21 to 25 which deal with prohibition and improvement notices. The Crown immunity exists because in the exercise of justice in the name of the monarch it is not constitutionally possible for one part of the Crown service to pursue another part of the service into the courts. Nevertheless, the HSE does apply a version of prohibition and improvement notices, called Crown Notices, when deficiencies are found on Crown property. Through the National Health Service (Amendment) Act 1986, Crown immunity for both food legislation and health and safety legislation have been removed from the Health Authority.

1.2.11 Subordinate legislation

This chapter has dealt briefly with the major legislation concerned with health and safety which is becoming progressively more proscriptive, i.e. it states the aims to be achieved but not how to achieve them. Supporting this major legislation is a vast and growing body of subordinate legislation, mainly in the form of regulations, which derive their authority from the main Acts. While not being completely proscriptive, these regulations, which tend to occur in particular areas of health and safety concern, point to possible routes to compliance while recognising that there may be other equally effective means for achieving the same goal.

The range of these regulations is continually developing and the list that follows the references gives some of the more generally applicable examples. Many of them are referred to in the texts of the following chapters where their relevant content is discussed. In the years following their introduction, many of the regulations have been amended.

References

1. Report of the Committee on Safety and Health at Work. 1970–71 (Robens Report), Cmnd 5034, The Stationery Office, London (1972)
2. Fife, I. and Machin, E.A. *Redgrave Fife and Machin Health and Safety*, Butterworth, London (1990)
3. Health and Safety Executive, *The Supply of Machinery (Safety) Regulations 1992*, SI 1992 No. 3073, The Stationery Office, London
4. Health and Safety Executive, *The Electromagnetic Compatibility Regulations 1992*, SI 1992 No. 2372, The Stationery Office, London
5. Health and Safety Executive, *The Gas Appliances (Safety) Regulations 1995*, SI 1995 No. 1629, The Stationery Office, London
6. Department of Trade and Industry, *The Lifts Regulations 1997*, SI 1997 No. 831, The Stationery Office, London
7. Health and Safety Executive, *The Construction (Design and Management) Regulations 1994*, SI 1994 No. 3140, The Stationery Office, London
8. Health and Safety Executive, *The Offshore Installations and Wells (Design and Construction etc.) Regulations 1996*, SI 1996 No. 913, The Stationery Office, London
9. Health and Safety Executive, *The Chemicals (Hazard Information and Packaging for Supply) Regulations 1994*, SI 1994 No. 3247, The Stationery Office, London
10. Health and Safety Executive, *The Health and Safety (Leasing Arrangements) Regulations 1992*, SI 1992 No. 1524, The Stationery Office, London
11. Health and Safety Executive, *The Health and Safety (Enforcing Authorities) Regulations 1998*, SI 1998 No. 494, The Stationery Office, London

12. Health and Safety Executive, *The Fire Precautions (Factories, Offices, Shops and Railway Premises) Order 1989*, SI 1989 No. 76, The Stationery Office, London
13. Health and Safety Executive, *The Management of Health and Safety at Work Regulations 1999*, SI 1999 No. 3242, The Stationery Office, London
14. Health and Safety Executive, *The Provision and Use of Work Equipment Regulations 1998*, SI 1998 No. 2306, The Stationery Office, London
15. Health and Safety Executive, *The Health and Safety (Display Screen Equipment) Regulations 1992*, SI 1992 No. 2792, The Stationery Office, London
16. Health and Safety Executive, *The Manual Handling Operations Regulations 1992*, SI 1992 No. 2793, The Stationery Office, London
17. Health and Safety Executive, *The Personal Protective Equipment at Work Regulations 1992*, SI 1992 No. 2966, The Stationery Office, London
18. Health and Safety Executive, *The Workplace (Health, Safety and Welfare) Regulations 1992*, SI 1992 No. 3004, The Stationery Office, London
19. Health and Safety Executive, *The Fire Precautions (Workplace) Regulations 1997*, SI 1997 No. 1840, Stationery Office, London
20. Health and Safety Executive, *The Health and Safety (Emission into the Atmosphere) Regulations 1983*, SI 1983 No. 943, The Stationery Office, London
21. Health and Safety Executive, *Employment Tribunals (Constitution and Rule of Procedures) Regulations 2001*, SI 2001 No. 1171, The Stationery Office, London
22. Health and Safety Executive, *The Health and Safety (Training for Employment) Regulations 1990*, SI 1990 No. 1380, The Stationery Office, London
23. Health and Safety Executive, *The Quarries Regulations 1999*, SI 1999 No. 2024, The Stationery Office, London
24. Health and Safety Executive, *The Adventure Activities (Licensing) Regulations 1996*, SI 1996 No. 772, The Stationery Office, London
25. Health and Safety Executive, *The Adventure Activities (Enforcing Authorities and Licensing Amendment) Regulations 1996*, SI 1996 No. 1647, The Stationery Office, London

List of some generally applicable regulations

The Highly Flammable Liquids and Liquefied Petroleum Gases Regulations 1972
The Safety Representatives and Safety Committee Regulations 1977
The Control of Lead at Work Regulations 2002
The Health and Safety (First Aid) Regulations 1981
The Control of Major Accident Hazards Regulations 1999
The Classification and Labelling of Explosives Regulations 1983
The Asbestos (Licensing) Regulations 1983
The Ionising Radiations Regulations 1999
The Control of Asbestos at Work Regulations 2002
The Electricity at Work Regulations 1989
The Pressure Systems Safety Regulations 2000
The Noise at Work Regulations 1998
The Health and Safety (Training for Employment) Regulations 1990
The Control of Explosives Regulations 1991
The Lifting Plant and Equipment (Records of Test and Inspection etc.) Regulations 1992
The Management of Health and Safety at Work Regulations 1999
The Provision and Use of Work Equipment Regulations 1998
The Manual Handling Operations Regulations 1992
The Workplace (Health, Safety and Welfare) Regulations 1992
The Personal Protective Equipment Regulations 1992
The Health and Safety (Display Screen Equipment) Regulations 1992
The Supply of Machinery (Safety) Regulations 1992
The Electromagnetic Compatibility Regulations 1992
The Control of Substances Hazardous to Health Regulations 2002
The Construction (Design and Management) Regulations 1994
The Gas Safety (Installations and Use) Regulations 1998
The Gas Appliances (Safety) Regulations 1995
The Chemical (Hazard Information and Packaging for Supply) Regulations 2002 (CHIP3)

The Reporting of Injuries, Diseases and Dangerous Occurrences Regulations 1995
The Food Safety (General Food Hygiene) Regulations 1995
The Health and Safety (Safety Signs and Signals) Regulations 1996
The Adventure Activities (Licensing) Regulations 1996
The Adventure Activities (Enforcing Authority and Licensing Amendment) Regulations 1996
The Gas Safety (Management) Regulations 1996
The Health and Safety (Consultation with Employees) Regulations 1996
The Work in Compressed Air Regulations 1996
The Carriage of Dangerous Goods (Classification, Packaging and Labelling) and Use of Transportable Pressure Receptacles Regulations 1996
The Construction (Health, Safety and Welfare) Regulations 1996
The Health and Safety (Young Persons) Regulations 1997
The Lifts Regulations 1997
The Confined Spaces Regulations 1997
The Fire Precautions (Workplace) Regulations 1997
The Diving at Work Regulations 1997
The Management and Administration of Safety and Health in Mines Regulations 1993
The Offshore Installations and Pipeline Works (Management and Administration) Regulations 1995
The Carriage of Dangerous Goods by Rail Regulations 1996
The Carriage of Dangerous Goods by Road Regulations 1996
The Lifting Operations and Lifting Equipment Regulations 1998
The Pressure Equipment Regulations 1999
The Railway (Safety Case) Regulations 2000
The Pollution Prevention and Control (England and Wales) Regulations 2000
The Health and Safety at Work (Application outside Great Britain) Order 2001
The Biocidal Products Regulations 2001
The Transportable Pressure Vessel Regulations 2001

Further reading

Selwyn, N., *Law of Health and Safety at Work*, Croners, London (1993)
Mahaffy and Dodson on Road Traffic, Butterworth, London, (loose-leaf)
Garner, *Environmental Law*, Butterworth, London, (loose-leaf)
Encyclopedia of Environmental Health, Sweet and Maxwell, London (loose-leaf)

Chapter 1.3

Influences on health and safety

J. R. Ridley

1.3.1 Introduction

Laws don't just happen.

They are a reaction to a perceived need, whether real or imaginary, and are aimed at protecting the individual and/or the community as a whole. But within the perceived need there are many trends and factors that influence attitudes towards the sort of action that needs to be taken. With time, situations and attitudes – both personal and political – change, making what was right yesterday not necessarily right today. These changes and developments are no more visible than in health and safety, an area on which a great deal of political, community and media attention has been focused and which can become very emotive issues. This chapter looks at some of the influences that affect the current laws, attitudes, interpretations and practices in health and safety.

1.3.2 The Robens Report[1]

There can be no doubt that the single largest influence on the organisation of and our approach to health and safety lies with the work of the Robens Committee and its subsequent report. Their remit gave them the opportunity to stand back and take an objective look at where health and safety was going – not only in the UK but in other major manufacturing countries. Their considerations were carried out in a political atmosphere that was changing from employer dominated to one in which the employee had a much greater say. Added to this, employees were no longer the down-trodden, exploited victims of a culture that blossomed in an environment of extremes of poverty and riches, a culture that had precipitated the passing of the first ever law to protect working people.

All the recommendations contained in the Robens Report were accepted by the government of the day and by the opposition parties. They have since been incorporated into health and safety legislation, notably the Health and Safety at Work etc. Act 1974 (HSW). Major changes brought about as a result include:

- all employees except domestic servants were brought within the scope of the Act and given a measure of protection while at work
- laws have been made more flexible to enforce and to comply with
- making laws less prescriptive and more proscriptive
- simplifying the process for making subordinate legislation
- increasing the powers of inspectors to reduce some of the bureaucratic and time-consuming enforcement procedures
- giving Ministers of State powers to approve codes of practice whose contents showed a means of achieving conformity with the laws
- accepting the concept of putting responsibility where authority lay, i.e. with the employer and of the need for a commitment at senior levels in an organisation
- recognising that employees were not the only ones who had a right to protection from the effects of the way work was carried out
- confirming the role that employees have to play in health and safety
- advocating a greater degree of self-regulation through the use of own rules and codes of practice.

It is over 25 years since the Roben's committee carried out its review of health and safety in the UK. In that time a great many changes have occurred in the health and safety field, changes in the attitudes of employers, employees and society as a whole, in the origins of laws and in the emphasis of legislative requirements. This raises a question of whether the time is approaching when a further review is needed to build on some of the radical changes initiated by the Roben's Report. Some relevant issues are considered in the following sections of this chapter.

1.3.3 Delegation of law-making powers

The original laws of the country, or dooms as they were then called, were made by the king. As the national administration became more complex, the law-making powers were delegated to, then finally taken over by, Parliament. Up until fairly recently, the laws were debated in depth by Parliament who incorporated a great deal of detail into their content so they stated not only what was to be achieved but how. Inevitably with the increasingly complex nature of the laws required, this proved a ponderous and time-consuming activity. So in the early part of the twentieth century the practice developed of Parliament delegating that part of its law-making authority concerned with the minutiae of compliance to a Minister of the Crown.

With the enormous strides in the development of technology that have been made over the past few decades the practice of delegating powers for making subordinate laws has become the norm in respect of health and safety. These powers are given only to the Secretary of State for the particular government department concerned (s. 15 HSW) who may require one of his Ministers to oversee the preparation of the necessary documents. As far as health and safety is concerned the drafting of proposals for subordinate legislation is normally passed to the HSC who, in turn, pass to the HSE the actual drafting work. Section 15 of HSW states '. . . the Secretary of State and that Minister acting jointly shall have power to make regulations . . .'. The interpretation of the word 'make' is important. It does not mean that they can put laws onto the Statute Book but only that they can prepare (make) the documents that state the requirements. To give the proposals the power of law needs the authority and approval of Parliament, which it gives through the procedure described below.

In the process of drafting its proposals, the HSE is required to consult with '. . . any government department or other body that appears . . . to be appropriate . . .' (s. 50 HSW). The procedure followed by the HSE is similar to that followed by Parliament with white and green papers (see section 1.15.3) where there may be consultation with the CBI, TUC and an industry with a particular interest in the subject. Thereafter public consultation takes place through the publication of a Consultative Document (CD) which outlines and explains the new proposals.

Opinions submitted are taken into account in the preparation of the final proposal which is then sent to the HSC for its approval. From there it goes to the Minister concerned who takes it to Parliament for their approval. Before that approval can be given, the proposal must lie in the House for thirty days during which time any MP can peruse it, raise objections and call for a debate on it.

At the end of the thirty days a vote is taken on whether to approve the proposal or reject it. There are two procedures for voting on a regulation 'made' by a Minister. The first for proposals of minor legal importance covering such administrative matters as the timing of the implementation of parts of an Act, or the approval of certain types of safety equipment where a *negative* vote is taken, i.e. if there are no objections to the proposal its approval is assumed and it becomes part of the law. The second procedure is followed for proposals that are of greater importance, such as those that may qualify or change part of an existing Act where a *positive* vote is required, i.e. there must be a majority vote in favour of the proposal for it to become part of the statute law. With subordinate legislation there is no need for royal assent since that is inherent in the authority contained in the substantive Act.

1.3.4 Legislative framework for health and safety

The foundations of health and safety legislation are the relevant Acts passed by Parliament on the subject – most noticeably HSW. But this particular Act covers such a wide range of subjects and activities that it needs supportive subordinate legislation to put flesh on its legislative skeleton. HSW incorporates clauses that empower the appropriate Secretary of State and/or the Minister to 'make' regulations and orders.

These regulations have the objective of detailing legislative requirements in the specific area at which they are aimed although there has been a trend – reflecting the requirements of EU directives – of being more proscriptive in the wording of the clauses or regulations than previously and of relying on annexes or schedules to identify areas that require particular attention. Conformity with the legislative requirements is then measured against compliance with particular official codes of practice or standards, whether BS or transposed harmonised (BS EN) standards.

To support the requirements, and to strengthen the arm of the enforcing authorities, HSW allows the HSC, subject to the approval of the Secretary of State, to make and approve Codes of Practice whether of their own drafting or of others. The fact of giving official approval to Codes of Practice (ACoPs) gives them a status of recognition in law such that they will be accepted by a court as documentary evidence of a means of conforming with the legislative requirements without the need to 'prove' the document should the defendant challenge it.

In the 'making' of Codes of Practice for approval there are statutory requirements that HSC/E should consult with *'any government department or other body that appears . . . to be appropriate . . .'* (s. 16(2) HSW). The use of ACoPs is becoming increasingly common as new regulations – reflecting EU directives – become more proscriptive. ACoPs lend greater flexibility to the making and amending of statutorily recognised standards – this is an important aspect that allows rapid changes to be made to keep pace with new hazards introduced by rapidly advancing technologies.

While ACoPs fill an important role they do not completely fill the gap which shows acceptable means for achieving conformity with legislative requirements – this is left to various advisory publications issued by HSC/E, – from the L series of booklets that explain the legislative requirements through the HS(G) guidance booklets to the series of technical Guidance Notes covering chemical, environmental hygiene, medical, plant and machinery aspects of health and safety.

In addition, a number of industries, often working with the HSC's Industry Advisory Committees (IACs), draw up codes of practice in respect of hazards particular to their operations. Where these are recognised or published by HSE they provide a statement of ways in which conformity with legislative requirements can be achieved.

1.3.5 Self-regulation

Self-regulation is an ideal to be aimed for where organisations set their own effective standards, make rules to ensure those standards are met and enforce adherence to those rules. The political mood of the 1970s and 1980s, looking for means to reduce the legislative burden on industry, turned to self-regulation as an answer. However, they failed to recognise that in the smaller companies, which comprise by far the greater number of employers, the top priority was economic survival in the face of increasing economic strictures, with health and safety as an also-ran. Furthermore, the attitudes and cultures to make such an innovation a success did not exist – nor was the political will sufficiently persistent to carry it through.

The move towards self-regulation reflects the political drift of the 1980s/90s to reducing the burden of imposed controls on employers and giving them greater freedom to act within their particular employment circumstances and environment. While many major manufacturing companies have practised this sort of control for years, it was from the consumer market that the motivation came. Those areas of industry and commerce that provide goods and services to major companies who supplied the consumer market were required to control the quality of goods to the standards demanded by major buyers. They had set their own standards and demanded it of their suppliers, in respect not only of quality but of health, safety and hygiene in the manufacture of those goods. This practice is likely to grow as the trend spreads to the manufacturing sector and more retail organisations demand quality assurance on goods they sell as 'own brand'.

Areas where self-regulation is included in legislation are seen in the requirements of both the Pressure Systems Regulations[2] and Lifting Operations and Lifting Equipment Regulations 1998[3] (LOLER) whereby employers and users decide what examinations need to be carried out and at what intervals according to the circumstances of use of the equipment.

For self-regulation to be a success requires:

- a commitment at the most senior level in an organisation
- a culture pervading all levels of the organisation that will accept only the highest standards (much as exists in successful quality assurance systems)
- a commitment by and the active co-operation of all levels
- the will to make the necessary resources available.

The advocation of self-regulation must not be a reaction to the excessive zeal or over-dogmatic enforcement of statutory requirements but must stem from a genuine desire to achieve high standards in ways that are pertinent to the particular organisation. The desire to implement schemes for self-regulation must be generated within the organisation. A move to encourage businesses in this direction can be seen in the HSC's campaign 'Good health is good business'.

The success of self-regulation depends to a large extent on the honesty and integrity of those following its regime. There is something of a parallel in the self-certification of machinery allowed under the Machinery Directive[4] but this requires the support of documentary evidence which can be examined.

1.3.6 Goal-setting legislation

Effectively this was first introduced by HSW which laid down broad objectives to be achieved rather than the very prescriptive nature of earlier legislation with its roots in a manufacturing-led economy. Now that manufacturing is no longer the major employer, and is facing severe economic and competitive strictures, the imposition of unrealistic controls is more likely to accelerate its demise than encourage its survival. To survive and expand requires a regime where controls are related to operating circumstances rather than to broad brush catch-all requirements. Legislation needs to recognise this and be more flexible in its ability to accommodate the enormous range of employment conditions and circumstances met today while still ensuring that acceptable standards are maintained. This requires a more flexible proscriptive approach which allows individual industries and employers to determine how best, within their trading and employment situation, they can achieve the desired end results.

Where industries and companies do agree and set standards, the role of enforcement – whether internally or by HSE inspectors – should be to ensure those standards are met and maintained.

The principle of goal setting permeates the making of EU health and safety directives and stems from a Resolution adopted by the European Council in June 1985 on a 'new approach to legislative harmonisation' whereby goal-setting objectives are outlined in broad terms in the main Articles of the directives and rely on supportive annexes to specify the particular areas requiring detail consideration. For directives concerned with work equipment, the annexes include a list of Essential Health and Safety Requirements (ESRs). In turn the matters covered by the annexes rely on harmonised standards to give evidence of conformity with the ESRs and hence the directive.

Goal-setting legislation cannot stand by itself but must rely heavily on associated requirements, in the form of Approved Codes of Practice, official guidance material, national and harmonised standards etc. to provide a base against which the degree of conformity can be assessed.

1.3.7 European Union

When the UK became a member state of the European Economic Community in 1972 it agreed to be bound by the legislative procedures of the Community (now the European Union (EU)). At that stage each of the Member States had the power of veto over any proposed EEC legislation with the result that very little progress was made towards unified laws. That changed in 1986 with the adoption of the Single European Act[5] (SEA) which brought in qualified majority voting (QMV), a system whereby each Member State is allocated a number of votes weighted according its population. *Table 1.3.1* shows the allocation of votes to the various Member States with a summed total of 87 votes. A majority vote (a minimum of 62) is needed for a matter to be adopted. This means that the UK lost its power of veto over directives and has become subject to the political inclinations and standards of its European neighbours with their cumulative majority vote.

The EU is a complex organisation which contains a number of salient institutions with various executive and advisory roles.

The *Council* is made up of the representatives of the governments of the 15 Member States. Normally these are Ministers of State with the Foreign Secretary being the UK's main representative. However, Ministers from other Departments meet to deal with specific matters such as finance, agriculture, transport etc. Twice a year the Heads of State meet for a *European Council* (sometimes referred to as the Summit). The Council is

Table 1.3.1 Allocation of 'weighted' votes to member states

Member State	*Number of votes*
France	10
Germany	10
Italy	10
United Kingdom	10
Spain	8
Belgium	5
Greece	5
Netherlands	5
Portugal	5
Austria	4
Sweden	4
Denmark	3
Finland	3
Ireland	3
Luxembourg	2
Total	87

Minimum number of votes needed for adoption of a proposal = 62

the supreme law-making body of the EU and is the only body that can adopt EU legislation. A great deal of background work is carried out between Council meetings by the *Committee of Permanent Representatives* whose members are the ambassadors to the EU of each member state.

The *Commission* is the executive body of the EU and consists of 17 members appointed by agreement between member governments. Commissioners must act independently of their national government and of Council. The Commission is answerable to the European Parliament and its detailed work is carried out by 23 Directorate Generals (DGs) each dealing with a specific aspect of EU business. Those most relevant to health and safety are:

DG III – Internal market and industrial affairs – covering the construction of work equipment for free movement in the EU.
DG V – Employment, social affairs and education – dealing with safety in the workplace and the safe use of work equipment.
DG XI – Environment, consumer protection and nuclear safety.

The Council and Commission are assisted by the *Economic and Social Committee* (EcoSoC) who must be consulted before decisions are taken on a large range of subjects. It is non-governmental and consists of 186 members representing the social partners plus particular sector interests. Additionally, preliminary proposals with a health and safety content are referred to the *Advisory Committee on Safety, Hygiene and Health Protection at Work* for their opinion.

The *Court of Justice*, which sits in Luxembourg, consists of 13 judges appointed by agreement of Member States. Its role is to hear cases concerning EU legislative matters. It does not interfere with national judiciary other than to advise where a national matter impinges on or clashes with EU law.

The *Court of Auditors* oversees the use of EU funds and ensures the proper use of such monies.

The *European Parliament* (EP) consists of 626 members, each elected by universal suffrage, who attach to EU-level political groups – there are no national sections. Detail consideration of specific matters is delegated to one of 18 standing committees whose rapporteurs report their findings to EP. There is an extensive consultation procedure between the Council, the Commission and EP on proposed legislation but EP seem set on increasing their involvement in EU law making.

A *European Agency for Health and Safety at Work* has been established to gather and disseminate information on health and safety matters to all member states. The Agency is located in Bilbao.

Each of these various institutions has a particular and separate function within the EU but work closely together to make an effective organisation.

1.3.7.1 EU legislation

Within the EU there are four basic types of legislation:

- *Regulations* which have direct applicability in Member States and take precedence over national laws. They arise mainly in respect of the coal and steel industries. They are rarely used for health and safety matters but have been used on transboundary environmental matters.
- *Directives* are the more usual EU legislation for health and safety issues. They do not have direct applicability but put an onus on Member States to incorporate their contents into national laws within a time scale specified in the directive.
- *Decisions* and *Recommendations* relate to specific matters of local concern and apply directly to the Members States at whom they are directed.

When a directive is adopted by the Council of Ministers, each Member State becomes committed to incorporate into its national laws the contents of the directive. This means in a number of areas – such as health and safety, employment, environment, finance etc. – the UK Parliament's power to decide the subject matter of the laws it passes is being eroded. While there are still domestic matters on which Parliament can decide to legislate, there is a growing body of laws whose subject matter and content have been dictated to Parliament. As the EU gets stronger and the eurocrats more bureaucratic the question arises of how much longer can the UK retain some measure of right of sovereignty in the making of its laws.

In the health and safey field, the EU approach to law making has polarised into two discrete areas. For new equipment, health and safety has been used as the criteria for setting equipment construction standards to allow free movement throughout the EU. This concerns free trade and is dealt with by Directorate General III. However, the use of work equipment – whether existing or new – is considered to be part of employment and is dealt with by Directorate General V. In both cases, early drafts of proposals are referred to the *Advisory Committee on Safety, Hygiene and Health Protection at Work* for its opinion. This committee is tripartite in composition with representatives from each Member State and can be a useful forum for influencing the final proposals in favour of national and sectorial positions.

One of the moves to create a free European market which allows free movement throughout the EU for all new machinery requires that the machinery conforms to the Machinery Directive[4]. Similar moves aimed at ensuring a greater degree of uniformity of working conditions in Member States requires comformity with the Workplace Directive[6] and Work Equipment Directive[7]. These directives were adopted in 1989 and had to be implemented in member states by 1 January 1993, the date when the European free market became established. In the UK the directives concerned with work equipment have been transposed into the Supply of Machinery (Safety) Regulations 1992 (SMSR) and the Provision and Use of Work Equipment Regulations 1998 (PUWER 98) which are considered in detail in later chapters.

1.3.8 European standards

Harmonised or 'EN' standards are those that have been approved by the European standard making organisations, Committé European de Normalisation (CEN) for standards of mechanical equipment and Committé European de Normalisation Electrotechnique (CENELEC) for standards of electrical equipment. They are given the prefix letters EN and in the UK can be recognised by the designation BS EN, e.g. BS EN 836, *Garden equipment – Powered lawn mowers – Safety.*

These standards are prepared by working parties whose members are representatives of the participating Member States – including members of EU, EFTA plus some from Eastern European and Middle East States. Representation at CEN and CENELEC is through the various national standards making bodies – in the UK, BSI.

There are parallel international standards making bodies, International Standards Organisation (ISO) for mechanical standards and International Electrical Commission (IEC) for electrical standards. These European and international standards making bodies are now working closely together to prevent duplication of effort and to expedite the preparation of standards.

1.3.8.1 Harmonised standards making procedures

In the UK, standards are developed by the British Standards Institution which remains the national standards making body but, with membership of the EU, procedures for making standards changed. In 1983 the Commission reached agreement with the two European standards making bodies that conformity with harmonised standards would be accepted as evidence of conformity with the directives. These two bodies, CEN and CENELEC, are sponsored by the Commission but act independently.

As soon as a subject is selected for a harmonised standard all Member States must cease individual work in the area. The work of drafting a harmonised standard is carried out by working groups of representatives of the Member State national standards making bodies. Final drafts are circulated to all members of CEN/CENELEC to vote for approval or rejection. A majority vote is required before the standard is adopted. Once adopted, the EN standard then applies in all EEA Member States and takes precedence over any national standard that covers the same subject.

Subjects for an EN standard are determined from a perceived need or by a proposal from a Member State that is considered to be of benefit to the EU as a whole. Where a Member State has a standard which they consider could beneficially apply across the EU they can submit it as the basis for an EN standard. BS 5304 'Safety of machinery'[8] was put forward and accepted on this basis. In the event, rather than keeping it as a comprehensive standard on the guarding of machinery, it has been split into component subjects each of which has become a separate harmonised standard. With the adoption of these EN standards, BS 5304 has had to be withdrawn as a British Standard and declared obsolete. However, because it is so comprehensive, well understood and widely used, it has been re-issued in a revised form – as a 'Published Document PD 5304'. This document does not have the status of a standard but does offer advice on basic principles of safeguarding for machinery. It contains an extensive cross-reference to European Standards, conformity with which is necessary for compliance with PUWER 98.

In their approach to developing harmonised standards, CEN and CENELEC have categorised standards into four types:

Type A – generic safety standards covering basic concepts, principles for design and general aspects that can be applied to all machinery.

Type B – dealing with one safety aspect or one type of safety related device that can be used across a wide range of machinery:

- type B1 cover particular safety aspects such as safety distances, surface temperatures, noise etc.
- type B2 cover safety related devices such as two hand controls, interlocking devices, pressure sensitive devices, guards etc.

Type C – giving detailed safety requirements for a particular machine or group of machines – machine specific standards.

The category status of harmonised standards is quoted in the introduction to each standard.

1.3.9 Our social partners

A phrase that has become part of the jargon of the EU and which effectively refers to the relationship between employer and employee. Its meaning has been enlarged from a definition of a simple relationship to encompass a very complex inter-relationship and in so doing its influence has grown enormously, particularly in the field of employment and health and safety. Employers' and employees' representatives have had a running dialogue for many years, going back to the days of the big unions and national wage negotiations. More recently big has become no longer beautiful and negotiations on conditions of work have become much more localised. But there is a continuing dialogue between employer and employees.

However, the attitudes, practices and relationships in our European partners with their influences through the European Commission and Parliament are resulting in changes to working standards in the UK. The EU works by consensus or through qualified majority rule recognising the standards of the less advanced Member States in the development of EU-wide standards. In effect, EU legislation sets the highest acceptable standard then allows derogation for those Member States that are less technicologically advanced. If the EU legislative requirements are less than those currently accepted in the UK, the question arises of how the UK can maintain its safety standards without its industries suffering economically from the resultant higher manufacturing costs.

1.3.10 Social expectations

Over the past two or three decades, the social attitude to work has undergone a considerable change. It has moved from one of being glad to have a job (and an income) and accepting the rough with the smooth where the rough could be very rough and result in horrendous injuries for which little or no compensation was paid. Also wages were such that there was little enough to barely survive on. Hours were long and the work demanding which left little time or energy for anything outside the work. There was little social life other than the men congregating in the pub or the women gossiping over the back fence. Rarely was there social entertainment for a man and his wife and, indeed, their family except on feast days and at the travelling fairs. Life centred on work and that was the sole object in life. However, with the considerable rise in incomes and living standards that have occurred since the end of the Second World War radical changes in social behaviour have occurred. Workpeople have the leisure time and are able to afford an extensive social life often making greater inroads on their time and energy than their paid employment. With these changes in circumstances has come a change in attitude towards the dangers experienced at work. Workpeople now expect to be able to enjoy their leisure time and demand that any dangers from the work process be removed and that employers provide conditions of work that do not put them – the workpeople – at risk.

1.3.11 Public expectations

Public expectations in health and safety are largely determined by the way in which a particular risk is viewed. The response to a hazard or risk, whether at individual, group or national level is significantly influenced by how it is perceived. To illustrate the point, each year many thousands of people are killed or injured on the roads and this fact rarely gets reported in the press. However, a multi-vehicle crash or a coach crash which causes several deaths immediately reaches the front page of the national press and television. In the area of health, whilst many people die from heart disease and cancers often related to smoking or excessive alcohol consumption, the media ignores them and focuses on the small number who die from the newly highlighted disease, Creutzfeldt–Jacob Disease (CJD). The occurrence of multi-death events or strange new diseases attract the media whose reporting treatment causes greater concern in the minds of individuals and the general public than the number of victims would seem to warrant.

Work by Slovic, Fischhoff and Lichenstein[9] shows that people are frequently unaware of the true risks from a specific hazard. For example, their study showed that whilst the actual number of deaths in the USA from botulism was less than 10 per year, it was estimated at between 100 and 5000 by the representative group of interviewees. Similarly the group estimated that deaths from stomach cancers at 5000 as against an actual figure of 100 000 per annum. The authors noted that biases in newspaper coverage closely matched the biases in people's perception. The question is, did the researchers report a natural bias or had the newspapers created one?

Further work by the same researchers indicated that different societal groups have different perceptions. In a survey, college students, members of a professional and business organisation and members of a women's political organisation all ranked nuclear power among the riskiest of 30 activities, whilst professional risk assessors placed it a lowly twentieth. Nuclear power is a very highly regulated activity and has become so because the media focus and public perception combined to provoke strong regulatory responses from the law makers.

Overall, the less that is known about a recognised risk, the more it is feared and the greater the anxiety about it. Conversely, the more we are familiar with a risk and feel in control of it, the less concerned we are about it. The concern arising from the perception of a risk can be at an individual, local or national level. When it surfaces at national level it frequently results in regulatory controls.

1.3.12 Political influences

These really operate at two levels: the grass roots level where the work people and electors put pressure on their Parliamentary representatives to make laws that provide protection against perceived dangers, conversely, politicians are anxious to appease the electorate and keep their votes so take up the rifles loaded by their constituents and fire the bullets. In some cases these bullets hit the target as Private Member's, Bills and partisan enactments are passed.

At other times an event or incident can occur whose result is so horrifying that even Parliament is jolted into action. Unfortunately there seems to be chronic shortsightedness among MPs with the result that in these circumstances any legislations passed tends to be aimed only at the circumstances and industry involved in the incident and to miss entirely the wider opportunity that the situation offers.

Finally there is the great annual party pilgrimage where the ruling party of the day decides in its infinite wisdom, and with an eye on political expediency, to introduce legislation that will increase its standing as a political party among marginal groups of voters. With luck, and it sometimes happens, the subject chosen will be in need of legislative control and the working community as a whole will benefit, but it is a random chance because the motivation behind it is fed by a desire to make political gain. Rationale does not seem to enter the issue when political standing is at stake.

1.3.13 Roles in health and safety

Early legislation put the responsibility for remaining safe at work very firmly in the hands of those who *operated* the machinery, so that if they had an accident it was their fault and they got no compensation. However, the pendulum swung very much the other way in the mid parts of the twentieth century with the onus being put firmly on the shoulders of the employer in their position of *controlling* what was done and how. But the pendulum is swinging back with the growing realisation of the positive role that employees have to play in health and safety. The catalyst for this oscillation seems to be *knowledge* which is being imparted through the extensive training programmes within industry and the higher level of education generally in the community at large. It remains to be seen how long it will be before and to what extent the judiciary will acknowledge and recognise that those with knowledge have an obligation to employ it in meeting statutory obligations.

Some of this has already been recognised in law through the obligations placed on manufacturers and suppliers to pass to their customers information about hazards and safe operating techniques for the equipment and substances that they supply. This obligation can be met relatively easily, and seen to be met, by preparing and issuing written instructions and data sheets. What is not going to be so easy to demonstrate is the extent of take-up of knowledge by an employee during training. Will NVQs and SNVQs have a role to play in providing documentary evidence of the knowledge of the workpeople? And to what extent will that documentary evidence of *qualification* be recognised by the courts as imposing a legal obligation on individual workpeople to work safely and follow working rules?

1.3.14 Safety culture

Anyone, who in the course of their work, has to visit a number of different workplaces very quickly acquires the facility for getting the 'feel' of the atmosphere in the workplace. That feel can be confirmed in the state of the workshops, offices and particularly the toilets which give an indication of the attitude of the workpeople. In any normal community there are inevitably one or two disgruntled individuals with genuine or imagined complaints against society. But when a whole community in a workplace exhibits this attitude the cause needs to be sought in a factor common to that workplace and that single common factor is often the manager, director or owner. It is a strange quirk of organisation that the attitude held in the board room, although there is no direct communication, inevitably manifests itself in the attitudes and behaviour of the shop floor or in the general office. This attitude or culture, which permeates the organisation, emanates from the highest level. Safety attitudes behave in the same way as water – they flow from the top down through the organisation. It is almost unknown for reverse flow to occur, i.e. for high health and safety standards on the workplace floor to influence attitudes in the boardroom.

The implications of this in health and safety are enormous. It's a bit like the adage 'look after the pennies and the pounds will look after themselves'. If the attitude at the top of an organisation is concerned with achieving high standards of health and safety that attitude will permeate the organisation and be measurable in the working areas with high levels of safety performance. Closely associated with this will also be high levels of job satisfaction with the spin-off of high quality of product and high output.

Thus the safety culture, which in itself cannot be quantified or evaluated as a function in absolute terms, can have great benefits to the organisation, its viability and to the people working in it. That safety culture can only be generated by the most senior people in that organisation – the board of directors – and this is probably the greatest contribution they can make towards the cause of high standards of health and safety.

1.3.15 Quality culture

There has been, in recent years, a recognition that to remain viable and to retain customers, the quality of product or service must be of an acceptably high standard. Much of the pressure for this has come from the need to remain competitive in a free market, but a great deal also stems from major purchasers, particularly in the retail and consumer durable markets, seeking greater market share through the quality of the goods and services they sell. They do this by ensuring within their own organisations high levels of service for the customer and by demanding from their suppliers a similar high quality in the goods they purchase. This latter point enables them to put enormous pressure on the manufacturer or producer to achieve the required standards. In many cases the purchaser will instruct the supplier in the techniques for achieving those quality standards. This move has been recognised nationally and internationally by the promulgation of standards[10,11]. These standards require systems to be in place that assure the quality of the items from the very start of manufacturing, and are based on the inculcation of a suitable attitude throughout all levels of the manufacturing organisation to achieve the goal.

So what is so different with health and safety? It requires the same attitude of mind and can bring similar benefits to the organisation. If the incentive is there for generating an attitude for satisfying a customer on quality, why should a similar attitude for protecting the most important and expensive asset in an organisation be so frequently lacking?

1.3.16 No fault liability

The concept of no fault liability is encompassed to a very low degree in the social security legislation[12] that ensures that anyone injured at work, subject to certain restrictions, can obtain a measure of financial payment while incapacitated. However, it is doubtful if that level of benefit is sufficient to sustain a standard of living equivalent to that enjoyed on full wages. In the event of ill-health or injury resulting from a work activity, compensation can still be sought from the employer in litigation with the surety that if damages are awarded payment will be guarateed under the compulsory insurance required by the Employer's Liability (Compulsory Insurance) Act 1969 and Regulations[13]. However, this can be a time-consuming and expensive process which most injured persons cannot afford unless supported by a union or other association.

In 1978 the Pearson Commission[14] made recommendations concerning the limited application of a system of no fault liability operating through a national insurance scheme. Because of sector interest pressure on the government of the day, nothing was done. However, schemes do exist in New Zealand and Canada which have been running for many years with varying degrees of success.

The question to be asked in respect of no fault liability, where individual employers contribute to a national scheme according to their performance, is whether the remoteness of the penalty (premium payment) from the events (the accidents and compensation payments) demotivates employers from taking actions necessary to remove the cause of the accident. This is a complex area since it inevitably involves humanitarian attitudes as well as the purely economic.

1.3.17 Risk assessments

A growing impact on health and safety has been an official recognition of the part that the assessment of risks can play in raising safety awareness and involving both workpeople and managers in health and safety issues. It is based on a technique that has, for many years, been an everyday tool of safety practitioners – hazard spotting tours. First formally incorporated into law in the original Control of Substances Hazardous to Health Regulations of 1988, risk assessments have now become a major plank of much health and safety legislation both in the UK and the EU.

While legislation requires risk assessments to be carried out, it does not state how or in how much detail although it does acknowledge that a little knowledge of the work being carried out and of health and safety matters is useful. Unfortunately the term 'risk assessment' has also become something of a 'buzz' phrase that slips easily off the tongue without being fully understood. In practice, assessing risks is just one relatively small step in a much larger hazard reduction process which basically has five main stages:

- identification of hazards;
- elimination of as many hazards as possible;
- assessment of risk from the residual hazards;
- implementation of measures to reduce to a minimum any likely ill effects from the residual hazards;
- monitoring the effectiveness of the precautionary measures.

The term risk assessment is frequently (and erroneously) used to describe the complete hazard reduction process.

Essentially the hazard reduction process is subjective with its effectiveness depending to a great extent to the experience, knowledge and attitude of the person carrying it out. It can result in a plethora of paperwork which may be seen as counter-productive in small companies with resultant reluctance to implement the technique. A 'check list' approach, as opposed to a risk-based approach, has been developed and appears popular with small to medium sized contractors and manufacturers who do not have the resources to carry out individual assessments. However, there is a danger that this stereotyped approach can lead to 'lip service' risk assessments being carried out from a manager's desk.

The concept of risk assessments as a means of forcing employers to get involved in safety is to be applauded, but there needs to be, by those who have to apply it, a much greater understanding of the technique and appreciation of its value before it can reach its full potential.

1.3.18 Conclusion

Activities, standards and legislation in the health and safety field are subject to an immense range of influences, at individual, company, community, national and international levels. Standards resulting from community and national influences tend to be imposed and can be resented.

Of all these influences, probably the most effective is in the area of company activities where a close relationship between employer and employee can generate immediate and long-lasting working methods of a high safety standard. This results from the attitude and culture within the organisation over which the organisation has some control. However, this is not a one-way arrangement but requires an ongoing close co-operation between employer and employee to ensure its continuing success.

References

1. HM Government, *Report of the Roben's Committee*, Cmnd 5034, The Stationery Office, London (1972)
2. *The Pressure Systems Safety Regulations 2000*, The Stationery Office, London
3. *The Lifting Operations and Lifting Equipment Regulations 1998*, The Stationery Office, London (1998)
4. European Union, Council Directive No. 89/392/EEC, *on the approximation of the laws of Member States relating to machinery* (Machinery directive), EU, Luxembourg (1989)
5. HM Government, *Single European Act 1986*, Cmnd 9758, The Stationery Office, London (1986)
6. European Union, Council Directive No. 89/654/EEC, *concerning the minimum safety and health requirements for the workplace*, EU, Luxembourg (1989)
7. European Union, Council Directive No. 89/655/EEC, *concerning the minimum safety and health requirements for the use of work equipment by workers at work*, EU, Luxembourg (1989)
8. British Standards Institution, BS 5304, *Safety of Machinery*. This standard has been overtaken by harmonised EN standards and has been withdrawn. However, it has been re-issued as Published Document 5304:2000, *Safe use of machinery* with the same content but enjoying only an advisory status. BSI, London (2000)
9. Slovic, P., Fischoff, B. and Lichenstein, S., *Perceived Risk: Psychological Factors and Social Implications*, Proceedings of the Royal Society, London, A 376, 17–34 (1981)
10. British Standards Institution, BS 9001, *Quality systems – Specification for design/development, production, installation and servicing*, BSI, London (1994)
11. British Standards Institution, BS ISO 14001 *Environmental management systems. Specification with guidance for use*, BSI, London (1996)
12. *Social Security (Industrial Injuries and Diseases) Miscellaneous Provisions Regulations 1986*, The Stationery Office Ltd, London (1986)
13. *Employer's Liability (Compulsory Insurance) Act 1969*, The Stationery Office, London (1969)
14. HM Government, *Report of the Royal Commission on Civil Liability and Compensation for Personal Injury* (The Pearson Report), The Stationery Office, London (1978)

Chapter 1.4

Law of contract

R. W. Hodgin

1.4.1 Contracts

1.4.1.1 Formation of contract

A contract is an agreement between two or more parties and to be legally enforceable it requires certain basic ingredients. It must be certain in its wording and consist of an offer made by one party which must be accepted unconditionally by the other (*Scammell* v. *Ouston*[1]; *Carlill* v. *Carbolic Smoke Ball Co.*[2]). This does not prevent negotiations taking place and alterations being made by both parties during the early stages of the discussions, but in its final stage there must be complete and clear agreement as to the terms of the contract (*Bigg* v. *Boyd Gibbins Ltd*[3]). The great majority of contracts need not be in writing and those made daily by the general public, buying a newspaper or food, clearly show this.

There must, however, be 'consideration' that flows from one party to the other. Consideration is the legal ingredient that changes an informal agreement into a legally binding contract. It is the exchange of goods for payment, of work for wages, of a journey for the price of a ticket that amounts to consideration (*Dunlop Pneumatic Tyre Co. Ltd* v. *Selfridge & Co. Ltd*[4]). It is possible for future promises, mutually exchanged, to amount to consideration. Thus X agrees to buy a car from Y on January 1st and payment will not be made until that date. The contract dates from the day that these promises are exchanged.

A potentially important change to contract law was brought in by the Contracts (Rights of Third Parties) Act 1999. Hitherto consideration needed to flow between the two parties to the contract. Third parties, i.e. those who were not party to the contract had no rights under the contract. Thus, if X agreed to carry out work for Y on the understanding that Y would pay a sum of money to Z once the work was completed, in the event of any default, Z could not sue Y because Z had not provided any consideration to the X–Y contract. The 1999 Act now remedies such a situation. It is necessary that the contract states that Z should acquire rights but, conversely, it is possible for the contract expressly to exclude Z's rights. The Act does not extend rights to a third party to enforce any term of a contract of employment against an employee, any term of a worker's contract against a worker (including a home worker) or any term of a relevant contract against an agency worker.

Two other essentials for a valid contract are that the parties must intend to enter into a legally binding agreement and both parties must have the legal capacity to make such a contract. Lack of intention is rarely a problem in reality, but if one party alleges that he never intended to enter into a contract, only the courts can say, having analysed the behaviour of the parties, whether or not a contract was created (*Balfour* v. *Balfour*[5]). Capacity to contract means that the parties must be sane, sober and over the age of 18, although there are obvious exceptions to the age requirement when it comes to contracts of employment; such employment contracts are now largely regulated by statute.

Although the limits of a contract should be reflected in what the parties to it have expressly agreed upon, it is possible for the courts, or for Parliament, to imply terms into the agreement. The most notable example of statutory implied terms is the Sale of Goods Act 1979 which among other things implies into contracts of sale a condition that goods shall be of a satisfactory quality. However this is only implied where the seller is selling in the course of business. Thus it would not apply to a private sale. The court's role in implying terms is a more difficult area of the law because the general approach adopted by the courts is one of non-intervention.

It is clear, however, that in certain circumstances they will adopt a more positive role particularly where the implied term is necessary to give business sense to the agreement (*Matthews* v. *Kuwait Bechtal Corporation*[6], *The Moorcock*[7]).

1.4.1.2 Faults in a contract

Despite the outward appearance of agreement there may be fundamental faults that will affect the validity of the contract. The parties are not always clear and precise in the language they use and it may be that the contract will be void, that is unenforceable, on the grounds of Mistake. This can arise in a number of ways. It may be that the subject matter of the contract is no longer in existence at the time the contract is made, e.g. where X appears to sell a machine to Y but earlier the machine had been destroyed in a fire at the factory.

It is possible that the parties have been negotiating at cross-purposes, e.g. where X intends to sell one particular machine but Y has in mind another machine at X's factory. In this situation the basic requirement of agreement is missing and no contract comes into existence.

A third possibility is where one party is mistaken as to the identity of the other contracting party and can prove that identity was crucial to his entering into the contract. It should be stressed, however, that the courts will not easily allow the mistaken party to avoid the contract for this would be an easy way for people to escape from contracts that look as though they are about to take a disastrous financial turn.

It may be that one party has been led into the contract by a Misrepresentation made by the other party, e.g. untrue statements about the capabilities of the machine. The remedies available in this situation vary depending on whether the misrepresentation was made innocently, negligently or fraudulently. Under the Misrepresentation Act 1967 where the statement was made innocently the party misled may ask for the contract to be set aside, but the court has the power to refuse this and instead grant damages. If the statement was made negligently then damages can be awarded and the contract may be set aside, but the court may decide that damages alone are sufficient and rule that the contract should continue. Lastly, the most serious misrepresentation is where it is fraudulent and here both damages and setting aside the contract will be ordered.

Even though the formation of the contract meets all the requirements it may still be declared to be an illegal contract and unenforceable. This is a complex topic but one example can be taken from the contract of employment. There are often restraint clauses in such contracts whereby an employer seeks to prevent an employee who leaves the firm from working for a rival company or setting himself up in competition. Such restraints are basically unenforceable because they are not in the public interest and are contrary to the employee's freedom to work. However, it is possible to enforce such a restraint if the employer can show that the wording of the clause was reasonable in scope and that he has some interests, such as trade secrets or customer lists, that need protecting (*Fitch* v. *Dewes*[8]). What the employer cannot do, however, is to seek to prevent competition that the ex-employee may threaten.

Similar restraints are often found in the sale of businesses. Part of the price of a business sale reflects the goodwill that the owner has built up over the years. The buyer obviously would not want the seller to start up in competition with him by opening a similar venture in a location which would pose a financial threat to the newly acquired business. The court approaches the problem in the same way as in employer–employee restraints. Basically the restraint will be struck out as being against the public interest unless the buyer can so word the restraint that the court regards it as reasonable in the circumstances. In both types of restraint the type of work or business, the length of time of the restraint and the geographical area of the restraint will all be taken into account by the court in deciding whether it is reasonable or not (*Nordenfelt* v. *Maxim Nordenfelt*[9]).

1.4.1.3 Remedies

In the vast majority of cases contracts are satisfactorily concluded with both sides completing their respective obligations. But when one party fails to do so and is in breach of contract then the question of remedies arises. The normal remedy is damages, or monetary compensation.

The aim of damages is to put the innocent party in the position he would have been in if the breach had not occurred (*Parsons* v. *B.N.M. Laboratories Ltd*[10]). It is not for him to profit from the wrongdoer's behaviour and there is in fact a duty on the innocent party to mitigate the loss wherever possible (*Darbishire* v. *Warren*[11]). The claim will also be limited to what the wrongdoer can reasonably have been expected to foresee would be the outcome of his breach. For instance where one party sends a piece of machinery to the other party for repair and the repairer is in breach of contract by not returning it by an agreed date, claims covering loss of production will be allowed if the repairer should have foreseen the likely losses caused by his delay (*Hadley* v. *Baxendale*[12]). The safest thing is to inform the repairer at the time the contract is made of the exact function the machinery plays in the manufacturing process so that he is aware of its importance (*Victoria Laundry (Windsor) Ltd* v. *Newman Industries Ltd*[13]).

Another possible remedy is *quantum meruit*. This arises where the innocent party has completed part of his contract but is prevented from continuing by the other party. His claim is then based on the amount of work he has completed up to that date (*Planche* v. *Colburn*[14]). However, this claim cannot be maintained by a party whose failure to complete is of his own doing. A builder cannot build half a garage and refuse to complete and yet claim for the work done. If the work is completed but badly, then the contract price must be paid less a deduction to compensate for the faulty work.

It is possible for the court to grant Specific Performance as a remedy whereby one party is ordered to complete his part of the contract. The remedy is discretionary and little used outside of land sales. It will not be granted where the contract is one of personal services, e.g. in a contract of employment. Although an industrial tribunal may order reinstatement of an employee following an unfair dismissal, such a remedy cannot be enforced against an unwilling employer and his refusal will merely be reflected in the compensation awarded to the former employee.

The above general discussion is obviously of the briefest nature. What follows is a closer look at specific contracts; contracts which depend for their content and form on legislation.

1.4.2 Contracts of employment

It is important to distinguish between contracts of service and contracts for services. The former describes the relationship between employer and employee while the latter is concerned with employing independent contractors to carry out certain specific tasks. Unfortunately it is not always easy to distinguish between the two and yet it is essential in order to determine the legal liabilities and responsibilities of the parties. This is particularly important in situations involving main contractors and sub-contractors. The wording of the contract can place responsibility on any party but care should be taken to set this out clearly in the various contracts. If this is done then the parties involved can cover their responsibilities by obtaining insurance.

In a contract of service it is said that a man is employed as part of the business; whereas under a contract for services his work, although done for the business, is not integrated into it but is only accessory to it (*Stevenson, Jordan and Harrison* v. *Macdonald and Evans*[15]; *The Ready-mixed Concrete (South East) Ltd.* v. *The Minister of Pensions and National Insurance*[16]). The distinction has serious repercussions on tortious liability for the general rule is that the employer is liable for the torts committed by his employees acting in the course of their employment, but he is not liable for the tortious behaviour of independent contractors. It must be stressed, however, that there are a number of exceptions to this basic rule. Even where an exception applies and the employer is liable to third parties, it may be that the contract will give the employer rights of reimbursement from the contractor.

1.4.3 Employment legislation

A contract of employment can be in any form, but the more informal it is the more difficult it may be to define its true scope.

Parliament enacted the first Contract of Employment Act in 1963, requiring that certain basic ingredients be expressed in writing. Governments since have been active in the area of employment law and much of the present law is to be found in the Employment Rights Act 1996 and the Trade Union Reform and Employment Rights Act 1993. The information to be communicated to the employee must be in writing and include the names of employer and employee, the date of commencement of employment, hours of work, pay, holiday entitlements, incapacity for work, sick pay provisions, length of notice which the employee must give and is entitled to receive, pension provisions and the employee's job title. Any changes in the terms of employment must be notified to the employee within one month of the change although they need not be retained by the employee.

Information regarding disciplinary rules and grievance procedure must also accompany the written particulars. These requirements are not, however, conclusive evidence of the terms of the contract of employment, but an employee can ask for the contract to be altered to correspond with the terms if he feels there are discrepancies. It is common also for the particulars to refer to other documentation, for instance, collective agreements, and by so doing to incorporate them into the Contract of Employment (*Systems Floors (UK) Ltd.* v. *Daniel*[17]). In all these the written agreement is persuasive but not necessarily conclusive evidence of the relationship between the parties.

The principal Act regulating employees' rights and employers' duties is now the Employment Rights Act 1996. This Act is a consolidation of earlier enactments and deals with such matters as pay, Sunday working, maternity rights, termination of employment and remedies for unfair dismissal.

Just a few of these points can be touched on in this chapter. An employee has a right to be paid in a situation where the employer has no work for him that day. To be eligible the employee must have been continuously employed for not less than one month. There will be no such entitlement if the lack of work is caused by industrial action nor will there be any entitlement if the employer has offered alternative and suitable work which has been refused.

The Act guarantees the employee's right of membership of trade unions and also payment for time off work while participating in certain union or public duties. Where an employee is given notice of dismissal by reason of redundancy he is entitled to take reasonable time off work to look for new employment or to make arrangements for training and he is entitled to be paid for such absences. Provisions are made for time off work to attend antenatal care and for payment to cover such periods. Rights to maternity leave and pay are set out in detail in Part VIII of the Act.

An employee who is suspended from work by his employer on medical grounds is entitled to wages for up to 26 weeks. Such suspension must arise from a requirement imposed by law or under a recommendation in a Code of Practice issued under HSW in relation to the Control of Lead at Work Regulations 2002, the Ionising Radiations Regulations 1999 and the Control of Substances Hazardous to Health Regulations 2002. There is no entitlement, however, if the employee is incapable of work due to disease or bodily or mental impairment.

An employer must give a minimum period of one week's notice to terminate employment of less than two years; one week's notice for each year of employment up to 12 years of continuous employment and not less than 12 weeks' notice for a continuous work period in excess of 12 years. An employee must give to his employer a minimum of one week's notice.

Part X deals with the question of unfair dismissal while Chapter II deals with employees' rights when unfairly dismissed. Of particular interest is s. 100 entitled 'Health and safety cases'. This states that where an employee has been designated by the employer to carry out activities in connection with preventing or reducing risks to health and safety at work and he is dismissed for carrying out those duties, then such dismissal will be regarded as unfair. A similar attitude is adopted in relation to a situation where the employee is a member of a safety committee and is dismissed for carrying out his functions. Where an employee is dismissed for refusing to return to a place of work on the grounds that he reasonably believes it to be a danger then any dismissal will also be regarded as unfair.

Several other pieces of legislation also play an important role in determining the relationship between employer and employee. The Equal Pay Act 1970 (see also the Equal Pay (Amendment) Regulations 1983 introducing the concept of equal value) introduces into all contracts an equality clause stating that where people are employed on broadly similar work with people of the opposite sex then any discrepancy in terms or conditions between them must be equalised upwards unless such differences can be explained on grounds other than a difference of sex. The Sex Discrimination Act 1975 as amended by the Employment Act 1989 and the Race Relations Act 1976 render it unlawful for a person to treat another less favourably on sexual or racial grounds. Any employment variations must be shown to be justified for non-sexual or non-racial reasons. A minimum wage has now been introduced by the National Minimum Wage Act 1998.

Apart from the above legislation certain other terms are implied into the employment contract, having been built up by court decisions over the years. These can be supplanted only by express provisions of an Act so stating.

The most important of the implied conditions are:

- that both parties exercise reasonable care in carrying out their duties under the contract. This means that the employer should provide a safe system of work including machinery, a safe place of work and skilled fellow employees;
- the employee must take care to act reasonably and not injure others (*Lister* v. *Romford Ice and Cold Storage Co. Ltd*[18]);
- the employee owes his employer a duty of fidelity which prevents him from working for a rival firm or divulging secret information (*Hivac Ltd* v. *Park Royal Scientific Instruments Ltd*[19]);
- the employee should co-operate fully with his employer to achieve the goal of the employment contract;
- an employer will not conduct his business in a manner likely to destroy or seriously damage the relationship of confidence and trust between him and his employee (*Malik* v. *BCCI*[20]).

1.4.4 Law of sale

The most common example of a contract is one for the sale and purchase of goods. While the basic common law rules and rules of equity still apply to such sales, many of the rules are to be found in the Sale of Goods Act 1979, as amended, which consolidates legislation that began in 1893. The Act covers the whole range of contract topics such as formation, terms, performance, transfer of ownership, rights of unpaid sellers and remedies.

Worthy of particular mention here are the implied terms found in ss. 12–15. Implied terms, as we have seen in the previous section, are terms which the courts will read into the contract where parties have failed to mention them. Because of the hesitancy of the courts in implying terms into contracts, the 1979 Act specifically sets out the important terms that must be read into the contract of sale. These terms provide the buyer with a certain basic protection against buying faulty or unsuitable goods.

Section 12(i) states that there is an implied term that the seller has good title to the goods and is therefore capable of passing true ownership to the buyer (*Rowland* v. *Divall*[21]).

Section 13 covers a situation where goods are sold by description. There is an implied term that the goods will correspond to the description given (*Beale* v. *Taylor*[22]).

Section 14 contains two important implied terms where the seller is selling in the course of a business, in contrast to a private sale. The first is that the goods must be of a satisfactory quality unless the seller has drawn the buyer's attention to the defect or the buyer has examined the goods and should have detected the faults before the contract was made (*Wilson* v. *Rickett, Cockerell & Co. Ltd*[23]). The second implied term is that where the buyer has expressly or by implication made known to the seller the particular purpose for which the goods are bought then such goods must be reasonably fit for such purpose (*Henry Kendall & Sons* v. *William Lillico & Sons Ltd*[24]).

Section 15 is concerned with sales by sample and implies a term that the bulk will correspond to the sample in quality; that the buyer will have a reasonable opportunity of comparing bulk with sample and that the goods will be free from any defect rendering them unsatisfactory which would not have been apparent on examination of the sample (*Ashington Piggeries Ltd* v. *Christopher Hill Ltd*[25]).

Certain transactions, i.e. exchange and barter, whereby goods change ownership may not fall within the definition of 'sale of goods'. Also, where a repair is carried out, the transfer of any new part itself is not regarded as a sale of goods.

To give protection to the new owner the Supply of Goods and Services Act 1982 was introduced whereby similar implied terms to those listed above under the Sale of Goods Act 1979 are incorporated into the contract. As the title of the Act suggests it also covers the service element of the contract. Thus, where the supplier is acting in the course of business there is an implied term that he will exercise reasonable care and skill. Where no time is stipulated for completion of the service then there is an implied term that it must be carried out within a reasonable time and that is determined by the facts of the case. Likewise, where no price is fixed by the contract then there is an implied term that a reasonable price will be paid.

This is an opportune moment to mention one of the most important pieces of consumer legislation in recent years, the Unfair Contract Terms Act 1977. The original Sale of Goods Act 1893, following the general principles of the common law contracts, permitted the parties to exclude themselves from legal liability for wrongful performance of their contractual duties by suitably worded clauses or notices. The 1977 Act, together with the 1982 Act, now forbids such clauses any legal recognition if the aim is to avoid liability for injury or death caused by negligence, for instance where an electrical appliance is faulty.

Where the clause is aimed at avoiding economic loss caused by one party to the other, for instance by selling unsatisfactory goods, the Act recognises two situations. If the seller sells in the course of business and the buyer is not buying in the course of business and the goods are of a type normally supplied for private use or consumption, then the Act prohibits the exclusion of the implied terms of the 1979 Act. Where the contract is not a consumer sale, then an exclusion clause will be valid but only if it can meet the test of reasonableness as laid down in the 1977 Act, the burden of proof resting on the party who wishes to utilise the exclusion clause (*George Mitchell (Chesterhall) Ltd.* v. *Finney Lock Seeds Ltd*[26]). Section 12 of the 1979 Act cannot be excluded in either category of sale. The 1982 Act adopts similar procedures. Somewhat surprisingly the

1977 Act does not apply to insurance contracts, an area where consumer protection would seem to be a necessary requirement.

The omission from the legislation was allowed in exchange for the insurance industry, through the Association of British Insurers, drawing up their own Statements of Insurance Practice. The purpose of such Statements is to give the private insured greater rights against his insurance company. The Statements do not, however, carry the force of law and represent only voluntary agreements on the part of those insurers who are members of A.B.I.

Also of great importance in terms of consumer protection is the Unfair Terms in Consumer Contracts Regulations 1999 which came out of an EU directive. As the title indicates, it is limited to consumer contracts only but it does apply to insurance contracts. It does not, however, apply to employment contracts. The determining factor is whether there is an 'unfair term'. This is defined as any term which, contrary to the requirements of good faith, causes a significant imbalance in the parties' rights and obligations under the contract to the detriment of the consumer. Such a situation will often arise where the consumer is faced with a written contract the contents of which were not negotiable.

1.4.5 Specialised legislation affecting occupational safety advisers

The responsibility of occupiers of land to those who enter the premises is to be found in the Occupier's Liability Acts of 1957 and 1984. The 1957 Act covers both tortious and contractual liability. The basic obligation is that the occupier owes a common duty of care to see that the premises are reasonably safe for the purpose for which the visitor has been permitted to enter (s. 2). It is possible, of course, for any contract that may exist between the occupier and the visitor to state a higher duty of care than that of s. 2. But under the Act it was possible for the occupier to exclude this basic duty by a suitably worded exclusion clause in a contract or by a notice (*Ashdown* v. *Samuel Williams & Sons*[27]). But the Unfair Contract Terms Act 1977 now prohibits any attempt to avoid liability for personal injury or death caused by negligence where the premises are used for business purposes and the injured person was a lawful entrant. Even under the 1957 Act where a party entered premises by virtue of a contract to which he was not a party, for example contractor's workmen, any exclusion of liability in that contract did not affect his rights against the occupier. He was owed the common duty of care unless the contract stated that higher obligations were owed, in which case he could then benefit from the higher duty (s. 3).

The Occupier's Liability Act 1984 makes two important changes to the law. The original Act made no reference to the duty of an occupier to a trespasser. The courts were left to evolve their own rules to cater for this category of person.

Section 1 of the 1984 Act states that an occupier owes a duty to take such care as is reasonable in the circumstances for the safety of a trespasser if he is aware or should have been aware of the existence of the danger and if he knows or should know that a trespasser may come within the vicinity of the danger.

Section 2 of the Act makes a change to the Unfair Contract Terms Act so that where an occupier allows someone to enter his business premises for purposes that are recreational or educational and not connected with the business itself then the occupier may rely on the use of an appropriately worded exclusion clause or notice.

Special reference should be made to two sections of the Health and Safety at Work etc. Act 1974 in a chapter on contract.

Section 4 is concerned with the various duties owed by those who have control of premises to those who are not their employees. Subsection (3) states that where such a person enters non-domestic premises by virtue of a contract or tenancy which creates an obligation for maintenance or repair or responsibility for the safety of or absence from risks to health arising from plant or substances in any such premises, then the person deemed to be in control owes a duty to see that reasonable measures are taken to ensure that such premises, plant or substances are safe and without risks to the health of the person entering. This section therefore would provide that safety standards be extended to someone who enters a cinema (contract) or enters a factory to inspect machinery (licensee) in addition to the other aim of the Act which is concerned with the safety of employees. It should be noted that liability is on the person who has control over premises or who can be described as an occupier and that case law shows that more than one person can be in that position (*Wheat* v. *Lacon & Co. Ltd*[28]).

The Consumer Protection Act 1987, schedule 3 amends and widens the scope of s. 6 of the Health and Safety at Work etc. Act 1974 which is concerned with the general standard of care and safety owed by manufacturers, designers, importers and suppliers of articles for use at work. Such persons must ensure, so far as is reasonably practicable, that the article is so designed and constructed that it will be safe and without risk to health at all times when being set, used, cleaned or maintained by a person at work. To meet these requirements there is a duty to carry out or arrange for such testing or examination as may be necessary in the circumstances. Also adequate information must be given to the person supplied about the use for which the article was designed. Revisions of earlier information must also be given. Similar obligations exist where it is a 'substance' rather than an 'article' that is being supplied. The major alteration here, however, is that the duties are not restricted to 'for use at work' but cover also when it is being 'used, handled, processed, stored or transported by a person at work'. It must be stressed that liability under the HSW Act 1974 is penal in character and civil remedies are only permissible where the Secretary of State introduces specific regulations.

However, there may be a contractual relationship between the supplier and the recipient of articles or substances referred to above. A remedy for breach of that contract may therefore be available. Subsection (8) of the HSW Act (as amended by the CP Act 1987) allows the originator of a defective article to escape liability if he has obtained a written undertaking from the person supplied that the person will take specified steps sufficient to ensure, so far as is reasonably practicable, that the article is safe and without risk to health when used, set, cleaned or maintained by a person at work. Subsection (9) also has the effect of pinpointing responsibility for contravening the standards imposed by s. 6 on the effective supplier of such goods when another person has in fact been the contracting party with the customer by virtue of a hire-purchase agreement, conditional sale or credit sale agreement. Usually the financing of these arrangements is carried out by means of finance houses. Contractually the goods are sold to the finance house who then in turn enters into his own contract with the customer. This subsection rightly seeks to ensure that the basic obligations of s. 6 remain with the originator of the faulty design, product etc., rather than allowing it to pass to the finance house. Similarly, the Health and Safety (Leasing Arrangements) Regulations 1992 extend to those who, purely as financiers, lease articles for use at work, the immunity from the duties of care imposed by s. 6 and leave the obligation of this section on the shoulders of the effective supplier rather than the ostensible supplier.

Further reading

For an introduction to the law of contract see:
Davies on Contract (Upex), 8th edn, Sweet & Maxwell, London (1999)
For more detailed coverage of employment law see:
Pitt, *Employment Law*, 4th edn, Sweet & Maxwell, London (2000)
Selwyn, N., *Law of Employment*, 11th edn, Butterworth, London (2000)
Cases and Material on Employment Law, 3rd edn, Blackstone (2000)

Reference cases

1. Scammell *v.* Ouston (1941) 1 All ER 14
2. Carlill *v.* Carbolic Smoke Ball Co. (1893) 1 QB 256
3. Bigg *v.* Boyd Gibbins Ltd (1971) 2 All ER 183
4. Dunlop Pneumatic Tyre Co. Ltd *v.* Selfridge & Co. Ltd (1915) AC 847
5. Balfour *v.* Balfour (1919) 2 KB 571
6. Matthews *v.* Kuwait Bechtal Corporation (1959) 2 QB 57
7. The Moorcock (1886–1890) All ER Rep 530
8. Fitch *v.* Dewes (1921) 2 AC 158
9. Nordenfelt *v.* Maxim Nordenfelt Guns and Ammunition Co. (1894) AC 535
10. Parsons *v.* B. N. M. Laboratories Ltd (1963) 2 All ER 658
11. Darbishire *v.* Warren (1963) 3 All ER 310
12. Hadley *v.* Baxendale (1854) 9 Exch. 341
13. Victoria Laundry (Windsor) Ltd *v.* Newman Industries Ltd (1949) 2 KB 528
14. Planche *v.* Colburn (1831) 8 Bing 14
15. Stevenson, Jordan and Harrison *v.* Macdonald & Evans (1951) 68 R.P.C. 190
16. The Ready-mixed Concrete (South East) Ltd *v.* The Minister of Pensions and National Insurance (1968) 1 All ER 433
17. Systems Floors (UK) Ltd. *v.* Daniel (1982) ICR 54; (1981) IRLR 475
18. Lister *v.* Romford Ice and Cold Storage Co. Ltd (1957) AC 535
19. Hivac Ltd *v.* Park Royal Scientific Instruments Ltd (1946) 1 All ER 350
20. Malik *v.* BCCI (1997) 3 All ER 1
21. Rowland *v.* Divall (1923) 2 KB 500
22. Beale *v.* Taylor (1967) 3 All ER 253
23. Wilson *v.* Rickett, Cockerell & Co. Ltd (1954) 1 All ER 868
24. Henry Kendall & Sons *v.* William Lillico & Sons Ltd (1968) 2 All ER 444
25. Ashington Piggeries Ltd *v.* Christopher Hill Ltd (1971) 1 All ER 847

26. George Mitchell (Chesterhall) Ltd. *v.* Finney Lock Seeds Ltd. (1983) 2 All ER 737
27. Ashdown *v.* Samuel Williams & Sons (1957) 1 All ER 35
28. Wheat *v.* Lacon & Co. Ltd (1966) 1 All ER 582

List of the Acts referred to

The Misrepresentation Act 1967
Contract of Employment Act 1963 (as amended)
Equal Pay Act 1970 and 1975
Sex Discrimination Act 1975 and 1986
Race Relations Act 1976
Sale of Goods Act 1893
Sale of Goods Act 1979
Unfair Contract Terms Act 1977
Occupier's Liability Acts 1957 and 1984
Health and Safety at Work etc. Act 1974
Supply of Goods and Services Act 1982
Consumer Protection Act 1987
Equal Pay (Amendment) Regulations 1983
Social Security Act 1986
Trade Union Reform and Employment Rights Act 1993
Unfair Terms in Consumer Contracts Regulations 1999, SI 1999 No. 2083
Employment Rights Act 1996
Contracts (Rights of Third Parties) Act 1999
National Minimum Wage Act 1998

Chapter 1.5

Employment law

R. D. Miskin updated by Amanda Jones

1.5.1 Introduction

Up until the end of the 1960s an employee in the UK had little or no legal protection so far as his employment was concerned. The employer had no duty to give the employee any specific form of contract or inform him of the basic terms of his employment. Thus, the employer had the power to dismiss virtually as and when he wanted and had no duty to give the employee the reasons for such dismissal, so the average employee, with only very few exceptions, had no continuity of employment nor any right to claim compensation when unfairly dismissed.

However, towards the end of the 1960s the then Government decided that an employee was entitled to be legally protected in the continuity of his employment, not to be unfairly or unreasonably dismissed and to be informed of the more important terms of his employment. There followed a number of Acts of Parliament implementing these rights and many others. It is the object of this chapter to consider the basic principles of such legislation and the rules, regulations and decided cases supporting it, but it must be appreciated that only an overall summary can be given. Where specific problems arise reference must be made to the relevant statutes.

1.5.2 Employment law

Employment law is governed by statute and the decided case law arising from those statutes. The most important is the Employment Rights Act 1996 (ERA) which consolidates earlier enactments relating to employment rights. It supersedes the Employment Protection (Consolidation) Act 1978, the Wages Act 1986 and parts of the Trade Union and Employment Rights Act 1993, many sections of which remain. However, it does not repeal the Equal Pay Act 1970, the Sex Discrimination Act 1975, the Race Relations Act 1976, the Transfer of Undertakings (Protection of Employment) Regulations 1981 or the Disability Discrimination Act 1995 but is the principal enactment to be taken into account when considering the rights of employees.

1.5.2.1 The Employment Rights Act 1996

This is 'An Act to consolidate enactments relating to employment rights' and deals mainly with the following areas:

(a) An employee's right to a written statement of the main particulars of his employment.
(b) Protection of wages.
(c) Guaranteed payments.
(d) Protection for shop workers and betting shop workers working on a Sunday.
(e) The right not to suffer detriment in employment.
(f) Time off work.
(g) Suspension from work on, *inter alia*, health and safety grounds.
(h) Termination of employment.
(i) The right not to be unfairly dismissed and remedies for unfair dismissal.
(j) Entitlement to redundancy payment.
(k) Position where the employer is insolvent.
(l) Definition of what amounts to a week's pay.

One of the main rights granted to an employee under this Act is that the employer must give a written statement of the particulars of his employment. Although it is often assumed that the written particulars form his 'contract' of employment this is not technically so. However, they amount to very strong evidence of the terms of a contract of employment and, from the purely practical point of view, they are the only so-called contract many employees receive. To prove that the terms are not of a contractual nature would be difficult.

The information must be given not later than two months after the commencement of employment and contain the following information:

(i) The names of both employer and employee.
(ii) The date when employment began.
(iii) The date on which the employee's period of continuous employment began.
(iv) The scale or rate of remuneration or the method of calculating it, whether such remuneration is paid weekly, monthly or at some other specified interval.
(v) Terms relating to hours of work.
(vi) Entitlement to holiday, including public holiday and holiday pay.
(vii) Payment during incapacity for work due to sickness or injury including sick pay provisions.
(viii) Pensions and pension schemes.
(ix) Length of notice to which an employee is entitled and is required to give to terminate his contract of employment.
(x) Job title and, where the employment is not permanent, the period for which it is expected to continue.
(xi) Place of work including details of any mobility clause and the employer's address.

(xii) Any collective agreements which affect the terms and conditions of employment.
(xiii) Details of necessity to work outside the UK if relevant.
(xiv) Details of any disciplinary rules that apply to the employee.
(xv) Name of person to whom complaints may be made if the employee is dissatisfied.

It should be noted that the employer may, so far as sickness, injury, pension schemes and collective agreements are concerned, specify the document or agreement in which the provisions are contained provided the employee has reasonable access to them to acquaint himself with their contents. An employee is entitled to an itemised pay statement and the employer should make no deductions from his wages unless the employee has specifically agreed to such deductions in writing. The exceptions to this rule are statutory deductions for income tax, National Insurance contributions and payments made under an Attachment of Earnings Order. This provides important protection for the employee and is firmly enforced by the Tribunals. Employment Tribunals procedures and practices are set out in the Employment Tribunal (Constitution and Rules of Procedure) Regulations 2001 and associated regulations. Other provisions of ERA are considered in the following sections.

1.5.3 Discrimination

1.5.3.1 Sex discrimination

1.5.3.1.1 The Sex Discrimination Act 1975

Section 1 states that a person discriminates against a woman if in any circumstances relevant to the purposes of any provision of the Act he:

(a) on the ground of her sex, treats her less favourably that he treats or would treat a man, or
(b) applies to her a requirement or condition which applies or would apply equally to a man but:
  (i) which is such that the proportion of women who can comply with it is considerably less than the proportion of men who can comply with it, and
  (ii) which he cannot show to be justifiable, irrespective of the sex of the person to whom it applied, and
  (iii) which is to her detriment because she cannot comply with it.

Section 1(1)(a) refers to direct discrimination where someone is treated differently because of his/her sex/marital status. It is best thought of in terms of comparative treatment. It should be noted that the complainant does not have to prove that the discrimination was intentional, only that it occurred.

The circumstances covered by s. 1(1)(b) are known as indirect discrimination and require that the employer has to prove his conduct was justifiable. Indirect discrimination occurs where an apparently neutral requirement or condition is applicable, but members of a certain sex are less able to meet the requirement.

Section 2 requires that men should receive equal treatment, but the vast majority of claims are from women. Section 6 concerns employment opportunities and says:

1 It is unlawful for a person to discriminate against a woman:
   (a) in the arrangements he makes for the purposes of determining who should be offered that employment, or
   (b) in terms on which he offers her that employment, or
   (c) by refusing, or deliberately omitting to offer her that employment.
2 It is unlawful for a person to discriminate against a woman employed by him:
   (a) in the way he affords her access to opportunities for promotion, transfer, training or any other benefits, facilities or services, or by refusing or deliberately omitting to afford her access to them, or
   (b) by dismissing her or subjecting her to other detrimental treatment.

Other parts of this section include further protection to a woman in employment.

There are exceptions where sex is a genuine occupational qualification and this is covered in s. 7 which provides:

1 In relation to sex discrimination:
   (a) s. 6(1)(a) or (c) do not apply to any employment where being a man is a genuine occupational qualification for the job, and
   (b) s. 6(2)(a) does not apply to opportunities for promotion or transfer to, or training for such employment.
2 Being a man is a genuine occupational qualification for a job only where:
   (a) the essential nature of the job calls for a man for reasons of physiology (excluding physical strength or stamina) or, in dramatic performances or other entertainment, for reasons of authenticity so that the essential nature of the job would be materially different if carried out by a woman; or
   (b) where the job needs to be held by a man to preserve decency or privacy because:
      (i) it is likely to involve physical contact with men in circumstances where they might reasonably object to it being carried out by a woman or
      (ii) the holder of the job is likely to do his work in circumstances where men might reasonably object to the presence of a woman because they are in a state of undress or are using sanitary facilities or

(ba) the job is likely to involve the holder doing his work or living in a private home and need to be held by a man because objection might reasonably be taken to allowing a woman:
   (i) the degree of physical or social contact with a person living in the home or
   (ii) the knowledge of intimate details of such a person's life or of the home.

These are the main exceptions to the general rule but it should be noted that there are others which apply.

In health and safety matters discrimination is allowed on health grounds and concerning both pregnancy and maternity. This occurred in *Page* v. *Freight Hire Tank Haulage Ltd*[1] where a woman lorry driver, who was of child bearing age, was prevented from driving a tanker lorry containing chemicals that could be harmful to a woman's ability to bear children.

1.5.3.1.2 Sex Discrimination Act 1986

This Act amends certain provisions of the 1975 Act and in particular makes reference to collective agreements, partnerships, employment in private households, rules of professional bodies or organisations, exemptions for small businesses and discrimination in training.

1.5.3.2 The Race Relations Act 1976

This Act is couched in almost exactly the same terms as the Sex Discrimination Act 1975 in that it provides in s. 1(1) that a person discriminates against another if in any circumstances relevant to the purposes of any provision he:

(a) on racial grounds treats that person less favourably than he treats or would treat other persons, or
(b) he applies to that person a requirement or condition which he applies or would apply equally to persons not of the same racial groups as that person but:
   (i) which is such that the proportion of persons of the same racial group as that person who can comply with it is considerably smaller than the proportion of persons not of that racial group who can comply with it, and
   (ii) which he cannot show to be justifiable irrespective of the colour, race, nationality or ethnic or national origins of the person to whom it is applied, and
   (iii) which is to the detriment of that other person because he cannot comply with it.

Race discrimination can be both direct and indirect, the latter being more difficult to recognise. Section 4 specifies that:

1 It is unlawful for a person, in relation to employment by him at an establishment in Great Britain, to discriminate against another:
  (a) in the arrangement he makes for the purpose of determining who should be offered that employment; or
  (b) in the terms on which he offers him that employment; or
  (c) by refusing or deliberately omitting to offer him that employment. It is also discriminatory if an employee is treated unfavourably in terms of employment, promotion, training, dismissal or is subject to any other detrimental treatment.

There are exceptions where genuine occupational qualifications are required and these are listed in s. 5 as:

1 In relation to racial discrimination:
  (a) s. 4(1)(a) or (c) do not apply to any employment where being of a particular racial group is a genuine occupational qualification for the job and
  (b) s. 4(2)(b) does not apply to opportunities for promotion or transfer to, or training for, such employment.
2 Being of a particular racial group is a genuine occupational qualification for a job only when:
  (a) the job involves participation in a dramatic performance, or other entertainment in a capacity for which a person of that racial group is required by reasons of authenticity; or
  (b) the job involves participation as an artiste or photographic model in the production of a work of art, visual image or sequence of visual images for which a person of that racial group is required for reasons of authenticity; or
  (c) the job involves working in a place where food or drink is (for payment or not) provided to and consumed by members of the public or a section of the public in a particular setting for which a person of that racial group is required for authenticity; or
  (d) the holder of the job provides persons of that racial group with a personal service promoting their welfare, and those services can most effectively be provided by a person of that racial group.

There are restrictions on advertisements which might reasonably be understood to indicate, or do indicate, an intention to racially discriminate.

1.5.3.3 The Disability Discrimination Act 1995

This Act makes it unlawful to discriminate against any disabled person in connection with employment, the provision of goods, facilities and services or the disposal or management of premises, makes provision for the employment of disabled persons and establishes a National Disability Council. The Act defines disability and disabled persons and in s. 4 makes it unlawful for an employer to discriminate against a disabled person:

(a) in the arrangements which he makes for the purpose of determining to whom he should offer employment;
(b) in the terms on which he offers that person employment; or
(c) by refusing or offer to deliberately not offering him employment.

It is also unlawful for an employer to discriminate against a disabled person whom he employs:

(i) in the terms of employment;
(ii) in the opportunities afforded for promotion, transfer, training or receiving any other benefit;
(iii) by refusing to offer, or deliberately withholding, any such opportunity; or
(iv) by dismissing him or subjecting him to detrimental treatment.

Section 5(1) states, *inter alia*, that the employer discriminates against a disabled person if:

(a) for a reason which relates to the disabled person's disability he is treated less favourably than others to whom that reason does not, or would not, apply; and
(b) he cannot show that the treatment in question is justified,
  (aa) he fails to comply with a duty under s. 6 imposed on him in relation to the disabled person; and
  (bb) he cannot show that this failure to comply with that duty is justified.

Section 6 deals with the duty of the employer to make arrangements to enable a disabled person to carry out his job properly. The provisions of s. 6 have important health and safety connotations and should be studied in conjunction with the Disability Discrimination (Employment) Regulations 1996 and the Code of Good Practice on the Employment of Disabled People[2] which deals, in ss. 4.2, 4.3 and 4.4, with safety aspects in particular.

Note should also be taken of the Equal Pay Act 1970 which is aimed at preventing discrimination in terms and conditions of employment between men and women. It is a complex Act to understand but its main principle is to ensure that where a woman is employed on like work with a man on the same employment she is entitled to the same terms of employment as a man. The most obvious claim by a woman under this Act is that she should be paid at the same rate as a man. Although the procedures for bringing such a claim are complex, there have been many Industrial Tribunal cases where this particular point has been argued.

1.5.3.4 The employment of children and young persons

The Management of Health and Safety at Work Regulations 1999 (MHSW) in reg. 1 give two important definitions; firstly, that a child is a person who is not over compulsory school age, and, secondly, that a young person is one who has not attained the age of 18.

They state that an employer who employs a young person shall, in carrying out a risk assessment take particular account of the following factors:

(a) the inexperience, lack of awareness of risks and immaturity of young persons;
(b) the fitting out and layout of the workplace and the work station;
(c) the nature, degree and duration of exposure to physical, biological and chemical agents;
(d) the form, range and use of work equipment and the way in which it is handled;
(e) the organisation of processes and activities;
(f) the extent of the health and safety training provided for young persons; and
(g) the risks from agents, processes and work listed in the annex to EU Directive 94/33/EC on the Protection of Young People at Work[3].

Prior to employing a child, an employer must provide a parent of the child with comprehensible and relevant information on:

(a) the risks to his health and safety identified by the assessment;
(b) the preventive and protective measures; and
(c) the risks notified to him by other employers sharing the same premises.

An employer must not employ a young person for work:

(a) which is beyond his physical and psychological capacity;
(b) involving harmful exposure to agents which are toxic, carcinogenic, cause heritable genetic damage, harm to the unborn child or which in any way chronically affect human health;
(c) involves harmful exposure to radiation;
(d) involves the risk of accidents which it may reasonably be assumed cannot be recognised or avoided by young persons owing to their insufficient attention to safety or lack of experience or training; or
(e) in which there is a risk to health from:
    (i) extreme cold or heat
    (ii) noise or
    (iii) vibration.

A young person, who is no longer a child, may be employed for work:

(a) where it is necessary for his training
(b) which is supervised by a competent person and
(c) where any risks have been reduced to the lowest level that is reasonably practicable.

1.5.3.5 Joint consultation

In October 1978, the Safety Representatives and Safety Committee Regulations 1977 (SRSC) came into effect and gave to those unions that were recognised in the workplace the right to appoint safety representatives. Those safety representatives were given certain functions and employers were required to give to the representatives, to enable them to perform their functions, time off work with pay for training and to carry out their functions, information necessary to fulfil their functions and allow them to carry out inspections of the workplace following accidents.

In 1989 the European Council adopted a directive no. 89/391/EEC (known as the Framework Directive) which contained a requirement for workpeople, whether union members or not, to be consulted about matters concerning their health and safety at work. The UK Government held that this was covered by the SRSC but a judgement by the European Court of Justice[4] in 1992 established the right of all employees to be consulted. This right was brought into effect in the UK by the Health and Safety (Consultation with Employees) Regulations 1996 as amended by the Employment Rights Dispute Resolution Act 1998.

These two Regulations effectively give the same rights and functions to safety representatives, whether union or employer appointed, and include the right:

1 to be consulted on:
   (a) the introduction of measures affecting health and safety
   (b) arrangements for appointing safety advisers
   (c) arrangements for appointing fire and emergency wardens
   (d) the health and safety information to be provided to employees
   (e) provisions for health and safety training
   (f) health and safety implications of new technologies
2 to be given sufficient information:
   (a) to carry out their functions
   (b) on accidents that had occurred but not:
      - where an individual can be recognised
      - if it could prejudice the company's trading
      - on matters subject to litigation
      - if it was against national security
      - if it contravened a prohibition imposed by law
3 to have time off work with pay:
   - to carry out their functions
   - to receive training
4 to carry out their functions which include:
   - making representations to the employer on hazards and incidents affecting his constituents' health and safety
   - being the contact with and receive information from HSE inspectors
   - investigating potential hazards and incidents affecting those he represents.

In addition, union appointed safety representatives have the right to:

- investigate complaints by those they represent
- carry out inspections of the workplace subject to the agreement of the employer
- attend meetings of the safety committee.

In both cases, complaints against the employer concerning refusal to provide, or allow time off for, training and for not paying for that time off are heard by an Industrial Tribunal.

1.5.3.6 Working time

The Working Time Regulations 1998, which derive from EU directive no. 93/104/EC[5], provide for a maximum working week of 48 hours. However, this can be extended by written agreement between the employee and employer. Night work is restricted to 8 hours in each 24 hour period and night workers are to have health assessments. Employers are required to keep records of the hours worked. Periodic rest times are specified as are the rest breaks to be taken if the working period is more than 6 hours ($4\frac{1}{2}$ hours for young persons). Workers (except those in agriculture) are entitled to 4 weeks paid leave each year.

Excluded from these requirements are workers in transport, trainee doctors, sea fishermen, police and armed forces and domestic servants. The Regulations allow the employer to vary working times to meet particular employment and trading circumstances.

1.5.4 Disciplinary procedures

Dismissal is dealt with in the following section, but a Tribunal, to find a dismissal fair, must be satisfied that the dismissal was reasonable in all the circumstances. In the majority of cases this entails the employer following his own disciplinary and grievance procedures. It is important that an employer should have formal disciplinary rules which should be communicated to each and every employee. It is a requirement of the Employment Rights Act 1996 that the written particulars of employment include the disciplinary rules that are applicable. Should such communication not have taken place the employer will not be able to rely on such rules, and a dismissal which might otherwise have been fair could be ruled unfair.

Acceptable procedures are outlined in the ACAS Code of Practice[6] which emphasise the importance of such rules by giving practical guidance on how to deal with issues involving disciplining and settling grievances. The rules should be set out clearly and concisely and be available to all employees.

To be effective the procedures should:

(a) be in writing,
(b) specify to whom they apply,
(c) be non-discriminatory,
(d) provide for matters to be dealt with quickly,
(e) provide for proceedings, witness statements and records to be kept confidential,
(f) indicate the disciplinary actions that may be taken,
(g) specify the various levels in the organisation that have the authority to take disciplinary action,
(h) provide for workers to be informed of the complaints against them and to have all relevant evidence before a hearing if possible,
(i) provide workers with an opportunity to state their case,
(j) allow workers to be accompanied,
(k) ensure that, except for gross misconduct, no worker is dismissed for a first breach of discipline,
(l) ensure that no action is taken until the case has been investigated,
(m) ensure that the workers are given an explanation for any penalty imposed, and
(n) provide a right of appeal and specify the procedure to be followed.

A record should be kept of any disciplinary actions taken against an employee for breach of the rules including lack of capability, conduct etc. and what disciplinary action was taken and the reasons supporting such action. The disciplinary procedures should be reviewed from time to time to ensure that they comply with the then practices of the employer. A written record should be kept of an oral warning to prove that it was actually given.

Many of these rules and procedures will incorporate items relevant to safety, health and welfare of the employees in that particular employment. The emphasis placed on particular aspects of safety and health will reflect the degree of risk or hazard faced by the employee in his daily work and what effect failure to follow these rules might have on the employees themselves, the environment or the continuing operation of the business. The onus is on the employer to draw up these rules and he may do this unilaterally but it is more prudent of him to consult the employees or their representative to obtain agreement to and acceptance of the various procedures before they are implemented.

The employer should ensure that, except for gross misconduct, no employee is dismissed for a first breach of discipline. Instead the employer should operate a system of warnings consisting of an oral warning, a first written warning and then a final written warning before dismissal is considered.

An employee at any disciplinary hearing must be informed of his right to appeal.

Under section 10 of the Employment Relations Act 1999, an employee is entitled to be accompanied at disciplinary or grievance hearings when they make a reasonable request to be accompanied. If an employer fails to allow an employee to be accompanied, a complaint may be presented to an Employment Tribunal. If successful, compensation up to two weeks pay (as defined by statute) may be awarded.

Until the case of *Polkey* v. *A.E. Dayton (Services) Ltd*[7] the courts tended to take the view that where employers did not follow their disciplinary procedures, but even if they had it would have made no difference to the outcome, then the dismissal was fair notwithstanding such failure. This principle was summarised by Browne Wilkinson, J. in *Sillifant* v. *Powell Duffryn Timber Ltd*[8] as follows:

> 'Even if, judged in the light of circumstances known at the time of dismissal, the employer's decision was not reasonable because of some failure to follow a fair procedure, yet the dismissal can be held to be fair if, on the facts proved before the Industrial Tribunal, the Industrial Tribunal comes to the conclusion that the employer could reasonably have decided to dismiss if he had followed fair procedure.'

The *Polkey* decision found that the one question an Employment Tribunal was not permitted to ask in applying the test of reasonableness was the hypothetical question of whether it would have made any difference to the outcome if the appropriate procedural steps had been taken. However, it was quite a different matter if the Tribunal was able to conclude that the employer himself, at the time of dismissal, acted reasonably in taking the view that, in the exceptional circumstances of the particular case, the procedural steps normally appropriate would have been futile, could not have altered the decision to dismiss and therefore could be dispensed with. In such a case the test of reasonableness may have been satisfied.

The *Polkey* decision makes it clear that a Tribunal will not err in law if it starts from the premise that breach of procedures, at least where they embody significant safeguards for the employee, will render a dismissal unfair. It is important that the employer follows his disciplinary procedures as closely as possible in the circumstances of any particular case.

1.5.5 Dismissal

Under s. 94 of ERA an employee has the right not to be unfairly dismissed. However, under section 95 an employee is dismissed if:

(a) the contract under which he is employed is terminated by the employer (whether with or without notice);
(b) he is employed under a contract for a fixed term and that term expires without being renewed under the same contract; or
(c) the employee terminates the contract under which he is employed (with or without notice) in circumstances in which he is entitled to terminate it without notice by reason of the employer's conduct.

An employee is dismissed for these purposes if:

(a) the employer gives notice to the employee to terminate his contract of employment; and
(b) at a time within the period of that notice the employee gives notice to the employer to terminate the contract on a date earlier than the date on which the employer's notice is due to expire and the reason for the dismissal is taken to be the reason for which the employer's notice is given.

In s. 97 'the effective date of termination' of employment is taken as:

(a) the date on which the notice expires whether the notice is given by the employee or the employer;
(b) the date on which termination takes effect if terminated without notice; and
(c) the date of expiry of the contract where it is a fixed term and is not being renewed.

An employee who has had his employment terminated is entitled to written reasons from the employer.

In determining whether the dismissal is fair or unfair, s. 98(1) requires the employer to show:

(a) the reason (or if more than one the principal reason) for the dismissal, and
(b) that it is either a reason falling within s. 98(2) or some other substantial reason sufficient to justify the dismissal of an employee from the position which he held.

A reason for dismissal is sufficient if:

(a) it relates to the capability or qualifications of the employee to perform work of the kind he is employed to do,
(b) it relates to the conduct of the employee,
(c) the employee is redundant, or
(d) the employee could not continue to work in the position which he held without contravention (either on his part or on the part of the employer) of a statutory duty or restriction.

The above are reasons upon which an employee can be fairly dismissed; however, s. 98(4) states that where an employer has fulfilled the requirements of subs. (1), the determination of the question whether the dismissal is fair or unfair:

(a) depends on whether, in the circumstances (including the size and administrative resources of the employer's undertaking), the employer acted reasonably or unreasonably in treating the reason as being sufficient for dismissing the employee, and

(b) shall be determined in accordance with equity and the substantial merits of the case.

It is important to understand the grounds upon which an employer can rely as having acted reasonably and fairly in the dismissal of an employee. The above reasons for dismissal are considered in the following sections.

1.5.5.1 Capability or qualification

Capability is defined in ERA as the employee's capability assessed by reference to skill, attitude, health or other physical or mental quality and such qualifications as any degree, diploma or other academic, technical or professional qualification relevant to the position the employee held. The two main classes of capability, or lack of it, are ill-health and the inability of the employee to carry out his duties in a reasonable and acceptable manner.

1.5.5.2 Ill-health

Ill-health falls into two categories. Firstly, where the employee is sick or incapacitated for one long period and, secondly, where he has regular short spells of illness which, added together, represent a lengthy period of absence. It is necessary to consider these two classes of illness separately as the legal position is different in each case.

1.5.5.2.1 Long-term illness

The leading case which sets out the main principles to support a fair dismissal for long-term illness is *Spencer* v. *Paragon Wallpapers Ltd*[9] in which the employee had been absent sick for approximately two months and the medical opinion was that he would return within another four to six weeks. The EAT held that in such cases the employer must take into account:

(a) the nature of the illness,
(b) the likely length of the continuing absence,
(c) the employer's need for the work to be done, and
(d) the availability of alternative employment for the employee.

Since all four criteria had been met, the dismissal was fair.

Consultation with the employee and investigation of the medical position by the employer would seem to be the two most important criteria. In *East Lindsay District Council* v. *Daubny*[10] the EAT stated that unless there were wholly exceptional circumstances the employee should be consulted and the matter discussed with him before his employment was terminated on the ground of ill-health.

1.5.5.2.2 Continuing periodic absences

In stark contrast to the above there have been several EAT cases where the employee has been dismissed for persistent absenteeism due to a succession of short illnesses. In *International Sports Company Ltd* v. *Thomson*[11] an employee was persistently absent for minor ailments that could not later be confirmed by medical examination. After review by the employer of her absence record and being given reasonable warnings she was dismissed. There had been no improvement in her attendance and the dismissal was held to be fair.

A factor in this decision was that there had been no medical investigation and that the employer would have been no wiser even if he had carried out such examination.

Further, it is essential for the employer to stick to his disciplinary procedures and give the appropriate warnings.

Ill-health caused by an employee's duties can lead to fair dismissal but an employer may be held not to have acted reasonably if reasonable steps had not been taken to eliminate the danger to health stemming from the job. If an employee claims he is absent for reasons of ill-health but the employer believes he is malingering it may be difficult for the employer to prove such. However, in *Hutchinson* v. *Enfield Rolling Mills Ltd*[12] the Tribunal was satisfied the employer had done so. They held that if there is evidence to suggest that the employee is, in fact, fit to work, despite his having a doctor's sick note, the employer can seek to rely upon that evidence to justify the dismissal.

1.5.5.3 Lack of skill on the part of the employee

Tribunals often find it difficult to decide whether a dismissal for incompetence is fair or unfair. What is clear is that it is not open to them to rely on their own view as to the employee's competence rather than that of the employer's. An employer has to show the he/she honestly believed that the employee was incompetent and that this belief was held on reasonable grounds.

The test therefore is the genuine belief of the employer based on the evidence that he has gathered to show that his view is a reasonable one. In such cases the employer must rely on his disciplinary procedures and give constructive warnings to the employee to give him an opportunity of improving his performance. Thus the employer should carry out a thorough evaluation of the employee's performance and discuss his criticisms with the employee personally, warn the employee of the consequence of there being no improvement and then give him reasonable opportunity to improve.

There are cases where an employer cannot follow the above procedure because of the seriousness of the consequences of further error. A pilot was dismissed and not given any further opportunities to improve when the company considered he was to blame when he made a faulty landing which caused considerable damage to the aircraft. In this case, the Court of Appeal specifically approved of the following statement made by Bristow, J.: 'In our judgement there are activities in which the degree of professional skill which must be required is so high, and the potential consequences of the smallest departure from that high standard are so serious, that one failure to perform in accordance with those standards is enough to justify dismissal. The passenger-carrying airline pilot, the scientist operating the nuclear reactor, the chemist in charge of research into the possible effects of, for example, thalidomide, the driver of the Manchester to London express, the driver of the articulated lorry full of sulphuric acid, are all in a position in which one failure to maintain the proper standard of professional skill can bring about a major disaster.'

Finally, the employer is entitled to dismiss an employee without warning where there is little likelihood of the employee improving his performance and his continuing presence is prejudical to the company's best interest. This is illustrated by the case of *James* v. *Waltham Holy Cross UDC*[13].

1.5.5.4 Misconduct

Misconduct in the place of work, or in certain circumstances outside it, is one of the major reasons for dismissal of an employee. It was defined by the Scottish EAT as 'Actings [*sic*] of such a nature, whether done in the course of employment or out of it, that reflect in some way on the employer–employee relationship.'

Discipline for misconduct falls into two main categories: firstly, the lesser transgressions should be dealt with under the employer's disciplinary practices, by way of warning and encouragement not to transgress again and, secondly, the more serious cases of gross misconduct, by instant dismissal. The employer should list in his disciplinary rules those acts that fall into the category of gross misconduct so that the employee is in no doubt whatsoever that by committing such act he renders himself liable to instant dismissal. The acts concerned vary from business to business but normally include:

- theft
- fraud
- deliberate falsification of records
- fighting
- assault on another person
- deliberate damage to company property
- serious incapability through alcohol or being under the influence of illegal drugs
- serious negligence which causes unacceptable loss, damage or injury
- serious acts of insubordination.

As well as referring to his disciplinary rules and procedures, an employer should refer to the contract of employment to ascertain what was required of the employee.

So far as criminal offences committed away from the place of work are concerned, the ACAS Code of Practice[6] makes it clear that these should not constitute automatic reasons for dismissal, but should be considered in the light of whether the offence in question makes the employee unsuitable for his or her type of work or unacceptable to other employees. If an employee secures employment by not disclosing a previous criminal conviction his dismissal on that ground is often fair provided he is not permitted to withhold such conviction under the Rehabilitation of Offenders Act 1974.

In cases of dismissal for misconduct it is essential that the employer has acted reasonably and fairly in all the circumstances. Although decided in 1978, the case of *British Home Stores* v. *Burchell*[14] still provides the basic guidelines to whether or not an employer has acted reasonably. The judgement in that case clearly sets out the steps an employer must take before dismissing an employee on the grounds of gross misconduct as:

1 belief in the employee's guilt,
2 having reasonable grounds for believing so, and
3 having carried out reasonable investigation to verify the grounds for sustaining that belief.

If the employer has followed these steps, then the Tribunal must uphold his decision although they may not necessarily have come to the same view themselves. Further, the standard of proof which an employer must meet is only that he should be satisfied on the balance of probabilities, and not beyond all reasonable doubt. These principles have been slightly eroded by subsequent legislation in that it is not essential for the last two elements to be proved but it will be very much in the employer's favour if he can do so.

There are a number of cases relating to misconduct which revolve around issues of health and safety in the workplace. The first of these is *Austin* v. *British Aircraft Corporation Ltd*[15] where the employer's attitude was considered unreasonable. Mrs Austin and her fellow employees were required to wear eye protection. Mrs Austin already wore glasses and the goggles provided were uncomfortable. However, she persevered for three months but eventually stopped wearing them. She raised the problem with her employers and the matter was put in the hands of the safety officer. Six months later nothing had been done so Mrs Austin resigned. The Tribunal hearing her case concluded that Mrs Austin had been constructively dismissed and was entitled to resign by reason of her employer's conduct.

The same principle applied in *Keys* v. *Shoefayre Ltd*[16] where the owner of a retail shop failed to take proper security precautions to protect his employees who worked in a shop in an area with a high crime rate that had suffered two armed robberies. Here it was held that the employer had failed to take reasonable care and provide a safe system of work and that Mrs Keys' resignation amounted to unfair constructive dismissal.

In the manufacture of tyres, part of the process emits dust and fumes that reports from America indicated might be carcinogenic. Negotiations resulted in face masks being provided as an interim measure until expensive capital equipment could be obtained which would improve matters, a step that was supported by the HSE. However, in *Lindsay* v. *Dunlop Ltd*[17] the employee was not satisfied with these precautions and delayed removing the tyres from the press until the fumes had dispersed. This seriously affected production and, following discussion with his union, the employer dismissed the employee. The Tribunal held that the dismissal was fair, a decision upheld by the Court of Appeal on the grounds that the employer had taken all reasonable steps in the circumstances.

1.5.5.5 Redundancy

The provisions in ERA regarding redundancy are both technical and difficult to understand but s. 139 states:

1 an employee who is dismissed shall be taken to be dismissed by reason of redundancy if the dismissal is wholly or mainly due to the fact that:
   (a) his employer has ceased or intends to cease:
      (i) to carry on the business for the purposes of which the employee was employed by him, or
      (ii) to carry on that business in the place where the employee was employed, or
   (b) the requirements of the business:
      (i) for the employee to carry out work of a particular kind, or
      (ii) for the employee to carry out work of a particular kind in the place where the employee was employed
   have ceased or diminished or are expected to cease or diminish.
2 For the purposes of subs. (1) the business of the employer together with the business or businesses of his associated employers shall be treated as one (unless either of the conditions specified in paragraphs (a) and (b) of that subsection would be satisfied without so treating them).

It is open to an employee to claim that the employer acted unreasonably in electing to make workers redundant. He may allege that his dismissal on this ground was unfair for two reasons. The first is that the method of selection was unfair and the second that within the meaning of s. 98(4)–(6) his selection was unreasonable. It is automatically unfair to select an employee for redundancy or any dismissal for:

1 a pregnancy or pregnancy related reason
2 a health and safety reason specified in s. 100
3 a reason related to the fact that they are protected shop workers specified in s. 101
4 being trustee of an occupational pension scheme specified in s. 102
5 being a representative or candidate to be such representatives as specified in s. 103

6 a reason relating to the assertion of a statutory right under s. 104
7 a reason connected with trade union membership or activities.

In *Williams* v. *Compair Maxam Ltd*[18] the EAT set out a general guideline as to the circumstances in which a selection for redundancy would be fair. These were that the employer should:

1 seek to give as much notice as possible;
2 consult the union as to the criteria to be applied in selecting the employees to be made redundant;
3 ensure that such criteria do not depend solely upon the opinion of the person making the selection but can be objectively checked against such things as attendance record, efficiency at the job, experience or length of service;
4 seek to ensure that the selection is made fairly in accordance with these criteria and consider any representation the union may make; and
5 see whether, instead of dismissing an employee, he could offer him alternative employment.

Whether or not a union is involved, a sensible employer will follow the above rules.

Over the years, the courts have varied in the importance they put upon consultation between the employer and the employee where redundancy is concerned. However, the present position is that consultation is of considerable importance and in *Polkey*[7] Lord Bridge said '. . . in the case of redundancy, the employer will normally not act reasonably unless he warns and consults any employee affected or their representative'. This does not mean that where no consultation takes place with the employee the redundancy is inevitably unfair but it certainly makes the employer's position more difficult to sustain. In some cases there is a statutory obligation to consult recognised trade unions over redundancies. A redundancy can also be rendered unfair by the failure of the employer to find alternative employment for the employee.

1.5.5.6 Contravention of an enactment

An employer can dismiss fairly where he can show that the employee could not continue to work without contravening a statutory enactment. The most common example is where an employee is disqualified from driving by a court and the principal part of his job is driving. In these circumstances the employer would have to consider either providing alternative transport or alternative employment. Other instances occur where an Authority is directed under the Schools Regulations not to employ a teacher because he is unsuitable or where an airline pilot cannot fly because an Air Navigation Order stipulates that he cannot do so unless the operator has satisfied himself that the pilot is competent to perform his duties.

1.5.5.7 Any other substantial reason

A dismissal that does not fall under one of the four potentially fair reasons may still be fair if it is 'for some other substantial reason' or SOSR. SOSR dismissals are often on the grounds of re-organisation necessary to protect business interests. The situation is primarily looked at from the employer's angle since, in re-organisation dismissals, it is their interests which are most important. It is difficult for an employee to challenge the employer's reasons as not being 'sound and good' as it is only the perceived advantages to the business at the time of the dismissal that need to be demonstrated.

1.5.5.8 Reasons for dismissal which are automatically unfair

Of particular importance to those involved in dismissals involving health and safety issues are subss. (1), (2) and (3) of s. 100 of ERA whereby:

1 An employee is regarded as having been unfairly dismissed if the principal reason for the dismissal was:
   (a) having been designated by the employer to carry out activities in connection with preventing or reducing risks to health and safety at work, the employee carried out, or proposed to carry out, such activities,
   (b) being a representative of workers on matters of health and safety at work or a member of a safety committee:
      (i) in accordance with arrangements established under or by virtue of any enactment,
      (ii) by reason of being acknowledged as such by the employer, the employee performed, or proposed to perform, the functions of a representative or a member of a safety committee,
   (c) being an employee at a workplace where:
      (i) there was no such representative or safety committee, or
      (ii) there was a representative or safety committee but it was not reasonably practicable for the employee to raise matters by those means, he brought to his employer's attention, by reasonable means, circumstances connected with his work which he reasonably believed were harmful or potentially harmful to health or safety,
   (d) in circumstances of danger which the employee reasonably believed to be serious and imminent and which he could not reasonably have been expected to avert, he left, or proposed to leave, or, while the danger persisted refused to return to, his place of work or any dangerous part of it, or
   (e) in circumstances of danger which the employee reasonably believed to be serious and imminent he took, or proposed to take, appropriate steps to protect himself and others from danger.
2 For the purposes of subs. (1)(e) whether steps which the employee took, or proposed to take, were appropriate is to be judged by reference to all the circumstances including, in particular, his knowledge and the facilities and advice available to him at the time.

3 Where the reason, or if more than one the principal reason, for the dismissal of an employee is that specified in subs. (1)(e), he shall not be regarded as unfairly dismissed if the employer shows that it was, or would have been, so negligent of the employee to take the steps which he took, or proposed to take, that a reasonable employer might have dismissed him for taking, or proposing to take, them.

The provisions of s. 100 create a number of difficult problems for the Tribunal in that they need to decide whether the employee used his rights reasonably or whether, deliberately or otherwise, he abused them. An important point is found in subs. (1)(e) where a union official might order individual employees to cease work until a potential hazard has been removed. The Act makes it clear that whether the steps taken by the official were appropriate or not must be determined by all the circumstances, including the facilities and advice available at the time.

Other examples of where an employee is regarded, prima facie, as having been unfairly dismissed include:

- pregnancy and connected reasons
- assertion of a statutory right
- shop workers and betting shop workers who refuse to work on Sundays
- being a trustee of an occupational pension scheme
- being an employees' representative
- matters relating to union membership or non-membership
- the transfer of an undertaking
- spent convictions.

1.5.5.9 Exclusions from the right to claim unfair dismissal

In order to claim unfair dismissal, the worker must have been employed for one year and be under the relevant retirement age. Certain categories, such as Crown employees, members of the armed forces and parliamentary staff among others, are excluded from claiming.

1.5.5.10 Rights of an employee who has been unfairly dismissed

An employee who has been found to have been unfairly dismissed is entitled to either compensation or reinstatement and re-engagement, the latter applying where the original job from which he was unfairly dismissed is no longer available. Section 114 of ERA defines an order for reinstatement as 'an order that the employer shall treat the complainant in all respects as if he had not been dismissed'. It follows from this that his original terms of employment once more apply and he is entitled to the benefit of any improvement, such as an increase in pay that he would have had if he had not been unfairly dismissed.

When making an order for reinstatement or re-engagement, the Tribunal must consider whether or not it is practicable to do so. If the employer fails to comply with such an order then the Tribunal must again consider the question whether or not it was practicable for him to comply with it. Where such an order is not complied with and the employer cannot show that it was not practicable for him to comply with it, then an additional or penal compensation can be ordered against the employer.

The other award a Tribunal can make is that of compensation. Such an award is made up of two factors: firstly, the basic award is the equivalent of the statutory redundancy payment the employee would have received if he had been dismissed for that reason. The second is the compensatory award for the financial loss which the employee has suffered. It falls under various heads which include, *inter alia*, loss of salary, future loss of salary, estimated fluctuations in earnings, future loss of unemployment benefit and loss of pension rights. A percentage of the award can be ordered to be deducted by the Tribunal if they feel that the employee has contributed in any way to his own dismissal. It is the employee's duty to mitigate his loss and he must be able to satisfy the Tribunal that he has sought other employment but without success.

1.5.6 Summary

The main purpose of employment legislation is to regulate the relationship between employer and employee and to determine the role and powers of trade union representatives in deciding the terms and conditions under which an employee has to work. It has become practice to include under the wing of 'industrial relations' anything that can affect the way in which an employee has to work, and in this respect safety has an important role to play.

This chapter has shown some of the ways in which decisions and actions taken for safety reasons can materially affect the employee's working conditions and, conversely, the ways in which employment legislation can affect safety issues. For the safety adviser to be able to perform his duties properly he must be aware of the wider implications of the recommendations he makes, particularly in the field of working conditions.

The law governing industrial relations is extremely complex and covers much more ground than it has been possible to cover in this chapter, but the most important of the statutory provisions have been covered briefly and some of the ways in which their application can affect the employer–employee relationship have been shown.

References

1. Page *v.* Freight Hire Tank Haulage Ltd (1980) ICR 29; (1981) IRLR 13
2. The National Disability Council, *Code of Good Practice on the Employment of Disabled People*, The Disability Council, London
3. EU Directive 94/33/EC *on the protection of young people at work*, EU, Luxembourg (1994)

4. European Court of Justice cases:
C382/92, *Safeguarding of employee rights in the event of transfer of undertakings*, Celex no. 692JO382
C383/92, *Collective redundancies*, Celex no. 692JO383, EU, Luxembourg (1992)
5. EU Directive 93/104/EC *concerning certain aspects of the organisation of working time* (The Working Time Directive), EU, Luxembourg (1993)
6. Advisory, Conciliation and Arbitration Service, Code of Practice No. 1, *Disciplinary and grievance procedures*, The Stationery Office, London (2000)
7. Polkey *v.* A.E Dayton (Services) Ltd (1988) IRLR 503; (1987) All ER 974, HE (E)
8. Sullifant *v.* Powell Duffryn Timber Ltd (1983) IRLR 91
9. Spencer *v.* Paragon Wallpapers Ltd (1976) IRLR 373
10. East Lindsay District Council *v.* Daubney (1977) IRLR 181
11. International Sports Co. Ltd *v.* Thomson (1980) IRLR 340
12. Hutchinson *v.* Enfield Rolling Mills (1981) IRLR 318
13. James *v.* Waltham Holy Cross UDC (1973) IRLR 202
14. British Home Stores *v.* Burchell (1978) IRLR 379
15. Austin *v.* British Aircraft Corporation Ltd (1978) IRLR 332
16. Keys *v.* Shoefayre (1978) IRLR 476
17. Lindsay *v.* Dunlop Ltd (1979) IRLR 93
18. Williams *v.* Compair Maxam Ltd (1982) ICR 800

Chapter 1.6

Consumer protection

R. G. Lawson

The expression 'consumer protection', and with it the notion of 'consumer law' first found expression in the Final Report of the Committee on Consumer Protection (Cmnd 1781) 1962[1], which led to the enactment of the Trade Descriptions Act in 1968[2]. This latter can fairly be regarded as the starting point of the modern law of consumer protection.

In more recent years much of the impetus for new consumer legislation has come from the UK's membership of the European Union (EU) as witnessed by the enactment in the UK of the Control of Misleading Advertisements Regulations 1988[3], the Consumer Protection Act 1987[4], the General Product Safety Regulations 1994[5] and the Unfair Terms in Consumer Contracts Regulations 1999[6], in each case derived from an EU directive. Further examples are the Control of Misleading Advertisements (Amendment) Regulations 2000[7], the Consumer Protection (Distance Selling) Regulations 2000[8] and the Stop Now Orders (EC Directive) Regulations 2001[9]. Each implementing in the UK EU Directives on, respectively, comparative advertising, distance selling and powers to take action against rogue traders.

1.6.1 Fair conditions of contract

It is central to any system of consumer protection that a potential customer is given only truthful and accurate information about the goods and services that he is wanting to buy. Even before Parliament had decided to intervene, the courts had already decided to allow a remedy where a contract had been induced by fraud or misrepresentation. Where a consumer has been duped into entering a contract through deception of the kind practised by some salesmen, he would be given the right to put an end to the contract and claim compensation for any loss which he may have suffered. This development in the courts was eventually confirmed by the Misrepresentation Act 1967[10].

Valuable though these controls were, they applied only to what is called the civil law, i.e. the law which regulates the relations between citizens. Where a consumer had been the victim of fraud or misrepresentation, the initiative lay entirely upon him to take remedial action. It was only with the advent of the Trade Descriptions Act that the criminal law came to the aid of the consumer in such cases.

1.6.1.1 False trade descriptions

The main feature of the Trade Descriptions Act 1968[2], is to penalise the use of false trade descriptions. Section 2 of the Act contains an exhaustive list of what constitutes a false trade description for the purposes of the Act. Anything not in the list is not a trade description for the purposes of the Act. The list includes any statement as to:

(a) quantity, size or gauge;
(b) method of manufacture, production, processing or reconditioning;
(c) composition;
(d) fitness for purpose, strength, performance, behaviour or accuracy;
(e) any physical characteristics not included in the preceding paragraphs;
(f) testing by any person and results thereof;
(g) approval by any person or conformity with a type approved by any person;
(h) place or date of manufacture, production, processing or reconditioning;
(i) person by whom manufactured, produced, processed or reconditioned;
(j) other history, including previous ownership or use.

Included within this list shall be matters concerning:
(i) any animal, its sex, breed or cross, fertility and soundness;
(ii) any semen, the identity and characteristics of the animal from which it was taken and measure of dilution.

Quantity is defined to include length, width, height, area, volume, capacity, weight and number.

This list can be summarised as saying that any statement about goods which when made can be either true or false is a trade description. This has meant that a statement made in relation to a bar of chocolate that it was of 'extra value' was not a trade description since it was such a vague kind of claim that no one could say of it that it was true or false: *Cadbury Ltd* v. *Halliday*[11].

While it is true to say that this part of the Act seems to have been used almost exclusively to control some of the more dubious antics of the second-hand car trade (convictions for turning back a mileometer have been particularly common) this is far from being entirely the case. One good example arose in the case of *British Gas Board* v. *Lubbock* (1974)[12]. A gas cooker was advertised as being ignited by a hand-held ignition pack. At the time the advertisement was shown, this was no longer true. The Board was prosecuted and convicted for making a false statement about the composition and the physical characteristics of goods.

Another example is the decision in *Queensway Discount Warehouses* v. *Burke* (1985)[13]. A wall unit was advertised in the national press. It was shown ready assembled. The advertisement was seen by a customer who later went to see the unit in store, where it was on display also ready assembled. The customer agreed to purchase the unit, but when it was delivered he found that it was in sections and that he had to assemble it himself. The advertisement was held to be a false trade description in that it gave a false description of the composition of the goods.

It is also possible under the Act for the description of goods to be accurate but still to give rise to an offence if that description is misleading. In *Dixons Ltd* v. *Barnett* (1988)[14] a telescope bore the clear statement that it magnified up to 455 times. This was true, but in fact the telescope had a maximum useful magnification of 120 times: beyond that, the image became less clear and became no clearer with higher magnification. The shop was convicted because, although the statement as to magnification was true, it was misleading.

Strict liability

An offence is committed under this part of the Act regardless of the absence of any blame on the part of the person making the false trade description. Its falsity is enough to secure the commission of an offence. This makes the offence one known as a 'strict liability' offence[15].

Due diligence defence

However, the Act does provide for what is called the 'due diligence' defence. This allows the defendant to escape conviction if he can show that he took all reasonable precautions and exercised all due diligence to avoid commission of the offence. The cases show that this is a very difficult defence to satisfy. In the case of *Hicks* v. *Sullam*[16] bulbs were falsely described as 'safe'. The bulbs had been imported from Taiwan. There were 110 000 in all. None had been sampled to test for their safety and no independent test reports were obtained. The defendant's agent in the Far East had checked the bulbs and had reported no defects. The court ruled that the defence had not been made out.

While most defendants failed to make out a defence, some do occasionally succeed. In one case[17], a supplier was charged under the Act with falsely describing jeans as being the manufacture of Levi Strauss. The jeans had been obtained by him from a business associate in Greece with whom he had dealt for a couple of years. They were sold to him for £1 to £2 less than normal wholesale price. He examined the goods and they appeared to him to be in order. It was held that the defence had been made out. Similarly, in *Tesco Supermarkets Ltd* v. *Nattrass*[18], the defence was made out when the defendants showed they had devised a proper in-store system for ensuring compliance with the Act, and that they had done all they reasonably could, despite its breakdown on the particular occasion, to ensure that the system was implemented by the staff.

1.6.1.1.1 Statements as to services, facilities and accommodation

Section 14 of the Trades Description Act makes separate provision in relation to false or misleading statements as to services, facilities or accommodation. It is an offence for any person in the course of trade or business to make a statement which he knows to be false, or recklessly to make a statement which is false, on any of the following:

- the provision of any services, accommodation or facilities;
- the nature of any services, accommodation or facilities so provided;
- the time at which, manner in which or persons by whom any services, accommodation or facilities are provided;
- the examination, approval or evaluation by any person of any services, accommodation or facilities so provided;
- the location of services, accommodation or facilities.

The statement of offence needs to assert only that the defendant is charged with recklessly making a false statement and in what way it is false. Cases have shown that to specify which subparagraph of s. 14(1) is contravened is unnecessary and likely to result in complications[19]. The interpretation of 'service' has been clarified in cases and means doing something for someone; while a facility means providing the opportunity or wherewithal for someone to do something for himself[20]. By guaranteeing a refund on the price of a book containing instructions for a gambling system, a party made a statement as to the nature of the services[21]. Similarly, in a timeshare presentation, the presenter also made a statement as to service to be provided[22].

The Act applies only to statements of fact and not to promises which cannot be said to be true or false when made: *Beckett* v. *Cohen*[23]; *R.* v. *Sunair Holidays Ltd*[24]. The Act, however, may extend to implied statements of present intention, means or belief: *British Airways Board* v. *Taylor*[25]. Statements about services provided in the past also fall within the Act: *R.* v. *Bevelectric*[26]. In *Roberts* v. *Leonard*[27] it was held that the provisions of s.1 of the Act applied to the professions and it can safely be assumed that this is now the case with s.14.

A statement is false if false to a material degree, and anything likely to be taken as a statement covering the matters referred to above would be false if deemed to be a false statement of the relevant matter [s.14(2)a and (4)].

The mental element

In contrast to the position with false trade descriptions, the offence created by s. 14(1) is conditional upon the party charged knowing that the statement was false or making it recklessly. A statement is made 'recklessly' if it is made 'regardless of whether it is true of false ... whether or not the person making it had reason for believing that it may be false' [s.14(2)b]. To prove recklessness, it is enough to show that the party charged did not have regard to the truth or falsity of a particular statement, even though it cannot be said that he deliberately closed his eyes to the truth or had any kind of dishonesty in mind: *MFI Warehouses Ltd* v. *Nattrass*[28].

Where a person has no knowledge of the falsity of a statement at the time of its publication in a brochure, but did when the statement was read by a complainant, the statement would be a false statement every time business was done on the basis of the incorrect brochure: *Wings Ltd* v. *Ellis*[29]. In theory, the due diligence defence applies equally to offences under s. 14, but, since this section requires knowledge or recklessness before an offence is committed, in practical terms the defence will not generally be applicable.

Disclaimers

The courts have also been prepared to allow the use of disclaimers to avoid conviction, but have insisted that the disclaimer will be effective only if it is as 'bold, precise and compelling' as the false description it is attempting to disclaim. This is laid down in *Norman* v. *Bennett*[30] where a car dealer sought to disclaim a false mileometer reading with the statement 'speedometer reading not guaranteed'. This was contained in the small print of the contract and was held to be ineffective. In contrast, it was held in *Newham London Borough* v. *Singh*[31] that a disclaimer was effective when it was placed over the mileometer and read 'Trade Descriptions Act 1968. Dealers are often unable to guarantee the mileage of a used car on sale. Please disregard the recorded mileage on this vehicle and accept this as an incorrect reading'.

In *R.* v. *Bull*[32], a statement in a sales invoice alongside an odometer reading stating that the reading had not been confirmed and must be considered incorrect was held to be a valid disclaimer, while in *R.* v. *Kent County Council*[33] a trader who sold counterfeit goods, but who posted disclaimers and advised customers that the goods were copies, was also held to have used a valid disclaimer.

Penalties

The penalties for breach of the Act depend on the court the case is brought in. Most cases are brought in magistrates' courts where the penalty is a maximum fine of £5000. More serious cases are brought in the Crown Court where the penalty is a fine of any amount, a maximum sentence of two years, or a combination of both. In addition, the Powers of Criminal Courts Act 1973[34] empowers the court to award compensation to consumers affected by the breach of the Trade Descriptions Act.

1.6.1.2 Pricing offences

The Consumer Protection Act 1987[4]. makes it an offence for a price to be indicated which is misleading as to the price at which any goods, services, accommodation or facilities are available. In *MFI Furniture Centres Ltd* v. *Hibbert*[35] the prosecution did not have to produce an individual consumer to whom a misleading price was given. It was held in *Toys R Us* v. *Gloucestershire County Council*[36] that an indication is not misleading when the goods carry a ticket price lower than that shown when the bar code is run through the till if the retailer has a policy of always charging the lower price.

A promise on the following line is very frequently to be seen in retail outlets: 'We won't be beaten on price. If you find exactly the same package cheaper in a local store within 7 days we will refund the difference'. In *the Link Stores Ltd v. The London Borough of Harrow*[37], an action arose when a purchaser of a mobile phone for £89.99 was refused a refund when he returned within the prescribed period for a refund after finding an identical set available at £69.99.

The defendants claimed that the customer had been told the promise was not going to be honoured but they were convicted by the Magistrates. This conviction was upheld on appeal to the Crown Court where the Judge said 'The damage had been done. It was done when [the consumer] made his purchase. Disabusing him now does nothing to repair the damage and is valueless in terms of consumer protection'. The key to an offence under this provision is that a supplier must have given an indication as to price which became misleading before a contract was entered into.

In the similar case of *DSG Retail Ltd v. Oxfordshire County Council*[38] which was taken to appeal, the High Court held that an offence could be committed under section 20 of the 1987 Act without evidence to prove that an indication applied to specific goods that were available at a specific price. DSG's price promise, although unqualified in its terms, was in application to be subject to conditions as to price. The High Court dismissed the appeal.

The Code of Practice

The novel feature of the Act is that it provided for the adoption by the Secretary of State for Trade and Industry, after approval by Parliament, of a code of practice which gave practical guidance to traders on how price indications should be given to avoid the commission of an offence. A Code of Practice for Traders on Price Indications[39] has been adopted.

It is important to understand that the Code is not binding on traders. A contravention of the Code is expressly said by the Act not of itself to give rise to an offence, but can be used as evidence that an offence in fact has been committed. Similarly, a trader who applies the Code cannot be entirely certain in law that his price indication is not misleading, although such compliance will again be evidence that his pricing is indeed not misleading. In practice, however, it will be very unusual for the presumptions raised by compliance with, or breach of, the Code to be displaced.

In a case on this point, Stanley Ltd advertised, on 2 April 1992, an occasional table for £7.99. On 14 October, it was reduced to £4.99 in a 'Style and Value' promotion. There was a point of sales notice on or near the tables stating 'Style and Value, Occasional Table, Chipboard, save £3, now £4.99 (in large numbers) was £7.99'. The tables continued to be advertised at this price until 12 March 1993. In the mean time, on 10 November 1992, the tables were advertised in the press as a 13 day event headed 'SUPER SAVERS 13 DAY EVENT MUST END TUES 24'. The advertisement stated in the top right-hand corner 'SAVE £3'. In the bottom of the right-hand side was a picture of the table and the body of the advertisement read 'OCCASIONAL TABLE WAS £7.99. NOW ONLY £4.99'.

Around 24 November, the 13 day event ended but, contrary to what the advertisement had said, the table continued to be priced at £4.99 and remained in the 'Style and Value' that it had previously been in before the 13 day event. On 27 November, a customer bought one of the tables at the advertised price of £4.99. On 21 December the Christmas sale began with the table still priced at £4.99 but with a slightly different point of sale notice headed 'SALE, ROUND OCCASIONAL TABLE, NOW £4.99 WAS £7.99'. On 30 December, the same purchaser returned and bought three more of the tables for £4.99. On 12 March 1993 the price of the table was increased to £7.99. Two informations were laid against the company alleging offences under the Act. The first arose from the newspaper advertisement of 10 November, the second from the point of sale notice. The court held that the comparisons made by the advertising were misleading and did not accept that the Code of Practice was ambiguous. The meaning of the Code was clear and, when construed in the context of the legislation, there was no problem in concluding what it meant[40].

Due diligence defence

A due diligence defence on the lines of that set out in relation to the Trade Descriptions Act applies.

Penalties

The penalty for misleading pricing is a maximum fine not exceeding £5000 where the case is prosecuted in a magistrates' court. More serious cases are taken to the Crown Court where the penalty is a fine of unlimited amount. There is no power to impose a custodial sentence for pricing offences. Under the Powers of Criminal Courts Act 1973 the courts may make compensation orders in favour of the victim of a pricing offence.

1.6.1.2.1 Price indications

As well as the general ban on misleading price claims discussed above, there are also enactments imposing positive duties as to price indications. The Price Indications (Method of Payment) Regulations 1990[41] apply to traders who give indications of price for goods, services, accommodation or facilities and who charge different prices for different methods of payment. The Regulations do not apply to motor fuel.

The Price Indications (Bureaux de Change) (No. 2) Regulations 1992[42] apply to any trader carrying on the business of selling foreign currency in exchange for sterling. Clear and accurate information must be given on the buying and selling rates, terms of business, commissions and other charges. Receipts must be given setting out the terms unless the transaction is made by machine.

The Price Marking Order 1999[43] applies to all products, but not services, offered by traders to consumers. A 'trader' is defined as any person who sells or offers or exposes for sale products which fall within their commercial or professional activity. The Order requires that the selling price, and where appropriate the unit price, of products be clearly displayed in sterling. The unit price is generally the price per kilogram, litre, metre, etc. for goods sold by quantity. All similar products should be priced using the same unit so consumers can easily compare prices. Similar conditions apply where products are prepacked in a constant quantity for sale in a small shop or from a vending machine. The Order requires that the prices be indicated in an unambiguous, easily identifiable and clearly legible manner and be placed in proximity to the products to which they relate such that customers do not have to ask for assistance to see them.

1.6.1.3 Truth in lending

The Consumer Credit Act 1974[44] requires that credit and hire advertising is not misleading. A building society offered 'low start' mortgages. While the normal period of a loan was 25 years, under the 'low start' arrangement a borrower would pay 1% interest for 6 months, 2% under the society's prevailing rate for the next 6 months and 0.5% less than the society's rate for the next year. Thereafter the rate would be the society's current rate. At the time of the advertisement, the current rate was 8.45%. The advertisement showed an annual percentage rate (APR) of charge of 1.1% variable. The prosecution claimed that the APR had not been calculated in accordance with the Consumer Credit (Total Charge for Credit) Regulations 1980[45]. These require that a calculation which provides for a variation dependent upon the occurrence of an event shall be based on the assumption that the particular event will not occur. The Regulations exclude from the definition of 'event' something which is certain to occur. The society argued that the Regulations required the APR to be calculated on the assumption that the initial interest rate of 1% would not change during the rest of the term of the mortgage unless the rate was 'certain' to change at the end of the low rate period. The court ruled that the assumption that the interest rate would not change in circumstances where it was certain to change was misleading. The test to apply was to ask whether there was a real or realistic possibility that at the end of the initial period the rate would be exactly 1%, or whether the chance of that was so remote that it should be disregarded. It held that this was so remote a chance that it could be ignored and the advertisement was, therefore, misleading[46].

The somewhat complicated provisions of the Consumer Credit (Advertisements) Regulations 1989[47] provide that a credit and hire advertisement must be in one of three categories, namely simple, intermediate or full. A simple advertisement will do little more than indicate the name of the advertiser and the nature of his business, while a full advertisement will often give an example of repayment terms and the APR.

1.6.2 A fair quality of goods and services

The Sale of Goods Act 1979[48] imposes a number of obligations on a seller of goods. First, he must provide goods which correspond to whatever description has been given; second, the goods must be reasonably (not absolutely) fit for their purpose, and lastly, the seller must provide the buyer with goods which are of 'satisfactory quality'. Prior to the amendment of the Act in 1994[49], the requirement was that the goods be of 'merchantable quality'. Goods are of satisfactory quality 'if they meet the standard that a reasonable person would regard as satisfactory, taking account of any description of the goods, the price (if relevant) and all other relevant circumstances'. 'Quality' refers to the state and condition of the goods and includes the following aspects:

- fitness for all the purposes for which the goods are commonly supplied
- appearance and finish
- freedom from minor defects
- safety and durability.

In the event of a breach of any of the foregoing provisions, the buyer will be entitled to reject the goods and to claim damages for any loss suffered. He will also be entitled to the return of the purchase price. If the goods had been bought by the use of a credit card and cost more than £100 but not more than £30 000, the Consumer Credit Act gives him the choice of bringing his action against the credit card company as an alternative to suing the seller.

The Sale of Goods Act only covers contracts of sale. Comparable duties are imposed in contracts which are closely related to sale under the provisions of the Supply of Goods (Implied Terms) Act 1973[50] (hire-purchase contracts) and the Supply of Goods and Services Act 1982[51] (contracts of hire or contracts where services and goods are supplied, such as the installation by a gas company of central heating).

Services

The Supply of Goods and Services Act requires that all who offer a service must provide that service with reasonable care and skill. It also states that if no time is agreed for the performance of the service, it must be carried out within a reasonable time, and if no charge for the service is agreed in advance a reasonable price must be paid. The Act expressly allows anyone providing a service to assume a stricter duty in respect of skill, care and time of performance.

1.6.3 Product safety

The general safety requirements are contained in the General Product Safety Regulations 1994[5] which make it an offence for any producer or distributor to market or supply a product which is not a 'safe product' or which is a 'dangerous product'. A 'dangerous product' is one which is not 'safe'. In turn a 'safe product' is one which, under normal and foreseeable conditions of use including duration, does not present any risk or only a minimum risk compatible with a high level of protection.

The general safety requirements do not apply to:

- second-hand products sold as antiques
- those supplied for repair or reconditioning before use provided that the buyer is so informed
- any product whose safety is already covered by specific EU legislation.

Producers are required to provide information to enable consumers to assess risks inherent in the product throughout its expected life and to take appropriate precautions. The producer should mark the product to identify batches, carry out sample testing, investigate complaints, keep records of feedback from retailers and have in place a product recall mechanism. Breach of the General Product Safety Regulations is a summary offence attracting a fine not exceeding £5000, a custodial sentence up to 3 months or a combination.

Consumer Protection Act 1987

Under s. 10(1) of this Act an offence is committed in relation to the supply, offer or agreement to supply, or the exposure or possession for the purpose of supply, of any goods when those goods fail to meet the general safety requirement. Section 10(2) defines the safety requirements as meaning that the goods 'are reasonably safe having regard to all the circumstances'. These circumstances include:

- the marketing, get up, use of any mark in relation to the goods and any instruction or warnings as to the keeping, use or consumption of the goods
- any published safety standards
- the existence of any means by which it would have been reasonable for the goods to be made safer.

In turn 'safe' is defined in s. 19(1) as meaning:

- that there is no risk or no risk apart from the minimum
- that none of the following will cause death or personal injury:
  - the goods
  - the keeping, use or consumption of the goods
  - the assembly of the goods
  - any emission or leakage from the goods or, as a result of the keeping, use or consumption of the goods, from anything else
  - reliance on the accuracy of any measurement, calculation or other reading made of the goods.

A person guilty of a breach of these general safety requirements is liable, on sentence, to a fine not exceeding £5000, a custodial sentence up to 6 months or a combination.

Since the General Product Safety Regulations came into effect, and because they include the disapplication of the general safety provisions in the Act in relation to products placed on the market by producers and distributors, those requirements of the 1987 Act have had diminishing importance. As a result, virtually all cases involving a breach of the general safety requirements are taken under the Regulations. However, the Act continues to apply where the supplier of the goods is not the person who first places the goods on the market.

1.6.3.1 Information exchange

Under the General Product Safety Directive if a Member State adopts emergency measures to prevent, restrict or impose specific conditions on the marketing or use, within its territory, of a product or product batch because of a serious and immediate risk to the health and safety of consumers, it must immediately inform the Commission, unless the effects of the risk do not or cannot extend beyond the territory of that Member State. The Commission will immediately inform the other Member States who will inform the Commission of any measures adopted to counter the risk. The Commission can intervene directly if it becomes aware through notification or information from the Member States of the existence of a serious and immediate risk from a product, and if:

- one or more Member States have adopted restrictions on its marketing, or withdrawn it from the market
- Member States do not agree on the measures to be taken
- The risk cannot be dealt with adequately under other specific EU legislation relating to the product, and
- the risk can only be eliminated by adopting Community measures.

The EU Council can then adopt a Decision, after consulting the Member States, requiring them to take certain temporary measures.

The Commission is supported in this area by a Committee on Product Safety Emergencies, which comprises representatives from the Member States and is chaired by the Commission. This committee is consulted on draft measures to be taken and must give its opinion, by weighted majority, within a time limit set by the chairman according to the urgency of the situation, but in any case in less than 1 month. The Commission will accept the measures recommended by the committee and submit them to Council for adoption. If the Council has not acted within 15 days, the Commission will act on the recommendations. Measures adopted under this procedure are only valid for 3 months but the period can be extended. Member States must implement the agreed measures within 10 days. The competent authorities of the Member States must allow the parties concerned an opportunity to give their views and will inform the Commission accordingly. In the UK, this system operates through the Consumer Safety Unit of the DTI. Pharmaceuticals, animals, products of animal origin and radiological emergencies are excluded from the rapid notification system since they are dealt with under separate EU legislation.

1.6.3.2 Product recall

Article 6(h) of the General Product Safety Directive requires EU Member States to have in place appropriate measures to ensure the effective and immediate withdrawal of a dangerous product and, if necessary, that product's destruction. No specific UK legislation has been introduced to implement this requirement, but it is dealt with by provisions of the Consumer Protection Act through prohibition notices, notices to warn, suspension and forfeiture orders (see section 1.6.3.3 below).

In the motor industry, the recall of motor vehicles is dealt with by a voluntary set of arrangements agreed between the Department of Transport and the industry which operates through the DVLA. Failure of a manufacturer to recall dangerous products can lead to an action against him in negligence: *Walton* v. *British Leyland UK Ltd*[52].

1.6.3.3 Notices and orders

The Consumer Protection Act 1987 empowered the enforcing authorities to issue various notices and orders.

1.6.3.3.1 Prohibition notices

These notices are addressed to individual traders requiring them to cease the supply of the specified goods which the Secretary of State considers to be unsafe. For example, a prohibition notice was issued to a shopkeeper requiring him to stop supplying certain rice cookers which were found to be unsafe. Notices have also been issued in respect of elastic sweet-like toys on the ground that they presented a potential choking hazard to children. Another case involved the withdrawal from the market of mercury soap used as a skin lightener.

1.6.3.3.2 Notices to warn

A notice to warn may be issued by the Secretary of State and require the person on whom it is served to publish, at his own expense, warnings in the form, manner and at times specified in the notice, that he supplies or has supplied unsafe goods. To date no such notices have been issued but it is not unusual for manufacturers voluntarily to advertise in the press that products identified in the advertisement are unsafe and to request their return for repair or refund.

1.6.3.3.3 Suspension notices

Local trading standards officers are empowered to issue suspension notices where they have reasonable grounds for supposing that there is a breach of the general safety requirement. The maximum duration of the notice is 6 months and during that period the person on whom it is served cannot deal in the goods without the consent of the local trading standards authority.

A person guilty of a breach of any of the above notices can be liable to a fine of up to £5000, a custodial sentence up to 6 months or a combination.

In *R. v. Liverpool City Council ex parte Baby Products Association*[53] following investigations into the safety of babywalkers, Liverpool City Council issued a press release. The applicants claimed the Council had acted unlawfully and, by acting as they did, deprived the companies concerned of rights and safeguards that Parliament had intended they should enjoy. The Council did have power to require the suspension of supply for a limited period, but it could only do so subject to conditions specified in the Act. Although the statutory procedure was 'relatively slow and cumbersome and difficult to apply effectively', it simply could not be ignored or circumvented.

1.6.3.3.4 Forfeiture orders

Local trading standard officers may apply, under s.15(2)b, for the forfeiture of goods on the grounds of a breach of either a general safety requirement, a prohition notice, a notice to warn or a suspension notice. Instead of ordering the destruction of the goods, the court may direct that the goods be released to a person nominated by the court provided that person only supplies them either to a trader whose business is that of repairing or reconditioning such goods or to one who receives them as scrap.

1.6.3.4 Safety regulations

Power is given by s.1(1) of the 1987 Act to the Secretary of State to make safety regulations whose purpose must be to ensure that goods are safe or that goods which are, or in the hands of a person of a particular description are, unsafe are not made available generally or to persons of that description. Also that appropriate information is, or inappropriate information is not, provided in relation to the goods. There is a considerable body of such regulations, many were made under previous legislation but are now treated as if made under the 1987 Act. Information about these regulations can be obtained from the Consumer Safety Unit, Department of Trade and Industry, Victoria Street, London SW1H 0ET.

1.6.3.5 Food and medicines

The Food Safety Act 1990[54] makes it an offence to sell food not conforming to the 'food safety requirements'. This is more narrowly defined in the general safety requirement discussed above. Food is deemed to breach the food safety requirement if it is unfit for human consumption, if it has been rendered injurious to health by any of the operations specified in the Food Safety Act, or if it is so contaminated that it would not be reasonable to expect it to be used for human consumption. Food is not covered by the general safety requirement laid down in the Consumer Protection Act, but it is within the general safety requirement contained in the General Product Safety Regulations.

The Medicine Act 1968[55] imposes strict controls on the manufacture and supply of medicinal products. In particular, most such products will require a 'product licence', while some will be available on prescription only.

1.6.4 Product liability

Part I of the Consumer Protection Act created what is called a system of 'strict liability' for defective products, allowing an injured person to sue without the need to prove negligence. It had long been regarded as anomalous that the person who bought defective goods could sue under the Sale of Goods Act for any injury caused without the need to prove negligence, whereas a non-purchaser (e.g. a bystander, a member of the purchaser's family or a person to whom the goods had been given as a gift) could only recover damages if he could prove negligence.

Part I of the Act now provides that damages can be recovered simply on proof that the product is defective, which means that its safety is not such as persons generally are entitled to expect. Liability is placed on the producer, which in this context is stated to include any person who 'own-brands' a product as though he were in fact the producer; and the first importer of the product into the EC. The actual supplier of the defective product can also be made liable, but this can only be where he has been asked by the injured party to name the actual producer and he fails to comply with the request or identify the party who supplied him with the goods within a reasonable time.

In *Worsley v. Tambrands Ltd*[56], the claimant had been using regular tampons during her menstrual period since the age of 15. The defendants were the manufacturers of the tampons. On 9 July 1994, the claimant began her period and inserted a Tampax regular tampon. The following day she became unwell with diarrhoea, vomiting, tiredness and lethargy and was diagnosed as having toxic shock syndrome (TSS). She brought an action seeking damages for personal injuries suffered as a result, contending that the warnings on the packet were defective within the Act because they did not contain sufficiently prominent or sufficient warnings to alert her to the association between TSS and the use of tampons. Finding in favour of the defendants, the court held that the warnings were provided from two sources, namely the information on the box itself and a detailed leaflet inside whose information was true and accurate and any warning was clear.

In *Richardson v. LRC Products Ltd*[57] an action was brought following the birth of a child notwithstanding the use of a condom. The female claimant contended that the cause of the fracture in the condom was ozone damage to the surface of the rubber condom while still in the factory. Finding for the defendants the court held that what a person was entitled to expect from a product depended on the product itself, the manner in which it was used, the purposes for which it was used and the warnings given. There was no evidence of a weakness in the system of testing used and the condoms were manufactured to the standard required.

In *Abouzaid v. Mothercare (UK) Ltd*[58], the respondent received an injury from a product called Cosytoes that had been purchased from the defendants in 1990. It had to be attached to a child's pushchair by elasticated straps which were joined by a metal buckle attached to one of the straps. In attempting to join the straps the buckle hit him in the eye causing severe injury. The question arose whether the product was defective within the 1987 Act. The Court of Appeal held that if the product was defective in 1999, it would also be defective in 1990 since there had been no material changes in public expectations in the meantime.

In *A & Others v. National Blood Authority & Others*[59] the claimants had been infected with hepatitis C in blood given to them in transfusions in the period from 1 March 1988 when the Consumer Protection Act came into force. They claimed that the blood was defective within the meaning of the Act and that the defendants were liable even though the virus had not been identified until May 1988 and a screening test not introduced until 1989. The court ruled that the Act related to the safety of the blood and not to whether tests were available or had been carried out.

The producer of a defective product will not be liable in every case. For instance, he will not be liable if he can show that the goods were not defective when supplied by him. Again, the producer of a defective component has a defence if he can show that he was following the instructions of the producer of the product which was to incorporate the component; or if the defect was due to the design of the end-product.

The most contentious defence contained in the Act is usually called the 'development risks' defence. Under this defence, the producer of a defective product will have a defence if he can show that the 'state of scientific and technical knowledge at the relevant time was not such that a producer of products of the same description might be expected to have discovered the defect if it had existed in his products while they were under his control'. The Act implemented the provisions of an EU directive under which the continuing existence of this defence was to be reviewed. That review has now taken place and has recommended the removal of this defence but no formal decision has been adopted.

In the Abouzaid case above Mothercare sought to rely on the 'development risk' defence since there was no record of a comparable accident at the time of supply. The defence failed because, as the court put it, the defect was present whether or not previous accidents had occurred.

A similar decision was made in *A & Others v. National Blood Authority & Others* where it was held that if there is a known risk, as there was in this case, then a producer would continue to produce and supply at his own risk. It would be inconsistent with the Act if a producer, in the case of a known risk, could continue to supply without responsibility on the basis that he could not identify the products in which the risk occurred. Once a defect became known by virtue of accessible information, infected blood products would no longer qualify for protection under s.4 of the 1987 Act. Known risks did not qualify for the defence even if unavoidable within the particular product.

The Act applies to damage to property in exactly the same way as it applies to personal injury. However, actions for damage to property cannot be brought unless the claim exceeds £275. If it does exceed that amount, then the whole of the loss can be claimed.

All claims for damage, whether to person or property, must be brought within three years, but no claims can be brought at all once 10 years has expired from the time of supply of the defective product.

1.6.5 Misleading advertising

In enacting the Control of Misleading Advertisements Regulations 1988[3], the UK adopted Council Directive 84/450. Under the Regulations, the Director General of Fair Trading may seek a court injunction against an advertisement where a complaint has been brought that it is misleading. In deciding whether to apply for a court injunction, he must first consider whether the advertisement in question has been the subject of complaint to 'established means' of dealing with such complaints. This is a reference to such bodies as the Advertising Standards Authority whose British Code of Advertising and Sales Promotion[60] represents the industry's efforts at self-regulation. The Office of Fair Trading (OFT) obtained an injunction in the case of *Director General of Fair Trading* v. *Tyler Barrett and Co. Ltd*[61] in which the court was presented with evidence that advertisements by this company were misleading and false in a number of respects. The injunction prevented the continuation of advertisements, in the form of telephone cold-calling and personal visits by representatives of the company to the premises of small local businesses offering to obtain business grants from the EU for a search fee of £350 plus 10% of any grant obtained. Clients were led to believe that the company would assist them in obtaining grant funding for a variety of business purposes. The availability of such grants and the likelihood of success in obtaining such funding were greatly exaggerated by the company. Clients were reassured that, should no grant be obtained for them, they would receive a full refund of their search fee.

The information given to the clients regarding the availability of grants was incorrect and what clients actually received was a standardised list of grant making bodies, most of which was wholly irrelevant to the client's needs. When clients realised the very limited nature of the service and that they had been deceived, they sought to obtain the refund without success. The information provided by the company to small businesses and its failure to provide a refund of the search fee had been the subject of many complaints to trading standards departments and other bodies, such as local business link offices. To comply with the injunction, Tyler Barrett and Co. Ltd and its director Peter Kemp will have to stop making misleading statements about what they offer to their clients. If they wish to get the injunction lifted, they will have to attend a further hearing.

Instead of taking court action, the OFT can obtain assurances from the advertiser that a particular course of advertising will stop. One advertiser sent unsolicited mailshots which used words like: 'Would you like more money in your pocket?', 'Too busy earning a living to make money?',

'Looking for a genuine chance to improve your lifestyle?'. Attached to them were handwritten self-adhesive notes bearing the comment 'Working wonders for me . . . knew you'd be interested' followed by the initial 'M'. The OFT considered the advertising was misleading in that it suggested personal recommendations. The advertiser agreed to give the Director General an undertaking that he would not put out advertisements which gave consumers the false impression that they were sent by, or a product was recommended by, somebody they knew[62].

The High Court granted a permanent injunction to the Office of Fair Trading (OFT) preventing the publication by Top 20 Ltd of certain advertisements making sweeping claims for the effectiveness of a Yummy Yum Yum Diet. In addition the company were restrained from publishing any advertisement for a diet or slimming plan which is in similar terms or is likely to convey a similar impression to any of the claims concerned. The Director General of OFT was asked to take proceedings by the Advertising Standards Authority after it had upheld complaints about the Yummy Yum Yum Diet and about other advertisements no longer published by Top 20 Ltd or associated companies[63].

Broadcast advertising

Under the Regulations, the Director General has no power in relation to advertisements carried on any television, radio or cable service. Where an advertisement is misleading, neither the Independent Television Commission nor the Radio Authority have power to seek an injunction, instead they can refuse to transmit the advertisment.

Under the Control of Misleading Advertisements (Amendment) Regulations 2000[7] the above powers now extend to comparative advertisements which must, *inter alia*:

- not be misleading
- ensure comparisons of features of goods and services are objective
- not create confusion in the market place between the advertiser and a competitor
- not denigrate or discredit trade marks, trade names or other distinguishing marks of a competitor
- not take unfair advantage of the reputation, trade mark, trade name or other distinguishing marks of a competitor
- not present goods or services as imitations or replicas of goods or services bearing a protected trade mark or name.

1.6.6 Exclusion clauses

At one time, particularly in the field of the sale of goods, the small print of the contract would often contain clauses, usually called 'exclusion clauses', which took away from the consumer the rights given him under such legislation as the Sale of Goods Act. The use of such clauses is now subject to the controls imposed by the Unfair Contract Terms Act 1977[64] and the Unfair Terms in Consumer Contracts Regulations 1999[6].

1.6.6.1 The Unfair Contract Terms Act 1977

In sales to a consumer, the Act does not allow the seller to avoid the obligations which are imposed on him by the Sale of Goods Act (see section 1.6.2 above). Even to try to avoid the obligation is a criminal offence under the provisions of the Consumer Transactions (Restrictions on Statements) Order 1976[65].

In the case of sales to other businesses, exclusion clauses will be effective, but only if they are reasonable. Similar constraints are imposed in relation to contracts where possession or ownership of goods passes, but the contract is not one of sale or hire-purchase.

The same Act also controls the operation of other types of exclusion clause when incorporated into a consumer contract or in a contract which is on written standard terms. The principle is that three types of clause which might be used in such a contract are valid only if they can be proved to be reasonable; such clauses are those which:

- seek to allow a person not to perform the contract;
- seek to allow him to provide a performance 'substantially different' from that which was reasonably expected; and
- allow the person in breach of contract to be free of all liability for his breach.

Suppose that a term in a holiday contract says that a person may have to share a room if the tour operator so decides instead of getting a single room that he has booked. This will be a clause seeking to provide a performance substantially different from that reasonably expected. Under the Act, this clause will not be valid unless the tour operator can prove it was reasonable. Where a clause is valid only if reasonable, the assumption laid down in the Act is that the clause is unreasonable until the contrary is proved.

1.6.6.2 The Unfair Terms In Consumer Contracts Regulations 1999

Unlike the Unfair Contract Terms Act, the Regulations do not automatically render any terms void, but the range of clauses dealt with is wider than in the Act. At the same time the scope of the Regulations is narrower because they only cover consumer contracts. The Regulations apply to all contracts made between an individual and a business which have not been individually negotiated.

A term is not regarded as 'individually negotiated' if it was drafted in advance without the individual's involvement. Notwithstanding that a specific term, or certain aspects of it, in a contract has been the subject of individual negotiation, the Regulations will apply to the rest of the contract if an overall assessment of the contract indicates that it was a pre-formulated contract.

Contracts must be in 'plain intelligible language'. If not, and there is doubt about a written term, the interpretation most favourable to the individual will prevail. An unfair term is not binding on the individual although the contract will remain binding if it can operate without the unfair term.

A term is unfair if, contrary to the requirements of good faith, it causes a significant imbalance of rights and obligations to the detriment of the consumer. A schedule to the Regulations includes what it calls an indicative and illustrative list of terms which may be regarded as unfair. These include terms requiring a consumer who fails to fulfil his obligations to pay a disproportionately high sum in compensation or which irrevocably binds a consumer to terms with which he had no real opportunity of becoming acquainted before the making of the contract. The Regulations go on to state that, provided it is in plain intelligible language, no assessment shall be made of the fairness of any term which defines the subject matter of the contract or which concerns the adequacy of the price or remuneration to be paid for the goods or services involved.

Complaints about unfair terms are dealt with by the OFT which can obtain undertakings and or seek injunctions in respect of the unfair terms. The Regulations extend to other bodies – including the Consumer's Association – the power to take action against those who put unfair terms into contracts but keep the obligation on the Director General of OFT to consider any complaint made to him about the fairness of any contract term drawn up for general use.

In *Director General v. First National Bank plc*[66], the Director General of OFT sought an injunction against the continued use of the clause:

> Time is of the essence in making all repayments to First National Bank (the Bank) as they fall due. If any repayment instalment is unpaid for more than 7 days after it became due, the Bank may serve a notice on the customer requiring payment before a specified date not less than 7 days later. If the repayment instalment is not paid in full by that date, the Bank will be entitled to demand payment of the balance on the customer's account and interest then outstanding together with all reasonable and legal costs charges and expenses claimed or incurred by the Bank in trying to obtain payment of the unpaid instalment or of such balance and interest. Interest on the amount which becomes payable shall be charged in accordance with condition 3, at the rate stated in paragraph D overleaf (subject to variation) until payment after as well as before any judgement (such obligation to be independent of and not to merge with any judgement).

The Director General submitted in the present case that the disputed clause operated in such a way in practice as to cause uncertainty and confusion among judgement debtors. The case was taken to the Court of Appeal who held: 'that the bank with its strong bargaining position, as against the relatively weak position of the consumer, has not adequately considered the consumer's interest in this respect. In our view, the relevant term in that respect does create unfair surprise and so does not satisfy the test of good faith, it does cause a significant imbalance in the rights and obligations of the parties by allowing the bank to obtain interest after judgement in circumstances when it would not obtain interest under the 1984 Act and the 1991 Order and no specific benefit to compensate the borrower is provided, and it operates to the detriment of that consumer who has to pay the interest'.

The OFT has established a special unit to deal with unfair terms and publishes regular bulletins. Anyone wishing to make a complaint should write to the Unfair Contract Terms Unit, Fleetbank House, 2–6 Salisbury Square, London EC4Y 8JZ; tel: 020 7211 8000.

1.6.7 Distance selling

The Consumer Protection (Distance Selling) Regulations 2000[8] apply to contracts for goods or services made under a scheme by a supplier who makes exclusive use of one or more means of distance communication. Distance communication effectively covers all forms of communication that do not involve face-to-face contact.

To protect the consumer the following information should be provided by the seller in writing and 'in good time' before a contract is made:

- the supplier's identity and, if advance payment is required, the supplier's address
- a description of the goods and services
- the price including VAT
- delivery costs if appropriate
- arrangements for payments, delivery or performance
- the cost of using the means of communication where calculated at other than the basic rate
- the existence and conditions of a right to cancel
- the period for which the offer or price remain valid
- if appropriate, the minimum duration of any permanent or recurring contract
- any rights of the supplier to provide substitutes of the same quality and price in the event of the contract goods or services not being available.

1.6.8 Stop now orders

These were introduced by the Stop Now Orders (EC Directive) Regulations 2001[9] and apply to domestic law implementing a number of EC Directives which are listed in the Regulations. Where there is an infringement of any of the terms of these Directives, the Director General of OFT or any 'qualified entity' may bring proceedings for an injunction. Qualified entity includes those organisations whose role is to ensure maintenance of standards in the public utilities and other nationwide services. Where a matter arises under these Orders, attempts must be made to resolve it, but if after 14 days it has not been resolved then proceedings may be started. There are provisions for cross-border co-operation within the EEA.

1.6.9 Consumer redress

Up until 1973 going to court in pursuance of a consumer claim could be a daunting and expensive business. It was then that a small claims or arbitration procedure was set up which operated through the County Court. Any claim within the County Court jurisdiction can be referred on application to an arbitration heard by the District or County Court judge or even by an outside arbitrator. Any such arbitration has the effect of a full County Court judgment, though it is usually heard in private by the arbitrator in an informal manner and without the normal rules of court procedure applying. If the sum claimed does not exceed £5000, it will go to arbitration automatically if either party desires. Above that limit, both parties will have to agree before the matter can use this procedure. An important feature of the arbitration system is that the rule as to costs has been considerably modified. The loser of a case normally will be asked to pay only a nominal sum. As a rule, he will not have to pay anything in respect of his opponent's legal fees.

References

1. Final report of the Committee on Consumer Protection, Cmnd 1781, The Stationery Office, London (1962)
2. *Trade Descriptions Act 1968*, The Stationery Office, London (1968)
3. *The Control of Misleading Advertisements Regulations 1988*, SI 1988 No. 915, The Stationery Office, London (1988)
4. *Consumer Protection Act 1987*, The Stationery Office, London (1987)
5. *The General Product Safety Regulations 1994*, SI 1994 No. 2328, The Stationery Office, London (1994)
6. *The Unfair Terms in Consumer Contracts Regulations 1999*, SI 1999 No. 2032, Stationery Office, London (1999)
7. *Control of Misleading Advertisements (Amendment) Regulations 2000*, SI 2000 No. 914, Stationery Office, London (2000)
8. *Consumer Protection (Distance Selling) Regulations 2000*, SI 2000 No. 2334, The Stationery Office, London (2000)
9. *Stop Now Orders (EC Directive) Regulations 2001*, SI 2001 No. 1422, The Stationery Office, London (2001)
10. *Misrepresentation Act 1967*, Stationery Office, London (1967)
11. Cadbury Ltd *v.* Halliday [1975] 2 All ER 226
12. British Gas Board *v.* Lubbock [1974] 1 WLR 37
13. Queensway Discount Warehouses *v.* Burke [1985] BTLC 43
14. Dixons Ltd *v.* Barnett [1988] BTLC 311
15. Alec Norman Garages Ltd *v.* Phillips [1984] JP 741
16. Hicks *v.* Sullam [1983] MR 122
17. Westminster City Council *v.* Pierglow Ltd (8 February 1994, unreported)
18. Tesco Supermarkets *v.* Nattrass [1972] AC 153
19. Regina *v.* Piper [1995] 160 JP 116
20. See: Newell *v.* Hicks [1983] 148 JP 308; Kinchin *v.* Ashton Park Scooters [1984] 148 JP 540; Dixons Ltd *v.* Roberts [1984] 159 JP 631
21. Ashley *v.* Sutton London Borough Council [1994] 159 JP 631
22. Global Marketing Europe (UK) Ltd *v.* Berkshire County Council Department of Trading Standards [1995] Crim LR 431
23. Beckett *v.* Cohen [1973] 1 All ER 120
24. Regina *v.* Sunair Holidays Ltd [1973] 2 All ER 1233
25. British Airways Board *v.* Taylor [1976] 1 All ER 65
26. Regina *v.* Bevelectric [1992] 157 JP 323

27. Roberts *v.* Leonard [1995] 159 JP 711
28. MFI Warehouses Ltd *v.* Nattrass [1973] 1 All ER 762
29. Wings Ltd *v.* Ellis [1985] AC 272
30. Norman *v.* Bennett [1974] 3 All ER 351
31. Newham London Borough *v.* Singh [1987] 152 JP 239
32. Regina *v.* Bull, *The Times*, 4 December 1993
33. Regina *v.* Kent County Council (6 May 1993, unreported)
34. *Powers of the Criminal Courts Act 1973*, The Stationery Office, London (1973)
35. MFI Furniture Centres Ltd *v.* Hibbert [1995] 160 JP 178
36. Toys R Us *v.* Gloucestershire County Council [1994] 158 JP 338
37. The Link Stores Ltd v. The London Borough of Harrow (21 December 2000 unreported)
38. DSG Retail Ltd v. Oxfordshire County Council (16 March 2001 unreported)
39. *Consumer Protection (Code of Practice for Traders on Price Indications) Approval Order 1988*, SI 1988 No. 2078, Stationery Office, London (1988)
40. AG Stanley Ltd *v.* Surrey County Council [1994] 159 JP 691
41. *The Price Indications (Method of Payment) Regulations 1990*, SI 1990 No. 199, The Stationery Office, London (1990)
42. *The Price Indications (Bureaux de Change) (No. 2) Regulations 1992*, SI 1992 No. 737, The Stationery Office, London (1992)
43. *The Price Marking Order 1991*, SI 1991 No. 1382, The Stationery Office, London (1991)
44. *The Consumer Credit Act 1974*, The Stationery Office, London (1974)
45. *The Consumer Credit (Total Charge for Credit) Regulations 1980*, SI 1980 No. 51, The Stationery Office, London (1980)
46. Scarborough Building Society *v.* Humberside Trading Standards Department [1997] CCLR 47
47. *The Consumer Credit (Advertisements) Regulations 1989*, SI 1989 No. 1125, The Stationery Office, London (1989)
48. *The Sale of Goods Act 1979*, The Stationery Office, London (1979)
49. *The Sale and Supply of Goods Act 1994*, The Stationery Office, London (1994)
50. *The Supply of Goods (Implied Terms) Act 1973*, The Stationery Office, London (1973)
51. *The Supply of Goods and Services Act 1982*, The Stationery Office, London (1982)
52. Walton *v.* British Leyland (UK) Ltd (12 July 1978, unreported)
53. R. *v.* Liverpool City Council ex parte Baby Products Association, *The Times*, 1 December 1999
54. *The Food Safety Act 1990*, The Stationery Office, London (1990)
55. *The Medicines Act 1968*, The Stationery Office, London (1968)
56. Worsley *v.* Tambrands Ltd [1999] 96 (48) LSG 40
57. Richardson *v.* LRC Products Ltd (2 February 2000 unreported)
58. Abouzaid *v.* Mothercare (UK) Ltd (21 December 2000 unreported)
59. A & Others *v.* National Blood Authority & Others, (26 March 2001 unreported)
60. *British Codes of Advertising and Sales Promotion 1999*, Advertising Standards Authority, London (1999)
61. Director of Fair Trading *v.* Tyler Barrett and Co. Ltd, (1 July 1997 unreported)
62. *Consumer Law Today*, December 1997
63. Office of Fair Trading press release PN21/99, 29 June 1999
64. *Unfair Contract Terms Act 1977*, The Stationery Office, London (1977)
65. *Consumer Transactions (Restrictions on Statements) Order 1976*, SI 1976 No. 1813, Stationery Office, London (1976)
66. *Director General of Fair Trading v. First National Bank plc* [2000] 2 All ER 759

Further reading

Lawson, R.G. *Exclusion clauses and unfair contract terms*, FT Law and Tax, London (2000)
Abbott, H. *Product safety*, Sweet and Maxwell, London (1996)
Wright, C. *Product liability*, Blackstone Press Ltd, London (1989)
Consumer Law Today, published monthly by Monitor Press

Chapter 1.7

Insurance cover and compensation

A. West

1.7.1 Workmen's compensation and the State insurance scheme

The first Workmen's Compensation Act was passed in 1897 (eventually consolidated in the Workmen's Compensation Act 1925) and, as an alternative to a workman's rights at common law, imposed on the employer an obligation to pay compensation automatically in the event of a workman sustaining an accident in the course of his employment. There was no requirement of fault, the legislation being introduced to provide compensation where the workman was injured in purely accidental circumstances with no blame attaching to anyone and resembled therefore an insurance scheme. The system was operated with recourse to the County Court in the event of any dispute arising and facilitated a cheap and relatively quick payment of compensation. The amount of compensation was expressed as a weekly sum and was based on the average wage earnings during the previous 12 months with the employer whereas at common law if successful in establishing liability a workman was awarded a lump sum by way of damages. The workman did, however, have to elect between claiming at common law or claiming under the Workmen's Compensation Act.

Following the decision in *Young* v. *Bristol Aeroplane Company Limited* [1944] 2 All ER 293 it became established that a workman was precluded from pursuing a claim at common law even where he did not know of his right to elect if he had in fact accepted weekly payments under the Workmen's Compensation scheme. The Workmen's Compensation insurance policies issued at that time indemnified the insured against his liability to pay compensation under the Workmen's Compensation Act, the Employer's Liability Act 1880 and the Factories Act 1846 or at common law in the event of personal injury to any employee arising out of and in the course of his employment.

The introduction of the State scheme by the National Insurance (Industrial Injuries) Act 1946 can be considered as a compromise between the complete abolition of the common law system with its requirement of proof of fault on the part of the employer and the differing opinions of the type of accident insurance which would be most desirable.

Various types of benefits are available under the State insurance schemes for industrial injuries and are payable in respect of any person who has suffered personal injury caused by an accident arising out of and in the course of his employment or where such person suffers from what is termed a prescribed disease with reference to certain industrial occupations which may give rise to that particular disease. The phrases 'accident' and 'arising out of and in the course of his employment' have given rise to much dispute over the years since their introduction. An accident has been defined as an 'unlooked for mishap or untoward event which is not expected or designed' and by definition may be distinguished from a process involving, for example, repetitive movements of the hand or wrist which may give rise to a disability such as tenosynovitis where it is difficult to identify any particular event causing injury as opposed to considering the series of events as a whole forming a process.

There are many cases involving the question whether an act of an employee arises out of and in the course of his employment especially under the State insurance scheme and while these are beyond the scope of this text they may be studied in detail elsewhere[1]. For a decision on the topic illustrating some of the problem areas see *Nancollas* v. *Insurance Officer* and *Ball* v. *Insurance Officer* [1985] 1 All ER 833.

An employee suffering from the effects of an accident at work or from a prescribed disease may be entitled to a range of benefits determined by the current Social Security Act and supporting Regulations. The benefits may include:

1 *Statutory Sick Pay (SSP)* – The Social Security and Housing Benefit Act 1982 introduced the concept of statutory sick pay payable by the employer for the first eight weeks of absence due to injury or sickness. From 6 April 1986 the period of payment was extended to 28 weeks. Payment is subject to taxation. While receiving SSP there is no right to claim incapacity benefit.
2 *Incapacity Benefit (IB)* – This is a contributory benefit and is paid when SSP has been exhausted in circumstances where SSP is not payable. There are currently three levels of payment:

   Short Term Incapacity Benefit payable at a lower rate for the first 28 weeks to those who do not qualify for SSP.
   Short Term Incapacity Benefit at the higher taxable rate payable for the period beyond the 28-week period of SSP payment.
   Long Term Incapacity Benefit payable if the incapacity runs beyond 52 weeks.

3 *Industrial Injuries Disablement Benefit (IIDB)* – where an employee becomes disabled as a result of an injury at work or as a result of one of the prescribed industrial diseases, then they should qualify for Industrial Injuries Disablement Benefit. The requirements for payment of benefit are broadly loss of physical or mental capacity as a result of an industrial accident or disease. This means some impairment of the power to enjoy a normal life and includes disfigurement even though this causes no bodily handicap. The impairment assessment is

expressed as a percentage subject to a maximum of 100%. Entitlement to benefit only arises where the degree of disablement arising from the loss of faculty is assessed at 14% or more although individual assessments can be aggregated. Payments for assessments between 14% and 19% will be paid at the 20% rate.

4 *Constant Attendance Allowance (CAA)* – A person in receipt of IIDB who is in need of daily care and attention and where the disablement is assessed at 100% may be entitled to CAA. This is a non-contributory benefit, is not income related and is currently paid at four levels.

5 *Disability Living Allowance (DLA)* – This is a non-contributory and non-income related benefit which is paid where someone is in need of help to look after themselves. It is paid at different rates depending upon the extent to which the disability affects the individual. There are two components of this benefit to take account of the care aspect and the mobility element.

5 *Reduced Earnings Allowance* – This benefit will provide help in circumstances where the person cannot earn as much as normal because of an accident or disease caused by work.

6 *Severe Disablement Allowance* – Claimants who do not qualify for incapacity benefit because of insufficient National Insurance contributions may be entitled to this allowance if they have been unable to work for 28 weeks provided they are assessed at 80% disabled unless under 20 when no assessment is necessary.

Section 22 of the Social Security Act 1989 made provision for the Department of Social Security to collect from those paying compensation for injury or illness, the amount of benefit paid to persons as a result of such injury or illness. Effectively, this entitles the government to repayment of any state insurance scheme payments made to those injured or ill where those persons are entitled to compensation following pursuit of a common law claim. This Act was revised and superseded by the Social Security (Recovery of Benefits) Act 1997, the effect of which was to expand the circumstances in which benefit can be reclaimed and restrict the entitlement of compensators to offset the benefits against damages. This is dealt with in more detail in section 1.7.4 on the quantum of damages.

7 *Pneumoconiosis etc. (Workers Compensation) Act 1979* – The purpose of this legislation is to provide compensation to sufferers (or their dependants) of certain dust-related diseases, who are unable to claim common law damages. To qualify for payment the employer where the dust exposure occurred must have gone out of business or there must be no realistic prospect of pursuing a court action. The diseases covered are:

- Diffuse mesothelioma
- Pneumoconiosis
- Diffuse pleural thickening
- Primary carcinoma of the lung (if accompanied by asbestosis or diffuse pleural thickening)
- Byssinosis.

1.7.2 Employer's liability insurance

Since 1 January 1972 it has been compulsory for employers to insure against their liability to pay damages for bodily injury or disease sustained by their employees arising out of and in the course of their employment. This was enacted by s.1(1) of the Employer's Liability (Compulsory Insurance) Act 1969 and failure to comply with the provisions of the Act by an employer renders him guilty of an offence and liable to summary conviction – s.5.

The Act contains a definition of the term 'employee' as including an individual who has entered into or works under a contract of service or apprenticeship with an employer whether by way of manual labour, clerical work or otherwise, whether such contract is expressed or implied or in writing – s. 2(1). Certain relatives of the employer are outside the ambit of the Act – s. 2(2)(a) – as are employees 'not ordinarily resident in Great Britain' – s. 2(2)(b).

The contract of insurance incorporates conditions compliance with which is itself a condition precedent to liability under the policy. Accordingly whilst an employer may incur liability to one of his employees, in the event of his failing to comply with a condition of the policy, for example failure to notify the insurer in reasonable time of an occurrence which may give rise to liability under the policy, the insurer may invoke non-compliance with the condition as a reason for refusing to indemnify the employer under the policy. In certain circumstances this could prejudice the injured employee's prospects of recovering damages. The Employer's Liability (Compulsory Insurance) General Regulations 1972 restrict the application of conditions in policies of insurance. The regulations do not, however, prejudice the rights of the insurer to recover from the policy holder sums which they have been required to pay by reason of application of the regulations. To ensure that employees are aware of the existence of the contract of insurance, ss. 5 and 6 of the Regulations deal with the requirement on the insurer to issue a certificate and the subsequent requirement on the employer for its display at his place of business in such a position as to be seen and read by every person employed whose claims may be the subject of indemnity under the policy.

The Employer's Liability (Compulsory Insurance) Act 1998 introduced new requirements, the main aspects of which are:

- The sum to be insured is raised from not less than £2 million to not less than £5 million.
- The prescribed wording on the certificates gives more information about the cover provided.
- Certificates of insurance are required to be retained for 40 years.
- A new power is given to authorised inspectors to require not just production of the current certificate but past certificates as well.

Policy Cover – the basic cover indemnifies the insured against liability at law for damages and claimant's costs and expenses in respect of bodily injury or disease caused during the period of insurance by any person under a contract of service or apprenticeship with the insured whilst employed in or temporarily outside the territorial limits which are normally Great Britain, Northern Ireland, the Isle of Man or Channel Islands and arising out of and in the course of his employment. In view of the increased use of subcontract labour and to clarify the position regarding temporary staff and others working for an insured under various schemes and arrangements, the definition of employee has now been extended to include persons supplied to, hired or borrowed by the insured in the course of his business.

The criteria by which 'arising out of and in the course of his employment' is established are different in relation to Employer's Liability insurance and the State insurance scheme, the latter incorporating a broader definition. For an illustration of this aspect see *Vandyke* v. *Fender* [1970] 2 All ER 335 concerning the question of which insurer, motor or employer's liability, should deal with a claim where a company provides a car for its employees to go to or from work and an accident occurs on the road.

A more recent example of these issues is *Smith* v. *Stages* [1989] 1 All ER 833. Two employees were sent by their employers to carry out work at a site some distance from the site at which they had previously been working. They were paid 8 hours pay for the travelling time in addition to the equivalent of the rail fare, although no stipulation was made as to the mode of travel. On returning from the site the vehicle crashed killing the passenger. It was held that the employers were vicariously liable for the negligence of the driver. Both men were acting within the course of their employment when returning to their ordinary residence after completing the temporary work as they were travelling back in the employers' time and were paid wages and not merely a travelling allowance.

With effect from 31 December 1992 the Motor Vehicles (Compulsory Insurance) Regulations came into force requiring all passengers to be covered by motor insurance, including liability arising out of and in the course of employment.

The majority of all Employer's Liability claims emanate from accidents on the 'factory floor' often involving injuries sustained through contact with dangerous moving machinery. The Employer's Liability policy is designed to indemnify the employer against his legal liability to pay damages to employees for injuries sustained in such circumstances. This liability may arise either from the employers' breach of certain statutory duties or from a breach of their common law duties to their employees where the injured person can prove that the breach was causative of the injury. Thus, employees are able to sue their employers following an injury received at work, with the main basis of an action being an allegation of a breach of the employer's duty of care to the employee. Many of the cases taken in the past have centred on a breach of the absolute duty imposed by s. 14 of FA – to protect employees from dangerous parts of machinery. The fact that compliance witnh that requirement would render the machine unusable did not absolve the employer from their duty. This principle is illustrated by *Frost* v. *John Summers and Son Limited* [1955] 1 All ER 870 where a grinding wheel was held to have been a dangerous part of machinery within the meaning of s. 14(1). This decision meant that any use of grinding wheels was illegal and was instrumental in bringing about special regulations (the Abrasive Wheels Regulations 1970) which established the conditions under which abrasive wheels could be operated without being in breach of the law.

The criterion applied in deciding whether a civil action can be founded on a breach of a statutory duty is whether the breach is of a specific requirement such as those relating to machinery that are contained in PUWER. Civil actions are not allowed for breaches of general requirements although, as a result of representations from the CEC, the HSC is preparing recommendations to remove the civil action exclusion clauses from certain of the general requirements contained in MHSW Regulations and the Fire Precautions (Workplace) Regulations 1997.

In a common law action centred on an alleged breach of statutory duty, if there has been a successful prosecution for the same breach, the claimant's case is, effectively, proved and the only matter to be resolved is the level of damages. However, if there has been no prosecution it is up to the claimant to prove the breach.

In an action for damages, a defence to a claim could be that providing protection was not 'reasonably practicable'. This term was defined by Asquith, LJ. In *Edwards* v. *National Coal Board* [1949] 1KB 712, [1949] 1 All ER 747 in that it *implies that a computation must be made in which the quantum of risk is placed in one scale and the sacrifice, whether in money, time or trouble, involved in the measures necessary to avert the risk is placed in the other; and that, if it be shown that there is a gross disproportion between them, the risk being insignificant in relation to the sacrifice, the person on whom the duty is laid discharges the burden of proving that compliance was not reasonably practicable. This computation falls to be made at a point of time anterior to the happening of the incident complained of.* Within the requirements of current health and safety legislation, the identification of a hazard and determination of the reasonable practicability of protective measures would emerge from a risk assessment.

Many of the employers' common law duties of care have now been incorporated into statute law through the HSW. These include the duties to:

1 take reasonable care that the place of work provided for the employee is safe (s. 2(2)(d) HSW);
2 provide sufficient safe and suitable plant (s. 2(2)(a) HSW);
3 maintain such equipment (s. 2(2)(a) HSW); and
4 provide a proper and safe system of work (s. 2(2)(a) HSW).

In addition an employer has an obligation to use care in the selection of fellow employees although this duty is less often encountered as a result of the development of the doctrine of vicarious liability whereby the employer will be liable for the negligent acts of his employees whilst acting in the course of their employment.

The circumstances in which someone is acting within the course of their employment appear to be inexorably widening. In May 2001 the

House of Lords in *Lister & Others v. Hesley Hall Ltd*[2] decided that a boarding house was liable for claims where there had been sexual abuse of young pupils by a warden who was one of the employees entrusted with the care of the boys. The decision overruled the earlier decision of the Court of Appeal in *Trotman v. North Yorkshire County Council* that assault fell outside of a deputy headmaster's course of employment. The impact of the House of Lords' decision is not confined to sexual abuse cases and will have far reaching consequences.

Any breach of these common law duties resulting in injury to an employee will give rise to liability against which the Employer's Liability policy may indemnify the insured in the event of damages being payable to the injured employee.

An insurer will on behalf of the employer, where applicable, raise a defence to a workman's claim. Various defences are available to him. These include the complete defences of:

1 *volenti non fit injuria* where the injured person has consented to run the risk,
2 'inevitable accident' where despite the exercise of reasonable care by the defendant the accident still occurred,
3 defences based on the Limitation Acts where the plaintiff fails to bring his action within the prescribed time limit, and
4 partial defences such as contributory negligence (see later text).

The defence of *volenti non fit injuria* has very limited application since the mere continuance in work that involves risk of injury does not imply acceptance of the risk of injury caused by the employer's negligence and this defence has rarely succeeded in circumstances of an injury to a servant by the negligence of his master. See, for example, *Bowater* v. *Rowley Regis Corporation* [1944] 1 All ER 465.

The onus of proving negligence or breach of statutory duty and that this failure was the cause of the accident rests on the plaintiff except where the facts of any accident are such that the accident would not have occurred without negligence. This is the doctrine of *res ipsa loquitur* whereby the defendant must prove that the accident could have occurred without negligence on his part, for example see *Scott* v. *London Dock Company* [1865] 3 H and C 596. For a more modern approach to this concept and a discussion of the problems involved see *Ward* v. *Tesco Stores* [1976] 1 All ER 219.

In the past, in contrast to the Public Liability policy, it was not usual to impose a limit of indemnity to the Employer's Liability policy. However, as from 1 January 1995 insurers have imposed a cap of £10 m per incident (a lower limit is usually provided for offshore risks). The Employer's Liability policy usually includes cover for all costs and expenses incurred with the insurance companies' consent and extends to include the cost of representation of the Insured at proceedings in a Court of Summary Jurisdiction arising out of an alleged breach of statutory duty resulting in bodily injury or disease which may be the subject of indemnity under the policy.

The phrase 'caused during the period of insurance' is designed to pick up the disease risk even where the symptoms do not become manifest until many years later. Insurers are increasingly finding themselves facing claims relating to events which took place many years ago, a situation brought about because of the relaxation in the time limit for bringing claims, in particular the introduction of the 'disapplying' provisions inserted into the 1939 Limitation Act by the Limitation Act 1975 and consolidated by the Limitation Act 1980. These developments are highlighted in the case of *Buck* v. *English Electric Co. Limited* [1978] 1 All ER 271 where the widow of a man who died of pneumoconiosis was allowed to continue her husband's action for damages for personal injuries against his former employers despite the lapse of some 16 years between the deceased's knowledge of the onset of the disease and proceedings being commenced. An insurer, however, will only indemnify the insured for that part of the damages relating to the period for which the risk was held and during which there was causative exposure to the process to which the disease is in part attributable.

Claims for damages for noise-induced hearing loss are a prime example of retrospective liability giving rise to substantial difficulties for liability insurers. Deafness was added in 1975 to the list of prescribed industrial diseases under the Social Security (Industrial Injuries) (Prescribed Diseases) Regulations 1975. However, the right to benefit was limited to deafness caused by exposure to specific noise producing machinery within the metal manufacturing and shipbuilding industries, also requiring an exposure of 20 years or more within that industry. The qualifying occupations have been extended by subsequent regulations now consolidated within the Social Security (Industrial Injuries) (Prescribed Diseases) Regulations 1985.

The first reported case of an employee succeeding in a damages claim against his employer for deafness was *Berry* v. *Stone Manganese Marine Limited* [1972] 1 Lloyd's Reports 182 although the law has developed since that case. In the case of *McGuiness* v. *Kirkstall Forge Engineering Limited* QBD Liverpool 22 February 1979 (unreported) the defendants were forgemasters and the plaintiff had worked for them for most of his working life operating a stamping press. The judge found that there was virtually no evidence that any employer in noisy industries was taking any steps at all to protect his workmen prior to the late 1950s and it was not until the late 1960s that anyone in the drop-forging industry began to show an interest in protecting workmen. The potential damage which might be caused by impact noise was not fully understood until the early 1970s and the judge concluded that the publication of the Ministry of Labour pamphlet *Noise and the Worker* marked the point where a reasonably careful employer ought to have become aware that, if his employees were exposed to a high level of noise, their hearing might be at risk and there were perhaps steps which could and should be taken to eliminate or at least reduce the danger.

Following the hearing in 1983 of a series of actions claiming damages for noise-induced hearing loss sustained whilst working in the ship building industry, it was established that an employer was not negligent at any given time if he followed a recognised practice which had been followed throughout industry for a substantial period, though that practice may not have been without mishap and at that particular time, the consequences of a particular type of risk were regarded as inevitable. Accordingly, 1963 marked the dividing line between a reasonable policy of following the same line of inaction as other employers in the trade and a failure to take positive action. After the publication of *Noise and the Worker*[3] there was no excuse for ignorance.

These cases also confirmed that claimants are only entitled to recover compensation for the additional detriment to their hearing caused during the period when the employers were in breach of their duty – see *Thompson* v. *Smiths Ship Repairers (North Shields) Limited* (1984) 1 All ER 881.

Some 26 years after the publication of *Noise and the Worker*[3], a comprehensive set of regulations was introduced to control the exposure of workers to the effects of noise – The Noise at Work Regulations 1989. These regulations, effective from 1 January 1990, require employers to eliminate or reduce noise exposures above prescribed levels subject to an overriding requirement to reduce, so far as is reasonably practicable, the exposure to noise of employees. What is reasonably practicable will vary according to the circumstances. An employer is required to weigh the quantum of risk against the money, time and trouble involved in remedying the problem and whilst he is not required to incur such cost as to make his business uncompetitive, the protection of the physical health of his employees must demand a high priority. Where it is not possible to reduce noise below the prescribed level protective equipment must be provided and the employee must wear it.

Technological and medical advances in recent years have increased the awareness of the possible relationships between diseases and working environments including contact with injurious substances and operation of machinery. Attempts are constantly being made to extend fields of potential liability. In 1980 a man who developed symptoms of vibration-induced white finger after working as a caulker/rivetter for many years failed in his claim for damages for personal injury against the Ministry of Defence as employers since in 1973 when the complaint arose little was known of the condition. See *Joseph* v. *Ministry of Defence* Court of Appeal Judgement 29 February 1980 – *The Times* 4 March 1980. Since that time knowledge of the condition has increased and vibration-induced white finger acquired in certain occupations has been introduced as a prescribed disease by the Social Security (Industrial Injuries) (Prescribed Diseases) Amendment Regulations 1985 with effect from 1 April 1985. There have now been a substantial number of claims brought by employees against their employers following development of the condition of vibration-induced white finger or hand/arm vibration syndrome (HAVS). In the case of *Heal* v. *Garringtons*, unreported, 26 May 1982, it was held that a workman had been exposed to excessive levels of vibration produced by a dressing tool used on a pedestal grinder.

In 1996, the Court of Appeal upheld the judgment in the case of *British Coal* v. *Armstrong and Others* (*The Times*, 6 December 1996, CA) which held that, in the light of the evidence, British Coal should have recognised by 1973 that the work undertaken by the claimants gave rise to forseeable risk of vibration white finger (VWF) and that they should have taken effective precautions to guard against the risk by 1976.

It is generally accepted that industry should have been aware of the risk of vibration-induced white finger from certain processes involving exposure of vibration-inducing equipment by 1976. This does not necessarily imply that an employer is liable from that date as the courts have shown a willingness to realise that an employer cannot modify processes overnight. In another unreported case – *Shepherd* v. *Firth Brown* 1985 – the judge allowed three years after the date of knowledge for the employers to modify an engineering process to reduce vibration.

1.7.3 Public Liability insurance

In addition to the statutory duty to insure against his legal liabilities to his employees an employer will usually insure against his liability to others. This liability may arise from his occupation of premises, the duty to visitors being governed by the Occupier's Liability Act 1957. In addition to the various statutory controls to eliminate the effects of pollution and environmental hazards under the Control of Pollution Act 1974 and the Health and Safety at Work Act 1974, and more recently the Environment Protection Act 1990, the common law has developed doctrines that impose strict liability for the escape of things likely to do damage should they be allowed to escape. See *Rylands* v. *Fletcher* [1861] 73 All ER Reprints No. 1. The occupier may even owe a duty to trespassers in certain circumstances, at least to act with humane consideration. This concept is of particular relevance to injuries to trespassing children, for example see *British Railways Board* v. *Herrington* [1971] 1 All ER 897. The duty an occupier of premises owes to persons other than visitors is now contained in the Occupier's Liability Act 1984. Public Liability insurance has been developed to indemnify the insured against this type of risk, the insurer providing cover against liability for injury to or illness of third parties (other than employees) and loss of or damage to third party property and including claimants' costs and expenses on the same basis as the Employer's Liability policy. It must be emphasised that for Public Liability policies to operate the occurrence must be accidental in origin, for example damage caused to plaster removed by an electrician to facilitate examination of wiring would not be covered. The injury or damage must also occur during the period of insurance and in connection with the business as defined in the policy although it is normally emphasised that the interpretation embraces the insured's legal liability arising from associated activities such as canteens, sports clubs, works fire service, medical facilities and the like.

The Public Liability policy will exclude liability arising out of the ownership, possession or use by or on behalf of the insured of a mechanically propelled vehicle, vessel or craft, the insurances of which are more properly the province of other policies. With regard to motor vehicles liability is often incurred by an employer where the driver of a vehicle who is acting as a servant or agent of the employer is negligent causing injury for which the employer is vicariously liable. However, the insurance against liability in respect of the death of or bodily injury to any person caused by or arising out of the use of a vehicle on a road is compulsory by virtue of the Road Traffic Act 1972, see part VI of the Act – Third Party Liabilities.

It is also not the intention of the insurer to provide cover against the insured's liability for damage to property belonging to or in the custody, possession or control of the insured which is more properly the province of material damage policies although often cover is extended in relation to the personal effects including motor vehicles of employees, but in each case legal liability for such damage must devolve on the employer before the policy cover operates.

As a result of an EEC Directive in 1985[4] relating to legal liability for defective products within Member States, the UK introduced the Consumer Protection Act 1987. This created a new civil liability for injury or damage caused wholly or partially by a defect in a product. However, the existing legal framework, under which a person could bring a claim for damages resulting from defective goods either by means of an action in contract or tort was retained.

Prior to the Consumer Protection Act a very limited form of strict liability existed in the form of statutory liability in contract arising from the direct supply of defective products. This is defined by the Sale of Goods Act 1893 as amended by the Supply of Goods (Implied Terms) Act 1973 (now consolidated into the Sale of Goods Act 1979) and the Unfair Contract Terms Act 1977.

The eventual consumer who sustains injury or damage may be able to succeed in an action in tort under the principle enunciated in *Donoghue (McAlister)* v. *Stevenson* [1932] All ER Reprints 1, if he is able to prove not only that a product was defective and it was that which caused the injury or damage but also that the defendant has failed in his duty of care. The defendant may raise various defences to the claim including contributory negligence or a defence based on the 'state of the art' whereby he asserts that he exercised all reasonable care in accordance with the present level of technological knowledge. This defence is also available to defendants in relation to claims brought under the Consumer Protection Act.

The Products Liability Insurance policy is designed to cover this type of risk, indemnifying the insured against his liability for bodily injury or illness to persons or loss of or damage to property caused by products sold, supplied or repaired by the insured although damage to the defective product itself is excluded.

1.7.4 Investigation, negotiation and the quantum of damage

Once a claim has been intimated by an injured person or by a solicitor on his behalf the insurer undertakes a detailed investigation into the circumstances of any accident prior to taking any decision regarding liability. Even before this stage is reached it is incumbent upon the insured to notify the insurer of any accident which may be the subject of indemnity under the policy. Some cases, for example fatal accidents, are serious enough to warrant immediate investigation to obviate the possibility of alteration or destruction of physical evidence and to ensure that the witnesses' evidence is secured before the facts become clouded through the passage of time. In fatal cases it is usual for an insurer to instruct solicitors to represent the insured at the inquest who will then report on the proceedings and where necessary obtain the depositions.

Any investigation will usually combine observation of the scene of any accident including the examination of any machinery or apparatus involved and the taking of detailed statements from witnesses, independent where possible. If litigation is in prospect full proofs of evidence may be obtained and particular regard paid to the demeanour of the individual in relation to the form and manner in which he is likely to reproduce his oral evidence in the court. Where both sides in an action produce technical and expert reports a judge will decide which opinion he is disposed to accept.

The landscape of civil actions for personal injury was transformed with the introduction of the Woolf reforms and the new Civil Procedure Rules on 26 April 1999. The first phase was the introduction of a unified code of civil procedure applicable to all civil courts thereby eliminating some of the unnecessary distinctions between County Court and High Court procedure.

Perhaps the most fundamental difference for personal injury practitioners was the introduction of the Pre-Action Protocol for personal injury claims which, for the first time, laid down rules and recommended practice governing behaviour of the parties prior to the commencement of proceedings.

The main aims of the protocol are to:

- encourage more pre-action contact between the parties;
- ensure better and earlier exchange of information;
- ensure better pre-action investigation by both sides;
- put the parties in a position where they may be able to settle cases fairly and without litigation;
- enable proceedings to run to the Court's timetable and efficiently, if litigation does become necessary.

The overriding aims are to encourage a 'cards-on-the-table' approach. For example, the use of a single expert is encouraged to avoid the necessity of both sides having to instruct their own and the Court hearing oral evidence from both before making a decision. The Courts treat the standards laid down in the protocol as the normal reasonable approach to conduct. If proceedings are subsequently issued, the Courts have the power to decide whether to impose penalties on the parties for non-compliance with the protocol.

With the aid of experts the insurer will assess the evidence and decide whether liability will attach to the insured. A condition in the policy stipulates that the insured themselves must make no admission of liability, even impliedly, without the consent of the insurers. Conversely insurers do not admit liability to a third party on behalf of their insured without prior consent. Repudiation of a claim will only be made after careful consideration of all of the evidence because litigation is both costly and uncertain in outcome.

The next stage is to assess the quantum of damage, in property damage cases often with the aid of loss adjusters and in personal injury cases with the assistance of medical experts. A medical examination will be arranged where the nature of the injury is sufficiently serious to warrant this expense and where possible the use of a single expert agreed with the claimant's representative is encouraged. Once medical evidence has been clarified the insurer will commence negotiations with a view to agreement of any amount to be paid in settlement of the claim.

The law of damages is complex and in a state of constant evolution. Consequently a full discussion and analysis is beyond the scope of this text. As a brief summary, damages may be classified in the following way (for a full analysis see McGregor on Damages[5] – and for up-to-date case law see Kemp[6]):

(1) Pecuniary loss

This may be subdivided into:

(a) Past losses – Included under this heading would be the claimant's net loss of earnings, medical expenses, nursing fees, damage to clothing, cost of repairs to property, all of which must have been reasonably incurred.

For accidents occurring or where a claim for benefit naming a disease was or is made on or after 1 January 1989 for which damages above £2500 were paid on or after 3 September 1990 the compensator could deduct all relevant State Benefits from the damages and repay them to the Department of Social Security by virtue of the Social Security Act 1989 and the Social Security (Recoupment) Regulations 1990.

On 6 October 1997 the Social Security (Recovery of Benefits) Act 1997 came into force, replacing all previous recoupment regulations. Under this Act, compensators must now pay to the DSS a sum equivalent to the amount of recoverable State Benefits paid during the relevant period, which is the period between the date of the accident (or in disease cases the date recoverable benefit is first claimed) and the date of settlement.

The compensator can then reduce the amount of compensation paid in respect of loss of earnings, past care costs and/or past loss of mobility, by way of a direct set-off against amounts payable to the DSS on a like for like basis.

Damages against which benefits are recoverable

Head of compensation	Benefit
Loss of earnings	Disability working allowance Disability pension Incapacity benefit Income support/unemployment benefit

	Invalidity pension and allowance Jobseekers allowance Reduced earnings allowance Severe disablement allowance Sickness benefit Statutory sick pay Unemployability supplement
Cost of care	Attendance allowance Care element of disability living allowance (DLA) Disability pension increase for constant attendance/exceptionally severe disablement allowance
Loss of mobility	Mobility allowance Mobility element of DLA

(b) Future losses – The court must attempt to predict the plaintiff's needs and the future costs thereof. If the plaintiff can show that his income will be substantially reduced in the future and this will result directly from the accident then this is a recoverable head of damages. In its simplest form it will be calculated by reference to the plaintiff's future earning capacity in relation to his notional pre-accident earnings and multiplied by the number of years over which the loss will exist, due allowance being made for the contingencies of life – see *Lim Poh Choo* v. *Camden and Islington Area Health Authority* [1979] 2 All ER 910.

In personal injury and fatal accident cases, the Courts will now use the Actuarial Tables compiled by the government's Actuary's Department to aid the calculation of lump sum compensation for future pecuniary loss.

(c) Loss of future earning capacity – Where the plaintiff has a disability but has returned to equally remunerative employment, compensation may be payable for the risk of loss of opportunity to earn in the future – see *Moeliker* v. *Reyrolle* WLR 4 February 1977.

(d) Loss of profit – In relation to some aspects of this head of damage see *Spartan Steel and Alloys Ltd* v. *Martin and Company (Contractors) Limited* [1972] 3 All ER 557 and *SCM (UK) Limited* v. *W.J. Whittle and Son Limited* [1970] 2 All ER 417.

(2) Non-pecuniary losses

Compensation for pain, suffering and loss of amenity falls into this category. This is awarded by way of general damages and the courts do not apportion individual amounts to each subdivision but merely make a global award. The potential value of any claim must be assessed by reference to previous awards falling within the same general category making due allowance for any individual distinguishing characteristics.

There are other heads of damages including loss of expectation of life and in particular the statutory entitlement of dependants of the deceased person under the Fatal Accidents Act 1976.

Any award or negotiated settlement should also take into account any reduction in the damages possible by virtue of the Law Reform (Contributory Negligence) Act where the plaintiff suffers damage partly as a result of his own fault. The criterion for the proportion of assessment is the degree to which the plaintiff has departed from the accepted norm as compared to the degree of culpability attached to the defendant. The statute itself refers to a reduction in damages 'to such extent as the court thinks just and equitable having regard to the claimant's share in the responsibility for the damage'. Contributory negligence is not always easy to establish. In particular, momentary inadvertence by an employee where the employer is in flagrant breach of his statutory duty will not suffice to mitigate damages, for example see *Mullard* v. *Ben Line Steamers Limited* [1971] 2 All ER 424. Although contributory negligence can amount to a significant degree of culpability it cannot equate to 100% – see *Pitts* v. *Hunt and Another* [1990] 3 All ER 344.

1.7.5 General

The role of the insurer extends beyond the mere limitations of indemnifying an employer against his liability for certain injury or damage. Accident prevention is of benefit to both the insurer and the insured because in the final analysis premiums are influenced by the claims cost ratio. The social benefits of accident prevention are of course impossible to measure in terms of the avoidance of personal suffering and financial loss. The insurers employ experienced surveyors whose job embraces risk reduction in a direct sense through their observation of potential hazards during surveys prior to the arrangements of Employer's Liability, Public Liability and Engineering insurances resulting in the making of recommendations to improve the risk to be insured.

References

1. Rideout, R. W., *Principles of Labour Law* – 5th edn, Sweet & Maxwell, London (1989)
2. Lister & Others *v.* Hesley Hall Ltd (2000) UKHL 22; 2 WLR 1311; HL
3. Department of Employment, Health and Safety at Work booklet No. 25, *Noise and the Worker*, HMSO, London (1974) (first published in 1963 – out of print)
4. EEC, *Directive on the approximation of the laws, regulations and administrative provisions of the Member States concerning liability for defective products*. Directive No. 85/374/EEC, Official journal No. 1210/29, Brussels (1985)
5. McGregor, Harvey, *McGregor on Damages*, 16th edn, Sweet & Maxwell, London (2001)
6. Kemp, D. A. M., *Quantum of Damages*, Vol. 2, *Personal Injury Reports*, Sweet & Maxwell, London (1989)

Chapter 1.8

Civil liability

E. J. Skellett

1.8.1 The common law and its development

The term 'the common law' means the body of case law of universal, or common, application formed by the judgements of the courts. Each judgement contains the judge's enunciation of the facts, a statement of the law applying to the case and his *ratio decidendi* or legal reasoning for the finding to which he has come. The judgements are recorded in the various series of Law Reports and have thus developed into the body of decided case law which we now have and which continues to develop.

The doctrine of precedent whereby an inferior court is bound to follow the judgement of a higher court ensures consistency in the law. Thus an earlier judgement of the Court of Appeal will bind a High Court or county court judge considering a similar situation and a decision of the House of Lords is binding on all inferior courts although the House itself is free to reappraise its previous judgements.

The common law is not a codified body of law clearly defined in its extent and limits. New law is being made all the time. Judges are asked to adjudicate on sets of circumstances which previously might not have been considered by the courts. Moreover a judge, in applying the established principles of common law to the facts he is considering, might well distinguish that particular case from earlier decided cases. Finally in determining whether in a case there has been compliance with standards such as that of 'reasonable care' the judge will of necessity approach the problem in the light of contemporary knowledge and thinking. Thus what is adjudged reasonable conduct in 1950, say, will not necessarily be adjudged reasonable in 1980. In these ways, judges bring up to date the body of common law and adapt and develop it in accordance with the standards and social principles of the era. Such changes are of course slow and gradual, but the common law is also subject to more drastic and immediate change by Parliament, examples being the Employer's Liability (Defective Equipment) Act 1969 and the Occupier's Liability Acts 1957 and 1984. Although Parliament thus exercises dominance over the common law, the statutes in their turn are interpreted by the judges following legal rules and principles already well established.

A feature of much litigation in respect of injury damage claims, particularly where technical issues have arisen, has been the practice of the contesting parties bringing their own expert witnesses to further their cause. These experts may put opposing views on the issue and the judge then has the onerous task of having to decide which is the more apposite, a process that can consume much court time and result in delays to the final judgement. This matter was considered by Lord Woolf in his report and a key element of the resulting reforms[16] is that the courts will appoint expert witnesses – except in major cases where each party will continue to produce their own expert witnesses. The emphasis of the reforms is to find an agreed settlement by conciliation so that the injured claimant can receive compensation more quickly.

1.8.2 The law of tort

This concerns the legal relationships between parties generally in the everyday course of their affairs, the duties owed one to the other and the legal effect of a wrongful act of one party causing harm to the person, property, reputation or economic interests of another.

The law of tort covers relationships generally, compared with the law of contract which applies where two or more parties have entered into a specific relationship between themselves for a specific purpose.

Three separate branches of the law of tort are trespass, nuisance and negligence, the latter being by far the most important and applying in particular to the field of an employer's liability for accidental injury to his employee.

1.8.2.1 Trespass

This is the oldest branch of the law of tort. An action for trespass is nowadays generally confined to the intentional invasion of a man's person, land or goods involving, for example, such civil claims for damages as those resulting from battery, assault, false imprisonment, unlawful entry onto the land of another. In the latter case, apart from legal action, direct action can be taken against the trespasser using reasonable force to regain possession against, for example, squatters or 'sit-in' demonstrators. It also includes claims for conversion, an intentional dealing with a chattel constituting a serious infringement of the plaintiff's right of possession.

1.8.2.2 Nuisance

There are two forms, private nuisance or public nuisance. An action for private nuisance lies only where there has been interference with the enjoyment of land and is appropriate where an occupier of land has acted in such a way as to harm his neighbour's enjoyment of his land. It need not be a deliberate interference and includes such cases as the emission of smoke, fumes or excessive noise. The interference must be sufficiently significant and must be unreasonable. In deciding if it is, the court will take into account all circumstances including the reason for the alleged nuisance, the locality (e.g. whether rural or industrial), the ordinary use of the land and the impracticability of preventing the nuisance.

The second classification of nuisance, public nuisance, constitutes a criminal offence as well as being an actionable wrong at civil law for which damages may be claimed for any injury or damage caused. Public nuisance relates to acts interfering with the public at large and includes, for example, obstruction of the highway, leaving open a cellar flap or leaving unlit scaffolding abutting onto the highway.

1.8.2.3 Negligence

A broad definition is careless conduct causing damage or injury to another.

Actions based upon the tort of negligence are far commoner than those based upon other torts. Distinctions are not exclusive. Very often the same facts can found an action both in negligence and nuisance. There are three elements necessary to establish a case in negligence:

1 that there is a duty of care owed by one party to the other,
2 that there has been a breach of that duty,
3 that the breach of duty has resulted in damage.

1.8.2.3.1 The duty of care

To whom is this owed? In the case of *Donoghue* v. *Stevenson*[1] this was defined as follows:

> 'You must take reasonable care to avoid acts or omissions which you can reasonably foresee would be likely to injure your neighbour.'

Neighbours are defined as:

> 'Persons who are so closely and directly affected by my act that I ought reasonably to have them in contemplation as being so affected when I am directing my mind to the acts or omissions which are called in question.'

There are no hard and fast rules as to who might or might not fall into this category, and this must be examined in each case. In some situations, the public at large may be owed a duty, for example by a motorist. In others, a duty is more closely defined. An employer owes a duty of care in tort to his employee, a manufacturer to the consumer, a solicitor to his client.

The standard of care owed

This requires an examination of the facts of the particular circumstances. The magnitude of the risk of injury and the gravity of the consequences of an accident must be weighed against the cost and difficulty of obviating the risk. A considered decision has to be made. Even though a risk may not warrant extensive precautions, the particular process, place or person may have features that make these vital. In *Paris* v. *Stepney BC*[2], for example, the House of Lords held goggles should have been provided for a one-eyed man doing work where there was a risk of metal particles striking the eye although the risk of this happening was such that for a man with normal sight it could be ignored. The question is put succinctly by Denning LJ in *Latimer* v. *AEC Ltd*[3]:

> 'It is a matter of balancing the risk against the measures necessary to eliminate it.'

The New Zealand courts give a convenient and simple approach to the issue in the case of *Fletcher Construction Co. Ltd* v. *Webster*[4]:

1 What dangers should the defendant, exercising reasonable foresight, have foreseen?
2 Of what remedies, applying reasonable care and ordinary knowledge, should he have known?
3 Was the remedy, of which he should have known, for the danger he should have foreseen, one he was entitled to reject as unreasonably expensive or troublesome?

1.8.2.3.2 Breach of duty

Once the existence of the duty of care which arises from the relationship of the parties concerned and its standard are established, one has to consider whether or not there has been a breach of that duty, and if so consideration can be given to the next question.

1.8.2.3.3 Res ipsa loquitur

This Latin maxim means literally 'the thing speaks for itself'. In other words the circumstances of the accident giving rise to the action are such as impute negligence on the part of the defendant, being an event which, if the defendant had properly ordered his affairs, would not have happened. If this plea by a plaintiff is accepted by the court then a presumption of negligence is raised against the defendants. In other words, effectively it is for the defendant to prove the absence of fault rather than for the plaintiff to prove fault. The defendant can set aside the presumption against him by:

1 Proof of reasonable care having been taken.
2 An alternative explanation for the accident which is equally probable and which does not involve negligence on the part of the defendant.

3 A complete analysis of the facts, i.e. the defendant laying before the court all the facts of the case and inviting full consideration of liability.

Illustrations of the application of this maxim are such cases as bricks falling from a bridge onto a person walking underneath or cargo falling from a crane onto an innocent passerby, i.e. where one would say that prima facie the accident could not have happened without someone's fault.

1.8.2.3.4 The resultant damage

The damage must result from the negligent act or omission and be caused by it. In other words it must be a direct consequence. Most cases of injury are straightforward but sometimes unexpected complications arise, as in the case of *Smith* v. *Leach Brain & Co. Ltd*[5] where a plaintiff was entitled to recover damages for cancer developing from a burn on the lip caused by molten metal. This was a direct result of the burn. However, the chain of causation must not be broken – there must not be a *novus actus interveniens*, i.e. an act of another party intervening between the defendant's breach and the loss, or a *nova causa*, i.e. an independent and unforeseeable cause intervening. For example, in *McKew* v. *Holland and Hannen and Cubitts Ltd*[6] it was held that a workman who had sprained his ankle and later fell down stairs when the ankle gave way, resulting in his breaking his leg, could recover from the original wrongdoer damages for the ankle injury but not for the fractured leg because he himself had been negligent for not holding on to the handrail. His negligence was held to constitute a *novus actus*.

If there are more than one possible causes of an injury, it is for the plaintiff to prove causation – *Wilsher* v. *Essex Health Authority*[7]. However, where a pedestrian was injured by one car then further injured by being thrown into the path of a second, it being impossible to say what proportion of injury was caused by each motorist, it was held that the plaintiff did not have to go so far as to prove the extent of injury caused by each – *Fitzgerald* v. *Lane*[8].

1.8.3 Occupier's Liability Acts 1957 and 1984

The 1957 Act defines the duty owed by the occupiers of premises to all persons lawfully on the premises in respect of:

> 'Dangers due to the state of the premises or to things done or omitted to be done on them. Section 1(i).'

The liability is not confined to buildings and has been held to include, for example, that of the main contractors retaining general control over a tunnel being constructed – *Bunker* v. *Charles Brand & Son Limited*[9].

Section 2 defines the standard of care, owed by the occupier to the persons lawfully on the premises, namely:

'A common duty of care to see a visitor will be reasonably safe in using the premises.'

Then by s. 2(3) 'A person present in the pursuance of his calling may be expected to appreciate and guard against any special risks ordinarily incidental to it, so far as the occupier leaves him free to do so'. In other words this class of visitor is expected to use his own specialist knowledge.

Under s. 2(4) 'A warning or notice does not, in itself, absolve the occupier from liability, unless in all the circumstances it was sufficient to enable the visitor to be reasonably safe'. Whilst the occupier could, under this section, avoid his liability by a suitably worded notice, this is superseded by the Unfair Contract Terms Act 1977, which provides that it is not permissible to exclude liability for death or injury due to negligence, by a contract or by a notice and this applies to a notice under s. 2(4) of the Occupier's Liability Act 1957. The 1957 Act made no provision for those outside this category of lawful visitors, i.e. contractors, invitees and licensees. The 1984 Act extended the classes of persons to whom the duty of care is owed to those exercising public and private rights of way, ramblers and trespassers. In the latter case the Act was directed to alleviate the position of the innocent, such as the young child or someone walking blithely unaware he had no right to be there, rather than the deliberate trespasser.

1.8.4 Supply of goods

In the normal course of obtaining goods, the purchaser can reasonably expect to be supplied with goods that are fit for the purpose for which he purchased them.

1.8.4.1 Manufacturers

They owe a duty of care to the consumer of their products independently of any rights the purchaser of their products may have under contract law, against the supplier to them of goods. Thus a consumer may be able to sue both his supplier and the manufacturer.

The leading case is *Donoghue* v. *Stevenson*[1] which established the principle, the House of Lords holding that someone who drank ginger beer from an opaque bottle, given her by a friend, and who became ill from the presence of a snail in the bottle was entitled to damages from the manufacturers if she could prove her case.

The manufacturer's duty is to take reasonable care in manufacture to ensure that the product is without defect and not liable to cause injury.

There is no liability on a manufacturer if there is the opportunity of intermediate examination particularly where this is expected, which it could not be in the case of a sealed opaque bottle. Nor for instance is a manufacturer liable to a workman injured by using defective goods the manufacturer supplied which an employer examines, sees are defective but decides to keep in use albeit only until they can be replaced.

1.8.4.2 Consumer Protection Act 1987

By s. 2, where damage is caused wholly or partially by a defect in a product, then producers, own-branders, importers and suppliers are liable for that damage.

Anyone damaged by a defective product has a right of action against those from whom they obtained the finished product or those involved in the production process. The Act does not cover liability for economic loss (even though recognised by the common law in *Junior Books Co. Ltd* v. *Veitchi*[10]) or damages below £275 or claims against repairers and second-hand dealers. Liability is non-excludable by contract, notice or otherwise.

The Act specifically makes it a defence that the product was supplied other than by way of the defendant's business, e.g. by gift. It also provides for a 'development risks' defence, i.e. that the defect was not one the defendant was aware of at the time, given the state of the scientific and technical knowledge then prevailing.

1.8.5 Employer's liability

An overall statement of the duty owed by an employer to his employees is that he must take such care as is reasonable for the safety of his employees. That duty is owned to each and every employee as an individual, taking into account his own weaknesses and strengths, and is owed wherever the employee may be in the course of his employment, on or off the employer's premises. It is a duty which the employer owes personally to the employee and the employer remains responsible for a breach of that duty even if he has delegated the performance of that duty to someone else, for example to a safety consultant who might have a separate liability. The same applies if he has put his employee to work under the order of another party – *McDermid* v. *Nash Dredging and Reclamation Co. Ltd*[11].

The employer can be held liable either directly for breach of his own duties or vicariously. Vicarious liability arises where an employee or an agent of the employer has acted negligently and caused injury to another employee. The employer is legally liable for the wrongful act or omission where it has been performed in his interests. However, he is not liable if the employee acts negligently on a frolic of his own independently of his employment. *Smith* v. *Crossley Bros Ltd*[12] illustrates this, where, as a joke, two apprentices injected compressed air into the body of a third and the employers were held not liable.

The employer's duty at common law can conveniently be considered under five heads. Obviously each will turn on the particular circumstances involving one or more of these elements and it is impossible to give more than general guidelines. The heads are:

1 system of work,
2 place of work,
3 plant and equipment,
4 supervision and/or instruction,
5 care in selection of fellow employees.

1.8.5.1 System of work

The employer is obliged to set up and operate a safe system of work, and it is a question of fact in every case what is safe. This includes such matters as the co-ordination of activities, the layout and arrangement of the way a job is to be done, the use of a particular method of doing a job. The employer is expected to plan and draw up an original method of operation which is safe and free, so far as possible, from foreseeable cause of injury. Regard will be held to established practice and absence of accident in assessing what is safe, but the court will still examine the practice to decide if it is safe. In *General Cleaning Contractors Ltd* v. *Christmas*[13] Lord Oakley said in his judgement:

> 'the common law demands that employers should take reasonable care to lay down a reasonably safe system of work.'

He continued that workmen even though experienced and competent to lay down a system themselves should not be expected to do so, making their decisions at their workplace where the dangers are obscured by repetition, compared with the employer who performs his duty in the calm atmosphere of a boardroom with the advice of experts.

1.8.5.2 Place of work

The employer is under a duty at common law to provide a reasonably safe place of work, relating to such matters as the provision of gangways clearly marked and free of obstruction, and the maintenance of floors and staircases. The duty is fulfilled through regular inspection of the workplace and keeping it in a safe state, free of hazard so far as reasonably practicable. It does not extend to protection from abnormal hazards which the employer could not reasonably have foreseen. For example, whilst in conditions of ice and snow, paths must as far as possible be sanded before the normal time for employees to arrive at the premises, if there is a sudden totally unexpected snowfall, the employer is not liable if paths are slippery or obstructed until he has had reasonable opportunity to remedy the situation.

The duty extends to any place at which the employee works whether belonging to his employer or not, but it will depend on the circumstances whether the employer should have inspected them before sending his employees to work there, and perhaps had steps taken to make them safer. For example, no court would suggest the employer of a plumber sent out to work at a private house should first send the foreman or supervisor to inspect the house unless the employer had prior knowledge of some particular feature of the premises which introduced added risk. In most cases involving factory or site accidents the relevant section of the Workplace Regulations 1992 or the Construction (Health, Safety and Welfare) Regulations 1996 will be pleaded in addition to the duty at common law.

1.8.5.3 Plant and equipment

The employer owes a duty to his employee to provide safe and proper plant and equipment which must also be suitable for the purpose to which it is put.

It is a far-ranging aspect of the employer's duty. In the first place the employer may have failed completely to provide equipment necessary for the safe performance of work, for example mechanical lifting equipment for a load too heavy to be manhandled.

Equipment supplied may be unsuitable for the particular function, or it may be the proper equipment but inadequately maintained or defective.

Consideration will be given in deciding if the employer is liable to the procedure followed for reporting and rectifying defects, routine maintenance, the issue of small items of plant and such like.

This aspect is relevant also to the question of whether an employer has provided protective equipment such as gloves, goggles and ear-muffs to reduce or prevent exposure to foreseeable risk of injury.

Where a claim for damages arises out of an accident in a factory, the appropriate sections of PUWER will be relied upon, for example relating to the guarding of machinery, in addition to the duty at common law.

The Employer's Liability (Defective Equipment) Act 1969 discussed later is relevant to this aspect too.

1.8.5.4 Supervision and/or instruction

An employer must take such care as is reasonable to ensure adequate and proper supervision over and instruction to his employees. What is reasonable must depend on the circumstances, including the complexity of the work to be done, the technicality of the equipment concerned and the age and experience of the workman. It must be obvious that if a young inexperienced man is set to work on a complicated machine or a complicated task where he can injure himself the employer will be held liable. It must not be thought, however, that an employer can leave even a senior experienced man to his own devices. Supervision and instructions are a matter of degree but always the courts will impute to the employer a superior knowledge of the dangers and risks in a work system with the consequent duty to supervise and instruct his employees.

1.8.5.5 Care in selection of fellow employees

This aspect of an employer's duty is of less significance since the employer will be held vicariously liable for the act of an employee who negligently injures another, which was not always so.

It is most relevant to the type of case where an employee indulging in horseplay or fighting has injured another and the man concerned has a history of such activities to the knowledge of the employer, who has taken no steps to dismiss him or prevent a recurrence.

1.8.6 Employer's Liability (Defective Equipment) Act 1969

Prior to the passing of this Act where a workman sued his employer in respect of injury caused by a defective tool or item of plant supplied by the employer to the employee it was a defence for the employer to prove that he did not know and could not reasonably have known of the defect and that he had exercised reasonable care when he obtained the item concerned, by going to a reputable manufacturer or supplier. This was the rule in *Davie* v. *New Merton Board Mills Limited*[14]. The Act changed the law and imposed liability on the employer where an employee was injured in consequence of a defect in equipment provided by his employer for the purpose of the employer's business, if the defect was attributable wholly or partly to the fault of a third party (whether identified or not). In other words the employer no longer has a defence if he provides defective equipment to his employee which results in injury and the defect was the fault of another party. This does not mean that the employer is without remedy against that other party. He is entitled to bring an action against the supplier in respect of the defective plant, but must be able legally to prove his case against the supplier. It is perhaps unnecessary to add that an employer is liable irrespective of the Act if it can be proved that the defect should have been found by the employer on inspection before being put into use or if an employer had caused or permitted his employee to keep in use defective items of plant.

1.8.7 Health and Safety at Work etc. Act 1974

Although ss. 2–8 of the Act impose general duties on parties including employers, failure to comply with the obligations imposed by the Act itself does not provide grounds for a civil claim. However, section 47(2) of the Act stipulates that an action can be based on a breach of Regulations made under the Act, unless the Regulation has a specific exclusion.

1.8.8 Defences to a civil liability claim

The first and obvious defence which may be raised is a denial of liability which may be based on a variety of grounds.

1 That the duty alleged to have been breached by the defendant was never imposed on him in the first place, for example in an employee's claim against his employer that the plaintiff was not an employee but was working for another company.
2 That the nature of the duty was different from that pleaded against the defendant.
3 That the duty owed was complied with and not breached.
4 That the breach of duty did not lead to the damage.
5 That the plaintiff was himself guilty of contributory negligence resulting wholly in the damage.

Secondly in the defence it may be pleaded that conduct of the plaintiff, constituting contributory negligence, caused and/or resulted in part in the damage he suffered and that any damages which might be payable to him should be reduced accordingly – the Law Reform (Contributory Negligence) Act 1945. By way of example, that he failed to see a hole into which he fell. Obviously such a consideration only comes into play in the event of a finding that the defendant is liable. The court will then assess the respective blameworthiness of the parties to decide whether there are grounds for finding the plaintiff partly to blame and, if contributory negligence is established, the court will determine the amount of damages the plaintiff would receive if he succeeded in full and then discount these by the proportion to which the plaintiff is himself found to blame.

Thirdly there is the situation where the accident is the fault not of the defendant sued but of some other party. If another party is blamed in the defence, the usual result is that they are joined in as a co-defendant by the plaintiff, and he sues both. However, if a defendant considers that if he is liable to the plaintiff, then he in turn is entitled to recover from someone else any damages he has to pay to the plaintiff in which case that person can be joined in the proceedings by the defendant as a third party. An example of the circumstances where there may be third party proceedings is one where an injured workman who has fallen into a hole at the place where he works sues his employer, who then brings in by third party proceedings the contractor who had left the hole unfenced. In such a case, the plaintiff would have to establish that the defendant was liable to him, and in turn the defendant would then have to prove his case against the third party. The third party will be liable only if the defendant is liable to the plaintiff.

Compare this situation with an action where the plaintiff sues more than one defendant, such as, in the example given above, both suing his employer and the contractor direct. The plaintiff might fail against both defendants, succeed against one or the other or succeed against both, the judge apportioning the degree of liability attaching to each defendant. In the case of *Fitzgerald* v. *Lane* the House of Lords held that where there were two or more defendants, the first consideration was whether the plaintiff had proved his case against the defendants, then the question of whether he was himself negligent and his damages should be reduced accordingly, and finally the apportionment of liability between the defendants themselves.

Thus a pedestrian who ran into the road and was hit by a car and then by another and who was held equally to blame with the car drivers, had his damages reduced by 50% on account of his own negligence. The car drivers' 50% share of the blame was then apportioned between them at 25% each.

1.8.8.1 Joint tortfeasors

Where two or more parties are responsible for breaches of duty leading to a single injury, i.e. the same damage, they are jointly liable as wrongdoers, in whatever proportion of fault is determined from the circumstances. The simplest example is where two vehicles collide, due to the fault of both drivers, injuring an innocent passenger. The passenger's claim may be enforced against either tortfeasor, who can, under the Civil Liability (Contribution) Act 1978, then claim contribution from the other to the extent of the other's liability.

Another illustration is where both the employer and another contractor engaged on work at the same building site are jointly liable for injury to employee. It must be noted that generally, an employer is not liable for the torts of an independent contractor, unless the work to be carried entails particular danger. Furthermore, an employer cannot get out of his liability for his employee's safety by delegating this to a contractor.

1.8.9 Volenti non fit injuria

Where a person has agreed either expressly or by implication to accept the risk of injury, he cannot recover damages for damage caused to him by that risk.

For this defence to succeed the person concerned must have had full knowledge of the nature and extent of the risk to be run and have accepted that risk of his own free will. Such a defence is available only in extremely limited circumstances in an action by an employee against his employer. In the case of *Smith* v. *Baker*[15] it was pleaded against an employee drilling rock in a cutting over whose head a crane lifted stones. The court held that although he knew of the danger and continued at work he had not voluntarily undertaken the risk of injury from a stone falling from the crane and hitting him.

Such a defence does not apply to an action for damages brought by a rescuer deliberately running risks to rescue someone who has been injured by dangers created by another. The case of *Baker* v. *T.E. Hopkins & Sons Limited*[16] confirms the entitlement of a rescuer to damages for injury in respect of that negligence.

1.8.10 Limitation

The Limitation Act 1980 stipulates that an action founded on tort shall not be brought after six years from the date when the cause of action accrued but an action for damages for personal injuries or death must be commenced within three years. Otherwise the actions are barred by the statute and the defendant can plead this as a defence. The three years start to run from the date of the accident or date of the plaintiff's knowledge if later. If the injured person dies within the three years, the period starts to run again from date of death or of his personal representatives' knowledge. The saving provisions of 'knowledge' are aimed primarily at the industrial disease cases where the accidental exposure almost invariably dates back many years before the effects of that exposure had developed and were known. There is a careful definition of what is meant by 'knowledge' in s. 11 of the Act.

The Act also permits an overriding discretion to the court to let in late claims where it is equitable or fair to do so. Furthermore in the case of someone under a disability, e.g. an infant or person of unsound mind, the three years do not start to run until the age of 18 or recovery.

1.8.11 Assessment of damages

Once liability is established the question for consideration is the amount of damages or compensation to be awarded. The object is to put the injured party as far as possible in the same position as before.

In an action for breach of contract or for debt, this amount will already have been defined in the dealing between the parties and is known legally as a liquidated claim. However, in an action for damages in respect of tort, where damage and/or injury has been caused, damages are called unliquidated, i.e. they will have to be calculated and assessed after the event giving rise to the claim.

These damages will comprise special damage and general damages.

1.8.11.1 Special damage

Special damage consists of heads of specific expenditure or loss as a result of the accident, damaged goods or loss of wages during time off work. In actions for personal injury, it consists primarily of the loss of wages and the figure recoverable is the net wage lost after deduction of income tax and national insurance contributions, i.e. the actual amount the plaintiff would have received in his pocket. He will be awarded both his total loss of wages during total incapacity from work and partial loss if by reason of continuing disability, as a result of the accident, he cannot do his full work or has to change to a lighter job and is thereby earning less. Credit is given for non-contributory payments by the employer such as sick pay. There are also offset against the loss of earnings claim, any tax refunds and unemployment benefit if, after having been certified fit to return to work following an accident, a man cannot return to his old job and is unable to get another. Redundancy payments under the Redundancy Payments Act 1965 as amended are deductible if attributable to the injury. In actions for damages for personal injury the Social Security (Recovery of Benefits) Act 1997 compels, a person making a compensation payment in consequence of an accident to obtain a Certificate of Total Benefit paid by the Department of Social Security and then to deduct from the compensation payment the amount of the benefit, accounting to the DSS for this.

1.8.11.2 General damages

General damages are those recovered to compensate for pain, suffering and loss of amenity resulting from an injury. Whilst there are no set tariffs there are published guidelines to assist in the assessment and ensure compatibility between awards made by judges and lawyers who also take into account decided cases involving similar injury. However, in calculating the appropriate sum to award, account is taken of such matters as the particular idiosyncrasies of the plaintiff, his age, occupation, hobbies and such like. The court also has regard to the effect of inflation on past awards or similar injuries. For example, the loss of a finger would attract higher damages for an employee who in his spare time was a skilled musician; similarly damages for an incapacitating leg injury to a keen and energetic sportsman would be higher than those for someone in a sedentary occupation with no active hobbies.

Where there is partial or complete incapacity for work continuing after the trial general damages also include a capital sum awarded for future loss of wages. A sum will, where appropriate, be awarded too for loss of opportunity on the labour market. This is intended to compensate for a permanent disability which a prospective employer may take into account in deciding whether to offer employment, compared with a candidate of equal competence who has no such disability.

Awards of damages generally are once and for all. However, there is an exception.

1.8.11.3 Provisional damages

Section 6 of the Administration of Justice Act 1982 introduced provisional damages for cases where there is a chance that some serious disease or serious deterioration in the plaintiff's condition will accrue at a later date. Appropriate cases include industrial disease claims where there may be a risk of the development of cancer or a malignant tumour in the future. Provisional damages are assessed ignoring that possibility. If it occurs then a further award may be made.

1.8.12 Fatal accidents

A cause of action in tort, save for defamation by or against a person, survives for the benefit of or to the detriment of the estate under the Law Reform (Miscellaneous Provisions) Act 1934.

On behalf of the estate, loss of earnings to date of death and general damages for pain and suffering during lifetime are claimable, without reference to any loss or gain to the estate resulting from the death.

Under the Fatal Accidents Act 1976 damages for loss of financial support can be claimed by or for the dependants. The definition of dependant is set out in the Act as amended by the Administration of Justice Act 1982 and includes spouse or former spouse, ascendants and descendants as well as adopted children and anyone living with the deceased as spouse, the latter subject to certain conditions.

Damages are calculated by the measure of actual financial loss. Thus the deceased's earnings will be established and the proportion expended on the dependant determined. This will then be multiplied by a number of years' purchase to allow for the length of time the deceased would have worked. A deduction will be made for capitalisation.

The Administration of Justice Act 1982 also introduced a claim for 'bereavement damages' under which a fixed sum is payable by way of damages – the amount is currently £7500 – for loss of a spouse and to parents for the loss of a child.

1.8.13 'No fault' liability system

Over the years the possibility of compensation being paid to victims of accidents irrespective of responsibility has been discussed and canvassed but not adopted. The attraction of such a system lies in the removal of the conflict between employer and employee over liability for the payment of damages and the consequent expense in time spent by the employer in detailed assessment of fault and in costs. However, such schemes still lead to dispute over the entitlement to compensation or the amount to be paid, such as those cases fought to establish entitlement to payment under the Workmen's Compensation Acts of the 1940s, later repealed. Many points of question remain to be answered, such as, how should such a scheme be funded? By the State, or by privately arranged insurance cover? Would it be practicable? Where would the limits be drawn, both as to the recipients of compensation and the nature of the compensation – damages for injury alone or including income loss? Injury caused solely by accident or including industrial disease and conditions due to the environment? Direct employees only or contractors too? What about road traffic casualties? If they too are included, is this not unfair to the victims of other accidents, such as those in the home?

References (cases referred to)

1. Donoghue *v.* Stevenson (1932) AC 562
2. Paris *v.* Stepney Borough Council (1951) AC 367

3. Latimer *v.* AEC Limited (1953) 2 All ER 449
4. Fletcher Construction Co. Ltd *v.* Webster (1948) NZLR 514
5. Smith *v.* Leach Brain & Co. Ltd (1962) 2 WLR 148
6. McKew *v.* Holland and Hannen and Cubitts Ltd (1969) 2 All ER 1621
7. Wilsher *v.* Essex Health Authority (1989) 2 WLR 557
8. Fitzgerald *v.* Lane (1988) 3 WLR 356
9. Bunker *v.* Charles Brand & Son Limited (1969) 2 All ER 59
10. Junior Books Co. Ltd *v.* Veitchi (1983) AC 520
11. McDermid *v.* Nash Dredging and Reclamation Co. Ltd (1987) 3 WLR 212
12. Smith *v.* Crossley Bros. Ltd (1951) 95 Sol. Jo. 655
13. General Cleaning Contractors Ltd *v.* Christmas (1953) AC 180
14. Davie *v.* New Merton Board Mills Ltd (1959) 1 All ER 67
15. Smith *v.* Baker (1891) AC 325
16. Baker *v.* T.E. Hopkins & Sons Ltd (1959) 1 WLR 966
17. *Civil Procedure Rules 1998* (emerging from the *Access to Civil Justice Report* by Lord Woolf – known as the Woolf Reforms), The Stationery Office, London (1998)

Further reading

Munkman, J., *Employer's Liability at Common Law*, 11th edn, Butterworth, London (1990)
Heuston, R.F.V. and Chambers, R.S., *Salmond on the Law of Torts*, 18th edn, Sweet & Maxwell, London (1981)
Kemp, D., *Damages for personal injury and death*, Oyez Publishing, London (1980)
McGregor, Harvey, *McGregor on Damages*, 15th edn, Sweet and Maxwell, London (1988)

Part II

The management of risk

In every activity there is an element of risk and the successful manager is the one who can look ahead, foresee the risks and eliminate or reduce their effects. Risks are no longer confined to the 'sharp end', the shop floor, but all parts of the organisation have roles to play in reducing or eliminating them. Indeed, the Robens' Committee recognised the vital role of management in engendering the right attitudes to, and developing high standards of, health and safety throughout the organisation.

A number of specialised techniques have been developed to enable risks to be identified, assessed and either avoided or reduced but there are other factors related to the culture of the organisation and the interrelationship of those who inhabit it that have a significant role to play. An understanding of those techniques and the roles and responsibilities of individuals and groups is a necessary prerequisite for high levels of safety performance.

Chapter 2.1

An introduction to risk management

J.E. Channing

2.1.1 Introduction

Reader, you are sitting in a chair beginning to read this chapter. There is a chance that you will never finish it. The chair may break and you may strike your head when you fall. A fire may break out in the room. The ceiling may collapse upon you. You are living with risks at the very moment you seek to understand how to control them! When considering risk management there are six lessons to be learnt:

1 Absolute safety is a chimera (a pious hope!)
2 We do have some ability to change the risk equation if we understand the hazard. For example, if you read this chapter in an open field you remove the hazard that a ceiling may fall on you or that a fire may trap you in your room.
3 However, you may expose yourself to a new hazard such as an attack from a swarm of bees so you clothe yourself in protective netting leaving so little of your skin exposed to the potential hazard that likelihood of receiving fatal stings is negligible. It is sometimes possible to alter the consequences should the hazard impact on you.
4 Some risks are so unlikely to occur that they can be accepted and lived with. You judge that the chair you are seated in is unlikely to collapse so you do not remove yourself from it and sit on the floor to read this chapter.
5 However, another reader may not be so confident in the chair and will change seating positions to the sofa. Other persons may view the same risk situation differently.
6 Even if you could change your seating position someone has decided that the classroom chairs are safe so you are compelled to read this chapter sitting on one. You may be constrained by circumstances to endure the risks you face and your attitude to them may change accordingly. Alternatively, your familiarity with equipment and work situations may influence your perception and tolerance of the risk.

The components of risk involve acceptance of its existence; understanding the hazard, the consequences, the likelihood of a hazard causing injury or damage; the perception of the risk and the tolerance of the risk by individuals or by a group.

It is not surprising therefore that the definitions of risk can be complex. The Royal Society Study Group report[1] offers the following definition:

> RISK is the probability that a particular adverse event occurs during a stated period of time, or results from a particular challenge.

The report continues with associated definitions:

> HAZARD is seen as the situation that in particular circumstances could lead to harm, where HARM is the loss to a human being (or to the human population) consequent on damage and DAMAGE is the loss of inherent quality suffered by an entity (physical or biological).
> RISK ASSESSMENT is the general term used to describe the study of decisions subject to uncertain consequences.
> RISK ESTIMATION is the first subdivision of Risk Assessment and includes the identification of outcomes, the estimation of the magnitude of the associated consequences of these outcomes, and the estimation of the probabilities of these outcomes.
> RISK EVALUATION is the second subdivision of Risk Assessment and is the complex process of determining the significance or value of the identified hazards and estimated risks to those concerned with or affected by the decision.
> RISK MANAGEMENT is the making of decisions concerning risks and their subsequent implementation and flows from Risk Estimation and Risk Evaluation.

The above definitions and the tenor of the report are based on an approach to the subject of risk from the standpoint of the natural scientist. The view of the social scientist had more influence in the subsequent report[2] that places greater emphasis on the perception of risk by individuals and the public at large, especially when overlaid by media involvement. Most managers do not have the time to consider what some will view as the esoteric components of risk. The Health and Safety Executive[3,4] offer the following definition of hazard and risk:

> HAZARD means anything that can cause harm (e.g. chemicals, electricity, working from ladders, etc);
> RISK is the chance, high or low, that somebody will be harmed by the hazard.

The assessment of risk can range from a profound intellectual exercise or a day-to-day practical activity. This chapter introduces the concepts and ideas of risk that subsequent chapters develop.

2.1.2 The components of risk

Risk can be subdivided into many elements. The major components are considered below.

2.1.2.1 Hazard

The definition of 'hazard' presented above has two elements. The first is that a hazard has within it the ability to *harm* a person. The second is that the existence of a hazard does not mean that harm will arise – a hazard only has to have the *potential* to harm. Identifying hazards is an ongoing process. There are everyday hazards associated with living – e.g. using gas as a fuel to cook food. There are unusual hazards that most people encounter only rarely – e.g. undergoing surgery.

There are hazards that will cause immediate harm if they are encountered. These are termed *acute* hazards and are usually recognisable to most people so there is rarely a need for them to be explained. For example, most people will understand that being struck by a moving vehicle will result in immediate harm. Other hazards may affect us but we do not experience immediate harm. An example is exposure to asbestos fibres which may be inhaled many times over many years before harm is caused to the body. These are termed *chronic* hazards. Some hazards can be both acute and chronic. Radiation in small repeated doses can cause chronic harm in the form of cancers. However, a large single dose can cause acute harm in the form of burns and poisoning.

Some hazards are caused by workplace exposure. Other hazards arise from a combination of workplace exposure and personal lifestyle. Two examples illustrate this point. Stress may arise at work (and usually does to some extent) and be quite tolerable to an individual. However, combine that with stress from the individual's personal life (such as undergoing a divorce or a bereavement) and harm to health can easily arise. The other example is musculo-skeletal injury, such as carpal tunnel syndrome, which may be experienced by a VDU operator using a keyboard all day. Combine the workplace activity with a hobby of surfing the net and it easy to see that this additional exposure increases the risk of wrist injury. In both these examples it is not easy to determine which of these activities are the causative factors and what their contributions are to the resulting harm. An activity that is not a hazard because it does not cause harm can become one under different circumstances.

Recognising hazards is not always straightforward or easy. If a check list of hazards is used, it should be reviewed periodically, preferably by different people so that there is a chance that what one reviewer misses another will identify.

2.1.2.2 Consequence

The harm that arises from a hazard is the *consequence* of it. It is important to identify the possible consequences before embarking on a hazard control strategy. The more serious the consequence the greater the need to control the hazard. If a hazard will result in serious injury or death then control of it becomes urgent. Hence there is an emphasis on guarding machines because they have sufficient energy to cause immediate serious harm to the machine operator. Where the consequence is likely to be a minor injury then a warning notice may suffice. For example, when a spill occurs in a supermarket, it is mopped up leaving the floor wet and slippery for a short period. A notice warning shoppers of the hazard is considered adequate. However, a shopper may slip on the damp floor and strike her head on a shelf and suffer serious injury. Consequences are not possible to predict with certainty.

Compliance with recommended standards does not confer immunity from risk. Where exposure limits are applied to an employee's exposure to toxic chemicals, the values in official lists of exposure limits are commonly interpreted as being safe levels, i.e. harm will only occur when there is exposure above the stated level. However, careful reading of the supporting information with the lists will reveal that the exposure level has been set at a point where *most* people *most* of the time will not be harmed. Clearly *some* people may be harmed. If it is known that a particular employee may be susceptible to a particular substance in use then preventative action must follow. The law recognises that below the listed exposure levels an amorphous group of employees does not need to be protected but where particular individual susceptibilities are known the employees must be protected. A good example is the case of a pregnant woman. Before pregnancy the body may not be harmed at all by exposure to low levels of a chemical agent. Once pregnant, however, the foetus may be harmed by the same level of exposure that would not harm the adult woman or any others in the group.

A further factor to consider under *consequence* is *who* may be harmed. It is not necessarily those who are immediately exposed to the hazard who are the only ones likely to be affected. For example, legionella is a form of virulent pneumonia. The virus causing it multiplies in warm water and causes harm when it enters the lungs, usually in the form of an aerosol. It can be fatal to people whose immune suppression system is depressed. The breeding grounds of legionella are cooling towers, air conditioning systems, shower heads in centrally heated hotel rooms, etc. The consequences of poor maintenance of these systems are the release into the atmosphere of contaminated aerosol vapours that affect not only the relatively healthy employees but also neighbours and passers-by and, in particular, people whose immune system is depressed.

Deciding what the consequences are likely to be and who may be affected by the hazards is not as straightforward as it might appear. It requires lateral thinking.

2.1.2.3 Likelihood

An important element of risk is the likelihood or probability that the hazard will cause injury. In its simplest form, probability can be considered as high, medium or low and for the majority of risk assessments this should prove adequate. The valuation of probability is subjective with the risk assessor drawing on his knowledge and experience to decide whether a risk should be rated high, medium or low. These ratings can be given a numerical value that when combined with similar numerical values given to the likely worst injury can give an indicative ranking or 'risk rating'.

Elegant techniques such as Fault Tree Analysis[5] can be applied which list all the elements and sub-elements which contribute to an incident. By placing against each element the probability of its occurrence a very detailed and accurate assessment of the likelihood of a hazard or fault causing harm can be made. Earlier work by Farmer[6] generated graphs of frequency against the number of deaths (a frequency–consequence or fC line) arising from the release of the isotope Iodine 131 from nuclear reactors. Much work has been done to put numerical values to what are regarded as acceptable probabilities of a hazard causing harm. Current probability values for ascceptable risks are:

- An acceptable risk of death for a single individual lies within the range of 10^{-3} to 10^{-4}.
- An acceptable risk of death for a group of individuals – a multi-casualty incident lies within the range 10^{-5} to 10^{-6}.
- A risk of death can be ignored completely if it exceeds 10^{-7}.

However, while useful guidance, these figures are controversial. As society's expectations of health and injury protection develop these figures are likely to be challenged.

Deciding the likelihood of a hazard causing injury or damage is difficult. Whatever system is chosen it should be remembered that human beings do make errors, whether they be risk estimators or those at risk, so any probability assessment carries with it a degree of uncertainty.

2.1.2.4 Perception

Understanding risk can assist in providing an insight into the factors that contribute to it. A rational analysis should be sufficient to convince everyone about the risk and hence the control measures that are needed. This may not be the case. Newby[7] comments that despite a huge increase in road traffic since the 1930s, the risk of an accident occurring involving pedestrians has been reduced by a factor of four. That argument does nothing to alter the response of a mother whose child has been injured in a road accident. Her perception of risk is decisive and the quantified assessment does not sway her or the immediate community. Visitors to an old established factory site keenly felt the risk from large trucks using the internal roads that were designed for the horse and cart. Employees were used to the risk and only occasionally voiced concern. They were more concerned about risks arising from hazards in the workplace. Their perception of the risk was dulled by their familiarity with it and the more immediate personal risks they faced elsewhere.

Different groups have a different perspective. Key studies on this topic have been reported by Slovic *et al*[8]. Their work, which is reviewed extensively in chapter 5.1 shows that:

- people are unable to estimate risk with accuracy, they are biased by media reports
- different groups will rank risks differently and ignore expert assessment
- people react more to risks that are relatively unknown to them where they feel they have little control over the consequences compared with known risks that kill thousands of people each year, over which they have some control.

People, as individuals and as groups, act in accordance with their own perception of the risk. Risk assessment and risk management processes will only be successful in reducing injury if there is widespread involvement of those who are potentially affected by the hazard. The safety professional may be the person who provides technical data and an initial ranking of risk. It is for a larger group to decide the appropriate responses.

2.1.3 Strategies to control risk

The management of risk is a strategic approach to health and safety that organisations must adopt in order to control the hazards that employees, contractors, community residents and others are exposed to. It requires more than just a focus on the hazard itself. The control of hazards requires organisational and administrative processes in order to be effective. Those processes need to be in place to influence the behaviours of directors, managers, supervisors and employees so that harm does not occur. They should also be bound together by a policy and their effectiveness established by measurement, review and audit. A structure to accommodate these processes is necessary if the risks from hazards are to be controlled. Its success is demonstrated when the hazard has been eliminated. Elimination is the first step in the risk control hierarchy.

2.1.3.1 Risk control hierarchy

A risk control hierarchy is a structured approach whereby for each hazard a set of action options is considered. The action that should be adopted is the one that gives the greatest degree of protection, not only to the operator but also to others who may be exposed to the hazard. The options in order of decreasing effectiveness are:

1. **Elimination** by the removal of the hazard itself to ensure that injury or damage will not occur. Many injuries occur each year from manually handling objects. By changing the work method and employing

mechanical handling, the hazards from manual handling are eliminated. Boxes of product being moved from one storage location to another do not add value to the business but do add injuries and costs. Removing, simplifying and streamlining handling operations can eliminate potential injury causing hazards.
2. **Substitution** is sometimes possible by utilising a less hazardous material instead of a more hazardous one. An example is in the replacement of benzene (MEL 3 ppm) that is a proven human carcinogen by toluene (MEL 50 ppm) which is in many cases as good a solvent. This option requires a good understanding of the hazardous properties of both materials to ensure that new or additional hazards are not introduced into the workplace by the replacement material.
3. **Reduction** in the risk faced can be achieved by reducing the quantity of materials held in the workplace. This was recognised in the HFL Regulations that allow only sufficient flammable materials for the day or shift to be kept in the work area.
4. **Personal protection** is the final option in the risk control hierarchy. This requires the issue to the exposed employee of equipment that will protect him only and may consist of a facemask, eye protection, safety shoes, bad weather clothing, etc. It must be seen as a last option after all the other options have been investigated and proved not feasible. The employee must be told of the hazards faced, be trained in the control measure in place and in the proper use of the equipment. Checks should be carried out periodically to ensure there is compliance with the rules associated with the use of protective equipment. The purchase of PPE may be a cheap option, but the infrastructure necessary to ensure that it is properly used and maintained may be onerous.

The risk control hierarchy should be applied to every identified hazard. The risk control method chosen need not be just one of the options but can be a combination of two or more. Consideration should also be given to the ease of use of the chosen control method and to the ease with which it can be defeated. For example, eliminating a hazard altogether is an option that nobody can defeat. By contrast, asking an employee to avoid harm by wearing a dust mask is a control measure that is easy to defeat – it is simply taken off when conditions become unbearably hot.

The usefulness of the risk control hierarchy is enhanced when it is a part of a comprehensive risk management process, that is, when it forms part of an established management process.

2.1.3.2 Risk management processes

The effective managing of risk has been central to the insurance industry's success. It involves assessing the financial risk and making provisions to reduce it or spread it. This process has been adopted and adapted to issues concerning health and safety at work. Early laws made tentative prescriptive attempts to manage risks, i.e. by controlling a specified hazard. However, recent legislation, such as HSW[9], has emphasised the control the risks through management processes, an approach that has been recently confirmed by the Health and Safety Commission[10]. The Act requires the management of risks by employers within the following framework:

- **Policy.** There must be a policy statement outlining the organisation's intentions in respect of health and safety.
- **Organisation.** There must be organisational arrangements in place to implement that policy where the roles and responsibilities of the office holders are stated.
- **Administration.** The necessary administrative arrangements, such as internal standards and procedures, must be in place to enable the policy's intentions to be carried out.
- **System.** A system must be established to monitor the effectiveness of the measures taken, to identify any shortcomings and indicate the corrective actions needed.

A central plank of modern health and safety legislation, both national and international, is the carrying out of risk assessments. In the UK, regulations such as COSHH[11] and the Management Regulations[12], recognise that risks can most effectively be identified through risk assessments but must be controlled within a comprehensive management framework that includes:

- having available health and safety expertise to assist and guide managers;
- the carrying out of suitable and sufficient assessment of risks to which employees and others may be exposed as a result of the work;
- a review of the risk assessment if there are significant changes in the work process;
- the keeping of records of risk assessments;
- the application of a risk prevention strategy based on the hierarchy of controls;
- arrangements to plan, organise, control, monitor and review risk control measures and record them.

Guidance from the HSE[13] suggests that risk management will be most effective within a framework that includes:

- **Policy** that outlines the intentions of the organisation towards health and safety.
- **Organising** the structures of the company such that responsibilities and relationships support the aims of the policy. The roles of departments and people with respect to health and safety should be documented so that they can be clearly understood. The guidance refers to the 4 C's of organising:
  - **control** by specifying and allocating responsibilities for safety within the organisation;

  - **cooperation** between managers, employees and their representatives in the organisation through the involvement of all levels in safety committees, problem solving of safety issues and incident investigation;
  - **communication** of information into and through the organisation – upwards, downwards, across, into and out of – by means of circulars, notice boards, emails and so on;
  - **competence** in the development of the skills needed at the various levels within the organisation.
- **Planning and Implementing** work activities to achieve the policy aims. In many ways these two activities are the core of the risk management process. They are concerned with taking the general objectives and drawing up specific detailed plans with assigned responsibilities to achieve them. It may be decided to work on a broad front looking at all facets of the organisation's activities or to look at a specific aspect that is giving cause for concern. For instance, the organisation may decide to focus on risks arising from stress in the workforce. It may arrange instruction on recognising and dealing with stress in the workplace and require department managers and their safety committees to focus on this risk. Within this overall plan the company may decide to draw up a detailed plan which could contain the following specific elements:
  - inform the work group of the objective
  - provide guidance on stress to members of the work group
  - develop a stress survey form
  - arrange for employees to complete the form and analyse the results
  - arrange feedback
  - make arrangements to deal with specific employee issue.
- **Measuring Performance** by setting time scales for achieving each step of the plan against which performance can be measured. This is an example of proactive monitoring – measuring the performance of a process designed to prevent a hazard causing injury. Reactive monitoring, on the other hand, is a review of lagging indicators such as accident and ill-health data. Both these monitoring techniques can be used to show how effective the risk prevention processes have been and may indicate where additional risk prevention action is needed.
- **Reviewing Performance** refers to a review of the extent to which the policy intention has been successfully achieved through the organising, planning, implementing and measuring phases. Its value lies in gaining an understanding of how well the management process to control the hazard has worked and what improvements need to be made to the risk control process.
- **Audit** is an overview across the organisation of the entire risk management process. It should consider whether the objectives are sufficiently comprehensive to cover all hazards and their effects on all those persons (internal and external) who may be affected. It should include the objectives and management style of the directors and their influence on the safety culture of the organisation. It can

use data from the review of performance to assess the strengths of the organisation and those aspects where improvement is necessary.

Where, in small organisations, this process is too cumbersome, guidance on a simpler process has been issued[4] which lists the following five steps to risk assessment:

Step 1: Look for the hazards
Step 2: Decide who might be harmed, and how
Step 3: Evaluate the risks and decide whether existing precautions are adequate or more should be done
Step 4: Record the findings
Step 5: Review the assessment and revise it if necessary.

Effective risk management can become complicated. It is not just about gaining an understanding of the hazards, their consequences, the probability of their occurring and how risks are perceived but it is also about establishing the correct management processes both in overall and detail terms to ensure the ongoing effective control of risks.

2.1.4 Risk management in the 21st century

Health and safety has developed out of the ashes of countless failures and injuries. There was a time in history when an injury was seen as a result of the inadequacy of the operator. Later, catastrophes were largely viewed purely in technology terms. Today both are seen as manifestations of management failure.

The work of social scientists shows that people tend to be more risk averse and are less tolerant of risks that are imposed upon them. They are especially nervous of unknown risks and this fear is fuelled by media interest in the topic of the day. There is a great tendency to apportion blame in the event of injury or ill health occurring. There are moves towards making directors take greater responsibility for health and safety in their organisations and being accountable for discharging that responsibility. The Turnbull report[14] arose out of financial malfeasance and was addressed primarily at financial risk management. This must include health and safety simply because a catastrophic loss due to safety mismanagement can severely cripple the organisation financially. The Health and Safety Commission[15] has produced guidance for directors containing five action points:

- *The board needs to accept formally and publicly its collective role in providing health and safety leadership in its organisation.*
  - This demands strong leadership, commitment and continuous improvement in health and safety performance.
- *Each member of the board needs to accept an individual role in providing health and safety leadership for their organisation.*
  - The key message of this statement is that each director should know and accept that part he plays in the company's risk management

efforts so that mismatches between the director's individual attitudes, behaviours or decisions and the organisations health and safety policy are avoided. Disagreements at board level undermine the workers' belief in the board's intentions and have an adverse effect on good health and safety practice.

- *The board needs to ensure that all board decisions reflect its health and safety intentions, as articulated in the health and safety policy statement.*
  - Consideration should be given to the health and safety implications of decisions concerning finance, staffing, investments in new plant, premises, processes or products and to the health and safety attitude of the companies with whom contracts are placed.
- *The board needs to recognise its role in engaging the active participation of its staff in improving health and safety.*
  - Employees at all levels should be involved in risk management and a partnership for prevention fostered throughout the organisation.
- *The board needs to ensure that it is kept informed of, and alert to, relevant health and safety risk management issues. The Health and Safety Commission recommends that boards appoint one of their number to be the 'health and safety director'.*
  - At board level there must be a focal point for health and safety matters. That role should review performance regularly, update priorities, check on the effectiveness of management systems, keep fellow directors informed of significant health and safety issues, report failures and their causes and actively address the implications of the 'system review'.

Waterman[16] suggests that directors need to undertake the following actions in order to demonstrate the proper discharge of their responsibilities:

- Ensure reports are made to the board on accident and ill health performance, take failure seriously and seek to correct it.
- Incorporate safety performance into staff appraisal systems to show that directors demand high levels of performance and commitment to succeed in health and safety.
- Integrate health and safety into staff development, training programmes and decision making.

The commitment of directors to achieving high levels of health and safety should be capable of being seen by employees at all levels in the organisation. 'Rubber stamp' commitment is quickly recognised and disgruntled employees may exercise their right to telephone the regulatory enforcement officers or even use the 'whistleblower' legislation[17] if internal objections are ignored or overridden.

The emphasis on risk management is set to increase. A practical response must be to ensure a comprehensive and integrated approach to risk management that starts in the boardroom and cascades down through the whole organisation, its people and its activities.

Risk is a complicated subject. It is easy to concentrate on the techniques used and the principles upon which they are based. It is possible to become embroiled in the mathematics of probability or the complexity of perceptions of risk. There is a temptation to become fascinated in the complexities of organisational structures. Amidst all the esoteric elements of the risk equation it should always be remembered that the objective is simply to prevent people from being harmed.

References

1. The Royal Society, *Risk Assessment, A Study Group Report*, The Royal Society, London (1983)
2. The Royal Society, *Risk: Analysis, Perception and Management*, The Royal Society, London (1992)
3. Health and Safety Executive, *Five Steps to Risk Assessment*, publication no. INDG163, HSE Books, Sudbury (1998)
4. Health and Safety Executive, Guidance publication no. HSG 183, *Five Steps to Risk Assessment – Case Studies*, HSE Books, Sudbury (1998)
5. Cox, S., and Coz, T., *Safety Systems and People*, Butterworth-Heinemann, Oxford (1996)
6. Farmer, F.R., Siting criteria – a new approach, Atom, No 128, pp. 152–170
7. Newby, H., Risk analysis and perception, the social limits of technological change, *Trans. Inst. Chem. Eng.*, 75, part B, 1997
8. Slovic, P., Fischoff, B., and Leichtenstein, S. Perceived risk: psychological factors and social implications. *Proc. R. Soc. London, A*, 376, pp. 17–34 (1981)
9. *Health and Safety at Work etc. Act, 1974*, The Stationery Office, London (1974)
10. Health and Safety Commission, *Revitalising Health and Safety*, HSE Books
11. *Control of Substances Hazardous to Health Regulations, 2002*, The Stationery Office, London (2002)
12. *Management of Health and Safety at Work Regulations, 1999*, The Stationery Office, London (1999)
13. Health and Safety Executive, Guidance publication no. HSG 65: *Successful health and safety management*, HSE Books, Sudbury (1997)
14. Turnbull, N., *Internal Control – Guidance for Directors on the Combined Code, report of the Internal Control Working Party*, Institute of Chartered Accountants in England and Wales, London (1999)
15. Health and Safety Executive, *Directors, responsibilities for health and safety*, HSE Books, Sudbury (2001)
16. Waterman, L., Directors in the dock?, *Safety and Health Practitioner*, October 2001
17. *Public Interest Disclosure Act 1998*, The Stationery Office, London (1998)

Chapter 2.2

Principles of the management of risk

L. Bamber

2.2.1 Principles of action necessary to prevent accidents

2.2.1.1 Introduction

The Ministry of Labour and National Service[1] postulated six principles of accident prevention in 1956 that are still valid today. These are:

1 Accident prevention is an essential part of good management and of good workmanship.
2 Management and workers must co-operate wholeheartedly in securing freedom from accidents.
3 Top management must take the lead in organising safety in the works.
4 There must be a definite and known safety policy in each workplace.
5 The organisation and resources necessary to carry out the policy must exist.
6 The best available knowledge and methods must be applied.

It would appear that these principles have only received legislative backing in more recent times – i.e. in the UK via the Health and Safety at Work etc. Act 1974, the Safety Representatives and Safety Committees Regulations 1977 and via the European Union, through the Management of Health and Safety at Work Regulations 1999 and the Health and Safety (Consultation with Employees) Regulations 1996.

Before a closer examination of principles of action necessary to prevent accidents is undertaken, there is a need to examine more closely what is meant by an accident.

2.2.1.2 What is an accident?

To start, consider the following axiomatic statements:

1 All accidents are incidents.
2 All incidents are *not* accidents.

3 All injuries result from accidents.
4 All accidents do *not* result in injury.

An early definition was propounded by Lord MacNaughton in the case of *Fenton* v. *Thorley & Co. Ltd* (1903) AC 443 where he defined an accident as 'some concrete happening which intervenes or obtrudes itself upon the normal course of employment. It has the ordinary everyday meaning of an unlooked-for mishap or an untoward event, which is not expected or designed by the victim.'

This definition refers to an event occurring to a worker that was an unlooked-for mishap having a degree of unexpectedness about it. However, taking into account the axiomatic statements above, this definition would seem to be somewhat narrow, as it is only concerned with accidents resulting in injury to employees.

From research of some 40 accident definitions from general, legal, medical, scientific and safety literature, it appears that the ideal accident definition should have two distinct sections: a description of the causes, and a description of the effects.

Causes should include: unexpectedness or unplanned events, multi-causality and sequence of events; while the effects should cover: injury, disease, damage, near-miss and loss.

Based on the research, the following definition is suggested: 'an accident is an unexpected, unplanned event in a sequence of events, that occurs through a combination of causes; it results in physical harm (injury or disease) to an individual, damage to property, a near-miss, a loss, or any combination of these effects'.

This definition requires recognition of a wider range of accidents than those resulting in injury.

2.2.2 Definitions of hazard, risk and danger

The HSE leaflet *Hazard and Risk Explained*[2] presents the definitions of 'hazard' and 'risk' in relation to the COSHH Regulations:

2.2.2.1 Hazard

The *hazard* presented by a substance is its potential to cause harm. Hazard is associated with degrees of danger, and is quantifiable.

2.2.2.2 Risk

The *risk* from a substance is the likelihood that it will cause harm in the actual circumstances of use. This will depend on: the hazard presented by the substance; how it is used; how it is controlled; who is exposed . . . to how much . . . for how long. Risk should be thought of in terms of 'chance-taking'. What are the odds – the probability – of an accident occurring? Risk can be taken after careful consideration, or out of ignorance. The result can be fortuitous or disastrous, or anything in between.

The link between *'hazard'* and *'risk'* must be understood. In terms of the COSHH Regulations, poor control can create substantial *risk* even from a substance with low *hazard*. But with proper controls, the risk of being harmed by even the most hazardous substance is greatly reduced.

Also, the Approved Code of Practice[3] for the Management of Health and Safety at Work Regulations 1999 clearly states:

(a) a hazard is something with the potential to cause harm (this can include substances or machines, methods of work and other aspects of work organisation);
(b) risk expresses the likelihood that the harm from a particular hazard is realised;
(c) the extent of the risk covers the population which might be affected by a risk; i.e. the number of people who might be exposed and the consequences for them.

Risk, therefore, reflects both the likelihood that harm will occur and its severity. Hence these factors should be taken into account when undertaking either qualitative or quantitative risk assessment.

Danger can be associated with situations where there is a distinct possibility of:

(a) Interchanges of energy above tolerance levels. Such interchanges can be equated to any form of matter, including animals, vegetables and inert objects. The interchanges in energy can be in the form of physical, chemical, biological or psychological energy.

    An example of an interchange of energy occurs when an employee is trapped by a moving part of a machine, e.g. a power press. If the press is inadequately guarded, and the employee is able to get his hand into the danger area, injury results as the energy interchange is above the tolerance level of his hand.
(b) An organisation's financial well-being being placed at risk because of deficiencies in management; deficiencies in design and/or production capabilities; deficiencies in product quality/conformance; inability to expand and/or change; lack of adequate human resources; poor financial stability; lack of market appreciation/penetration; lack of awareness of cultural/social responsibilities; and failure to meet legal obligations.

    An example of a risk to an organisation's well-being is the serving by an enforcement officer (e.g. HSE Inspector) of a prohibition notice when a risk of imminent danger exists. The prohibition notice stops the risky process, machine, department, or – in some cases – the whole business. As it is not possible to insure against the consequential financial loss arising from such a stop notice, the loss must be financed from within the organisation's profitability. If profit margins are tight, then an overall loss situation may result, thus threatening the overall financial viability of the organisation as a whole.

2.2.3 Risk management

Risk management may be defined as the eradication or minimisation of the adverse effects of the pure risks to which an organisation is exposed.

Pure risks can only result in a loss to the organisation, whereas with speculative risks, either gain or loss may result.

An example of a pure – or static – risk concerns a build-up of combustible material in the corner of a large distribution warehouse. If a source of ignition is present in the vicinity, then the risk of fire spread is greatly enhanced by the build-up of combustible material, thus posing the threat of a large loss of stock caused by fire. There will also be consequential loss resulting from the fire to consider, e.g. loss of profit on goods in stock; loss of market share etc.

An example of a speculative – or dynamic – risk concerns commodity purchasing. A company – speculating to accumulate – buys in quantities of a key raw material at price £x/tonne, as the price is favourable, hoping that a cost saving will be made, as the price is likely to increase in the future. The speculative risk may result in gain or loss, as the price of the raw material could continue to fall, after the purchase price has been agreed with the supplier. Alternatively, the price may rise to £(x + y)/tonne, thus achieving the anticipated saving.

It should be borne in mind that the division between 'pure' and 'speculative' risks is not absolute. For example, the speculative risk of operating a machine without an adequate guard has a number of pure risks associated with it – injury; death; prohibition notices; prosecution; fines; loss of profit etc.

The principles of a risk management programme are: risk identification, risk evaluation and risk control. These three principles have been enshrined in recent health and safety legislation – e.g. Lead, Asbestos, COSHH, Noise, Manual Handling and Display Screen Equipment – and have brought risk management strategies and legislative compliance together.

Indeed, the Management of Health and Safety at Work Regulations 1999[4] have now ensured that risk identification, evaluation – i.e. assessment – and control become the cornerstone of all health and safety management systems.

2.2.3.1 Risk management – role and process

The role[5] of risk management in industry and commerce is to:

1 consider the impact of certain risky events on the performance of the organisation;
2 devise alternative strategies for controlling these risks and/or their impact on the organisation; and
3 relate these alternative strategies to the general decision framework used by the organisation.

The process of risk management involves: identification, evaluation and control.

Risk identification may be achieved by a multiplicity of techniques, including physical inspections, management and worker discussions, safety audits, job safety analysis, and Hazop studies. The study of past accidents can also identify areas of high risk.

Risk evaluation (or measurement) may be based on economic, social or legal considerations.

Economic considerations should include the financial impact on the organisation of the uninsured cost of accidents, the effect on insurance premiums, and the overall effect on the profitability of the organisation and the possible loss of production following the issue of Improvement and Prohibition Notices.

Social and humanitarian considerations should include the general well-being of employees, the interaction with the general public who either live near the organisation's premises or come into contact with the organisation's operations – e.g. transportation, nuisance noise, effluent discharges etc. – and the consumers of the organisation's products or services, who ultimately keep the organisation in business.

Legal considerations should include possible constraints from compliance with health and safety legislation, codes of practice, guidance notes and accepted standards, plus other relevant legislation concerning fire prevention, pollution, and product liability.

The probability and frequency of each occurrence, and the severity of the outcome – including an estimation of the maximum potential loss – will also need to be incorporated into any meaningful evaluation.

2.2.3.2 Risk assessment

This topic has been discussed elsewhere in this book. The concept of risk assessment within a risk management framework is the combination of risk identification and risk evaluation mentioned above.

Risk assessment has been a legal requirement within the EU for some time, but in all its guises it is merely the means to an end. The 'end' is the timely implementation of workable control measures designed to either eliminate or reduce risks to an acceptable or insignificant level. There are many established risk assessment systems. The HSE have developed their *Five Steps to Risk Assessment*[22] which is largely qualitative in nature. The BSI have included a quantitative risk assessment model in their guide to occupational safety and health management systems[23]. Chapter 2.4 provides an alternative risk assessment model.

Whichever system of risk assessment is used, the main aim is to get control measures in place as soon as is practicable after the assessment has been completed.

2.2.3.3 Risk control strategies

Risk control strategies may be classified into four main areas: risk avoidance, risk retention, risk transfer and risk reduction.

1 Risk avoidance
This strategy involves a conscious decision on the part of the organisation to avoid completely a particular risk by discontinuing the operation producing the risk and it presupposes that the risk has been identified and evaluated.

For example, a decision may be made, subject to employees' agreement, to pay all wages by cheque or credit transfer, thus obviating the need to have large amounts of cash on the premises and the inherent risk of a wages snatch.

Another example of a risk avoidance strategy – from the health and safety field – would be the decision to replace a hazardous chemical by one with less or no risk potential.

2 Risk retention
The risk is retained in the organisation where any consequent loss is financed by the company. There are two aspects to consider under this heading: risk retention with knowledge, and risk retention without knowledge.

(a) With knowledge
This covers the case where a conscious decision is made to meet any resulting loss from within the organisation's financial resources. Decisions on which risks to retain can only be made once all the risks have been identified and effectively evaluated.

(b) Without knowledge
Risk retention without knowledge usually results from lack of knowledge of the existence of a risk or an omission to insure against it, and this often arises because the risks have not been either identified or fully evaluated.

3 Risk transfer
Risk transfer refers to the legal assignment of the costs of certain potential losses from one party to another. The most common way of effecting such transfer is by insurance. Under an insurance policy, the insurer (insurance company) undertakes to compensate the insured (organisation) against losses resulting from the occurrence of an event specified in the insurance policy (e.g. fire, accident etc.).

The introduction of clauses into sales agreements whereby another party accepts responsibility for the costs of a particular loss is an alternative risk transfer strategy. However, it should be noted that the conditions of the agreement may be affected by the Unfair Contract Terms Act 1977 and the interpretation placed on 'reasonableness'.

4 Risk reduction
The principles of risk reduction rely on the reduction of risk within the organisation by the implementation of a loss control programme, whose basic aim is to protect the company's assets from wastage caused by accidental loss.

The collection of data on as many loss producing accidents as possible provides information on which an effective programme of remedial action can be based. This process will involve the investigation, reporting and recording of accidents that result in either injury or disease to an individual, damage to property, plant, equipment, materials, or the product; or those near-misses where

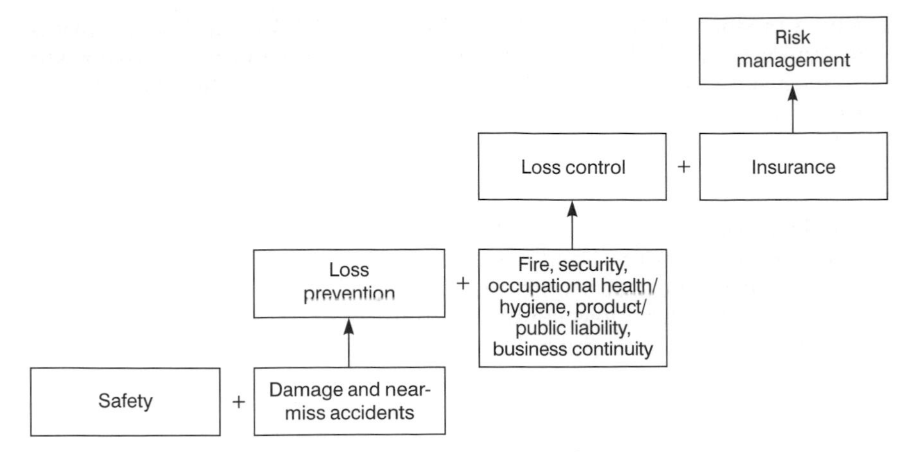

Figure 2.2.1 Pictogram showing development from safety to risk management.

although there has been no injury, disease or damage, the risk potential was high.

The second stage of the development towards risk reduction is achieved by bringing together all areas where losses arise from accidents – whether fire, security, pollution, product liability, business interruption etc. – and co-ordinating action with the aim of reducing the loss. This risk reduction strategy is synonymous with loss control.

Loss control (or risk reduction) may be defined as a management system designed to reduce or eliminate all aspects of accidental loss that lead to a wastage of an organisation's assets.

As the emphasis on the economic argument increased, the technique of loss control has become more closely allied to financial matters, and in particular insurance.

The bringing together of insurance (risk transfer) and loss control (risk reduction) was the final stage in the development of the new discipline of risk management.

This logical, progressive development from safety to risk management may be presented in pictogram form as in *Figure 2.2.1*.

2.2.4 Loss control

2.2.4.1 Introduction

Loss control may be defined as a management system designed to reduce or eliminate all aspects of accidental loss that may lead to a wastage of the organisation's assets. Those assets include manpower, materials, machinery, methods, manufactured goods and money. Loss control is based mainly on the economic approach to accident prevention, and loss control management is essentially the application of sound management techniques to the identification, evaluation and economic control of losses within a business. It has been shown in section 2.2.3.2 that loss control is synonymous with risk reduction, so the practical techniques associated with each stage of the process – i.e. identification, evaluation and control – are closely related.

Bird and Loftus[6] state that loss control management involves the following:

1 The identification of risk exposure.
2 The measurement and analysis of exposures.
3 The determination of exposures that will respond to treatment by existing or available loss control techniques or activities.
4 The selection of appropriate loss control action based on effectiveness and economic feasibility.
5 The managing of the loss control programme implementation in the most effective manner subject to economic constraints.

2.2.4.2 Component parts of a loss control programme

The component parts of a loss control programme may be considered in terms of protecting one or more of the organisation's assets from accidental loss, and will generally include: injury prevention (safety); damage control; fire prevention; security; industrial health and hygiene; pollution; product liability; and business interruption.

Injury prevention is concerned directly with the protection of the manpower asset within an organisation. To a lesser extent, it is indirectly concerned with the protection of the money asset, as a reduction in the number of injuries should result in a reduction in both the insured and the uninsured accident costs. In certain injury prevention programmes, the protection of the manpower asset is extended outside the factory, via the inclusion of off-the-job safety, road safety and home safety.

The following specific areas need to be examined in an injury prevention programme: safety policies; safety training; safety audits; identification of hazards; accident reporting and investigation; safe systems of work; machine and area guarding; housekeeping; personal protective equipment; and legislative compliance.

Damage control is directly concerned with the protection of assets comprising machinery, materials and manufactured goods from accidental loss before they reach the customer. Indirectly, this leads to the protection of the money asset through the elimination of repetitive damage and associated repair/replacement costs. Also, there may be some indirect protection of the manpower asset if damage causes and injury causes are similar. Essentially, damage control is an extension of the injury reporting and prevention programme to encompass also those accidents which result in damage only to plant, property, equipment and materials. The practices and techniques of damage control are considered in depth by Bird and Germain[7], and Bird and Loftus[8].

Incident recall[9,10] is another technique that can be utilised in a damage control programme to gain information about near-miss accidents.

Essentially, the incident recall technique may be used to identify unsafe acts, unsafe conditions, non-compliance with safe systems of work, and near-miss accidents by following a *confidential* interviewing procedure to a stratified random sample of employees. Each interviewee is asked to recall and report verbally any of the above-mentioned situations in which he was involved or has knowledge. Details of near-miss accidents are then obtained to enable remedial action to be taken *before* further similar accidents result in both damage and injury.

Fire prevention may be considered to be a special aspect of damage control in that it protects the machinery, materials and manufactured goods assets. It also protects the manpower asset, since fire can cause injury as well as damage and because fire damage is a very costly item, it is indirectly protecting the money asset as well. The Association of British Insurers has stated that the total cost fire insurance claims in the UK for 1999 was £1.2 bn, up 18% on 1993 figures.

From a practical viewpoint, consideration should be given to aspects of fire prevention techniques, methods of fire control, firefighting and extinguishment, fire protection including fixed equipment (e.g. sprinklers etc.), storage and handling of flammable liquids, fire safety of employees, means of escape, evacuation drills and procedures, explosion potential, handling, storage and use of explosives, electrical installations, disaster contingency planning and the requirements of the Fire Precautions Act 1971 and the Dangerous Substances and Explosive Atmospheres Regulations 2002 and, more recently, the Fire Precautions (Workplace) Regulations 1997.

Security protects the materials, methods, manufactured goods and money assets. Its inclusion in a loss control programme is primarily based on economic considerations, as any breaches of security that result in losses of the organisation's assets may not be considered by the organisation to be accidental in nature.

However, certain safety hazards arise because of lack of security: for example, the high risk involved when employees are sent to collect wages; or the potential risk to children and others when a factory or construction site is not physically secure in terms of preventing unauthorised access.

Hence, indirectly, a system of security can improve the overall safety of the organisation, as well as directly protecting the assets mentioned above. Security – in the form of locked doors – can, however, sometimes conflict with the protection of the manpower asset, especially if the locked doors are the emergency fire exits.

The loss control programme should examine such areas as the physical security of premises; cash collection, handling, and distribution; theft and pilfering; vandalism; storage of valuable and attractive items; sabotage; industrial espionage; and the control of confidential data and methods. Consideration should also be given to defensive techniques, such as stocktaking, accounting/auditing checks, and the use of mechanical/electrical safeguards. Aspects of computer security should also be included in this area, especially in the light of the increasing number of cases being reported that have involved some form of fraud, embezzlement or espionage utilising the computer.

Occupational health and hygiene is concerned with the protection of the manpower asset from the effects of occupational diseases – i.e. long-term accidents and other adverse conditions associated with the industrial environment. Indirectly, this also protects the money asset, as an improvement in health and hygiene within a factory should lead to a reduction in the incidence of occupational diseases, and hence a reduction in the associated costs.

Specific areas for consideration include: noise, dusts, gases, vapours, corrosives, toxic materials, radioactive materials, ventilation, heating, lighting, humidity, environmental monitoring, biological monitoring, health checks, general and personal hygiene, health education, counselling, and employment/pre-employment medicals. Compliance with the COSHH Regulations is vitally important to ensure a healthy workforce as is consideration of physical and mental stressors, as well as the more obvious chemical hazards.

Pollution control/environmental protection in all its aspects is concerned not only with the environment within the factory, but also the environment outside and around the factory. Control of air, ground and water pollution protects the manpower asset directly and the money asset indirectly. Adverse publicity resulting from an organisation causing some form of pollution would initially harm the organisation's image, and perhaps harm it economically. Persistent breaches of one or other of the Acts dealing with pollution can ultimately lead to prosecution and fines with further adverse publicity, both at local and national level.

Special attention needs to be paid to the Control of Pollution Act 1974 that deals with the control of noise as a pollutant, in addition to the other areas of air, ground and water pollution. All possible areas and types of pollution should be identified as part of the loss control audit. With the advent of the Environmental Protection Act 1990[11] and BS 7750:1992: Environmental Management Systems[12] – now BS EN ISO 14001: 1996[20] – the environmental aspects of the loss control programme have increasingly gained in importance in recent years. Indeed, many health and safety practitioners now have the added responsibility to provide advice on environmental matters.

Product liability extends the protection to all consumers of the organisation's products or services, and is therefore primarily concerned with the protection of the money asset. This asset may suffer accidental losses directly because of increased insurance premiums that reflect large compensation payments, or indirectly because of adverse publicity which is detrimental to the organisation's image.

Practical considerations in this area should centre on the development of a product safety strategy that is in keeping with both occupational (section 6, Health and Safety at Work etc. Act 1974) and consumer (Consumer Protection Act 1987) product safety legislation, codes of practice and guidance notes.

Areas to be considered should include: products loss control policy; products loss control committee chaired by senior manager and comprising designers, R and D personnel, manufacturing and production management, safety advisers, quality control, servicing, sales and advertising, and distribution; product safety incorporated at the design and R and D stages; written quality control/assurance procedures, i.e. BS 5750: 1987[13] – now BS EN ISO 9001 2000[21]; role of sales, marketing, advertising, distribution and servicing personnel in product safety; complaints system; and products recall system.

Business interruption or continuity further extends the loss control strategy to take account of the fact that time is money, and, as such, any loss of production or service is detrimental to the overall profitability of the company. Hence, business interruption is primarily concerned with the protection of the money asset. Indirectly, however, it serves to maintain the assets of machinery, materials, manufactured goods and methods.

A programme to prevent business interruption can include planned lubrication, planned preventive maintenance, condition monitoring, statutory inspections, machinery replacement programmes, availability of key spares, identification of key machines, processes, areas, personnel etc. within the organisation, continued supply of raw materials, minimisation of production bottlenecks, and highlighting dependencies on specific items of plant, suppliers, customers, personnel, computer systems, and/or public utilities (e.g. gas, electricity).

2.2.4.3 Loss control management in practice

The aforementioned areas – from injury prevention to business interruption – require to be co-ordinated within one senior management function (possibly risk management) in order to ensure a rational and concerted approach to the problem of eliminating or reducing the costly accidental losses that can occur within an organisation. These so-called operational losses inevitably lead to an erosion of profit margins, and also adversely affect the overall performance of the organisation.

This co-ordinated management role is crucial to the success of any loss control programme. The senior manager responsible for programme implementation should have authority to make decisions and take action without the need to seek day-to-day approval for his decisions. He should report on a regular (monthly) basis to the main board on the implementation of the loss control programme within the organisation. Without the backing and commitment of the most senior executives, it is doubtful whether a programme can be successfully introduced.

An effective programme of loss control (risk reduction) not only leads to a more profitable situation, but will also greatly assist legislative compliance, and will result in a reduction in the total number of accidents within the organisation's operations.

2.2.5 Degrees of hazard

An awareness of the differing degrees of hazard to people will enable appropriate control measures to be developed and implemented.

Immediate physical danger can manifest itself through very short-term injury accidents – e.g. hand amputation in a power press; person falling from a height. The result of immediate physical danger – if it goes uncontrolled – will inevitably be *immediate physical injury*. The enforcing agencies use the phrase 'risk of imminent danger' or 'risk of serious personal injury' in connection with the issuing of prohibition notices – a legal control measure designed to reduce the risk of immediate physical danger.

Long-term physical danger is more cumulative or chronic than acute or short term. Cumulative back strain caused by poor kinetic handling techniques is an example of *long-term physical injury*.
Immediate chemical danger may be caused by strong acids and alkalis being poorly stored and handled, thus leading to a risk of skin contact and corrosive burns – i.e. *immediate chemical injury*.

Long-term chemical danger is again chronic or cumulative – e.g. lead poisoning or exposure to asbestos fibres. The result is some form of occupational disease – i.e. *long-term chemical injury*.

Immediate biological danger may be caused by the presence of contagious diseases or via genetic manipulation. The result is again some form of occupation disease or illness.

Long-term biological danger is usually cumulative in nature, for example noise-induced occupational deafness.

Immediate psychological danger is linked to short-term trauma – e.g. a disaster at home or work; social problems – domestic illness etc. This may result in a loss of concentration, abruptness with work colleagues, and other short-term stress-related symptoms.

Long-term psychological danger may be linked to fears connected with fear of failure, unemployment/job security, or lack of career direction and motivation. The symptoms are similar to those described above, but often only become apparent over a longer timescale.

2.2.6 Accident causation models

2.2.6.1 Sequence of events – domino theory

The 'Domino Theory' attributed to Heinrich[14] is based on the theory that a chain or sequence of events can be listed in chronological order to show the events leading up to an accident:

event a → event b → event c → accident → effect

Each event may have more than one cause, i.e. be multicausal.

Heinrich states that the occurrence of an injury accident invariably results from a completed sequence of factors culminating in the accident itself. He postulates five factors or stages in the accident sequence, with the injury invariably caused by the accident, and the accident in turn the result of the factor that immediately precedes it.

The five factors or stages in the sequence of events are:

(a) ancestry and social environment, leading to
(b) fault of person, constituting the proximate reason for
(c) an unsafe act and/or mechanical hazard, which results in
(d) the accident, which leads to
(e) the injury.

Heinrich likens these five stages to five dominoes standing on edge in a line next to each other, so that when the first domino falls it automatically knocks down its neighbour which in turn knocks down its neighbour and so on. Removal of any one of the first four will break the sequence and so prevent the injury.

In fact, Heinrich suggested that accident prevention should aim to remove or eliminate the middle or third domino, representing the unsafe act, mechanical or physical hazard, thus preventing the accident.

During accident investigations, in addition to asking 'What action has been taken to prevent recurrence?', the investigator needs to be aware of the chain of events leading up to the accident, and tracing it back. Similarly, on safety audits and inspections, when the risk of an accident has been identified, possible event chains should be investigated and action taken to remove potential causes.

2.2.6.2 An updated domino sequence

Bird and Loftus[15] have extended this theory to reflect the influence of management in the cause and effect of all accidents that result in a wastage of the company's assets. The modified sequence of events becomes:

(a) lack of control by management, permitting
(b) basic causes (personal and job factors), that lead to
(c) immediate causes (substandard practices/conditions/errors), which are the proximate causes of
(d) the accident, which results in
(e) the loss (minor, serious or catastrophic).

This modified sequence can be applied to all accidents, and is fundamental to loss control management.

2.2.6.3 Multiple causation theory

Multicausality refers to the fact that there may be more than one cause to any accident:

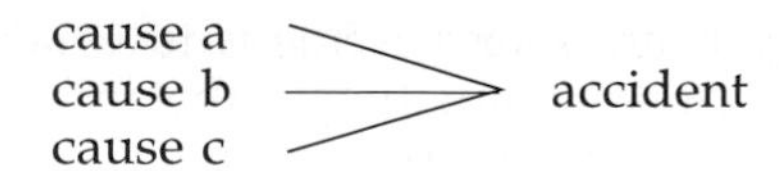

Each of these multicauses is equivalent to the third domino in the Heinrich theory and can represent an unsafe act or condition or situation. Each of these can itself have multicauses and the process during accident investigation of following each branch back to its root is known as 'fault tree analysis'.

The theory of multicausation is that the contributing causes combine together in a random fashion to result in an accident. During accident investigations, there is a need to identify as many of these causes as possible. In reality, the accident model is an amalgam of both the domino and multicausality theories.

Petersen has compared and contrasted both theories and gives an example[16] which illustrates the comparative narrowness of the domino theory in relation to the multicausality theory and concludes that this has severely limited the identification and control of the underlying causes of accidents.

The theory of multicausality has its basis in epidemiology. Gordon[17] points out that accidental injuries could be considered with epidemiological techniques. He believes that if the characteristics of the 'host' (accident victim), of the agent (the injury deliverer), and of the supporting 'environment' could be described in detail, more understanding of accident causes could be achieved than by following the domino technique of looking for a single cause only. Essentially, Gordon's theory is that the accident is the result of a complex and random interaction between the host, the agent and the environment, and cannot be explained by consideration of only one of the three.

2.2.6.4 Failure modes and effects

This technique involves a sequential analysis and evaluation of the kinds of failures that could happen, and their likely effects, expressed in terms of maximum potential loss.

The technique is used as a predictive model and would form part of an overall risk assessment study.

2.2.6.5 Fault tree analysis

Fault tree analysis is an analytical technique that is used to trace the chronological progression of factors (events) contributing to the accident situation, and is useful in accident investigation and as a predictive, quantitative model in risk assessment. Again, the principle of multicausality is utilised in this type of analysis. (A fuller treatment on fault tree analysis is given at section 2.4.6.)

2.2.7 Accident prevention: legal, humanitarian and economic reasons for action

2.2.7.1 Introduction

In order to get action taken in the field of accident prevention, safety advisers have the three fundamental lines of attack on which to base their strategies for generating and maintaining management activity in this area. These three reasons for accident prevention make use of the legal, humanitarian and economic arguments respectively. An optimum accident prevention strategy for a particular organisation would involve a combination of the three, because they are interrelated and probably reinforce one another.

2.2.7.2 Legal reasons for accident prevention

The legal argument is based on the statutory requirements of the HSW, FA and other related legislation.

The HSW imposes a general duty on employers to ensure, so far as is reasonably practicable, the health, safety and welfare of all his employees. The term 'reasonably practicable' involves balancing the cost of preventing the accident against the risk of the accident occurring. Thus, economic considerations need also to be taken into account.

PUWER lays down more specific statutory requirements which impose a minimum but absolute standard of conduct on the employer.

Any breach of the statutory duties imposed by either of the aforementioned Acts can result in the employer being involved in criminal proceedings. The penalties under the Health and Safety at Work Act include unlimited fines and imprisonment for up to two years, for prosecution on indictment. On average, 20 directors, managers, supervisors, employees are individually prosecuted per annum. A number of individuals have been given custodial sentences under, for example, asbestos, machinery safety and gas safety legislation. The maximum fine on summary conviction for certain offences is currently (December 2000) £20 000 with the maximum for other offences being £5 000.

The safety adviser can therefore reason via the legal argument for accident prevention on the basis that the employer should avoid attracting a prosecution.

The economic argument is also relevant here, because of the fines that may be imposed as a result of statutory breaches, and also because of the impact of Improvement and Prohibition Notices in terms of uninsurable consequential loss arising out of enforced cessation of work.

The image of the company or organisation is also likely to be tarnished as a result of adverse publicity received in connection with any prosecution for breaches of statute. Loss of company image has predominantly economic disadvantages, usually because of the loss of good will or other intangible and invisible company assets, which in turn indirectly leads to a loss of business.

2.2.7.3 Humanitarian reasons for accident prevention

The humanitarian reason for accident prevention is based on the notion that it is the duty of any man to ensure the general well-being of his fellow men. This places an onus on the employer – the common law duty of care – to provide a safe and healthy working environment for all his employees.

An illustration of this occurs in the case of *Wilsons and Clyde Coal Co. Ltd* v. *English*[18], where Lord Wright said that 'the whole course of authority consistently recognises a duty which rests on the employer, and which is personal to the employer, to take reasonable care for the safety of his workmen, whether the employer be an individual, a firm or a company and whether or not the employer takes any share in the conduct of the operations'.

There is some overlap here between common and statute law, as the Health and Safety at Work Act places a general duty on an employer to ensure, so far as is reasonably practicable, the health, safety and welfare of his employees.

The safety adviser is therefore able to argue – via humanitarian reasoning – that it is immoral to have a process or machine which may injure employees, and he can stress the possible outcome of such dangers in terms of pain and suffering.

2.2.7.4 Economic reasons for accident prevention

The fundamental reason for utilising the economic argument in the promotion of accident prevention is the fact that accidents cost an organisation money. However, in order to press the economic argument, knowledge is needed of the costs to the organisation of all types of accident.

Essentially, there are two types of accident costs – the insured costs, and the uninsured costs.

The insured (or direct) costs are predominantly covered by the Employer's Liability insurance premium, which to all intents and purposes is the direct accident cost to the majority of organisations.

The uninsured (or indirect, hidden) costs of accidents should also be established. Bamber[19] developed a list of uninsured costs which is considered to be objective, and which will readily be accepted by operational management as being costs associated with accidents:

1. Safety administration and accident investigation.
2. Medical and treatment.
3. Cost of lost time of injured person.
4. Cost of lost time of other employees.
5. Cost of replacement labour.
6. Cost of payments to injured person.
7. Cost of loss of production and business interruption.
8. Cost of repair to damaged plant.
9. Cost of replacement of damaged materials.
10. Other costs – e.g. photographs, transport, accommodation, wage details, fees etc.

The above list of costs should be utilised in the calculation of the total accident costs to the organisation, to enable senior management to gauge the relative impact of such costs, by comparing them with other business costs.

An HSE document[24] demonstrates that accident costs can amount to 37% of annual profits, 8.5% of tender prices or 5% of the running costs of the organisations studied.

The safety adviser is therefore able to reason – via the economic argument – that accident prevention may well be cost-effective. But the organisation is reducing pain and suffering by having an effective system of accident prevention, as well as saving money. Thus, the economic argument gives support to both the legal – via the use of economic sanctions – and the humanitarian arguments. In order to achieve maximum co-operation in any programme of accident prevention, use should be made of an amalgam of all three arguments, i.e. legal, humanitarian and economic. However, from a motivational point of view, it is the economic argument that has the greatest impact with directors and senior management.

References

1. Ministry of Labour and National Service, *Industrial Accident Prevention*, Report of the Industrial Safety Sub-Committee of the National Joint Advisory Council (1956)
2. Health and Safety Executive, *Hazard and Risk Explained – Control of Substances Hazardous to Health Regulations 1988* (COSHH), Leaflet No. IND(G)67(L), HSE Books, Sudbury (1988)
3. Health and Safety Commission, Legal Series booklet No. L21, *Management of Health and Safety at Work Regulations 1999: Approved Code of Practice and Guidance*, HSE Books, Sudbury (1999)
4. *Management of Health and Safety at Work Regulations 1999*, The Stationery Office, London (1999)
5. Carter, R. L. and Doherty, N., *Handbook of Risk Management*, 1.1–06, Kluwer-Harrap, London (1974)
6. Bird, F. E. and Loftus, R. G., *Loss Control Management*, 52, Institute Press, Longanville, Georgia (1976)
7. Bird, F. E. and Germain, G. L., *Damage Control*, American Management Association, New York (1966)
8. Ref. 6, pp. 93–138
9. Ref. 6, pp. 215–246
10. Bamber, L., Incident recall – a (lack of) progress report, *Health and Safety at Work*, **2**, No. 9, 83 (1980)
11. *The Environmental Protection Act 1990*, The Stationery Office, London (1990)
12. British Standards Institution, *BS 7750:1992 Specification for Environmental Management Systems*, BSI, London (1992)
13. BS 5750: Parts 1–6:1987, *Quality systems*, British Standards Institution, London
14. Heinrich, H. W., *Industrial Accident Prevention*, 4th edn, 13–16, McGraw-Hill, New York (1959)
15. Ref. 6, pp. 39–48
16. Petersen, D. C., *Techniques of Safety Management*, 2nd edn, 16–19, McGraw-Hill, Kogakusha, USA (1978)
17. Gordon, J. E., The epidemiology of accidents, *Amer. J. of Public Health*, **39**, 504–515 (1949)
18. Wilsons and Clyde Coal Co. Ltd *v.* English, (1938) AC 57 (HL)
19. Bamber, L., Accident prevention the economic argument, *Occupational Safety and Health*, **9**, No. 6, 18–21 (1979)

20. British Standards Institution BS EN ISO 14001: 1996, *Environmental management systems – Specification with guidance for use*, BSI, London (1996)
21. British Standards Institution BS EN ISO 9001: 2000, *Quality systems*, BSI, London
22. Health and Safety Executive, publication no INDG 163 *Five Steps to Risk Assessment*, HSE Books, Sudbury (1994)
23. British Standards Institution, BS 8800:1996, *Guide to occupational health and safety management systems*, BSI, London (1996) [Note: this standard has been overtaken by OHSAS 18002:2000, *Occupational health and safety management systems* which in turn is likely to become ISO 18002]
24. Health and Safety Executive, Guidance booklet HSG 96. *The costs of accidents at work*, HSE Books, Sudbury (1997)

Further reading

Heinrich, H.W., Petersen, D. and Roos, N., *Industrial Accident Prevention – A Safety Management Approach*, 5th edn, McGraw-Hill, New York (1980)

DeReamer, R., *Modern Safety Practices*, John Wiley & Sons Inc., New York (1958)

Bird, F.E. and Loftus, R.G., *Loss Control Management*, Institute Press, Loganville, Georgia (1976)

Petersen, D.C., *Techniques of Safety Management*, 2nd edn, McGraw-Hill, Kogakusha, USA (1978)

Hale, A.R. and Hale, M., *A Review of the Industrial Accident Research Literature*, Committee on Safety and Health at Work: Research Paper, The Stationery Office, London (1972)

Crockford, G.N., *An Introduction to Risk Management*, Woodhead-Faulkner, Cambridge (1980)

Carter, R.L. *et al.*, *Handbook of Risk Management*, Kluwer Publishing Ltd, Kingston-upon-Thames (1997–1998)

Health and Safety Executive Publication No. HS(G)65 *Successful Health and Safety Management* (2nd ed), HSE Books, Sudbury (1997)

British Standards Institution BS 8800: 1996, *Guide to occupational health and safety management systems*, BSI, London (1996)

Chapter 2.3

Risk management: organisation and administration for safety

J. E. Channing

2.3.1 Introduction

The identification and control of hazards does not occur in a vacuum. The methods and techniques of risk assessment may be well understood but they will be far from effective in reducing incidents and injury unless they are deployed within an effective management framework. Risk assessments result in changing the actions and behaviour of those affected. The behaviour change is sometimes mistakenly thought of as affecting shop floor employees only but directors, managers and supervisors are also affected. If risk assessments are to be effective the organisation as a whole needs to understand its roles and responsibilities. This factor is evident in the *Health and Safety at Work etc. Act, 1974*[1] where section 2 (2) requires the employer to provide safe systems of work, information, instruction, training and supervision. This can best be achieved through a proper organisation and administration.

The above steps are necessary to implement the safety policy requirements of the Act (Section 2 (3)). The point is further endorsed in an HSE publication[2] which outlines the need for:

- effective health and safety policies to send a clear direction for its organisation to follow;
- effective management structures and arrangements for delivering the policy;
- planned and systematic approaches to implementing the health and safety policy through an effective health and safety management system;
- performance to be measured against agreed criteria to reveal when and where improvement is needed.

Thus, to derive the greatest benefit from safety activities, and particularly risk assessments, the organisational and administrative arrangements within the company should be clearly understood by all involved.

For any organisation to function effectively and successfully it is necessary that those who constitute that organisation understand the goals of the enterprise and identify with them. They should know where they fit into the executive structure and must be competent and confident in the work they have to do. This applies to all the many facets of the enterprise's activities, whether production, financial, administrative or safety. There needs to be an understanding of the influences, both internal and from outside, that bear on the success of the component parts and the organisation as a whole. There should be an appreciation of the conflicts that can arise, their causes as well as the techniques that can be used to defuse the situation.

Similarly there needs to be an understanding and recognition of the informal structures and relationships that occur within the more formal imposed organisation structure. These informal arrangements very often 'oil the wheels' of commerce and keep the organisation running smoothly and effectively. While a firm structure is necessary to ensure a consistent direction of the efforts of the enterprise, that structure must not be so rigid that it cannot adjust to changes in the trading, economic, legislative and other aspects of the operating environment over which the enterprise has no control. This chapter looks at the various relationships and organisations that occur within an enterprise and which can materially affect its success not only in serving its customers and the community but also the satisfaction it gives to, and the safety it provides for, the people it employs.

The chapter also examines the administrative processes that contribute to the safe and successful running of the organisation. The organisational structure and administrative processes are linked because together they shape the behaviour of the enterprise in the entirety of its activities. A successful enterprise achieves its goals by constantly reviewing and refreshing its structure and processes. In so doing it guides the behaviour of those who work in it towards the desired goals. Many organisations have formal structures, determined rationally by the senior executives and often displayed as charts or 'organograms' showing the functions that are considered essential for the effective and smooth running of the organisation. Equally important but very rarely committed to paper are the informal working groups established socially by members within the organisation.

Understanding all these aspects is as important in achieving successful health and safety programmes and maintaining the control strategies they generate as any other part of the business enterprise.

2.3.2 Organisation structure models

An overall management structure needs to be established in order to achieve successful health and safety performance and the management of risks. An HSE guidance booklet[2] recommends a cohesive management process. An international health and safety management standard, ISO 18001, is in preparation and which follows the general structure of the quality standard[3] and the environment standard[4]. An existing UK standard[5] is a precursor which offers a scheme for integrating the elements of the international systems with the HSE guidance booklet. Both refer to the need for effective organisational and administrative processes.

2.3.2.1 Formal organisation structures

For an enterprise to succeed it needs to have some sort of organisation and the most common is hierarchical with authority flowing through definite channels from top management to the workpeople. A typical hierarchical organisation chart is shown in *Figure 2.3.1* and for it to be effective certain positions must be vested with power to exercise their authority to direct and control the activities of the organisation. How that power is exercised will be determined by the culture of the organisation, whether it is authoritarian or bureaucratic.

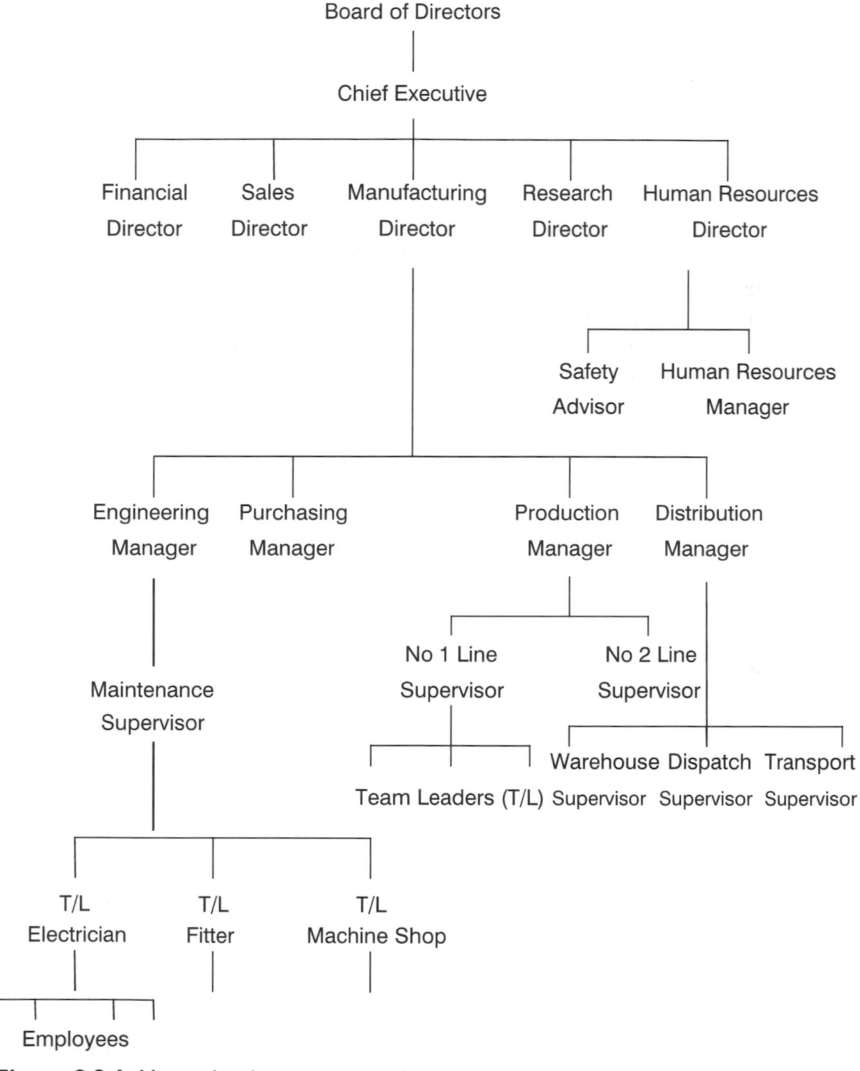

Figure 2.3.1 Hierarchical organisation chart

Organisation structures of the sort shown in *Figure 2.3.1* vary with the size and complexity of the enterprise. As profit margins are cut, there is a tendency to remove intermediate layers of management and supervision to reduce costs. The rationale is based upon the concept that it is only the shop floor operative who actually 'adds value' to the product or in providing a service. Other (higher) members of the management and supervisory hierarchy are engaged in planning, organising and monitoring. These tasks are seen as 'burdens' because they do not directly 'add value' but do add cost. The increasing use of electronic information technology is making some of the administrative tasks performed by managers and supervisors redundant allowing these roles in the organisation to be dispensed with.

An example of a 'de-layered' or flatter organisation structure is given in *Figure 2.3.2*. In this example there are just three levels between shop floor employees and the Chief Executive compared with five levels in the previous structure. The reduction in layers has occurred at lower manager and supervisor levels of the organisation.

As a further move at cost cutting, some organisations have 'outsourced' whole parcels of their operations. The original enterprise continues to assemble the final product or provide the service but employs only key people to fulfil essential administrative functions, the non-core functions being undertaken either by subcontractors or by people hired on contract. This type of arrangement applies more easily to engineering manufacture than it does to line production. There are moves for organisations to create ways of focusing only on the essential functions of their operations with the aim of having as few direct employees as possible. In parallel with this trend in organisational change, information technology is permitting more people to work from home. Employees in sales, purchasing, systems support, etc., are all able to do much of their work from home. While still on the direct payroll, they are removed to a significant degree from the day-to-day control of their supervisor or manager. This arrangement exhibits the features of the 'delayered' or flatter organisational structure shown in

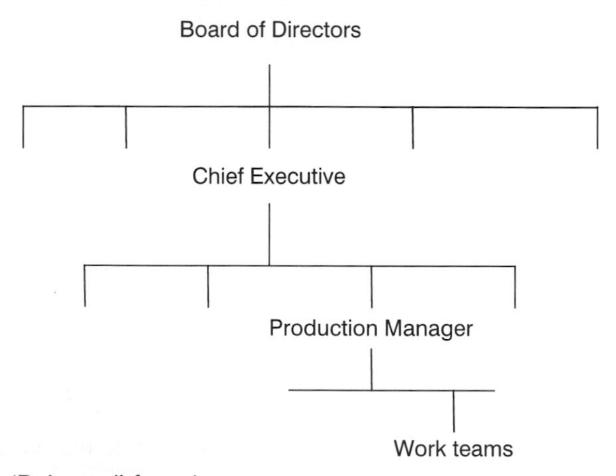

Figure 2.3.2 'Delayered' formal structure

Figure 2.3.2. It should be noted that the employer owes to those employees working from home the same duties as he does to those who attend the place of work. By the same token, he owes similar duties to contracted staff who work in his premises.

2.3.2.2 Informal organisation structure

Within any formal organisation will be found whole networks of informal organisations based on personal relationships, social needs, personal allegiances and sometimes a desire to be helpful in by-passing the formal organisation. These informal organisations are rarely committed to paper. They come into existence to serve a perceived need of those involved and may remain for many years or can disappear when the need is satisfied.

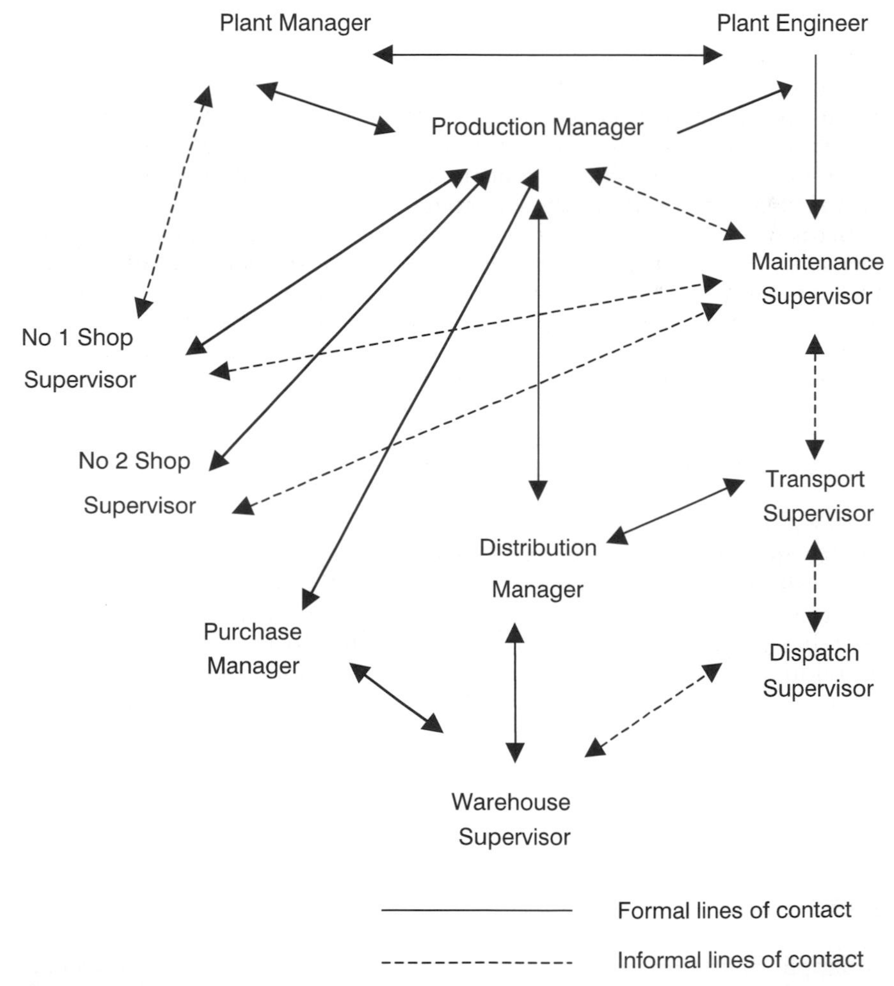

Figure 2.3.3 Diagram of workplace contacts

An informal organisation may use the formal structure but create unofficial lines of communication to circumvent failures and weaknesses in individuals that make up the formal organisation in attempts to achieve the enterprise's objectives. This is shown in *Figure 2.3.3* which demonstrates the actual relationships which could occur where the Production Manager has been running affairs for some time because of the indifferent health of the Plant Manager. The Production Manager has gathered a group of key personnel who recognise him as 'next-in-line' after the Plant Manager and acknowledge his authority. However, the Plant Engineer, who happens to be the Managing Director's son-in-law, does not acknowledge the Production Manager's authority and endeavours to exert his own authority by resisting the Production Manager. As a result the Plant Engineer is often by-passed when maintenance work needs to be done and indeed the maintenance foreman refuses to take work from him since he, the foreman, only acknowledges the Production Manager as his superior.

The informal structure serves to improve communication and to develop non-official roles. It has an important part to play in the resolution of conflicts between roles and positions. Questions raised in the formal organisation often elicit answers that rely on the informal structure. Questions such as 'what's the best way to get this information?' or 'who can get this done?' produce answers that cut right across the formal structure – but get the job done.

Occasionally there is tacit recognition of the informal structure by the formal since the informal level allows for practical interpretation of rules and procedures that can otherwise place restrictions on achieving enterprise targets. This flexible interpretation of the formal rules may be to the common organisational good, but may also be detrimental where they concern safe working practices.

2.3.3 Roles and responsibilities

Each person in an organisation should be confident of the role they have to play and aware of the degree of executive responsibility they bear. This responsibility must be limited to the extent of the control they have been authorised to exercise. While the company, as a legal entity, carries the overall legal responsibility and is answerable to the law of the land, individual employees carry executive responsibilities and are answerable to the Chief Executive or owner of the company. Executive responsibility can be delegated, legal responsibility cannot.

Factors to be taken into account when considering the extent of control exercised by an individual person are the *authority* and the *power* that the person has.

2.3.3.1 Authority

A simple definition of authority is 'legitimate power' and Max Weber[6] saw authority and its legitimacy as central to the question of organisational structure. He maintained that authority exists when instructions are obeyed in the belief that they are legitimate. In other words that they are justified and that obedience is the appropriate response. Weber classified authority into three kinds:

- rational legal authority based upon rules and procedures and with bureaucracy as its purest form,
- charismatical authority where the personality of the individual predominates, and
- traditional authority founded on respect for custom and practice.

These three authority types are not mutually exclusive but can appear in any organisation at any hierarchical level.

Authority often has at its base the rules and regulations that the organisation has drawn up to regulate its affairs, but for the authority to be effective relies on a number of presumptions:

- that the rules exist and are accepted by those they are aimed at,
- that the rules are relevant to the particular circumstances,
- that the rules will be obeyed by every member of the organisation including those exercising the authority, and
- that the authority is vested in the office and not the individual.

A question that has frequently been debated but never resolved is the relationship between authority and responsibility. Which should come first? If authority is vested in an individual does responsibility follow or does the holding of responsibility grant the taking of authority? Common sense suggests a balance has to be struck if internal conflict is to be avoided and an organisation is to operate at its most effective.

2.3.3.2 Power

Power has been described as the capacity to influence others to do that which they would not have done voluntarily and is defined by C. Wright Mills[7] as 'the capacity to make and carry out decisions even if other people resist.' Power can derive from authority vested in a person by the organisation, the control of a desired product, such as money, or by virtue of special knowledge or expertise.

Kaplan[8] says that there are at least three dimensions of power which are of practical importance: *weight*, *domain* and *scope*. He describes weight as the ability of an individual to affect the probability that another individual will act in a certain way under certain circumstances; domain as the span or number of individuals or groups influenced, and scope as the range of outcomes over which power can be exercised.

A safety adviser in a factory may have the power to cause a manager to abandon a particular production method or process immediately on his advice, or at the other extreme so little power that his advice is completely ignored. The scope of his power may extend to authorising production procedures within the factory but not extend to influencing basic safety matters that affect employees in their leisure pursuits, thus his domain encompasses the factory premises but does not extend outside it.

The power referred to above is 'power over' – the exercising of control over the actions of others, but it is arguable whether this is as effective use of power as 'power with' – the concept of two people pulling in the same direction exerting greater force than one person trying to drive another.

Each person may be seen to have a degree of power and authority. At the lowest levels of an organisation an individual may exercise formal power and authority only over his own individual actions. However, by his actions and behaviour, he may influence another employee, i.e. through the use of informal or personality power.

Employees whose position places them in higher levels in the organisation exercise power through the formal recognition of their position which carries the appropriate authority. With this elevated level of control comes increased responsibility. An issue for people who hold higher positions in an organisation is the extent to which they should exercise control over particular and specific tasks delegated to subordinates. For example, it is not reasonable for an engineering manager to provide specific instructions and personally oversee a qualified electrician re-wiring a 220 V circuit. In most circumstances, the engineering manager will have discharged his responsibility by ensuring the electrician chosen for the job is competent in terms of training and experience. If, however, the task involved a 11 kV circuit, then much more vigorous steps need to be taken for the manager to discharge his responsibilities. These could include establishing a safe procedure of work including ensuring a risk assessment is carried out, the work people are competent and experienced in the particular work, ensuring isolation from live conductors and counter-signing the permit-to-work. The key skill of the manager is to be able to identify when the hazard and the risk demand specific action and ensure it is taken. Only then can the responsibility be discharged in a proper manner.

In more complex circumstances there must still be one person in overall charge although he may delegate responsibility for a number of functions, including safety. When he does delegate, he must ensure the people concerned are competent and experienced in the area of responsibility they have been given.

Roles and responsibilities are usually considered in relation to the position within the formal organisation structure. In recent years there has been a trend towards self-managed workgroups which have a much greater degree of control over the planning, organisation and carrying out of the work.

2.3.4 Work groups

Work groups may be defined as collections of individuals interacting with each other in the pursuance of a common work-related task or goal and who, for this purpose, are dependent upon each other. Important characteristics of groups are:

- the existence of standards of expected behaviour or norms
- the social experience of working together
- having a group identity
- the sharing of common goals
- the social approval given by other group members to those who meet the group norms.

There may be a presumption that the goals of the work group are the same as those of management, but this is not necessarily the case.

Work groups, as a special kind of social phenomenon, have been studied by many observers. The best known of these studies is the Hawthorne Experiment[9] which started in the early 1920s and ran for 12 years. Initially it monitored work and behaviour responses to varying physical conditions. Later in the Bank Room study, the experiments investigated work norms where Mayo and his fellow workers found that the act of monitoring workers' behaviour influenced that behaviour (the Hawthorne Effect), and that the social environment within the groups must be considered equally with the physical conditions as factors governing production rates. Mayo's work stimulated the birth of the human relations movement as opposed to the mechanistic scientific management school of thought of Taylor and Gilbreth.

The size of a work group may be determined by the tasks to be undertaken but other factors can be its role in the formal organisation, the social atmosphere and the geographical spread of the work. Work groups may extend their activities to beyond the work place when there are strong social ties and common social interests between the members.

Within a work group there may be two levels, the 'primary' group whose relations are personal and informal and the larger 'secondary' group which is established within the formal structure and where the role relationships predominate. Primary groups tend to be small in size dividing into smaller units as they grow in size so maintaining the personal contact and satisfying the social needs of its members.

Membership of a work group carries with it obligations in the form of pressure to conform as well as the benefit of rewards. The need to remain within a work group stimulates a desire to conform to group norms that can be so strong as to change individual attitudes and beliefs. The group itself may determine its own boundaries and membership often through an informally recognised leader or leaders. Where the work group is management-organised there is a risk of the inclusion of an informally unacceptable member – a sort of bad apple in the barrel – that can disrupt the whole group.

Consultation with informal work groups by management in order to gain acceptance of changing work patterns has become a recognised work practice. As such, work groups have been trained in risk assessment techniques in the hope of stimulating group as well as the individual action. A typical example of this practice is the establishment of Quality Circles. Changes devised or approved by the group have been found to have more ready acceptance among members. Formal recognition of the autonomy of the work groups has been advocated from time to time in an attempt to improve job satisfaction and output, but there is some evidence that this technique is not totally successful since it relies on management and group goals being identical. An autonomous group may seek to redefine its goals in ways which may not be to the benefit of the organisation, and are unacceptable to the management. The need to maintain a balance of interests within the organisation may be the most significant constraint on the development of work groups within it.

The ability of a work group to be effectively involved in health and safety depends upon the structure of the group and its maturity as a team. In a formal traditionally structured work group which is given specific operating instructions, the supervisor will take the lead role in ensuring the safety of his group, satisfy himself that the workers are familiar with procedures, are properly trained and comply with rules and regulations.

However, with the move towards flatter organisational structures, there is a trend to delegate administrative responsibilities to as low a level as possible so that much of that work – both production and safety – is carried out by those employed on production itself. There is only minimum supervision from senior managers who act in the role of coaches rather than supervisors. In this type of organisation the team has a large degree of autonomy. The team leader will co-ordinate and oversee the activities of the team and allocate jobs to various members. The members (or they may be sub teams) will plan the detailed work (in co-operation with the central planning department), develop their own training programmes (with the human resources department), keep track of safety legislation and standards (with the safety department) and so on. High levels of productivity and safety performance have been achieved by work teams organised on these lines. While each work team is responsible for the health and safety of its members, the overall legal responsibility for compliance with statutory requirements still rests with the employer.

2.3.5 Organisational theory

The character of organisations has attracted the interest and study of researchers. Taylor developed scientific management theories and focused on the production aspect of work considering people much like machines. They were given precise documented tasks to do in a set time. The tasks had to be performed repeatedly with efficiency being monitored by 'work study engineers' who sought ways to cut minutes and seconds off each task. People were employed for their 'brawn' not their 'brain'. Manual dexterity alone was valued. Employees were at risk from endlessly repetitive tasks placing strain on the same muscle groups of the body. Later other researchers began to examine people at work from a different perspective.

Maslow developed motivation theories based upon a Hierarchy of Human Needs[10] which is reproduced in *Figure 2.3.4*. He concluded that the best motivated and most productive workers are those whose work allows them to fulfil the needs at the top of the Hierarchy.

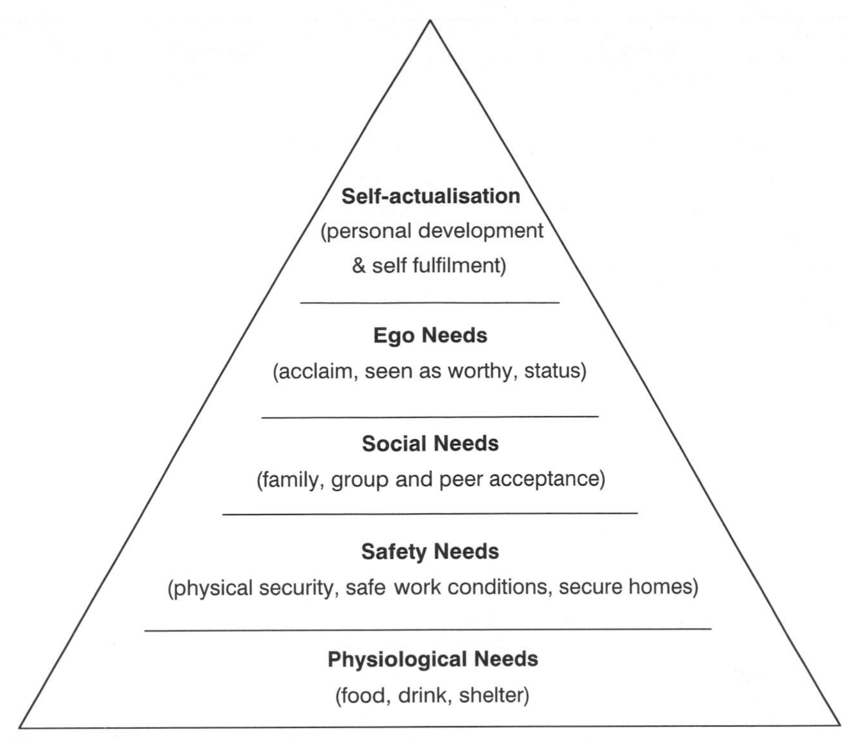

Figure 2.3.4 Maslow's Hierarchy of Human Need

McGregor[11] defined two sets of assumptions about human nature using them to explain how people influence the behaviour of others and in particular how managers view their employees. Theory X was a 'carrot and stick' approach to achieving high levels of productivity. Workers either were 'bribed' or 'threatened', and frequently a mixture of both, to get them to achieve the work required. Specific targets for work were set and additional money paid if these targets were exceeded. Conversely a failure to meet the standard targets would result in disciplinary action. Theory Y, on the other hand, assumed people were not inherently lazy and materialistic but eager to achieve goals and take pride in their activity. A more participative style of organisation, based on high standards and expectations placed on employees, resulted. A typical list of Theory X and Theory Y characteristics are produced in *Figure 2.3.5*.

Herzberg[12] developed a theory of job motivation based upon two dimensions – *Hygiene* and *Motivation*. Hygiene factors, covering such matters as poor company policies, poor supervision and poor working conditions, made employees unhappy in their work. Addressing hygiene factors reduced job dissatisfaction, but contrariwise while eliminating the dissatisfier factors, it did not produce a state of positive satisfaction. To achieve this latter aim required a completely different set of Motivating factors which included achievement, recognition for

Theory X (traditional)	Theory Y (potential)
people are naturally lazy	people are naturally active
people work mostly for money and status	people seek satisfaction from work and pride in achievement
people expect and depend on direction from above	people close to the situation are capable of self-direction
people need to be pushed or driven	people need to be managed and assisted

Figure 2.3.5 Adapted from McGregor Theory X and Theory Y assumptions about people

Hygiene factors (contribute to job dissatisfaction)	Motivating factors (contribute to job satisfaction)
Company policy and administration	Achievement
Supervision	Recognition for achievement
Work conditions	Interesting work
Pay	Task responsibility
Relationship with peers	Professional advancement
Security	Personal growth

Figure 2.3.6 Hygiene and Motivator factors (adapted from Herzberg)

achievements, interesting work and responsibility. *Figure 2.3.6* lists some of the Hygiene and Motivator factors.

The work of Maslow, McGregor and Herzberg has led to organisations being seen as socio-technical systems. An understanding of the management style in an organisation is critical if risk assessments are to be effective and high safety performance achieved. If the manager exercises control in a Theory X fashion with autocratic tendencies then employees are unlikely to respond to involvement in risk assessment processes. Maslow's and Herzberg's theories suggest that safety is a basic expectation of employees and a poor safety record is a major dissatisfier.

Involving employees to make improvements in the pursuit of safety excellence is, however, more difficult to achieve and is discussed in another chapter.

2.3.6 Organisational techniques

In the working community there are a number of techniques available that, used in various combinations, will assist in ensuring the achievement of the enterprise goals through enlisting the co-operation of the workforce, as individuals and as groups. The techniques work on the premise of involving work people to the greatest extent consistent with maintaining discipline and control.

2.3.6.1 Risk assessment and administrative processes

The objective of a risk assessment is to identify hazards and formulate actions that will ensure injury is avoided. This process is a daily living experience for all human beings. Natives of South American jungles walking through forests are aware of the hazards they face and take precautions to avoid harm from them. In cities, automobiles are driven with care and circumspection to avoid road accidents. In both cases, the individual is aware of the risks and applies controls to avoid injury from them. This is done mentally in real time, drawing upon their training and experiences to make the correct behavioural decisions to ensure their safety and survival. Risk assessments in occupational activities use administrative devices to achieve the same end result. The purpose of the administrative techniques is to bring some formality to the day-to-day behaviour so that:

- hazards are identified;
- control strategies are formulated and documented;
- training is given to those at risk in the implementation of control strategies;
- actions necessary to implement the control strategies are completed;
- hazards and controls are periodically reviewed.

These administrative techniques rely on documentation, consultation and meetings for their successful implementation.

Employers are primarily responsible for ensuring risk assessments are carried out and implementing controls to prevent the identified hazards from causing harm. Unions may disagree with the means by which the employer achieves this objective and may advocate alternative ways. However, the employer must make the final choice because it bears the ultimate legal accountability in the event of injury. It is no defence to say '. . . that was what the union wanted!'.

2.3.6.2 Administration and documentation

Formal risk assessments must be written down and recorded. Risk assessment documents should exhibit the following features:

- Clear identification of the hazards being addressed. Listing the hazards associated with the process enables a reviewer to see if any have been missed. It also allows others in the future to see if newly identified hazards in the process can be controlled by the original controls. This is important as new technical and scientific information, emerges (e.g. newly identified risks from an existing chemical).
- Identification of the risk assessment process being followed. Several formal risk assessment processes are discussed in other chapters. Each has merits. The documentation should clearly reveal which process is being employed so that its relevance to the current operations can be evaluated. For example, a typical job hazard analysis is not appropriate to assess risks arising from a machine interlock.
- Identification of the actions to be followed to avoid the hazards identified. This should be accompanied with time limits within which the actions must be completed.
- Assessment of the residual hazards, i.e. those that cannot be eliminated, and the means used to award priorities for actions.
- The system whereby the risks from the residual hazards are reduced to a minimum.
- Arrangements for monitoring the actions agreed.
- Identification of the person(s) who carried out the risk assessment. Risk assessors must be trained and experienced in the type of work covered by the assessment. It seems to be a fact of human nature that requiring people to sign their name to an assessment heightens the degree of responsibility they bring to the task. Identifying the assessors also permits an auditor to check that the assessors have received suitable training.
- Management sign-off to accept the assessment and implement the controls identified.
- The document should bear a date and number so that it can be identified and reviewed periodically. The review process is best performed by different assessors to ensure an independent review with greater objectivity.

2.3.6.3 Meeting structure

In any organisation, meetings convene to share information, establish goals, set objectives, allocate objectives to participants and monitor progress in meeting the objectives. The effectiveness of meetings depends upon several factors. These include:

- the purpose of the meeting is understood by the attendees;
- the attendees are the persons necessary to have an effective meeting;
- the meeting agenda has been pre-published and attendees come prepared;
- the chairperson is experienced at running meetings;
- people's comments are listened to and their opinions respected;
- disagreements are voiced and resolutions are sought;
- action-based decisions are made and allocated.

While these and other factors contribute to successful meetings they are essential to achieve successful safety meetings. Safety meetings should be chaired by a person with management authority because of the primary duty placed on management for safety. Supervisors and employee representatives should be in attendance. The meeting should be conducted in a spirit of co-operation and partnership by all. Sometimes in safety meetings emotions will rise when an employee concern is not shared to the same extent by management. This situation can arise from a different perception of the hazard being discussed or a perceived tardiness in response by management to a hazard which is acknowledged by them. The attendees at the meeting should remember that it is often those outside of the meeting who are at risk and highly charged emotions within a meeting may not assist them! Resolving conflicts which arise are key skills for all attendees and especially the chairperson.

Since 1979 legislation[13] in the UK has given the right to recognised trades unions to appoint safety representatives and require a safety committee if one does not exist. Where there are no recognised unions in the workplace subsequent legislation[14] requires managers to consult with their employees. It also gave elected safety representatives additional entitlements.

Safety Committees should have certain permanent agenda items which arise at each meeting supplemented by additional items of immediate or local concern. Permanent agenda items can include:

- actions completed since the previous meeting;
- actions outstanding from the previous meeting;
- incidents occurring since the previous meeting;
- hazards identified since the previous meeting;
- new safety regulations, standards and information;
- risk assessments performed since the previous meeting and the control measures proposed;
- member's items (Note: a member should not be allowed to raise an item with the Safety Committee until the supervisor of the area concerned has had an opportunity to deal with the matter. Only if no actions results may the matter be raised).

Periodic agenda items may include:

- a review risk assessments in the workplace and identifying those which need review;
- safety training plans;
- an annual review of the safety performance of departments and the company.

Healthy organisations do not limit discussions on safety to the safety meeting, but will make safety a topic at general meetings. For example, a morning production meeting may review any health and safety issues that have arisen in the last 24 hours. If major shutdowns are planned the health and safety implications must be included in the plans. On construction sites, daily meetings should include safety matters and an effective means of consulting with the employees of the various contractors should be in place. This is a requirement of the Construction (Design and Management) Regulations[15].

By these administrative means health and safety as a subject, with risk assessment as a core element, can be woven into the fabric of industrial and commercial life in the same way as are costs, satisfying the customer and quality.

2.3.7 Culture

The manner in which an enterprise deploys its resources is in reality a reflection of its culture. 'Culture' can be defined in many and various ways. The following definition is given in an HSE publication[16]:

> The safety culture of an organisation is the product of individual and group values, attitudes, perceptions, competencies, and patterns of behaviour that determine the commitment to, and the style and proficiency of, an organisation's health and safety management. Organisations with a positive safety culture are characterised by communications founded on mutual trust, by shared perceptions of the importance of safety and by confidence in the efficiency of preventative measures.

The publication goes on to list five organisational factors which tend to characterise enterprises with a positive safety culture. These factors are:

- Senior management commitment demonstrated by the perceived priority given to safety and the resources devoted to it.
- Management style that is cooperative and humanistic as opposed to autocratic and dictatorial.
- Visible management activity, including shop floor walkabouts and personal communication.
- Good communications horizontally and vertically in an organisation with an emphasis on sharing experiences, perceptions and especially an ability to share and learn from incidents.
- Balance between health and safety and operational goals so that both are achieved without compromise of either.

This list echoes other attempts to identify the elements which constitute an effective safety culture. The Confederation of British Industries lists:

- Leadership and commitment from the top
- Acceptance of health and safety as a long-term strategy requiring sustained effort
- A policy statement with high expectations
- Health and safety treated as a corporate goal
- Line management responsibility
- Ownership at all levels

- Realistic and achievable targets
- Thorough incident investigations
- Consistent behaviour against agreed standards
- Prompt remedy of deficiencies
- Adequate and timely information.

Both lists show that it is easier to list the characteristics and behaviours required than it is to define 'culture'. It is the organisation and administrative procedures that delivers these behaviours and their ability to do so should be assessed against the characteristics listed.

2.3.8 Potential problems

While good organisational and administrative arrangements are necessary to ensure risk assessments are undertaken and implemented effectively, problems can arise which jeopardise the risk assessment process in particular and health and safety in general. These problems can arise in three major areas, bureaucracy, conflict, and loss of focus.

2.3.8.1 Bureaucracy

This term has come to be used to describe what are felt to be the worst features of contemporary organisation and conjures up visions of over-regulation, inflexible procedures, 'red tape', disinterest in the customer and accountability to a 'faceless' committee. However, Weber considers that 'bureaucracy has a crucial role in our society as the central element in any kind of large scale administration' but in its most rational form depends upon rules, procedures and authority to achieve its control. He suggests it has the following characteristics:

- specialisation between positions;
- hierarchy of authority;
- a system of rules even extending to the recruitment of new members;
- impersonality; and
- written records of administrative acts, decisions and rules.

A bureaucratic organisation can be thought of as one that aims to maximise its efficiency in administration. Claims that a bureaucratic organisation offered benefits from cost reduction, precision, impersonality, inflexibility, etc., may owe more to the informal staff relationships, and practices than to the organisation itself. However, it must be recognised that elements of bureaucratic organisation can probably be found in parts of most medium and large organisations.

The benefits, however, can become liabilities. This occurs when the fabric of bureaucracy becomes more important than the purpose of bureaucracy. It is possible for organisations to spend many hours in committee developing exemplary risk assessment procedures with carefully detailed paperwork but lose sight of the fact that the purpose is to identify hazards and implement controls to prevent those hazards causing harm. Bureaucratic organisations will assiduously set targets for the number of risk assessments to be completed within a set timescale and staff groups will spend hours writing detailed procedures specifying how employees should work safely. Unfortunately, however, little attention is paid to the practical implementation of these plans which should occur if the results of this work are to prevent injury.

Examples of the bureaucratic mind set are revealed by its response to audit. Many organisations audit themselves for safety. A typical audit will generate a list of 'non-conformances' against internal, national or international standards. Success is measured by the division having fewest 'non-conformances'. The focus of management then becomes one of how to 'close the gaps'. This is a 'compliance attitude' which shows that management is simply reacting to the auditor's evaluation and uses the best performers as the target to aim for. It is satisfied when the auditor's criticisms have been dealt with and fails to realise that the performance being achieved is measured against the auditor's opinion rather than accepted standards. Management and work groups that have the 'pursuit of excellence' as their intention respond to audits more positively and seek to address the underlying deficiencies which generated the 'non-conformances' in the first place.

2.3.8.2 Conflict

Within any organisation, people have their own ideas about priorities for themselves and for the organisation which not infrequently conflict with the 'official view'. Many of the individuals in an organisation are likely to be subject to conflicting demands upon their time, energy and their principles not only in their work where they may play a number of roles but also in their private lives. Conflicts can arise within and between individuals, groups, departments and organisations.

A side effect of conflict is stress which can occur whenever an individual is put in a position of having to attack or defend. As stress builds up so equanimity is eroded and the propensity to argue, disagree or openly oppose grows with the risk of an escalation of potential conflict.

Conflicts arise whenever there are differences between individuals or groups and other individuals or groups and it can be between those at the same level or at different levels. At the individual level, there can be a reaction to not being consulted about a matter that materially affects the individual or resentment when the reason for working in a particular way is not understood or has not been explained.

Often the cause of the conflict is either obscure or not appreciated by those taking entrenched positions such as occurs in the case where a union official insists on representing a group of members with whom he has previously had little contact and without fully investigating the reason for the conflict, antagonising not only the employer but often the members he purports to represent. Similarly, where the supervisor is not given the support by management in resolving a relatively minor difference on the shop floor but where management (often the Personnel Manager) insists on handling the affair without fully appreciating the points at issue and finishes up with a full-blown failure-to-agree and a major industrial relations problem. This can discredit the supervisor and antagonise the workforce, both of which militate towards further conflicts.

Conflicts can stem from the different ways in which individuals or groups believe that affairs should be run. This can be seen in the broad differences between political parties over the allocation of national resources, in the differing views on how a social club should be organised and, within an organisation, the different views on whether, for example, promotion should be on the basis of merit or seniority. Again, employees may be disenchanted with the way their tasks are organised because the planned way does not match their natural way of working.

Perhaps the more frequent, but less disrupting, conflicts are those between one individual and another. However, these can escalate where there are strong allegiance ties with other individuals who rally to support the contesting parties. Refusal by an individual to conform to group standards of thought or behaviour can result in pressure to do so or isolation – 'being sent to Coventry'. Between groups, demarcation of jobs, pay scale and differentials and the threat of redundancy with the competition for diminishing job opportunities are fruitful sources of inter-group conflicts. Demands for more say in corporate decision making at their roots pose questions over the use of authority and power by management and individuals.

Conflicts can arise between organisations which compete for shares in a fluctuated market where the creation and the removal of jobs is at stake. Also organisations that exercise control over others in the way they perform their tasks can have important results in the workplace. Typical of the latter is the effect of those who enforce statutory regulations where unnecessarily expensive safety controls and procedures can be insisted upon with consequent adverse effect both on operator earnings and on the profitability of an operation.

Conflict sources may be inter-personal (a clash of personality or the frustration of an ambition), based on fact (overtly bad production planning), unjust exercise of authority or philosophical involving a clash of beliefs or aims.

Conflict in safety can arise when there is a different perception of hazards between the management, employees and their representatives, and/or the enforcement officers. It can also arise from slowness to address and resolve safety issues. Another cause is the perceived allocation of liability or civil liability when injury occurs. The adversarial nature of the litigation process is not conducive to the pragmatic allocation of liability and can cause resentment in both the claimant and the defendant This situation may be amended by the reforms brought in as a result of the report by Lord Woolf[17].

2.3.8.3 Loss of focus

Organisations and the administrative processes that they put in place should be living and dynamic. They need to respond to workplace changes. It occurs all too often that committees, meetings and documentation processes remain unchanged year in and year out without thought to the developments that are occurring in the work environment and in the risks associated with work. It is possible for an organisation's administration to have a life of its own virtually independent of all else and with a tendency simply to perpetuate itself. This occurs when it loses focus on the original reason for its existence. The organisation exists to achieve the goals of the enterprise in a safe and effective manner. The administrative processes exist for the same reasons. The 'customers' of both are the people who buy the product or service provided, the employees, and the general public who may be affected. All expect safety and the freedom from risk. When an organisation ceases to think of the people affected by its activities, it has lost its safety focus. Risk assessment is a form of critical self-appraisal and is critical in maintaining safety focus. Third party safety audits also provide a vital safeguard against a loss of safety focus.

2.3.9 The role of specialists in the organisation

Many organisations employ specialists to assist them in meeting their health and safety responsibilities. These specialists may be employees or consultants brought in to help the organisation meet its safety obligations. It is important that their advisory role is understood and that they are not used as a check on line management. Those in control must always be accountable for safety in their area of responsibility. As specialists, they should have no executive authority and their role should be seen as providing a '3A' service – Advice, Assistance, and Assessment.

Part of the *advice* provided should include bringing to the organisation's attention new legislation, standards and hazards which may be relevant to the organisation's activities. In order to fulfil this task, the specialist needs to keep abreast of statutory and technical developments, which can be achieved through contacts with professional and regulatory bodies and with trade associations. It is often incumbent upon the specialist to interpret this information and apply it to fit the culture of his/her client's organisation. The specialist should have technical and communication skills to enable him/her to present advice to managers and employees in an understandable way. The wording of Regulations can be somewhat convoluted and clear interpretation of them in as jargon-free manner as possible is essential.

The specialist should *assist* by providing members of the organisation with the ability and skills to enable them to carry out the practical work involved in safety activities. The specialist should train employees to undertake:

- general risk assessments involving a general survey of the workplace to identify hazards and initiate the appropriate remedial actions;

- specific risk assessments dealing with particular matters such as manual handling, display screen equipment, hazardous substances, etc.;
- simple occupational hygiene monitoring such as noise measurements using a noise level meter and the preparation of noise contours;
- routine training in the use of personal protective equipment.

In these activities the specialist's role will be as trainer to impart the skills. It is important that the specialist monitors the trained employees work and makes arrangements to refresh the skills imparted from time to time.

Assessment is the third aspect of the specialist's role and involves the observation and monitoring of the overall ability of the organisation to fulfil its health and safety obligations. This part of the role is far wider than an audit of technical compliance with legislation. It requires examination of other facets of the organisation's activities insofar as they affect health and safety performance including:

- the efficacy of the organisation's structure in accommodating the various safety activities;
- the effectiveness of the organisation in meeting safety targets;
- the usefulness of the documentation;
- the activity and impact of safety committees;
- the training programmes.

Each of these disparate roles of the specialist, although interlinked, stand in their own right and should be reported on separately to the company. In this, the specialist can confirm in writing any recommendations that earlier may have been made verbally and so reduce the possibility of misunderstanding of interpretation.

2.3.10 Conclusion

Attitudes are the cornerstones of health and safety. The risk assessment is a major technique but however excellent the techniques employed, and however competent the people using them, the safety goals will not be achieved without a vibrant and attuned organisation backed by sound and sensible administrative practices.

References

1. *Health and Safety at Work etc Act 1974*, The Stationary Office, London (1974)
2. Health and Safety Executive, Booklet No. HSG 65, *Successful Health and Management*, HSE Books, Sudbury (1997)
3. British Standards Institution, BS EN ISO 9001: 1994, *Quality Systems: Model for quality assurance in design, development, production, installation and servicing*, BSI, London (1994)
4. British Standards Institution, BS EN ISO 14001: 1996, *Environment management systems – Specifications with guidance for use*, BSI, London (1996)

5. British Standards Institution, OH SAS 18002:2000, *Occupational health and safety management systems – Guidelines for the implementation of OHSAS 18001*, BSI, London (2000)
6. Weber, M., *The Theory of Social and Economic Organisation*, Free Press, New York (1964)
7. Mills, C. Wright, *The Sociological Imagination*, Oxford University Press (1959)
8. Kaplan, A., *Power in perspective*, in Khan, R.L. and Boulding, E. (eds) *Power and Conflicts in Organisations*, Tavistock Institute, London (1964)
9. Mayo, E., *The Social Problems of an Industrial Civilisation*, Routledge, London (1949) reprinted in Pugh, Ed., *Organisation Theory*, chapter on *Hawthorne and the Western Electric Company*, Penguin Modern Management Text (1971)
10. Maslow, A.H., *Motivation and Personality*, 2nd edn, New York, Harper & Row (1970)
11. McGregor, D., *The Human Side of Enterprise*, McGraw-Hill (1961)
12. Herzberg, F., *The Managerial Choice*, Irwin (1976)
13. *Safety Representatives and Safety Committee Regulations 1977*, The Stationery Office, London (1977)
14. *The Health and Safety (Consultation with Employees) Regulations 1996*, The Stationery Office, London (1996)
15. Health and Safety Executive, publication No. HSG 234, *Managing health and safety in construction, Construction (Design and Management) Regulations 1994 Approved Code of Practice and Guidance*, HSE Books, Sudbury (2001)
16. Health and Safety Executive, publication HSG 48, *Reducing error and influencing behaviour*, HSE Books, Sudbury (1999) ISBN 0 7176 2452
17. *Civil Procedures Rules 1998* (emerging from the Access to Civil Justice Report by Lord Woolf – known as the Woolf Reforms), The Stationery Office, London (1998)

Chapter 2.4

Risk management: techniques and practices

L. Bamber

2.4.1 Risk identification, assessment and control

2.4.1.1 Introduction

As discussed in section 2.2.2.2, the risk from a hazard is the likelihood that it will cause harm in the actual circumstances in which it exists.

Essentially, the technique of risk management involves:

1 identification
2 assessment
3 control (elimination or reduction).

Within the workplace, operational management at all levels has a responsibility to identify, evaluate and control risks that are likely to result in injury, damage or loss. Part of these responsibilities should involve implementation of a regular programme of safety inspections of the work areas under their control. These inspections should include physical examinations of the workplace – i.e. the nuts and bolts – and also the systems, procedures, and work methods – i.e. the organisational aspects.

The process of risk management has been briefly outlined in section 2.2.3.1. The following sections (2.4.1.2–2.4.1.4) consider the practical application of the techniques in the workplace.

2.4.1.2 Risk identification

Within an organisation, there are several ways by which risks may be identified. These include:

1 Workplace inspections.
2 Management/worker discussions.
3 Independent audits.
4 Job safety analysis.
5 Hazard and operability studies.
6 Accident statistics.

Workplace inspections are undertaken with the aim of identifying risks and promoting remedial action. Many different individuals and groups within an organisation will – at some time – be involved in a workplace inspection: directors, line managers, safety adviser, supervisors and safety representatives. The key aspect is that results of all such inspections should be co-ordinated by one person within the factory, whose responsibility should include (a) monitoring action taken once the risk has been notified, and (b) informing those persons who reported the risk as to what action has been taken.

The vast majority of workplace inspections concentrate on the 'safe place' approach – i.e. the identification of unsafe conditions – to the detriment of the 'safe person' approach – i.e. the identification of unsafe acts.

Heinrich states that only 10% of accidents are caused by unsafe mechanical and physical conditions, whereas 88% of accidents are caused by unsafe acts of persons. (The other 2% are classed as unpreventable, or acts of God!).

Hence for workplace inspections to be beneficial in terms of risk identification and accident prevention, emphasis *must* be placed on the positive safe person approach, using techniques such as:

- managing by walking about (MBWA)
- safe visiting – talking to people
- catching people doing something right (not wrong)
- positive behavioural reinforcement
- one-to-one training/counselling sessions,

as well as the more traditional safe place approach which tends to be more negative as it evokes fault finding and blame apportionment at all levels within an organisation – i.e. catching people doing something wrong and penalising them for it.

Workplace inspections tend to follow the same format but are given many different names including: safety sampling, safety audits, safety inspections, hazard surveys, etc. Certain of the above are discussed below but all have the same aim – namely risk identification.

Management/worker discussions can also be useful in the identification of risks. Formal discussions take place during meetings of the safety committee with informal discussions occurring during on-the-job contact or in conversations between supervisor and worker. The concept of incident recall[1,2] is an example of management/worker discussion.

Indeed, incident recall has in effect been given legal status via Regulation 14 of the UK Management of Health and Safety at Work Regulations 1999 which requires employees to highlight shortcomings in systems and procedures – i.e. hazards, defects, damage and near-miss accidents, unsafe conditions and unsafe activities. This requirement emanated from the EU Framework Directive and should, therefore, be reflected in national legislation of all EU Member States.

In all cases, however, the feedback element is important from a motivational viewpoint. The risk identifier must be kept fully informed of any action taken to prevent injury, damage or loss arising from the risk he has noted.

Independent audits can also be used to identify risks. The term 'independent' here refers to those who are not employees of the organisation, but who – from time to time – undertake either general or specific workplace audits or inspections. Such independent persons may include:

1 Engineer surveyors – insurance company personnel undertaking statutory inspections of boilers, pressure vessels, lifting tackle etc. They are employed by the organisation as 'competent persons'.
2 Employers' liability surveyors – insurance company personnel undertaking general health and safety inspections in connection with employers' liability insurance.
3 Claims investigators – insurance company personnel investigating either accidents in connection with injury or damage claims under insurance policies.
4 Insurance brokers personnel – risk management or technical consultants undertaking inspections in connection with health and safety, fire, or engineering insurance as part of client servicing.
5 Outside consultants – undertaking specific investigations on a fee-paying basis. For example, noise or environmental surveys may be commissioned, if the expertise is not available within the organisation. Trade associations may be of assistance in this area.
6 Health and Safety Executive – factory (and other) inspectors undertaking either general surveys or specific accident investigations.

Again, with all the above there is a need to co-ordinate their independent findings to ensure that action is promptly taken to control any risks identified. It is quite possible that with the advent of the Turnbull Guidance on Corporate Governance[31], independent occupational safety and health audits may well become a legal requirement for publicly listed companies.

Job safety analysis is another method of risk identification. A fuller discussion of this method is presented below (see section 2.4.2).

Hazard and operability studies are useful as a risk identification technique, especially in connection with new designs/processes. The technique was developed in the chemical process industries, and essentially it is a structured, multi-disciplinary brainstorming session involving chemists, engineers, production management, safety advisers, designers etc. critically examining each stage of the design/process by asking a series of 'what if?' questions. The prime aim is to design out risk at the early stages of a new project, rather than have to enter into costly modifications once the process is up and running.

Further information on Hazop studies may be found in the Chemical Industries Association's publication on the subject[3].

Accident statistics will be useful in identifying uncontrolled risks as they will present – if properly analysed from a causal viewpoint – data indicative of where control action should have been taken to prevent recurrence. Ideally, an analysis of *all* injury, damage and near-miss accidents should be undertaken, so that underlying trends may be highlighted and effective control action – both organisational and physical in nature – taken. It should be borne in mind that the use of accident statistics is classed as reactive monitoring, whereas the use of audits and inspections is classed as active (or proactive) monitoring.

2.4.1.3 Risk assessment

Once a list of risks within a company has been compiled, the impact of each risk on the organisation – assuming no control action has been taken – requires assessment, so that the risks may be put in order of priority in terms of when control action is actually required, i.e. immediate; short term; medium term; long term on the basis of a ranking of the risks relating to their relative impact on the organisation. Such an assessment should take account of legal, humanitarian and economic considerations (as outlined in section 2.2.7).

The fundamental equation in any risk assessment exercise is:

Risk magnitude = Frequency (how often?) × Consequence (how big?)

In general:

- Low-frequency, low-consequence risks should be retained (i.e. self-financed) within the organisation. Examples include the failure of small electric motors, plate-glass breakages, and possibly motor vehicle damage accidents (via retention of comprehensive aspects of insurance cover).
- Low-frequency, high-consequence risks should be transferred (usually via insurance contracts). Examples include explosions, and environmental impairment.
- High-frequency, low-consequence risks should be reduced via effective loss control management. Examples include minor injury accidents; pilfering; and damage accidents.
- High-frequency, high-consequence risks should (ideally) be avoided by managing them out of the organisation's risks portfolio. If this appears to be an uneconomic (or unpalatable) solution, then adequate insurance – i.e. the risk transfer option – *must* be arranged.

A quantitative method of risk assessment – which takes into account the risk magnitude equation discussed above – considers the frequency (number of times spotted); the maximum potential loss (MPL) – i.e. the severity of the worst possible outcome; and the probability that the risk will actually come to fruition and result in a loss to the organisation.

From this type of quantitative assessment, a list of priorities for risk control can be established, and used as a basis to allocate resources.

Quantitative risk assessment is a complex and hotly debated subject. Practitioners use techniques such as Event Tree Analysis or Fault Tree Analysis to give estimated failure rates to key actions in the sequence of events. An example from everyday life might be the probability of stopping a motor vehicle before it struck an object. The outcome is the combination between the driver applying the brake in time and the reliability of the braking mechanism. Numbers are put on these two required events based upon historical data or informed opinion. The likelihood of the brake mechanism failing will be more accurately known (especially for a new vehicle) than the driver's reaction time. The latter depends on an array of personal factors (driving ability, reaction time, etc.) and environmental factors (weather, state of the road, etc.). For these reasons quantitative techniques need to be used with knowledge of their limitations. They are helpful techniques and have value in giving an insight into the relative importance of the factors involved.

A simpler form of quantitative risk assessment which has been used by the author is set out below. It takes into account frequency, MPL and probability using the formula:

$$\text{Risk rating} = \text{Frequency} \times (\text{MPL} + \text{probability})$$

In the above formula, frequency (F) is the number of times that a risk has been identified during a safety inspection.

Maximum Potential Loss (MPL) is rated on a 50-point scale where, for example:

multifatality	– 50
single fatality	– 45
total disablement (para/quadraplegic)	– 40
loss of eye	– 35
arm/leg amputation	– 30
hand/foot amputation	– 25
loss of hearing	– 20
broken/fractured limb	– 15
deep laceration	– 10
bruising	– 5
scratch	– 1

Probability (P) is rated on a 50-point scale where, for example:

imminent	– 50
hourly	– 35
daily	– 25
once per week	– 15
once per month	– 10
once per year	– 5
once per five or more years	– 1

Consider an example where the risk to be assessed has been identified once during an inspection. The MPL (worst possible outcome) was considered to be the loss of an eye with the probability of occurrence of once per day.

Thus, for this risk, the rate is:

$$\begin{aligned} \text{RR} &= \text{F} \times (\text{MPL} + \text{P}) \\ &= 1 \times (35 + 25) \\ &= 60 \end{aligned}$$

This risk rating figure should then be compared to a previously agreed risk control action guide, such as:

Risk rating	*Urgency of action*
Over 100	Immediate
80–100	Today
60–79	Within 2 days
40–59	Within 4 days
20–39	Within 1 week
10–19	Within 1 month
0–9	Within 3 months

These action scales should be drawn up by individual organisations, taking into account both the human and financial resources available for risk control.

In our example, the risk rating was found to be 60, hence control action to eliminate (or reduce) the risk should be taken within two days.

The above scales, example and action guide serve only to illustrate the principles involved, and – because of resource constraints – may not be generally applicable for practical use in all organisations. However, it does enable insights to be gained in order to prioritise risks, decide the order in which they should be addressed and the amount of money that should be allocated for risk elimination or control.

2.4.1.4 Risk control

The four risk control strategies – avoidance, retention, transfer and reduction – have been discussed in section 2.2.3.2.

The bulk of the risks identified by regular safety inspections will require some form of risk reduction (or avoidance) through effective loss control management.

The control of risks within an organisation requires careful planning, and its achievement will involve both short-term (temporary) and long-term (permanent) measures.

These measures can be graded thus:

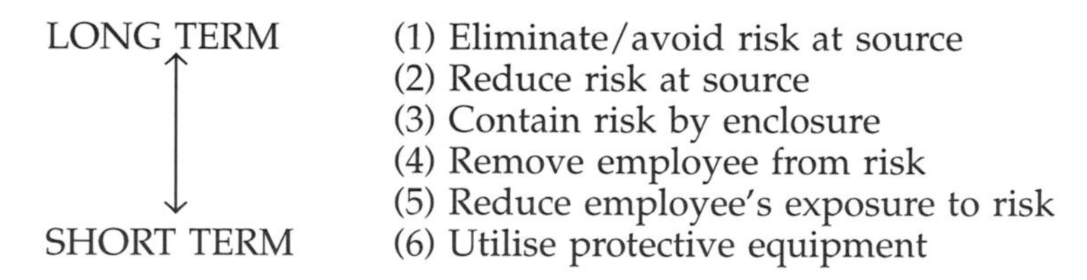

LONG TERM
SHORT TERM
(1) Eliminate/avoid risk at source
(2) Reduce risk at source
(3) Contain risk by enclosure
(4) Remove employee from risk
(5) Reduce employee's exposure to risk
(6) Utilise protective equipment

The long-term aim must always be to eliminate the hazard at source, but, whilst attempting to achieve this aim, other short-term actions – for example, utilisation of protective equipment – will be necessary. This list indicates an 'order-of' priority for remedial measures for any risk situation.

Various techniques are available to control risks within the workplace.

Mechanical risks may be engineered out of the process, or effectively enclosed by means of fixed guarding. Alternative forms of guarding involve the use of interlocked guards, light-sensitive barriers or pressure-sensitive mats. Trip devices and other forms of emergency stops may also be incorporated.

Risks from the working environment may be controlled by effective ventilation systems, adequate heating and lighting, and the general provision of good working conditions.

Chemical risks may also be controlled by effective ventilation, regular monitoring, substitution of material, change of process, purchasing controls, and the use of protective equipment.

A necessary corollary of risk assessment is the establishment of safe systems of work and training for the workforce to make them aware of the risks in their work areas, and of the methods for the control of such risks.

2.4.2 Job safety analysis

2.4.2.1 Job safety analysis – procedure

Job safety analysis (or job hazard analysis) is an accident prevention technique that should be used in conjunction with the development of job safety instructions; safe systems of work; and job safety training.

The technique of job safety analysis (JSA) has evolved from the work study techniques known as method study and work measurement.

The method study engineers' aim is to improve methods of production. In this they use a technique known as the SREDIM principle:

*S*elect (work to be studied);
*R*ecord (how work is done);
*E*xamine (the total situation);
*D*evelop (best method for doing work);
*I*nstall (this method into the company's operations);
*M*aintain (this defined and measured method).

Work measurement is utilised to break the job down into its component parts and, by measuring the quantity of work in each of the component parts, make human effort more effective. From experience standard times have evolved for particular component operations and these enable jobs to be given a 'time'.

Job safety analysis uses the SREDIM principle but measures the risk (rather than the work content) in each of the component parts of the job under review. From this detailed examination a safe method for carrying out each stage of the job can be developed.

The basic procedure for job safety analysis is as follows:

1 Select the job to be analysed. (SELECT)
2 Break the job down into its component parts in an orderly and chronological sequence of job steps. (RECORD)

Job Safety Analysis Record Chart		
Job title:		Date of job analysis:
Department:		Time of job observation:
Analyst/reviewer:		
Description of job:		
Accident experience:		
Maximum potential loss:		
Legal requirements:		
Relevant codes of practice/guidance notes/advisory publications:		
Sequence of job steps	Risks identified	Precautions advised
Suggested safe system of work:		
Suggested review date:		
Suggested job safety instructions:		
Suggested training programme:		
Signed:		Date:
Department:		Function:

Figure 2.4.1 Job safety analysis record chart

3 Critically observe and examine each component part of the job to determine the risk of accident. (EXAMINE)
4 Develop control measures to eliminate or reduce the risk of accident. (DEVELOP)
5 Formulate written and safe systems of work and job safety instructions for the job. (INSTALL)
6 Review safe systems of work and job safe practices at regular intervals to ensure their utilisation. (MAINTAIN)

From a practical viewpoint, this information can be recorded on a job safety analysis chart of the sort shown in *Figure 2.4.1*.

This is a typical job safety analysis chart. The detailed format will depend on the process and company and should be adapted to suit.

Criteria to be considered when selecting jobs for analysis will include:

1 past accident and loss experience;
2 maximum potential loss;
3 probability of recurrence;
4 legal requirements;
5 the newness of the job; and
6 the number of employees at risk.

The ultimate aim must be to undertake JSA on *all* jobs within an organisation.

Once the job has been selected, the next stage is to break it down into its component parts or job steps. On average, there will be approximately 15 job steps; if more than 20, then the job under study should be subdivided; if less than 10, then a bigger slice of the job should be analysed. Each job step should be one component part of the total job where something happens to advance by a measurable amount the doing of the work involved. The breakdown should be neither too general nor too specific. An example of such a job breakdown is given on p. 236.

From the above, it may be seen that each job step has been systematically analysed for its component risk factor. For each identified risk factor a control action has been developed.

The third column – Control action – becomes the Job Safety Instructions, and forms the basis of the written safe system of work.

2.4.2.2 Job safety instructions

Once the individual job has been analysed, as described above, a written safe system of work should be produced.

The purpose of job safety instructions is to communicate the safe system of work to employees. For each job step, there is a corresponding control action designed to reduce or eliminate the risk factor associated with the job step. This becomes the job safety instruction which spells out the safe (and efficient) method of undertaking that specific job step.

Such job safety instructions should be utilised in as much job safety training both formal (in the classroom) and informal (on the job contact

Example of a job breakdown – changing a car wheel

Job step	*Risk factor*	*Control action*
1 Put on handbrake	Strain to wrist/arm	Avoid snatching, rapid movement
2 Remove spare from boot and check tyre pressure	Strain to back	Use kinetic handling techniques
3 Remove hub cap	Strain; abrasion to hand	Ensure correct lever used
4 Ensure jack is suitable and is located on firm ground	Vehicle slipping. Jack sinking into ground	Check jack
5 Ensure jacking point is sound	Vehicle collapse	Consider secondary means of support
6 Jack up car part-way, but not so that the wheels leave the ground	Strain; bumping hands on jack/car	Avoid snatching, rapid movements
7 Loosen wheel-nuts	Hands slipping – bruised knuckles. Strain	Ensure spanner brace in good order. Avoid snatching, rapid movements. Use gloves
8 Jack up car fully in accordance with manufacturer's advice	Strain; bumping hands on jack/car	Avoid snatching, rapid movements
9 Remove wheel	Strain to back. Dropping onto feet	Use kinetic handling techniques. Use gloves (if available) to improve grip
10 Fit spare	Strain to back	Use kinetic handling techniques
11 Tighten wheel-nuts	Hand slipping – bruised knuckles. Strain	Use gloves. Avoid snatching, rapid movements
12 Lower car	Strain; bumping hands on jack/car	Avoid snatching, rapid movements
13 Remove jack and store in boot, together with removed wheel	Strain to back	Use kinetic handling techniques
14 Retighten wheel-nuts	Hand slipping – bruised knuckles	Use gloves. Avoid snatching, rapid movements
15 Replace hub cap	Abrasion to hand	Use gloves
16 Ensure wheel is secure, prior to driving off		Check wheel and area around car

sessions) as possible. All managers and supervisors concerned should be fully knowledgeable and aware of the job safety instructions and safe systems of work that are relevant to the areas under their control.

From a practical viewpoint, job safety instructions should be listed on cards which should be (a) posted in the area in which the job is to be undertaken; (b) issued on an individual basis to all relevant employees; and (c) referred to and explained in all related training sessions.

2.4.2.3 Safe systems of work

Safe systems of work are fundamental to accident prevention and should: (a) fully document the hazards, precautions and safe working methods, (b) include job training, and (c) be referred to in the 'Arrangements' section (part 3) of the Safety Policy.

Where safe systems of work are used, consideration should be given in their preparation and implementation to the following:

1. Safe design.
2. Safe installation.
3. Safe premises and plant.
4. Safe tools and equipment.
5. Correct use of plant, tools and equipment (via training and supervision).
6. Effective planned maintenance of plant and equipment.
7. Proper working environment ensuring adequate lighting, heating and ventilation.
8. Trained and competent employees.
9. Adequate and competent supervision.
10. Enforcement of safety policy and rules.
11. Additional protection for vulnerable employees.
12. Formalised issue and proper utilisation of all necessary protective equipment and clothing.
13. Continued emphasis on adherence to the agreed safe method of work by all employees at *all* levels.
14. Regular (at least annually) reviews of all written systems of work to ensure:
    (a) compliance with current legislation,
    (b) systems are still workable in practice,
    (c) plant modifications are taken account of,
    (d) substituted materials are allowed for,
    (e) new work methods are incorporated into the system,
    (f) advances in technology are exploited,
    (g) proper precautions are taken in the light of accident experience, and
    (h) continued involvement in, and awareness of the importance of, written safe systems of work.
15. Regular feedback to all concerned – possibly by safety committees and job contact training sessions – following any changes in existing safe systems of work.

The above 15 points give a basic framework for developing and maintaining safe systems of work.

2.4.3 System safety

2.4.3.1 Principles of system safety

A necessary prerequisite in connection with the study of system safety is a working knowledge of the principles of safe systems of work and job safety analysis. Also an appreciation of how hazard and operability studies[3] can be used will be of assistance.

System safety techniques have primarily emanated from the aviation and aerospace industries, where the overriding concern is for the complete system to work as it has been designed to, so that no one becomes injured as a result of malfunction.

Therefore, system safety techniques may be applied in order to eliminate any machinery malfunctions or mistakes in design that could have serious consequences. Thus, there is a need to analyse critically the complete system in order to anticipate risks, and estimate the maximum potential loss associated with such risks, should they not be effectively controlled.

The principles of system safety are founded on pre-planning and organisation of action designed to conserve all resources associated with the system under review.

According to Bird and Loftus[4], the stages associated with system safety are as follows:

1 The pre-accident identification of potential hazards.
2 The timely incorporation of effective safety-related design and operational specification, provisions, and criteria.
3 The early evaluation of design and procedures for compliance with applicable safety requirements and criteria.
4 The continued surveillance over all safety aspects throughout the total life-span – including disposal – of the system.

System safety may therefore be seen to be an ordered monitoring programme of the system from a safety viewpoint.

It may be seen that the system safety approach is very closely allied to the risk management approach. Indeed, the logical progression of system safety management techniques has been incorporated into many risk management processes, and also to other linked disciplines such as total quality management and environmental management systems.

2.4.3.2 The system

The system under review is the sum total of all component parts working together within a given environment to achieve a given purpose or mission within a given time over a given life-span.

The elements or component parts within a system will include manpower, materials, machinery and methods.

Each system will have a series of phases, which follow a chronological pattern; the sum total of which will equate to the overall life-span of the system. These phases are: conceptual phase, design and engineering phase, operational phase, and disposal phase:

1 The conceptual phase considers the basic purpose of the system and formulates the preliminary designs and methods of operation. It is at this stage that hazard and operability studies should be undertaken.
2 The design and engineering phase develops the basic idea from the conceptual phase, and augments them to enable translation into practical equipment and procedures. This phase should include testing and analysis of the various components to ensure compliance with various system specifications. It is at this stage that job safety analysis should be undertaken.
3 The operational phase involves the bringing together of the various components – i.e. manpower, materials, machinery, methods – in order to achieve the purpose of the system. From a practical viewpoint, it is at this stage that safe systems of work should be developed and communicated.
4 The disposal phase begins when machinery and manpower are no longer needed to achieve the purpose of the system. All components must be effectively disposed of, transferred, reallocated or placed into storage.

2.4.3.3 Method analysis

There are many methods of analysis in use in systems safety including:

1 Hazard and Operability Study[3]
This analytical method has been discussed above.
2 Technique of Operations Review[5]
This analytical technique or tracing system directs system designers and managers to examine the underlying and contributory factors that combine together to cause a failure of the system. It is associated with the theory of multicausality of accidents.
3 Gross Hazard Analysis
This analysis is done early in the design stage, and would be a part of a 'Hazop' (hazard and operability) study. It is the initial step in the system safety analysis, and it considers the total system.
4 Classification of Risks
This analysis involves the identification and evaluation of risks by type and impact (i.e. maximum potential loss) on the company. A further analysis – Risk Ranking – may then be undertaken.
5 Risk Ranking
A rank ordering of the identified and evaluated risks is drawn up, ranging from the most critical down to the least critical. This then enables priorities to be set, and resources to be allocated.
6 Failure Modes and Effects
The kinds of failures that could happen are examined, and their effects – in terms of maximum potential loss – are evaluated. Again this analysis would form part of an overall Hazop study.

7 Fault Tree Analysis
Fault tree analysis is an analytical technique that is used to trace the chronological progression of factors contributing to the accident situation, and is useful not only for system safety, but also in accident investigation. Again, the principle of multicausality is utilised in this type of analysis.

2.4.4 Systems theory and design

The word 'system' is defined in the Oxford Dictionary as 'a whole composed of parts in orderly arrangement according to some scheme or plan'. In present day parlance, we tend to think of 'systems' as connected with computers. However, the word is used in a wider sense in Operational Research to imply the building of conceptual and mathematical models to simulate problems and provide quantitative or qualitative information to executives who have to control operations, e.g. a maintenance system, a system governing purchase and use of protective clothing, a training system etc.

In this chapter, only an outline can be given of the concepts underlying systems theory and the theory will be presented mainly as an aid to clear thinking. The mathematical techniques associated with quantifying it can be found in textbooks of operational research.

The essential components of a systems model are goals or objectives, inputs, outputs, interactions between constituent parts of the system (e.g. storage, decision making, processing etc.) and feedback.

The stages in establishing and using a systems model are:

1 Define the problems clearly.
2 Build a systems diagram (including values).
3 Evaluate and test the system using already solved problems to check that the model gives the correct answer.
4 Use the model on new problems.

If we take as an example the provision of cost-effective machinery guards, we might produce a diagram such as *Figure 2.4.2* to indicate some of the factors affecting the process.

Such a conceptual type of model shows not only the sequence of events taking place, but further highlights feedback (fb on model) which informs management whether or not legal requirements are being satisfied. Besides the legal and technical considerations the diagram shows that the new guard could upset previously agreed incentive earnings and lead to conflict between management and unions, which in turn may lead to work stoppage and delays. The aim of the organisation (the system within which the subsystem is embedded) is to satisfy its customers and this can only be achieved by consistent output both in terms of quantity and quality. It can be seen that the fitting of a relatively insignificant machine guard can affect wider areas of the company's operations. Systems diagrams can direct the attention of those who are responsible for the effective running of

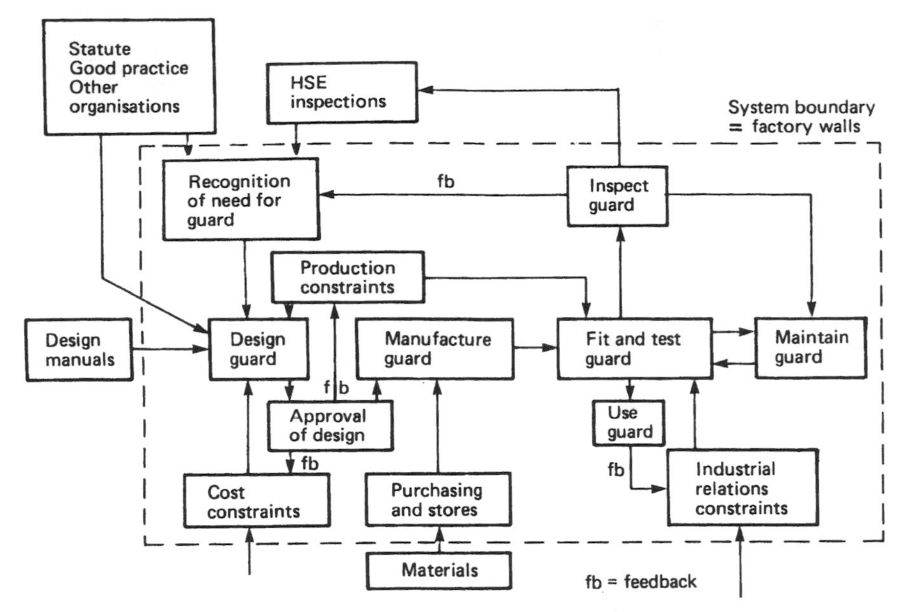

Figure 2.4.2 Systems diagram of the provision of a guard

a company to possible interactions, between either individuals or groups, inside or outside the company which could lead to conflicts and hence work disruption, long before it actually occurs, thus enabling suitable provisions to be made.

It is a useful exercise to consider how the safety adviser fits into the above system and which activities he should be involved in.

The example shows a system boundary drawn to correspond to the boundary of the organisation. The fact that there are inputs and outputs across this boundary indicates that the system is an open one. Closed systems have no transactions across the system boundary. Consider a simple example, which compares these two main types. A steam engine's speed is controlled by a valve which controls the supply of steam. If the valve is adjusted by an attendant (an outside agent), the system is open, whereas if the valve is controlled by a governor responsive to the engine speed, the system is a closed one.

The system boundary could be drawn at various levels – e.g. in the guarding example it could be drawn at the level of the department in which the machine is located, the works, the company, or the country (in the last case there might be inputs across the boundary (frontier) of materials, designs or EC regulations which would still make it an open system). For a full systems analysis and model it is usually necessary to produce a hierarchy of diagrams showing the total system, main subsystems and subsubsystems etc.

System diagrams sometimes only contain the hardware or technical elements as in *Figure 2.4.3* of a simple diagram of a car.

This is a very incomplete system diagram as it leaves out human control. A complete model or sociotechnical system should include both

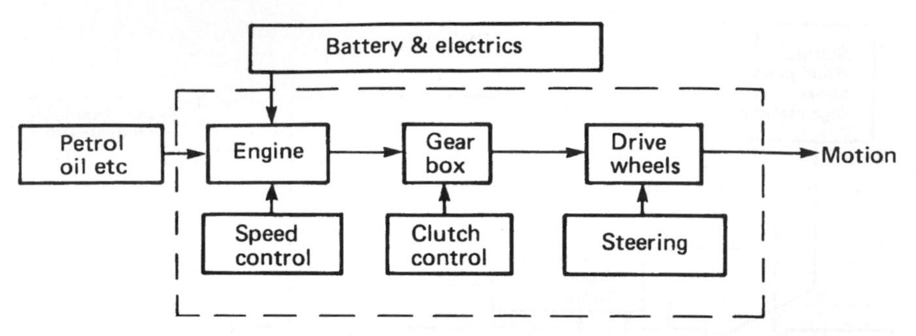

Figure 2.4.3 System diagram of a car

technical and human aspects (both desired and undesired, e.g. vandalism, sabotage etc.). Only in this way can the mode of operation or breakdown of the whole system be investigated and if necessary redesigned. A useful exercise is to add the human element to the car system above, or to devise a complete sociotechnical systems diagram for a company.

Accidents can be modelled as breakdowns of systems. The individual as a system set out in Chapter 2.7 is one example. Another which illustrates a fatal forklift truck accident in a warehouse is given in *Figure 2.4.4.*

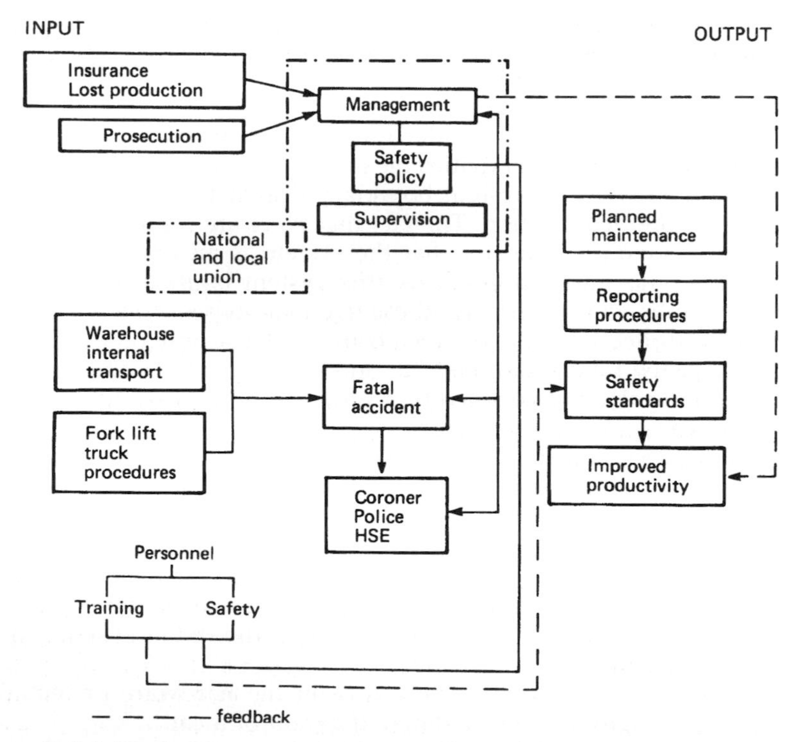

Figure 2.4.4 Systems diagram showing an accident and the environment in which it occurred

2.4.5 System safety engineering

System safety engineering has been defined[6] as an element of systems engineering involving the application of scientific and engineering principles for the timely identification of hazards and initiation of those actions necessary to prevent or control hazards within the system.

It draws upon professional knowledge and specialised skills in the mathematical, physical and related scientific disciplines, together with the principles and methods of engineering design and analysis to specify, predict, and evaluate the safety of the system.

An examination of system safety engineering methodology requires the consideration of two basic and interrelated aspects, namely system safety management and system safety analyses.

System safety management provides the framework wherein the findings and recommendations resulting from the application of system safety analysis techniques can be effectively reviewed and implemented.

System safety analyses employ the three basic elements of identification, evaluation, and communication to facilitate the establishment of cause. System safety analyses provide the loss identification, evaluation and communication factors and interactions within a given system which could cause inadvertent injury, death or material damage during any phase or activity associated with the given system's life-cycle.

Examples of system safety analyses include: routine hazard spotting; job safety analysis; hazard and operability studies; design safety analysis; fault-tree analysis; and stimulation exercises using a computer.

2.4.6 Fault tree analysis

2.4.6.1 Introduction

Fault tree analysis is a technique that may be utilised to trace back through the chronological progression of causes and effects that have contributed to a particular event, whether it be an accident (industrial safety) or failure (system safety).

The fault-tree is a logic diagram based on the principle of multicausality that traces all the branches of events that could contribute to an accident or failure.

2.4.6.2 Methodology

In constructing a fault-tree to assist in cause analysis, firstly the event – the accident or failure – must be identified. Secondly, *all* the proximate causes (contributory factors) must be investigated and identified. Thirdly, each proximate cause (i.e. each branch of a contributory factor) must be traced back to identify and establish all the conceivable ways in which each might have occurred. Each contributing factor or cause thus identified is then studied further to determine how it could possibly have happened, and so on, until the beginning or source of the chain of events has been highlighted for each branch of the fault-tree.

Certain standard symbols are used in the construction of a fault-tree; some of the more common are:

The rectangle identifies an event or contributory factor that results from a combination of contributory factors through the logic gates.

The 'AND' logic gate describes the logical operations whereby the co-existence of *all* input contributory factors are required to produce the output event or contributory factor.

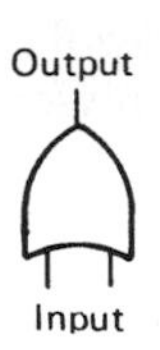

The 'OR' logic gate defines the situation whereby the output event or contributory factor will exist if *any* one of the input contributory factors is present.

Examples of fault-tree analyses are presented in Bird and Loftus[7] and Petersen[8].

By tracing back in this way the causes of accidents during accident investigation, a clearer and more objective assessment may be made of *all* the contributory factors, and hence more effective preventive action may be taken to ensure that there is no recurrence.

2.4.7 Probabilistic risk assessments

A probabilistic risk assessment consists of the following stages:

(a) the identification of undesired events and the mechanisms by which they occur (WHAT IF?),
(b) the likelihood – probability – that these undesired events may occur (HOW OFTEN?),
(c) the consequences of such an event once it occurs (HOW BIG?),
(d) a calculated judgement as to the significance of (b) and (c) (SO WHAT?) which may or may not lead to,
(e) the taking of control action.

Stage (a) – Identification of undesired events

Primarily, this stage involves the use of HAZOP studies (see section 2.4.1.2). Hazops are usually carried out at various stages of a system's cycle:

- conceptual design stage
- detailed design stage
- operational stage.

Stage (b) – Likelihood of undesired events
This stage involves an estimation of the probability of whether the undesired event is likely to occur or not.

Probabilities can be established mathematically, based upon the probable failure rates of individual components within the system. Data on individual components may be obtained from manufacturers' reliability statistics or quality assurance information. Specific failure rate data for individual items can also be obtained from reliability data banks such as that operated by the United Kingdom Atomic Energy Authority's (UKAEA) System Reliability Service.

Aspects such as maintenance schedules, condition monitoring, replacement criteria and human reliability/failure should also be taken into account.

An ideal technique for summating these individual probabilities to obtain the overall probability of the event occurring is Fault-Tree Analysis (see section 2.4.6), which is in essence a logic diagram with the event at the top of the tree.

Stage (c) – Consequences of undesired event occurring
Initially a hazard analysis (HAZAN) is undertaken so as to ascertain the magnitude of the potential problem and its potential for harm to the people, plant, process and the public.

A subsequent risk analysis will then go on to examine the actual consequences – worst possible case considerations – and express them in quantifiable terms. This then enables Stage (d) to be performed.

Stage (d) – Is the risk of the event occurring significant?
The output from Stage (c) may be expressed in the form of individual risk or of societal risk. Individual risk is the probability of death to an individual within a year (e.g. 1 in 10^4 per year). Societal risk is the probability of death to a group of people – either employees or members of the general public – within a year (e.g. a risk of 500 or more deaths of 10^{-8} per year).

Societal risks are usually given as fatal accident frequency rates (FAFRs).

The fatal accident frequency rate is defined[9] as the number of fatal injury accidents in a group of 1000 in a working lifetime (10^8 hours).

In making judgements to enable decisions concerning control action (Stage (e)) to be made, use is made of (published) risk criteria. These criteria are expressed in the form of numerical risk targets and they provide a yardstick for decision-makers against which to judge the significance of estimated risks. Generally two forms of risk criteria – as indicated above – are used:

- employee risks (on site)
- public risks (off site).

Stage (e) – Taking control action
If the calculated risk criteria figure is above the agreed accepted (published) figure, then control action is necessary. The amount by which

the calculated figure is higher than the agreed figure is useful in setting priorities, i.e. the greater the difference, the higher is the priority for control action.

If the calculated risk factor is below the accepted figure, then the safety provisions of the system may be considered to be adequate, and hence no further control action is required.

For employees in the UK chemical industry the FAFR is approximately 3.5. An acceptable target would obviously be below this, at approximately half, i.e. 1.7.

For the general public, FAFRs are rare. However, it has been suggested[10] that from an individual risk viewpoint as involuntary risks expose members of the public to a risk of death of about 10^{-7} per person per year, then industrial activities should not increase this figure. Hence, a risk criteria of less than 10^{-7} is acceptable.

2.4.8 Health and safety in design and planning

2.4.8.1 Introduction

The consideration of health and safety aspects at the design and planning stages of new projects, buildings, plant and processes is vitally important in order to ensure that health and safety are built in, rather than bolted on.

It is therefore essential that all engineers, designers, and architects receive education and training in such matters, so as to ensure that relevant legislative and technical factors appertaining to health and safety are taken into account at the design and planning stage of all new projects.

Certain risk identification (e.g. Hazop) and risk evaluation techniques (e.g. Hazan, fault-tree analysis) that may prove useful in this regard have been discussed above.

2.4.8.2 Project design

It is imperative that assessment and control of all new projects take health and safety aspects into account at the earliest – and at all – stages of a project's development.

The project originator should ensure that the project is appraised from a health and safety viewpoint, and it should not be allowed to proceed until it has been approved by a safety adviser. Ideally the safety adviser should be involved at all stages of the project's design and planning, so that specialist guidance and advice may be incorporated as necessary.

The risks associated with new projects may include: use of hazardous substances; insufficient product data; faulty electrical equipment; poor access/egress; poor ergonomics; noisy equipment; poorly guarded machinery; imported equipment/materials; lack of risk assessment; lack of training/awareness on behalf of management, supervision and employees; poor environmental control; inadequate emergency procedures; inadequate maintenance considerations; poor construction methods; little or no consideration of waste disposal/demolition.

The whole life-cycle of the project – from inception to ultimate disposal – should be considered at the design stage.

2.4.8.3 Project design: health and safety action plan

When health and safety is considered at the design stage, the following action plan should ensure that risks are designed and engineered out of the system *before* they are able to cause injury, disease, damage or loss:

- Ensure advice on health and safety is made available to the project team/ originator.
- Ensure a Hazop-type brainstorming meeting of key personnel associated with the development of the project is held to identify risks and establish control actions. (The list of risks in section 2.4.8.2 may serve as a checklist.)
- Ensure that suitable written safe systems of work are prepared and communicated to all concerned – i.e. develop the 'software' to go with the 'hardware'.
- Ensure that all aspects of the project comply with relevant legislative and technical standards.
- Ensure that all personnel concerned with the project receive necessary health and safety training.
- Ensure that suitable emergency procedures are developed.
- Ensure that the project commissioning procedure involves approval by the safety and health adviser at all stages of the project's development.

2.4.8.4 Project commissioning

Once the project has been approved, practical aspects of supply, installation, commissioning, use and ultimately disposal follow on. As stated in section 2.4.8.2, health and safety should already have been considered at the design stage of the project.

From a legislative viewpoint, the supply, installation and commissioning aspects are covered by s. 6 of the Health and Safety at Work etc. Act 1974, which requires manufacturers, suppliers, installers etc. of articles and substances to ensure that they are safe and without risks to health when set, used, cleaned maintained, stored, transported and disposed of.

In order to ensure safe commissioning of new plant and equipment, a three-part plant acceptance system should be utilised:

Part one – provisional safety certificate (for test purposes)
– this enables only design/engineering personnel to undertake testing, once approved by safety adviser and project originator.

Part two – trial production run
– this enables production employees to become familiar with the new equipment under the supervision of the project originator/safety adviser and enables any previously unforeseen risks to be engineered out at the man/machine interface.

Part three – final certification
– once all testing and production trials have been satisfactorily completed, the plant/equipment is handed over to production management.

By involving a multidisciplinary team – including a safety adviser – at all stages of the design, planning and commissioning process, the risk of having to provide additional – and more costly – safeguards *after* the plant is in full use is minimised.

2.4.9 Quality, Environment, Safety and Health Management Systems (QUENSH)

2.4.9.1 Introduction

Established quality assurance procedures provide a sound basis for the development of systems for health and safety management[11]. The introduction of the HSE's publication *Successful Health and Safety Management* (revised 1997) clearly states[19] that many of the features of effective health and safety management are indistinguishable from the sound management practices advocated by proponents of quality and business excellence.

This logic has more recently been extended into the realms of environmental management systems via BS 7750: 1994[25]/BS EN ISO 1400: 1996[26] and also into occupational health and safety management systems via BS 8800: 1996[27]. Indeed, Annex A to BS 8800: 1996 clearly demonstrates the commonality of approach between BS 8800 and BS EN ISO 9001: 2000[28] for quality management systems. Also BS 8800 postulates two routes towards occupational health and safety management systems: route 1 via HS(G)65[19]; route 2 via BS EN ISO 14001: 1996[26].

Hence the QUENSH approach towards integration is gathering momentum.[29,30]

2.4.9.2 Quality systems

The original British Standard on quality systems[12] – BS 5750: 1987 – has now been superseded by BS EN ISO 9001: 2000[28]. This tells suppliers and manufacturers what is required of a quality-orientated system from a practical viewpoint. It identifies basic disciplines and specifies procedures and criteria to ensure customer requirements.

Within the context of ISO 9001, quality means that the product is fit for the purpose for which it has been purchased, and has been designed and constructed to satisfy the customer's needs.

The Standard sets out how an effective and economic quality system can be established, documented and maintained.

The Standard considers that an effective quality system should comprise: management responsibility; quality system principles; quality system audits; quality/cost considerations; raw material quality control; inspection and testing; control of non-conforming product; handling, storage, packaging and delivery; after sales service; quality documentation and records; personnel training; product safety and liability; and statistical data/analyses.

2.4.9.3 Quality and safety

Although the Standard does not explicitly refer to 'people safety', there are obvious parallels to be drawn between the quality systems approach and health and safety management.

Indeed, the management systems described in ISO 9001 are as applicable to health and safety management as they are to product liability risk management[11]. ISO 9001 is concerned with the achievement of quality, which is measurable against specific criteria. It lays down systems which demonstrate achievement against these specified criteria.

One of the benefits of an effective quality system is to minimise the risk of product liability claims and losses.

In the case of product liability risk management, the specified criteria of performance are:

- Compliance with the relevant consumer safety and consumer protection legislation, e.g. in the UK this would include s. 6 of the Health and Safety at Work etc. Act 1974 (as modified by the Consumer Protection Act 1987).
- Compliance with all other relevant statutory provisions, especially the Management of Health and Safety at Work Regulations 1992.
- The ability to adhere to all product contract conditions.
- The minimisation of defective products.
- The maximisation of health and safety benefits to the consumer.

This parallels very closely the perceived criteria of an effective health and safety management system, namely:

- Compliance with the relevant occupational safety and health legislation, regulations, codes of practice and standards.
- The ability to adhere to the common law duty of care, and relevant aspects of employment contract conditions.
- The minimisation of risks likely to cause injury or disease.
- The maximisation of health and safety benefits to employees, third parties, and the general public.

From the above, it may be seen that the application of quality systems to the management of health and safety at work has distinct benefits, especially when consideration is given to the tremendous overlap between the two subject areas. Overlap examples include: policies; systems and procedures; standards; documentation – records; training (including record keeping); statistical analyses – causal, numerical; accident/complaint investigations; audits/inspections (internal and external); and the taking of remedial control action.

Hence effective quality systems management will greatly enhance the management of health and safety, and will lead to an overall improvement in the level of safety performance.

2.4.10 Use of data on accidents

In addition to the use of qualitative and quantitative accident data when identifying and evaluating risk (see sections 2.4.1.2 and 2.4.1.3), there is a need to consider the relative occurrence of different accident types, in order that effective accident control measures may be implemented throughout an organisation.

Accidents, whether they result in injury, damage, disease or loss, need to be controlled. Similarly those that have no end result – i.e. the near-misses – should be considered for control action.

To enable an accident control system to be developed, it is necessary that *all* accidents are reported, recorded, investigated and analysed, so that after remedial action has been decided, plans can be drawn up to prevent a recurrence. The most important question to be asked in any accident investigation is: 'What action has been taken to prevent a recurrence?'

The collection of accident data on a much broader base to facilitate the planning of control action has been undertaken by a number of researchers and one of the most widely applied accident ratios is that propounded by Bird[13] in 1969.

The Bird study is generally depicted in triangular form:

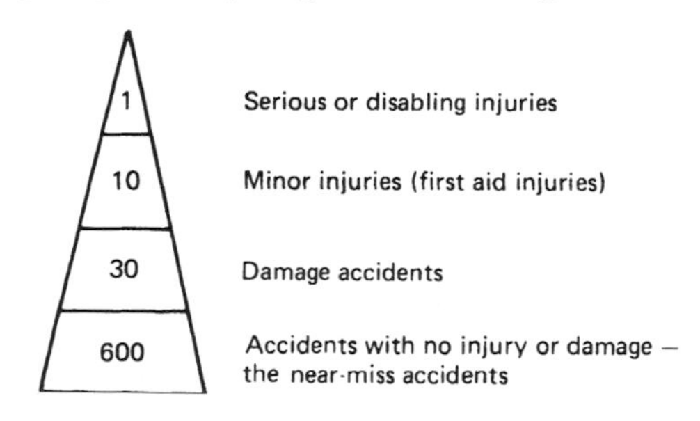

From this it can be seen that the majority are either 'near-misses' or damage only accidents, so that any accident control programme, for its greatest effect, must concentrate much of its attention on these two areas. Related techniques of damage control and incident recall are dealt with in sections 2.2.4.2 and 2.4.12 and also by Carter[14].

Accident data are only one form of risk information which assists in the identification, assessment and control of risks. Other forms of risk information include: reliability data; epidemiological studies; mortality/morbidity data; vulnerability analysis; results of audits/inspections; insurance claims data; and – probably the best – personal experience. Sources of risk information are listed in the Further Reading section of this chapter (see especially the *Handbook of Risk Management* by Carter *et al.*).

2.4.11 Maintenance systems and planned maintenance

2.4.11.1 Maintenance – risk management aspects

From a risk management viewpoint, health and safety aspects concerning all maintenance operatives should be considered and planned for at the design stage of a new project or process.

Aspects such as: safe access; operations at unguarded machinery; breaking into pressurised systems; blanking off; inerting; hot work; installation; setting up; inching; dismantling; and demolition, should have been taken into account.

Effective maintenance systems that take into account health and safety aspects serve not only to increase the lifetime of key plant and equipment, but also ensure safer and more cost-efficient operation.

Indeed, a planned maintenance system is essential to the continued safe operation of a range of plant and equipment, as well as meeting both general and specific legislative requirements – e.g. statutory inspections.

Plant and equipment requiring such regular inspection and planned maintenance include: boilers; pressure vessels – air receivers; power presses; cranes; lifting equipment – chains, pulleys; ventilation systems; pressurised systems; access equipment – ladders; scaffolding; electrical equipment – portable electric tools; office equipment; emergency eye-wash points and showers; fire alarm systems and fire points; fire extinguishers; fixed firefighting appliances/systems; stairways, walk-ways, gangways; in addition to all key items of production plant and equipment.

2.4.11.2 Planned maintenance

It is generally accepted that a system which only allows for the maintenance/repair of plant on a crisis or breakdown basis is not cost-efficient, especially in the longer term.

Hence any form of planned maintenance – whether it be planned lubrication at one end of the scale, or sophisticated techniques such as condition monitoring at the other end – will improve both safety and plant integrity/reliability.

A system of planned maintenance may be introduced progressively, commencing with limited maintenance routines only on key items of plant and equipment. This may then be extended to a wider range of plant, utilising more complex routines, until such time as a 100% planned maintenance system has been implemented. As with quality systems, maintenance records and histories are essential to ensure smooth operation.

When a total (100%) system of planned maintenance has been implemented, then, by definition, any breakdown that occurs must be accidental in nature. Hence such accidental breakdowns – i.e. damage accidents – should be reported and investigated (see section 2.4.12 for fuller discussion).

2.4.12 Damage control

2.4.12.1 Introduction

The technique of damage control involves the systematic reporting, investigation, costing and control of damage accidents within an organisation.

2.4.12.2 Reporting

The reporting system for damage accidents within an organisation should, as far as possible, be the same as that for injury accidents. The fundamental differences between injury and damage accidents are that in the former, the victim (the injured person) starts the reporting chain by advising his supervisor or first-aider of his injury.

However, in the case of damage accidents, the victim is an inanimate object, physically incapable of reporting the accident. Therefore, there is a need to develop a system by which accidental damage can be highlighted and clearly identified.

It would appear that the situations likely to give rise to accidents can be divided into three main categories:

(i) Fair wear and tear – if we define this as the failure of an item in service after a lifetime of proper use, then, in theory, such a failure will be eliminated by a total (100%) system of planned maintenance and condition monitoring. In such a situation any failure will be an accident, i.e. unexpected, unplanned.
(ii) Malicious damage – this will need to be identified and dealt with separately.
(iii) Accident damage – it is this final category that we are dealing with in this section.

There will, of course, be many occasions in practice where it will be necessary to determine whether damage arises from (i) or (iii) above and the opinion of all relevant staff will need to be sought.

2.4.12.3 Investigation

In order to ensure that prompt action is taken to prevent the damage accident recurring, a detailed investigation should be undertaken. This should aim to establish the sequence of events, and to identify the combination of causes of the damage accident.

It should be remembered that the prime aim of any accident investigation is prevention, rather than blame. As with injury accidents, the investigating team should generally comprise the relevant supervisor and/or manager, together with the safety adviser and safety representative (if appointed).

To aid the investigation it is suggested that a detailed accident investigation report form is utilised.

2.4.12.4 Costing

The prime aim in costing damage accidents is to obtain an objective measure of severity. This will in turn assist in the setting of priorities and hence the allocation of resources and the overall planning of the damage control programme.

Information regarding the cost of all accidents – both injury and damage – may also be used to motivate management action in accident prevention.

2.4.12.5 Control

During the investigation the question 'What action should be taken to prevent a recurrence?' should be clearly answered. This should ensure that the correct remedial action is promptly taken.

The findings of the investigation should be clearly communicated to the manager responsible for instigating the control action.

To gain the full value from a control system, it is important that findings be communicated to all management within the organisation who may be faced with similar problems. This will greatly assist in the elimination of the causes of potential accidents before either injury or damage results.

2.4.13 Cost-effectiveness of risk management

2.4.13.1 Accident costing

The three fundamental arguments that may be employed to promote action in the field of risk management concern the legal, humanitarian, and economic aspects (see section 2.2.7).

For the vast majority of organisations, there will be few or no quantitative data relating to the economic facets, and particularly to the costs associated with accidents. In order to be able to demonstrate cost-effectiveness (or cost benefit), there is therefore a need to be able to quantify the cost of all losses associated with accidents.

An investigation of the costs of occupational accidents for the UK as a whole[15] has considered the costs under two headings: resource costs covering lost output, damage to plant, medical treatment and administrative costs, and subjective costs relating to pain and suffering of the victim and his family.

	Resource costs £	*Subjective costs £*	*Total costs £*
Fatality	71 500	38 000	109 500
Serious injury	1 800	2 750	4 550
Temporary disabling diseases	1 600	2 750	4 350
Minor injuries	300	150	450

However figures relating to the national situation are often not pertinent to individual factories and departments. Some figures obtained in May 2001 for the uninsured costs of three types of accident[16] were:

Lost time injury	£2148
Non-lost time injury	£ 34
Damage	£ 144

The above figures – in isolation – do not have much impact, but must be related to the total costs of accidents in an individual factory or organisation. When these accident costs are compared with other business costs, such as production, sales and distribution, a true indication of the drain on financial resources can be appreciated. The following hypothetical case is presented as an example:

Consider Factory Z in 2000; the data given below apply:

Number of lost time injury accidents	=	30
Number of non-lost time injury accidents	=	750
Number of damage accidents (estimated)	=	780
Employers' liability premium (£1.10% wages)	=	£90 000
Number of employees	=	1000

When the average uninsured accident cost figures are applied to the above data, the following result:

Uninsured cost of lost time injury accidents			
	30 × £2148	=	£64 440
Uninsured cost of non-lost time injury accidents			
	750 × £34	=	£25 500
Uninsured cost of damage accidents	780 × £144	=	£112 320
Total uninsured accident cost		=	£202 260

Adding to this the Employer's Liability insurance premium (£90 000), the total accident cost becomes £292 260, or approximately £292 per employee per year.

Of particular concern is the impact of such costs on the overall profitability of the organisation. Although the total cost of accidents within a company may be relatively small, it can amount to approximately 2% of the annual running costs, and represents a direct drain on profits.

The table below illustrates the sales necessary to cover these accident costs.

Accident costs £	*If your organisation profit margin is*		
	1%	*3%*	*5%*
1 000	100 000	33 000	20 000
10 000	1 000 000	330 000	200 000
100 000	10 000 000	3 300 000	2 000 000

Any reduction of these costs that may be made through a cost-effective risk management programme will lead both to a safer and more profitable organisation.

The use of cost figures presented in this way will be more meaningful to managers and executives and is likely to stimulate their motivation to reduce the number of accidents.

An alternative – more specific – method of relating accident and wastage costs to costs of production is presented in the following example:

On a construction site, the number of facing bricks lying around was estimated by random counting to be 1300, valued at £455.

The cost in terms of profit on turnover (of 6.3%) was:

$$\frac{\text{Cost} \times 100}{\text{Profit on turnover}} = \frac{455 \times 100}{6.3\%} = £7222$$

i.e. a turnover of £7222 was required to pay for the bricks – i.e. to break even.

The cost expressed as a percentage of the total contract price (of £1 650 000) was:

$$\frac{\text{Turnover to pay for bricks} \times 100}{\text{Contract price}} = \frac{£7220 \times 100}{£1\ 650\ 000} = 0.43\%$$

Such exercises may be undertaken for all types of accident and wastage situations, to enable the costs involved to be judged in relative – rather than absolute – terms.

2.4.13.2 Cost benefit analysis

Cost benefit analysis techniques have been developed in recent years, as decisions concerning risk management have been made on a cost versus risk basis – i.e. so far as is reasonably practicable.

Indeed, all proposed legislation has to pass the cost/benefit test at the consultative stage, before being allowed to pass on to the statute books. Although, in the majority of cases, the benefits of proposed legislation are not accurately quantified, nevertheless the qualitative benefits are listed, and these may be seen to outweigh the costs. Further discussion on the cost/benefit of specific proposed legislation may be found in relevant Health and Safety Commission Consultative Documents which contain a review of the costs and benefits associated with proposed changes in legal requirements.

To undertake a cost benefit analysis, answers to the following questions are required:

- What costs are involved to reduce or eliminate the risk?
- What degree of capital expenditure is required?
- What ongoing costs will be involved, e.g. regular maintenance, training?
- What will the benefits be?
- What is the pay-back period?
- Is there any other more cost-effective method of reducing the risk?

The cost factors associated with poor risk management have been discussed in sections 2.4.7 and 2.4.13.1 and include both insured and uninsured elements.

The benefit factors should initially be listed, and should always be quantified, where possible, so that the pay-back period can be established. Some benefits are easier to quantify than others.

Benefits of effective risk management include:

- few claims resulting in lower insurance premiums,
- less absenteeism,
- fewer injury and damage accidents,
- better levels of health,
- higher productivity/efficiency,
- better utilisation of plant and equipment,
- higher morale and motivation of employees,
- reduction in cost factors.

The costs are then balanced against the benefits (both qualitative and quantitative) and then an objective decision may be made on whether to allocate resources to the project or not. This will usually be based on the length of the pay-back period. Most health and safety projects will generally have a pay-back period of between three and five years – i.e. medium-term rather than short term.

2.4.14 Performance evaluation and appraisal

2.4.14.1 Introduction

In the vast majority of company health and safety policies the health and safety responsibilities of line and functional management are clearly laid down, together with – in some cases – the mechanism by which the fulfilment of these responsibilities will be monitored.

Indeed, the Accident Prevention Advisory Unit (APAU) have produced three excellent publications[17–19] which provide additional guidance and discussion on the aspects of policy management, implementation and monitoring.

2.4.14.2 Financial accountability and motivational theory

However, the most effective way to fix accountability for health, safety and indeed risk management responsibilities is by financial accountability of directors and managers.

This is borne out when consideration is given to the use of the legal, humanitarian and economic arguments for health and safety at work (section 2.4.7).

Maslow[20] related his theory of motivation to human needs. He suggested five sets of goals which are usually depicted as a progression or hierarchy:

Self actualisation (Achievement; Doing a good job)
↑
Esteem (Status; Approval)
↑
Love (Social)
↑
Safety (Security)
↑
Physiological (Sustenance)

The logic is that once the lower needs are well satisfied, the individual is motivated to attain satisfaction at the next higher level, and so on up the hierarchy.

From a health and safety viewpoint, therefore, the humanitarian argument tends to operate in terms of doing a good job, caring for people, and being well thought of and accepted, i.e. the higher end of the hierarchy: esteem/self actualisation.

The legal argument tends to motivate via the safety (security) need – the real (if remote) threat to physical security – and the esteem (approval) need of others. This motivation is in the middle of the hierarchy.

However, the economic argument derives its motivational impact from the fact that the goals involved tend to be generally lower in the hierarchy – i.e. the safety (security) need and, in certain cases, the physiological need. The security need involves the manager's performance in his job –

the ability to effectively carry out his responsibilities whilst keeping within budgetary constraints. Hence adverse performance could result in a decrease in or loss of the next salary increment or merit rise, as part of the performance appraisal exercise, thus directly threatening the security needs. In times of economic recession, this failure to achieve satisfactory performance could even affect the manager's position in the company and his ability to maintain employment, thus threatening the physiological needs.

Hence it would appear that financial accountability is the key.

2.4.14.3 Use of accident costs

However, the present system of accounting for health, safety, accident prevention and risk management operating in most of industry does not attempt to make line management financially accountable for accidents and uninsured losses, and very little use is made of economic arguments in stimulating management interest in risk management. Any arguments put forward rely mainly on legal and humanitarian considerations which, in some instances, fail to convince management that there is a need for risk management, at least beyond compliance with statutory duties.

One method that may prove useful involves the use of budgetary control which would introduce economic accountability into the field of accident prevention. Such measures would involve the reorganisation of the existing accounting procedures in most companies in order to overcome the lack of accountability for accident prevention and risk management.

When an accident occurs within a factory department, the cost of the accident usually is absorbed into the running costs of the factory as a whole, and will not be itemised on the departmental balance sheet. Nor will many of the uninsured costs be paid for from the departmental manager's budget. Furthermore, the insured cost – i.e. insurance premiums, such as employer's liability, will generally be paid from a central fund, usually administered by Head Office.

However, when the company safety adviser or factory inspector recommends safety measures such as guarding for machinery, the cost is usually charged against the departmental manager's budget, though it is very unlikely that it would be itemised as an accident prevention cost.

Thus under accounting systems currently employed in many companies, accident costs are not charged to departmental managers' budgets, whereas accident prevention is. Hence the departmental manager has no economic motivation to undertake any accident prevention; rather the reverse.

A positive economic motivating factor for encouraging accident prevention may be introduced by interchanging the budgetary system. For each accident – injury or damage – that occurs within a department, a charge is made against that department. Any accident prevention expenditure that is required within the department is financed from a central fund subject to approval by the risk manager or safety adviser. The result – as far as the departmental manager is concerned – is that it is costly to have accidents but not to prevent them.

Thus line management become accountable for the accidents occurring within their areas of control. At the end of the financial year, when the budgets are drawn up, a realistic allowance for accidents will be set within the budget as a target for the manager to achieve. Failure to achieve the agreed target would adversely reflect on performance and should be taken into account at job performance appraisal interviews. This allowance will form an integral part of the management plan as with budgetary control in other business areas.

However, the number – and hence the cost – of accidents budgeted for should be less than the previous year so that reduction of accidents becomes part of the management plan. The reorganised system would bring about the necessary economic accountability and would make full use of the knowledge and data obtained in establishing what accidents were costing the company in financial terms.

Once the costs of accidents have been established, the reorganised budgetary system can be implemented. The charges to be made against the departmental manager's budget can then be calculated and allocated on a monthly basis. The departmental manager might receive a monthly report giving information on the costs of accidents and accident prevention expenditure. This would enable him to plan any action necessary to maintain or improve the level of safety within his department. Also it would facilitate decision-making in connection with the allocation of scarce resources. Any deficiencies in the current programme would be highlighted in cost terms rather than by a frequency rate – a measure of safety increasingly questioned by safety advisers.

On their own, the legal and humanitarian arguments for risk management may not be sufficient to achieve a reduction in accidents and other losses. The addition of economic accountability – through accident costing – should greatly assist in reducing losses arising from accidents and ill-health at work.

2.4.15 Loss control profiling

General aspects of loss control are discussed in section 1.4. Loss control profiling is one of the major evaluation and control techniques associated with loss control management. The technique of profiling has formed the basis for a number of proprietary auditing systems such as International Safety Rating System (ISRS), Complete Health and Safety Evaluation (CHASE) and Coursafe.

Between 1968 and 1971, Bird[21,22] designed a loss control profile to quantify management's efforts in this area. He considered 30 areas of management activity that are connected either directly or indirectly with the reduction of loss.

These 30 areas are:

1 Management involvement and policy making.
2 Professional competence of loss control manager.
3 Technical experience of loss control manager.

4 Aptitude and talents of loss control manager.
5 In-depth accident investigations.
6 Plant and facility inspection.
7 Laws, policies, standards.
8 Management group meetings.
9 Safety committee meetings.
10 General promotion through the use of posters, banners, signs.
11 Personal protection.
12 Supervisory training.
13 Employee training.
14 Selection and employment procedures.
15 Reference library.
16 Occupational health and hygiene.
17 Fire prevention and loss control.
18 Damage control.
19 Personal communications.
20 Job safety analysis.
21 Job safety observations.
22 Records and statistics.
23 Emergency care and first aid.
24 Product liability.
25 Off-the-job safety including on the road and at home.
26 Incident recall and analysis.
27 Transport including managers and salesmen driving cars.
28 Security.
29 Ergonomic applications.
30 Pollution and disaster control.

Each of the 30 areas needs to be evaluated to pinpoint where action is necessary to improve the organisation's control of losses. The evaluation should be undertaken by trained personnel using the technique of asking a series of questions related to each of the 30 areas. Up to 500 questions may be required to cover the 30 areas.

Thus, the first stage in loss control profiling is to develop the list of questions that relate specifically to the organisation under review.

The answer to each question is rated on a scale ranging from 0% for a bad to 100% for a good response. An evaluation of each of the 30 areas is then calculated by taking the average percentage of those answers relating to a particular area.

A percentage figure of 25% or less indicates those areas where immediate action needs to be taken. A percentage figure between 25% and 50% indicates those areas where there is a need for improvement within the near future. A percentage figure between 50% and 75% indicates those areas in which an acceptable level has been achieved, but in which there is still room for improvement. A percentage figure of 75% or more indicates those areas where the organisation is operating at optimum performance, but which have to be monitored to ensure that this performance is maintained.

The results of such evaluations can be presented graphically in the form of a horizontal bar chart where each subject area is shown on a separate line. Those areas giving cause for concern, i.e. the short lines, are immediately highlighted.

Fletcher and Douglas[23] and Fletcher[24] developed Bird's original ideas on profiling and formulated their own detailed evaluation questionnaire in which the answer to each question was rated on a six-point scale, ranging from 'fully implemented and fully effective' (score 5) to 'nothing done to date' (score 0). The scores of each question in a subject area are then summated, and the value is expressed as a percentage of the maximum attainable score. Loss control profiles are then constructed and utilised in a similar manner to that described above.

Once the losses – both actual and potential – have been evaluated, and a loss control profile developed, then – and only then – can a definite action programme of loss control be planned and implemented.

This would be based on assessing the deficiencies highlighted by the loss control evaluation and profile, then initiating a programme of work to make good those deficiencies.

Annual profiles may be undertaken to assess progress made, and also to ensure that all areas under review are maintained at an acceptable level.

References

1. Bird, F. E., and O'Shell, H. E., Incident recall, *National Safety News*, **100**, No. 4, 58–60 (1969)
2. Bamber, L., Incident recall – a (lack of) progress report, *Health and Safety at Work*, **2**, No. 9, 83 (1980)
3. Chemical Industries Association Ltd, *A Guide to Hazard and Operability Studies*, Chemical Industries Association Ltd, London (1977)
4. Bird, F. E. and Loftus, R. G., *Loss Control Management*, 464, Institute Press, Loganville, Georgia (1976)
5. Ref. 4, pp. 165–171
6. Ref. 4, p. 474
7. Ref. 4, p. 493 et seq.
8. Petersen, D. C., *Techniques of Safety Management*, 2nd edn, p. 174 et seq., McGraw-Hill, Kogakusha, USA (1978)
9. Kletz, T. A., *Hazop and Hazan: notes on the identification and assessment of hazards*, Institution of Chemical Engineers, Rugby (1983)
10. Kletz, T. A., Hazard analysis – its application to risks to the public at large (Part 1), *Occupational Safety & Health*, **7**, 10 (1977)
11. Industrial Relation Services, A systems approach to health and safety management, *Health & Safety Information Bulletin* No. 168, 5–6, Industrial Relations Services, London (1989)
12. BS 5750: 1987, *Quality systems*, British Standards Institution, London (1982)
13. Bird, F. E, *Management Guide to Loss Control*, 17, Institute Press, Atlanta, Georgia, (1974)
14. Carter, R. L., The use of non-injury accidents in risk identification, 4.6–01–4.6–05, *Handbook of Risk Management*, Kluwer Publishing Ltd, Kingston-upon-Thames (1992)
15. Morgan, P. and Davies, N., Cost of occupational accidents and diseases in GB, *Employment Gazette*, 477–485, HMSO (Nov. 1981)
16. Ref. 14, pp. 6.4–01–6.4–12
17. Health and Safety Executive, *Managing Safety, Occasional Paper Series No. OP3*, HSE Books, Sudbury (1981)
18. Health and Safety Executive, *Monitoring Safety, Occasional Paper Series No. OP9*, HSE Books, Sudbury (1985)
19. Health and Safety Executive, Publication No. H5(G)65, *Successful Health and Safety Management* (2nd edn.), HSE Books, Sudbury (1997)

20. Maslow, A. H., A theory of human motivation, *Psychological Review*, **50**, 370–396 (1943)
21. Ref. 13, pp. 151–165
22. Ref. 4, pp. 185–197
23. Fletcher, J. A. and Douglas, H. M., *Total Loss Control*, 113–154, Associated Business Programmes, London (1971)
24. Fletcher, J. A., *The Industrial Environment – Total Loss Control*, 18–122, National Profile Ltd, Willowdale, Ontario, (1972)
25. British Standards Institution, BS 7750: 1994, *Specification for environmental management systems*, BSI, London (1994)
26. British Standards Institution, BS EN ISO 14001: 1996, *Environmental management systems – Specification with guidance for use*, BSI, London (1996)
27. British Standards Institution, BS 8800: 1996, *Guide to occupational health and safety management systems*, BSI, London (1996)
28. British Standards Institution, BS EN ISO 9001: 2000, *Quality systems*, BSI, London (2000)
29. Fishwick, L. and Bamber, L., Common Ground – Practical ways of integrating the environment into your health and safety programme (Part 1), *Health and Safety at Work 1996*, **18**, 2, pp. 12–14
30. Fishwick, L. and Bamber, L., Common Ground – Practical ways of integrating the environment into your health and safety programme (Part 2), *Health and Safety at Work 1996*, **18**, 3, pp. 34/35
31. *Internal Control: Guidance for Directors on the Combined Code on Corporate Governance* – The Turnbull Guidance, Institute of Chartered Accountants in England and Wales, London (1999)

Further reading

Diekemper and Spartz, A quantitative and qualitative measurement of industrial safety activities, *J. Amer. Soc. Safety Engrs*, **15**, No. 12, 12–19 (1970)

Fine, W. T., Mathematical evaluation for controlling; hazards, *J. Safety Research*, **3**, No. 4, 57–166 (1971)

Chemical Industries Association Ltd and Chemical Industry Safety and Health Council, *A Guide to Hazard and Operability Studies*, Chemical Industries Association Ltd, London (1977)

Petersen, D.C., *Techniques of Safety Management*, 2nd edn, McGraw-Hill, Kogakusha, USA (1978)

Bird, F.E. and Loftus, R.G., *Loss Control Management*, Institute Press, Loganville, Georgia (1976)

Heinrich, H.W., Petersen, D. and Roos, N., *Industrial Accident Prevention – A Safety Management Approach*, 5th Edn, McGraw-Hill, New York (1980)

Dewis, M. *et al.*, *Product Liability*, Heinemann, London, (1980)

Carter, R.L. *et al.*, *Handbook of Risk Management*, Kluwer Publishing Ltd, Kingston-upon-Thames (1992)

Health and Safety Executive, *Report – Canvey: an investigation of potential hazards from operations in the Canvey Island/Thurrock area*, HSE Books, Sudbury (1978)

Health and Safety Executive, Advisory Committee on the Safety of Nuclear Installations (ACSNI), ACSNI Study Group on Human Factors, *2nd Report – Human Assessment: A Critical Overview*, HSE Books, Sudbury (1991)

Health and Safety Executive, Publication No. H5(G)65, *Successful Health and Safety Management* (2nd edn.), MSE Books, Sudbury (1997)

Chemical Industries Association, *Guidance on Safety, Occupational Health and Environmental Protection Auditing*, Chemical Industries Association, London (1991)

Health and Safety Commission, Publication L21, *Management of Health and Safety at Work, Approved Code of Practice: Management of Health and Safety at Work Regulations 1999*, HSE Books, Sudbury (1999)

Croner's Management of Business Risk, Croner, CCH Group Ltd., Kingston-upon-Thames (2000)

Chapter 2.5

The collection and use of accident and incident data

Dr A. J. Boyle

2.5.1 Introduction

Although the title of this chapter refers to accidents and incidents, there is no general agreement about how accidents and incidents should be defined. For this reason, the chapter begins with a discussion of the various types of data which might be included in these two categories and the practical implications of this discussion.

The remainder of the chapter is divided into the following main sections:

1 The collection of accident and incident data and the systems which have to be in place if accident and incident data are to be collected and recorded effectively.
2 The main uses of accident and incident data such as the techniques for learning from the analysis of aggregated accident and incident data, using trend analysis, comparisons of accident and incident data, and epidemiological analyses.
3 Lessons to be learnt from individual accidents and incidents by means of effective investigations.
4 The relationship between accident and incident data and risk assessment data. In addition, there is an appendix dealing with the UK legal requirements to notify accidents causing injuries of particular types and levels of severity.
5 The use of computers with accident and incident data.

2.5.2 Types of accident and incident data

A commonly used distinction between accidents and incidents is that accidents have a specific outcome, for example injuries or damage, while incidents have no outcome of this type, but could have had in slightly different circumstances. In the UK, the HSE uses the following definitions[1]:

- *Accident* includes any undesired circumstances which give rise to ill-health or injury; damage to property, plant, products or the environment; production losses or increased liabilities.
- *Incident* includes all undesired circumstances and 'near misses' which could cause accidents.

However, it is preferable to think of accidents and incidents as part of a single, much larger, group of undesired events or circumstances which varies in two main dimensions.

1 Qualitative differences of actual and potential outcomes, for example injuries, ill-health and damage.
2 Quantitative differences in outcomes, for example 'minor' injuries, 'major' injuries and damage.

Each of these dimensions will now be considered in more detail.

2.5.2.1 Qualitative differences

Table 2.5.1 shows some of the main categories of outcome and examples of each.

To a certain extent, the divisions shown in *Table 2.5.1* are artificial and some examples of overlaps are given below:

- A customer complaint may be about product safety.
- A spillage may result in injury, ill-health, and asset damage.
- An injury accident can also involve asset damage and damage to the environment.
- Some injuries, such as back injury, can result in chronic illness.

Table 2.5.1 The main types of accident and incident data

Quality	*Environment*	*Injuries*	*Health*	*Asset damage and other losses*
Customer complaints	Spillages	Injuries to employees at work	Sickness absence	Damage to organisation's assets
Product non-conformances	Emissions above consent levels	Injuries to others at work	Chronic illness	Damage to other people's assets
Service non-conformances	Discharges above consent levels	Injuries during travel	Sensitisation	Interruptions to production
		Injuries at home		Damage arising from unsafe products
		Injuries arising from unsafe products		Losses from theft and vandalism

However, despite these divisions being artificial, it is traditional to keep them separate, with different specialists dealing with particular types of outcome. This chapter will continue this tradition by restricting discussion to the following categories:

- Damage to people, including mental and physical damage, damage which occurs instantaneously (mostly injuries) and damage which is caused over a longer period of time (mainly ill-health).
- Damage to assets, including assets of the organisation, and assets belonging to other people which are damaged by the organisation's activities.

However, the principles described apply equally well to all categories and there should be little difficulty in generalising them if required.

There is one further point on qualitative differences. The preceding discussion has assumed that there is either an outcome (accidents) or no outcome (incidents). However, there is a type of outcome which does not fit either of these categories and that is the creation of a hazard. For example, people can quite safely lay cables across a walkway, but they have then created a hazard for themselves and others. It is not usual to deal with this type of outcome as part of a discussion of accidents and incidents but there appears no good reason for this exclusion and hazard creation will be discussed at relevant points in this chapter.

2.5.2.2 Quantitative differences

It is well known that accidents vary in severity, ranging from minor injuries through major injuries and ill-health to fatalities and catastrophic damage.

There is a relationship between the severity of the outcome and the frequency of the outcome. As the seriousness of the outcome increases, the frequency of that outcome decreases. This means that there are many more minor injuries, and cases of minor ill-health, than there are fatalities. In addition, there are many more 'near misses' than there are minor injuries or cases of minor ill-health. The sort of relationship which exists between frequency and severity is illustrated in *Figure 2.5.1*.

The relative numbers in this sort of relationship are not important, what is important is that it is recognised that there is a continuum from near miss to fatality and that definitions such 'minor', 'three day' and 'major' are arbitrary points on this continuum.

There have been various studies which have put numbers to the different categories of outcome and these are usually referred to as 'accident triangles'. A typical accident triangle is shown in *Figure 2.5.2*.

The figures given in *Figure 2.5.2* are from a study by Bird (1969), but this type of study goes back to 1931 (Heinrich). What this sort of diagram is intended to show is that for every major injury there are increasingly larger numbers of less serious losses. However, accident triangles of this type can be misleading because it is possible to have damage-only accidents which are very serious indeed in financial terms.

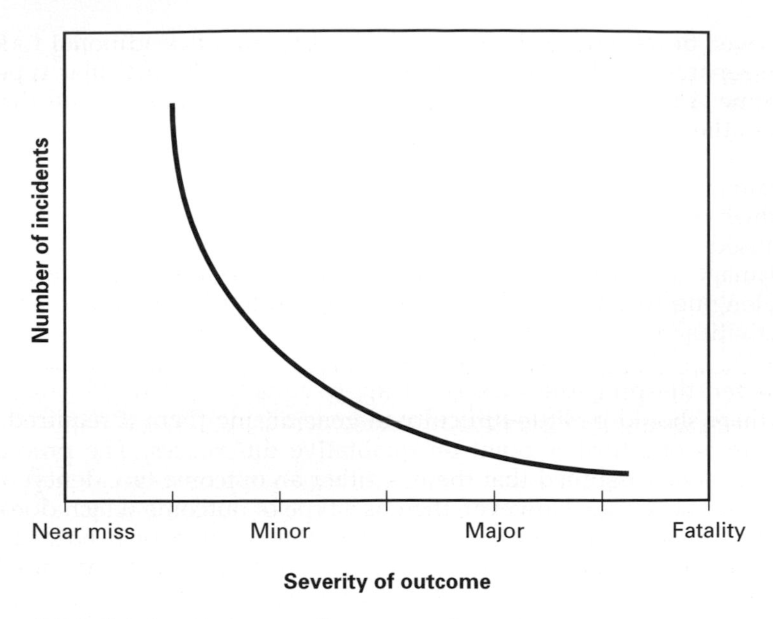

Figure 2.5.1 Relationship between frequency and severity

Since 1931, there have been various versions of the accident triangle, with different incident categories and different numbers. Several such studies are reported in the HSE publication[2], and these are summarised below:

	'Over 3 days'	*'Minor'*	*'Non-injury'*
Construction	1	56+	3570+
Creamery	1	5	148
Oil platform	1	4	126
Hospital	1	10	195

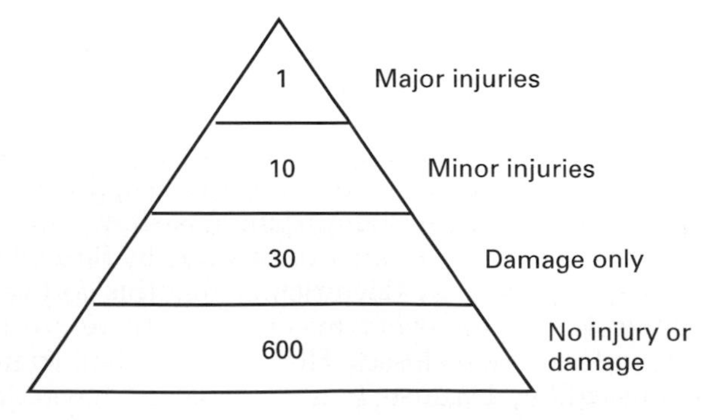

Figure 2.5.2 Accident triangle

These results demonstrate that the ratios for different outcomes differ from industry to industry and this is likely to be due to the different ranges of risk involved. Note, however, that they continue the conflation of the qualitative and quantitative dimensions.

2.5.2.3 Practical implications

The preceding discussion has demonstrated that accidents and incidents are subsets of a much wider category of events and that there is no universally agreed definition of the subsets involved.

This means that the first stage in any work on accident and incident data should be a clear definition of the particular subset which will be used. To be of practical value, the definition should deal with the following:

1 The nature of the outcomes to be included, for example injury, sickness or damage to assets and, in particular, whether hazard creation will be considered as an outcome.
2 The severity of the outcomes to be included. This can be difficult since it will usually involve specifying a more or less arbitrary point on a continuum and a decision will have to be made on, for example, near misses.
3 The population to be covered. For example, will it be restricted to the organisation's personnel and assets or will contractors and members of the public be included?

2.5.3 Collection of accident and incident data

It is assumed for the purposes of this section that the collection of accident and incident data involves three main stages:

1 Ensuring that accidents and incidents are reported. Unless individuals know what accidents and incidents to report, and how to report them, the relevant data are unlikely to be collected.
2 Checking for non-reporting of the types of accident or incident which should be reported and recorded.
3 Recording details of the accidents and incidents which are reported. If the data collected on accidents and incidents are to be of real value, they have to be recorded accurately.

However, before starting a discussion it is necessary to consider some problems with terminology and, in particular, the way 'accident reporting' and 'incident reporting' are used to mean a number of different things.

If we consider the chronology of an accident and its aftermath, we have the following identifiable stages:

1 The person who sustains the injury, or someone else, *reports* that an accident has happened, usually to a supervisor or manager.

2 The person to whom the accident is reported, makes a written record of the salient points, usually on an *accident report form* or *accident record form*.
3 The accident is investigated and, if it is sufficiently serious, is *reported* to the relevant national authority, for example the Health and Safety Executive in the UK.
4 The person who investigates the accident writes a *report* on his or her findings, to which are added any suggestions for remedial action.
5 The person who investigates the accident *reports back* to those involved in the outcome of the investigation and the action to be taken.

Since it is not always obvious from the context which of these uses of 'report' is the relevant one, the following terminology will be used in this chapter:

- *Accident report*. The report made by the person who sustains the injury, or someone else on their behalf.
- *Accident record*. The written record of the salient points, usually on an *accident record* form.
- *Accident notification*. The notification of an accident to the relevant national authority, for example the Health and Safety Executive in the UK.
- *Investigation report*. The written report on the findings of an accident investigation, together with any suggestions for remedial action.
- *Feedback*. The reporting back to those involved on the outcome of the investigation and the action to be taken.

2.5.3.1 Ensuring accidents and incidents are reported

In general, the less serious an incident is, the less likely it is to be reported. It is extremely difficult in most organisations to 'cover up' a fatality or major injury, but minor injuries often go unreported. However, there are various things which can be done to improve reporting and these are described below.

Have a 'user friendly' system. Reporting and recording systems which are too onerous for the quantity of data to be collected will not be used. For example, using 'major' accident forms for collecting information on 'minor' accidents or incidents will discourage reporting of minor accidents and incidents because the amount of effort required is not perceived as being commensurate with the seriousness of the accident.

Emphasise continuous improvement. The reasons for collecting the data (continuous improvement and prevention of recurrence) should be clearly stated and repeated often.

Avoid a 'blame culture'. If accident or incident reports are followed by disciplinary action or other minor forms of 'blame', people will stop reporting.

Demonstrate that the data are used. If people who have to report and record cannot see that use is being made of their efforts, they will stop making the effort.

Always give feedback. It is not always possible or necessary to take action on a report, but there should be feedback to the people concerned explaining the action being taken or reasons for the lack of action.

The practicalities of implementing these various points will vary from organisation to organisation but any weaknesses in accident and incident reporting systems can usually be identified quite easily in the course of a straightforward review of the systems against the criteria listed above.

2.5.3.2 Checking for non-reporting

Where it is important to have an accurate measure of the occurrence of a particular category of accident or incident, checks should be made that all of the relevant accidents or incidents are being reported. Three methods of carrying out such checks are described below:

1 Interviews with people who are likely to have experience or knowledge of the relevant accidents or incidents. People are more willing to talk about accidents or incidents they did not report if they are confident that there will be no adverse consequence as a result of their revelations. They are even more willing to talk about accidents or incidents other people did not report, if they know it will not result in adverse consequences for the people being identified. A skilled interviewer who has carried out an appropriate sample of interviews should be able to make a reasonably accurate assessment of the proportion of accidents or incidents which is going unreported.
2 Inspections of locations and people. The simplest example in this category involves inspecting plant and equipment for damage and comparing the findings from the inspection with the most up-to-date damage records. A similar approach can be used for minor injuries by, for example, inspecting people's hands, checking for dressings, cuts, grazes and burns and then comparing the inspection findings with the injury records. However, this approach can engender resentment and should be undertaken with care. In some organisations, the dressing for minor injuries are a characteristic colour, easily recognised even from a distance. The use of this type of dressing makes inspections for minor injuries much easier.
3 Cross-checking one set of records with another. The usefulness of this type of technique will depend on the records available and their accuracy, but possible cross-checking includes the following:

- Where there are records of what is taken from a first aid box, these can be checked against injury records to see whether everyone who has made use of the first aid box has reported an injury.

- Where there are records of plant and equipment maintenance, these can be checked against records of accidental damage to plant and equipment to see whether all of the relevant repairs which have had to be carried out have been recorded as accidents.
- Where records of 'cradle to grave' or 'mass transfer' are available for particular chemicals or substances, these can be used to check whether unexplained losses of chemicals or substances have appeared as, for example, accidental spillage records.

Any, or all, of the above techniques can be used to check the adequacy of reporting and to ensure 'good' data are available for analysis.

2.5.3.3 Recording details of accidents and incidents

Details of accidents and incidents are usually recorded on some type of form and the design of this form can have a marked influence on what gets reported. When designing for recording accident data the following points should be taken into account:

- The form should require only those data which it is reasonable to expect people will be willing to record. Many forms are designed to cover all of the eventualities of a major injury with boxes for whether the accident has been notified, next of kin, and a range of other details. People are then expected to use this form to record details of a cut finger! It could be argued that there is no need for a form for serious accidents since they will all be investigated in detail and an investigation report written. Even if a form is considered necessary for serious accidents, it should be used only for the purpose for which it was designed, and a separate form designed to use for the recording of less serious accidents and incidents.
- The form should require only those data which it is reasonable to expect people to be competent to provide. Many forms include spaces for such things as 'root cause of accident' and 'suggestions for risk control measures' and expect them to be filled in by people without the competences to provide accurate data. In the worst cases, these data are then analysed as though they had the same validity as the data it is reasonable to expect will be accurate, such as time of incident and part of body injured. There are two solutions to this, either omit from the form items requiring judgement, or provide the necessary competences.
- Ensuring completion of the form should be the responsibility of people at an appropriate level in the organisation. It is reasonable to expect work people to report minor injuries, near misses and hazards but it is not necessarily the case that these people are willing or able to record the details necessary for effective analysis.

There are various ways of meeting the requirements listed above but they all depend on well-designed forms and well-thought out systems for reporting and recording. A suitably designed accident report form may also be accepted by insurers as notification of a claim.

2.5.4 Legal requirements to notify accidents and incidents

Accident notification requirements are specific to a particular country and readers outside the UK should identify the requirements of their local legislation, or requirements imposed through other means. A detailed study of those reporting requirements will be necessary for safety practitioners and those responsible for reporting. The requirements for accident reporting in the UK are summarised in the appendix at the end of this chapter.

2.5.5 The use of accident and incident data

2.5.5.1 Introduction

In this section a more detailed look is taken at how accident and incident data can be used to learn from what has gone wrong in the past so that risk control measures can be implemented or improved. There are three main aspects:

1 Measuring whether performance is improving or deteriorating using trend analysis.
2 Making comparisons using accident and incident data.
3 Learning from accident and incident occurrence by using epidemiological analysis.

2.5.5.2 Trend analysis

By making continuous measurements of the numbers of accidents and incidents, it is possible to make comparisons of performance in different time periods and compare one with another, that is, carry out trend analysis over time. However, any such analysis can be influenced by changes other than changes in the effectiveness of the safety management. For example, if an organisation is reducing the amount of work it does it is likely that the number of accidents will decrease, whether or not there are any changes in safety management practices. This is an obvious, but important, point. If a press operates one million times, or delivery drivers drive one million miles, there is a certain scope for accidents. If a reduction in work halves the number of press operations, or the number of miles driven, the scope for accidents is reduced. Similarly, if the amount of work being done is increasing, we would expect the number of accidents to increase.

Because numbers of accidents can be influenced by these sorts of changes, the trend analysis will be dealt with in two stages. First, assuming that there is a steady state, with no relevant changes, this will allow study of the basic techniques without undue complication. Second, the techniques required to take into account the sorts of changes described above will be considered.

2.5.5.2.1 Trend analysis with a steady state

The most straightforward method of trend analysis is to plot the numbers of accidents or incidents against a suitable measure of time. However, it is also possible to plot on a graph not just the numbers of accidents and incidents but also some measure of severity. Typical examples of this sort of measure include days lost through sickness, litres of lost fuel, cost of damage repair, etc. Typical time measures include monthly, quarterly and annually. *Figures 2.5.3* to *2.5.6* show examples of these sorts of plots.

One practical problem with graphs is that the more detailed they are, the more difficult it is to judge the trend 'by eye'.

Compare the two graphs shown in *Figures 2.5.7* and *2.5.8*.

Although using the same data for both *Figures 2.5.7* and *2.5.8*, it is easier to see from the quarterly plot that there appears to be a slight downward trend. Generally grouping data in this way 'smooths out' variations and makes trends easier to identify. One technique for smoothing data, the

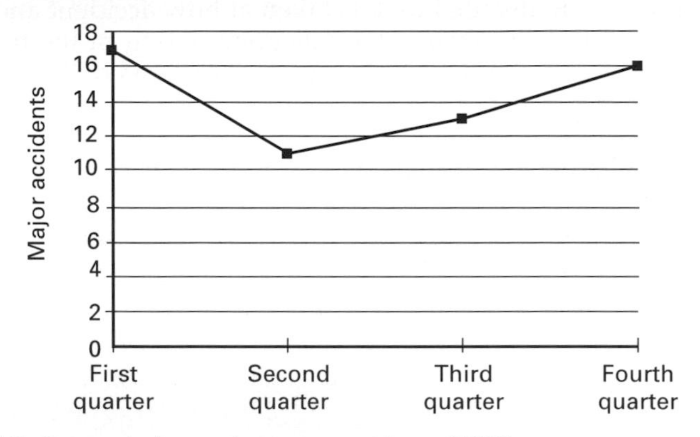

Figure 2.5.3 Quarterly figures for major accidents (1998)

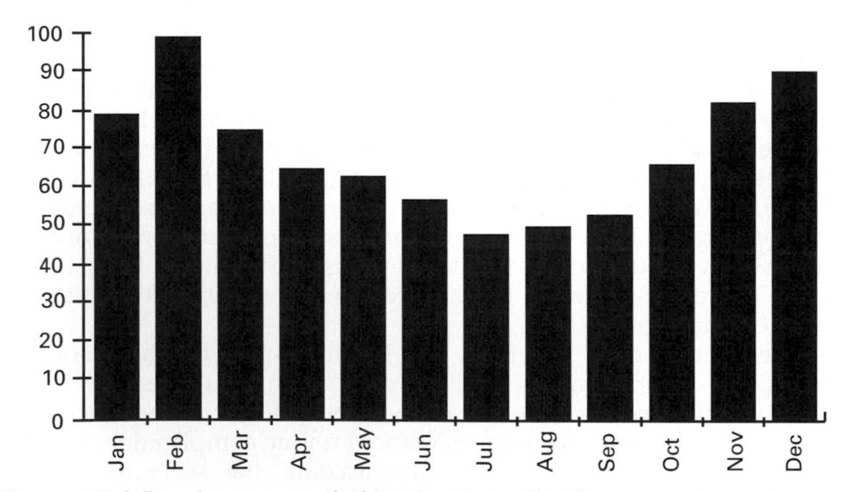

Figure 2.5.4 Days lost per month through sickness (1998)

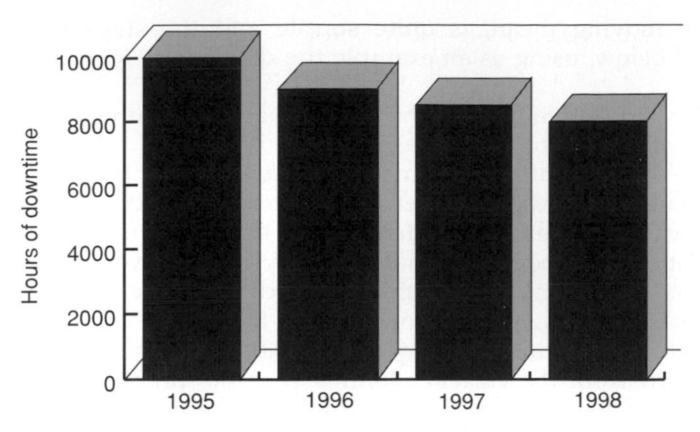

Figure 2.5.5 Hours downtime by year (1995 to 1998)

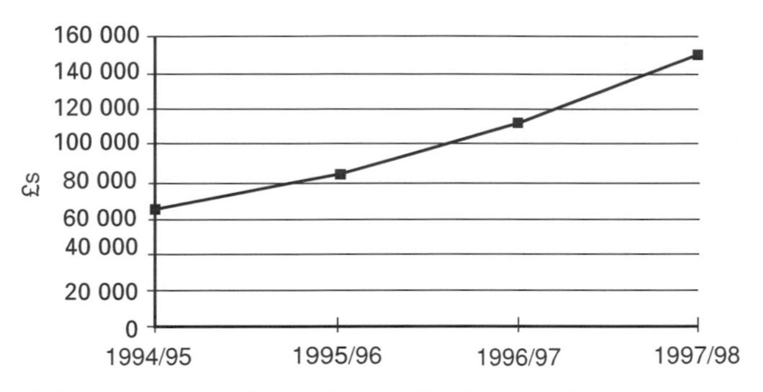

Figure 2.5.6 Damage costs – financial years 1994/95 to 1997/98

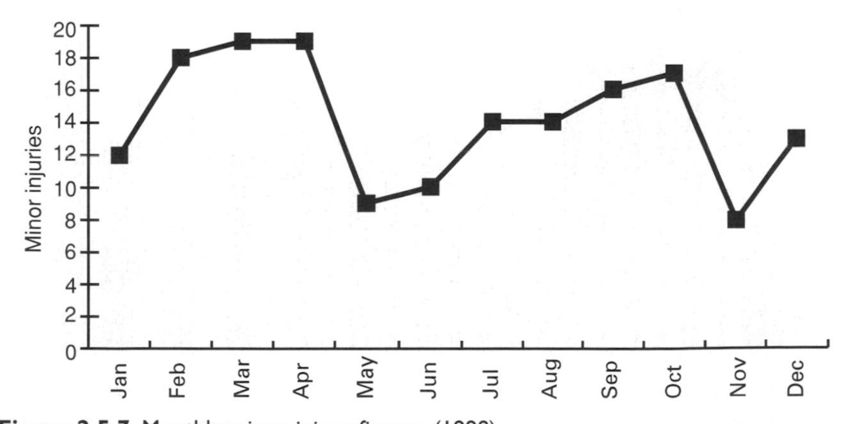

Figure 2.5.7 Monthly minor injury figures (1998)

quarterly moving mean, is quite simple and the steps required are described below, using as an example the data from *Figure 2.5.7*.

- For the first two months of the year, the accident numbers are plotted on a month-by-month basis as in *Figure 2.5.7*.
- For the third month, the numbers of accidents for January, February and March are added together and the result divided by three to give the mean number of accidents and this is plotted[1]. While this calculation may be more familiar as giving the 'average' number of accidents per month, it should be noted that there are a number of different averages, only one of which is the mean.
- For the fourth, and subsequent months, the quarterly mean is calculated from the current month's plus the previous two months' figures divided by three and the results of these calculations plotted.

The quarterly moving mean for the data in *Figure 2.5.7* is illustrated in *Figure 2.5.9*. Note that it is usual to plot both the actual monthly figures as well as the moving mean, and this has been done in *Figure 2.5.9*.

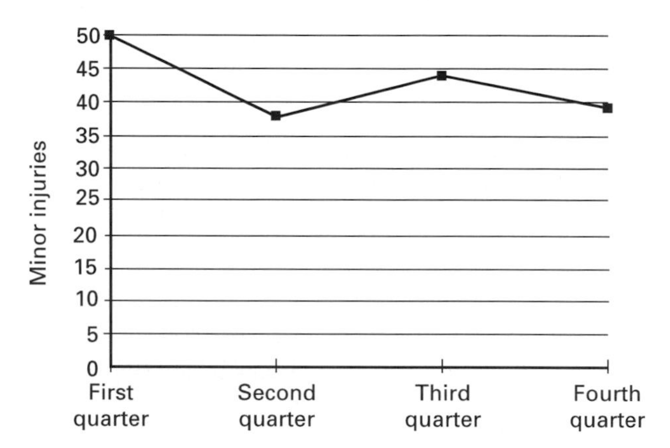

Figure 2.5.8 Quarterly minor injury figures (1998)

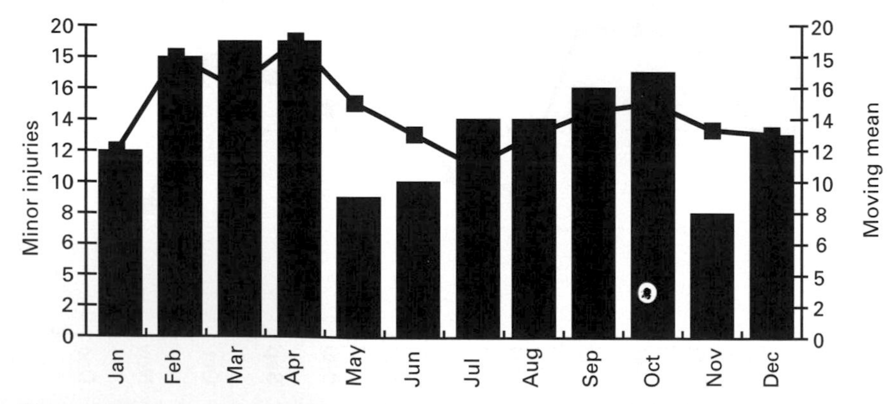

Figure 2.5.9 Monthly accidents and quarterly moving mean (1998)

Whichever method of trend analysis is used, a check should be made that any change in direction is more than a random fluctuation.

Suppose that in a particular year there were 100 accidents in a company and that in the following year the company proposed to carry out the same amount of work with no changes which would affect risk. In these circumstances, we would expect around 100 accidents in the year following the one for which records were available. Note that we would not expect exactly 100 accidents, but *around* 100 accidents. If there were 99 accidents or 101 accidents we would be able to say that this was due to random fluctuation and, more generally, anything between say 95 and 105 accidents could also be random.

The difficulty arises when the number of accidents reaches 85 or 90. Are these numbers due to random fluctuation, or is someone doing something which is improving risk control and influencing the accident numbers? Statisticians refer to fluctuations in numbers which cannot reasonably be attributed to random fluctuation as 'significant' when they may make statements like: 'There is only a 5% chance that the improvement in accident performance is due to random fluctuations', or 'This deterioration in accident performance would have occurred by chance in only 1% of cases'.

The working out of the significance of fluctuations in numbers has practical importance in the more advanced techniques of loss management since we can only draw valid conclusions when we know whether or not particular fluctuations in numbers are significant. For this reason, it is valuable to have some idea of the significance of fluctuations and trends. One way of doing this is to use historical accident data and work out upper and lower limit lines, based on the mean of these data. If we used this technique on the data shown in *Figure 2.5.9*, we could draw up a chart for 2002 which would look like the one shown in *Figure 2.5.10*.

As the monthly accident figures for 2002 become available, they are plotted on the chart in the usual way. Monthly numbers of accidents which are within the limit lines are defined as random fluctuations. Only if the number of accidents is above the upper limit line, or below the lower limit line, is the fluctuation considered significant.

Using this type of upper and lower limit line has practical advantages since it can prevent resources being expended on attempts to reduce

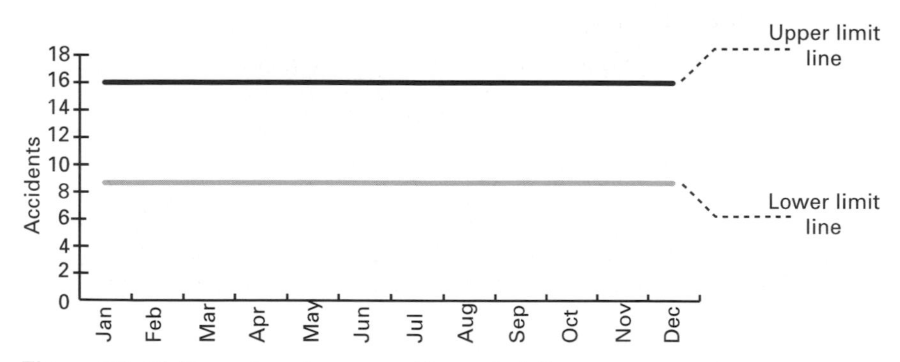

Figure 2.5.10 Illustration of upper and lower limit lines

increases in accidents which are purely random. While it might be argued that no resources spent on attempts to reduce accidents are wasted, resources are always limited and it is preferable to use them where there is good statistical evidence that they will do the most good.

The details of the calculations required for upper and lower limit lines, and related statistical techniques such as confidence limits and control charts, are beyond the scope of the present chapter, but details can be found in references 4 to 7 inclusive.

2.5.5.2.2 Trend analysis with variable conditions

So far, for the sake of simplicity, we have assumed that everything has remained stable in the organisation. In the real world, however, things rarely remain the same for any length of time and we need methods of trend analysis which can take this into account.

In an ideal world, we would be able to measure changes in risk in an organisation and hence determine how well the risk was being managed. For example, the measures would enable us to say such things as 'despite a 50% increase in risk due to additional work being done, the accidents increased by only 25%', or 'there was a 10% reduction in risk because of the new machines and work procedures, but accidents increased by 5%'.

Unfortunately, it is rarely possible to measure risk in this sort of way so what we have to do in practice is to find some proxy for risk which we can measure and use instead. Two such proxy measures in common use are numbers of people employed and numbers of hours worked which are used to calculate two accident rates.

- *Incidence rate.* This index gives the number of accidents for 1000 employees and is used to take into account variations in the size of the workforce:

$$\text{Incidence rate} = \frac{\text{Number of accidents} \times 1000}{\text{Number employed}}$$

- *Frequency rate.* This index gives the number of accidents for every 100 000 hours worked and takes into account variations in the amount of work done, and allows for part-time employees.

$$\text{Frequency rate} = \frac{\text{Number of accidents} \times 100\,000}{\text{Number of hours worked}}$$

There are, however, a number of problems with these accident rates.

- *Terminology.* Although the versions given above are in general use, there is no universal agreement as to the basic formula. A rate cannot be interpreted unless the equation on which it was based is known.
- *Definitions.* There is no general agreement on what constitutes an 'accident', with some organisations basing their rates on only major

injuries, while others use both major and minor injuries. Similarly, there is no general agreement on what constitutes an employee – incidence rates can be reduced by employing more part-time people! Hours present similar problems, with different types of hours having significantly different types of risk. For example, 'working' time, when the risk is high, versus 'waiting' time, when the risk is low. Also, 'staff' do not normally book their time although they can face the same risks as hourly paid employees.
- *Multipliers*. There is no general agreement on which multipliers should be used and it is normal to select one that suits the particular organisation.

In general, the value of accident rate figures depends on the quality of the data on which they are based and the honesty of the person preparing them. Quoted rates should always be treated with caution until the basis of the calculation has been determined.

Using *incidence* and *frequency rates* enable sensible trend analyses to be carried out during periods when changes are being made in the organisation that affect the number employed or the amount of work being done and, used properly, they can provide useful safety information.

These rates also enable us to make comparisons between one organisation and another, or between different parts of the same organisation, i.e. enable comparative analyses to be made.

2.5.5.2.3 Comparisons of accident data

It is only possible to make valid comparisons when there is some measure of the risk being managed. When the numbers employed or the hours worked are taken into account, these are only a proxy for risk and are used because we can measure them, rather than because they are good indicators of risk.

Considering two organisations, each with a frequency rate of 100, this could be because:

- The organisations have roughly equal levels of risk and are managing them equally effectively.
- One organisation has high levels of risk and is managing them well, while the other organisation has low levels of risk and is managing them badly.

This should be borne in mind when making, or interpreting, comparisons of accident data since it is a fundamental weakness of such comparisons. In general, a comparison will be valid only to the extent that the risk levels in the organisations being compared are equal.

Having dealt with this caveat, the types of comparison which can, with reason, be made are:

- *Comparisons between parts of the same organisation*. In theory, these are the simplest and potentially most accurate comparisons. This is because the measurement of risk, the definition of what has to be reported, reporting procedures, and methods of calculation are all under the organisation's control and can be standardised. However, the value of this comparison depends on the effectiveness of the reporting, which may not be consistent throughout the organisation.

- *Comparisons between one organisation and another.* Industries in the same sector can compare accident data one with another, assuming that they are willing to do so. In the UK, for example, there are national associations for particular industry and service sectors which provide a forum for comparing accident data. More formally, there have been moves recently to include accident data in 'benchmarking' exercises where organisations compare various aspects of their performance with those of their competitors.
- *Comparisons between an organisation and the relevant industry or service sector.* Some trade organisations publish aggregated data on accidents for their industry or service sector giving, for example, the 'average' frequency and incidence rates for a particular year. Examples of these sorts of accident data for the UK are published by the HSE annually[8]. The HSE's Epidemiology and Medical Statistics Unit also produces statistics on occupational ill health.
- *Comparisons between countries.* Where appropriate data are available, comparisons can be made between accidents in one country and another, either for the country as a whole, or by industry or service sector. However, there are major variations in accident reporting procedures between countries so that comparisons of this type should be made with great care.

A particular problem with all of these comparisons is that there is no consistency about what constitutes an 'accident' and it should be remembered that this was one of the problems with any comparison of incidence and frequency rates. One way of improving comparisons is to calculate a rate which takes into account the severity of the accidents, i.e. the number of days lost per accident, to give the mean duration rate:

$$\text{Mean duration rate} = \frac{\text{Number of days lost as a result of } x \text{ accidents}}{x \text{ accidents}}$$

This mean duration rate can be used in trend analysis in the same way as other rates. A disadvantage of it is that it can give a misleading picture since it can show a decrease when the numbers of days lost is increasing, i.e. more accidents but fewer days lost per accident.

For this reason, some organisations use an alternative severity rate:

$$\text{Severity rate} = \frac{\text{Number of days lost as a result of accidents}}{\text{Number of hours worked}}$$

The final point to make on comparisons is that the rates described above should, when the relevant data are available, be used in conjunction with each other. This is because they do not necessarily give the same result, as is illustrated, using simplified data, in *Table 2.5.2*.

Table 2.5.2 Comparisons using incidence, frequency and severity rates

	A	*B*	*C*	*D*
Number of accidents	100	80	60	20
Numbers employed	100	40	60	20
Incident rate	1000	2000	1000	1000
Hours worked	10 000	8000	3000	2000
Frequency rate	1000	1000	2000	1000
Days lost	100	80	60	40
Mean duration rate	1	1	1	2
Severity rate	0.01	0.01	0.02	0.02

2.5.5.2.4 Accidents and incidents as a measure of risk

Accurate accident and incident data will provide a measure of what has gone wrong in the past, and allow comparisons over time (trend analyses) and comparisons between one organisation and another. What these data will not do, even if they are accurate, is to provide a measure of risk.

Information on the number of accidents gives us very little information about risk. Two organisations can have the same number of accidents because one is managing high levels of risk very well, while the other is managing low levels of risk very badly. Alternatively, because risk is probabilistic, two organisations with the same levels of risk can have widely different numbers of accidents because one was 'lucky' and the other was not.

True levels of risk in an organisation can only be determined accurately using appropriate risk assessment methodologies, details of which will be found elsewhere in this book. However, more detailed discussion of the relationship between accident and incident data and risk assessment data will be found in section 2.5.8 of this chapter.

2.5.6 Epidemiological analysis

2.5.6.1 Introduction

The techniques of epidemiological analysis were first applied to the study of disease epidemics and historical example will be looked at by way of illustration to show how epidemiological techniques can be applied to accident and incident data.

Typhoid plague was a major cause of death in cities for many years. No one knew what caused the plague but many doctors looked for patterns in where the epidemics occurred. This was done on a trial and error basis with different people looking at where plague victims lived, what they ate, and the work they did. Eventually it was discovered that plague epidemics were centred around certain wells from which the city dwellers of those days obtained their drinking water. It was also found that closing these wells stopped the spread of the plague in those areas. Although no one knew why the wells, or the water from them, was causing the plague, they had found an effective way of stopping the plague spreading. In fact, it was many years before the water-borne organisms responsible for plague infection were identified.

This example illustrates the essential elements of epidemiological analysis. It is the identification, usually by trial and error, of patterns in the occurrence of a problem which is being investigated. These patterns can then be analysed to see whether causal factors can be identified and remedial action taken.

Epidemiology is used to identify problems which would not be apparent from single incidents. For example, if accidents occurred more frequently at a particular type of location, the records provide a guide to where investigation will be most fruitful and cost effective, although they provide no information on the possible causes.

2.5.6.2 Techniques of epidemiological analysis

Epidemiological analysis is only possible when the same type of information (data dimension) is available for all (or a substantial portion) of the accidents being analysed. Typical data dimensions include location and time of the accident or incident, the part of the body injured in an accident and the nature of the injury.

The simplest form of epidemiological analysis is *single dimension analysis*. This involves comparing incidents in the population on a single data dimension, for example time of occurrence or nature of injury. The analyst would look for any deviation from what would reasonably be expected. For example, if work is spread evenly over the working day, we would expect times of injuries also to be spread evenly. Where peaks and troughs are found in accident occurrences, these should be investigated. The analysis is slightly more complicated when an even spread is not expected as the analyst has to carry out preliminary work to determine the expected spread.

The analyst will look for both over-representation and under-representation when carrying out the analysis. Both should be investigated, over-representation because it suggests that there are risks which are being managed poorly, under-representation since it suggests either a degradation in the reporting and recording system, or particularly effective management of risk from which others might learn.

The principles and practices described above for single dimension analysis can also be applied to two or more dimensions analysed simultaneously, this is referred to as *multi-dimensional analysis*. This type of analysis can identify patterns which would not be apparent from analysing the data dimensions separately and examples include part of body injured analysed with department, and time of day analysed with nature of injury.

Full-scale epidemiological analysis of a set of data will involve analysis of all of the single data dimensions separately and analysis of all of the possible combinations of these single dimensions. For this reason, epidemiological analysis is a very time consuming process and where more than a trivial number of data are involved, the only practical approach is to use a computer. Suitable software for epidemiological analysis is described later in the chapter.

The epidemiological analysis merely identifies patterns in data distribution, it does not give information on why these patterns are occurring. This can only be determined by appropriate follow-up investigations and these are dealt with in the section on accident investigation.

2.5.6.3 Epidemiological analysis with limited data

The fact that the detailed data described earlier as necessary for full-scale epidemiological analysis does not prevent the techniques being applied to information that had already been gained.

Valuable results can often be obtained simply by tabulating accident data for the past two or three years and looking for patterns in accident occurrence. It is also worth trying to discover if there were no accidents for particular places, times, people, etc. since this can provide clues on non-reporting or effective risk control measures.

2.5.7 Accident investigation

2.5.7.1 Introduction

Accident investigations can be carried out for a number of reasons, including:

- Collecting the information required for reporting the accident to the enforcing authorities.
- Establishing where the fault lay.
- Obtaining the information required to pursue, or defend, a claim for damages.
- Obtaining the information necessary to prevent a recurrence.

In theory, a thorough investigation will result in the collection of the information required to satisfy all of these purposes but, in practice, this is rarely the case. If, for example, the primary purpose is to collect the information required for accident notification then the investigation is usually stopped when the relevant information has been collected, whether or not this information includes that required for the prevention of a recurrence. When the primary purpose is to establish where the fault lay, if this is allowed to extend to who was responsible, there may be an additional problem in that the investigation may become adversarial, that is, the investigators are on one 'side' or the other, for example the employer's 'side' or the injured person's 'side'. This can lead to biases in data collection with, for example, information which does not support a particular investigator's 'side' being ignored or not recorded.

The ideal investigation is, therefore, one which is neutral with respect to fault and has the primary purpose of obtaining the information necessary to prevent a recurrence.

In all accident investigations of this type there are two types of information to collect:

- Information about *what* happened which is usually factual and has limited scope for interpretation, for example the date and time of the incident, and what caused the injury, damage or other loss.
- Information about *why* it happened is concerned with the causes of the incident. It is more difficult to identify and more open to interpretation.

This distinction between 'what' and 'why' corresponds with the terminology used elsewhere to make roughly the same distinction. Typical terms include:

- *Immediate or proximate* causes are the direct causes of the injury, damage or other loss.
- *Underlying or root* causes are the reasons why the accident or incident happened.

These terms are used throughout the remainder of this chapter.

Collecting information about what happened is the essential first step in an investigation and must be completed before considering why it happened.

2.5.7.2 Collecting information on what happened

The two main sources of information are observation of the accident site and interviews with those involved (the injured person, witnesses, those who rendered assistance and so on). Observation of the site is fairly straightforward but interviewing is a skill which has to be learned. There are a number of key points to be followed for good interviewing.

2.5.7.2.1 Interviewing for accident investigations

There are three important aspects of interviewing which have to be considered:

- Coverage
- Keeping an open mind
- Getting people to talk.

(a) *Coverage*

This aspect of interviewing deals with the nature and amount of information which has to be collected, how to decide when all the relevant information has been obtained and how to avoid collecting information which is of no value?

What is relevant and valuable will, of course, depend on the purpose of the investigation and as a general guide, coverage should include all the information necessary to enable a decision to be made about the appropriate remedial action. However, in this first stage of the investigation, the purpose is to establish a clear idea of what happened. The information required falls into two categories:

1. Information which is common to all types of incident and which is best dealt with by using a pro-forma containing spaces for the information required. The accident record form used for this purpose should include information which gives:

   - Details of the incident – e.g. time, date and location.
   - Details of person injured – e.g. names, age, sex, occupation and experience.
   - Details of the injury – e.g. part of body injured, nature of injury (cut, burn, break etc.), the agent of injury (knife, fall, electricity etc.), and time lost.
   - Details of asset or environmental damage – e.g. what was damaged, nature of damage, and the agent of damage.

   It is this type of information which is best used for the sorts of analyses discussed earlier since it is common to all incidents and can, therefore, be used for trend, comparison and epidemiological analyses.
2. Other information has to be recorded as a narrative and space for this should be included on the accident record form. However, it is often necessary for this brief summary to be supplemented by a more detailed investigation report.

(b) *Keeping an open mind*

One of the main difficulties during an investigation is avoiding assumptions about what has happened. The greater the experience of the type of site involved, the nature of the work and the people, the more likely is it that assumptions will be made. There is always the possibility that an investigation will result in a summary of what was thought likely to have happened, rather than what actually happened.

To avoid making assumptions questions should be asked about all aspects of what happened, even if the answer is known. Perhaps even especially when confident of what the answer will be!

Making assumptions can lead to forming an inaccurate picture of what happened, which in turn can have serious implications if it leads to suggestions for remedial actions which are wholly inappropriate. Where possible remedial action is identified early in the investigation, this is a warning sign that too many assumptions may have been made.

(c) *Getting people to talk*

Interviewees will volunteer information more readily if a rapport can be established and maintained with them. Rapport is the term used to describe the relationship between people which enables a ready flow of conversation without nervousness or distrust. A wider range and more accurate information can be collected when a rapport has been established with the people being interviewed. There are no techniques which will guarantee that rapport is established, but the guidelines listed below will make it more likely:

(i) *Interview only one person at a time.* It is difficult to establish rapport with two or more people simultaneously since each will require different responses. This may not be possible in some circumstances, for example if the person interviewed requests that a representative attends. In these circumstances, the status of any attendees should be clearly established at the start of the interview including whether they are just observers, will be answering questions on the interviewee's behalf or whether they will be entitled to interrupt.

(ii) *Have only one interviewer at a time.* 'Board' or 'panel' interviews should be avoided since they require the interviewee to communicate with more than one person, and this is rarely successful. Note, however, that there are many circumstances where it may be necessary for more than one person to be involved in the interview. For example, the employee's representative may wish to be involved. In these circumstances, the interviewer should lead the interview and invite the second interviewer or representative to ask questions at an appropriate point. This procedure should be explained to the interviewee and his representative at the start of the interview so that it has a minimal effect on rapport.

(iii) *Introduce yourself and explain the purpose of the interview.* Do this even if you have already been introduced by someone else. The interviewee will gain confidence if he or she knows who you are and why the interview is taking place. Emphasise that the primary purpose of the interview is the prevention of a recurrence and that action will be taken on the results of the investigation.

(iv) *Check the interviewee's name and the part they played in the incident.* This may sound obvious but checking before the interview can save embarrassment later on. Confusion can arise when, for example, more than one person has been injured, where more than one accident has occurred in the same area, or where other interviews are in progress for some different purpose, for example work study.

(v) *Start the interview on the interviewee's home ground.* The idea is to start the interview with things which are familiar to the interviewee and hence establish a rapport, then move on to the details of the accident. This is helped by beginning the interview at the interviewee's place of work and talking about their normal job before moving on to discussion of the accident.

It is important to establish rapport before moving on to collect detailed information. If this is not done, the interview may degenerate into a series of stilted questions and one word answers. This can also happen if rapport is not maintained and there are a number of things which will help maintain rapport:

(vi) *Prevent interruptions.* Make sure the interview is not interrupted. Interruptions come from other people and an effective way of preventing this is by choosing a suitable place for the interview where interruptions are unlikely. However, the interviewer can interrupt the interview by stopping the interviewee to ask questions. In general it is best to let the interviewee talk and ask any questions when he or she gets to a natural break in their story.

(vii) *Use open questions rather than closed questions.* Open questions are ones which cannot be answered with 'yes' or 'no'; closed questions are ones which can be answered with a 'yes' or 'no'. For example, 'What was the noise level like?' is an open question, 'Was it noisy?' is a closed question. In general, closed questions should be used only to check on specific points already made by the interviewee.

(viii) *Avoid multiple questions.* For example, a question such as 'Can you tell me what everyone was doing at the time?' is better asked as a series of questions starting with 'Can you tell me who was there at the time?' and then a single question about what each of them was doing. Asking multiple questions is likely to result in only part of the question being answered.

(ix) *Keep your manner positive and uncritical.* Interviewees will form an opinion of your manner based on what you say and on your body language. Avoid expressing your views and opinions during the interview, especially if these are critical of what the interviewee has done or not done. Similarly, avoid such obvious signs of lack of interest as not listening, yawning or looking at your watch.

2.5.7.2.2 Recording the interview

It is essential that written notes are taken during an interview for a number of reasons:

(a) *So that what has been said is not forgotten.* Most people believe that their memory is much better than it really is. Few people can remember all the relevant facts raised during even a short interview.

(b) *So that there can be no confusion over what different people have said.* In most investigations more than one person will have to be interviewed and unless notes are made of each interview it is unlikely that who said what will be remembered, especially if there is a delay between the interviews and writing the report.

(c) *So that the interviewee's narrative is not interrupted.* The importance of not interrupting was mentioned earlier. It is a help in avoiding this if questions are written as they occur ready to be asked at a later and more suitable time. This means that the interview is not interrupted and the points to be raised are not forgotten.

Making notes during the interview is difficult at first but it is a skill, and like all skills can be learned with practice. This skill should be practised whenever possible, and the following should be borne in mind:

(i) *Timing*. Wait until rapport has been established before starting to take notes. Establishing rapport is difficult enough without the added distraction of note taking.
(ii) *Agreement*. Always tell the interviewee that notes will be taken and get their agreement to this.
(iii) *Content*. Make notes of everything that is said. Even parts of what the interviewee says that seem irrelevant should be recorded. Their relevance should be judged when all the information has been collected, from this and other interviews.
(iv) *Take your time*. Note taking shows the interviewees that what they are saying is of interest. They do not consider it an interruption and are usually happy to wait while notes are made.
(v) *Review*. At the end of the interview go over the notes with the interviewee checking that what has been written down is an accurate record of what has been said.

2.5.7.3 Collecting information on why things happen

Once what has happened in an accident has been clearly established, the reason why it happened (the causes) can be investigated. There are various approaches to ensuring adequate coverage of possible accident causes and three options are described below:

1 One or more of the models of human error, such as the one devised by Hale and Hale[9], are summaries of the ways in which human beings think and act and, in particular, how failures in thinking and acting can result in errors. Familiarity with models of this type will help structure an interviewer's approach to the human error aspects of the accident or incident.
2 The Domino Theory provides a succinct description of how the organisational aspects of accident and incident causes link with individual losses, and how human errors can be the result of organisational arrangements. Familiarity with this theory and its variants will help the interviewer avoid too narrow a concentration on the role of the injured person to the exclusion of broader organisational issues.
3 The approach described in an HSE publication[10] is particularly useful for those organisations which have adopted the HSE's Safety Management System since it facilitates the identification of accident and incident causes in terms of weaknesses in the existing Safety Management System.

The next section describes, in outline, how one of these approaches, the Domino Theory, can be used as a means of identifying more accurately the required remedial actions.

2.5.7.3.1 The Domino Theory

There are various versions of the Domino Theory and the one illustrated in *Figure 2.5.11* is a generalised version. The basic idea behind the Domino Theory is that individual errors take place in the context of organisations

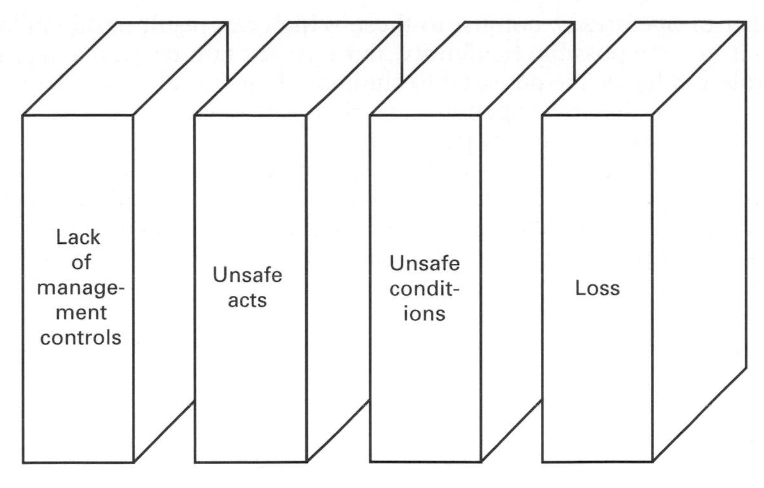

Figure 2.5.11 Generalised Domino Theory

and a useful concept for illustrating them is a series of dominoes standing on end.

If one of the dominoes to the left of the Loss domino falls, it will knock over those to the right and a loss will occur. For example:

- Lack of supervision (management control) results in a situation where oil can be spilt and not cleared up.
- An unsafe act occurs, spilling oil and not clearing it up.
- An unsafe condition results, a pool of oil on the floor.
- A loss occurs when someone slips on the oil, falls and breaks an arm.

When we investigate the loss, we can identify unsafe conditions, unsafe acts and lack of management controls and establish causes for these, as well as causes for the loss. Continuing the example:

- Possible causes of a person slipping on a patch of oil might be not looking where they were going, or not wearing appropriate footwear.
- Possible causes of not clearing up spilled oil might be lack of time, or not seeing it as part of the job.
- Possible causes of spilling oil might be working in a hurry, inappropriate implements or a poor method of work.
- Possible causes of poor management control might be excessive pressure for production (resulting in hurrying), lack of funding for proper implements, or insufficient attention to designing appropriate systems of work.

The further the cause of the accident is to the left of the dominoes, the greater the implications for lack of management control. By inference it follows that lack of appropriate systems of work may apply to a large number of operations, not just to those which can result in oil spillages. Thus it may be possible to identify and remedy failures in management controls and hence the potential to eliminate large numbers of losses. The usefulness of the investigation can, therefore, extend beyond simply preventing that accident happening again.

One way of doing this is to look systematically at what each of the dominoes represents, determine which one started the fall and concentrate investigation in that area. But it is important to remember that there is rarely a single function or action that causes a particular domino to fall, rather there are a number of reasons which contribute to the fall. There is a need to continue to ask why things happened until all of these contributory causes have been identified. The oil spillage example used earlier illustrates this.

Possible reasons for the person slipping on the oil were that he was not looking where he was going and that he was wearing inappropriate footwear. The question 'Why' should be asked about each of these to see whether further useful information can be obtained. For example, not wearing appropriate footwear could be because:

- he did not know he should be wearing special footwear
- he did not know which type of footwear was appropriate
- the appropriate footwear was uncomfortable
- the appropriate footwear was too expensive
- and so on.

The different answers to these 'why?' questions will have different implications for remedial action so it is important to establish the reason 'why' before making any recommendations. A similar technique should be applied to the other dominoes and again this can be illustrated using the oil spill example.

Unsafe condition. The possible reasons for not clearing up the oil spillage were lack of time and not seeing it as part of the job. Asking 'why?' about the lack of time could produce the following types of answer:

- management pressure
- piece work
- wanted to get home
- understaffing
- and so on.

Again, the remedial action suggested will depend on the answer obtained. There is little point in suggesting that people take time to clear spillages if management are continuing to insist on giving production priority.

Unsafe act. Possible reasons for the spillage of oil may include using inappropriate implements and using an inappropriate system of work. Asking 'why?' about the inappropriate system of work might produce the following types of answer:

- no one has prepared a system of work
- the people who do the work do not know about the system of work
- the recommended system of work is impractical
- the recommended system of work is out of date
- and so on.

As before, whichever reason is identified, it should be followed up so that any remedial action suggested is as relevant and practical as possible.

Lack of management controls. Possible reasons for the lack of management controls were the pressure for production, lack of funding and failure to produce written systems of work. Asking 'why?' about written systems of work might produce the following types of answer:

- no one knows it is necessary
- no one has the time
- no one has the skills
- no one has clear responsibility
- and so on.

It should be noted that as the basic investigation moves from the loss domino to the lack of management controls domino, a wider range of people will have to be interviewed. The injured person, for example, is unlikely to have the required information on lack of management controls. He or she can probably tell you about the effects of lack of controls but is unlikely to know the reasons why the controls are not in place.

Identifying who should be interviewed in the course of an investigation and knowing which questions to ask are matters of experience and practice and, as with the other skills aspects of accident investigation, they should be practised whenever possible.

Note also that if a manager is conducting an investigation into an accident within the area of another manager's control, a conflict of interests may arise. The person who should be implementing management controls may have a tendency to avoid going into details of the weaknesses in management control as thoroughly as might be required. In these circumstances it may be preferable to hand the investigation over to, or seek the assistance of, a neutral investigator.

Safety professionals have a related problem when there are requests from managers to take part in investigations. In general this is to be encouraged since it increases management involvement in safety matters, but it should be explained to these managers that they may have to be interviewed as part of the investigation if lack of management control is identified as an underlying cause.

2.5.7.4 Writing investigation reports

It is not always necessary to prepare a formal written report of an accident investigation, but where it is, the techniques of good report writing should be followed. Key points on report writing are given below.

It is preferable to prepare a draft report since this provides an opportunity to check that nothing has been omitted from the investigations. In particular, that the information is available for:

- making any statutory or other notifications
- making reasoned suggestions on measures for preventing recurrence, and
- any other tasks, e.g. completing insurance claims.

Some people find that drafting reports is best done with techniques such as system diagrams, system maps and flow charts and it is worth experimenting with these techniques to find out their suitability.

The first question to ask about a final report is whether or not it is necessary. Answers should be based on whether there is an audience, who they are and what this audience needs from the report. In many cases, a detailed draft report is adequate as a record, and as a basis for justifying remedial actions.

The following points should be covered where a final report is required:

(i) *Good signposting*. Any report, but especially a long one, will be difficult to read and action if the various sections are not clearly identified. If the report is intended for more than one audience, the sections relevant to particular audiences should be clearly identified.
(ii) *Separate fact and opinion*. Facts should be unarguable, opinions can, and should, be debatable. It is good practice to keep the two separate.
(iii) *Base opinions on the facts*. Conclusions should not be drawn which cannot clearly be supported by the facts presented, nor should conclusions be drawn which do not take all of the relevant facts into account.

2.5.7.4.1 Feedback of investigation results

The relevant results of investigations, including any recommendations for remedial action, should be fed back to all of the people who were involved in the investigation.

If this is not done, there may be the following detrimental effects:

- Subsequent investigations will be more difficult, and less information will be given, because people will have seen no results from helping with earlier investigations.
- Credibility will be damaged since people will have been told that the investigation is to prevent recurrence and they have received no instructions on the action to take.

Even though the results from a particular investigation indicate that no action needs to be taken, the results and the reasons for taking no action should be fed back to those who were involved.

2.5.7.5 Learning from minor incidents and near misses

It is often the case that only the more serious incidents are considered worthy of investigation. The rationale for this is usually that investigations take time and, therefore, cost money so that they are only worth doing when there has been a significant loss. However, various researchers have demonstrated that there is no relationship between the causes of accidents and the seriousness of the outcome and that, for example, minor injuries have the same range of causes as major injuries.

It follows from this that as much can be learned from investigating individual minor incidents and near misses as can be learned from investigating individual major injuries. Since it is also the case that there are many more minor incidents than major incidents, investigation of minor incidents gives us many more opportunities to learn from what has gone wrong.

Since there are so many minor incidents, we are left with the practical problem of the time required for adequate investigation of them all. There are two ways of dealing with this problem.

1 Provide managers with the competences to carry out proper investigations so that the required work is spread among a number of competent persons.
2 Identify patterns in minor incident occurrence and investigate groups of minor incidents which are likely to have related causes. How this pattern identification is carried out was described in section 2.5.6 on epidemiological analysis.

Assuming that managers can be trained in investigation techniques, the first option is to be preferred. However, the second option can provide an acceptable alternative and it should be used as a backup when managerial investigations are in place.

2.5.7.6 Advanced investigation techniques

Effective observation and interviewing will be adequate for the majority of investigations and, when combined with the discipline imposed by good report writing, should ensure appropriate recommendations. However, there will be certain accidents and incidents which, because of their complexity, require the use of more advanced techniques. A full description of these techniques is beyond the scope of this chapter but key points are as follows:

- Complex accidents may have many 'sites'. For example, the root cause of a road traffic accident may have occurred in a design office (for the car or for the road) many years before, and many miles away from, the fatal crash.
- The amount of information required to describe effectively a complex accident is likely to be beyond the scope of succinct narrative summary

so that some type of formal collation technique is to be preferred. A number of such techniques are available but Events and Causal Factors Analysis (ECFA) is the most straightforward and most generally useful.
- The production of effective recommendations for the prevention of recurrence is unlikely to be straightforward in these complex accidents and incidents so that techniques such as Fault Tree Analysis (FTA) may be required to analyse the causal sequences. In addition, creative thinking techniques such as brainstorming and systems thinking may be required to generate a suitable range of recommendations.

More extended discussion of these advanced techniques are given by Boyle[11].

2.5.8 Accident and incident data and risk assessment data

There are two types of accident and incident data to be considered, the aggregated data used for trend and epidemiological analyses and the data on single accidents and incidents collected during investigations. Each of these data types is dealt with separately.

2.5.8.1 Aggregated accident and incident data

As has already been mentioned accident and incident data do not provide a measure of risk. This is because the number of accidents and incidents depends on three factors:

(i) the underlying level of risk,
(ii) how well the risk is controlled, and
(iii) the operation of chance.

Risk assessment techniques are intended to estimate the first of these factors, the underlying level of risk, and should determine the number and the nature of the accidents and incidents which would occur if there were no risk control measures. However, there are two limitations with the techniques currently available.

1 The techniques are generally restricted to the assessment of a single activity or group of activities and there are no recognised methods for the aggregation of risks across an organisation. This is why comparisons between organisations are still based on proxies for risk such as numbers employed and hours worked.
2 The techniques are based on probabilities which many people find difficult to understand. In essence, the fact that an accident happens does not mean that the risk assessment was incorrect. For example, if it is correctly calculated that there is a very low likelihood of a multiple fatality, the fact that the multiple fatality occurs does not necessarily

mean that the estimate of likelihood was incorrect. Rather, it is the third of the factors listed above, i.e. the operation of chance. What is required in the longer term are numerical techniques for risk assessment which identify the underlying level of risk and the extent to which risk control measures will reduce the risk. It will then be possible to predict the number of accidents and incidents that will occur by chance and this can be compared with the numbers of accidents and incidents that do occur. It may then also be possible, by examining the accident and incident data in more detail, to determine whether any problems are due to an underestimate of the underlying level of risk or a failure to select or implement appropriate risk control measures. These are discussed by Boyle[11].

2.5.8.2 Data on single accidents and incidents

The investigation of each accident and incident should include a review of the relevant risk assessment or risk assessments. This review should include checks on the following:

- That the risk assessment has been carried out, reviewed at appropriate intervals and adequately documented. This is, in effect, a check on the operation of the risk assessment element of the Safety Management System and, where weaknesses are identified, suitable corrective action should be instigated.
- That the estimates of likelihood and severity on which the risk calculation was based were realistic, given the information available at the time. Again, this is a check on the risk assessment element of the Safety Management System but it is checking how well the risk assessment was carried out, not just whether it was carried out.
- That any recommendation for risk control measures would, had they been implemented, have effectively controlled the risk. Risk assessments should include a calculation of the extent to which the recommendation will reduce risk since if it is not possible to demonstrate a risk reduction, the recommendations are pointless.
- That any recommendations for risk control measures have been implemented and effectively maintained. Different organisations use different procedures for the implementation and maintenance of risk control measures but whatever procedures are used they should be checked.

All of these checks should be made in the context of the information available to the assessors at the time of the risk assessment since the purpose of the checks is to identify weaknesses in the current procedures for risk assessment and risk control. Once this has been done, the relevant risk assessment(s) can be reviewed in the light of the new information arising from the accident or incident investigation and, if necessary, the risk assessment can be revised. However, it should always be remembered that the occurrence of an accident or incident is not, *per se*, a demonstration that the risk assessment was incorrect.

2.5.9 The use of computers

2.5.9.1 Introduction

This section consists of a brief description of the sorts of computer software which are available for the recording and analysis of accident and incident data, and for a range of related data handling tasks. The criteria to be used in selecting software are also briefly discussed.

2.5.9.2 Hardware and system software

There are many types of computer (usually referred to as hardware) but the most common type is the personal computer (PC), either in its desktop form, or as a portable ('laptop' or 'notebook'). This discussion will, therefore, be restricted to software available on PCs.

Before any application program can be run on a PC it has to be equipped with system software. This software does a number of things but essentially it is an interface between the hardware and any application program to be run. The major practical value of system software is that people who write, for example, statistical programs do not have to produce a different version for each different type of hardware. Instead, they write a program for a particular type of system software. The most common system software is Windows in its various versions and this discussion will be restricted to software packages which run under Windows. However many of the points made will also apply to other systems software.

2.5.9.3 The nature of programs

The sorts of programs discussed all operate in essentially the same way. Each one provides a framework, or shell, into which data can be put and, for the present purposes, the programs can be classified according to the types of data they accept. The main categories are as follows:

- *Free format text, diagrams, pictures, tables, etc*. These data types are all accepted by programs such as word processors, desktop publishing packages and presentation packages.
- *Structured alphanumeric data*. These data types consist of mixed letters and numbers in a highly structured format of records and fields. Database programs accept these data types, including the specialised database programs used for specific purposes such as accident and incident recording.
- *Questions and answers*. This is a subcategory of the structured alphanumeric data but because it has special relevance to health and safety it is dealt with separately. Packages for active monitoring, audit, attitude surveys and measuring safety culture accept these types of data.

- *Numeric data*. Spreadsheets are the most common programs for numeric data but these types of data are also used by the specialised statistical packages.

There is always an overlap between programs, for example word processors will do elementary calculations. However, all programs are designed to deal primarily with a single data type. Specific programs are dealt with after some general points.

In theory, there could be one computer program which did everything but, in practice, the more a computer program does, the more difficult it is to learn and use. For this reason, program authors compromise in two main ways:

1 *Reducing functions*. This involves limiting the number of things thc program encompasses, for example the sorts of calculations that can be done using a word processor, or the level of word processing that can be done using a spreadsheet.
2 *Reducing flexibility*. This involves limiting the data the program will accept, or the number of things which can be done with these data. For example, any database program can be used for accident and incident recording but database programs are difficult to learn. A program designed solely for accident and incident data, although it is less flexible, should be much easier to learn and use.

However, the link between functionality and flexibility, and speed of learning and ease of use, depends on the skill of the software designer. Some very limited programs are badly designed and are difficult to learn and use, while some very powerful programs are relatively easy to learn and use.

2.5.9.4 Free format text programs

The main programs in this category include word processors and presentation packages.

So far as the present purposes are concerned, the primary use for these packages is for getting messages over to other people, either as a written report or as a presentation.

The key point to consider when selecting suitable software of this type is whether it will accept data directly from the other packages being used. Having to retype data, particularly numeric data, is tedious and error prone, and it is preferable to have a word processor and presentation package which will read data directly from the output of the other packages in use.

It is prudent to select for general use well-known packages such as Word for Windows (word processor) and Powerpoint (presentation package) since authors of other software are likely to ensure that the output from their programs will be compatible. However, there are a number of specialised packages which are of particular relevance to investigations since they facilitate the preparation of the diagrams used for ECFA, FTA and other related techniques. While these diagrams can be prepared in, for example 'Word', the work required is extensive for all but the simplest diagrams.

2.5.9.5 Structured alphanumeric data

The main programs in this category include general databases and databases designed for use with specific types of data such as accident records.

General databases such as Access have a very wide range of functions and are very flexible. However, they are difficult to use without some programming experience or the willingness to devote time to learning how to use them.

There are two separate stages in the use of general databases:

1 Setting up the database so that it will do the recording and analysis required. If, for example, a general database is to be used to record and analyse accident and incident data it would be necessary to set up the fields for recording such things as name of person injured, time of injury and number of days lost. This is specialised work requiring a high level of skill.
2 Entering data into the framework created in step 1. This requires a lower level of skill but, unless step 1 has been carried out properly, it will be highly error prone. For example, step 1 should include building in automatic checks on the data being entered with appropriate error messages when incorrect data are entered.

Because of the high levels of skill required to set up general databases for specific uses, it is not usually worthwhile for health and safety professionals to learn the skills required. What normally happens is that the health and safety professional specifies what is required and then hands over the work of setting up the database to the IT professionals who then produce a program which looks like a specific database when it is being used for data input and analysis.

Specific databases are available for a wide range of uses including the recording and analysis of data on accidents, risk assessments and various test results such as audiometry and LEV tests. Several different versions of each database type, which differ in function, flexibility and price, are available on the market.

The key selection strategies for these types of databases involves two main elements:

1 Being clear about what data are to be recorded and what analyses are to be carried out. Software suppliers will try to convince potential purchasers that their program does what is required, but this is not always the case. On the other hand, purchasing new software should be taken as an opportunity to review what is being

done by way of recording and using data since there is little point in computerising a poor paper system.

2 Looking to the long term. Many program demonstrations are carried out with just a few records on a highly specified computer and they appear fast and easy to use. Ask to see demonstrations involving the sort of computer you have with the numbers of records there will be in the system in two to three years' time. Some programs may be so slow as to be unusable.

The health and safety trade press carries advertisements for these types of specific databases and it is easy to get further information simply by contacting the suppliers.

2.5.9.6 Questions and answers

The main uses for programs of this type are the recording and analysis of active monitoring data, audit data, and data from surveys such as attitude or safety culture surveys. The strategy for the selection of these programs includes the points already made about specific databases, plus the following:

- *Flexibility of the question set*. Some programs are supplied with a set of questions which cannot be altered, while others can be supplied in a form which allows users to put in their own questions 'from scratch', or tailor a set of questions provided with the program. Fixed questions are fine so long as they exactly meet an organisation's requirements, but this is not often the case.
- *Use of more than one question set*. Some programs allow the use of only one set of questions (fixed or tailored) for all analyses while others allow the use of as many different sets of questions as may be required. The latter type of program is to be preferred when, for example, there is a wide range of risks and it is preferable to avoid asking people a lot of questions which do not apply to them.
- *Analysis options*. Some programs have very limited analysis options while others provide a range of alternatives. An important point to note is the extent to which the program allows 'labelling' of the answers to a particular set of questions. For programs designed for auditing, it may be adequate to have one label for each set of questions, usually the location which was audited. However, for attitude and safety culture surveys a range of labels will be required including, for example, department, level in the management hierarchy, length of time with the company and age.

2.5.9.7 Numeric data

Programs for numeric data are similar to databases in that they are split into general programs, i.e. spreadsheets, and programs which are designed to do specific things with numeric data, i.e. statistical packages.

So far as spreadsheets are concerned, the principles of their use and selection are the same as for general databases, although people in general tend to be more familiar with spreadsheet use.

There is a range of statistical packages available ranging from cheap and easy to use packages which will do most of the basic statistical tests to expensive, 'heavy weight' packages suitable only for the professional statistician. However, none of these packages will compensate for poor statistical technique. Easy, accurate calculation of confidence limits are of no value if incorrect types of confidence limits are being used.

2.5.9.8 Choosing software

Summarising the steps to take in choosing appropriate software of any type:

- Know the hardware and system software to be used since this will put restrictions on which programs can be used.
- Know exactly what is to be achieved by using the software. However, always take the opportunity to review the extent of the recording and analysis being carried out since the availability of software may make it possible to do more than is currently being done.
- Check what relevant software is on the market. This is probably best done by reading the health and safety trade press, or one of the many computer magazines.
- Get a demonstration of the software under conditions which match those under which it will be used. Many software houses will supply 'demonstration versions' which can be tried out on the computer setup to be used.
- Do a cost benefit analysis on the options available. It is unlikely that any package will exactly meet your requirements but remember that having a program written is likely to be several orders of magnitude more expensive than buying one 'off the shelf'. A decision may have to be made as to whether being able to do exactly what is required is worth the extra cost.

References

1. Health and Safety Executive, Guidance Book No. HSG 65, *Successful health and safety management*, HSE Books, Sudbury (1997)
2. Health and Safety Executive, Guidance Book No. HSG 96, *The costs of accidents at work*, HSE Books, Sudbury (1997)
3. Health and Safety Executive, Legal Series Book No. L 73, *A Guide to the Reporting of Injuries, Diseases and Dangerous Occurrences Regulations 1995*, HSE Books, Sudbury (1999)
4. Moroney, M.J., *Facts from Figures*, Penguin Books (1980)
5. Shipp, P.J., *The Presentation and Use of Injury Data*, British Iron and Steel Association. No date. (Out of print, but copies should be available through interlibrary loan services.)
6. Siegel, S.S., *Non-parametric Statistics for the Behavioural Sciences*, McGraw Hill (1956)
7. Whaler, D.J., *Understanding Variation – the Key to Managing Chaos*, SPC Press,

8. Health and Safety Commission, *Health and Safety Commission Annual Report, Statistical Supplement*, HSE Books, Sudbury (published annually)
9. Hale, A.R. and Hale, M., Accidents in perspective, *Occupational Pschology*, **44**, 115–121 (1970)
10. Appendix 5 of reference 1
11. Boyle, A.J., *Health and safety: Risk management*, IOSH Services Ltd., Sudbury (2000)

Appendix. UK requirements for reporting accidents and incidents

This appendix summarises the UK requirements for reporting accidents and incidents. It is only a summary and detailed study of the Regulations is essential for safety practitioners and those responsible for reporting accidents.

The Reporting of Injuries, Diseases and Dangerous Occurrences Regulations 1995 (RIDDOR) with its supporting guide[3] place duties on employers and the self-employed to report certain incidents which occur in the course of work. These reports are used by the enforcing authorities to identify trends in incident occurrence on a national basis. The reports also bring to the attention of the enforcing authorities serious incidents which they may wish to investigate. Reports must be made by the 'responsible person' who, depending on circumstances, may be an employer, a self-employed person, or the person in control of the premises where the work was being carried out.

The methods of reporting depend on the type of incident. For an incident resulting in any of outcomes listed in the table the relevant enforcing authority must be notified by the quickest practicable means, usually by telephone.

This notification must be followed by a written report within 10 days using Form F2508, details of which are given below. If they wish, the enforcing authorities can make a request for further information on any incident.

Dangerous occurrences are, in general, specific to particular types of machinery, equipment, occupations or processes and knowledge of the relevant incidents is necessary to ensure proper reporting. Some examples are given in the second part of the table illustrating the range of incidents involved.

An accident, other than one causing a major injury, which results in a person 'being incapacitated for work of a kind which he might reasonably be expected to do . . . for more than three consecutive days (excluding the day of the accident, but including days which would not have been working days)' is referred to as a 'three day' accident and is only required to be notified by a written report.

Fatalities, major injuries, dangerous occurrences and three day accidents have to be reported on Form F2508. The main requirements for information on this form are:

- Date and time of the accident or dangerous occurrence.
- For a person injured at work, full name, occupation and nature of injury.

Table of incidents to be reported by the quickest practicable means

Fatalities

Major injuries as listed below:

Fractures (other than finger, thumb or toe).

Amputations.

Dislocations of shoulder, hip, knee or spine.

Loss of sight (temporary or permanent).

Chemical or hot metal burn to the eye or any penetrating injury to the eye.

Electric shock or burn leading to unconsciousness, or requiring resuscitation, or admittance to hospital for more than 24 hours.

Any injury leading to hypothermia, heat induced illness or to unconsciousness, or requiring resuscitation, or requiring admittance to hospital for more than 24 hours.

Loss of consciousness caused by asphyxia or by exposure to a harmful substance or biological agent.

Either of the following conditions which result from the absorption of any substance by inhalation, ingestion or through the skin: (a) acute illness requiring medical treatment; (b) loss of consciousness.

Acute illness which requires medical treatment where there is reason to believe that this resulted from exposure to a biological agent or its toxins or infected material.

Dangerous occurrences. Specified incidents involving:

Lifting machinery (includes fork-lift trucks) – collapse, overturning, or failure of any load bearing part.

Pressure systems – failure of any closed vessel or associated pipework, where the failure has the potential to cause death.

Freight containers – failure of container or load bearing parts while it is being raised, lowered or suspended.

Overhead electric lines

Electrical short circuit leading to fire or explosion resulting in plant stoppage for more than 24 hours, or with the potential to cause death.

Explosives

Biological agents

Malfunction of radiation generators

Breathing apparatus

Diving operations

Collapse of scaffolding

Train collisions

Wells (NB not water wells)

Pipelines

Fairground equipment

Carrying of dangerous substances by road

Collapse of building or structure

Explosion or fire which results in stoppage or suspension of normal work for more than 24 hours where the explosion or fire was due to the ignition of any material.

Escape of flammable substance the sudden uncontrolled release, inside a building, of e.g., 100 kg or more of a flammable liquid or 10 kg or more of a flammable gas. If in the open air, 500 kg or more of flammable liquid or gas.

Escape of substances in any quantity sufficient to cause the death, major injury, or any other damage to the health of any person

- For a person not at work, full name, status (e.g. visitor, passenger) and nature of injury.
- Place where incident happened, brief description of the circumstances, date of first reporting to the relevant authority and method of reporting.

If an employee dies within one year as a result of an accident the employer has to inform the enforcing authority as soon as he learns of the death, whether or not the accident had been reported originally.

If a person at work suffers from an occupational disease and his or her work involves one of a specified list of substances and activities, the

Table of some examples of occupational diseases and associated activities

Diseases	*Activities*
Cataract due to electromagnetic radiation	Work involving exposure to electromagnetic radiation (including heat).
Cramp of the hand or forearm due to repetitive movements	Work involving prolonged periods of handwriting, typing or other repetitive movements of the fingers, hand or arm.
Beat hand, beat elbow and beat knee	Physically demanding work causing severe or prolonged friction or pressure on the hand, or at or about the elbow or knee.
Hand arm vibration syndrome	Work involving a specified range of tools or activities creating vibration.
Hepatitis	Work involving contact with human blood or blood products, or any source of viral hepatitis.
Legionellosis	Work on or near cooling systems which are located in the workplace and use water, or work on hot water service systems located in the workplace which are likely to be a source of contamination.
Rabies	Work involving handling or contact with infected animals.
Tetanus	Work involving soil likely to be contaminated by animals.
Tuberculosis	Work with persons, animals, human or animal remains or any other material which might be a source of infection.
Poisoning by specified substances including mercury and oxides of nitrogen	Any activity.
Various cancers	Various activities.
Occupational dermatitis	Work involving exposure to a range of substances.
Occupational asthma	Work involving exposure to a range of agents.

responsible person must send a report to the relevant enforcing authority as soon as he learns of the disease. The table on p. 301 gives some examples of the diseases and associated activities listed in Schedule 3 of RIDDOR. Notification of occupational diseases is normally by Form F2508A, the main requirements of which are:

- Date of diagnosis of disease.
- Name and occupation of person affected.
- Name and nature of disease.
- Date first reported to the relevant authority, and method of reporting.

Copies of F2508 and F2508A are contained in the guide[3].

- Records of reportable incidents must be retained for at least three years. This can be as photocopies of the Forms F2508 and F2508A or the data can be kept on computer when registration under the Data Protection Act will be necessary.

RIDDOR covers:

- employees
- the self-employed
- those receiving training for work
- members of the public, pupils and students, and other people who suffer injuries or diseases as a result of work activities

but does not cover:

- patients who die or are injured while undergoing medical or dental treatment
- some incidents on board merchant ships
- death or injury where the Explosives Act applies
- death or injury as a result of escapes of radioactive gas
- cases of agricultural poisoning
- where the incident is reportable under the Road Traffic Act.

Serious incidents as defined have to be recorded and reported to the relevant authority. However, there is also a requirement to record details of less serious incidents, for example minor injuries, although these do not have to be reported to an authority. This requirement is imposed by the Social Security (Claims and Payments) Regulations 1979 but it does not apply to all employers. However, where it does apply, these less serious incidents can be recorded either in an Accident Book (BI 510) or on an organisation's own form or forms. In either case, the minimum information which must be recorded is:

- Full name, address and occupation of injured person.
- Date and time of accident.
- Place where accident happened.

- Cause and nature of injury.
- Name, address and occupation of person giving the notice, if other than the person injured.

Where an organisation is using its own form, additional information for internal use can be recorded. As with the serious incidents, if minor injury records are kept on computer, registration under the Data Protection Act will be required.

Chapter 2.6

Practical safety management: systems and techniques

J. E. Channing

2.6.1 Introduction

Societies exist and work as their members formulate rules by which to live. The foundations of the rules are either religious or ethical and develop slowly over many years to reflect the changing culture and values of the particular society. Breaking the rules incurs censure and punishment. They are carefully scripted by highly educated legal minds, communicated by the written word and interpreted by judges. Most citizens grow to appreciate the general principles of the rules (or laws) without knowing the intricate legal details. If, however, an individual citizen is accused of disobeying the laws, then the details become important. In these circumstances another highly trained legal mind defends the accused citizen by arguing over the written words of the law in front of a judge who has to interpret their precise meaning and intent and decide whether or not a contravention has occurred. This process generates many laws of an intricate and confusing nature which extend to cover health, safety and the environment. These laws address, *inter alia*, hazards that need to be controlled, some of which are obvious and some are not. The responsible citizen needs to comply but may not have the time to read and understand the complexity of the requirements. The objective of this chapter is to provide insights into the techniques and processes that can be used to control the health, safety and environmental risks effectively and sensibly while complying with legal requirements. It is often forgotten that the legal objective is simply to prevent people from being injured or suffering ill-health from the activities of the enterprise. A confusing jargon has emerged full of 'risk assessments', 'safe systems of work' and 'reasonably practicable options'. The straightforward approach – 'how can we be hurt and what can we do about it?' – has been put aside.

Yet as people live longer and their expectations of good health increase it is inevitable that complexity from ever more subtle risks to our well-being increases.

The conundrum facing many managers is to find practical ways of dealing with these issues without becoming a fully trained lawyer. The problem facing the safety professional is to utilise the hazards and the legal requirements and create everyday tools that the manager and the work group can use. The solutions will vary with the size and nature of the enterprise. A small marketing operation which just uses computers and telephones to operate its business has fewer and different risks from a supermarket business. Other enterprises may be larger and encompass many different types of operations so that each manager must consider hazards which are general to the whole business as well as ones which are special to the part of the business under local control. This situation can arise on larger mixed occupancy sites where, for example, chemical storage facilities are adjacent to the sales operation.

2.6.2 Legal obligations

The responsibility for managing the safety of employees lies with the owners of the enterprise and their appointed agents, usually the managers of the workplace. This obligation is a constant feature of legislation throughout the world. In the UK it is encoded in the Health and Safety at Work etc. Act 1974 (HSW) and reinforced in the Management of Health and Safety at Work Regulations 1999 (MHSW). This basic obligation is also explicit or implicit in all other regulations and is a reflection of each citizen's common law duties.

2.6.2.2 Common law

Lord Maugham articulated the duties under common law in the court case between *Wilsons and Clyde Coal Co. v. English*[1] when he said:

> 'In the case of employment's involving risk it was held that there was a duty on the employees to take reasonable care, and to use reasonable skill, first, to provide and maintain proper machinery, plant, appliances, and works; secondly, to select properly skilled persons to manage and superintend the business, and, thirdly to provide a proper system of working.'

The above statement was published in 1937. In the same case Lord Wright quoted a previous judgement by Lord McLaren in *Bett v. Delmeny Oil Co.* in 1905[2], which said:

> 'The obligation is threefold, the provision of a competent staff of men, adequate material, and a proper system and effective supervision.'

The duties of employers, thus well established in common law, became encoded in criminal law in the 1974 Act where these duties are applied to the extent that is 'reasonably practicable'. The meaning of this phrase was summarised in a common law case by Lord Asquith in his judgement in *Edwards v. National Coal Board*[3]. He said:

> ''Reasonably practicable' is a narrower term than 'physically possible', and seems to me to imply that a computation must be made by the owner in which the quantum of risk is placed on one scale and the sacrifice involved in the measures necessary for averting the risk (whether in money, time or trouble) is placed in the other, and that, if it be shown that there is a gross disproportion between them – the risk being insignificant in relation to the sacrifice – the defendants discharge the onus on them. Moreover, this computation falls to be made by the owner at a point of time anterior to the accident.'

The phrase and the interpretation once more summarised the common law duty of care to take 'reasonable care'. How they have been applied to different accident situations is to be found in many legal publications such as *Munkman's Employer's Liability*[4]. Common themes emerge where factors such as the nature of the hazard, the obviousness of the hazard, the potential consequence to the employee as well as the cost of the control measures must be considered.

2.6.2.3 Statute law

Now that these common law duties are encoded in HSW, failure to comply is a criminal offence. Particular regulations extend those duties by requiring risk assessments to be undertaken as a means of determining the appropriate control measures for specific hazards. The Management Regulations (MHSW) encompass far more, requiring that risk assessments be undertaken for all workplace risks not otherwise referred to in specific regulations.

The practical problem faced by employers in meeting the legal requirements is twofold:

- The employer must identify hazards, undertake risk assessments and institute control measures to protect employees and others who may be affected by the enterprise.
- The employer must achieve these objectives as part of a recognisable management system which is capable of audit by enforcement officers.

2.6.3 Generic safety management

2.6.3.1 Management models

The law is a crude management model for safety. It defines a goal (a hazard which it seeks to control) and sets forth a set of criteria to achieve that control. It also motivates employers to apply those controls by threatening punishment for failures to do so. The safety policy can also be seen as another attempt to apply a management model to health and safety.

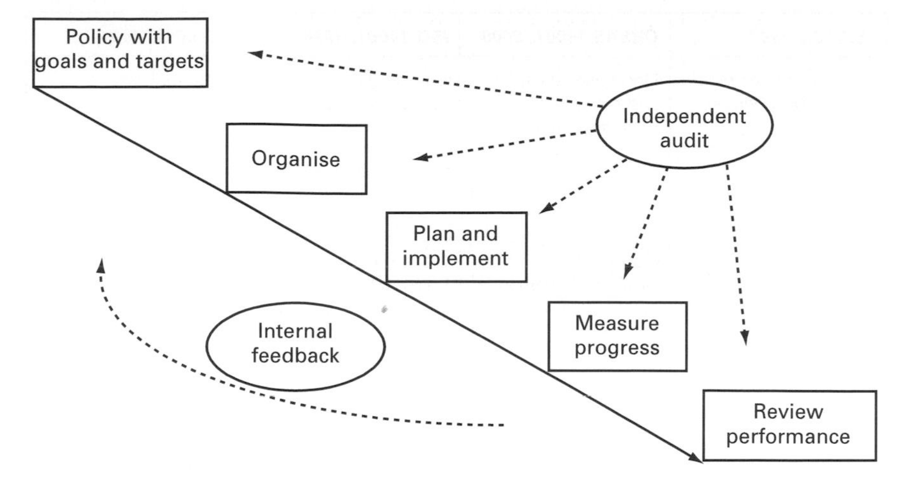

Figure 2.6.1 Schematic diagram of HSE management model

More recently models for health and safety have emerged which are far more user friendly to the hard pressed manager. The HSE's publication 'Successful Health and Safety Management'[5] provides one of many models to manage health and safety issues within an enterprise. It outlines a system based upon establishing a policy with targets and goals, organising to implement it, setting forth practical plans to achieve the targets, measuring performance against the targets and reviewing performance. The whole process is overlaid by auditing. A schematic representation of this process is given in *Figure 2.6.1*.

Another similar OHSAS model[6] chooses as its starting point an Initial Status Review by an independent auditor and the steps that follow are:

1 The occupational health and safety policy.
2 Planning.
3 Implementation and operation.
4 Checking and corrective action.
5 Management review.

The above steps are to be considered within an overall framework of 'continuous improvement'.

This model has been structured to fit with the international quality management system standard BS EN ISO 9000[7] and the equivalent environmental management system standard BS EN ISO 14001[8]. Applying the OHSAS model therefore offers the opportunity for companies who have opted for the quality management system to integrate environment and health and safety into one comprehensive management process.

Figure 2.6.2 interrelates the elements of these management standards with the HSE's guidance[5].

The quality standard, which was the first in the series of management system standards, is an uncomfortable fit with the safety and environ-

HSG 65: 1997	*OHSAS 18001: 2000*	*ISO 14001: 1996*	*ISO 9001: 1994*
Successful Health and Safety Management	*Occupational health and safety management systems*	*Environmental management system*	*Quality system*
Chapter 2: Effective health and safety policies	System requirements, general requirements and policy		
	Clause 4 OH&S management system elements	Clause 4 Environmental management system requirements	Clause 4 Quality system requirements
	Clause 4.1 General requirements	Clause 4.1 General requirements	Clause 4.2.1 (part) General
	Clause 4.2 OH&S policy	Clause 4.2 Environmental policy	Clause 4.1.1 Quality policy
Chapter 3: Organising for health and safety	Structure, responsibilities, training, communication and documentation		
	Clause 4.3.4 OH&S management programme	Clause 4.3.4 Management programme	Clause 4.2 Quality system
	Clause 4.4.1 Structure and responsibility	Clause 4.4.1 Structure and responsibility	Clause 4.1 and 4.1.2 Management responsibility and organisation
	Clause 4.4.2 Training, awareness & competence	Clause 4.4.2 Training, awareness & competence	Clause 4.18 Training
	Clause 4.4.3 Consultation & communication	Clause 4.4.3 Communication	No equivalent
	Clause 4.4.4 Documentation	Clause 4.4.4 Documentation	Clause 4.2.1 (part) General
	Clause 4.4.5 Document & data control	Clause 4.4.5 Document control	Clause 4.5 Document and data control
Chapter 4: Planning and implementing	Hazard identification, risk assessment and control, legal requirements, objectives, targets, operational control		
	Clause 4.3.1 Hazard identification, risk assessment & control	Clause 4.3.1 Environmental aspects	Clause 4.2 Quality system
	Clause 4.3.2 Legal requirements	Clause 4.3.2 Legal requirements	No equivalent
	Clause 4.3.3 Objectives	Clause 4.3.3 Objectives and targets	Clause 4.2 Quality system

HSG 65: 1997	*OHSAS18001:2000*	*ISO 14001:1996*	*ISO 9001:1994*
Successful Health and Safety Management	*Occupational health and safety management systems*	*Environmental management system*	*Quality system*
	Clause 4.4.6 Operational control	Clause 4.4.6 Operational control	Clause 4.2.2/4.3/4.4/ 4.6/4.7/4.8/4.8/4.9/ 4.15/4.19/4/20 Detailed operational requirements
	Clause 4.4.7 Arrangements for emergencies	Clause 4.4.7 Arrangements for emergencies	No equivalent
Chapter 5: Measuring performance	Performance measurement, inspection, incident/non-conformance investigation, preventive actions and records		
	Clause 4.5.1 Performance measurement and monitoring	Clause 4.5.1 Monitoring and measurement	Clause 4.10/4.11/4.12 Detailed requirements for testing and inspection
	Clause 4.5.2 Accidents, incidents and preventive actions	Clause 4.5.2 Non-conformances and corrective action	Clause 4.13/4.14 Non-conformances and corrective action
	Clause 4.5.3 Records	Clause 4.5.3 Records	Clause 4.16 Records
Chapter 6: Auditing and reviewing performance	Audit and management review		
	Clause 4.5.4 Audit	Clause 4.5.4 Management system audit	Clause 4.17 Internal audit
	Clause 4.6 Management review	Clause 4.6 Management review	Clause 4.1.3 Management review

Figure 2.6.2 Alignment of HSG 65 with Safety, Environment and Quality Standard

ment management system standards. Managers should be aware that regulatory enforcement officers are likely to be influenced more by their own (HSE's) guidance than by accreditation to a safety management system awarded by commercial organisations.

2.6.3.2 Local safety management assessment systems

Many enterprises have chosen this route and based their safety management systems on the quality standard. The chemical industry has linked its pre-existing Responsible Care programme[9] to the ISO quality and environmental standards[7, 8].

Regulatory bodies increasingly seek a robust formal management system and managers can expect to be asked, during a site visit by inspectors, to explain how they manage health and safety. It is therefore important that systems are developed which integrate with the responsibilities and the business needs of the enterprise.

The management system for health and safety needs to be structured in a similar way to the financial system of controls in a company. This establishes budgetary targets based upon its previous performance and its future aspirations. The overall targets are then broken down to area budgets. Each level of management is given defined targets to achieve and is measured against them. Progress is reviewed regularly at each management level and the whole financial system is independently audited.

The parallel health and safety system needs to define the safety risks and to establish the effectiveness of the control measures taken and to monitor performance at each operational level in the organisation. A consideration of accident data and an audit of performance can provide this information and enable an improvement target for the year to be set.

The management systems described above are high level models and will need to be supported by more detailed programmes. However, they offer little guidance on the process of implementation beyond setting standards and requiring audits and assessments.

In this chapter, an assessment refers to the identification of risks in a local area whereas an audit covers health and safety across the whole organisation with a view to determining its overall health and safety condition.

The local management system consists of the operational checklists that evolve out of the generic model. An example for a telephone call service centre is given in *Figure 2.6.3*.

2.6.4 Implementing a regulation within a safety management system

An essential component of any safety management system is a mechanism for recognising and implementing new regulations into the working environment. Once identified, the regulation should be broken down into simple discrete components. Choosing the MHOR and the associated Guidance[10] as the example, the process can be applied as follows.

2.6.4.1 Defining the key steps

In the Regulations, the sequencing of the subsections may be logical from a legal viewpoint, but from a practical manager's point of view they need to be rearranged. They require the employer to assess the risks from manual handling and to develop methods of work that reduce these risks to a minimum. Employees are required to follow the work method.

ITEM	STATUS ✓ OR ×	COMMENT
1. Are aisles and exits kept clear?		
2. Are trailing leads contained within safety ducting?		
3. Are chairs adjustable and in sound condition?		
4. Can employees adjust their workstations to avoid fatigue and strains?		
5. Are computer screens free from glare?		
6. Do operators take work breaks at suitable intervals?		
7. Is lighting appropriate for all workstations?		
8. Are employees using telephone headsets not the handsets?		
9. Are workstations tidy?		
10. Is the air conditioning effective for all areas?		

Figure 2.6.3 Example of local management system checklist for a call centre

Step 1 There is no requirement in the Regulations for a list of tasks to be generated. However, planning a compliance programme is difficult without knowing the scope of the task. The first step is to list jobs with a manual handling component together with the specific tasks within those jobs. Groups of similar tasks can be put together for generic assessments. For example, the task of loading supermarket shelves with produce would be one assessment – separate assessments would not be required for stacking baked beans one moment and sugar the next! However, a separate assessment may be required for loading Christmas turkeys into a chest freezer due to the greater weight of the product and the bending component of the task. In effect an informal assessment is being undertaken whilst compiling generic assessments so that like-for-like risks are grouped together.

Step 2 is to undertake the formal risk assessment, as required by reg. 4(1)(b)(i) possibly using the proforma outlined in Appendix 2 of the HSE guidance.

Step 3 is to identify the agreed control measures (or 'remedial actions' in the HSE proforma). It may have emerged from the study that the manual handling task can be eliminated. If so the employer has been able to comply with reg. 4(1)(a). This conclusion, however, may not have been reached until the assessment study has been completed.

Step 4 implements the control measures which have been identified by the assessments. Employers are required by reg. 4(1)(b)(iii) to give employees information on the weight of each load. This can be achieved in different ways. For loads of constant weight it may only be necessary to put the information in a local procedure. However, enterprises which supply a range of goods to a variety of customers may need to state the weight on each package. For off-centre loads, the packaging should indicate the heavier side.

Step 5 is a procedure for reviewing the assessments whenever there is reason to suspect the existing one is invalid, or there has been a significant change in the operation. This step complies with reg. 4(2).

Step 6 implements reg. 5 which requires employees to adopt the safe system of work established by the employer as a result of the assessment. This can be achieved through training to make employees aware of the risks, the proper working methods for the safe performance of the task.

Decoding the regulations into practical and easily understood steps is just the commencement of the implementation process.

2.6.4.2 Organising the implementation process

In establishing the above steps line managers may be best placed to generate the initial list of jobs and tasks. They will need an understanding of ergonomics and of the regulations. The risk assessments could be undertaken by line managers or team leaders since they usually have a better appreciation of the tasks undertaken in both normal and abnormal circumstances. It becomes their responsibility to find less risky alternatives and hence gain ownership of the whole process. Furthermore, they fulfil a basic tenet of safety law in that the employer, through his line managers, has responsibility for the safety of his employees. Some companies use safety advisers, specialist trainers or medical specialists to undertake the assessments. There are, however, considerable benefits in keeping as much of the assessment and implementation activity 'in the line' rather than offloading them onto a specialist.

The safety adviser's role is to assist in the construction of the implementation process and training those in the team who will undertake the various assessment tasks. Training provided to people to implement a particular regulation should include:

- An explanation of the general structure of law and regulations and where any particular regulation fits within the overall process.
- The reasons why the regulation came about and why the existing controls were deemed inadequate. These could be:
  - a significant public incident has occurred which demonstrates the inadequacy of current legislation (e.g. a chemical plant disaster);
  - injuries continue to occur at a persistent level in spite of current controls (e.g. musculo-skeletal injury from manual handling);
  - the government is implementing a European Directive to harmonise laws across the region.
- A description of how people can be harmed if the hazard is not controlled. For example, the majority of manual handling injuries occur to the lower back, the shoulder, or the wrist/forearm. Knowing this helps to focus on the risk factors.
- Information on the circumstances that contribute to injury such as the causative factors in manual handling injuries that include lifting, twisting, repetitive actions, or posture while moving objects of weight.
- A demonstration that compliance can generate feedback and be applied to give low cost practical solutions to particular problems.
- An appreciation of the need for a programme timescale and of the process to be used in measuring the progress of implementation.

A benefit of using line employees to implement regulatory controls is that the training they are given to help them undertake the task results in a well-educated workforce with benefits for the entire safety programme.

2.6.4.3 Measuring the progress

'What gets measured, gets done' was a phrase coined by Peters[11]. The phrase sums up the belief that managers respond best to measurement systems, often numerical, to chart progress and thereby ensure plans reach completion. A measurement matrix can be applied to implementing regulations once the steps have been established. An example of a measurement matrix for implementing the manual handling operations risk assessment is presented in *Figure 2.6.4.*

The matrix has several important features. Some steps are measured so that a 'No' answer attracts a zero score whilst a 'Yes' answer gains the maximum score. Other steps gain a graduated score according to the percentage of the task completed. Another important feature is that not all steps are equally weighted. In general the higher weightings are given to the more important steps or those requiring the greater workload. In the example, step 4 has the heaviest weighting as this is viewed as the single most important step to achieve a safe manual handling workplace. Adjusting the weighting in this manner encourages a focus on the essential elements of complying with the Regulations.

As the implementation plan proceeds, a score for each step is calculated by multiplying the level achieved by its weighting. A copy of the chart can be provided to senior managers each month from which the pace of

MANUAL HANDLING REGULATIONS COMPLIANCE CHART														
STEP	LEVEL ACHIEVED											SCORE	WEIGHT	OVERALL SCORE
score values	0	1	2	3	4	5	6	7	8	9	10			
1. List of jobs and tasks generated	no										yes		5	
2. % risk assessments completed	0	10	20	30	40	50	60	70	80	90	100		20	
3. List of control measures agreed	no										yes		15	
4. % of control measures implemented	0	10	20	30	40	50	60	70	80	90	100		25	
5. procedure to mark loadweight	no										yes		5	
6. Review process in place	no										yes		10	
7. % employees trained	0	10	20	30	40	50	60	70	80	90	100		15	
8. Training packages available	no										yes		5	
													TOTAL SCORE=	

Figure 2.6.4 Measurement matrix for the implementation of MHOR

progress can easily be seen. This can assist in decisions regarding the allocation of resources to maintain progress.

In a multi-departmental enterprise, each area should have its own matrix with summary scores published to spur the laggards to catch up. It acts as a simple but effective motivational and behaviour-shaping tool.

Guidance on the measurement of health and safety[15] observes that companies measure their financial performance in terms of positive outcomes whereas for health and safety they use a negative measure – injury and ill health data – which are measures of failure. Success in health and safety boils down to the absence of adverse (failure) outcomes. The guidance recommends a 'basket of measures or a balanced scorecard' to provide indicators of the health of the organisation and early detection of deterioration in safety performance before it becomes overtly manifest. *Figure 2.6.5*, which is based on the five elements for successful safety management advocated by the HSE[5] provides a simple list of 20 criteria for measuring the success of management.

2.6.5 Safety management and housekeeping

Managers often target 'housekeeping' as an area which needs improvement. Many accidents occur because of poor housekeeping in the form of uncleared spillages, overstacked shelves etc., which indicate a lack of control in the workplace. The achievement of high housekeeping standards is also a key indicator of good safety performance in the area because it is a manifestation of the attitudes that prevail amongst those working there. Good housekeepng requires a well-structured work process, discipline in execution, and motivated employees. For this reason an HSE inspector may use the workplace condition to indicate possible weaknesses in the safety management systems.

Scoring techniques can be used to improve and maintain housekeeping in the work area and an example of a checklist is provided in *Figure 2.6.6*. The checklist identifies the expectations clearly and precisely and should be phrased in such a way that full compliance gains the maximum points. Questions can range from the physical conditions in the workplace to the practical knowledge of the employees. Approaching housekeeping in this way has several benefits. Firstly, the area's requirements are stated clearly. Secondly, because the questions are clear and concise, each employee in the work area can complete the checklist on a rota basis. Thirdly, the checklist can be changed periodically.

Changing the checklist contents needs to be considered carefully so that the desired behaviours and standards are developed and improved. Leaving an item off a well-established checklist may lead to it not receiving attention in the future. On the other hand, simply adding more and more items may make the whole checklist unwieldy. If this occurs, a split checklist may be necessary with each part completed in alternate weeks.

A final point to consider relates to the scores themselves and whether to use 'points' or 'percentages'. In general 'points' are preferred because

ELEMENT	TEST QUESTION	RESPONSE (Y/N)
1. Policy	1.1 Does the safety policy reflect the current organisation? 1.2 Does the policy specify responsibilities of those in the organisation? 1.3 Is the policy pertinent to the hazard burden? 1.4 Are employees aware of the policy and the roles, responsibilities and arrangements to make it effective?	
2. Organising	2.1 Does a system exist to control health and safety? 2.2 Do safety committees exist at appropriate levels in the organisation? 2.3 Do consultation arrangements promote effective co-operation and participation of all relevant stakeholders in the organisation and others who may be affected? 2.4 Are the communication channels effective in distributing all the types of health and safety information? 2.5 Are the people appointed to specific roles and tasks trained and qualified to a sufficient competence level?	
3. Planning and Implementing	3.1 Does the organisation and its parts have plans with objectives? 3.2 Do the plans have specific deliverables and suitable time scales? 3.3 Do the plans address the identified hazards? 3.4 Do the plans ensure risk assessments are carried out and controls implemented? 3.5 Are individuals who are responsible for delivering specific elements named?	
4. Measuring Performance	4.1 Does the organisation measure progress against the plan elements each quarter? 4.2 Is progress published to employees? 4.3 Are explanations provided to employees if progress is slow?	
5. Audit and Review	5.1 Are there arrangements to audit the entire health and safety programme periodically? 5.2 Does a senior manager/director review health and safety performance? 5.3 Do senior managers go outside the organisation to gain an objective view of its performance?	

Figure 2.6.5 A 20-point list to measure successful safety management

Housekeeping Checklist – *Computer Suite*		
Item	*Maximum score*	*Actual score*
1 Emergency fire escape routes are unobstructed (*2 routes, 10 points per clear route*)	20	
2 First aid boxes contain correct contents (*2 boxes, 4 points per box*)	8	
3 Work areas clear of trailing leads in walkways (*deduct 2 points per trailing lead*)	10	
4 Electrical leads in area visually inspected and show no sign of mechanical damage (*deduct 4 points per damaged lead*)	12	
5 Computer workstation meets requirements of Regulations (see supplementary list) (*2 workstations, deduct 4 points per defect*)	20	
6 Workstation users have made correct adjustments for their own use (*2 workstations, deduct 4 points per defect*)	20	
Score maximum =	90	actual =

Figure 2.6.6 An example of a scoring checklist for housekeeping

new items to the checklist add more to the 'maximum points horizon' without diminishing the value of the points scores for the existing items. 'Percentages', where 100% is the goal, do not achieve this, thereby diminishing the relative worth of an existing item on the checklist.

Techniques that link housekeeping to safety, productivity and quality have become prominent. One example is known as '5 S's'[16]. Based upon five Japanese words (*Figure 2.6.7*), it is a tool which encourages employees to improve their own working conditions and reduce waste in all its forms. The essential philosophy of the process is to engage employees in a formal structured process that is within their control but is aligned with business objectives, of which health and safety performance is but one.

- 'Seiri' refers to the need to separate and retain only those things that are necessary for the tasks undertaken by the workgroup. It requires the identification and removal of superfluous materials and process steps. It directly relates to safety performance insofar as many accidents occur from non-added value activity. Examples of tasks which add cost and hazards without adding value include storage and transportation of product in warehouses.

Japanese Term	English Equivalent	Meaning	Examples
Seiri	Tidiness	Organisation	Discard rubbish Everything in place
Seiton	Orderliness	Neatness	Items close to hand Documents nearby
Seiso	Cleanliness	Cleaning	Individual cleaning responsibility
Seiketsu	Clean-up	Standardisation	Transparency of storage
Shitsuke	Discipline	Self-discipline	Daily commitment to the programme

Figure 2.6.7 Overview of five S's terminology

- 'Seiton' refers to workplace orderliness and is basically about efficiency. This part of the process addresses the question of how quickly an employee can get the items needed for the task. It requires that just sufficient materials are kept in a convenient specifically allocated place. An example would be to keep tools mounted on a pegboard rather than haphazardly placed in a drawer. Safety benefits arise from locating the pegboard nearby and at a suitable height to reduce manual handling hazards.
- 'Seiso' relates to cleanliness. It requires that everyone keeps their own work areas clean. Communal areas can be allocated to individuals or responsibility rotated between individuals in the workgroup. Each employee is encouraged to see their workplace through the eyes of a visiting dignatory – 'would I be proud for the Prime Minister to visit my own work area?' The impact on safety is to remove hazards such as those which result in slips or trips from dirty floors. It also prompts questions such as 'why do oil leaks recur from this machine?' and prompts preventive action. Organisations advanced in Seiso do not employ cleaning staff!
- 'Seiketsu' embraces continually and repeatedly cleaning up to maintain the work area's tidiness, orderliness, and cleanliness. It is achieved by 'visual management' using colour coding to highlight quickly a breach of standards. For example, walls are painted with light coloured paint which quickly shows the presence of chemical dust should a bag filter malfunction. Some organisations have attired employees in white overalls to show when the workplace or individual habits are creating dirt and mess.
- 'Shitsuke' means applying discipline to do in the right way the things that need to be done. The emphasis is on creating good work habits and it is successful when the self-discipline of all employees is seen as pride in the work area. It is particularly important for safety where regular checks on safety systems need to be undertaken to prevent failure. Reliance on the individual employee to inspect, check and record conscientiously is critical.

The 5S's approach is more about attitude and culture and it elevates humble housekeeping into a core value.

2.6.6 Assessment techniques

The models reviewed at the beginning of this chapter set an overall framework for managing health and safety issues. Once the system is established there is a requirement for it to be assessed. The role of an assessor can vary. One aspect is to review the management system and procedures to ensure that they can operate as a cohesive structure. It does not need a professional understanding of health and safety to perform this task. Another approach that might be taken is to use a qualified health and safety adviser who checks the system and is able to judge whether safety in the workplace is being achieved. Whichever approach is used the assessment report should list positive features of the system as well as listing 'non-compliances', 'negative findings' or 'corrective action requests' according to the terminology used. The benefit of this approach is that it requires an assessment by a third party.

But there are weaknesses. Even an enterprise which, by any objective standards, is good in terms of health and safety performance may still be given a list of non-conformances. Also the items listed in successive assessments become of increasingly less importance, even trivial, and may reflect the assessor's opinion rather than a requirement of a regulation or internal standard.

A typical response to a third party health and safety assessment is to see it as a target to be aimed for and to do just sufficient to satisfy the assessment. This response can prevail where managers fail to embrace health and safety as central to their function.

Assessments may be undertaken once or twice a year and may follow a 'sampling' technique whereby a few specific areas are examined in depth and conclusions are drawn which can be extrapolated to cover the whole organisation. An assessment may seem like an examination where the result is always 'can do better'! Whilst assessments are necessary and can be beneficial to an organisation they are often approached with apprehension because they are seen as punitive where the options lie between being beaten with a big stick and beaten with a small one.

2.6.6.1 The positive approach to health and safety assessments

Techniques have been developed to make health and safety assessments an ongoing and central element of a manager's function. Section 2.6.4 demonstrated how a regulation can be translated into an assessment with a numerical format so that progress towards implementation can be given a score. The same principle of scoring can be applied to embrace all regulatory requirements once they have been implemented with other health and safety aspects of the workplace. *Figure 2.6.8* shows a simple set of questions which can be used to maintain compliance with MHOR. Each question attracts a points score as indicated.

Audit question	Points available	Points scored
1 Have tasks in the area been reviewed in the last year to identify those with a manual handling component?	10 if 'yes'	
2 State the % of manual handling tasks with formal written risk assessments	50 points for 100% compliance and pro rata	
3 On what % of tasks has the risk been reduced by the control measures to as low as reasonably practicable?	50 points for 100% compliance and pro rata	
4 What % of employees know where the risk assessments are kept and have been made familiar with them?	50 points for 100% compliance and pro rata	
5 Have steps been taken to inform employees of load weights?	20 if 'yes'	
6 Have assessments been reviewed within the last 3 years?	30 if 'yes'	
7 What % of employees have undergone training in handling methods in the last 5 years?	30 points for 100% compliance and pro rata	
8 Is there a procedure in place to review assessments if significant change occurs?	10 points for 'yes'	

Figure 2.6.8 Example of audit questions for maintaining compliance with manual handling operations regulations

The scoring concept can be applied to each element of the management systems. *Figure 2.6.9* is an example of a question set to manage the performance of safety committees.

The questions are designed to be straightforward and easily understood by everyone. This makes it possible for any employee or a small group of employees to monitor the performance of the system.

The scoring principles for the question sets should give fewer points for simple administrative tasks and greater points for actions by the people involved.

In the safety committee example, the mere existence of a safety committee attracts few points. Furthermore the attendance of the manager outscores the attendance of the safety adviser and employee representatives because the manager has the greater responsibility and plays the key role in safety issues in the workplace.

This is an example of the technique of 'shaping' which can be used to guide the contents of a set of practical actions aimed at improving health and safety performance in the area.

Audit question	Points available	Points scored
1 Does the area have a joint safety committee?	10 for 'yes'	
2 Has the committee met at least every 3 months in the last year?	10 per meeting up to 40	
3 Does the area manager attend the meetings?	20 per attendance	
4 Does the safety adviser attend the meetings?	10 per attendance	
5 Do employee representatives attend the meetings?	10 per meeting up to 40	

Figure 2.6.9 An example of audit questions for safety committees

This approach can be applied across the range of health and safety hazards which exist in the work area but it does raise issues of the scores given to dissimilar tasks. For example, one question on fire safety may concern the maintenance of suitable fire escape routes. The administrator of the assessment process must decide how many points should be given to that question compared to a question on labelling of chemicals. There is no absolute answer. Nor need there be. What is important is that each assessment is repeated on a regular basis. Increasing scores for particular areas indicate improving safety standards.

The safety adviser has a key role in establishing a system for ongoing assessments and for deciding on the questions and their scores. Software packages are available and may be preferable to paper systems in some organisations[14].

Scoring systems of this type permit managers to undertake a self-audit of the performance of their areas of responsibility. However, rather than carry out a comprehensive audit of all areas of responsibility occasionally, greater benefit may be derived by assessing one aspect only each month.

The following section headings have been used with success in some organisations:

1 Health and safety management and administration.
2 Fire, loss and emergencies.
3 Investigation and monitoring.
4 Chemicals and substances.
5 Environment and waste.
6 People.
7 Systems of work.
8 Machinery, plant and equipment.
9 Product safety.

Each section itself can be split into separate parts. For example, section 6 on 'People' can be split into:

6.1 Standard operating procedures.
6.2 Permits-to-work and lock-off systems.
6.3 Manual handling operations.
6.4 Repetitive work.
6.5 Large vehicles.
6.6 Lone workers.

One section should be completed each month over an agreed period of, say, 9 months. In the 3 months following the final assessment the safety adviser can review and agree the scores, thereby providing an independent validation of the self-assessment. It also allows the manager to review his performance and act to improve his score before submitting the final performance score for the year to his director.

With ongoing assessments an opportunity can be taken to alter or add to the question set for the following assessment. The amendments to the assessment enable the enterprise to incorporate newly implemented regulations into the package. It also allows an examination of its performance and identification of the improvements to be made. In one organisation, it was observed that a safety committee in one area was less than effective because it didn't have a procedure for closing actions. A procedure was implemented and additional questions (see *Figure 2.6.10*) were added to the existing assessment form. As the questions set applied to all areas of the organisation the deficiency observed in one area was able to be addressed in all areas.

Audit question	Points available	Points scored
6 Is a list of actions produced with agreed timelines for completion?	20 for 'yes'	
7 What % of actions were completed in the last year within time?	100 points for 100%	

Figure 2.6.10 Additional audit questions on safety committees

The objective of any assessment process is to reduce accidents at work.

Figure 2.6.11 shows data from an organisation using the self-assessment system outlined above. Graphs of this sort can demonstrate correlations between lost time accident levels and self-assessment scores. A manager who takes the right actions in a structured approach to health and safety should produce an improved accident performance.

2.6.7 Proprietary audit systems

Several proprietary audit systems are available. The International Safety Rating System[12] was one of the first comprehensive safety audit

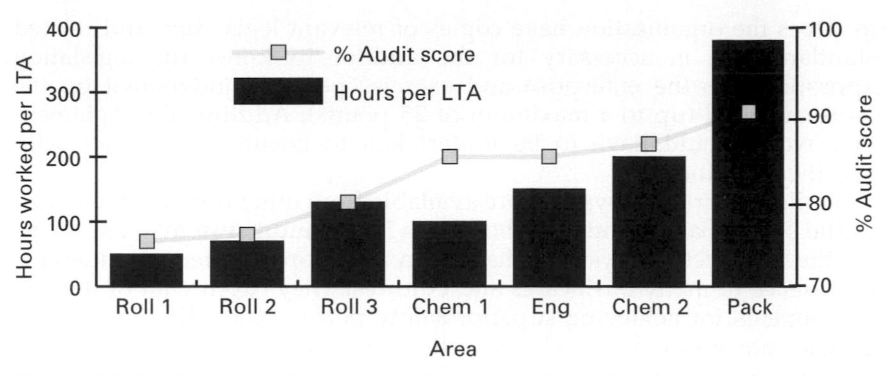

Figure 2.6.11 Correlation between accident and audit performance

systems to be developed. It categorises health and safety issues into 20 management elements and includes a 'Physical Conditions' inspection. From the score obtained an enterprise can be judged to have achieved one of five standard levels or one of five advanced levels. Achievement at the highest level demands a high score in over 600 questions in the management elements and high performance in 'Physical Conditions' (loosely equated to housekeeping). The assessment is undertaken by a third party auditor who has been trained in the system. A successful application of the system also requires that managers are educated to understand its jargon and objectives. Using the system means that comparisons can be made between companies in different business sectors and even in different countries. It has been developed over many years to give organisations comprehensive audit processes.

The elements measured by the system are:

Leadership and administration
Management training
Planned inspections
Task analysis and procedures
Accident/incident investigation
Task observation
Emergency preparedness
Organisational rules
Accident/incident analysis
Employee training
Personal protective equipment
Health control
Programme evaluation system
Engineering controls
Personal communication
Group meetings
General (safety) promotion
Hiring and placement
Purchasing controls
Off-the-job safety

Disadvantages of the system include the scores given to each question and the 'pass marks' required for each level which are set by the system and do not allow flexibility to meet a particular emphasis on which the enterprise wishes to focus. Another problem arises in the application to enterprises that use a different culture and jargon from that employed by the system. Finally, the system does not address specific national legislative requirements since it does not set out to focus on particular national laws but seeks a position above them by asking questions such as 'Does the organisation have copies of relevant legislation and related standards?' It is necessary for the auditor to know the legislation appropriate to the enterprise and exercise personal judgement on the score awarded (up to a maximum of 25 points). Additional supplementary work would have to be undertaken to ensure compliance with specific regulations.

Similar proprietary systems are available from other organisations such as the British Safety Council who offer a 5-star audit system[13]. The scope of the proprietary systems has been developed after considerable experience in many businesses and cultures. They capture all of the key components for achieving superior safety performance. If an enterprise chooses an internally developed self-assessment route, then keeping a proprietary system as a reference and guide for future developments may be useful.

Computer-based systems such as 'CHASE'[14] can provide preloaded audit questionnaires or blank formats into which the organisation's own audit questions can be written. The benefits of writing the audit questions within the organisation is that it will reflect the culture and jargon of the workplace, but the process of setting it up can be time consuming.

2.6.8 Safety systems and incidents

The previous sections deal with safety management from a proactive standpoint. Audit and assessment are essential but they do tend to follow pre-designed checklists or protocols. Many auditors under time pressure or through inexperience will limit their assessment to the prescribed list of topics. When incidents occur an opportunity is presented to do more than identify the proximate causes. Gaps in the audit system itself that allowed the causes to arise and had not been corrected can be identified. In practice investigations tend to seek out a 'root cause' which is often identified as a defect in the work system giving the impression that all would have been well if the work system had been properly designed in the first instance. Such an analysis does not generate the complete list of lessons from the incident.

A better approach is to re-focus the investigation process by asking the question '*who* could have done *what* to avoid, help to avoid, or mitigate the incident?' Normally there are three controlling minds to a workplace incident:

- the manager who controls the system;
- the supervisor who controls the day-to-day workplace, and
- the employee who controls the activity.

Each of the above could have contributed to avoiding the incident or reducing its consequence. By examining these three aspects a greater insight into the deficiencies of the safety system can be gained and more improvement opportunities identified. *Figure 2.6.12* provides an example of this approach.

The maintenance supervisor sends Joe, the mechanic, to fix a leak coming from a capped pipe elbow. The capped elbow is located 9 feet high on a wall and had been left after the removal of a pump which was once required for the production process. Joe takes a step ladder but he finds that he cannot get to the nearest point because pallets of product have been stored underneath. Joe places the ladder as close as he can so that he can stretch to reach the leaking joint. As he applies force to the wrench, the ladder moves away from underneath him. He falls and badly bruises his shoulder and ribs. At the inquiry, it emerges that a permit-to-work was not issued as required by the Maintenance Supervisor and the Production Supervisor was not informed (who would have looked at the job and removed the pallets to give safe access). The inquiry noted that the capped pipe elbow could and should have been removed when the pump was removed during the process changes made two years previously.

Actions within the managers control:
Remove leaking joint
Review permit-to-work training
Organise ladder safety training

Actions within the supervisors control:
Check location of pallet storage
Check job locations before sending people out to work

Actions within the employees control:
Request permit-to-work if forgotten
Do not place oneself in unsafe positions
Choose the correct tools for the job

Figure 2.6.12 Alternative to 'single root cause' incident model

Once the various alternatives that could wholly or partly have contributed to avoiding the incident have been identified then preferred solutions can be chosen. In the example given in *Figure 2.6.12* the system cause was the 'non-removal' of the leaking joint. It would be appropriate to remove it whatever the expense or inconvenience if it leaks flammable solvent and does so regularly. However, if leakage is rare and when it occurs only small amounts of water drip from it, and if it is awkward or difficult to remove, then the 'permit-to-work' option may be more appropriate. In either event, some training in the permit procedure and in ladder safety should be undertaken. The incident investigation approach suggested promotes more creative wider ranging thinking than traditional prescriptive root cause analysis.

2.6.9 Learning organisations

A key element of successful health and safety management is the ability of the organisation to learn from its experiences. Arrangements should be made so that proactive data (e.g. from audit results) and reactive data (e.g. from incidents) in one area of an organisation are shared with other areas in the organisation. In this way, areas have the opportunity to learn from the experiences of others and to consider whether they have similar deficiencies in their systems and processes. The availability of electronic mail systems greatly facilitates the rapid sharing of such information. The information format should provide:

- a summary description at the opening of the document;
- highlights of the key improvement opportunities;
- an opening screen containing all the essential information;
- attachments with greater detail for those receiving areas that need more information.

Sharing 'failure' information such as this is a feature of mature and confident organisations. It means that their first intention is to learn from their experiences, improve their safety performance and indicates that they operate in a 'no-blame' environment. It also means that their desire to improve outweighs their fear of any consequences that may arise from publishing information across their organisation. Included in this process of sharing data should be any comments made by visiting regulators or external third party auditors.

Many organisations find that incidents repeat themselves after a passage of time either in the same location or in another location. This is a particular challenge. It arises because organisations *per se* do not have memories – it's the employees who remember, and as they change jobs or leave and are replaced, their memories, knowledge and experiences are lost. It is important to record and file reports of incidents, remedial actions and their effectiveness for future reference. Particular attention should be paid to ensuring that the safety management system retains key learning points that have been developed with the passage of time. This can be achieved by:

- turning incidents into training packages for new entrants;
- adding explanatory notes to procedures explaining why a certain action is necessary;
- issuing company newsletters or bulletins referring to past experiences and the lessons to be learnt.

Learning from external sources is also important. Major incidents are usually investigated and reported by the regulatory authorities and their reports published. They are always worth reading and can provide lessons for organisations that operate in different business sectors. Commercial health and safety journals often carry a summary of incidents and court cases which again can highlight for other organisations areas that need to be checked. Learning from other organisations about their health and safety processes can best be achieved through a 'benchmarking' visit.

2.6.9.1 Benchmarking

Benchmarking is an increasingly important tool in safety management. The term 'benchmarking' describes a concept of comparison with and learning from other companies. There is a misconception that benchmarking of safety performance and processes should only take place with companies who are perceived to be superior, i.e. who set the 'benchmark'. However, by undertaking a benchmarking exercise the enterprise is able to obtain a better understanding of its own safety processes. This aspect alone brings benefits before it seeks to identify alternative processes and quality gaps in comparable processes observed in other companies.

A natural starting point for benchmarking exercises is to focus on the elements contained in their audit process. Features of benchmarking studies include:

- identification of the elements of the safety system
- how the elements fit together
- how management leads and monitors the safety system
- what topics are subject to formal risk assessments, who does them, and what is their quality
- how employees are involved and motivated to play a full part in safety.

From the above it is clear that safety benchmarking is less about a comparison of accident statistics and much more about comparing management processes and their effectiveness. Benchmarking should not be the sole preserve of the safety adviser but should involve line managers who can gain considerable benefit from it. The process compels them to focus their attention on both the subject of safety and their own performance. Safety advisers benefit because it enables them to formulate new improvement strategies which have been proved elsewhere and can be recommended to managers with confidence.

2.6.10 Safety management systems in small organisations

Small organisations are required to meet the same legal standards as large organisations. They usually have fewer, if any, specialist support staff and need to find ways to manage the hazards appropriate to their activities. This needs to be done systematically and a good starting point is to examine four aspects of the operations:

- Facilities – which includes premises, plant and employees.
- Products – which refers to the articles the enterprise produces for customers.
- Services – the service elements which the enterprise provides to customers.
- Contractors – the use made of contractors in any of the enterprise's activities.

Each of these aspects should be examined with a view to:

- identifying the hazards involved;
- identifying the different groups of people who might be harmed;

FACILITY	PRODUCT	SERVICES	CONTRACTOR
– Fire safety – Office and office equipment – Chemicals and substances – Machinery – Visitor safety – Electrical safety – Storage – Radiations – Noise – Employee communication etc.	– Information on its use – Composition of product – Maintenance of product – Transportation of product – Storage of product – Disposal of product etc.	– Safety while working on client's premises – Lone worker safety – Home working – Working hours – Travelling between locations etc.	– Selection criteria – Work methods – Materials used – Onsite hazards – Offsite hazards – Communication on safety issues and performance etc.

Figure 2.6.13 Example of safety management system topics for small organisations

- identify the possible circumstances that could arise to cause the harm and how likely they are to occur;
- examining the measures that currently exist to prevent harm occurring and to decide if they are adequate or if additional controls are necessary;
- making a record of the findings with the date and the name of the person who carried out the examination;
- making arrangements to review the findings periodically (at least once a year).

Figure 2.6.13 lists topics which fall within each aspect.

Small organisations generally find checklists a valuable aid in assessing their standard of compliance. Periodically they may need to seek professional assistance to ensure that the safety standards achieved are adequate for the hazards faced and that there are no glaring omissions in the scope of their checklists that need to be addressed.

2.6.11 Conclusion

Safety legislation has been on the statute book for many years but the development of safety management as a subject is a recent innovation. It is now recognised that the levels of safety expected by employees and the general public cannot be achieved without utilising safety management strategies. So much has been written about the subject that it is sometimes forgotten that conventional management techniques are just as applicable to safety as to any other aspect of business. In particular the emergence of quality management processes is highly relevant. The exact application of each element of a cohesive practical safety management system depends on the area of business and the aspirations of the organisation itself.

Effective safety management does not occur by chance. It arises out of a clear understanding of obligations and the application of considered strategies and techniques. Managing safety does not exist in isolation from other aspects of the business. An enterprise may change direction in terms of its business plan, its products or markets which in turn result in increases or reductions in the number of employees, relocation of premises and adopting changing technologies. In all these circumstances the obligations to the safety of its employees, customers and the public at large remain but it is only the style and techniques applied that differ. Recognising changing operational environments and adjusting the management techniques to suit are essential to improving safety. The principles outlined in this chapter remain constant. The techniques described offer flexibility. Together they offer a pragmatic approach to safe working for the busy manager.

References

1. Wilsons and Clyde Coal Co. Ltd. *v.* English (1938) AC 57 (HL)
2. Bett *v.* Dalmey Oil Co. (1905) 7F (Ct of Sess.) 787
3. Edwards *v.* National Coal Board (1949) 1 KB 704, (1949) 1 All ER 743
4. Munkman, J., *Employer's Liability at Common Law*, 12th edn. Butterworths, London (1996)
5. Health and Safety Executive, Guidance booklet no. HSG 65, *Successful health and safety management*, HSE Books, Sudbury (1997)
6. British Standards Institution, OHSAS 18002: 2000, *Occupational health and safety management systems*, BSI, London (2000)
7. British Standards Institution, BS EN ISO 9000, *Quality systems. Specification for the design/development, production, installation and servicing*, BSI, London (1994)
8. British Standards Institution, BS EN ISO 14001, *Environmental management systems. Specification with guidance for use*, BSI, London (1996)
9. Chemical Industries Association, *Responsible Care*, CIA, London
10. Health and Safety Executive, Legal publication no. L23, *Manual Handling. Manual Handling Operations Regulations 1992, Guidance on the Regulations*, HSE Books, Sudbury (1998)
11. Peters, T.J. and Waterman, R.H., *In Search of Excellence*, Harper & Row, New York (1982)
12. Bird, F.E. and Germain, G.L. *Practical Loss Control Leadership*, International Loss Control Institute, Loganville (1986)
13. British Safety Council, *5 Star Health and Safety Management System Audits*, BSC, London
14. HASTAM, *CHASE*, HASTAM, Birmingham
15. Health and Safety Executive, Guidance publication, *A guide to measuring health and safety performance*, www.hse.gov.uk/opsonit/perfmeas.htm
16. Pojasek, R.B., Five S's: A tool that prepares an organisation for change, *Environmental Quality Management*, pp. 97–103, Autumn 1999

Chapter 2.7

The individual and safety

Andrew Hale

2.7.1 Introduction: What does this chapter try to do?

The history of safety revolves around the way in which we see the role of the individual in accident causation and prevention. Fundamentalist religions in many ages and countries have seen accidents as the punishment for sins and prevention as the need to pray and live a God-fearing life. The 1930s saw the flowering of the theory of accident proneness[1], which sought the major causes of accidents in innate or learned characteristics of individuals, and prevention in careful selection, job placement and early training. There is still an obsession with blame in some companies and in sensational media headlines. The fact that the blame still focuses most often on the last person to touch the system (the operator, driver, pilot) indicates that elements of the accident prone view of individuals still haunt us. Ergonomics, from the 1950s onwards, changed our focus to the human–technology interface, rather than the individual alone[2]. Since the 1980s there has been an increasing interest in the behaviour of managers and supervisors as individuals vital in the control and management of safety. We have come to see individual behaviour in relation to safety as something conditioned by the whole social and organisational environment, and something which can be managed. The turn of the century has seen organisational safety culture as the fashionable topic for study and concern. Now we are trying to influence and steer the collective attitudes and beliefs of a whole department or organisation in a direction favourable to safety[3]. *Figure 2.7.1* illustrates these changing views of accident cause and prevention.

From each of these many different viewpoints individual behaviour is always crucial to the attainment of safety performance, but the way we have seen it has been different. Sometimes the individual has been seen as *the problem*, to be eliminated as much as possible from workplaces and the vicinity of dangerous technology. Sometimes the individual has been seen as our only *guarantee of safety*, to be carefully selected, groomed, consulted and placed centrally in our attention and our culture of safety. As we might expect, there is some truth in all these viewpoints; the trick is knowing which view works in which situations.

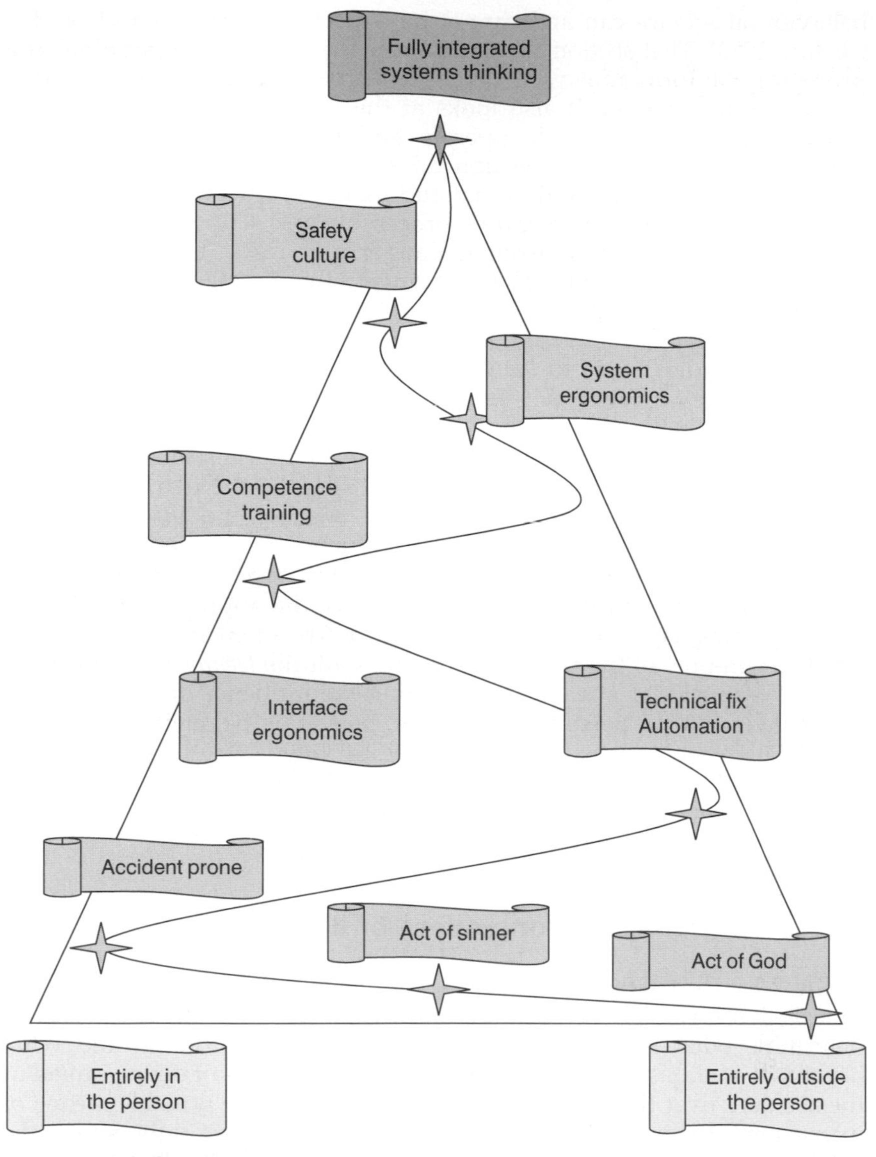

Figure 2.7.1 Changing perceptions of accident causes and the role of humans

This chapter tries to place all of these different views of the role of the individual in safety performance into the broad context of system safety (the peak of the triangle in *Figure 2.7.1*). It provides some basic grounding in the ideas, theories and models of behavioural science, which can illustrate and illuminate the complexity of the individual and help to explain why we always tend to have only a partial view of that role.

It starts off by positioning the individual in the process of development of accidents (section 2.7.2). Then it gives some basic ideas about what behavioural science can and cannot be expected to explain and predict (section 2.7.3). That section also introduces the idea of the individual as a processor of information, which is the main model and metaphor used in the rest of the chapter. It also looks at the different levels of conscious processing at which people operate and the different strengths and weaknesses at each level. Section 2.7.4 goes through an information-processing model of the individual and looks at how good people are at the different aspects of perception, processing and action, when it comes to understanding and controlling danger and risks. The final section (2.7.5) looks at the individual and change. How can the behaviour of a person be changed and how does an individual respond to change?

In the paragraphs above the word 'individual' has been consequently used. The reader needs to keep constantly in mind that managers are individuals too. We will not be talking just about how to control and influence the behaviour of the workforce at the sharp end of safety. Managers and safety staff who think that behavioural science is there to help them control their workforce are still stuck in the patriarchal ideas characterised by the concept of accident proneness. It is just as, or even more, important to understand and influence the individual in the boardroom, the designer's office, the finance department or the regulatory body. The individual is also often not the appropriate entity for understanding behaviour or influencing it. We often need to think of the 'three-in-one' of the *person* interacting with the *technology* to perform a *task*. We must also always keep in mind the influence of social groups on behaviour. One person defines him/herself as belonging to several social or peer groups and is therefore interested in pleasing or impressing them. These can be work, family or social groups, whose attitudes, beliefs and experience strongly condition the behaviour of each and every one of us.

2.7.2. Individuals as controllers of danger

Figure 2.7.2 shows the process of development of an accident from the first design ideas of a technology or workplace. These design decisions determine what is the normal, or accepted work situation and what hazards it contains. The central boxes show the process of development of the accident over time. The boxes on the left show the possible means of intervening to arrest the process. The boxes on the right show the learning process we should go through, as individuals and as an organisation, based on the events. In-built control measures, either technical or procedural, keep the danger under control in the normal situation. Deviations from this controlled state, or failures to predict scenarios which can lead to harm, set the process of accident development in motion. Detection and recovery can arrest the process and restore the state to 'normal'. Once control is irrecoverably lost, actions can still protect the most valuable and vulnerable system elements, rescue them from harm and speed their recovery.

This model illustrates where people have an influence on the accident process. Those who see the individual operator as the source of problems

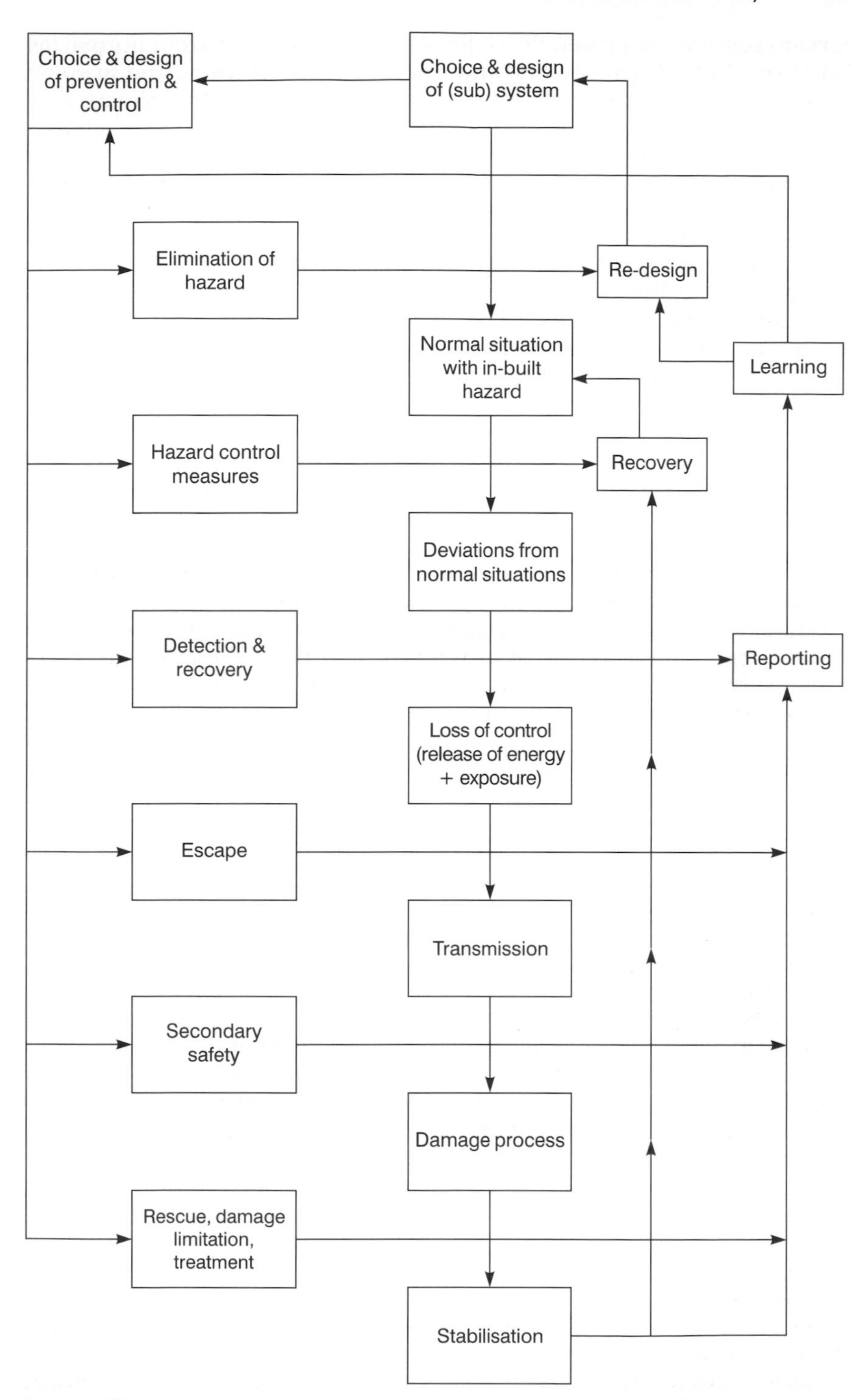

Figure 2.7.2 Accidents as deviation processes

tend to concentrate on people as the causes of deviations from normal, as what sets the accident process in motion. This is certainly true in some accidents. People do deviate from rules and safe procedures. We need to understand why and know how to discourage that deviation when it is inappropriate. However, we also need to emphasise the positive side of human behaviour. Sometimes it is necessary to deviate from rules to preserve safety, or to improvise when there are no rules. People are still far better than machines or computers at doing this, especially in unpredicted (or unpredictable) situations. People also play an overwhelmingly positive role in two other aspects. They can, as designers of the 'normal situation', predict hazards and deviations in advance and carry out their designs to prevent them or enable their recovery. Designers are responsible for the content of all the boxes on the left-hand side of the figure and need to know enough about human behaviour to predict and support the role of people in all those boxes. The overwhelmingly positive side of human behaviour is, however, its ability, in the vast majority of cases, to detect and recover the accident process before it results in harm. People are past masters at error recovery. This usually more than compensates for their great ability to create errors. If we try to eliminate people from systems by automation, we can live to regret that we have also eliminated them as agents of recovery.

Ergonomics has taught us that we should never expect to eliminate human error completely. Anyone who conscientiously logs all the errors they make will come up with several, if not tens, per hour. If your rate of typing errors is anything like mine, it can even run into hundreds per hour. Most errors have few consequences; most are corrected on the spot. The advent of word processors has proved once again that facilitating the correction of errors, rather than forcing people to type correctly first time, can enormously speed the process of typing, especially for the non-professional typist. This emphasises that the support of error recovery can be an option preferable to error elimination. If we try to automate the human out of the system by replacing people with the machines and computers we now have, we end up leaving the human to do the tasks the automation cannot do. Unfortunately these prove to be the supervisory tasks, to intervene when the machine cannot cope, which are the very tasks humans can no longer do if they have been taken 'out of the loop' of normal control of the technology. We then get the worst of both worlds[4]. The objective of this chapter is therefore to provide enough insight into human behaviour to see where behaviour can be so guided as to eliminate risks and where it can be supported to recover from incipient danger.

While the idea of accidents and errors as deviations from the normal and the designed is still a powerful and useful concept, we should not overemphasise it. It encourages the idea that safety is purely a matter of defining the one right way to do things, teaching that to people and then keeping them on this straight and narrow road. This underestimates the difficulty of defining safe ways of dealing with all conceivable situations, and particularly the impossibility of conceiving what all dangerous situations could be. It is also a far too negative view of safety, which equates it with policing people and preventing them doing what they want, in a paternalistic sort of way. Jens Rasmussen[5] has provided a more challenging analogy in talking about the way organisations are pushed by competing pressures towards unsafe areas on the edge of controllability. We can adapt his metaphor and apply it to understanding individual behaviour in relation to danger. This is like steering a course in partly unknown waters, under pressure from many motives, which are always only partially compatible with safety. The task is continuously to stay off the reefs which represent the potential accidents. If we can stay within a certain boundary, we will not come close to the reefs and only very rare obstacles in the open waters will harm us. But sailing very close to the reefs can have great advantages for other objectives, such as convenience, speed, production, etc. The trick is to find a course which allows us constantly to measure the distance to the reefs, but gives us time to manoeuvre away from them if danger increases. This is steering by the boundaries of what is safe enough, rather than sticking strictly to the middle of the channel, as far as possible away from the reefs. Such a concept offers more room for manoeuvre and for individual choice than an idea of strict standards and no deviation. We can also still use the steps in *Figure 2.7.2*, if we replace the 'deviations from normal' with 'approaches towards safe limits'.

2.7.3 Behavioural science and the human information processor

2.7.3.1 What is behavioural science?

Behavioural science has four main aims: to *describe*, to *explain* and to *predict* human behaviour, with the objective of *influencing* it. The safety adviser is interested particularly in behaviour at work and more particularly in the behaviour of people in situations which may endanger their health or safety. Even describing behaviour in such situations is not always easy. People who are observed may not behave as they normally would. Observers, or individuals trying to explain afterwards what they did in a situation, may describe what they expected to see or do, rather than what someone actually did. Explaining behaviour requires theories about why it happens. We need to get to this level of understanding in order to understand behaviour in accidents and to decide how to design hardware and organisations which will be useable by people and complement their skills. What we really want to do is to predict in detail how decisions on selection, training, design and management will influence the way people will behave in the future. If we can do that, we can modify our decisions and influence the ultimate behaviour. This is a very severe test of theories about individual behaviour, and psychology is often not far enough advanced as a science to withstand such scrutiny. In this chapter the aim is to describe in broad terms what is known and can be used to help us understand and guide human behaviour.

A human individual is far more complex than any machine, and when individuals are placed together in groups and organisations the interactions between them add many times to the complexity which needs to be understood. Individuals are also extremely adaptable. They change their behaviour as they learn and if they know that they are being observed. They may change it in different social situations, behaving in front of their friends or work colleagues in ways they would be embarrassed to do before their parents. Each individual is to an extent unique because of their unique experience. Because of all this the behavioural scientists' task can be seen to be daunting indeed. What they try to do is to understand the patterns and the influences in order to simplify the complexity. The explanations and predictions of behavioural science therefore have wider margins of error than those which can be offered by engineers or even by doctors. Statements made about behaviour will usually be qualified by words such as 'probably' or 'in general'. Individual exceptions to the predictions will always occur.

Because of its limitations behavioural science is dismissed by some as being no more than common sense dressed up in fancy language. Everyone thinks they are an expert on human behaviour, and they are partially right. All individuals must have some ability to explain and predict the behaviour of themselves and others, or they would not be able to function effectively in the world. However, the most common way for non-experts to try to understand another's behaviour is to think how you would behave yourself in those circumstances. People forget how broad the range of individual differences is, and so how poor this comparison will often be. Most individuals' explanations and predictions are, therefore, quite often proved wrong. Behavioural science used in a systematic and rigorous way can always improve on unaided 'common sense'.

Behavioural science commonly works by developing models of particular aspects of human behaviour. These models are inevitably simplifications of real life, in order to make it comprehensible. The models are frequently analogies drawn from other branches of knowledge and can reflect in their history the history of technology. We used to represent the brain as a telephone exchange; we now routinely compare it to a computer. Different behavioural scientists may use different analogies, or divide up the complexity in different ways. This, to some extent, explains why there sometimes appear to be parallel and incompatible theories about the same aspect of human behaviour. Analogies are powerful and useful, but they have limitations which must always be acknowledged. They can never be perfect descriptions of the way that an individual functions, and will be useful only within their limits. In the sections which follow some models will be described and used to explain particular aspects of behaviour. Readers are urged to use them, but with care.

2.7.3.2 The relevance of behavioural science to health and safety

Here are some of the questions relevant to a safety practitioner which behavioural science can help to answer:

- What sort of hazards will people spot easily, and which will they miss?

- Are there vulnerable times of day for errors and accidents?
- Can you predict what sorts of people will have accidents in particular circumstances?
- Why do people ignore safety rules or fail to use protective equipment and what changes can be made in the rules or equipment to make it more likely that people will use them when they should?
- What sorts of beliefs do supervisors and managers have about what causes accidents and how does that affect how they try to manage them?
- If people understand how things harm them, will it make them take more care?
- Can you frighten people into being safe?
- What will motivate a line manager to spend more time on safety?
- What knowledge and training do people need to cope with emergencies?
- What dangers arise, or are prevented when people work in teams?
- How do company payment, incentive and promotion schemes affect people's behaviour in the face of danger?
- How can training help people to take care?
- When are committees better than individuals at solving health and safety problems?
- What constitutes a good set of attitudes and beliefs, which make safety a central goal of an organisational culture?

The list of questions can go on almost indefinitely. Before studying behavioural science it is a valuable exercise to draw up a list of questions relevant to your own workplace, which you hope more knowledge of behavioural science will help you to solve. See how many of the questions you have answered, or reformulated, by the end of your study. That is a good test of your study course, and of this book as a part of it.

2.7.3.3 The human being as a system

A common model used in behavioural science, and in the biological and engineering sciences, is the 'systems' model. Systems are defined as organised entities which are separated by distinct boundaries from the environment in which they operate. They import things across those boundaries, such as energy and information; they transform those inputs inside the system, and export some form of output back across the boundaries. Open systems are entities which have goals or objectives which they pursue by organising and regulating their internal activity and their interchange with their environment. They use the feedback from the environment to check constantly whether they are getting nearer to or further away from their objectives. *Figure 2.7.3* shows a generalised system model.

Such models can be applied to a single cell in the body, to the individual as a whole, to a group of individuals who are working together, and to an organisation such as a company.

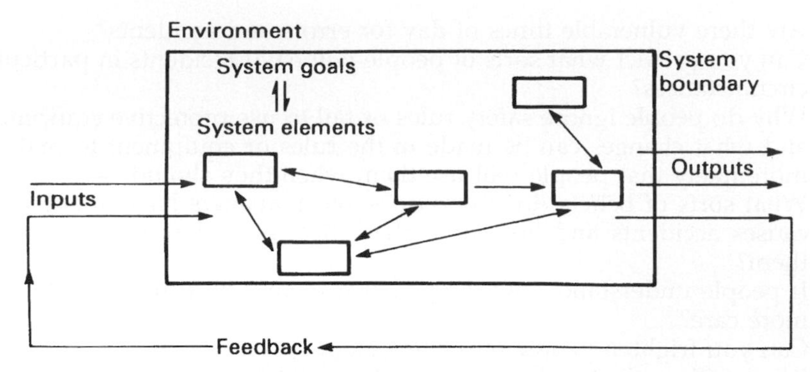

Figure 2.7.3 Simplified system model (adapted from Hale and Glendon[6])

Figure 2.7.4[7] considers the human being as a system for taking in, processing and acting on information. The system objective we consider in this chapter is the avoidance of harm to the person or to others. Accidents and ill-health can then be conceived as damage which occurs to the system when one or other part of this information processing fails. The human factor causes of accidents can be classified according to which part of the system failed. In section 2.7.4 we will apply this model directly to understanding behaviour in the face of danger, but it is useful first to give some basic insights into the different aspects of human functioning, particularly of goals and motivation.

2.7.3.4 Some basic facets of human information processing and action

2.7.3.4.1 Goals, objectives and motivation

Any understanding of human behaviour must start with an attempt to describe the goals and objectives of the human system. Individuals have many goals. Some such as the acquisition of food and drink are innate. Others are acquired, sometimes as means of achieving the innate goals, and sometimes as ends in themselves, for example the acquisition of money, attainment of promotion, purchase of a house, etc. Some are short term, e.g. food at dinnertime; others are much longer term, e.g. earning enough for retirement. In some cases the short- and long-term goals may be in conflict. For example, a person may fail to check equipment before starting work in order to satisfy the short-term goal of getting the job done as fast as possible, as a result jeopardising the long-term goal of preserving his own health and safety.

Not all goals are consciously pursued, either because people may not want to admit even to themselves that they are pursuing a particular goal, or because the goal is so basic that it has been built into the person's behaviour and no longer requires any conscious attention.

An individual's goals can be conceived of as vying with each other to see which one will control the system from moment to moment. People

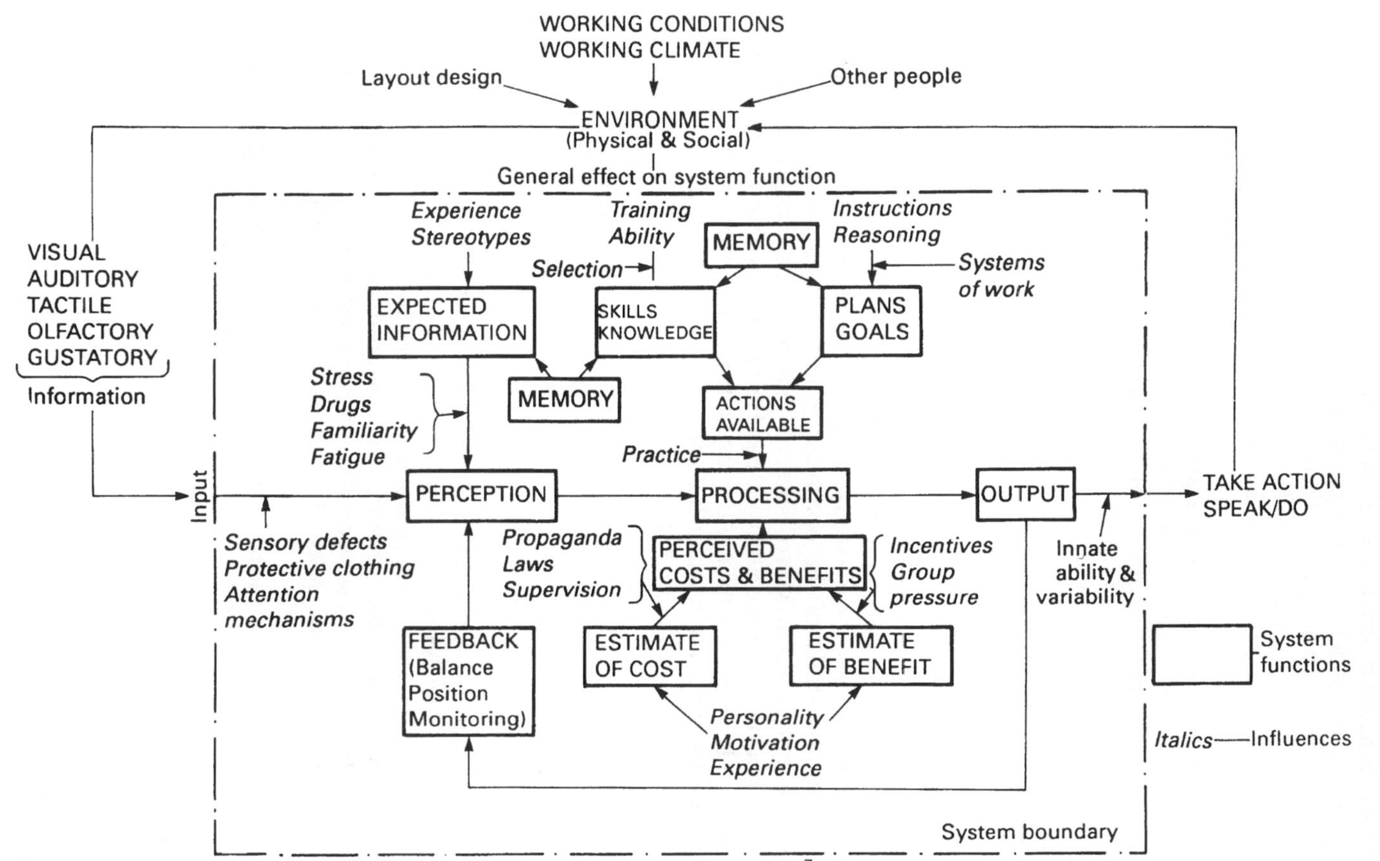

Figure 2.7.4 Systems model of human behaviour (adapted from Hale and Hale[7])

will therefore show to some extent different and sometimes contradictory behaviour from day to day, and certainly from year to year, depending on which goal is uppermost at the time. However, people will also show consistency in their behaviour, since the power of each of their goals to capture control of the system will change only slowly over the sort of time periods which concern those interested in behaviour at work.

Many theorists have written about motivation, particularly motivation at work. They have emphasised different aspects at different times as being dominant motives, and have frequently tried to give a hierarchical ordering of their importance. At the start of the 19th century Taylor[8] divided people into two groups: potential managers who were competent at and enjoyed planning, organising and monitoring work, and the majority of the workforce who did not like those activities but preferred to have simple tasks set out for them. Taylor considered that, once work had been rationally organised by the former and the latter had been trained to carry it out, money was the main motive force to get more work out of them. His ideas of scientific management encouraged the development of division of labour and the flow line process, work-study and the concentration on training, selection and study of the optimum conditions for work. Later work, such as that by Elton Mayo[9] showed that this was much too simple a view. His studies in the 1930s at the Hawthorne works of the Western Electric Company in Chicago led to the realisation that people were not automata operated by money, but that they worked within social norms of a fair day's work for a fair day's pay. It showed that they were responsive to social pressure from their peers, and to interest shown in them by the company. This led to a new emphasis on the role of the supervisor as group leader, rather than as autocrat, and also to a greater emphasis on building group morale. Maslow[10] looked at the motivation of people who were successful and satisfied with their work. He found that there was always an important element of achievement, self-esteem and personal growth in their descriptions of their behaviour. He put forward his theory of the hierarchy of needs (*Figure 2.7.5*) to express this concept of growth. He postulated that the homeostatic needs had to be satisfied before the growth needs would emerge. Although this hierarchy has not been subjected to rigorous scientific confirmation, it is broadly borne out by research studies, at least in Western capitalist countries.

Modern motivation theory tries to incorporate what is valuable from all of the earlier theories, and recognises that there are individual differences in the strengths of different motivations both between individuals and over time in the same individual. As far as possible, incentives need to be matched to the individual and the situation (the job of human resources management). It is also recognised that the human system is more complex than many early theories postulated, and that expectations play a strong part in motivation[11]. In other words the force of a motivator is dependent on the sum of the value of the reward and the expectancy that a particular behaviour will lead to the reward. If someone perceives that it will take a great deal of effort to

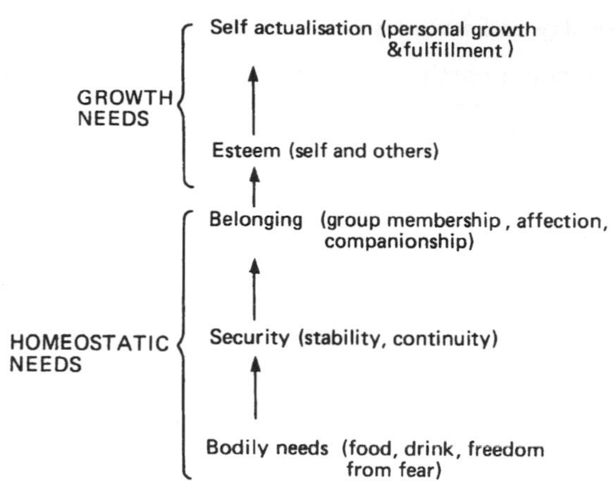

Figure 2.7.5 Hierarchy of needs (after Maslow)

gain any increase in reward, or that the reward does not appear to be dependent upon how much effort is actually put in, their behaviour will not be influenced by that reward.

The unique combination of goals and behaviour which represents each individual's adaptation to the environment in which he finds himself is one definition which is given to the word personality (see 2.7.3.4.2). Thus, those who habitually place a high value on their need for acceptance by people around them are called gregarious or friendly, whereas those people who habitually subordinate their need for approval by peers to their goal of achieving high status in the organisation, we call ambitious.

It may be thought axiomatic that the preservation of the self (i.e. of safety and health) would be one of the basic goals of all individuals. Clearly this is not a goal of all people at all times, as the statistics of suicides must indicate*. However, we can assume that most failures to achieve that goal are because individuals do not perceive that their safety is immediately threatened, and so other goals which the individual has are given priority over the one of self-preservation. If risks are perceived to be small and gains great, then individuals are willing to trade off a slight increase in (long-term) risk for a bigger short-term gain in speed or comfort. It also seems that the majority of people are optimists in respect of risk. They think that situations will stay under control, particularly if they themselves are the ones who can influence the risk. Hence, typically three-quarters of drivers asked will say that they are safer than the average driver†.

* In the case of suicide, murder or self-sacrifice some other goal supersedes, but we do not deal with those cases here. Freudian theorists have speculated that accidents can be unconscious attempts to punish oneself, which override self-preservation. However, we do not find this a useful concept when dealing with behaviour at work.

† In case the paradox here is not clear, the figure should be no higher than 50%!

2.7.3.4.2 Personality and attitudes

Personality is formed partly from innate characteristics, inherited genetically, partly by what happens in the critical years of maturation and partly by subsequent experience. Since each individual will have been subject to a unique mixture of all of these factors, the result is that no two individuals will be entirely alike in the combination of characteristics which make up their behaviour. No two people will perceive the world in quite the same way. No two individuals will react in quite the same way to the same circumstances confronting them. To predict with certainty how any one individual will behave in a particular set of circumstances would require a complete knowledge of all the factors which had gone to make up that person, and that we never have. However, the position is not entirely hopeless since there is enough common ground in individual responses to most circumstances to make predictions worthwhile. That common ground within one person is labelled personality; where it is common ground between people in a group we label it norms or group attitudes.

The study of personality is an area of psychology which has spawned many parallel and conflicting theories. One style of theory tries to explain where personality comes from and classifies people into 'types' or groups based on differences in personality development; other theories merely classify the end result and measure existing differences (trait theories). A typical example of the latter is Cattell's trait theory[12]. From extensive research based upon the responses to questionnaires on their beliefs and preferences by many thousands of individuals, Cattell produced a list of 16 personality factors (see *Table 2.7.1*). The factors are envisaged as 16 dimensions on which an individual's position can be plotted to produce a profile which describes that unique individual. Since someone can score from 1 to 10 on each scale, these 16 scales provide 10^{16} unique character combinations or personalities, which is more than the total number of human beings who have ever walked the earth since Homo sapiens evolved.

Table 2.7.1 Cattell's 16 personality factors

1	Reserved, detached, critical	Outgoing, warm-hearted
2	Less intelligent, concrete thinking	More intelligent, abstract thinking
3	Affected by feelings, easily upset	Emotionally stable, faces reality
4	Humble, mild, accommodating	Assertive, aggressive, stubborn
5	Sober, prudent, serious	Happy-go-lucky, impulsive, lively
6	Expedient, disregards rules	Conscientious, persevering
7	Shy, restrained, timid	Venturesome, socially bold
8	Tough-minded, self-reliant	Tender-minded, clinging
9	Trusting, adaptable	Suspicious, self-opinionated
10	Practical, careful	Imaginative
11	Forthright, natural	Shrewd, calculating
12	Self-assured, confident	Apprehensive, self-reproaching
13	Conservative	Experimenting, liberal
14	Group-dependent	Self-sufficient
15	Undisciplined, in self-conflict	Controlled, socially precise
16	Relaxed, tranquil	Tense, frustrated

There have been attempts to relate personality types to accident rates, notably to define 'risk takers'. In section 2.7.4 this is briefly discussed under the heading of accident proneness.

While personality is the underlying core of relatively unchanging behavioural consistency in a person, we can consider attitudes as rather more superficial manifestations. Attitude is sometimes defined as 'a tendency to behave in a particular way in a certain situation'. Underlying this definition is one of the thorniest problems in psychology, the consistency between what people say they believe or will do and what they actually do. As with personality many theories abound in this area. We illustrate just one. In their theory Fishbein and Ajzen[13] define:

Attitude. Attraction to or repulsion from an object, person or situation. Evaluation, e.g. liking rock climbing, favouring trades unions, avoiding unproven new machinery, etc.
Belief. Information about an object, person or situation (true or false) linking an attribute to it, e.g. that machine guards are a hindrance to production, that accidents are caused by careless workers.
Behavioural intention. People's beliefs about what they will do if a given situation arises in the future, e.g. that they will use a safety belt when driving on a motorway, or that they could find the exits to the building fast enough to escape a fire.
Behaviour. Actual overt action, e.g. actually wearing your seat belt or evacuating the building.

All these are linked in Fishbein and Ajzen's theory as shown in *Figure 2.7.6.*

As an example, someone may believe that breathing apparatus is uncomfortable and dislike it (which may feed back to beliefs by making that person hypercritical of the comfort of any new apparatus). This may result in resistance to wearing it, but, knowing that it is a company rule (norm) the person will hurriedly put it on (behaviour) when the safety adviser walks by (trigger). If this happens many times he may find it is not so bad after all and there will be feedback which will change the beliefs.

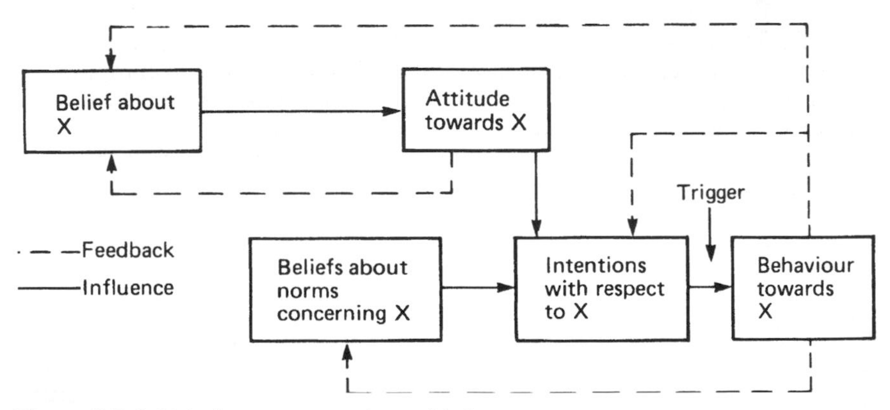

Figure 2.7.6 Links between attitudes and behaviour

2.7.3.4.3 From inputs to outputs

Figure 2.7.4 shows the process of taking in information from the environment, processing it and taking actions. The general effects of the environment will be discussed in section 2.7.4.6. In section 2.7.4.2 we consider the question of individual differences in accident susceptibility because of the way an individual functions at each of these steps.

Feedback and monitoring loops. The most important thing to say about human behaviour and information processing is that it is a closed loop process. It is purposive and not purely reactive. We formulate intentions and objectives, scan the world to find information useful for reaching them and make plans to steer ourselves towards them. We then take action and monitor whether it takes us in the right direction to achieve what we want. If not, we adjust our behaviour. Attempts to influence behaviour are therefore always attempts to modify an ongoing process. They neither operate in a vacuum, nor start with a blank sheet. Unless we succeed in changing the goals and objectives a person is seeking to achieve, we must therefore always expect that pressure to change behaviour will be met with resistance or will only be partly vectored into the direction people will steer their own behaviour.

These feedback and monitoring loops operate at all levels. A basic physiological one is the 'proprioceptive' or 'kinaesthetic' sense which transmits information from the muscles and joints to the brain, informing it about their position in space, and their orientation one to another. This drives the sense of balance and position which can be disorientated by rapid movement, resulting in motion sickness. Other loops partly inside and partly outside the brain monitor where we have got to in a task. Still longer loops drive our learning.

Perception is a process which is also purposive and selective. It is not like a camera taking a snapshot of a whole scene. People register some aspects of the situation very rapidly, but ignore or overlook others which are not relevant to their goals at the time. These attention mechanisms can be crucial in spotting hazards or warning signals. Perception is also strongly influenced by expectations. We sometimes see what we expect to see, or rather we accept evidence from our senses much more readily if it matches what we expect. In this way we can be fooled by situations which have something in common with what we expect to happen and overlook vital differences. The control room operators at Three Mile Island did that in misdiagnosing the problem there.

Expectancy can be seen as a mental model of the real world which has been built up from experience over an individual's lifetime. This can form a very 'real' alternative for perception to direct input from the world itself. Every one living in an industrial society knows what a motor car looks like and can conjure up a mental picture of one comparatively easily. This means that, when confronted with a particular car in the real world, there is no need to take in all of the details, which are already on file in the brain. People can concentrate upon only those characteristics which differentiate this car from the 'standard' car of their mental picture, e.g. its colour, or make, or its driver. Again there is a cost: we can make errors. For example, in the UK we are so used to expecting the driver to be sitting on the right of the car, that we may not see that this particular car is from abroad and that the person on the left is the one driving.

Machines, processes, people, and whole situations are stored in the brain and can be recalled at will, like files from the hard disk of a computer. This cuts down enormously on the amount of information about any scene which an individual need take in order to perceive and understand it. The reliance upon expectation is an essential mechanism in skilled operation and many tasks would take a great deal longer to carry out if this was not so. Thought therefore needs to be given to ways in which reality can be made to fit people's simple models. Standardisation of machine controls, layout of workplaces, colour coding and symbols, etc. are all designed to achieve this result, as are codes of rules such as the Highway Code or Plant Operating Procedures. But standardisation has a hidden snag. The more standard things normally are, the more likely are exceptions to trap someone into an error. So, once they have given rise to strong expectations, standards and rules must be enforced 100% to avoid this danger. Any circumstances which are unclear or ambiguous (e.g. fog, poor lighting) or where an individual is under pressure of time, is distracted, worried, or fatigued, will encourage expectancy errors. In extreme cases individuals may even perceive and believe in what are in fact hallucinations.

Memory. The storage facility of the human system is the memory. The memory is divided into two different types of storage, a long-term, large capacity store which requires some time for access, and a short-term working storage, which is of very small capacity and rapidly decays, but can be tapped extremely rapidly.

The short-term memory is extremely susceptible to interference from other activities. It is used as a working store to remember where one has got to in a sequence of events, for example in isolating a piece of equipment for maintenance purposes. It also stores small bits of information between one stage of a process and another, such as the telephone number of a company between looking it up in the directory and dialling the number.

Long-term memory contains an abundant store of information which is organised in some form of classification. Any new information is perceived in terms of these categories (closely related to expectations) and may be forced into the classification system even when it may not fit exactly. In the process it can become distorted. This process probably also results in specific memories blurring into each other, with the result that the wrong memory may be retrieved from the store when it is demanded.

People are not able to retrieve on any one occasion all the things which they have stored in their memory. There are always things which they know, but cannot recall, and which 'pop out' of store at some later stage. They are there, but we have forgotten where we put them. This sort of limitation can frequently be overcome by recalling the circumstances in which the original memory was stored, or by approaching it via memories which we know were associated with it. Unavailability of memories may be crucial in emergency situations where speed of action is essential. A technique for overcoming unavailability is to recall and reuse the memories (knowledge and skills) at regular intervals. Refresher courses, emergency drills, and practice sessions all perform this function. However, one unwanted side effect of constant recall and reuse of memories is that they may undergo significant but slow change. When the memory is unpleasant, or shows the individual in a bad light, it is extremely likely that distortion will occur at each recall and it will be these that will be remembered rather than the original story. Testimony following an accident is notoriously subject to such distortion. People can quite genuinely remember doing what they should have done (the rule) rather than the slip they actually made. They can also construct memories as they go over the story time and again. 'I must have done X, if Y happened', can unconsciously become 'I did X and Y happened'.

Decision-making, or processing, has been the subject of an enormous amount of research. Much of it has been normative, often carried out by economists or management and policy theorists, and studies how decisions *should ideally* be reached. These decision theories often assume perfect knowledge of all alternative courses of action and their consequences and then calculate trade-offs and optimum courses of action based on utility functions. The theories can be modified to allow preferences such as 'minimax', which tries to minimise the occurrence of extreme (maximum) consequences, rather than optimising across all decisions. This is an attempt to model the well-known risk aversion which many people show in decisions they make in practice. However, even the most sophisticated normative theories do not predict or match normal human decision-making very well. Natural decisions are far from rational in the economists' sense of the word, which may explain why economists are so poor at predicting the behaviour of markets and economies, which are a combination of many people's naturalistic decisions interacting with each other.

Human decision-making is influenced by a combination of many factors. All alternatives are never fully known, or even looked for, and all of the consequences are not worked out in detail. This is usually too complex a process for people to carry out in their heads in the time available, and they do not take the time to do it exhaustively even when they could. They set limits to what they take account of, a situation we call 'bounded rationality'. These limits are set by experience and expectation and are subject to many biases. Reason[14] discusses these in relation to human error in his classic book. In many instances people reduce conscious decision-making to a process of following rules. We call these rules *habits*, which are pre-programmed sequences of decisions stored for use in routine circumstances (see also section 2.7.3.5). Decision-making is also strongly influenced by beliefs which may have only a tentative link to reality. We can see this at work in gamblers' betting behaviour[15], their belief that they can see regularities in the random behaviour of a roulette wheel, or can influence the outcome of a throw of the dice by blowing on them. It is also to be found in the gross overestimates of the time saved by (possibly dangerous) speeding and overtaking on the roads and the unquestioning acceptance of the efficacy of some safety measures, which may have no objective basis in the scientific literature[16].

Action. Once the decision to act has been made, the remaining limitations on the human system are those of its capacity to act, e.g. its speed, strength, versatility. Humans differ from machine systems in that their actions are not carbon copies of each other, even when the individual is carrying out the same task again and again. The objective may be unchanging, but the system adapts itself to small changes in body position, etc. to carry out a different sequence of muscular actions each time to achieve that same result. This use of constantly changing combinations of muscles in coordination is an essential means for the body to avoid fatiguing any particular muscle combination.

In all human actions there is also a trade-off between the speed of an action and its accuracy. Speed can be improved by reducing the amount of monitoring which the brain carries out during the course of the action, but only at the cost of reducing the accuracy.

2.7.3.5 Levels of behaviour and types of error

The description of behaviour given up to now has blurred a distinction which is vital in understanding the types of error which people make. This distinction has arisen from the work of Rasmussen[17] and Reason[14]. They distinguish three levels of behaviour, which show an increasing level of conscious control:

1 Skill-based behaviour in which people carry out routines on 'automatic pilot' with built-in checking loops.
2 Rule-based behaviour in which people select those routines, at a more or less conscious level, out of a very large inventory of possible routines built up over many years of experience.
3 Knowledge-based behaviour where people have to cope with situations which are new to them and for which they have no routines. This is a fully conscious process of interaction with the situation to solve a problem.

As a working principle we try to delegate control of behaviour to the most routine level at any given time. Only when we pick up signals that the more routine level is not coping, do we switch over to the next level (see *Figure 2.7.7*). This provides an efficient use of the limited resources of attention which we have at our disposal, and allows us to a limited extent to do two things at once. The crucial feature in achieving error-free operation is to ensure that the right level of operation is used at the right time. It can be just as disastrous to operate at too high a level of conscious control as at too routine a level.

Each level of functioning has its own characteristics and error types, which are described briefly in the following sections.

2.7.3.5.1 Skills and routines

The moment-to-moment decision-making about what action to take next in order to cope with and respond to the environment around the

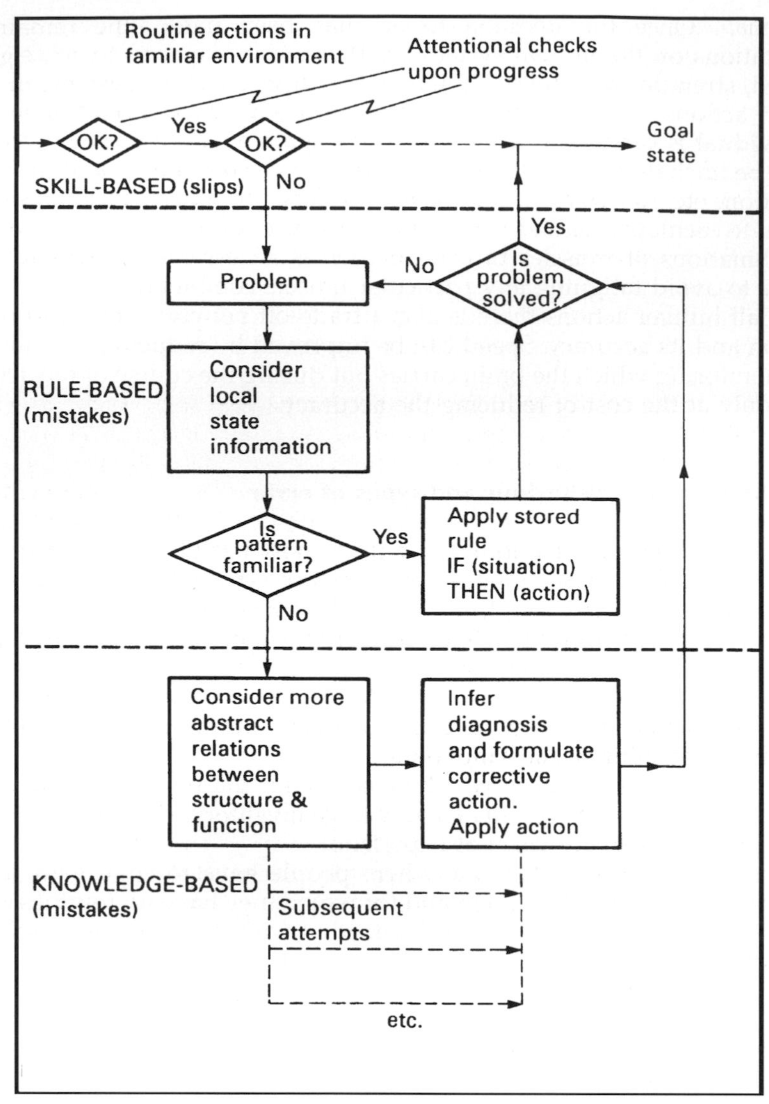

Figure 2.7.7 Dynamics of generic error-modelling system[14]

individual has to be funnelled through a narrow capacity channel which can only handle small numbers of items at one time. This limited capacity can be used to best effect by grouping actions together as packages or habits which can be set in motion as one, rather than as separate steps. Such habits form the basic structure of many repetitive skills, for example signing one's name, loading a component into a machine, changing gear in a car. Such grouping of activities does, however, carry with it the penalty that, once initiated, the sequence of actions is difficult to stop until it has run its course. Monitoring is turned down low during the routine. The packaging of actions in these chunks also places a greater premium on correct learning in the first place, since it becomes very difficult to insert any new actions into them at a later stage. All routines consist of a number of steps which have been highly practised and slotted together into a smooth chain, where completion of one step automatically triggers the next. Routine dangers which are constantly or frequently present in any situation are (or should be) kept under control by building the necessary checks and controls into the routines as they are learned. The checks still require a certain amount of attention and the relatively small number of errors which occur typically at this level of functioning are ones where that attention is disturbed in some way. The four examples below demonstrate this point.

1 If two routines have identical steps for part of their sequence, it is possible to slip from one to the other without noticing. This nearly always occurs from the less frequently to the more frequently used routine. An example is driving up to your normal workplace rather than turning off at a particular point to go to an early meeting in another building. Almost always these slips occur when the person is busy thinking about other things (e.g. making plans, worrying about something, under stress).
2 If someone is interrupted half way through a routine they may return to the routine at the wrong point and miss out a step (e.g. a routine check). Alternatively they may carry out an action twice (e.g. switching off the instrument they have just switched on, because both actions involve pushing in the same button).
3 If the environment in which the routine is practised changes without the person noticing, or that change slips the mind, the routine can be used or persist in the wrong situation. This can result in injury, for example if someone steps off a loading platform at the point where the steps always used to be, without remembering that they have recently been moved.
4 The final problem at this level is that routines are dynamic chains of behaviour and not static ones. There is a constant tendency to streamline them with practice and to erode steps which appear unnecessary. The most vulnerable steps are the routine checks for very infrequent problems in very reliable systems (e.g. checking the oil level in a new car engine).

Many of these errors occur because the boundary between skill-based and rule-based activity has not been correctly respected.

These sorts of error will be immediately obvious in many cases because the next step in the routine will not be possible. The danger comes when the routine can proceed apparently with no problem and things only go wrong much later. You can still start your car with a low oil level and find yourself driving down the fast lane of the motorway when the oil warning light comes on. The cure for the errors does not lie in trying to make people carry out their routines with more conscious attention. This will take too long and so be too inefficient, and will be subject over a short time to the erosion of the monitoring steps. The solution lies to a great extent with the designer of the routines (and so of the apparatus or system) to ensure that routines with different purposes are very different, so that unintended slipping from one to the other is avoided. Where this is not possible extra feedback signals can be built in to warn that the wrong path has been entered by mistake (see section 2.7.4.2).

The second line of defence is to train people thoroughly so that the correct steps are built into the system, and then to organise supervision and monitoring (by the people themselves, their work or reference group, and supervisors or safety staff) so that the steps do not get eroded.

2.7.3.5.2 Rules and diagnosis

When the routine checks indicate that all is not well, or when a choice is needed between two or more possible routines, people must switch to the rule-based level. Choice of a routine implies categorisation of the situation as 'A' or 'B' and choice of routine X which belongs to A or Y which belongs to B. This is a process of pattern recognition. This is analogous to computer program rules of the form IF . . ., THEN . . .

The errors which people make at this level are linked to a built-in bias in decision-making. We all have the tendency to formulate hypotheses about the situation which faces us on the basis of what has happened most often before. We then seek evidence to *confirm* that diagnosis rather than doing what the scientific method bids us and seeking to disprove the hypothesis. This means that people tend to think they are facing well-known problems until they get unequivocal evidence to the contrary. The Three Mile Island accident was a classic case of misdiagnosis. The operators persisted with a false diagnosis for several hours in the face of contradictory evidence until a person coming on shift (and so without the perceptual set coming from having made the initial diagnosis) detected the incompatibility between symptoms and diagnosis.

The solution lies in aiding people to make diagnoses more critically (e.g. defensive driving courses, permit-to-work systems) and in supporting their decisions with warnings about signals they may have missed (e.g. checking critical decisions with a colleague or supervisor before implementing them).

2.7.3.5.3 Knowledge and problem solving

When people are facing situations they have no personal rules for, they must switch to the fully interactive problem-solving stage. Here they have to rely upon their background knowledge of the system and the principles on which it works, in order to derive a new rule to cope with the new situation. There are meta-rules for problem solving which can be taught (see section 2.7.5.2). Besides these there is the creativity and intelligence of the individual and the thoroughness of their training in the principles underlying the machine or system. Errors at this level can be traced to:

1 Inadequate understanding of these principles (inadequate mental models).

2 Inadequate time to explore the problem thoroughly enough and to explore the consequences of different courses of action.
3 The tendency to shift back to rule-based operation too soon and to be satisfied with a solution without checking out the full ramifications it has for the system.

The first two are typical errors of novices, the last more of the expert. Experts are by definition the most capable of functioning at this knowledge level, but also the people who need to do so least often, because they have learned to reduce most problems to rules. They may also become less willing to accept that there are situations which do not fit their rules. Almost all experts overestimate their own expertise.

2.7.4 Individual behaviour in the face of danger

Hale and Glendon[6] combined the insights from the information-processing model with the theories of the three levels of functioning[14,17] and other sources[18], into a model of individual behaviour in the face of danger (*Figure 2.7.8*). Their model allows us to discuss a number of practical issues of how to influence human safety behaviour, e.g. through task design or training.

Danger is always present in the work situation (as in all other situations). The only question is how great is the danger and is it increasing, or could it suddenly increase in the foreseeable future. That is indicated at the top of the figure, which is the starting point for this discussion. The task of the individual is to keep danger under control, to avoid errors which provoke an increase in danger and to detect and avoid or recover from danger increasing from other reasons. Much of this activity occurs by more or less routine reaction to warning signals. Only occasionally do people in their normal work situations need actively to contemplate danger. However, these occasions are vitally important when they do occur. Examples of such activities are:

1 Designers making decisions about machine or workplace design, plant layout, work procedures, etc. They need to predict the actions of the people who will use these products and the hazards which will arise in use.
2 Operators, safety committees, safety advisers and inspectors carrying out hazard inspections, safety audits and surveys, who need to seek out hazards or shortcomings in preventive measures.
3 Policy makers in industry and government deciding whether a level of risk associated with a technology or plant location is to be accepted. Members of the public assessing whether that policy decision is acceptable to them.
4 Planners designing emergency plans for reacting to disasters.

Such decisions and activities are all largely carried out at the knowledge-based level and the borders with the rule-based level.

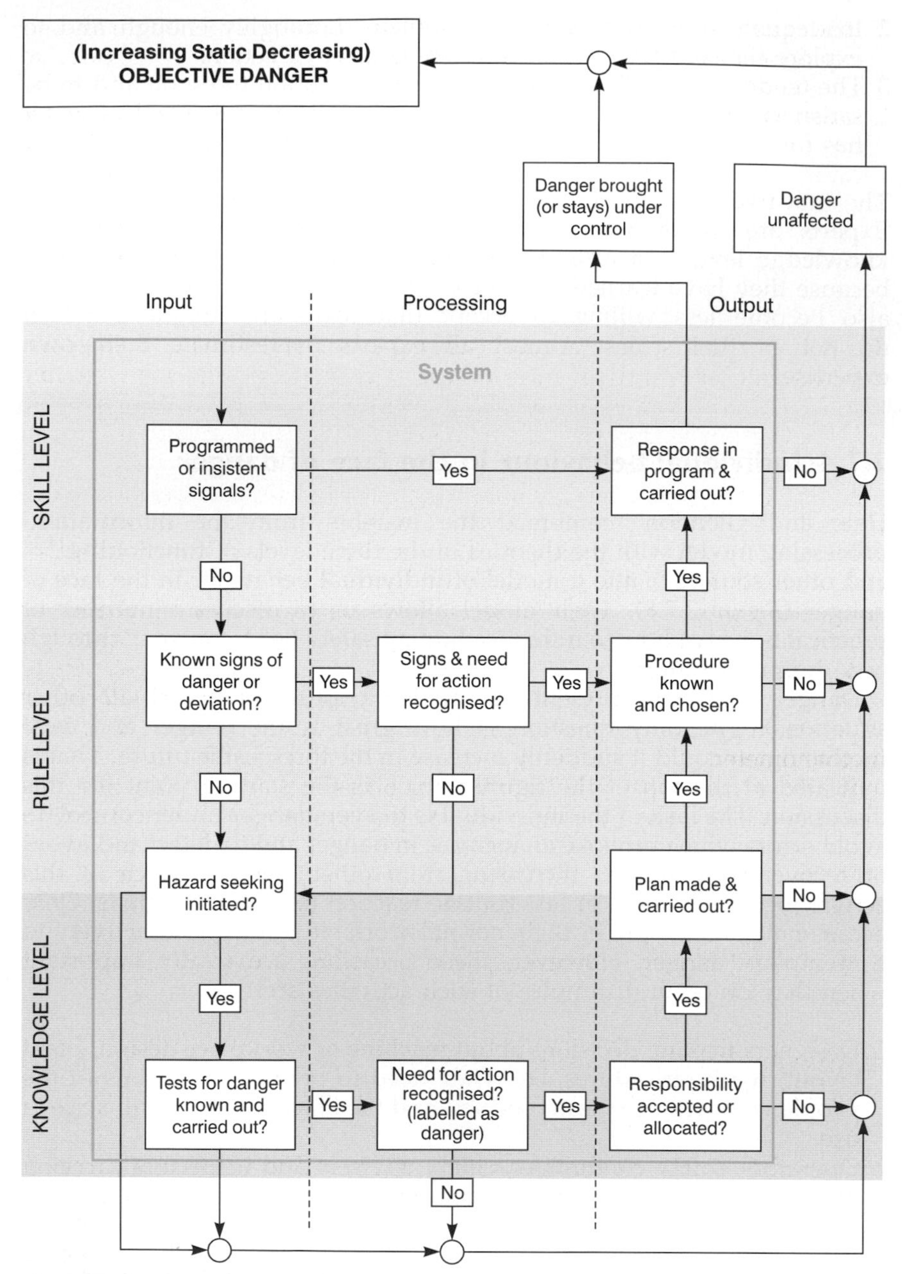

Figure 2.7.8 Individual behaviour in the face of danger[6]

2.7.4.1 Perception and hazard detection

This covers the four steps joined by arrows at the left-hand side of *Figure 2.7.8*. Hazard detection is important in three situations:

1 Emergency situations where the signals of danger are so clear and insistent that they lead to an instant, pre-programmed response to escape or to control the danger.
2 Other situations in which danger is known to be possible, but is not always present. Here an obvious warning can alert people to the danger and trigger the correct response to keep it under control.
3 In all other situations in which the warnings about danger are not obvious, people will only detect danger if they go looking for it and we need to know what initiates hazard seeking and what makes it successful.

Surprisingly little scientific study has been made of how hazard detection and recognition operates in any of these situations and what alerts people to the presence of danger. What follows is a summary of the available information[6]. It starts with some general information about perception and then takes each of the three situations above in turn.

Information gets into the human system through the sense organs. Hazards which are not perceptible to the senses will not be noticed unless suitable alarms are triggered by them or warnings given of them, or people go intelligently in search of them. Examples are odourless, colourless gases such as methane, X-rays, innocuous looking chemicals which are in fact carcinogens, ultrasound, or hazards in the dark. The canary falling off its perch in the mine because of its greater sensitivity to methane was an early example of a warning device, subsequently superseded by the colour change in a safety lamp flame and now by the methanometer.

If any of the senses are defective the necessary information may not arrive at the brain at all, or may be so distorted as to be unrecognisable. Some sensory defects are set out in *Table 2.7.2*. Limitations similar to sensory defects can also be 'imposed' by some of the equipment or clothing provided to protect people against exposure to danger, e.g. safety goggles, gloves or ear defenders. These sensory defects can be overcome by taking more care elsewhere in the behavioural model. This

Table 2.7.2 Some sensory defects

Sense	*Natural and 'imposed' sensory defects*
Sight	Colour blindness, astigmatism, long and short-sightedness, monocular vision, cataracts, vision distortion by goggles and face screens
Hearing	Obstructed ear canal, perforated ear drum, middle ear damage, catarrh, ear plugs or muffs altering the sound reaching the ear
Taste and smell	Lack of sensitivity, genetic limitations, catarrh, breathing apparatus screening out smells
Touch senses	Severed nerves, genetic defects, lack of sensitivity through gloves and aprons
Balance	Ménièrés disease, alcohol consumption, rapid motion, etc.

means, for example, that people with poor sight or hearing do not necessarily have more accidents[2]. They often learn to avoid situations which would be critical in this respect. If that choice is not open, however, they may be caught out.

The sense organs themselves have a limited capacity for receiving and transmitting information to the brain. The environment around us always contains far more information than they can accept and transmit. We try to cope with this limited capacity by using a switchable attention filter.

The brain classifies information by its source and type. It is capable of selecting on a number of parameters those stimuli that it will allow through a filter into the system. The setting of the filters on each sensory mode is partly conscious and partly unconscious. The main visual attention mechanism is the direction of gaze which ensures that the stimulus from the object being looked at is directed to the most sensitive part of the retina (fovea) where it can be analysed in detail. The rest of the field of view is relegated to the less sensitive parts of the retina. In normal activity this centre of focus is shifted constantly in a search pattern which ranges over the field of view until an object of interest is picked out. The senses can be tuned to seek out a particular facet such as a defect in a machine or component, provided that we know in advance what characteristics to tune it to. This ability is known as 'perceptual set'. People can also tune their hearing sense to pay attention to strange sounds coming from a particular part of a machine, while ignoring all others. Perceptual set can also show longer term settings which produce differences between individuals because of their interests and their experience. Safety advisers notice hazards because they are interested in them and used to finding them; motor cycle addicts spot a Bonneville in a crowded street, where others might not even notice that there was a motor bike.

Inputs which do not vary at all are usually not particularly useful to the system, e.g. a constant noise or smell, a clock ticking, the sensation of clothes rubbing on the skin. The filter alters over time to exclude such constant stimuli from consciousness. As soon as they change, however, e.g. the clock stops, the filter lets through this information and we notice it. In this way we are alerted very efficiently to 'something wrong' and we can go in search of it. This seems to be one of our main hazard detection devices.

These selective attention mechanisms are extremely efficient and invaluable in many tasks. But any mechanism which is selective carries with it the penalty that information which does not conform to the characteristics selected by the filter will not get through to the brain, however important that information is. People can be concentrating so hard on one task that they are unaware of other information. Hence someone can fall down a hole because they were walking along staring at some activity going on in the opposite direction. Pre-setting the filter can also lead to false alarms; searching a list for the name Jones we can sometimes be fooled by James. The maintenance fitter can expect to see a particular type of fault and jump to the conclusion it is there, based on just a few of the necessary symptoms. The cost of a rapid response is an increase in errors.

Expectations are one of the main bases for setting the attention filters. In that case we may see what we expect and not what is there. This is most often the case with situations which go against population stereotypes, for example a machine on which moving a lever downwards turns it off. We see the lever in the down position and 'see' it as 'on'. Other population stereotypes are red for danger and stop, clockwise turns the volume up or shuts the valve. These examples are very widely shared, but in other cases stereotypes for one population may contradict those held by others, e.g. you turn the light on by putting the switch down in Britain but by pushing it up in the USA and parts of Europe. Designs which do not match expectations can trap people into making errors.

In some circumstances these false expectations may result in little more than annoyance and delay. In other cases they may be a prelude to physical damage or injury. For example, a machine operator may reach rapidly towards his pile of components without looking and gash his hand on the sharp edge of one of them which has fallen off the pile and is nearer than he expected. The truck driver may drive rapidly through the doors which are reserved for trucks without looking or sounding a warning, because no one is supposed to be there, only to find that someone is using the truck doors as pedestrian access.

2.7.4.1.1 'Flight or fight' responses

The body has built-in danger detectors, which trigger instant reaction. All extremes of heat, cold, loud noise, rapid movement, strong smells, smoke or irritant chemicals in the lungs are pre-programmed to set off the body's fight or flight responses. Whether we respond adequately to such stimuli depends partly on how extreme they are, but has everything to do with the programming of the response and little to do with any perceptual problems. In extreme danger, such as a large fire, only some 15% of people seem to respond with rational alertness and rapidity. Another 15% seem to freeze and become totally passive, while the rest show impaired alertness and fall back on well-learned behaviour and routines, which may or may not be appropriate for the situation. Hence customers in a shop or disco will tend to try to find the way back to the entrance they normally use, ignoring closer fire exit signs. People may stop to collect their belongings before leaving, as they routinely do at the end of the working day. Car drivers will tend to brake and steer in an imminent collision, as they do to avoid more common and less serious situations. It takes an enormous amount of training and practice to get over these conservative responses and give people the full armoury of necessary responses and the capacity to use them appropriately. An effective response to emergencies in situations where a large or shifting population may be present, therefore, depends upon training at least a small number of people to that level and making sure they can influence and control those who have not been trained. This applies to sports stadia, shops, theatres and discos, hotels and many workplaces with high staff turnover.

Some workplaces, such as textile mills, steel rolling mills or high rise construction sites have the same effect on people the first time they see or hear them as the emergency situations mentioned above. They are scared by the noise, heat and fast moving machinery or by the heights. Yet people who have worked there for some time are quite happy there and do not regard them as overly dangerous. This illustrates a learning process, which is vitally important in understanding hazard detection and recognition. We can learn that the insistent danger signals do not actually mean danger. As a person learns where the specific dangers in such workplaces lie, the general fear response gets replaced by a much more sophisticated sense of hazard. The individual learns that it is quite possible to approach seemingly horrific dangers quite closely as long as the last vital barriers are not breached. The sense of danger is tempered by the knowledge of how to control the danger. For this reason experienced workers can do things and enter situations which the novice cannot. The question is always whether the novice realises his lack of control and so avoids the danger, or copies too rapidly the behaviour of the experienced worker without having the control skills necessary. The research evidence[2] is that there is a large peak in accidents in the first few days, when novices get caught out by completely unknown problems. Then there is a rapid decline in accidents. However, the sense of danger seems to decline faster than the control ability increases in many situations, resulting in another small peak in accidents some days or weeks later.

2.7.4.1.2 Responding to warnings and error signals

If the danger signals are not so insistent that they demand response, the reaction will depend upon a more conscious assessment of the warning signs. These may be the learned specific warnings which are already present (see 2.7.4.1), or they may be artificially introduced warnings. These can range from alarms which go off when the danger is present (a fire alarm), to general notices alerting people to the fact that danger may be present ('Beware of the dog').

The following criteria can be used for the design and placing of warnings[19]. They should:

- be present only when and where needed,
- be clearly understandable and stand out from the background in order to attract attention,
- be durable,
- contain clear and realistic instructions about action,
- preferably indicate what would happen if the warning is not heeded.

Warnings should preferably not be present when the hazard is absent, otherwise people will soon learn that it is not necessarily dangerous in that area. They will then look for further confirmatory evidence that something really is a problem before taking preventive action. The philosophy of 'if in doubt put up a warning sign' is counterproductive, unless an organisation is prepared to go to great lengths in enforcing it even in the face of the patent lack of need for the precautions at some times.

If an alarm goes off and there proves to have been no danger, there will be a small, but significant loss of confidence in it. If false alarms exceed true ones, the first hypothesis an individual will have when a new alarm goes off is that it is a false one. Tong[20] reports that less than 20% of people believe that a fire alarm bell going off is a sign that there really is a fire. The rest interpret the bell in the absence of other evidence as a test, a faulty alarm or a joke. The recognition of the presence of fire is therefore often delayed, and the first reaction to warnings of a fire is often to approach the area where it is, in order to check whether there really is a problem, rather than to move away.

Because warnings must be understood quickly and sometimes under conditions of stress, there can never be too much attention to their ease of comprehension. The language used must be consistent, whether verbal or visual (e.g. a red triangle round a sign must always mean a warning, a red circle a prohibition). The language must be taught. Written warnings must take account of the reading age of the intended audience and the proportion of illiterates or foreign nationals with a poor understanding of the language. A word such as 'inflammable' should, for example, be avoided because many misunderstand it to mean 'not flammable' (by analogy with 'inappropriate' or 'incomprehensible').

A special category of warnings are those applicable in routine tasks, which need to bring people up short with an insistent indication that their routine has gone off on the wrong track. In routine tasks we normally only conduct minimal monitoring. Designers should improve upon the availability of information about deviations in such tasks and consciously build into their designs feedback about the actions which the individual has just taken, and their consequences. Examples of such feedback which helps error detection are the following:

- displays on telephones which show the number you have just keyed in, so that you can check before dialling it;
- a click or bleep when a key is pressed hard enough to enter an instruction on a keyboard;
- commands echoed on a visual display as they are entered on a keyboard; and
- the use of tick boxes on a checklist to indicate the stage in the check which has been reached.

2.7.4.1.3 Inspection and hazard seeking

If there are no immediately relevant warnings, people will not usually go in search of hazards. There may be a general level of alertness to unusual signals, but it will not be high. Entry into novel situations may trigger some hazard seeking. There may be a strong influence of personality or experience here. Those with a distrust of technology or with experience of dangers in the past may be more inclined to go in search. Recency effects may be important. After reports of a horrific hotel fire on the television, we may all check the fire exits of the hotel we stay in that week, but the effect usually fades by the next month. In general, if we want people to search for hazards in such circumstances we have to instruct them to do so, or train them to make it a routine. During such workplace safety inspections the people concerned are, therefore, already alerted to the possibility that there are hazards present, and are actively seeking them. But to seek is not always to find. Untrained inspectors characteristically miss hazards which have one or more of the following features:

1 Not detectable by the unaided eye, but requiring active looking in, behind or under things, rattling guards or asking questions about the bag of white powder in the corner.
2 Transient; e.g. most unsafe behaviour which can only be discovered by asking questions and using the imagination.
3 Latent; i.e. contingent upon other events, such as a breakdown, a fire, or work having to be done by artificial light. Again, only 'what if' questions can uncover these.

Too many people think that inspection rounds can be passive walks, 'just keeping the eyes open'. This is absolutely not the case. Inspection must be an active and creative search process, of developing hypotheses about how the system might go wrong. It requires the allocation of time and mental resources. Checklists can help to make the search systematic and to avoid forgetting things, but they should not be allowed to become a substitute for active thinking. It often helps to get people in from other work areas, or with other backgrounds, who see things through new eyes and spot hazards which the regular workforce have ceased to see.

2.7.4.1.4 Predicting danger

Prediction at the design stage is an extension of the problem of inspections, made more difficult because there may be no system in existence, comparable to the one envisaged in the design, from which to learn. Imagination and creativity are therefore relevant attributes for the risk analyst in addition to both plant knowledge and expertise in behavioural sciences.

There are large individual differences in how good people are at imagining the creative misuse that operators will make of their systems. Those who are good are known as 'divergent' thinkers. There is some evidence that people who gravitate towards the sciences, maths and engineering are more 'convergent' in their thinking, and tend to be more bound by experience and convention, than those who choose social sciences and the arts. They may therefore be less able to anticipate the more unusual combinations of events which could lead to harm. Designers are, anyway, rather inclined not to want to think about how people might use (or misuse) their beloved designs in the real world. They tend to think normatively: this is how it should be used, and so, this is how it will be used. This suggests the need for teamwork in design and hazard prediction.

Risk assessment techniques such as HAZOP, design reviews and fault and event trees are systematic methods to guide and record the process of creative thinking about risks. They are also often used to quantify the chance of failure. That step is only legitimate if we are certain that all modes of failure have been identified. That is particularly difficult with human-initiated failures because people can act (and fail) in so many more ways than hardware elements.

A systematic task analysis[21] is essential for a good prediction of human error. The data for such an analysis come partly from logical analysis of what **should** happen and partly from observation of what **does** happen when people carry out the tasks. Such job safety analysis techniques are dealt with in chapter 2.4 of this book.

Task analysis forms the basis for the use of techniques for error prediction[22]. They all depend upon one or other sort of checklist. For example each sub-task can be subjected to the following standard list of questions to specify what would happen if these types of error occurred and how such an error could happen (c.f. HAZOP):

- sub-task omitted
- sub-task incorrectly timed:
  - too soon
  - too late
  - wrong order
- inadequate performance of sub-task:
  - input signals misinterpreted (misdiagnosis)
  - skills not adequate
  - tools/equipment not correctly chosen
  - procedure not correct
  - inappropriate stop point
    - too soon
    - too late
    - not accurate enough
    - quality too low
- routine confusable with other routines.

Other checklists have been produced linked to the models of Reason[23] and Rasmussen[17] or derived from ergonomics[24].

2.7.4.1.5 Knowledge of causal networks

Hazard detection has been shown above to be dependent on the mental models people have of the way in which events happen and systems develop. If these mental models are incomplete or wrong they can lead to inappropriate behaviour in the face of hazards. Interview studies[6] show that such problems frequently occur, particularly in relation to occupational disease hazards.

Examples of such significant inaccuracies are:

- Men sawing asbestos cement sheets who said that they only wore their face masks when they could see asbestos particles in the air. (Yet it is the microscopic, invisible particles which are the most dangerous because they are in the size range which can penetrate to the lung.)
- Wearers of ear defenders who fail to incorporate the notion of time weighted average exposure in their concept of what constitutes

dangerous noise. Hence they fail to realise that 'just taking the ear-muffs off for a few moments to let the ears breathe' in high noise areas can negate much of their protective effect.
- Misconceptions about the link between posture and musculo-skeletal damage such as considering postures as 'relaxed' and therefore good when they show the body and notably the spine in a slumped position (which in fact puts extra load on the back muscles to stabilise the spine in that position).

The examples above are of causal links quite close to the actual harm. When we look at perceptions of causal links further back in the chain, we find even greater distortions and individual differences in attribution of cause. This is particularly clear when we look at people's attribution of causes to accidents that have happened to them or others. They find it very hard to construct logical and complete causal trees, which satisfy the test that each effect needs a set of causes which are necessary and sufficient to produce it. This means that fault trees are often constructed with missing branches and people checking them do not notice the gaps. People also tend to mention as causes only those things they can envisage as being modifiable[25]. They often fail to mention design features of the workplace or machine as causes, if they see them as 'given'. In contrast, human behaviour is almost always seen as modifiable (I could have done X differently), and hence gets mentioned more often as cause.

We need to pay considerable attention to causal perceptions in training. They will determine both how seriously people take the risks (how imminent, nasty or probable they consider them: section 2.7.4.2), and who they see to be responsible for taking action (section 2.7.4.3). If the prevailing attitude among both managers and shop floor is that accidents are inevitable, because their causes are too varied always to predict them, we have an organisational culture that will never excel at safety and always be willing to shrug its shoulders at a continuing trickle of accidents. Both initial training and the discussion of accidents and incidents which have happened need to go deeply into what these perceptions are and how appropriate they are.

2.7.4.2 Labelling as dangerous: reactions to perceived risk

Perceiving a hazard and assessing its seriousness are very closely related processes. They are separated here in order to discuss them more clearly. However, in practice the two steps are iterative. We only see something as a hazard if we see it as (potentially) out of control. The law may require us first to make an inventory of all possible hazards and then to assess their likelihood (risk). In fact we do it the other way around. We have already applied a cut-off in doing a risk assessment, which has excluded for us 'non-credible' accidents.

The reactions of different groups to risks they perceive has been the subject of much research in the past few decades[6,26,27]. Much of it has concentrated on decisions about siting hazardous plants or developing technologies such as nuclear power. A main focus has been the question of 'acceptability of risk'. This term has led to much confusion because it implies that people are, or should be content, or even actively happy with a particular risk level. The word 'tolerated'[28] gives a better assessment of the situation, since it carries with it an idea that the opportunity to do something about the hazard is a relevant factor in any decision. There is also overwhelming evidence that people do not consider the risk attached to an activity or technology in isolation from the benefits to be gained from it[29]. Therefore no absolute 'acceptable' level for a wide range of different hazards can be meaningful, since the benefits which go with them will vary widely.

A clear distinction has emerged from the research between threats to personal safety, threats to health and threats to societal safety[30]. The factors which people use in assessing each of them appear to weigh differently, and this is likely to be related to the sort of action which people perceive they can take against the differing threats. For example, moving house or changing jobs will solve the threat to an individual's safety from a particular chemical plant, but will do nothing for the threat to societal safety from that same plant.

Despite these differences there appear to be common factors which people use to make assessments of danger and to apply a label to a situation indicating that something must be done about it. What varies between types of hazard and between responses to different questions is the weighting given to the different factors. The research has used two basic approaches. Either to ask people directly what they think about hazards and how they react to them, or to measure actual behaviour in respect of different hazards. The first is called 'expressed preference' research, the second 'revealed preference'.

One clear result of expressed preference research is that people use a more sophisticated assessment process in judging risk decisions than just considering probability of harm. They also consider a wide range of other factors, which can be grouped under the following headings:

1 Whether the victim has a real choice to enter the danger or not, or to leave it once exposed.
2 Whether the potential for harm in the situation is under the control of the potential victim or another person, or outside any human control.
3 The foreseeability of the danger.
4 The vividness, and severity of the consequences.

2.7.4.2.1 Choice to enter and leave danger

People generally think of those who choose to engage voluntarily in dangerous activities like rock climbing, rallying or caving, as people who realise the nature of the hazards and have accepted their own responsibility to control them. The problem of accidents or disease is then seen as their affair, and not a matter of concern for society. On the other hand if there is no choice about exposure to the danger, e.g. whether a chemical plant or nuclear waste dump is built near your village, far higher demands on the level of safety are made. The situation is, however, seldom clear-cut. Can, for example, the choice of a person to take a job on a construction site in an area of high unemployment be called a voluntary acceptance of risks associated with that job? Hazards frequently come as part of a package with other costs and benefits. In the early years of the industrial revolution workers were deemed to have accepted voluntarily the hazards of the job that they accepted. Therefore they were deemed liable for their own accidents. Now both society's view and the law have changed. The employee is not considered to accept occupational hazards voluntarily, unless there is talk of some gross deviation from normal carefulness. Perceptions of what is voluntary do change, but the dimension remains important in determining how serious we consider a safety problem.

The demand for increased safety levels is also stronger if the risks and benefits are not equitably shared and one group profits from the risk exposure of another. The great public concern about risks of genetic damage from radiation, genetically manipulated organisms or teratogens seems partly explainable in terms of the threats they bring to unborn generations who have no choice in the matter and no share in benefits occurring now.

If we look at risk perception for those taking part in an activity, there is some evidence that dangerous activities which are voluntarily chosen are positively valued just because of their finite element of danger. Mountaineers choose to attempt climbs of increasing difficulty as their skill increases, finding the old ones tame. There is an element here of testing the degree of control which one has over a situation to check that it is real. The element of apparent loss of control is one of the attractions of fairground rides such as the 'wall of death'. But the fascination seems to go further than this. Greater danger, such as in war or time of disaster, is associated in the minds of survivors with greater group friendliness, shared emotions, sense of purpose and competence which makes that danger in retrospect positively valued, or at least willingly accepted.

2.7.4.2.2 Controllability

Perhaps the most important factor in taking a risk seriously is our judgement of whether it is controlled. The largest element here seems to be the feeling of personal control. Those who believe themselves knowledgeable about and in control of a dangerous situation, even where the magnitude of the consequences is potentially great, show little fear or concern about it. Laboratory chemists can appear blasé about handling toxins, steel erectors about walking across narrow planks high above the ground. This also applies to our attitudes towards the safety of others. Research has shown[31] that construction site supervisors often consider that the site hazards are under the control of skilled contractors. Hence they do not personally concern themselves with prevention, even when they see that that control is not being fully exercised and even when their bosses tell them they are responsible for all that goes on on the site. Similarly, workers in a plant are much less concerned about the hazards from it than those who live nearby but do not work there.

If the assessment of personal control is such an important factor in evaluating hazards, it is very important that the assessment is accurate, and that people do not believe they are in control when they are not. But there is ample proof that people can have illusions of great control where none or less exists. Svenson[32] quotes a number of examples from the field of driving. For example between 75% and 90% of drivers believe themselves to be better than average when it comes to driving safely. Similarly 88% of trainees in cardiopulmonary resuscitation felt confident after an interval of several months to perform it, while only 1% actually performed adequately. Experts always tend to be overconfident in their expertise. This is particularly dangerous when specific knowledge, for example of a theoretical nature, about a process is taken to mean control over the whole activity involving that process. This can be a serious source of overconfidence in skilled personnel such as research chemists or toolroom personnel, most of whose accidents in fact come from the everyday hazards of the machinery or the laboratory which have little to do with their speciality.

When people have no personal control they may place their trust in others to keep the situation safe. Again it is a question of whether the assessment is accurate and whether the trust is justifiably placed. The work of Vlek and Stallen[29] suggests that one of the clusters of beliefs characterising those who oppose large nuclear, transport and chemical plant developments is personal insecurity and lack of trust in those controlling the technology. The situation is made worse by the spectacle of experts disagreeing violently with each other about the safety issues of the developments in question.

In the field of health promotion the concept of control has also been shown to be important. One of the main thresholds to be crossed before people will act to change their own behaviour is to admit that they personally are susceptible to the health threat, for example, from smoking, alcohol, drugs, or heart disease, i.e. that they have lost control. The other side of this coin is the need to believe in the efficacy of the preventive action before it will be adopted. This means believing that the proposed action would restore the lost control; that giving up smoking would reduce the risk of cancer and heart disease, that wearing the protective ear-muffs would reduce the hearing loss and so on. The opportunity to prove for oneself the effectiveness of protective devices is therefore important in persuading people to wear them.

2.7.4.2.3 Foreseeability

Foreseeability is a word familiar from the case law of the English legal system relating to health and safety. It has been defined by judges as being what the 'reasonable man' would expect to happen given access to the current state of knowledge at the time of making a decision. It is used to draw a dividing line between situations in which people should have taken action to prevent an accident, and those for which it is not reasonable to hold them liable.

At an individual level hazard detection is limited by what is foreseeable or foreseen. However, if people cannot foresee exactly what may happen, but suspect that it may still go wrong they will be afraid. If this feeling goes hand in hand with the belief that there could be severe consequences and that the person is powerless to do anything, the reaction may be extreme. Evidence that a new and unknown technology like genetic engineering is not as much under control as previously thought would therefore have a profound effect on people's beliefs, shifting them rapidly from indifference to strong opposition. This is approximately the effect which Chernobyl had on nuclear power.

2.7.4.2.4 Vividness, dreadfulness and severity

The most recent accident or tragedy weighs heavily in the minds of people when they are asked about priorities for prevention, but this may fade rapidly. On a more permanent basis, people have reasonably consistent ratings of how nasty particular types of injury or disease are[33]. For example cancer is greatly feared, an eye is worth more than a leg and some injuries such as quadriplegia and brain damage are consistently rated as worse than death.

An important element in memorability is 'kill size', the number of people who either actually do, or potentially could get killed in an incident. This is why people tend to be more concerned about the safety of air or train travel, where accidents tend to cause multiple deaths, and much less about road safety which usually kills people one at a time. The cumulative numbers of deaths in a year and the probability of death per trip have a much smaller influence on the assessment of seriousness.

2.7.4.2.5 Risk scales and probability

With such a complex of factors determining the reaction of both individuals and society to risk it is not surprising that no simple scale such as Fatal Accident Frequency Rate can capture its essence. Managers and planners may wish to reduce decision-making to a tidy consideration of probability times cost of harm (usually deaths). They may even wish to label as irrational any opposition to this definition of risk. But this is no more than one powerful group putting an emotive label on something they seek to oppose. A better approach is to treat the factors for what they are, namely the basic elements which must be influenced if we wish to change behaviour. If you want someone to use a safety device, you must convince them that the danger is foreseeable, unpleasant and avoidable, that the safety device is effective and that they can choose how to use it.

There is an additional problem with the use of probability in the definition of risk. For the average person probability is not a concept that comes naturally. This is an idea which will be readily accepted by anyone who has tried to learn the fundamentals of statistics. Most normal people have little need for accurate probability judgements and little practice in making them. Individuals normally only rate whether they think things will remain under control. Despite this it is surprising how good the correlation is between measured probability of particular types of accident and subjective assessments by the general population. The major bias is that the subjective scale is compressed and foreshortened in relation to the objective one, typically spanning only three orders of magnitude instead of six. Very rare risks are treated as non-existent, slightly less rare risks may be overestimated and common ones are underestimated. Some hazards are raised in the order of probability; typically those which receive much media coverage.

The framing of questions about probability and of statements about risk can strongly affect people's responses to them. It is more effective as an argument to get people to be vaccinated to tell them that a vaccine offers total protection against one strain of disease that accounts for half of a given sickness, than to tell them that the vaccine offers 50% protection against the sickness. The presence of the word 'total' in the message gives the illusion of certainty. Framing information about road accidents in terms of the probability of accidents over a lifetime (probability of death is about 0.01 – and of disabling injury about 0.33) is much more effective in getting people to wear seat belts than quoting the probability over one trip (probability of death is about 1 in 3.5×10^{-6}, and of disabling injury about 1 in 10^{-5}).

2.7.4.3 Responsibility for action

Even if people can see a danger and appreciate the need for action, they may not act because they think they cannot or should not. This may be because it is seen as someone else's job or responsibility. This attitude is found among supervisors who are not willing to tell skilled workers how to avoid risks in their job[31]. Social pressures determining what is or is not acceptable behaviour may discourage people from warning others because of the fear of being told to mind one's own business, or of being thought to be interfering.

The major factor at this stage will be the way in which people view the courses of action open to themselves and others to influence the danger. Again the crucial role of the mental models of cause and effect are clear. If I believe as a supervisor that accidents to my staff are caused by their own carelessness and lack of attention to rules I will only think in terms of selection, training and discipline as actions. If I believe that the machine design is such that nobody can be expected to concentrate 100% of the time to avoid injury, I shall give attention to redesign or guarding as well. Biases in the way people allocate responsibility for accidents or prevention are therefore of vital importance and should be the subject of training. Such biases can be summarised as follows[6]:

1 When people are looking at their own future behaviour they think that they can exercise more control than is usually the case. Hence they accept great (even too great) responsibility to act to control the situation and to prevent future accidents.
2 When people personally suffer an accident, they are inclined to attribute it too much to the force of external circumstances rather than to personal responsibility.

3 When people observe others' behaviour they grossly underestimate the effect that the situation has in determining it; hence they overestimate the control that others have over what they do, and blame them unfairly for their accidents. This can lead to a reluctance to intervene in situations to warn, instruct or help people.

These biases arise when there is some ambiguity in a situation which allows for more than one interpretation. Such occasions are most frequent in rapidly changing situations, and when people are trying with hindsight to reconstruct an accident of which they may have been a witness or about which they have merely heard reports. Putting the three biases together goes some way towards explaining the inactivity in accident prevention in a number of situations. Designers must consider hazards to others (the users). They tend to overestimate users' ability to look after themselves, and so underestimate the need to build in safeguards. Supervisors place the onus for avoiding accidents on the victims and not on themselves; while their bosses think that it is the supervisors' job and so shuffle off their own responsibility for the climate of rules and priorities which they create. If managers and designers sit round a table with workers from the shop floor who are talking about their own accidents (and so are subject to the first two biases), a great measure of agreement is possible. Both sides will agree that the main onus lies on the worker to prevent accidents. However, this agreement does not lead to any action, as both sides will also tend to believe that everything is under control and nothing needs to be done except to be alert more of the time and to take more care – an impossible discipline to realise in most working situations.

2.7.4.4 Decision-making and action in the control of danger

This section covers the right-hand side of *Figure 2.7.8* in respect of the plans and procedures for controlling the danger and the actions which follow from them. The factors of importance here are more straightforward than those on the left-hand side of the figure. The bases for human decision-making have been discussed in section 2.7.3.

People can simulate in their mind the results of different possible courses of action before they make any decision about which course to choose. This skill is an immensely valuable one because it allows some courses of action to be rejected without ever trying them, on account of the unpleasant consequences which we correctly predict. However, as a skill it depends upon knowledge of how factors interact and the ability to manipulate many factors together in the mind. The latter is related to intelligence, and to the amount of practice in using the skill. The world is always so complex that we can only take account of a limited set of factors and considerations. This bounded rationality protects us from waiting too long before acting, but like all limits and choices, it can exclude vital considerations, which trap us later. The question is, therefore, whether people are good at considering risks in their decisions. We have already seen that people may be overconfident of their ability to control risks and hence estimate them too low in their decisions. A greater problem, however, seems to be the fact that we do not succeed in being creative enough in imagining the breadth of possible consequences of the range of decisions to be taken. We are also too reluctant to go back on decisions, once made, when there is evidence that things are not going as expected. We do not see the new evidence as proof of a wrong decision, but as new, unrelated facts, which have to be made sense of, *given the decision*. People are creative enough, and many situations unclear enough, that it is frequently possible to find a plausible set of reasons why the original decision or diagnosis can be maintained and the new facts reconciled with it.

Individuals are also not entirely logical in the way in which they make decisions. The value assigned to particular outcomes, such as amount of effort saved, money earned, approval obtained or withheld by colleagues and superiors, etc. is a subjective one. It will be influenced by the personality of the individual making the decision, and by experience of the way in which previous decisions have turned out.

The rate and efficiency of mental processing are also limited by the level of arousal of the brain. At low arousal levels performance is poor, rising with increased arousal to an optimum and then falling with further increases. This change in arousal corresponds roughly to a movement from drowsiness, through optimum co-ordinated performance, to the confused activity resulting from overanxiety and panic.

In the case of rule- and skill-based decisions, the whole process of choice is also much less conscious, and, in the latter case, we may not even be aware of making it. These decisions are, therefore, even more bounded than the ones at knowledge-based level. In order to make these decisions as good as possible, the influences we have are:

- Training and experience, particularly explicit reflection on, and conscious learning from the experience.
- Built-in checks on vital decisions involving risks, in which we arrange or require that someone checks the decision with a colleague or superior before taking action. There are some examples of building software programs to do this questioning role.
- Limited possibilities of using automation or guarding to block certain actions in certain circumstances, or to take actions over. Examples are interlocks on machines which prevent access when the machine is under power or in motion, or automatic train protection systems which apply the brakes if a train passes a red signal. The latter example is one which leaves the human in the loop to take the decision, but intervenes in clear emergency situations to cope with human failures. We can contrast this with automatic shutdown systems, which take the decision-making away from the person and take them out of the loop. The person then has to intervene if the automatic system fails, an almost impossible task if this occurs very infrequently (e.g. once every few years)[4].

The major problem in the last step under this heading is whether the necessary actions have been correctly learned and are available when needed. The latter point is particularly crucial in emergencies, such as evacuation, plant shut down and first aid treatment. Considerable investment in refresher training, sometimes on simulators, is needed to keep rarely used skills available enough to be deployed accurately when they are wanted. Even simple tasks like evacuating a building will not go successfully if not practised. The experience of real-life fires and bomb warnings demonstrates this regularly. There is always more chaos than is expected, even when escape routes are clearly marked and there is trained staff present to lead and direct those not trained. Lack of training always shows itself most in situations where the simple rules and procedures unexpectedly go wrong; e.g. where escape doors turn out to be blocked, or where vital evacuation orders fail to get given (as in the Piper Alpha disaster).

The speed with which complex skills decline with lack of use is often underestimated. For example, the statutory period for refresher training in first aid will not preserve such complex skills as cardiopulmonary resuscitation, which decline to dangerous levels within 2 months if not practised.

2.7.4.5 Error correction

The preceding paragraphs have taken us once round the loop of *Figure 2.7.8*. Safe behaviour is a constant process of going round the loop. This means we have many chances to correct errors and to stay safe. Only if we continuously do it wrong will the danger get out of control and catch us. This is only likely to happen if we spend too much time responding unquestioningly to the work and its environment. To be fast and efficient, we must operate for a great percentage of time at the skill- and rule-based levels. However, we need to organise that we spend sufficient time, at sufficiently frequent intervals questioning and rechecking that automatic behaviour. This 'creative mistrust' of routines and accepted ways of working is an essential ingredient of a safe organisational culture. How often 'sufficiently frequent' is and how long 'sufficient time' is will depend on the activity, its dynamic and the turnover of people carrying it out. It is certain, however, that even well-learned routines of machine operation, manual handling and computer use can erode and change subtly, but significantly, in a matter of months. This gives an indication of how often a work group or individual needs to audit its own behaviour, or have an outsider check it.

2.7.4.6 General effects on performance and safety

All of the above behaviour has been described in isolation from the overall effects of the external environment on behaviour and the issue of the general safety performance of the individual. This section summarises some of the main issues under those headings.

2.7.4.6.1 Effects of the environment and degradation of performance

General environmental conditions such as noise, glare and lighting level, dust and fumes, social environment, etc. will influence the factors which have been described above. Noise and high temperature both have an effect on the arousal level. Noise increases it, heat decreases it, and both have an effect on the accuracy of detection of information and the speed of processing it. These physical environmental factors are dealt with in the chapters on Occupational Health and Hygiene. The effects of fatigue and the social environment upon individual behaviour are sketched here very briefly, but a detailed discussion is outside the scope of this chapter.

The performance of the human system is only at an optimum within certain environmental limits. As part of the price of its sensitivity and flexibility the human system is susceptible to the influence of a very large range of factors which can affect its performance. Unlike machines, human beings show a slow and often subtle degradation of performance over a wide range of environmental conditions, but arrive at a total breakdown only comparatively rarely. This means that individuals can maintain some sort of functioning long after they have passed the peak of their performance, but it also blurs the point at which they should be required to stop in order to avoid errors. Regulations and good practice on, for example, working hours for coach drivers, hospital doctors and others must wrestle with the problem of making these black and white decisions at some point on the continuum of shades of grey.

Performance is degraded under the following types of situations:

1. Working for too lengthy a period, which produces **fatigue**. Muscular fatigue results from overloading of individual muscle groups, either through static loading to maintain posture, or through awkward or repetitive dynamic loading. In addition to limiting working hours, the cure to this problem lies in the design of workplaces to minimise static working load and to allow for the utilisation of the most efficient muscle groups, and the opportunity to rest muscle groups by shifting posture. General mental fatigue is characterised by an increase in the length and variability of reaction time, especially for decision-making. This leads to an increase in errors and a tendency to neglect peripheral aspects of tasks such as checking routines. These effects can be demonstrated in most tasks after periods of uninterrupted working of between 10 and 50 minutes, depending upon the task load. Rest pauses of one or two minutes interposed when the performance begins to fall from its optimum level are sufficient to restore functioning to its former level. If performance is allowed to carry on without a break until more obvious signs of degradation have appeared, then proportionately longer rest pauses are required for complete recovery. If the task is machine paced, or strong motivation from pressure of work or incentive bonus schemes prevents natural breaks, artificial breaks should be introduced in order to maintain performance at an optimum level. The recovery produced by rest pauses becomes less as the length of working hours increases. Old research[2] showed very clear effects on safety of working days longer than 10 hours. These

tended to be jobs with physically heavy workloads. However, effects on mental performance of such long periods can also be shown.

2. Working at times of day when body mechanisms are not functioning efficiently and **diurnal** or **circadian rhythm** is disturbed[34]. Around 10% of the working population spend some time working on a night shift, and a larger proportion are on shift work, outside the period 7.00 a.m. to 7.00 p.m. In addition, an increasing number of company staff fly regularly across time zones, as do airline employees on a far more frequent basis. Body systems follow a cyclical variation in activity which is linked to the 24 hour light–dark cycle, known as the diurnal or circadian rhythm. The difference between performance at the peak and the trough of the curve is of the order of 10%, which is as significant as the degradation in performance caused by a blood alcohol level at the legal limit or by approximately two hours' loss of sleep on the previous night. If people work at night, rhythms are thrown into some disarray and take time to begin to adjust. Adjustment begins to be apparent after 2–3 days, and goes on increasing up to a period of about 14 days provided that the person continues both to live and work on a night-time schedule, and does not return to day-time living over a weekend. Even after two weeks the curves have not fully reversed, but have flattened out. If the shift is caused by flying to a new time zone, the adaptation is much faster, and will become complete, because the signals of light/dark and social behaviour all point in the same direction.

   Night workers have the additional disadvantage that they are trying to sleep when the rest of the world is awake and making a noise. Hence their sleep is far more disturbed than that of day workers. Night and evening workers also suffer from a major disruption to social life. This can result in conflicts and stress generated within the family. Studies of night workers show that they tend to have a higher incidence of gastro-intestinal disease such as ulcers, and nervous disorders. There is no clear evidence that the physical and health effects on women and young persons are greater than on men. The original reasons for the introduction of the ban on night work for protected persons were as much for its supposed moral dangers as for its health effects.

3. Lack of stimulation resulting in lowered arousal (**boredom**). This is typical on supervisory jobs in control rooms, security functions, gatekeepers, etc. where people are waiting for things to go wrong before they have a primary task. It can result in people seeking other distractions to keep them occupied. They are then 'out of the loop' when their intervention is needed. This problem can only be tackled by careful task planning which gives them tasks to do which keep them in the loop, combined with excellent alarm systems to arouse them for interventions.

4. Working under conditions of conflict, threat, both physical and psychological, or conditions which threaten the body's homeostatic or coping mechanism and cause **stress**[35,36]. There is extremely wide

variation in the reaction of different individuals to the same stressor. In the case of physiological stressors this variation is largely a case of differences in physical or physiological tolerance. In the case of psychological stressors there is the important intervening variable of the perception of the stressor and the degree to which it is seen as a threat to valued goals. Individuals have different abilities to deploy coping responses when faced with a stressor which will not go away. The psychological symptoms of stress are disturbed concentration, impaired memory, impaired decision-making, tension and aggression, sleep disturbance, and, in severe cases, mood changes. There are a number of coping strategies which can be adopted by individuals. They can withdraw from the source of the stress either physically by leaving their job or going absent, or psychologically by lowering their ambitions, e.g. ceasing to fight for promotion when promotion prospects are blocked. Companies can remove some of the unnecessary demands of the workplace[37], e.g. by work restructuring or by providing greater support through discussion groups, meetings, or counselling services. Finally, there has been some success in bolstering the individual's own resources for countering stress through the teaching of relaxation techniques, through counselling and psychotherapy. A coping strategy open to the employing organisation is to identify people who would appear to be susceptible to stress and to redeploy them into jobs where the demands are lower.

2.7.4.6.2 Individual differences in accident susceptibility

Research into individual differences in accident rate (accident proneness) has a long and complex history littered with mistaken conclusions from invalid methodology and poor experimentation[1,2]. It set out to discover whether there were stable differences in accident susceptibility when individuals were subjected to equal hazards (both in type and length of exposure). It was established very early in the research that both age and experience were correlated with differences in accident susceptibility (see *Figure 2.7.9*) The exact shape of the graphs will vary from job to job.

Job-related experience appears to be the most relevant to accident rate[38] although the effects of number of years in the industry and of number of hours worked on a specific task (where the job involves a number of tasks) can also be demonstrated. Effects of advancing age can be found in their own right, but these tend to be in the form of the type of accidents and errors occurring (more slips and lapses and less mistakes with increasing age), rather than in numbers. Older people also adjust their work rhythm and change their relationship to the work group, which can influence their vulnerability to accidents, either positively or negatively.

The relationship of physical and anthropometric differences to accident susceptibility has also been shown in many specific tasks. For example, colour blindness can be a danger where hazard perception depends on colour discrimination. Extremes of height, reach and slimness of arms, wrists or fingers can result in individuals being able to reach into danger areas around or through guards. Susceptibility to epilepsy, bronchitis or eczema can be problems on jobs involving moving machinery, dust and

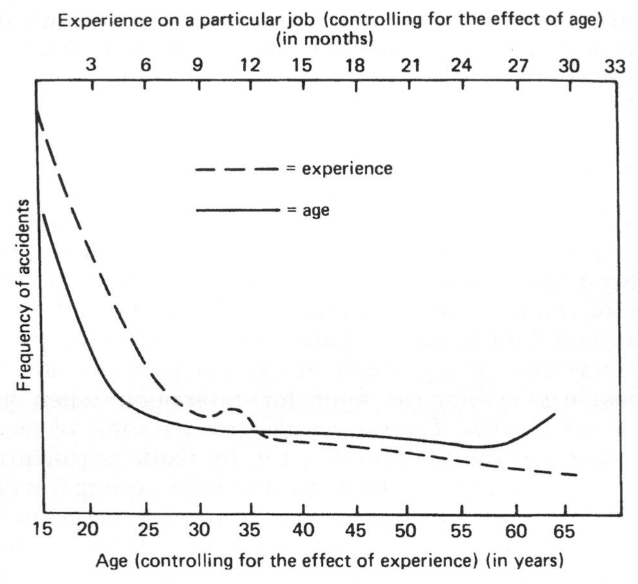

Figure 2.7.9 Distribution of accidents by age and experience (derived from Hale and Hale[2])

oils respectively, etc. Research on sex and ethnic differences in accident liability often shows apparent differences in gross accident rates, but these almost always turn out on closer examination to be differences in risk exposure (i.e. immigrant workers and men tend to be found more often in the dirtier and more dangerous jobs). In driving there is, however, evidence that men have more serious accidents than women per kilometre driven.

Research on the relationship of other factors to accident susceptibility has produced few clear-cut results; personality factors, intelligence, co-ordination and attention skills and many other characteristics have been studied but the correlations produced have usually been quite low and have been specific to the job or task studied. In the area of driving accidents insurance companies clearly work on the basis that certain groups have more or less accidents and adjust their premiums accordingly. However, much of this difference is accounted for by the factors discussed above and by the difference in distance driven. Accident proneness as an explanation for accidents or a basis for safety policy is therefore usually unprofitable and only helps to reinforce a blame culture instead of a problem-solving one. It is hardly ever possible to use selection criteria to keep those who will have many accidents out of the company. It is also doubtful if it is sensible to monitor individual minor incident or accident rates in extreme circumstances to identify poor adaptation of an individual to a job. Such poor adaptation can usually be picked up much earlier from other signals.

We should not underestimate the importance of individual differences in behaviour and their importance for safety. However, we should look for the reasons for the differences, and the means of influencing them in factors such as training, experience, social pressures and group cultures, all of which we can influence and manage.

2.7.5 Change

Underlying the above discussion of human characteristics, limitations and differences has been the notion of change. The human system is in dynamic equilibrium with its physical, social and cultural environment, constantly adjusting itself to changes in that environment while it pursues its goals and objectives. The environment changes as society and technology change and as family responsibilities and job type and location change. At the same time the individual changes, grows up, matures and grows old, learns new skills, forgets old knowledge and acquires new goals. Change is intrinsic in the human conditions. It is, therefore, a mistake to think of solutions to behavioural problems in health and safety in terms of changing a person from one stable state to another. The problem is better seen as one of prodding and guiding behaviour along one path out of the many possible ones and trying to stabilise it in a new equilibrium which will maintain its essential characteristics while it carries on adapting to other changes in the world. An appropriate analogy for coping with change is steering a sailing ship in a turbulent ocean. This gives a very different picture to the idea of picking up a piece of plasticine, remoulding it and setting it down again, a picture many managers have.

In the following sections the various processes of change are briefly presented. First the 'natural' processes of growing up and maturation are sketched, which lead to individual differences in personality, motivation, knowledge and skill. Then the process of individual learning is outlined, which must form the basis for the systematic acquisition of experience in training courses. Then there is a short section on individual responses to change, which also introduces the concept of risk homeostasis. Finally this section reviews briefly the ways of changing individual behaviour through design, information and communication, training, rules and culture. More detail on a number of specific techniques for behavioural change is to be found in chapter 2.6.

2.7.5.1 Growth and maturation

Many of the physical characteristics of individuals are strongly influenced by genetic factors, for example hair and eye colour, height, physical build, and body dimension, although the last three are also influenced by environmental factors such as nutrition. With skills such as language, the evidence now seems to be that there are critical stages in maturation in which they have to be learned. For language that is the years from birth up to 5 or 6, when a child can even learn two languages almost perfectly. If learned after the age of about 7 the second language will never be so effortless and accentless as the first. When it comes to factors such as intelligence or personality there is far greater argument about the importance of genetic and maturational factors, compared to learning from the environment, in determining the final measured characteristic in an individual. With attitudes we arrive at a characteristic which is largely determined by learning and experience, though some fundamental attitudes can be so firmly learned in childhood, that they can only with difficulty be changed later. This debate about the influence of genetics, maturation and learning is important wherever we are trying to change behaviour through education or by changing social or company policy, since what is genetically or maturationally determined is likely to be more or less unchangeable.

Some changes during growth are irreversible. The newspaper headlines of the last decades are eloquent witnesses to some of the more dramatic effects occurring between conception and birth which can produce individual differences. Drugs such as thalidomide have massive deforming effects, as do diseases such as German measles contracted during critical phases of pregnancy. Alcohol and cigarette consumption in pregnancy has also been shown to have less severe but widespread effects on the foetus.

The opportunities provided by the family, and the goals and the motivations which are learned because of the behaviour which is rewarded during childhood have a great effect on the attitudes, skills and abilities shown in maturity. During later childhood and adolescence school, teachers and peer groups take over from the immediate family as the major influence on the development of personality and attitudes. The rate at which personality and attitudes change slows down once the age of 20 is reached, but working environments and social groups in adulthood still have a formative and changing influence. Individuals tend to seek out social groups which suit their personality and match their attitudes, so reducing the pressures on themselves to change. However, when there is a mismatch the individual will be changed by the group to a greater or lesser extent, depending upon the importance that the individual attaches to acceptance by the group.

Some of the factors which have been mentioned above are often drawn together and labelled as *cultural* factors if they are influences that are shared by a defined group of people, whether in a particular country, region, social class, age, company or occupational group. Thus certain attitudes towards risk taking may be shared by members of one adolescent culture, which are markedly different from those of other age groups. An organisational culture which favours good safety management has become an issue of increasing concern in the last decade and will be dealt with in section 2.7.5.4.5.

2.7.5.2 Learning

This section deals with the psychological principles of learning. A number of authors[39] have classified learning processes into a hierarchy of levels, building from the simplest stimulus response learning to the most complex processes of research and problem solving. Three levels of learning are particularly relevant, because they link with the three levels in *Figure 2.7.7*:

(1) Stimulus response learning (skill-based).
(2) Concept and rule learning (rule-based).
(3) Problem solving (knowledge-based).

2.7.5.2.1 Stimulus response learning

This is the building block which is particularly important in all 'habits' and repeated invariant sequences of behaviour. The sequences are built up by a process called conditioning. This has three essential elements:

(1) Evoking the correct response when the stimulus is presented; by trial and error, explanation or demonstration.
(2) Reinforcing correct responses. This can best use rewards, such as praise for correct performance, or latch onto the motives of achievement and interest by feeding back information of how well the task is being learnt or how near to the objective the learner has approached. It is also possible to use punishment for incorrect performance, but this can engender resentment and a desire to escape or take (subtle) revenge.
(3) Practise of the correct sequence which establishes the response more and more firmly until it occurs without conscious effort.

Once sequences of action of this sort have become established it is extremely difficult to add significant steps to them or subtract steps from them. They have to be largely dismantled and painstakingly relearned, as any golfer who has changed his grip can tell you. As second best, individuals must be trained consciously to break into the chain at the appropriate point in order to carry out the missing response action or to avoid the bad habit. It is therefore important that in safety all the necessary steps are built into the sequence at the learning stage. For example, if the response of donning protective goggles is built into the sequence of setting up and starting an abrasive wheel, it will become an automatic part of that habit.

The sequences of actions will become less automatic if they are not used regularly. Behaviour in response to infrequent emergencies will therefore not be available unless it is practised in the interim.

2.7.5.2.2 Concept and rule learning

A fundamental human characteristic is the tendency to think about and classify objects, experiences or ideas into mental categories or 'concepts', which share things in common.

Concepts are built up from experience and each person's set of concepts will therefore differ slightly. Thus, for example, the concept 'dangerous' to one individual may contain hundreds of items which include any situation which is new or strange (such a person we might label 'nervous'). For another the same label might contain very few items and leave out some which should be there, such as noisy discotheques or bottles of weed killer kept in the larder.

Acquiring concepts depends on amassing examples which have elements in common until the individual arrives at a tentative definition of the boundaries of the group of objects or ideas. This process can be abbreviated by defining the concept for the person. It must then be consolidated by providing examples which fall inside it, and outside it, gradually reducing the difference between these positive and negative examples until the boundaries become clearly defined. People continue even then to test out and modify the boundaries of the concept. It is therefore very useful to have brainstorm sessions at intervals with managers or work groups in which they recalibrate their concepts of danger in a particular workplace: what are credible risks, especially those falling under the heading of deviations of disturbances?

Concept learning is appropriate whenever someone is expected to recognise a new stimulus as belonging to a particular category, so that they can respond to it even though they have never met it before. It is also necessary whenever someone is expected to follow a rule. For example the rule 'all hazards must be reported to a responsible person' requires that individuals should learn three concepts. First a correct concept of what is a hazard: only such things as a very near miss when something drops from a shelf right next to you, or also unusual combinations of events which you can imagine remotely causing a problem at some stage? Secondly it needs a correct concept of who are responsible persons: a senior colleague, the safety adviser, your boss? Finally it needs an adequate idea of what reported means: a casual mention, a formal verbal report, or a formal written statement?

2.7.5.2.3 Problem solving

Where someone is faced with a situation which they have never met before, and they cannot clearly place it into an existing concept category, they are faced with the need to produce a solution new to them. Problem solving is a creative process which relies on the basic building blocks described above and consists of re-ordering them and re-interpreting them.

Learning to solve problems can be aided by teaching the steps in systematic problem solving:

1 To recognise the problem area and define it in broad terms.
2 To explore all the possibilities for solving the problem.
3 To analyse all the facts available to determine whether or not subsequent problems will follow from the solution suggested.
4 To choose the best possible solution and implement a plan of action for introducing it.

Some techniques for creative thinking (step 2) can also be taught, e.g. brainstorming. A problem is presented to a group of people who are encouraged to throw out ideas, sparked off by each other, no matter how wild. These ideas are recorded by the group leader on a board for all to see. It is important that all ideas be put forward without either interruption or criticism because cross-fertilisation and the building up on ideas produced by others will generate many new ideas. After a suitable period, the ideas noted can be discussed in open forum as part of the evaluation stage to obtain the best solution to the problem from the large pool of ideas so formed.

2.7.5.3 Stimuli for and resistance to change

A stimulus for the individual (or indeed for the organisation) to change will be the perception that his or her adaptation to the environment is no longer as close as desired. If the failure in adaptation is not perceived there will be no acceptance by the individual that change is needed. In that case attempts to impose change will be met with the sort of resistance which is characterised by the remark 'We have done it this way for 50 years and it's been OK; why change now?' This conservatism in attitudes and beliefs seems to become more marked with advancing age, perhaps because there is more past experience to call on for support in rejecting the need for change.

Social groups are frequently bastions against change. If a number of people can be found to share the view that things are all right as they are, they reinforce each other's view of the world and unite to resist change. This can lead to 'group-think' in which a powerful group totally ignores even strong signals that they are wrong in their judgements and decisions. This is something which has been found to occur under emergency conditions or conditions of strong external threat. Even major risks can then be negated.

Those wishing to promote change have the task of convincing individuals, groups and organisations that change is needed because the old adaptation is no longer appropriate. This process is least difficult when something dramatic such as an accident demonstrates that old methods were not safe, or there is a new law, a new machine, a new boss or a new job, all of which self-evidently require change. The alert safety adviser always has plans available for necessary changes awaiting such an opportunity.

It is the slow changes in environment and individuals which often go undetected, and hence slowly increasing risks may not be responded to. Examples are changes in technology, in social mores or in attitudes to work. Small changes may accumulate in plant through a series of modifications, each of which is too small individually to make a significant difference to risk. Gradual increases in the size and power of fireworks over a number of years led, for example, by small steps to a disastrous accumulation of explosive potential at one Dutch fireworks company in the centre of a town. It finally exploded in a fire in 2000, flattening a complete suburb and killing 22 local inhabitants in the blast. This alerted the company, the local authority and the government to the fact that, with hindsight, these creeping changes had been going on for many years, with too little attention and too much local acceptance of each small change. If these changes are to be brought to the notice of individuals or organisations it is often necessary to dramatise them in order to get through the conservatism or blinkered vision which is failing to see them.

Finally we must mention the phenomenon of risk compensation or homeostasis. Wilde[40] defines this as the desire of individuals to operate at a more or less constant level of risk, which is not zero. He has collected very convincing evidence that people respond in many situations to a lowering of the risk, by changing their behaviour so that it increases again. The lack of effect on safety of the introduction of anti-blocking brakes on cars is one of the best examples of this compensation. Drivers with ABS tend to drive faster and brake later and harder and not to use the ABS to increase the safety margin. Their accident rate stays roughly the same. In mixed traffic, some of which does not have ABS, another effect occurs, which effectively transfers the accidents from the front of the car with ABS to the back, when the car behind without ABS cannot stop fast enough. Wilde postulates that compensation occurs in almost all situations and is positively driven by a target risk level (in other words that there is a homeostatic mechanism at work). His opponents accept that compensation, i.e. change in behaviour, occurs, but only when the risk changes are perceivable and only when there is something else to be gained by the change. In other words they suggest that the change is not driven by actively seeking a particular level of risk, but by trading off the reduction of risk against another gain, such as speed, comfort, production, which is valued more highly.

Whatever the mechanism underlying the effect, we need to take it seriously. Any obvious change in risk levels which does not also succeed in changing the value of risk, will trigger some compensation. Our only defence is to anticipate that and to try to design and steer the change so as to make it difficult or unattractive to trade off the safety gain against some other goal. Wilde also underlines that any lasting change in safety behaviour must always aim to increase the motivation to be safe and the rewards for acting safely. This touches on a fundamental debate about successful safety strategies. Authors[41,42] have long advocated a 'safe place' strategy over and above a 'safe person' one. This was initially argued from a situation in which companies were too inclined to exhort their employees to be safe despite the manifest dangers of their working places, rather than spending money to try to make accidents impossible through good design. Such a priority is unexceptionable for relatively high levels of risk. However, the research discussed in this section must bring us to the conclusion that there comes a level of risk which is seen as so low by those involved with it, that they do not respond positively to any further lowering. At that point a further implementation of the safe place strategy will only work if it is preceded by a strong dose of safe person influence to change that implicit risk acceptance.

2.7.5.4 Methods of change

Change can be brought about by:

(1) Changing the physical environment of the individual, so that different behaviour is triggered by the design and unsafe behaviour is made less likely or less easy. The chapter on ergonomics in this book deals with this in depth. Hence the treatment here is brief.
(2) Providing new information to an individual which shows that the existing adaptation is not as close as the individual thought it was, or as it could be. This requires successful communication.
(3) Changing the knowledge and competence of the individual to perform all the steps in *Figure 2.7.8*.
(4) Changing the rules which the company provides and enforces. This overlaps with both the former and the next category, since it partly sets out knowledge of how to act safely and partly makes clear what will be rewarded or punished.
(5) Changing the motivation, goals and objectives which an individual is seeking, or the rewards for achieving it, so that old, unsafe behaviour does not bring the rewards it did, or new, safe behaviour brings greater rewards. This whole complex has come to be given the title of changing the safety culture.

A basis for any process of change is that the current situation has been clearly understood, the desired situation defined and the plans for reaching the second from the first have been carefully made. Task analysis and job safety analysis are essential ingredients in this process[21,22].

2.7.5.4.1 Changing the design and the environment

Under this heading come modification to physical work design and layout, new machinery and work methods and changes in allocation of jobs to people. Ergonomists have been saying for two generations that design is the biggest influence on use. The operator or user has too often in the past been saddled with the impossible task of recovering the inadequacies of the designer. Machines which operate in ways which do not fit expectations, routines which are easily confusable, and tasks which have been left to the operator only because technology cannot yet take them over, are all examples of accidents planted like time-bombs in the system. Users are to be congratulated that they manage for so much of the time to operate safely despite them. The increase in feedback from users to designers and the heavier emphasis on designers' liability which have occurred in the last decades (e.g. in the European Directives on Products, Machines and Temporary and Mobile Workplaces) are welcome signs that this tolerance has reached its limit. Designers have to think not only about how people should operate their designs, but also how they might be tempted to use and misuse them because of the design. This substitutes for the old normative way of thinking a new predictive, creative and proactive one. It emphasises that designers both go through *Figure 2.7.8* themselves and must understand how the constructor, manufacturer, installer, user, maintainer, cleaner, modifier and disposer of their design go through it too.

Designers should also be humble enough to realise that their designs are not eternal and that operators have a need to modify or adjust their workplace and not have to operate within rigid constraints which are not perfectly attuned to operating conditions. Therefore enough information and training must be provided to the operators to allow them to oversee the room for modification and not fall into unsuspected traps. In addition, predictable modifications which will lead to danger (like removing guards or defeating safety interlocks) should be made as difficult as possible.

Designers should not think that people are automata. There are differences both between and within individuals. People will never be as consistent in their response as hardware components. Therefore designs must be error tolerant and make error recovery as easy as possible. Nor should designers respond to human error with an unthinking push to automate the individual out of the system as much as possible. That is a recipe for creating residual monitoring tasks which are boring and unsatisfying. It will also result in the loss of skill and insight to such an extent that the operator cannot intervene effectively when the hardware fails.

Finally designers should not have unrealistically high hopes of the effectiveness of their hardware solutions. People will always adapt to system changes by altering their behaviour, sometimes trading off increased safety margins against other gains; e.g. straightening out roads with dangerous curves results in an increase in traffic speed; more reliable hardware results in less spontaneous checks of its functioning. This risk compensation should be anticipated by the designer, who should design against such trade-offs.

2.7.5.4.2 Giving information

Under this heading fall the provision of information about danger and communication about the state of any task or activity involving risk, or the performance on it, so that everyone knows what the situation is. The following chapter describes a number of techniques under this heading in detail. Other topics which fall under this heading are communication at shift changeovers, or at handovers between operations and maintenance. Permit-to-work systems are one method to make this process systematic. Another approach is to improve communication skills so that people are more open to influence (e.g. sensitivity training). The success of all these endeavours will depend upon the credibility of the source of information and the ability of that person or organisation to organise and put over information. Some general considerations about communication are given here, but specific applications are left to other chapters.

Communication is the process whereby one person makes his ideas, feelings and knowledge known to others and learns in exchange about

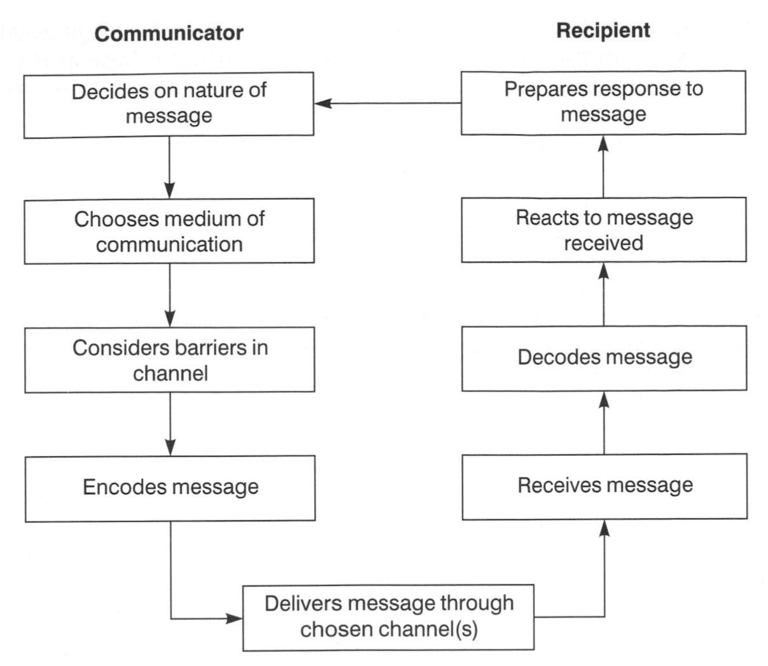

Figure 2.7.10 Steps in communication

theirs. It is therefore a two-way process which depends crucially on both clear sending and receipt of the message (*Figure 2.7.10*)

The first essential is to know with whom you are communicating (directors, line managers, accountants, the workforce, the HSE Inspector). Then the precise objectives of the message must be planned (what change is wanted? what are the precise obstacles? must the message succeed with more than one group?). The message must be coded in terms appropriate for the audience. It must latch onto their way of thinking, priorities and language. It must not use jargon they do not know. To do all of this it is necessary to think about the subject of the communication from the point of view of the receiver. The message must be conceived from that viewpoint, cover the disadvantages of the change from that viewpoint and how they will be overcome and finally spell out the advantages of the change from that viewpoint. On the basis of this planning the communication medium can be chosen.

1 **Face to face communication** has the advantage that it allows feedback and adjustment of the message based on the response. It is also friendlier and less formal. However, it is less easy to control because it is interactive. It is important to remember that it consists of two elements, verbal and non-verbal. The words seem dominant and must indeed be chosen appropriately and put over clearly, but the non-verbal clues can either reinforce or destroy their effect. The tone of voice can indicate boredom, the stance friendliness, hand movements nervousness. The very different effect of messages over the telephone

and face-to-face demonstrate the effect of the non-verbal. It is excellent practice for communicators to listen to themselves on tape and watch themselves on video to see and correct these elements of their style.

2 **Written communication** allows for much more complex messages to be sent and understood because they can be reread and carefully weighed. It also forms a permanent record for future reference.
3 **Visual communication** allows for very rapid transmission of the relations between things in one glance. It can therefore have great power and emotional impact, but it may be less easily controllable. There is just as much a language of pictures which must be learned by both parties to the communication, as there is a language of words.
4 **Electronic communication**. One of the most dramatic changes in the last decade has been the rapid advance of electronic mail, internet and computer use. This has enormously speeded up communication and expanded the options for presentation, but it has also increased the problems of overload. Electronic communication has the advantages that it is less limited in time and space than face-to-face or telephone communication. E-mail can also be very informal. Well-designed internet and computer-based information is, however, costly and time-consuming to produce.

2.7.5.4.3 Training[16]

The principles of learning were set out in section 2.7.5.2. Safety training needs to cover all the aspects of behaviour covered in *Figure 2.7.8* and in section 2.7.4. The value of safety training is undisputed, despite the fact that there is remarkably little literature evaluating its effectiveness[16]. The success of training is usually strongly determined by how actively the trainees can and do participate. Skills training almost always incorporates such active participation. It is much harder to achieve this with knowledge training, but active exercises, group discussions and case studies and application of knowledge to problems from the trainee's own company in reports can increase it. It is essential to plan some evaluation of performance at the end of the training to prove that it has had its desired effect.

2.7.5.4.4 Safety rules[43,44]

We should be suspicious of anyone who claims that safety is merely a matter of laying down and enforcing rules. It can never do any harm to define clearly and as exhaustively as possible how the system should operate to overcome all known hazards. This can form the 'instruction book' for the technology or company and can prove to auditors that the company has thought deeply about its risks and their control. However, we should not think that people working in the company will be able to use such a detailed rule book, any more than you frequently read the instruction book for your video or computer from cover to cover. For daily use rules need to be much less detailed and presented more clearly, or thoroughly learned. Enforcement of a voluminous and detailed rule book is difficult to achieve. This approach can seem to take away all individual freedom and control over the work. It will only work where danger is very evident and it can be guaranteed that application of the rules will always result in safety. Even then it will work only with difficulty if following the rules is also not the easiest and most obvious way of doing the job.

The following extract from a study of rules is typical:

50 railway workers were asked about safety rules governing work on and near railway tracks:

- 80% considered that the rules were mainly concerned with pinning blame.
- 79% thought there were too many rules.
- 77% found the rules conflicting.
- 95% thought that work could not be finished on time if the rules were all followed.
- 85% found it hard to find what they wanted in the rule book.
- 70% found the rules too complex and hard to read.
- 71% thought there was too little motivation to follow rules.
- Not one could remember ever having referred to the rule book in a practical work situation.

Rules are subject to exceptions and to erosion. Safety manuals and safety laws tend to be full of complex specifications with many 'if . . ., then . . .' clauses which are perfect if followed, but which are too complex to remember. Execution of all the checks to see which sub-clause applies in any one case would often take too long in practice. Such rule books only serve to assuage the consciences of the rule makers. After an accident they can establish exactly who should have done what and so who was to blame. The existence of such a complex edifice of rules is a signal that the system is inwardly sick and in urgent need of redesign to incorporate behavioural rules into either training or hardware design. Ideally design should precipitate the right action, and articulated, written rules are only necessary where the way someone would expect to have to operate in a given situation is not in fact correct.

This conflict between establishing rules and leaving the flexibility to cope with exceptions and with changes can be seen at all levels in safety. It is reflected in the arguments about rigid central specification in laws and standards in contrast with enabling frameworks with objectives and the freedom for each company to comply in the way it wishes. It can be seen at the level of the company where operating managers are keen to reduce problems to fixed rules as fast as possible, in order to be able to get on with production. Safety departments have a task here to act as the protagonists of continual revolution in the firm. Safety rules need to be written with the involvement of those who must follow them. They also need to be updated at regular intervals from the critical experience of those same people.

The critical factor in rule making is to ensure the participation of those who will carry out the rules in the process of making them. An extreme version of this, which occurs in organisations called in the literature[45] 'High Reliability Organisations', is that the working group is left entirely to develop its own specific working methods and rules, within a very strong culture of safety. Through constant on-line discussion, revision of rules in the light of shared experience, constant checking of each other's behaviour and openness to mutual criticism, these organisations can even run with very few written rules. The social interaction and active concern with safety makes them unnecessary. New people are very rapidly indoctrinated and trained. The archetypal application of this approach has been described in the US Marine aircraft carriers. The principles are now being applied elsewhere in modified form[46].

2.7.5.4.5 Safety culture[47]

Safety culture became a fashionable word in the last decade of the 20th century. There is a great deal of confusion about its precise definition[3]. A definition which tries to cater for all of these confusions is the following:

> Safety culture is the attitudes, beliefs and perceptions shared by natural groups as defining norms and values, which determine how they act and react in relation to risks and risk control systems.

We would actually do better to refer to 'the culture of an organisation which affects safety', rather than 'a safety culture'. There are only a few rare companies where safety as a value is really so central that we can consider them as having a safety culture. Most companies have values which are at odds with safety to some extent, preferring production or cost-saving above risk reduction. The essence of the values which would make safety central are the following, cited from reference 3.

- The importance which is given by all employees, but particularly top managers to safety as goal, alongside and in unavoidable conflict with other organisational goals; e.g. whether actions favouring safety are sanctioned and rewarded even if they cost time, money or other scarce resources.
- Which aspects of safety in the broadest sense of the word are included in that concept, and how the priority is given to, and felt between the different aspects.
- The involvement felt by all parties in the organisation in the process of defining, prioritising and controlling risk; the sense of shared purpose in safety.
- The creative mistrust which people have in the risk control system, which means that they are always expecting new problems, or old ones in new guises and are never convinced that the safety culture or performance is ideal. If you think you have a perfect safety culture, that proves that you haven't. This means that there must be explicit provision for whistleblowers. A role for health and safety staff in very good organisations may be as a professional group constantly questioning and seeking the weak points in the prevailing culture.

- The caring trust which all parties have in each other, that each will do their own part, but that each (including yourself) needs a watchful eye and helping hand to cope with the inevitable slips and blunders which can always be made. This leads to overlapping and shared responsibility.
- The openness in communication to talk about failures as learning experiences and to imagine and share new dangers, which leads to the reflexivity about the working of the whole risk control system. If coupled with a willingness only to blame in the case of unusual thoughtlessness or recklessness, this can drive a responsible learning culture.
- The belief that causes for incidents and opportunities for safety improvements should be sought not just in individual behaviour, but in the interaction of many causal factors. Hence the belief that solutions and safety improvement can be sought in many places and be expected from many people.
- The integration of safety thinking and action into all aspects of work practice, so that it is seen as an inseparable, but explicit part of the organisation.

Attempts to instil such a culture will always take a long time, since they will have to change the basic assumptions on which the organisation works. There are many ways to do this. Under this heading come education, media and advertising campaigns to build and change 'images'. However, these are only the instruments. What drives them is a long-term change in management beliefs and practice, leading to consistent management standards, insistent example from top managers and opinion leaders, and an openness to learning and improvement. This may need to be supported by broader changes, even in national culture and law. These methods of change are usually long-term, and are often poorly understood. They operate by training people to look at, question, and so develop, their own goals. They also expose people to different opportunities and chances for achievement, and present them with examples of what are labelled 'acceptable' and 'unacceptable' behaviour for people to copy. Above all they depend on the investment of much time from top management to inculcate these values all through the organisation. In essence the culture described is also a participative one, in which all members of the workforce are valued for their unique contribution to the whole.

2.7.6 Conclusion

This chapter has surveyed in broad sweeps a huge area of knowledge and study. It has only been able to do so to a limited depth. On each topic there are books written to take the interested reader further. The reference list and the list for further reading attempts to indicate some books to do this. It is hoped that the picture which has been painted here gives enough structure to show that behaviour is predictable and can be influenced and managed, but not without the willing acceptance and participation of those being managed. People need to be supported in the tasks they are good at, which keep a vast array of risks under control already. They need to be protected from being required to carry out tasks they are not good at. Above all they need to be encouraged to see the control of safety as a shared task that needs to be made explicit and discussed and improved on at regular intervals, so that it can then be performed as smoothly and automatically as possible most of the time.

References

1. Shaw, L. and Sichel, H., *Accident Proneness*. Pergamon, Oxford (1971)
2. Hale, A.R. and Hale, M., A review of industrial accident research literature, Committee on Safety and Health at Work: Research Paper, The Stationery Office, London (1972)
3. Hale, A.R., Cultures confusions. Editorial to a special issue on safety culture and climate. *Safety Science*, **34**, 1–14 (2000)
4. Bainbridge, L., The ironies of automation. In Rasmussen, J., Duncan, K.D. and Leplat, J. (eds), *New Technology & Human Error*. London, Wiley, pp. 271–283 (1987)
5. Rasmussen, J., Risk management in a dynamic society: a modelling problem. *Safety Science*, **27**(2/3), 183–213 (1997)
6. Hale, A.R. and Glendon, A.I., *Individual Behaviour in the Control of Danger*. Elsevier, Amsterdam (1987)
7. Hale, A.R. and Hale, M., Accidents in perspective, *Occupational Psychology*, **44**, 115–121 (1970)
8. Taylor, F.W., *Principles of Scientific Management*, Harper & Row, New York (1911)
9. Mayo, E., *The Social Problems of an Industrial Civilisation*, Routledge & Kegan Paul Ltd, London (1952)
10. Maslow, A.H., *Motivation and Personality*. Harper, New York (1954)
11. Porter, L.W., Lawler, E.E. and Hackman, J.R., *Behaviour in Organisations*. McGraw-Hill, Kogushawa, Tokyo (1975)
12. Cattell, R.B., *The Scientific Analysis of Personality*, Penguin Books, London (1965)
13. Fishbein, M. and Ajzen, I., *Belief, Attitude, Intention and Behaviour – an Introduction to Theory and Research*. Addison Wesley, Reading, MA (1975)
14. Reason, J., *Human Error*. Cambridge University Press (1990)
15. Wagenaar, W.A., *Paradoxes of Gambling Behaviour*. Lawrence Erlbaum, Hove (1988)
16. Hale, A.R., Is safety training worthwhile? *J. Occupational Accidents*, **6**(1–3), 17–33 (1984)
17. Rasmussen, J., What can be learned from human error reports, in Duncan, K., Gruneberg, M.M. and Wallis, D.J. (eds), *Changes in Working Life*. Wiley, Chichester (1980)
18. Surry, J., *Industrial Accident Research*, Department of Industrial Engineering, University of Toronto (1969)
19. Lehto, M.R. and Miller, J.M., *Warnings: Fundamentals, Design & Evaluation Methodologies*. Fuller Technical Publications. Ann Arbor Michigan. (1986). Special issue of *Safety Science on Warnings & Risk Communication*. DeJoy, D.M. and Wogalter, M.S. (eds), **16**(5/6), (1993)
20. Tong, D., The application of behavioural research to improve fire safety. *Proc. Ann. Conf. Aston Health and Safety Society*, Birmingham (1983)
21. Kirwan, B. and Ainsworth, L.K. (eds), *A Guide to Task Analysis*. Taylor & Francis, London (1992)
22. Kirwan, B., *A Guide to Practical Human Reliability Assessment*. Taylor & Francis, London (1994)
23. Reason, J.T. A Framework for Classifying Errors, in Rasmussen, J., Leplat, J. and Duncan, K. (eds), *New Technology and Human Error*. Wiley, New York (1986)
24. Feggetter, A.J., A Method for Investigating Human Factors Aspects of Aircraft Accidents and Incidents. *Ergonomics*, **11**, 1065–1075 (1982)
25. Weegels, M., *Accidents involving consumer products*. Doctoral thesis. Faculty of Industrial Design. Delft University of Technology (1996)

26. Lowrance, W., *Of Acceptable Risk: Science and the Determination of Safety.* W. Kaufmann, Los Altos, CA (1976)
27. Royal Society, Risk assessment, a Study Group Report, London (1983)
28. Health and Safety Executive. *The Tolerability of Risk from Nuclear Power Stations*, HSE Books, Sudbury (1992)
29. Vlek, C. and Stallen, P.-J., Judging risks and benefits in the small and in the large. *Organisational Behaviour and Human Performance*, **28**, 235–271 (1981)
30. Starr, C., Social Benefit versus Technological Risk. *Science*, **16**, 1232–1238 (1969)
31. Abeytunga, P.K., *The Role of the First Line Supervisor in Construction Safety: the Potential for Training*. PhD thesis, University of Aston in Birmingham (1978)
32. Svenson, O., Risks of Road Transportation in a Psychological Perspective. *Accident Analysis and Prevention*, **10**, 267–280 (1978)
33. Green, C.H. and Brown, R.A., The perception of, and attitudes towards, risk: Preliminary Report: E2, Measures of Safety. Research Unit, School of Architecture, Duncan of Jordanstone College of Art, University of Dundee (1976)
34. Waterhouse, J.M., Minors, D.S. and Scott, A.R., Circadian rhythms, intercontinental travel, and shiftwork. In Ward Gardiner, A. (ed.), *Current Approaches to Occupational Health, 3*, Wright, Bristol (1987)
35. Cox, T., *Stress*, Macmillan Press, London (1978)
36. Theorell, T., Psychosocial factors in the work environment. In Brune, D., Gerhardsson, G., Crockford, G.W. and D'Auria, D. (eds), *The Workplace: Fundamentals of Health, Safety and Welfare*, Vol. 1, pp. 158–186. ILO, Geneva (1997)
37. Burke, R.J., Organisation-level interventions to reduce occupational stressors. *Work and Stress*, **7**, 77–87 (1993)
38. Powell, P.I., Hale, M., Martin, P. and Simon, M., *2000 Accidents*. National Institute of Industrial Psychology, London (1971)
39. Gagné, R.M., *The Conditions of Learning*. Holt, Rinehart and Winston, London (1970)
40. Wilde, G.J.S., *Target Risk*. PDE Publications, Toronto (1994)
41. Atherley, G.R.C.A., *Occupational Health and Safety Concepts: Chemical and Process Hazards*. Applied Science Publishers, London (1978)
42. Culvenor, J., Driving the science of prevention into reverse, *Safety Science*, **27**(1), pp. 11–83 (1997)
43. Hale, A.R., Safety rules OK? Possibilities and limitations in behavioural safety strategies. *J. Occupational Accidents*, **12**, 3–20 (1990)
44. Hale, A.R. and Swuste, S., Safety rules: procedural freedom or action constraint?, *Safety Science*, **29**(3), 163–178 (1998)
45. Roberts, K.H., New challenges in high reliability research: high reliability organisations. *Industrial Crisis Quarterly*, **3**, 111–125 (1998)
46. Bourrier, M., Elements for designing a self-correcting organisation: examples from nuclear plants. In Hale, A.R. and Baram, M., *Safety Management: the Challenge of Change*. Pergamon, Oxford (1998)
47. *Safety Science*, Special issue on safety culture and climate (2000)

Further reading

Primary texts for further reading are:

Hale, A.R. and Glendon, A.I., *Individual Behaviour in the Control of Danger.* Elsevier, Amsterdam (1987). This chapter is in great part a summary of the material covered there in great detail. It also contains detailed references for still deeper reading.

Glendon, A.I. and McKenna, E.F., *Human Safety and Risk Management*. Chapman Hall, London (1995). In many ways an updated version of Hale and Glendon, with a somewhat different structure and more oriented to basic psychological studies.

Many of the following texts have an overlapping coverage of subject matter. The reader should therefore select from among them. The brief notes attached will guide that choice.

Brune, D., Gerhardsson, G., Crockford, G.W. and D'Auria, D. (eds), *The Workplace: Fundamentals of Health, Safety and Welfare*. ILO, Geneva (1997). A multi-author work covering a much wider range of topics, but with some useful chapters on working environment and safe behaviour.

Canter, D., *Fires and Human Behaviour.* Wiley, Chichester (1980)
A good review of work on the specific topic of reactions to fire.

Coleman, J.C., *Introductory Psychology,* Routledge & Kegan Paul, London (1977). Written with medical and nursing students in mind. The individual chapters are by different experts. Covers almost the full range of the subjects in these chapters.

Cohen, J. and Clark, J.H., *Medicine, Mind and* Man, W.H. Freeman & Co., Reading, MA (1979). A parallel text to Coleman also written for students of health sciences.

Feyer, A.M. and Williamson, A., *Occupational Injury: Risk Prevention and Intervention.* Taylor & Francis, London (1998). An edited collection of papers on many of the aspects covered in this chapter.

Hoyos, C.G. and Zimolong, B., *Occupational Safety and Accident Prevention: Behavioural Strategies and Methods.* Elsevier, Amsterdam (1988). A parallel text to Hale and Glendon written somewhat more from the viewpoint of safety management.

Hale, A.R. and Hale, M., *A Review of Industrial Accident Research Literature,* Committee on Safety and Health at Work: Research Paper, HMSO, London (1972). A brief review of the literature up to 1972 on human factors in accident causation. Valuable source of further references for the older studies.

Hollnagel, E., *Cognitive Reliability and Error Analysis*: CREAM. Elsevier, Oxford (1998). A book drawing on human factors work in the major hazards industry, with a clear structure and excellent referencing.

Powell, P.L., Hale, M., Martin, P. and Simon, M., *2000 Accidents,* National Institute of Industrial Psychology, London (1971). Summary report of a four year field study of accident causes. Good overview of the priorities in the field.

Rasmussen, J., Duncan, K. and Leplat, J. (eds), *New Technology and Human Error.* Wiley, Chichester (1987). A very valuable book of readings of both theory and practice in human error assessment and control.

Reason, J., *Human Error.* Cambridge University Press (1990). An excellent book setting out the theories of a very influential researcher. It goes much more deeply into the psychological mechanisms behind error.

Stammers, R.B. and Patrick, J., *The Psychology of Training,* Methuen Essential Psychology E3, London (1975). Short text covering the main psychological approaches and insights into the subject.

Wilde, G.J.S., *Target Risk.* PDE Publications, Toronto (1994). A thorough review of the risk homeostasis literature. A book to make you think about the effectiveness of safety interventions aimed at human behaviour.

Readers wishing to keep up-to-date with research on this topic will find research and review articles in scientific journals such as *Safety Science, Applied Ergonomics* and to an extent in *Work and Stress.* Road traffic safety papers are to be found in the *Journal of Safety Research* and in *Accident Analysis and Prevention.*

Chapter 2.8

Risk management and behaviour modification

J. E. Channing

2.8.1 Introduction

The objective of law is to regulate the behaviour of the citizen. Consequently, changing the behaviour of people at work to improve their safety is a feature of many aspects of legislation. The most obvious elements include the requirements for systems of work, the provision of information to employees, and requirements for training. These legal obligations have been in place for many years and are key elements of the Health and Safety at Work etc. Act 1974. In practice this approach alone has not eradicated accidents at work. This should be no surprise. The existence of written procedures and attendance at training courses are not particularly effective at gaining the correct behaviour from people preoccupied with their jobs and private thoughts when a hazardous situation suddenly arises.

From another perspective most businesses find that the majority of their accidents at work do not arise out of their specific technology or business activity. They arise mostly from everyday events such as slips, trips, falls and handling accidents. Yet trying to focus the workforce upon reducing these apparently trivial and unglamorous accidents is difficult. Paradoxically it is essential to attack all types of accidents. Focusing on some types of accident yet tolerating others is illogical. If a person slips the most likely outcome is a bruise, yet with a minor change in circumstance it could be fatal. An accident can be seen as a 'loss of control' and the consequences cannot be predicted with certainty. An approach which treats these 'everyday' accidents as unacceptable also promotes an attitude that prevents the obviously serious or catastrophic accidents. Such an approach also prevents the insidious chronic conditions which cause ill-health – such as musculoskeletal, skin or lung disorders – from occurring and does not tolerate superficial testing of safety critical trip systems whose failure could be catastrophic. This is a 'zero tolerance' approach to accidents.

Achieving a zero tolerance position requires a change of culture in the workplace and of the attitudes of people working in it. Terms like 'culture' and 'attitude' are easily understood in general terminology but quite difficult to develop into practical safety programmes. The relationship between attitude and behaviour is the subject of ongoing research but it can be argued that where a positive attitude toward safety

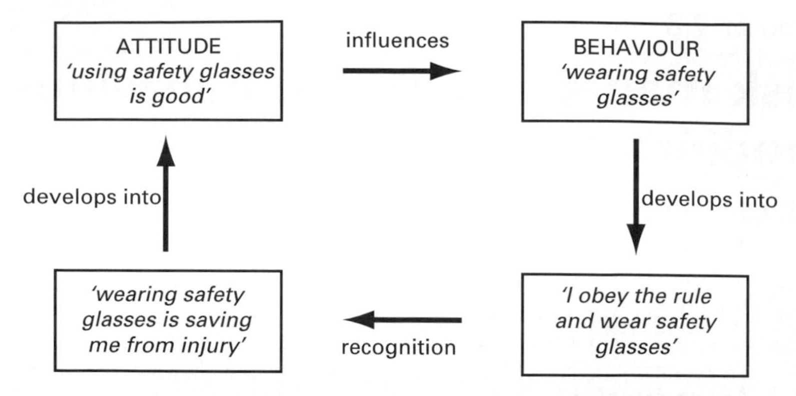

Figure 2.8.1 Attitude and behaviour: the wearing of eye protection

exists correct safety behaviour occurs. A simple model linking attitude to behaviour is presented in *Figure 2.8.1*. This example examines the attitude and behaviour towards the wearing of eye protection. People who normally do not need to wear spectacles in everyday life often find that it is uncomfortable to wear safety glasses at work. They only wear them because it is the rule in the workplace although they recognise that the rule is applied to save them from injury. Over time the action of wearing safety glasses develops into an attitude that wearing them is 'good'. This attitude change can be seen when the same people begin to wear eye protection when they do jobs at home.

The model implies that attitude and behaviour are linked such as to influence and reinforce each other. Psychologists began to look at behaviour as a subject itself rather than as merely an indicator of internal states of mind (i.e. attitudes) following early work by Skinner.[1] This approach has led to behavioural analysis and behavioural modification. By focusing on behaviours, accidents can be prevented. In turn this reinforces an attitude toward a safe work environment and a culture of zero tolerance.

The UK Health and Safety Executive believe that an individual's actions at work depend upon a number of human factors which they define as: *'Human factors refer to environmental, organisational and job factors, and human and individual characteristics which influence behaviour at work in a way which can affect health and safety'*[2]. They suggest that human factors can best be understood by considering three aspects.

- Organisational and management aspects:
  - Poor work planning
  - Lack of safety systems
  - Poor response to previous incidents
- Job aspects:
  - Poor equipment design
  - Poor instruction
  - Poor work conditions

- Individual aspects:
  - Low skill levels
  - Demoralised, bored employees.

This chapter considers how behaviour modification processes can assist to control risks and reduce injury.

2.8.2 Behaviour modification for employees

Behaviour shaping is a function of management. Employing people to undertake tasks for the benefit and prosperity of the enterprise for which they receive a reward (an income) is itself behaviour shaping.

Behaviour changing programmes, however, seem most effective when feedback occurs which shows the positive consequences of the safe behaviour. Typical areas of work where behaviour modification to improve safety can be successful include the wearing of personal protective devices, the proper handling of materials, the use of safe working methods around dangerous machines, and housekeeping. Researchers such as Komaki *et al.*[3,4,5] and Suzler-Azaroff[6] consider that the highlighting of consequences when the desired safety behaviour occurs stimulates the adoption of safe work practices. They also promote the idea that feedback when the desired behaviour occurs is itself a motivational strategy.

A study by Nasanen and Saari[7] examined positive feedback as applied to housekeeping. They looked at how feedback on its own, without a target goal being set, improved housekeeping performance. An improvement in housekeeping was achieved by simply publishing data to the work group. The employees were only aware of the key practices used to measure housekeeping performance and were told the percentage score achieved by the independent observers. Although the study did not address accident performance it found that by focusing on housekeeping and providing feedback on performance, accidents were reduced significantly below those caused by poor housekeeping. The study suggested that the factors that caused a response to improve poor housekeeping also worked to reduce accidents even where housekeeping was not a contributory factor.

2.8.2.1 The performance management approach

Management gurus have been active in exploring techniques which will improve performance of groups and individuals to achieve business goals. One approach has been termed 'Performance Management'[8], which considers four responses to a behaviour. These responses are termed 'positive', 'negative', 'punishment' and 'extinction'.

In 'Positive Reinforcement' the individual receives something that is wanted or valued after the proper behaviour is completed. Reinforcements of this type encourage the behaviour to be repeated in the future.

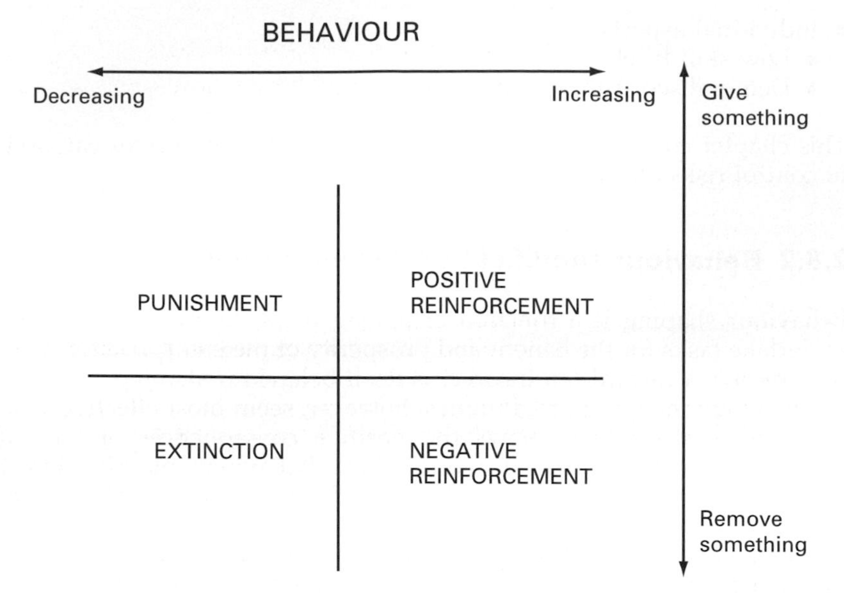

Figure 2.8.2 Summary of behaviour reinforcers

'Negative Reinforcement' encourages a desired behaviour when the consequence is *removed*. People will work to avoid certain outcomes such as reprimand, suspension or dismissal. They will choose to repeat a behaviour which avoids or escapes this sort of negative outcome, thereby making the required behaviour more likely to be repeated. 'Punishment' reinforcers aim to decrease the likelihood that the behaviour will be repeated. This type of reinforcement employs the giving of an unacceptable response such as criticism or the allocation of undesired work as a means of reducing the recurrence of the undesirable behaviour.

'Extinction' is a type of consequence in which an outcome desired by an individual is removed following a behaviour and is withheld each time

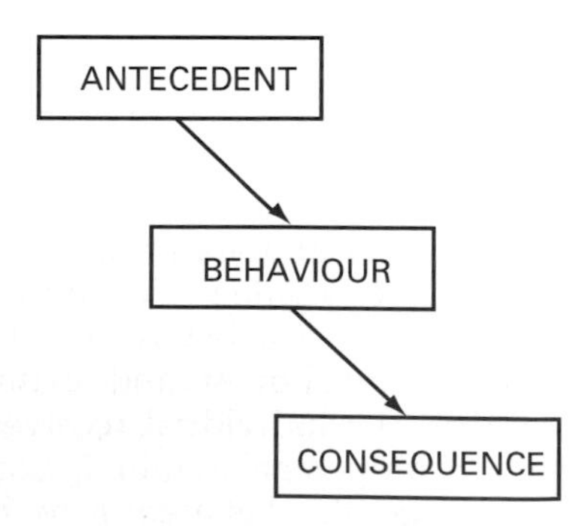

Figure 2.8.3 The ABC model

Why you do what you do when the telephone rings!

In 'ABC' terms the sequence of events when the telephone rings is as follows:

- Antecedent – the telephone rings
- Behaviour – you answer the telephone
- Consequence – you talk to the caller

However, if the phone rings regularly around the time when the children have returned from school and your experience is that in most cases the calls are for the children, you as the parent soon desist from the Pavlovian response of answering it. Instead you call the children to answer it!

Whilst the antecedent still occurs (the telephone rings) your behaviour to it has been conditioned by the anticipated consequence (the caller will want to talk to the children).

Figure 2.8.4 An example of antecedents, behaviours and consequences

that behaviour occurs with the intention of reducing the occurrence of the behaviour. *Figure 2.8.2* summarises these behaviour reinforcers.

In everyday life, both in the family and at work, all four types of reinforcement are used. Positive reinforcement is generally viewed as being the most effective motivator to achieve work-related and safety goals since it is often 'free' because a few words of praise or encouragement may be all that is required. Furthermore it often outlasts the presence of the manager or supervisor who gives it, thus making the (safe) behaviour more likely to continue.

This approach has also been applied to safety situations by Krause *et al.*[9]

The key concept is that behaviours are mostly shaped by the expected consequences of that behaviour rather than by anything else. The model put forward is outlined in *Figure 2.8.3*.

An 'antecedent' is an event which initiates a visible behaviour. A 'consequence' is the outcome of that behaviour. Whilst both antecedents and consequences have an effect on behaviour, the consequences are more powerful in exerting control over and directly influencing behaviour. Antecedents, on the other hand, control behaviour indirectly, largely because they serve to predict the consequences. An example of this theory is to be found in *Figure 2.8.4*.

Further work in this area has shown that there are a number of features which make consequences stronger behaviour modifiers to groups or individuals than others.

The first feature is *Timing*. A consequence that follows on quickly from a behaviour is far more effective than one which occurs after a delay (i.e. later).

The second feature is *Reliability*. A consequence that with certainty will follow a behaviour is more effective than one which may or may not follow that behaviour.

The third feature is the *Nature* of the consequence. When the individual or group feel they gain from the consequence, i.e. it is positive, the effect

Antecedent	Behaviour	Consequence	s/l	c/u	+/–
Eagerness to use the workstation	Failure to adjust the computer station before use	Saves time	s	c	+
		Musculoskeletal injury	l	u	–
Inadequate training on the need to adjust the workstation		Eye strain	l	u	–
Lack of awareness of the chronic injury potential		No immediate ill effects	s	c	+
Anticipation of zero consequence		Not seen by colleagues to be 'fussy'	s	c	+

s/l – soon/late c/u – certain/uncertain +/– – positive/negative

Figure 2.8.5 ABC analysis of computer stations

is more powerful than when they lose, i.e. a negative consequence. There are more problems with this feature than the others.

A positive feature to one person may not be seen as such by another. It can be dependent upon national or local culture. Some people may respond to a simple 'well done' but others may only respond to something far more tangible such as a gift. Problems can also arise if one person's behaviour receives a different level or quality of response from another. For these reasons the positive consequence should be consistent and an appropriate token rather than a chancy lottery win! Notwithstanding this difficulty, a positive response to a behaviour does stimulate a repeat of the behaviour.

When a consequence is imbued with these three characteristics, viz. 'soon', 'certain', and 'positive', it is an effective motivator to achieve the required behaviour. In contrast a consequence which is 'late', 'uncertain', and 'negative' is a weak motivator for achieving the desired behaviour but it is not totally insignificant. At least the behaviour has been recognised and not ignored!

Equally, applying two or even just one feature to the consequence has intermediate levels of influence on the behaviour.

The whole point of a behavioural analysis is to identify consequences which will reinforce the behaviour that is wanted.

The starting point for applying this theory in the workplace is to identify a specific behaviour and analyse it. *Figure 2.8.5* looks at behaviour common to many workplaces, namely the failure by the operator to adjust a computer workstation for individual use. The analysis proceeds by first listing the possible reasons why the current behaviour should occur as it does. These are the antecedents. The

Antecedent	*Behaviour*	*Consequence*	*s/l*	*c/u*	*+/–*
Understanding of injury potential	Computer workstation adjusted by user before use	Observation and comment by supervisor	s	c	+
Training in use		Expectation of colleagues	s	c	+
Expectation of comment by supervisor		Observation and comment by colleagues	s	c	+
Expectation of ongoing reminders to use workstation correctly					

s/l – soon/late c/u – certain/uncertain +/– – positive/negative

Figure 2.8.6 Revised ABC analysis of computer workstations

consequences of the behaviour are also listed and analysed to establish the features they possess.

It can be seen that the avoidance of musculoskeletal injury or eyestrain are weak consequences because they are 'late', 'uncertain' and 'negative' in nature. The other consequences arising from not adjusting the workstation are 'immediate', certain', and 'positive' to the individuals and thus reinforce the unwanted behaviour. Whilst it is possible to debate each component the overall balance of consequences is disproportional and in favour of reinforcing the unwanted behaviour, namely the failure to adjust the chair height, or alter the screen tilt, or draw the blinds so as to prevent glare from windows etc.

The next step in the process is to state in precise terms the observable behaviours which are desired. Then the antecedents and consequences which influence the desired behaviour can be added. This is demonstrated in *Figure 2.8.6.*

The antecedents are likely to focus around education and training followed by ongoing reminders. The consequences would include comments and intervention by supervisors and fellow workers. For the consequences to be effective, supervisors would need regularly and frequently to observe and comment on the individual's use of, and performance at, the workstation. The comments should be positive and approving when the desired behaviour has occurred. Commenting only when the desired behaviour has not occurred is far less effective.

Reducing accidents in the workplace requires that the performance management approach is applied to all unsafe behaviours. Clearly this is a mammoth task that should be approached systematically. It can be achieved either by an analysis of the accident data (i.e. using historical data) or by using Job Hazard Analysis of tasks undertaken (i.e. using

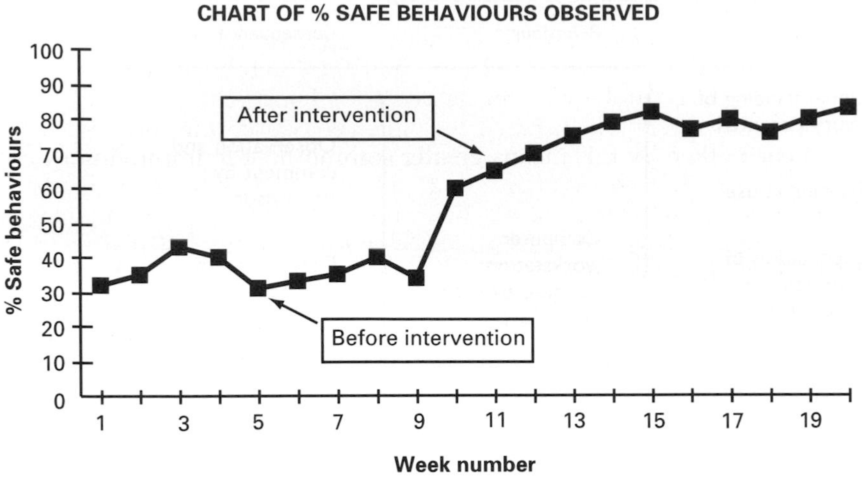

CHART OF % SAFE BEHAVIOURS OBSERVED

Figure 2.8.7 Typical effect of behaviour intervention process in the workplace

predictive data). In either case the analysis is looked at from a behavioural perspective.

Using the accidents analysis data, the first step is to group it by task and by area, e.g. 'fork-lift truck accidents in the distribution area'. The second step is to examine how each accident occurred and to identify the significant behaviours which contributed to the accident. This list is variously termed the 'critical behaviour list' or the 'key behaviour list' for fork-lift truck accidents in the area. The third step is to analyse each of the critical behaviours identifying their antecedents and consequences, and the features of each consequence (the 'ABC' analysis). The fourth step is to state the desired behaviour which would avoid the accident and to give it consequences which will reinforce the use of the desired consequences. At this stage this analysis is complete.

Job hazard analysis begins by examining the task and listing the desired behaviours to accomplish it safely. Each desired behaviour is examined and the necessary antecedents and consequences added.

The analysis is the first part of the programme which then has to be implemented. In practice this means that a number of observers have to be trained. Their task is to understand the safety critical behaviours for the workplace and to become skilled in identifying them.

The observers audit the work area recording the number of safety critical behaviours observed and the number of unsafe behaviours observed to produce a 'percentage safe behaviour' score as follows:

$$\% \text{ Safe behaviour score} = \frac{\text{number of safe behaviours observed}}{\text{total number of behaviours observed}}$$

This data is plotted and posted in the workplace so that the workgroup is encouraged to work toward a rising trend. A typical graph of the results is shown in *Figure 2.8.7*.

The strength of the process is more than an analysis of actions in the workplace and observations of activity. It works best where the employees take a leading role in managing and implementing it. Employees thus undertake the analysis of behaviours, add the consequences necessary to achieve the safe behaviours and subsequently audit each other. By this means greater commitment to improving safety occurs. In addition the workgroup often know how a job is actually done (as opposed to what the procedure for it says) and is in the best position to draw up the list of desired safety behaviours in the first place and to monitor compliance with it.

When accidents occur they are analysed to see what might have gone wrong. It may be that the safety critical behaviours were not identified correctly in the first place. Alternatively the analysis might reveal that the consequence modifiers are ineffective and have to be rethought.

2.8.2.2 The structural feedback approach

Performance management requires that consequences are first analysed and then restructured to encourage the preferred (safe) behaviours. The chart demonstrating the improvement in the percentage of safe behaviours is a feedback tool demonstrating the gains made.

Other work by Cooper *et al.*[10,11,12,13] places greater emphasis upon the feedback process. In particular they stress that publicly displaying a chart showing how well, or otherwise, a group of employees is doing in relation to the areas of safety in which improvement is sought is itself a very powerful agent for change. Consequently it becomes important that feedback charts are posted prominently and are regularly updated. In order to achieve this it is necessary that managers adopt a particular role, namely:

1. Champion the behavioural process and inform his workpeople of it and of his support for it.
2. Encourage employees to become active in the process especially as observers.
3. Allow employees the time to be involved in the training and meetings needed for goal setting.
4. Allow each observer one observation session each working day. An observation session should last no longer than 20 to 30 minutes.
5. Be committed to attend goal setting sessions with the observers thereby demonstrating his support.
6. Praise employees who work safely.
7. Encourage employees to reach the safety goals.
8. Arrange for senior managers to visit the workplace each week to encourage the safety improvement effort.

The observers, who are members of the workgroup, commence their training by analysing local accident data. They identify contributory factors for each accident and subdivide them into observable behaviours or situations which are safe or unsafe. These observable data form the basis of a checklist. Emphasis is placed upon gaining agreement from the workforce that the items that form the checklist of behaviours are valid. This is an important step as the workforce is assessed and scored against the list that has been generated. The process of gaining agreement is itself a type of feedback which seeks to gain involvement of and ownership by employees of the safety programme.

Scoring takes the conventional form of making observations in the workplace of safe and unsafe behaviours to generate a 'percentage safe behaviour score'. The data are charted and posted visibly in the workplace. Feedback of the data is not the only emphasis. The employees are asked by their observers to establish their own goals and subgoals against which the performance is measured

2.8.2.3 Behaviour observation and counselling techniques

Any behaviour modification technique must involve an interaction with people. As an accident can occur at almost any time and the consequences in terms of injury outcome are not predictable, concentrating on a list of identified safety critical behaviours can have the following limitations:

1 The critical behaviour list may be incomplete.
2 Behaviours may appear on the list as a result of a perception of, rather than an analysis of, an actual risk. This is more likely if the list has been compiled from a job hazard analysis.
3 As the size of the list grows to encompass more behaviours (a result of ongoing accident experience and the desire to eradicate all accidents by adding more safety critical behaviours) the whole system can become unwieldy because too may behaviours are included in the observation process.
4 The very existence of a list may limit the focus of employees and observers to only those behaviours which are on the list. This becomes a more significant problem if observers are under pressure to complete a quota of observations per week or per month. Under these circumstances the objective can alter subtly from one of using the technique to reduce accidents to becoming merely an exercise in completing a checklist. The resultant quality fall-off which takes place can undermine and discredit the entire effort.

Other approaches have been developed such as the DuPont Safety Training Observation Program ('STOP') or their similar 'Safety Management Audit Programme'[14]. In both these programmes the approach tends to be less analytical in defining prescribed unsafe behaviours with a different emphasis that requires a management top-down approach in which one level of manager, having been taught the process, subsequently teaches the next subordinate level. The emphasis is upon observation of employee behaviour and immediate counselling of the observed employees. Implementation of the process is through members of line management from team leaders to senior managers. Each undertakes a workplace safety behaviour audit to an agreed schedule. For example, a team leader of a large workgroup may be expected to undertake a daily audit, middle managers may do an audit each week, and senior managers and directors an audit each month. The training they receive assumes a degree of knowledge of the workplace and the hazards it contains. This is not unreasonable given that many managers will have several years' experience and knowledge of the work areas. Furthermore they are not necessarily expected to know in detail how safe working on each job should be achieved. They are expected, however, to recognise how injury might occur.

The training emphasises the skill in observing people as they work and learning to approach and discuss safety with them in a constructive manner. This applies to employees who are observed working safely as well as those working unsafely. In the former case discussion can commend the safe behaviour and be widened to encompass other tasks the employee might do, seeking out any safety concerns arising from them. The very fact that a person in authority is discussing safety issues with the employee is of great importance in raising awareness and commitment to accident-free working. Thus in these programmes, there is greater emphasis on observation and immediate intervention than on observation, completion of a checklist, and the posting of a chart in the workplace. Nevertheless as employees do voice their concerns they have expectations that remedial measures will be taken. Feedback in this case often takes the form of a list of actions identified from the audits and a rolling calculation of the percentage completed.

2.8.2.4 Behaviour modification and the lone worker

Behaviour modification is most easily applied where large groups of people work in a systematic activity. Examples of applications include factory production lines, call centres, packaging and assembly lines and large construction sites. In these situations the tasks can be easily identified and the safe behaviours required to perform them without injury specified. Other jobs, where employees work alone and away from the direct control of the supervisor present different challenges. An example is maintenance tasks that require employees to work away from the workshop. In these situations specific risks can arise that cannot be identified until the job is underway. For example, maintaining a pump on a workbench is far simpler than performing the same task while the pump remains in its original location two miles away from the workshop. Typical of the problems that can arise are:

- Access to the pump is restricted.
- Bolt fixings may be corroded and the use of a blow torch to free them may not be permitted in the area.
- Particular internal parts of the pump may need additional maintenance work due to unexpected wear and tear, but the spare parts are at the workshop and the mechanic must make a decision to use the existing part believing 'it will last until next time' or stop production while the part is replaced.

The individual employee can rarely be kept under observation as these decisions and actions are taken. Guidance from the Health and Safety Executive[15] recommends the use of:

- structured incident reviews where the assessor lists contributing causes and seeks clarification and comment from employees;
- workforce questionnaires that seek to capture the perception of employees in identifying which of a list of eighteen management issues warrant improvement.

Approaches such as this are indirect behaviour modification and can be effective where there is extensive employee participation with feedback provided on progress made. Communication is critical to success. The objective is to address and influence the behaviour of the remote maintenance worker so that he can deal with the hazard at the time when it becomes a serious risk.

2.8.2.5 Behaviour modification and employee involvement

The application of behavioural techniques to improve the control of risks in normal employment have usually occurred as a result of a senior management initiative. It occurs as a dictate from on high and is imposed on the workforce. A typical response from the workforce is three-fold:

- it is seen as a management lay-on by middle managers and supervisors adding greater burden to their (the operator's) busy working lives;
- it is viewed with scepticism by employees who have seen initiatives come and go over past years ('flavour-of-the-month' syndrome);
- managers, supervisors and employees alike do not expect the initiative to last beyond the lifespan of the current senior management team or until something else comes along to distract their attention.

The result of these attitudes is that a minimum commitment is made to the initiative until senior management show by continuous example that the initiative is here to stay. A significant new safety initiative is likely to require at least two years of operations to convince middle managers and supervisors and four years to convince employees that the initiative is seriously intended!

In order to gain acceptance of behaviour change initiatives, but also to promote a wider safety culture, the involvement of employees in a partnership is recommended. A Health and Safety Executive publication[16] suggests that workforce involvement can improve performance in an organisation which:

- does not involve the workforce in determining company policy;
- does not treat the workforce or its representatives as equal partners in the health and safety committee;
- does not allow employees to set the health and safety agenda during meetings; and

Consultation model	Serious injury rate per 1000 workers
No union recognition and no joint committee	10.9
Worker representation but no joint committee	7.3
Joint committee but no trade union safety representatives	6.1 to 7.6
Full union recognition and joint committee	5.3

Figure 2.8.8 Type of consultation and serious injury rates

- does not involve workers in writing safe operating procedures.

If these and other aspects show that workforce involvement is limited, it is unlikely that a positive health and safety culture exists.

The concept of partnership is promoted in a union publication[17] that quotes from a study by Reilly *et al.*[18] which shows that in-depth consultation with employees reduces serious injury rates. Data are reproduced in *Figure 2.8.8*.

2.8.2.6 Refreshing behaviour modification processes

Employee behaviour modification processes, as with all other human processes, can become stale and ineffective. A significant proportion of the safety improvement comes from the early interaction between the observer and the observed. If this process becomes superficial then few improvements in safety performance can be expected. Refreshing the whole intervention process periodically is essential and can be achieved by:

- training new observers;
- using trained observers from one department in an adjacent department;
- changing the mix of observers to include managers, supervisors and employees in rotation;
- revising the checklist of critical behaviours. If some behaviours always remain on the critical list then generate two lists and mix the observation sequence;
- revising the checklist and discussing the changes in the light of incident experience from within the area and from appropriate external data.

2.8.2.7 Generic behaviour modification model

Most studies have focused upon modifying the behaviour of production operators to improve safety performance. While variations in detail exist, the basic generic behavioural process consists of:

- **Specify/know** the behaviours that are necessary for safe working.
- **Observe.** The observation step can take the form of a general observation of an employee's behaviour and counselling in the appropriate target behaviour. Alternatively the observer can refer to a previously established critical behaviour checklist and score the number of safe and unsafe behaviours observed.
- **Intervene and discuss.** The observer discusses with the individual their personal safety performance. This step can take several forms. The discussion can simply be a comparison of their performance against the critical behaviour checklist, safe behaviours and an emphasising of the recommended advice on the immediate corrective action, that should be taken to remedy unsafe situations.
- **Follow-up action.** A review of the data are collected and the action necessary to ensure safe operator behaviour.
- **Feedback** can take a number of forms ranging from a chart showing the percentage of safe and unsafe behaviours to a rolling list of open and closed corrective actions.

2.8.3 Behaviour modification for managers and supervisors

Wrong behaviour occurs at all levels. The narrow focus on front line employees behaviour has been criticised by labour unions[19] because it infers:

- only front line employee behaviour causes accidents;
- the employee is to blame;
- that a blame environment drives safety problems underground;
- that capital expenditure is not authorised because incidents would be avoided if employees did not make errors.

Figure 2.8.9 provides an illustrative example that demonstrates how several levels and functions of management as well as front line employees can contribute to the occurrence of an unsafe situation. It is therefore important that behaviour modification processes are applied to people in management and supervisory roles.

The generic behaviour modification model can be applied to managers and supervisors as illustrated in *Figure 2.8.10*.

The observer of a supervisor's behaviour can be the direct manager or a third party, such as a safety professional. A procedure that has proved successful comprises a list of fifty behaviours that align with the overall safety initiative for that year (*Figure 2.8.11*) with the results plotted on a 'radar screen' chart (*Figure 2.8.12*). Measurement occurred in several different ways which included:

- personal self-assessment by the individual manager and supervisor;

OPPORTUNITIES FOR ERRORS LEADING TO A PIPE RUPTURE WITH CONSEQUENTIAL LOSS OF OUTPUT	
The research chemist	The chemist recommended in his report a minimum operating temperature for the new compound but failed to emphasise that it froze below that temperature and expanded as it froze.
The design engineer	The designer did not allow for extremes of temperature and failed to specify adequate heat tracing for a heat exchanger bypass line.
The construction contractor	Because the bypass pipe was awkward to get at, did not lay the trace heating evenly along its length.
The supervisor	Wrote into the operating instructions turning on the trace heating when the temperature of the incoming compound dropped below a specified temperature. He did not incorporate a check to ensure the compound was flowing.
The plant operator	Neglected to check that the trace heating was on and that the compound was flowing.
The maintenance mechanic	Failed to report or repair damage to the trace heating caused during earlier maintenance.
The plant manager	Delayed activating the trace heating system to save energy but failed to recognise the cooling effect of a cold spell.
The corporate director	Cut the plant budget causing staff shortages that prevented all the safety checks being carried out, particularly to the trace heating system.
Result	Compound temperature dropped and it froze, expanded and fractured the bypass pipe. Process shut down for 6 hours while a repair was effected. No injuries but much lost production.

Figure 2.8.9 Error opportunities at different organisational levels (adapted from Lorenzo[20])

TASK	EXAMPLE
Specify the supervisors behaviour that is required	To hold a weekly toolbox talk on the previous weeks safety issues
Observe that the behaviour is occurring	Check/attend/sample the toolbox talks
Intervene to commend the activity or counsel if improvements are needed	Provide comment to the supervisor on the content and the impact of the toolbox talk
Follow-up on actions that the observer has undertaken	If the supervisor needs training in running group meetings ensure it is arranged
Feedback on overall performance periodically	Discuss the safety behaviours at the annual performance review

Figure 2.8.10 Generic behavioural check for managers and supervisors

BEHAVIOR DIMENSION A	BEHAVIOR DIMENSION B	BEHAVIOR DIMENSION C	BEHAVIOR DIMENSION D	BEHAVIOR DIMENSION E
Management Commitment, Involvement and Leadership	Training and Education	Employee Involvement	Hazard Identification and Analysis	Hazard Prevention, Elimination and Control
☐ I have an HSE policy, supported by annual goals and objectives, which is written and shared with my direct reports	☐ The rules for Personal Protective Equipment (PPE) in my area are posted and the safety signage is unambiguous	☐ My areas have safety committees which meet at least quarterly	☐ My area investigates all significant incidents immediately (whether a person is injured or a serious near-miss occurs)	☐ I ensure that individuals responsible for correcting hazards are identified and are set clear timelines to achieve the corrective action
☐ I have reviewed performance against my policy, goals and objectives with my direct reports and my HSE staff this quarter	☐ My area has a process to review work procedures with all employees regularly (not just the HSE people) and to ask them if changes are necessary	☐ I chair my safety committee	☐ A line manager or supervisor always leads the incident reviews, not an HSE staff person	☐ I keep track of the corrective actions in my area
☐ My safety policy states what I expect the line organization to do and what I expect the HSE staff organization to do	☐ I have attended the Basic Law training class	☐ My safety committee is comprised primarily of line people, with a minimum of HSE staff in attendance	☐ I make clear my expectation that an injured person must go to Medical even if they also wish to see their private physician	☐ I visit work areas and make it a point to comment on hazards I have identified and safety improvements achieved

☐ I energize the HSE support function to behave and act as a 'coach' rather than a 'do-er'	☐ I have attended an employee absence Case Management training class	☐ All actions arising from my safety committees have agreed-upon actions and timelines	☐ My area uses the new accident model form during incident reviews to brainstorm all possible contributions to an incident	☐ I recognize and reward work areas when they achieve safety performance milestones
☐ I use HSE issues to demonstrate a leadership style that is co-operative, participative, and inclusive as distinct from an autocratic or adversarial style	☐ I have assisted in delivering the 'Incident 2001' training class to my direct reports	☐ The safety meeting minutes are freely available. Departmental level safety committee minutes are posted on local bulletin boards	☐ My area conducts incident reviews in a positive, open, no-blame environment to ensure all views and opinions are discussed	☐ The area's safety data is rolled up by the line organization and not by an HSE staff person
☐ I make clear an expectation of zero incidents because 'professional' individuals and 'professional' workgroups do not have incidents	☐ My HSE staff person has reviewed my personal safety training plan	☐ Each safety bulletin board has an assigned owner who updates it frequently	☐ I fulfill my monthly Behaviour audit goals	☐ My area has an Emergency Preparedness program which has been tested within the last 12 months
☐ I promote an expectation of personal responsibility and accountability to avoid incidents in my work group	☐ My area has a safety training plan for each employee	☐ As a manager, I encourage participation by seeking my employees' input and actively listen to their concerns	☐ My area uses data from incident reviews to focus the Behaviour audits on topics of concern	☐ During this quarter, I have undertaken a spot-check on a key maintenance or inspection procedure (e.g., lock-out/tag-out) that is critical to control a hazard in my area

Figure 2.8.11

BEHAVIOR DIMENSION A	BEHAVIOR DIMENSION B	BEHAVIOR DIMENSION C	BEHAVIOR DIMENSION D	BEHAVIOR DIMENSION E
☐ I promote feelings of ownership among team members for their collective safety	☐ My area has a process to review all workplace changes so that safety issues are addressed	☐ I communicate my expectation of an overall 'partnership' in safety matters that creates an equal responsibility between the manager, the supervisor, or the employee	☐ The work groups in my area get feedback on incidents and audits	☐ I insist that hazards and safety incidents are reported to me by the line organization, not the HSE staff person
☐ I seek each day an opportunity to verbally and personally reinforce the safety values with my direct reports	☐ I encourage my HSE staff to increase the training they give to my managers, supervisors, and employees to do the activities that the HSE staff currently do	☐ I use my HSE representative to review and guide in improving the areas 'partnership in safety'	☐ I use incidents and behaviour audits to promote personal responsibility for the 'system' (manager's behaviour), the 'workplace' (supervisor's behaviour), and the 'work activity' (employee's behaviour)	☐ I set a personal example by following local safety rules and procedures
☐ Safety is one of the top 3 agenda items at my group and individual meetings with my direct reports	☐ Each person in my area has received the Safety Calendar for 2001	☐ My area exhibits a high standard of 'orderliness' knowing that to achieve it requires a high standard of workplace discipline and motivation	☐ In my area, less significant incidents are investigated and reported at a local level	☐ I always intervene if another person is behaving in a way that could result in injury

Figure 2.8.11 *(Continued)*

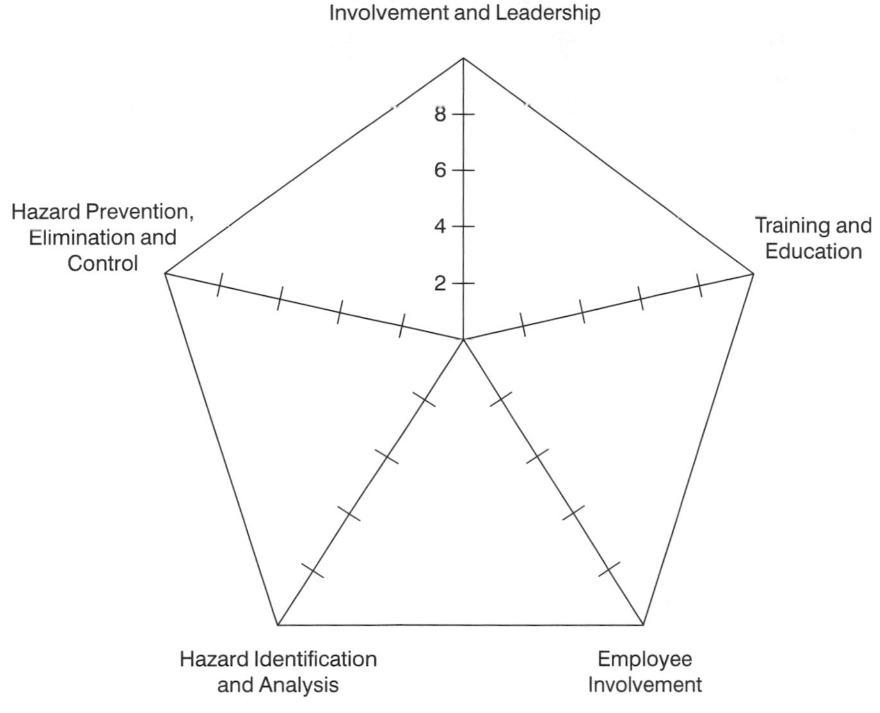

Figure 2.8.12 Safety leadership radar screen

- personal self-assessment which was reviewed with the direct superior;
- the divisional management team set the minimum behaviours expected with achievement against or beyond these targets being recognised in the annual performance review;
- direct reports of the manager or supervisor scoring the performance of the senior manager.

The particular method selected must depend on the maturity, confidence and interpersonal skills of the people involved. The fifty behaviours are organised into five dimensions each containing ten desirable behaviours. In this example, the behaviours are not ranked in any order and each behaviour bears the same value. The number of behaviours that a manager or supervisor demonstrates in each dimension is plotted along the appropriate axis of the radar screen. It is then possible to see at a glance what behaviours are being achieved and what behaviours should be undertaken. The radar screen was chosen in one organisation for a particular reason – that other production data was also displayed in the same format. Consequently 'safety' was seen as an integral part of the manager's or supervisor's work in the same way that production was.

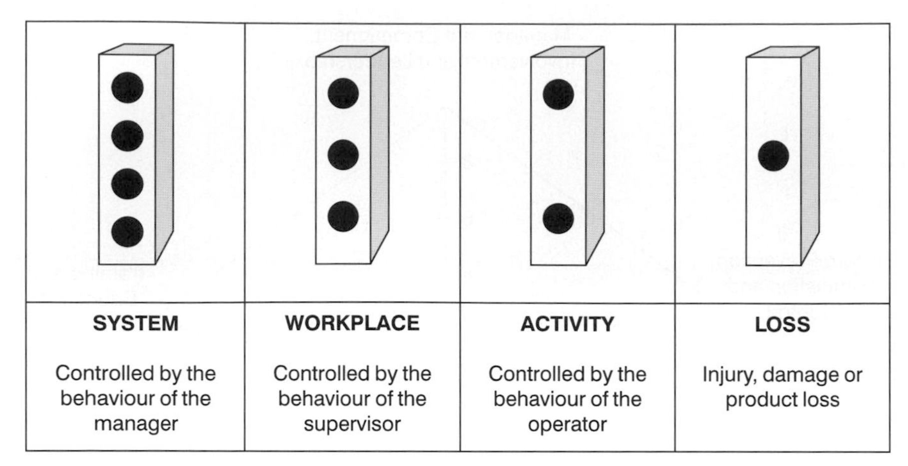

Figure 2.8.13 Domino incident model related to behavioural activity

2.8.4 Applying behaviour concepts to incident investigation

Incident investigation is more effective when it is structured to engage people in a way that promotes a sense of personal responsibility. This can be achieved by modifying the well known domino accident model as illustrated in *Figure 2.8.13*

In a normal application of the domino incident model, the user works back from the 'loss' to identify the 'root cause'. The process by its very terminology seeks to identify a single causative factor (seldom is the analysis referred to as a 'root cause' analysis) and usually focuses upon system deficiencies. Applying a behavioural perspective is achieved by asking a different question that changes the focus of the investigation. The question asked is *who could have done what to prevent the incident occurring or reduce the likelihood of it occurring, or reduce the consequence when it occurred*. The investigation process should first focus on the managers' behaviours (the system controllers), then the supervisors' behaviours (the workplace controllers) and finally the activity controller(s) (the operators'). If this sequence is reversed it is easy to give an impression of trying to blame the operator. Once all the possible contributory factors have emerged, suitable remedial actions can be chosen according to the circumstances, the risks and the costs. Several actions can usually be identified and the behavioural responses addressed through the behavioural audit process.

When carried out within a mutually supportive culture, the process promotes greater personal responsibility, accountability and commitment to improve.

2.8.4.1 Stress and using the incident model

A previous chapter used an example of Joe falling from a stepladder (see *Figure 2.6.12*). That analysis was undertaken using the Incident Model

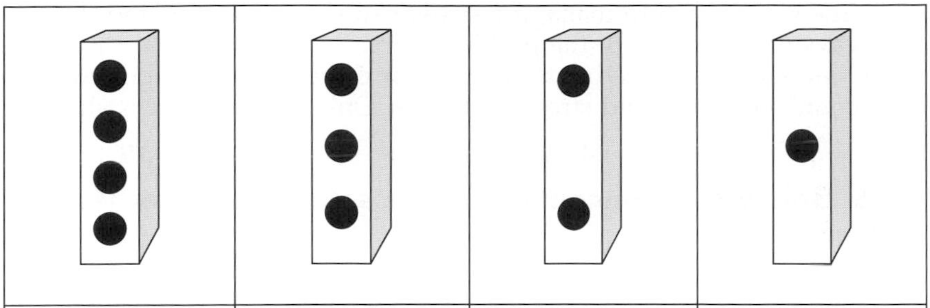

SYSTEM	WORKPLACE	ACTIVITY	LOSS
Controlled by the behaviour of the manager	Controlled by the behaviour of the supervisor	Controlled by the behaviour of the operator	Injury, damage and product loss
Possible contributory causes to stress			**Factor**
Are staff levels balanced with the task burden?	Did the employee work too many hours?	Did the operator take allocated breaks?	Ill health and product loss
Is sufficient equipment available for the tasks?	Was equipment located conveniently in the workplace?	Is the operator familiar with the use of the equipment?	Frustration with own inability to cope
Is there a policy on stress that has been communicated?	Has the supervisor discussed the stress policy with the operator?	Did the operator adhere to the stress policy?	Not recognising or being aware of being stressed
Does the stress policy or HR policies indicate how employees can get help for any stress issues whatever the source?	Did the supervisor discuss operator well-being in general?	Did the operator recognise the value of a stress policy? Is there stress from personal problems?	Operator not sympathetic to admitting stress

Figure 2.8.14 Behavioural incident model applied to a stress incident

and it was discussed in terms of the *organisational* responsibilities for health and safety.

The example provided in *Figure 2.8.14* relates to the condition of stress and demonstrates how the model can be applied to chronic health and safety issues. It shows how stress incidents can be analysed in such a way to influence the behaviours of managers, supervisors and operators.

Stress has been the subject of much recent research and comment. Life itself is not without stress but the addition of work-related stress can contribute to a situation where the individual is emotionally and psychologically overcome. Stress affects the individual's decision-making ability and also his actions (behaviours) and can lead to acute accidents. Research by the Health and Safety Executive[21] indicates that a risk assessment approach to managing stress can stimulate new and innovative ways of dealing with the problem.

The model can be used with a checklist of behaviours expected from each contributor to avoid stress problems. Only a partial list is shown.

2.8.5 Behaviour concepts and the safety management system

In chapter 2.6 the elements of safety management systems were discussed. Systems as systems can amount to no more than a list of sterile procedures and general expectations that deliver little improvement. In order to put 'flesh on the bones' of a system thought must be given to the behaviour that is necessary to support it and the system must be constructed empathetically with those behavioural aims. A safety management system based on the model suggested by the Health and Safety Executive[22] can be designed to generate the practical behaviours that make the system effective. An example is provided in *Figure 2.8.15*.

The process demonstrates the following features:

- expectations are specific;
- expectations have an assigned owner;
- expectations are measurable.

The whole system must be visible to the operator workgroup and their representatives. The process is most effective where the workgroup play a large role in assessing performance. In large organisations the system elements and behavioural expectations can be established and measured level by level down the chain of command. Each manager or supervisor should ask their work group to monitor its performance and report upwards.

2.8.6 Risk, behaviour, leadership and commitment

Changing peoples' behaviour to avoid hazards and reduce risks can become a transactional process and increasingly ineffective. A *transactional* style of leadership can be defined as a rule-following approach that is mainly concerned with achieving a task. By contrast, a *transformational* leadership style is mostly focused upon people and is more effective in changing behaviours. An HSE research report[23] quotes a study[24] which concluded that a transformational leadership style had a strong positive impact on the safety performance of those individuals who generally were otherwise less committed to safety. Some differences between a transactional and transformational leadership style are listed in *Figure 2.8.16*.

The behavioural processes that have been described in this chapter can be applied with either a transactional or a transformational style. For example if a behavioural checklist becomes a rule-based observation and intervention activity, then it is being applied in a transactional manner. In

HSG65 ELEMENT	SYSTEM REQUIREMENT	MEASURE
POLICY	1. The board to review and reissue the policy every 3 years	A policy less than three years old
	2. Each board member to ensure that the policy is reviewed in cascaded discussions throughout the organisation	At least 80% of a sample of employees to confirm their knowledge of the new policy within 3 months of its issue
	3. Independent auditors to assess that the policy covers all risks etc.	An audit report to be reviewed at the senior safety committee
ORGANISING	1. Each job will have specified safety responsibilities which are reviewed every 3 years	Job responsibilities to be reviewed/checked by the safety department
	2. All work procedures will be reviewed every three years	Work procedures to be reviewed by the workplace representative
	3. Safety committees at appropriate organisational levels will meet bi-monthly etc.	Records of meeting minutes to be kept under review
PLANNING AND IMPLEMENTING	1. Employees will receive a copy of the safety plan	Operator reaction to be sampled by the safety reps
	2. Suitable training will be undertaken for key plan personnel etc.	The safety department to specify the training and ensure delivery of it
MEASURING PERFORMANCE	1. Safety committees will monitor their area's part in the overall plan	Data from safety committees to be summarised for the board
	2. Incidents will be reviewed and key findings shared widely etc.	Safety department to monitor compliance
AUDIT & REVIEW	1. The safety department will provide a review of plan progress and incidents each quarter	A report will be sent to the board, members of the management team and safety committee members
	2. An external audit will be performed every 3 years etc.	– the audit will be available to employees

Figure 2.8.15 Example of behaviour shaping associated with a safety management system

TRANSACTIONAL **LEADERSHIP STYLE**	*TRANSFORMATIONAL* **LEADERSHIP STYLE**
Managers command	Managers involve
Managers 'have the answers'	Managers ask employees for answers
Managers resist change – conform to 'same old way'	Managers seek out and consider new ideas – prepared to change
Managers communicate one-way	Managers listen and encourage everyone to communicate
Managers ignore failings	Managers right wrongs

Figure 2.8.16 Features of transactional and transformational leadership styles

these circumstances behavioural processes will be less successful. A study by Griffin *et al.*[25] concluded that safety performance improved where the manager adopted a supportive and caring style. This is not surprising because a good manager will seek to explain and understand a poor or 'at-risk' behaviour before applying a workplace safety rule. Specific rules, generally speaking, do not cover every situation in and nuance of workaday life. Different levels of manager have different impacts. O'Dea's study[24] indicates that senior managers influence the general climate and expectations for proactive safe behaviour while middle managers and supervisors influence the adherence to local procedures and rules. Simard *et al.*[26] found that low accident rates occurred where the supervisor adopted a participatory leadership style whereas higher accident rates occurred where a hierarchical style was used. A study of safety on construction sites[27] concludes that management commitment is vitally important. Duff *et al.*[28] noted that the best performing construction sites in their study were those where managers showed their commitment.

Leadership style and committed application at all levels, but especially from managers and supervisors, are essential to the success of behaviour modification programmes. Experience has shown that success occurs where senior managers create the fertile soil in which safety professionals can plant appropriate seeds for supervisors to tend and bring to full bloom. A flowery metaphor, but apposite!

2.8.7 Behaviour modification processes: the hazards

Behaviour modification processes can fail to have the aimed-for impact for a number of reasons including:

1 Behaviour processes have not been placed within the context of the wider safety management system but have been viewed as an alternative to it.
2 The behaviour modification process designed does not include managers or supervisors, only front-line operators.

3 The cost and time commitment has been underestimated especially in regard to initial training, refresher training, and employee time taken in observation and intervention.
4 There is a lack of willingness to make a sustained commitment over several years. Behaviour modification processes are seen as a 'quick fix' to boost safety performance.

For success, the organisation must be ready to adopt a behaviour modification process and the factors which indicate this is so include:

- Accident levels have stopped falling.
- Management is frustrated at the lack of improvement in safety performance and is willing to embrace a new initiative.
- Poor improvements in safety performance have resulted from capital expenditure.
- Incident analyses indicate that behaviour modifications by employees could have avoided many of the accidents.

It must be understood that behaviour modification processes are inappropriate where failure will result in a serious or fatal injury. Human beings by their very nature are not perfect – they fail. If the consequence of a single failure is so serious, the risk must be controlled by other means.

2.8.8 Behaviour and safety culture

A positive safety culture is often seen as the endpoint of the systematic efforts to improve safety. Once attained any hazards will be controlled in a coherent, supportive, constructive, even happy environment. But what is a positive safety culture? There are many answers. The Confederation of British Industry[29] lists eleven features:

1. Leadership and commitment from the top which is genuine and visible. This is the most important feature.
2. Acceptance that it is a long term strategy which requires sustained effort and interest.
3. A policy statement of high expectations and conveying a sense of optimism about what is possible, supported by adequate codes of practice and safety standards.
4. Health and safety should be treated as other corporate aims, and properly resourced.
5. It must be a line management responsibility.
6. 'Ownership' of health and safety must permeate at all levels of the workforce. This requires employee involvement, training and communication.
7. Realistic and achievable targets should be set and performance measured against them.
8. Incidents should be thoroughly investigated.

9. Consistency of behaviour against agreed standards should be achieved by auditing and good safety behaviour should be a condition of employment.
10. Deficiencies revealed by an investigation or audit should be remedied promptly.
11. Management must receive adequate and up-to-date information to be able to assess performance.

Research by Pidgeon[30] implies three major features constitute a good safety culture:

- the existence of procedures and rules reinforced by high expectations of compliance;
- attitudes toward safety that are constructive and positive;
- an ability, capacity and willingness to consider and learn from experience from within and without the organisation.

A technique for assessing where the organisation stands in respect of safety behaviours at any point in time is to ask those most exposed to the hazards – the operators. The deliberations in the boardroom lead to decisions that can affect the health and safety of employees. If asked, employees can indicate the impact of these high level decisions and show whether their intentions and objectives are being achieved. This reflects the 'culture in being'.

Understanding employees' perceptions and opinions can be considered as a 'reality check' and can help to focus on concerns that they see as impediments to good safety performance. Psychologists such as Stanton and Glendon[31] have developed structured survey tools for this purpose. Another is available from the HSE[32]. This latter publication asks 71 questions covering the following 10 factors:

1 Organisational commitment and communication.
2 Line management commitment.
3 Supervisor's role.
4 Personal role.
5 Workmates' influence.
6 Competence.
7 Risk-taking behaviour and some contributory influences.
8 Some obstacles to safe behaviour.
9 Permit-to-work.
10 Reporting of accidents and near misses.

Using the tool requires that managers, supervisors and their operators each provide their opinion on these 10 topics. This permits the views that managers hold to be compared with the views of other levels in the organisation. The survey can be undertaken in other areas of the company and at other dates enabling comparison across an organisation and over a period of time.

Measurement is only the first step in using survey techniques. Collected data, including accident and incident data, should be used for

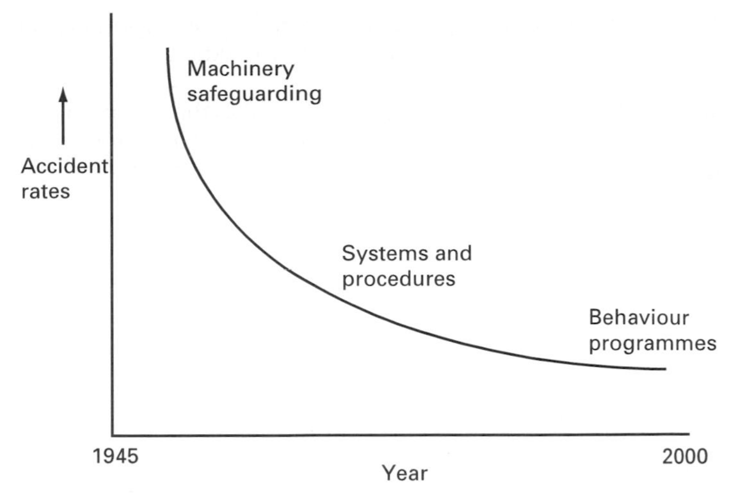

Figure 2.8.17 Safety emphasis and post-war accidents trends

discussion with operators. Utilising the data in this way is as important as getting it in the first place. The involvement of operators in this way raises expectations – operators expect more from managers and managers expect a greater contribution from operators.

The process measures opinions, and it also generates a focus on, and commitment to, improved safety performance, behaviours that eventually produce a positive safety culture.

2.8.9 Conclusion

The experience of many established companies is reflected in *Figure 2.8.17*.

Accidents have been decreasing over the last fifty years but are tending to reach a trough. In early post-war years the emphasis remained on physical safeguarding. This gave way to the systems-of-work approaches in the 1970s which continues today in the form of risk assessments. The behaviour-based approach is becoming increasingly important as a technique to improve safety at work for two reasons. First, the other approaches have been implemented but there still remains room for considerable improvement. Second, the nature of work is changing. The emergence of service sector employment and the increasing freedom from a conventional workplace, which modern communication and computer systems allow, have caused written procedures and close supervision to become less easy to apply as an effective means of exerting control. Effective safety measures will emerge from better education and training and the implementation of motivational theories to change attitudes and behaviour.

Whilst an understanding of the links between attitude, behaviour and the consequences of the behaviour are still developing, enough is known to establish that certain key elements of a behaviour approach are more likely to yield success. These key elements include:

1 Management commitment. This should take the form of providing encouragement, being supportive and offering coaching to the workgroup. A 'command and control' approach is not a recipe for lasting success.
2 Workforce involvement. It is essential that the employees, who themselves suffer the accidents, are thoroughly immersed in the process. Their involvement is crucial because they are the people who know the unsafe acts that are committed and the reasons for them. Clearly the workgroup themselves are in the best position to know the antecedents and consequences which operate and how they can be adjusted to promote safer working.
3 Effective feedback is essential and should be twofold in content. First, there should be immediate interaction with the employee being observed irrespective of whether safe or unsafe working is noticed. Second, there should be feedback to the work group as a whole for comparison with the agreed checklist of safety critical behaviours. This can give a behaviour score which can be compared with goals agreed by the work group. The result can be used to measure progress and can be posted on a chart in the workplace.

Behavioural processes may have limitations as to their effectiveness. Most examples come from areas where the hazards are acute and understood and the behaviour is observable, such as the wearing of personal protective equipment. It may be harder to apply these techniques to work systems such as 'permits-to-work' or where the risk is low in probability but high in consequence and where precautions are taken but effects not immediately visible. An example might include the maintenance of pressure systems where poor work may not become visible until an explosion occurs much later.

The objective of a behaviour-based approach is to change work habits for the better. There is evidence to show that success does result from such programmes. However, they require considerable resources and commitment which might not always be forthcoming. Little work has been undertaken to show if improvements in performance continue to accrue or can even be maintained when resources are scaled down, for example, by undertaking less frequent audits. It may be that the imposed effort can be scaled down because attitudes have effectively and permanently altered in favour of ongoing safe working. The evidence from major companies which have reputations for continuously superior safety performance seems to be that the *formality* can be reduced but there must be an ongoing management focus. In practical terms this involves being seen to be committed to safe working, walking the workplace, coaching employees in safety behaviours, and encouraging their full participation into what is a stated major workplace value, namely an ongoing reduction in accidents and ill-health at work. When this point is attained, a culture change has been achieved.

References

1. Skinner, B.F., *The Behaviour of Organisms*, Appleton-Century-Crofts, New York (1938)
2. Health and Safety Executive, Guidance publication no. HSG 48. *Reducing Error and Influencing Behaviour*, HSE Books, Sudbury (1999)
3. Komaki, J., Barwick, K.D. and Scott, L.R., A behavioural approach to occupational safety: Pinpointing and reinforcing safe performace in a food manufacturing plant. *J. Appl. Psychol.*, **63**: 434–445 (1978)
4. Komaki, J., Heinzman, A.T. and Lawson, L., Effect of training and feedback: component analysis of a behavioural safety program. *J. Appl. Psychol.*, **65**: 261–270 (1980)
5. Komaki, J.D., Collins, R.L. and Penn, P., The role of performance antecedents and consequences in work motivation. *J. Appl. Psychol.*, **67**: 334–340 (1982)
6. Suzler-Azaroff, B., The Modification of Occupational Safety Behaviour. *J. Occupational Accidents*, **9**: 177–197 (1987)
7. Nasanen, M. and Saari, J., The effects of positive feedback on housekeeping and accidents at a shipyard. *Journal of Occupational Accidents*, **8**: 237–250 (1987)
8. Daniels, A.C. and Rosen, T.A., *Performance Management: Improving Quality and Productivity through Positive Reinforcement*. Performance Management Publications, Inc., Tucker, Georgia (1987)
9. Krause, T.R., Hidley, J.H. and Hodsen, S.J., *The Behaviour-Bases Safety Process*. Van Nostrand Reinhold, New York (1990)
10. Cooper, M.D., Makin, P.J., Phillips, R.A. and Sutherland, V.J., Improving safety in a large, continuous shift, production plant using goal setting and feedback: benefits and pitfalls. Brit. Psychol. Soc. Annual Occ. Psychol. Conference, Brighton, Jan. 3–5 (1993)
11. Cooper, M.D., Goalsetting for safety. *The Safety and Health Practitioner*, November 1993, 32–37.
12. Cooper, M.D., Implementing the behaviour based approach, a practical guide. *The Safety and Health Practitioner*, November 1994, 18–23.
13. Cooper, M.D., Phillips, R.A., Sutherland, V.J. and Makin, P.J., Reducing accidents using goal setting and feedback: A field study. *J. Occ. & Org. Psychol.*, **67**, 219–240, (1994)
14. Safety Training Observation Program. E.I. du Pont de Nemours and Company, Wilmington, Delaware, 1989.
15. Health and Safety Executive, Guidance publication, *Improving Maintenance: a Guide to Reducing Human Error*, HSE Books, Sudbury (2000)
16. Health and Safety Executive, Guidance publication no. HSG 217, *Involving Employees in Health and Safety*, HSE Books, Sudbury (2001)
17. The Trades Union Congress, *Partners in Prevention: Revitalising Health and Safety in the Workplace*, Trades Union Congress, London
18. Reilly, Paci and Hall, *British Journal of Industrial Relations*, **33**(2): June (1995)
19. Howe, J., *Warning, Behavior Based Safety Can Be Hazardous To Your Health And Safety Program!*, Union of Automotive Worker, International Union, September 1993
20. Lorenzo, D.K., *A Manager's Guide to Reducing Human Errors*, Chemical Manufacturers Association, USA (1990)
21. Health and Safety Executive, Contract Research Report no. CRR 435/2002, *Interventions to Control Stress at Work in Hospital Staff*, HSE Books, Sudbury (2002)
22. Health and Safety Executive, Guidance publication no. HSG 65, *Successful Health and Safety Management*, 2nd edn, HSE Books, Sudbury (1999)
23. Health and Safety Executive, Contract Research Report no. CRR 430/2002, *Strategies to Promote Safe Behaviour as part of a Health and Safety Management System*, HSE Books, Sudbury (2002)
24. O'Dea, A. and Flin, R., Site managers, supervisors and safety in the offshore oil industry, Academy of Management symposium, Canada (2000)
25. Griffin, M., Burley, I. and Neal, A., The impact of supportive leadership and conscientiousness on safety behaviour at work. Academy of Management symposium, Canada (2000)
26. Simard, M. and Marchand, A., Workgroup's propensity to comply with safety rules: the influence of micro-macro organisational factors. *Ergonomics*, **40**(2): 172–188 (1997)
27. Health and Safety Executive, Contract Reasearch Report no. CRR 299/1999, *Improving Safety on Construction Sites by Changing Personal Behaviour*, HSE Books, Sudbury (1999)

28. Duff, A.R., Robertson, I.T., Phillips, R.A. and Cooper, M.D., Improving safety by the modification of behaviour. *Construction Management and Economics*, **12**: 67–78 (1994)
29. Confederation of British Industry, *Developing a Safety Culture*, CBI, London (1991)
30. Pidgeon, N.F., Safety culture and risk management in organisations, *Journal of Cross-Cultural Psychology*, **22**: 129–140.
31. Stanton, N. and Glendon, I., *Safety Culture Questionnaire*, Griffith University, Australia and University of Southampton, England, 1996 (Private communication)
32. Health and Safety Executive, *Health and Safety Climate Survey Tool* (Diskette), HSE Books, Sudbury (1998)

Part III

Occupational health and hygiene

In his work, the safety adviser may be called upon to recommend measures to overcome health problems that have been identified by the doctor or nurse. Part of his duties may include the identification of processes and substances that are known to give rise to health risks and advising on the procedures to be followed for their safe use.

The advice he can give will be more pertinent if the safety adviser has an understanding of the nature of the substance and the manner in which it affects the functioning of the human body.

This Part explains the functions of the major organs of the body, considers the characteristics and hazards of a range of commonly used substances and processes and discusses the techniques that can be employed to reduce the effects of those risks on the health and well-being of the workpeople.

Chapter 3.1

The structure and functions of the human body

Dr T. Coates

3.1.1 Introduction

Occupational medicine is that branch of medicine concerned with health problems caused by or manifest at work. Some health problems, although not caused by the job, may be aggravated by it.

A knowledge of the structure and functioning of the organs and tissues of the body is of value in the understanding of occupational illness and injury.

Some substances are particularly liable to damage certain organs; e.g. hydrocarbon solvents may affect the liver, cadmium may damage the lungs or kidneys and mercury may affect the brain.

A brief description of anatomy and physiology is given below and more details may be obtained from textbooks on the subject.

3.1.2 History

Although many hazards of work were well recognised in ancient times, very little was done to prevent occupational disease. Mining was a dangerous unpleasant occupation performed by slaves. The latter were expendable and the frightful conditions in which they worked may have been a deterrent to slaves on the surface!

By the second century AD some miners were using bladders to protect themselves from dust inhalation. (Apart from armour and shields this is probably the first example of protective clothing worn at work.)

Little is known about occupational diseases in the dark ages but by the sixteenth century there was extensive mining for metals in central Europe and several accounts of associated diseases. The year 1556 saw the publication of a work of 12 books on metal mining by a mining engineer and doctor called Agricola. The latter part of book VI was devoted to the diseases of miners. Agricola advised the use of loose veils worn over the face to protect the miner against dust and ventilating machines to purify the air.

Eleven years later another doctor with an interest in mining published a work on diseases of mining and smelting. Paracelsus was physician to an Austrian town and local metallurgist. He used several metals including lead, mercury, iron and copper to treat diseases. He described the signs and symptoms of mercury poisoning and recommended the use of mercury in treating syphilis. When challenged that some of his drugs were poisonous he replied 'All things are poisons, for there is nothing without poisonous qualities. It is only the dose which make things a poison'.

In 1700 a book on trade diseases was published by an Italian physician by the name of Bernardino Ramazzini. He based the book on personal observations in the workshops of Modena where he was professor of medicine and on the writings of earlier doctors. Ramazzini was the first person to advise that physicians should ask specifically about the patient's occupation when diagnosing illness.

The development of the factory system saw the rapid movement of people from the country to the towns with consequent disruption of family life. Large numbers of workers and their families housed near the factories resulted in overcrowding, poor housing and poor sanitation. At work, people suffered appalling injury and disease and worked very long hours until eventually the pressures of humanitarians such as the Earl of Shaftesbury promoted legislation which improved working conditions and reduced the hours of employment of workers in factories, mines and elsewhere.

During this time Charles Turner Thackrah, a doctor from Leeds, wrote a book about occupational diseases in his native city which was the first such work to be published in the UK. But this was 1832 and his work raised little interest, but did influence the House of Commons on future factory legislation.

The Factories Act of 1833 saw the appointment of Factory Inspectors and the need for doctors to certify that a child appeared to be at least nine years of age before being employed in textile mills. When birth certification was introduced in 1837 the assessment of children's ages became unnecessary. In 1844, the Factory Inspectors appointed Certifying Surgeons and by 1855 they were required to investigate industrial accidents and to certify that young persons were not incapacitated by disease or bodily infirmity.

By the mid nineteenth century the Registrar General had amassed a great deal of statistical information about occupational disease. Dr E.H. Greenhow of St Thomas' Hospital showed from these figures that much of the chest disease in certain areas of the country was due to the inhalation of dust and fumes at work.

In 1895, poisoning by lead, phosphorus and arsenic and cases of anthrax became notifiable to the Factory Inspectorate. Certifying surgeons examined workers in match factories, lead paint works, trinitrobenzene explosive factories and india-rubber factories using the vulcanising process which involved carbon bisulphide. The widespread occurrence of 'phossy-jaw' in phosphorus workers and lead poisoning gained much publicity and provoked the appointment in 1898 of Dr Thomas Legge as the first Medical Inspector of Factories.

Legge devoted the next 30 years to investigating and preventing occupational disease. His book *Industrial Maladies* was published posthumously in 1934.

By 1948, the Certifying Surgeons had become 'Appointed Factory Doctors' and numbered over 1800. They examined young people under the age of 18, investigated patients suffering from notifiable diseases and carried out periodic medical examinations on people employed in specific dangerous trades. The Appointed Factory Doctor system was replaced by the Employment Medical Advisory Service in 1972. This service, the nucleus of which was formed by the medical branch of the factory inspectorate, gives advice to employers, employees, trade unions and others on medical matters related to work.

Occupational Health Services in private industry were slow to develop and although there were rare instances of medical services at work even before the industrial revolution the first Workman's Compensation Act of 1897 was the first real stimulus which provoked employers to seek medical advice in their factories. At that time, the main reason for such appointments was to help protect the firm against claims for compensation. Exposure to hazards in munitions factories in World War I initiated many new medical and nursing appointments and increased the number of trained first aiders. Although the depression of the 1920s reversed the trend, interest returned in the 1930s and 1935 saw the founding of the Association of Industrial Medical Officers with some 20 members. This organisation grew into the Society of Occupational Medicine with a current membership of almost 2000 doctors.

A new surge of growth in Occupational Health Services occurred in World War II. The large factories were required to have their own doctors. After the war medical services grew but slowly. Many larger industries developed comprehensive medical services with X-ray, laboratory and other facilities. Some smaller factories shared medical services with their neighbours in schemes set up by the Nuffield Foundation.

In 1978, the Royal College of Physicians of London established a Faculty of Occupational Medicine as an academic centre for the subject. The Faculty has established criteria for the training and examination of specialists in the field and has a membership of over 1700.

Meantime, occupational health nursing has developed as an important aspect of health at work. Many factories with no occupational health physician employ one or more occupational health nurses. The first such nurse was employed by Colemans of Norwich in 1877. The Royal College of Nursing has formed a Society of Occupational Health Nursing for members employed in industry and commerce and provides training courses for those engaged in this branch of nursing. The House of Lords produced a report on Occupational Health and Hygiene Services in 1984. The report recommended development of group services which would benefit the smaller companies and suggested a Government-financed fund administered by HSE to initiate such services.

In the past decade the National Health Service has developed occupational health services for its own staff. These services are organised by individual NHS Trusts rather than on a national basis but many of them are extended to local authorities and local industry.

3.1.3 The functions of an occupational health department

These fall into clinical and advisory categories.

Health assessments

1 Pre-employment and other medical examinations, e.g. employees returning from sickness or those changing jobs.
2 Examination of people exposed to specific occupational hazards.
3 Treatment of conditions on behalf of the hospital or general practitioner. This may include physiotherapy or rehabilitation for which purposes a physiotherapist may be employed.
4 Emergency treatment of illness or injury occurring at work.
5 Immunological services, e.g. vaccination of overseas travellers, tetanus prevention, influenza prevention. Hospital workers require protection against hepatitis and tuberculosis.

Advisory services

1 The study and prevention of occupational disease.
2 Advice on problems of medical legislation and codes of practice.
3 Advice on medical aspects of new processes and plant.
4 The study of sickness absence.
5 Advice on the reduction or prevention of common non-occupational diseases such as alcoholism and the effects of smoking.
6 Advice to employees prior to retirement.
7 The preparation of contingency plans for major disasters at the place of work.

Nurses have an important part to play in these activities and much of the clinical treatment of patients is in their hands. In the UK nurses may be State Registered (SRN) or Registered General Nurse (RGN) with 3 years' training or State Enrolled (SEN) with 2 years' practical training. Full-time and part-time training courses in occupational health nursing are run by the Royal College of Nursing at various centres. The RGN may obtain a diploma in occupational health nursing after an examination. The SEN may take part in one of the courses which will help her carry out her duties in this field of nursing. As most nurses in industry and commerce lack full-time medical advice the need for formal training in the subject is very clear.

3.1.4 Overseas developments

Not all EC countries have introduced legislation on occupational health. In France, for example, there is no law requiring treatment services but pre-employment medical examinations are mandatory.

Holland and Belgium require medical services in companies of over a specified size. In Germany, doctors trained in occupational health must be employed in factories as must safety advisers, and in a number of other European countries the major concerns have occupational health services.

Some countries use the factory as the site for a medical centre which provides clinical services for workers and their families as well as making available similar medical facilities to those provided by many factory medical departments in the UK.

In the USA the National Institute for Occupational Safety and Health (NIOSH) determines standards of occupational health and safety at work and organises training and research facilities. The Occupational Safety and Health Act 1970 applies to workers in industry, agriculture and construction sites and requires that employers must provide a place of work free from hazards likely to cause death or serious harm to employees[1].

Elsewhere, occupational nursing arrangements generally reflect the national emphasis placed on occupational health.

3.1.5 Risks to health at work

The main hazards are of three kinds, physical, chemical and biological, although occupational psychological factors may also cause illness.

3.1.5.1 Physical hazards

Noise, vibration, light, heat, cold, ultraviolet and infrared rays, ionising radiations.

3.1.5.2 Chemical hazards

These are liable to occur as a result of exposure to any of a wide range of chemicals.

Ill-effects may arise at once or a considerable period of time may elapse before signs and symptoms of disease are noticed. By this time the effects are often permanent.

3.1.5.3 Biological hazards

These may occur in workers using bacteria, viruses or plants or in animal handlers and workers dealing with meat and other foods. Diseases produced range from infective hepatitis in hospital workers (virus infection) to ringworm in farm labourers (fungus infection).

3.1.5.4 Stress

This may be caused by work or may present problems in the time spent at work. Work related stresses may be due to difficulties in coping with the amount of work (quantitative stress) or the nature of the job (qualitative stress).

3.1.6 Occupational hygiene

In 1959, the American Industrial Hygiene Association defined occupational hygiene as 'that science and art devoted to the recognition, evaluation and control of the environmental factors or stresses arising in or from the workplace which may cause sickness, impaired health and well-being or significant discomfort and inefficiency among workers or among citizens of the community'[2].

The British Occupational Hygiene Society was founded in 1953 'to provide a forum in which specialist experience from many different but related fields could be exchanged and made available to the growing number of occupational hygienists at both national and international level, and to encourage discussion with other managerial and technical professions'. The Society holds frequent conferences and meetings.

The British Examining Board in Occupational Hygiene was set up by the British Occupational Hygiene Society in 1968 to examine candidates to well-defined professional standards. This Board has been superseded by the British Institute of Occupational Hygienists. Similar organisations exist in most advanced industrial nations.

The first stage in the practice of good occupational hygiene is to recognise the potential or manifest hazard. This may result from an inspection of the process in question or may be suggested by symptoms and signs of disease in the operatives. Ideally the potential risk should be considered at the planning stage before plant is installed.

The next stage is to quantify the extent of the hazard. Measurements of physical and chemical factors and their duration must be related to levels of acceptability and the likelihood of injury or disease arising if the hazard is allowed to continue. These measurements often involve the use of sophisticated measuring devices which must be calibrated and used very carefully in order to produce meaningful results.

For smaller firms or small isolated units in larger organisations, the person who carries out a limited range of tests may benefit from attendance at a short course leading to a Preliminary Certificate in Occupational Hygiene at a College of Further Education. Full-time specialists in Occupational Hygiene will require a professional qualification in the subject. The need to meet the requirements contained in the Control of Substances Hazardous to Health Regulations 1999 has increased the role of occupational hygienists, both full-time and part-time.

Having assessed the dangers of the process, the final stage is to decide how best to control the hazard. This may require some radical modification of plant design, special monitoring devices which will warn of increasing danger or the need for protective devices to be used by plant operators.

In deciding on appropriate ways of dealing with such problems, the occupational hygienist will often require the co-operation and understanding of the occupational physician, nurse, safety adviser, personnel officer and line manager in order to achieve his aims.

The degree of involvement and co-operation of advisers in the medical, nursing, engineering and safety fields will vary from one problem to another. Failure to achieve adequate health and safety measures may be due to lack of understanding or co-operation between advisers in the various disciplines or failure to influence line management.

The appointment of safety representatives under the Health and Safety at Work Act 1974 has focused attention on the part played by all who are involved in occupational health, safety and hygiene. Advice on prevention and safety measures is less likely to be ignored but more likely to be challenged than in the past. It is vital that the adviser's opinions can withstand challenge and are seen to be fair and unbiased.

3.1.7 First aid at work

The obligations placed on an employer to provide first aid facilities are contained in the Health and Safety (First Aid) Regulations 1981[3]. These Regulations recognise that the extent to which first aid provision is required in the workplace depends on a range of factors including hazards and risks, the size of the organisation, the distribution of the workforce, the distance from various emergency services and the extent of the occupational health facilities provided on site. When doctors and/or nurses are employed this can be a factor in determining the number of first aiders needed.

An assessment of the risk to employees' health and safety is required under MHSW and this should include an assessment of first aid requirements. If the risk is small, an easily identified fully equipped first aid materials container and an *appointed person*, trained to deal with emergencies, may be all that is required. Short courses lasting about 4 hours are available for training such people to cope with emergencies and the trainers do not require HSE approval.

In areas of greater risk, the employer needs to provide better services. In such areas a first aider should be available to give first aid immediately after an accident has occurred. Where the process includes the possibilities of gassing or poisoning, the first aider may need special training to deal with these specific risks. Adequate first aid rooms and equipment may be required, especially in areas of high risk such as the chemical industry and on construction sites.

Details of all first aid treatments should be recorded. The record may be made in the statutory accident book (BI 150) or in a record system developed by the employer. The local emergency services should be notified of all sites where hazardous substances are used.

3.1.7.1 Number and qualifications of first aiders

These should be available in sufficient number to be able to give first aid rapidly when the occasion demands. Where more than 50 people are employed a first aider should be provided unless the assessment can justify other facilities. For example, a small organisation with only minor hazards may require only an appointed person. On the other hand, serious hazards may warrant a first aider in each hazardous area. A detailed guide to the assessment of first aid needs is given in an Approved Code of Practice[4] which includes a table indicating the recommended number of first aiders.

First aiders must hold a currently valid certificate of competence in first aid at work. The HSE approves first aid trainers and provides information on the availability of local courses. Certificates obtained after basic training last for 3 years but refresher courses should be attended on a regular basis particularly leading up to the renewal of certificates. Where special training has been given for specific hazards, the standard certificate may be endorsed to confirm that such training has been received.

3.1.7.2 First aid containers

The contents of a first aid container should match the assessed needs of the work area, the lower the risk, the more basic the contents. It is the responsibility of managers to ensure that they provide sufficient number of containers, reflecting the extent and location of the hazards, and that the contents of each container are appropriate to the risks faced in the work area covered. Tablets and medication should not form part of their contents.

First aid containers should be sufficiently robust to protect the contents and should be clearly identified by means of a white cross on a green background. The containers should contain only first aid items, including a guidance leaflet[5], and should be kept properly stocked. This may require the holding of back-up first aid supplies. The approved Code of Practice[4] suggests the minimum content for a container in a low risk work area. First aid containers should be located near hand washing facilities. If peripatetic workers have to visit areas without first aid cover, they should be provided with a small travelling first aid kit.

3.1.8 Basic human anatomy and physiology

Anatomy is the study of the structure of the body. Physiology is concerned with its function. Although the various organs which make up the body can be studied individually it is important to remember that these organs do not function independently but are interrelated so that if one part of the body is not functioning properly it may upset the health of the body as a whole.

An organ like the stomach or the brain contains structures within it such as arteries, nerves and other specialised components. These components, which contain cells of a similar kind, are referred to as tissues. So we have nervous tissue, arterial tissue, muscular tissue and so on making up specialised organs which have a specific function. (The stomach, for example, is concerned with the first stage in the digestive process.)

The cells which make up the tissues and organs are so small that they are invisible to the naked eye. Under the microscope a cell consists of a mass of jelly-like material called protoplasm held together by a surrounding membrane. The shape and function of the cell vary according to the tissue of which it is composed and depend upon the job which the cell is required to do. For example, the nerve cell has long fibres capable of conducting electrical impulses while some cells in the stomach wall produce hydrochloric acid. Cells in the thyroid gland produce a chemical which influences other body cells. With these varying roles it is not surprising that cells differ from one another in appearance.

Although the human body is composed of many million cells, the work of each one is controlled so as to serve the body as a whole. If this coordination is lost, some cells can grow rapidly relative to others and the result may be disastrous. This sort of cell behaviour occurs in cancer when a group of cells may grow rapidly invading adjacent tissues.

Each cell is a sort of miniature chemical factory. It takes in food and converts it into energy to perform work. The sort of work carried out depends on the type of cell, e.g. locomotion (muscle cells), oxygen transport (trachea, lungs, blood vessels, red blood cells). Energy is also needed to repair the wear and tear of body cells. We refer to the chemical processes which convert food into energy as metabolism.

3.1.8.1 Foodstuffs

The energy needed to perform work and to maintain body temperature is provided by oxygen and various foods. Any diet which maintains life must contain six basic ingredients in a digestible form. These are as follows:

1 Proteins.
2 Carbohydrates.
3 Fats.
4 Salts.
5 Water.
6 Vitamins.

Proteins are composed of large complicated molecules which contain atoms of carbon, hydrogen, oxygen, nitrogen and often sulphur. They are made up of simpler substances called amino acids which form the basic structures of body cells. Foods such as meat, milk, beans and peas contain protein. This is broken down in the digestive process into its constituent aminoacids which are then realigned in a new pattern to make human proteins.

As the name suggests, *carbohydrates* are composed of carbon, hydrogen and oxygen. Sugars and starches are common examples of carbohydrate. *Fats* are used as reserve foodstuffs and insulate the body thus protecting it against cold. Carbohydrates and fats are a ready source of heat energy.

Various *salts* including those of sodium, iron and phosphorus are obtained from food such as milk and green vegetables. They are needed for the formation of bone, blood and other tissues.

Water is a vital constituent of all cells and a regular intake is essential to maintain life. Small quantities of a range of chemicals known as *vitamins* are also needed for healthy existence. The absence of a vitamin leads to a deficiency disease such as scurvy which occurs when vitamin C is absent from the diet. It is available in fresh vegetables, oranges and lemons. Vitamin D is formed by the action of ultraviolet light on a chemical in the skin (7-dehydrocholesterol), and is present in milk and cod liver oil. Absence of this vitamin may lead to rickets.

As well as containing the substances listed above an adequate diet must provide enough calories to satisfy metabolic requirements. This will depend on the physical demands of the person's work and hobbies as well as on his stature.

3.1.8.2 Digestion

When food is taken into the body much of it is in a form which cannot be used directly by the tissues as its chemical structure is too complicated. The larger molecules of food therefore need to be broken down into simpler molecules. This process takes place by chemical action and occurs in the digestive tract (*Figure 3.1.1*) which is a long tube of varying dimensions which starts at the mouth from where the food passes to the gullet, the stomach, the small intestine and finally the large intestine.

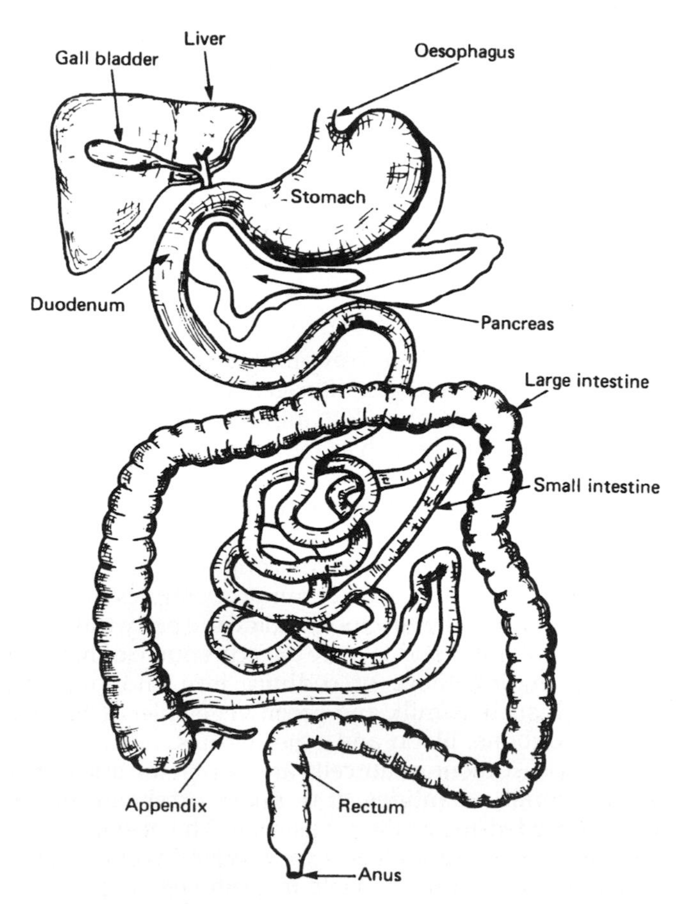

Figure 3.1.1 Diagram of digestive system

Alcohol is absorbed in the stomach and water is absorbed in the large intestine but the majority of energy-containing foods are absorbed in the small intestine. The products of digestion pass through the walls of the digestive tract into blood vessels and thence to the liver. This is a very large organ situated in the upper right side of the belly cavity below the diaphragm. It is made up of many units of cells arranged around blood vessels in a circular fashion. Sugar from the digestive tract is taken to the liver where it is changed into a chemical called glycogen which is stored in the liver cells. Proteins are broken down in the liver and form urea as a waste product which is then excreted via the kidneys and is the main chemical constituent of urine.

Old red blood corpuscles are removed from circulation by the liver which retains iron from them for later use. The liver is also responsible for the manufacture of bile which assists digestion and which is stored in the gall bladder adjacent to the liver.

Many poisons are dealt with by the liver which attempts to render the poison less toxic (detoxification) before it is excreted. Sometimes the poison damages or destroys liver cells but fortunately the liver has such a large reserve of cells that a great deal of damage is necessary to affect its function adversely. Liver damage may result from certain types of industrial poisons as well as from excessive consumption of alcohol.

3.1.8.3 Excretion

Just as a motor car needs to get rid of exhaust gases so the human body has to dispose of waste materials left over from metabolic chemical reactions. Special organs are involved in the process of excretion. Water and urea are disposed of by the kidneys, although some water is lost via the skin. Solid waste leaves the body through the bowel after water has been extracted in the large intestine.

The voiding mechanisms are of paramount importance when the body is affected by poisonous materials and may be by vomiting, diarrhoea or by being excreted in the urine. Sometimes the passage of a poison through the body may leave a trail of destruction in its wake and result in permanent liver or kidney damage.

3.1.8.4 The respiratory system

The foodstuffs absorbed from the digestive tract are converted into energy. In order to produce this energy the body cells require oxygen just as a motor car engine needs oxygen in order to function. This converting process generates carbon dioxide which, in large quantities, is poisonous and must be got rid of.

During the process of respiration oxygen is transferred from the air to the body cells and carbon dioxide is disposed of in exhaled air. Because body cells function at different rates their oxygen requirements vary from one tissue to another. If brain cells are starved of oxygen for more than four minutes there is little prospect of recovery of intellectual function. Other body cells can do without oxygen for longer periods of time.

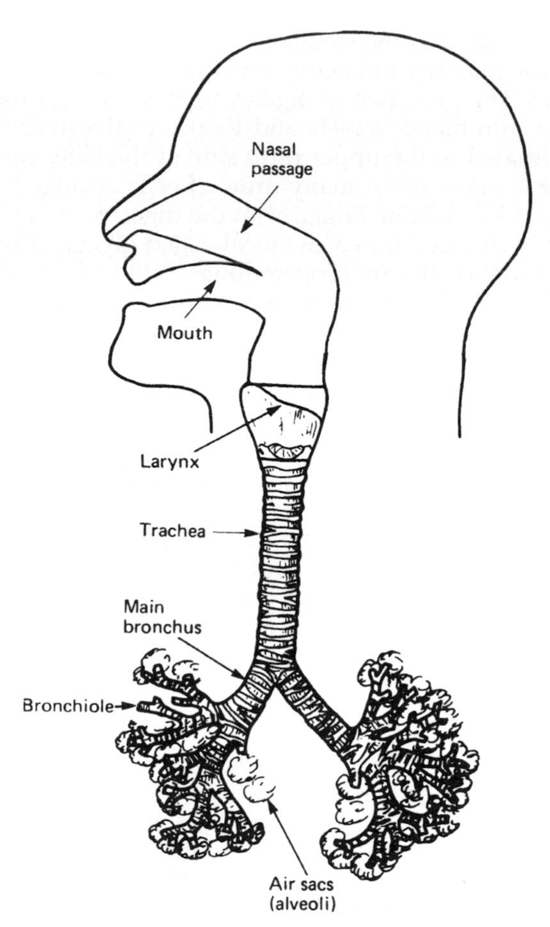

Figure 3.1.2 Respiratory system

A great deal of oxygen is needed to provide the requirements of all the body cells. The apparatus which fulfils this requirement is the respiratory tract (*Figure 3.1.2*) which is made up of the nose and mouth, the throat, the larynx (voice box) and trachea (windpipe), the bronchi and the lungs. The respiratory tract is lined by a wet shiny membrane which contains mucous secreting cells that keep the walls moist. Other cells are fringed with little hairs or cilia which by moving in one direction can evict towards the mouth particles of dust which have entered the airways. Sometimes the quantity of material which has entered the airway is greater than the cilia can cope with. The lungs then eject collections of particles by the mechanism of coughing. Primary filtration of air entering the respiratory tract occurs in the nose but many of the smaller particles enter the air passages where some hit the walls of the bronchi and are rejected by the cilia but others go on to reach the lungs.

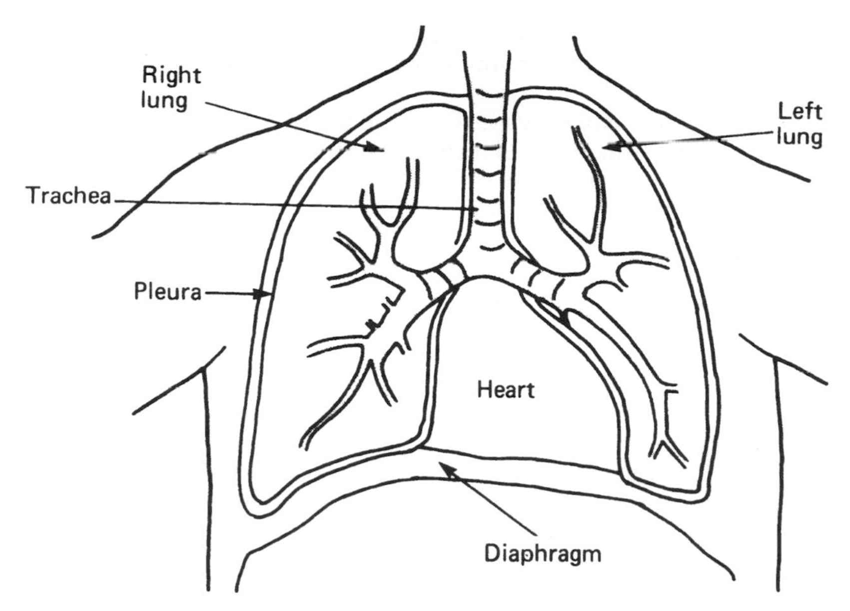

Figure 3.1.3 Diagram showing lungs and heart within chest

The respiratory tract is rather like an inverted tree. The windpipe is the trunk and the two main bronchi (one for each lung) the main branches. Each main bronchus then subdivides into smaller branches (bronchioles). All these tubes are held open by rings of cartilage which prevent their collapse when subject to suction when breathing in. The smallest tubes end in air sacs or alveoli which have very thin walls. These walls allow oxygen to pass into the small blood vessels with which they are surrounded and carbon dioxide to pass in the opposite direction from the blood into the air sac.

The blood vessels lining the air sacs carry the oxygen-containing blood to the left side of the heart from whence it is pumped via the arteries to all parts of the body. When the tissues have used the oxygen to carry out the metabolic processes described earlier carbon dioxide is produced which like oxygen is dissolved in the blood. The carbon dioxide is carried back to the lungs via the veins and the right side of the heart.

The lungs are surrounded by a tough layer of smooth membrane called the pleura which if it becomes inflamed gives rise to a condition known as pleurisy. Blue asbestos fibres can irritate the pleura to produce a type of cancer called a mesothelioma.

The lungs are located in the chest cavity in a space limited by the ribs, breastbone, backbone and diaphragm (*Figure 3.1.3*). The latter is a dome-shaped sheet of muscle which separates the chest and belly cavities. When the diaphragm moves downward the dome shape is flattened and at the same time the ribs move upwards, thus increasing the volume of the chest cavity creating a suction which draws air into the lungs and then into the air sacs (about 20% of air is oxygen). This movement is called inhalation. When the diaphragm expands the chest cavity contracts and the elastic recoil of the lungs forces air in the opposite direction (exhalation).

If respiratory movement ceases (e.g. due to electric shock damaging the breathing mechanism) artificial respiration is needed at once. This may be carried out by breathing into the casualty's mouth (mouth to mouth resuscitation) or by exerting pressure on the casualty's chest, thereby forcing air out of the lungs and allowing the recoil of the chest wall to draw air into the lungs.

Divers and caisson workers may suffer from pain in the joints if they return to the surface too quickly after working in deep water or under elevated air pressure. In these conditions, air contained in the body tissues, which was dissolved under high pressure, is released to form air bubbles (mostly nitrogen gas) in the joints and elsewhere to produce unpleasant symptoms. 'The bends' is the name of the illness resulting. Mild cases affect the elbows, shoulders, ankles and knees. As the illness develops, the pain increases in intensity and the affected joint becomes swollen. Serious cases of the bends may affect the brain and/or the spinal cord. In cases of brain damage the patient may suffer visual problems, headaches, loss of balance and speech disturbances. Spinal cord damage may cause paralysis of the limbs, loss of sensation, pins and needles and pain in the shoulders and/or hips. The problem is obviated by reducing the rate of change of pressure to which the workers are subjected to a level at which the bubbles of gas do not form.

3.1.8.5 The circulatory system

This consists of the heart, the arteries, the veins and the smaller blood vessels which permeate all tissues of the body (*Figure 3.1.4*). The heart is a muscular pump divided into left and right sides. The left side is larger and stronger than the right since it has the bigger job to do in pumping to all parts of the body via the arteries oxygen-containing blood which it has received from the lungs. The blood then takes carbon dioxide from the tissues and carries it back to the lungs through the veins via the right side of the heart.

Each side of the heart (*Figure 3.1.5*) has two chambers, an auricle or intake chamber and a ventricle or delivery chamber which are separated by valves which ensure that the blood travels in one direction only from auricle to ventricle. A further set of valves ensures that the blood being ejected cannot flow back into the ventricles each time the heart contracts.

The muscles of the heart are less dependent on the brain than the muscles which cause body movement. Cutting the nerves to a muscle in the leg, for example, will stop that muscle working. Cutting the nerves to the heart will not stop the heart since it is not entirely controlled by nervous impulses coming from outside the organ as is the case with voluntary muscle. The heart has a dual nerve supply. One nerve supply increases the heart rate, the other reduces it. The rate of the heart beat,

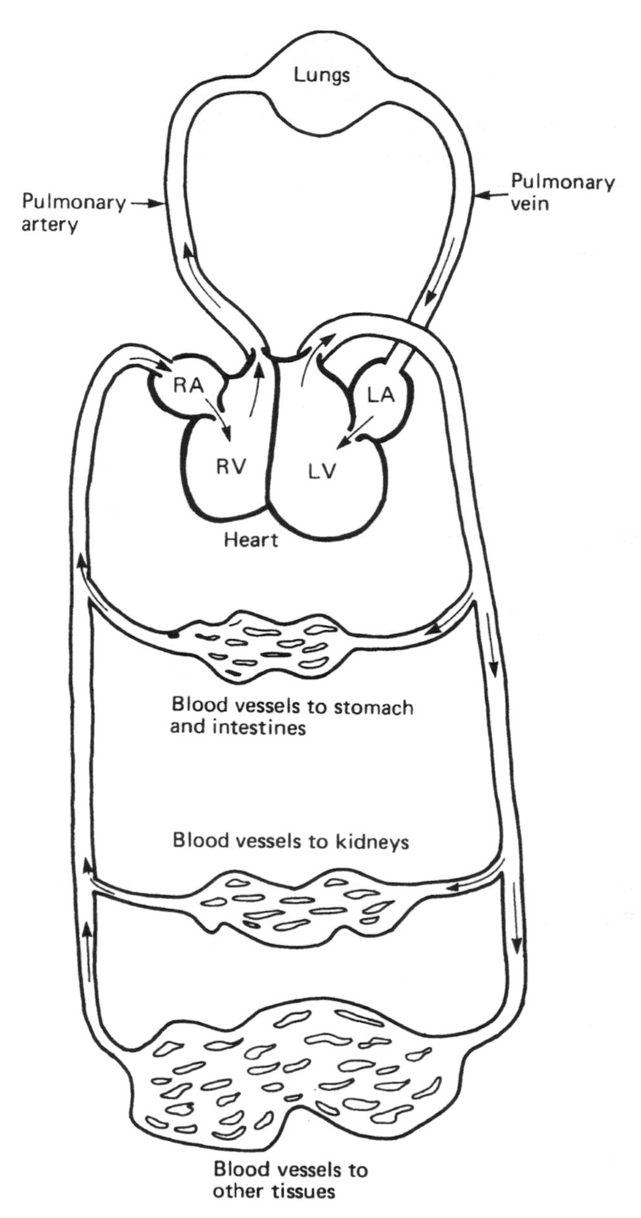

Figure 3.1.4 Diagram showing circulation of blood (RA – right auricle; LA – left auricle. RV – right ventricle; LV – left ventricle.)

which in the healthy adult at rest is about 70 beats per minute, may be slowed by stimulating the vagus nerve and increased by stimulating the sympathetic nerves. The latter are activated when people are afraid so that more blood is sent to the muscles and the person is keyed up for action.

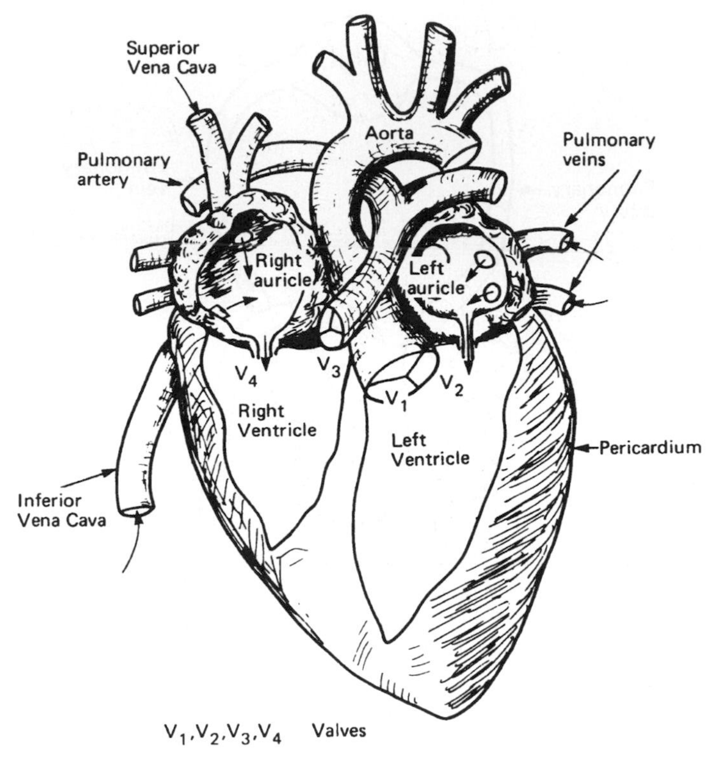

Figure 3.1.5 Diagram of human heart

Electric shock may affect the heart rate either by stimulating the nerves to the heart or by interfering with the conduction of electricity to the heart muscle. This may stop the heart or change the rhythm thus affecting its efficiency as a pump. Fibrillation of the heart muscle may occur as a result.

In addition to its oxygen-carrying capacity, the blood also carries nutrients to the tissues and removes waste products. It enables heat generated to be dissipated and its white cells defend the body when attacked by viruses and bacteria. The blood accounts for about one-thirteenth of body weight and in the average adult amounts to about 5 litres (12 pints).

It is composed of a straw-coloured fluid (plasma) and cells of two different colours – red and white. The red blood cells which account for the colour of blood are made up of minute circular discs which contain red pigment (haemoglobin) with which the oxygen and carbon dioxide transported in the blood combine temporarily. Haemoglobin also combines very readily with carbon dioxide.

The white blood cells are somewhat larger than the red blood cells but fewer in number. There is only one white cell for every 500 red cells. Several kinds of white cell exist which are mobilised when the body is infected by germs and viruses and attempt to destroy them.

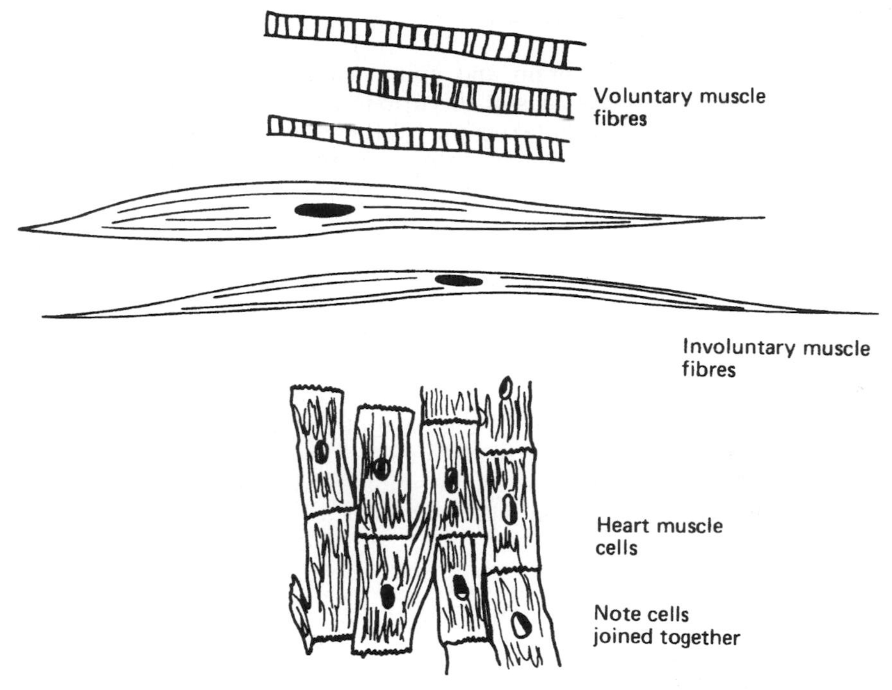

Figure 3.1.6 Structure of three types of muscle

3.1.8.6 Muscles

One of the characteristics of all animals is movement. In man, movement is brought about by the contraction of muscles. There are three varieties of human muscle (*Figure 3.1.6*). Voluntary muscles which can be moved at will are made up of many fibres which are covered by transverse stripes when seen under the microscope. The muscle is often attached to bones (hence the term skeletal muscle) by connective tissue fibres which collectively form a tendon. When the muscle contracts, the bones to which it is attached are drawn together producing movement of one part of the body relative to another part.

Involuntary muscle does not have the characteristic striped appearance of voluntary muscle. It exists in the digestive tract, the walls of the blood vessels and in the respiratory and genito-urinary apparatus. Involuntary muscle is controlled automatically by the autonomic nervous system.

The third type of muscle is that found in the heart. This shows some striated fibres under the microscope but is not under voluntary control and is not entirely dependent on its nerve supply. A 'pacemaker' within the heart muscles produces, at a rate appropriate to the body's need for oxygen, electrical impulses which cause the heart muscles to contract thus producing heart beats.

3.1.8.7 Central nervous system

This is made up of the brain and spinal cord. The brain is a highly developed mass of nerve cells at the upper end of the spinal cord. The largest part of the brain is taken up by the two cerebral hemispheres. These receive sensory messages from various parts of the body and originate the nerve impulses which produce voluntary movements. The layers of grey tissue (known as the cerebral cortex) overlying the cerebral hemispheres are covered in folds. This tissue is concerned with the intellectual function of the individual. The various parts of the cortex of the brain are associated with specific activities. For example, there are centres concerned with speech, hearing, vision, skin sensation and muscle movement.

The cerebellum is the part of the brain concerned with balance and complicated movements. Part of the base of the brain is involved with emotional behaviour (the hypothalamus).

The portion of the brain nearest the spinal cord contains centres which control the rate of respiration and heart beat.

3.1.8.8 The special senses

These specialised organs measure environmental factors such as light and noise and facilitate communication with other human beings.

The nerve cells within these organs pass their messages to the brain which then interprets them and determines appropriate action.

3.1.8.9 The eye

The eyeball (*Figure 3.1.7*) is a globe 25 mm (1 inch) in diameter which is made up of a transparent medium (the vitreous) through which light is

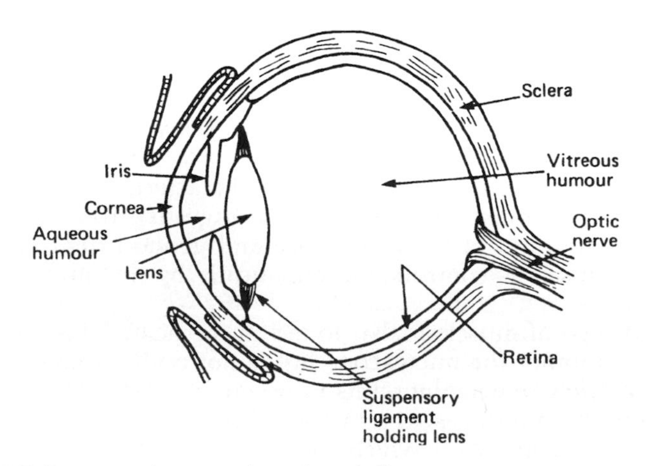

Figure 3.1.7 Diagram of section through eyeball

focused by a lens onto a light sensitive layer (the retina). The front of the globe (the cornea) is transparent. After light rays have entered the eye they pass through a fluid (the aqueous humour) and the lens which changes in shape in order to focus on to the retina light rays from objects at varying distances.

If the path of light rays is interrupted, e.g. by a foreign body or by an opacity of the lens (cataract), the vision may be distorted and/or diminished.

Light produces changes in the cells of the retina which transmit electrical impulses to the visual cortex at the back of the brain. The movements of the eyeball are controlled by six muscles in each eye which are carefully synchronised with those of the other eye. Imbalance of these muscles may give rise to double vision or a squint. Temporary muscle imbalance may result from exposure to toxic materials or alcohol.

Burns of the eye may result from chemical splashes and exposure to ultraviolet radiation as can occur with electric welding. It is vital that chemical burns be treated at once by irrigation with copious quantities of running water.

3.1.8.10 The ear

This organ is concerned with hearing and with orientating our position in space. The portion concerned with hearing consists of three parts (*Figure 3.1.8*). Sound travels through the ear canal to the eardrum which is a membrane stretched across the canal and separating it from the middle ear. Vibrations of the eardrum are produced by the sound waves passing along the ear canal.

These vibrations are transmitted through the middle ear by three tiny bones known as the 'ossicles', being the hammer (malleus), anvil (incus) and stirrup (stapes). The hammer bone is fixed to the eardrum and the stirrup to another membrane (the oval window) which separates the middle and inner parts of the ear. The section of the inner ear which receives sound waves is shaped like a snail's shell (the cochlea) and contains strands of tissue under varying tensions. These strands or hairs vibrate in response to sound waves of particular frequencies which have entered the inner ear from the bones of the middle ear and produce nerve impulses in the auditory nerve which are then transmitted to the cortex of the brain. It is at this point that the signals are received as sound of a certain pitch, intensity and quality.

Various factors may interfere with the transmission of sound impulses. Normally, the pressure on either side of the eardrum is equal but when a difference occurs, as with airline passengers who fly when suffering from a cold, temporary hearing impairment can be experienced. Infection of the middle ear may occur and this may result in thickening and scarring of the eardrum. Some unfortunate people suffer an inherited form of deafness in which the ossicles develop damage and are unable to transmit sound.

The inner ear is a very sensitive part of the hearing mechanism and may be damaged by prolonged loud noise. Usually, the frequencies around 4000 hertz (cycles/second) are first affected but the damage can

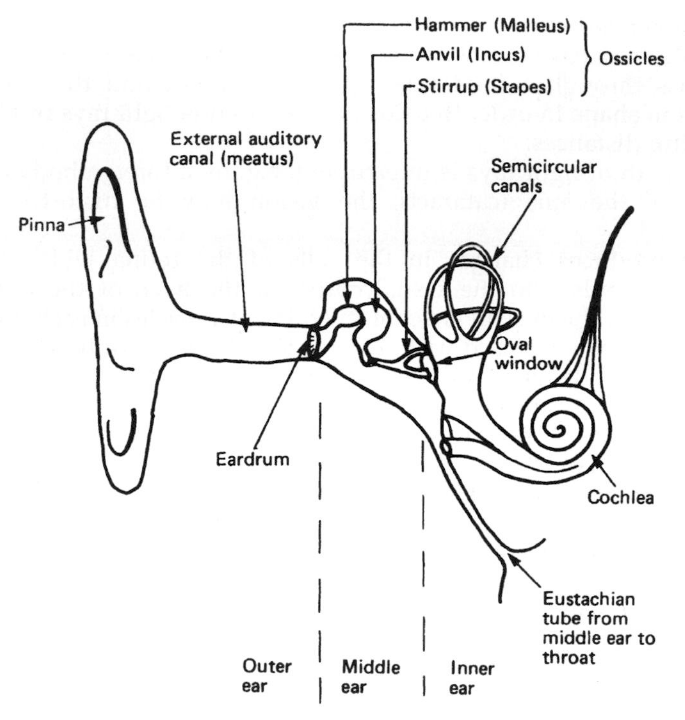

Figure 3.1.8 Diagram of human ear

extend to other frequencies as well as becoming more pronounced. In addition to this, deafness is associated with the ageing process and is more noticeable in males (presbyacusis).

A balance mechanism is also situated in the inner ear. This is composed of three semicircular canals which are at right angles to each other. Inside each canal are specialised nerve endings.

Moving the body into an unbalanced position stimulates the nerve endings in one of the canals in each ear and results in an urge to return the body to a normal balanced posture.

3.1.8.11 Smell and taste

The lining of the inside of the nose contains special cells which are capable of detecting chemicals in the air. Nerve fibres pass from these cells into the skull and connect with the brain.

The sense of smell may be an important safety factor. A cold may diminish or remove the facility. Hydrogen sulphide gas smells of bad eggs. Continuing exposure to increasing concentrations of this gas saturates the nerves concerned with smell so that the person exposed to this substance may be unable to smell it even if the concentration increases further.

The sense of taste originates when chemical stimulation of the taste buds occurs. These are collections of cells concentrated in certain areas of the tongue. The sides, tip and rear third of the organ have the most taste buds. The back of the tongue most readily detects bitterness and the tip sweetness.

3.1.8.12 Hormones

These are chemicals which act as messengers provoking action in some distant part of the body. They are produced by various hormone or endocrine glands. For example, the thyroid gland is a gland situated in the front of the neck which produces the chemical thyroxine. Too much thyroxine produces a rapid pulse and an overactive jumpy person. Too little thyroxine may result in a slowing of the pulse and too slow a rate of metabolism with the face becoming swollen and the skin dry and aged; the hair becomes coarse and falls out (myxoedema).

The suprarenal glands are two small glands situated above the kidneys. They produce a number of hormones including adrenaline and cortisone. Adrenaline is released in conditions causing fear or anger and makes the muscles in the artery walls contract. This increases the blood pressure, and consequently the supply of oxygen to the muscles, so that an animal is ready to meet a confrontation by either 'fight or flight'. This is not always an appropriate reaction for human beings in present day stressful situations where they cannot fight or run away.

Cortisone is released from the adrenal cortex at times of stress. It delays physical fatigue by increasing the ability of muscles to contract and has a euphoric effect on the brain which may give added confidence at a stressful time.

Stress is an engineering term describing the force applied to an object and the resulting deformity is referred to as strain. It has become customary to refer to the result of applying such force or pressure on human beings as 'stress'.

The effects of long-term stresses on the human body and mind are not clearly understood. Certain so-called 'stress diseases' such as asthma, duodenal ulcer and coronary heart disease may be aggravated at times of stress but a direct cause/effect relationship is difficult to prove.

Nevertheless, people in stressful situations, whether caused by work or by non-occupational factors, may be more liable to accidents.

Another hormone-producing organ, the pancreas, has two important functions. Its secretions flow into the digestive tract where the gland's products are involved in the digestion of carbohydrates. A different secretion which passes straight into the blood stream is a chemical called insulin. Without insulin the body is unable to use available carbohydrates as a source of energy and has to obtain energy from the breakdown of body tissues. The person whose pancreas is unable to make enough insulin for his needs is diabetic. Diabetes is a condition which can be treated by replacing the insulin deficiency and by careful dietary control.

The sex glands also produce hormone secretions which determine the growth of beard hair and the deep voice of the male and breast development in the female. Other endocrine glands include the parathyroid glands which are concerned with calcium metabolism and the pituitary gland which controls the other hormone glands. The pituitary gland situated at the base of the brain has two lobes: the front one controls growth in children while the rear lobe secretions cause contraction of the muscles of the womb and increase the output of urine. The part it plays in regulating the other endocrine glands has been referred to as 'direction of the endocrine orchestra'.

3.1.8.13 The skin

This is the largest organ in the body (*Figure 3.1.9*) and performs a variety of functions. Its most obvious purpose is a protective one. The superficial layers of cells keep out chemicals and germs as well as acting as a physical barrier. If the physical pressures on certain cells of skin are considerable, the tissues may be much thickened, for example on the soles of the feet.

Four different kinds of sensation may be appreciated through the skin, namely heat, cold, touch and pain.

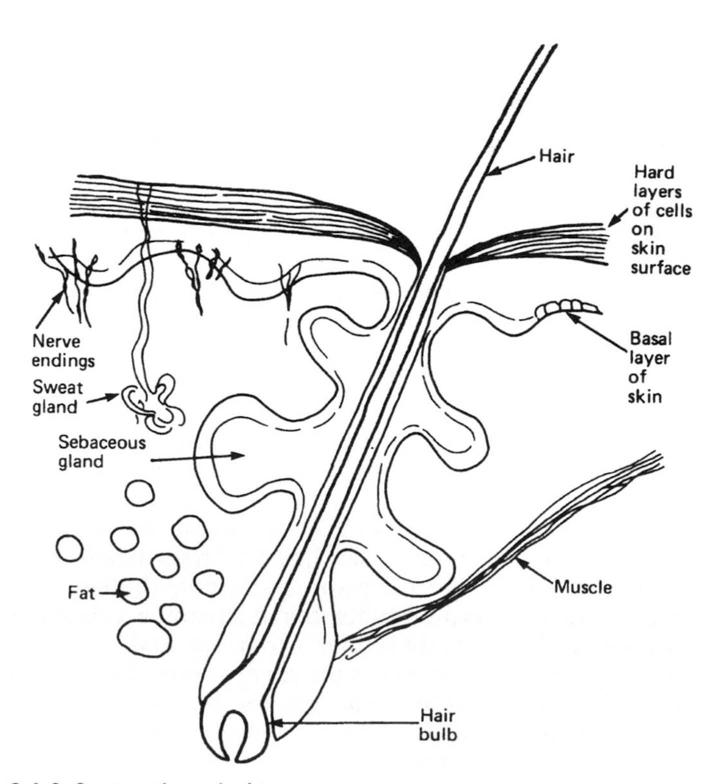

Figure 3.1.9 Section through skin

The skin is not uniformly sensitive to all these stimuli, some parts being more sensitive than others; for example, the tips of the fingers are more sensitive than the back of the hands. Apart from our awareness of touch and temperature on the surface of the skin, we are also aware of the position of our arms and legs even when our eyes are closed. This is because special nerve endings from the muscles return nerve impulses to the brain which is provided with information on the relative positions of various parts of the body. Water and various salts are lost through the skin as sweat. This may be important to people working in very hot environments such as in deep mines who may sweat profusely and as a result of the loss of salt suffer severe cramps in the muscles.

The human being is able to keep his body temperature fairly constant whatever the range of ambient temperature. This is because heat loss from the skin may be increased by sweating, followed by evaporation of the sweat and increase in the size of the blood vessels in the skin which encourages more heat to reach the skin surface.

In a cold environment the blood vessels in the skin contract and the muscles under the skin produce the phenomenon of shivering in an attempt to generate heat.

The skin also secretes sebum which is a waxy material produced by special glands. This may help protect the skin surface from attack by foreign substances. Exposure to various chemicals including solvents may remove such secretions and predispose the skin to attack by germs.

3.1.9 Cancer and other problems of cell growth

Cells reproduce in order to replace other cells which are continually being lost, e.g. on the surface of the skin by wear and tear or by damage such as that caused in a wound.

Sometimes, cells do not develop as nature intended and a variety of abnormal cells may be formed, some of which may endanger life. Some examples of maldevelopment are given below:

Aplasia The tissues may fail to develop. This happened in the unfortunate children whose mothers had taken the drug thalidomide during their pregnancies. In some of these babies, the limbs were only partly developed (hypoplasia).

Hyperplasia The organ or tissue has an increased number of cells. When the cells increase in size this is termed hypertrophy. The latter condition may occur in the muscles of a leg when the other has been amputated and the remaining limb has to work harder.

Metaplasia This process involves changes in the types of cell present in a tissue often from a complicated cell to a more simple one. Long lasting irritation from chemicals may bring about this change which is often seen in the lining of the air passages of smokers.

Mutagenesis This occurs when any abnormal changes which may have arisen in a group of cells are passed on to the next generation as the original genetic material within the cell has been damaged.

Teratogenesis This refers to congenital abnormalities which occur in the new-born as a result of damage to the genetic material or to the developing embryo.

Neoplasia This is the process in which there is a mass of new tissue (a tumour) which is made up of abnormal cells which cannot stop growing. Malignant tumours are more likely to occur in tissues which are frequently producing new cells.

The chemical DNA which carries the coding mechanism which determines the structure and function of new cells is liable to be damaged by some chemicals. New malformed cells may form and grow in an uncontrolled manner producing more damaged cells of the same type. Malignant tumours consist of undifferentiated cells which grow into surrounding tissues. Benign tumours are slower to develop and their cells look more like normal cells.

Cancer cells may spread and 'seed' themselves elsewhere in the body. From these new sites further secondary cancerous growths may develop, hastening the destructive process which may well end in the victim's death. Neoplasms may spread via the blood stream or along lymph vessels.

3.1.10 The body's defence mechanisms

The various systems of the body have a range of mechanisms for dealing with foreign materials, germs and viruses that enter the body.

Particles entering the trachea or smaller airways may be ejected by the reflex action of coughing. Irritating substances entering the nose will provoke sneezing. The lining of the nose and upper airways produce a glairy fluid (mucus) on which particles may land in which case they are dealt with by small moving hairs (cilia) which propel them towards the throat and nose. However, the smallest particles may be inhaled into the lungs where they can cause damage. Gases may be absorbed from the air sacs (alveoli) and enter the blood stream which conveys them to tissues where harm can be caused if the gases are toxic.

Foreign bodies on the surface of the eye may cause weeping (lachrymation) which can flush out some of the particles but not those that have become embedded in the eye. Tears will also dilute irritating chemicals.

The immune system provides a wide variety of defence mechanisms. Bacteria entering the blood stream are engulfed in a type of blood cell (the macrophage), a process known as phagocytosis. Macrophages in the lungs attack other foreign bodies as well as bacteria. They may also lodge in organs, such as the liver and spleen, where they eat any organism that invades their area.

A second type of white cell, the polymorphonuclear leukocyte (PMNL) is more common than the macrophage and makes up 60% of all white blood cells. These cells are attracted to sites of infection by the chemicals that are released by invading bacteria. Special cells, known as B lymphocytes, produce antibodies that are chemicals that can attach themselves to the surface of various germs thus rendering them ineffective, but any particular antibody can only bind to one bacterium. T cytoxic (killer) cells can kill cells directly. They attacked cells filled with viruses before the virus can reproduce and are involved in the defence of the body against cancer cells.

When the body is severely affected by invading organisms or irritant chemicals or burns, inflammation of the affected tissue occurs. The signs and symptoms of the inflammatory process include heat, redness, swelling, pain and loss of function of the affected part of the body. These reactions result in an increased blood supply to the affected part where the beneficial effects of the immune system's cells are enhanced. Severe inflammatory damage may be followed by scarring of the affected tissues when collagen fibres form in the damaged area. These fibres are important in wound healing where they strengthen the damaged tissues. However, scarring in the lungs may reduce the efficiency of that organ and scarring within the eye may reduce visual acuity.

3.1.11 Factors determining the effect of substances in the body

Poisonous materials may enter the body via the mouth and be absorbed from the food tract, via the airways and lungs or through the skin.

The quantity of, and duration of exposure to, a given substance are the most important factors in determining the severity of effects on an individual. Specific chemicals can cause damage to particular organs (target organs), e.g. solvents may cause damage to the liver and kidneys. If these organs have already suffered previous damage from some other disease process, they will be less able to cope with further insult from occupational hazards.

Another factor in determining the toxicity of a substance is its solubility in body fluids or the ease with which it becomes attached to tissues within the body. Carbon monoxide owes its high toxicity to its ability to attach to molecules of haemoglobin, the pigment inside red blood cells. Some materials cause disproportionate damage by synergistic effects. Both smoking and asbestosis (the presence of asbestos particles in the lungs) are well-recognised causes of lung cancer but there is a far higher risk of this disease in smokers with asbestosis than one would expect by adding together the risk from each hazard.

Some chemicals act as 'sensitisers' provoking an unpleasant response, usually in the lungs or the skin, from a very small quantity of the offending substance. This process is caused by a response of the body's immune system to foreign material.

In the case of particle inhalation, their size and shape may determine whether they are retained in the lungs or are rejected. Long thin needle-like fibres, such as asbestos, may be more readily retained than rounder particles, and may cause more physical irritation.

Some materials have an immediate adverse effect on surrounding tissue. This so-called acute effect is seen when strong acids or alikalis are in contact with body tissues. Other substances take much longer, perhaps weeks, months or years, before adverse effects are noticed, e.g. cancers. These are termed chronic effects which, unfortunately, tend to be irreversible changes that often do not respond well to treatment.

3.1.12 The assessment of risk to health

The hazards of specific substances are described in textbooks on toxicology and the details of hazardous substances used in industrial processes may be obtained from the suppliers' safety data sheets. However, the employer is responsible for carrying out assessments of the risks that may arise from the use of those substances in the particular circumstances of his own plant.

Various systems of the body can be monitored to identify the presence of harmful substances so that people exposed to a known hazard may be removed from contact with the offending agent before clinical damage ensues. Examples of such monitoring include sampling of the blood and urine for chemicals, lung function tests to assess the efficiency of the lungs and the measurement of visual and hearing acuity. When the functions of specific organs are being measured it is important to note that non-occupational disease or previous occupational exposure may account for some or all of the damage revealed by such tests.

References

1. US Public Law 91–596. The Occupational Safety and Health Act 1970
2. Gardner, A. Ward (Ed.), *Current Approaches to Occupational Medicine*, 205, John Wright & Sons, Bristol (1979)
3. Health and Safety Executive, *The Health and Safety (First Aid) Regulations 1981* (SI 1981 No. 917), HMSO, London (1981)
4. Health and Safety Executive. *First aid at work. The Health and Safety (First Aid) Regulations 1981. Approved Code of Practice*, Publication No. L74, HSE Books, Sudbury (1997)
5. Health and Safety Executive, *Basic advice on first aid at work*, Publication No. IND(G)215L, HSE Books, Sudbury (1997) (Single copies free)

Further reading

Croner's Handbook of Occupational Hygiene, loose-leaf publication with regular updates, Croner Publications Ltd, Kingston-upon-Thames

Gill, F.S. and Ashton, I., *Monitoring for Health Hazards at Work*, Blackwell Science, Oxford (2000)

Harrington, J.M., Gill, F.S., Aw, T.C. and Gardiner, K., *Occupational Health Pocket Consultant*, Blackwell Science, Oxford (1998)

Raffles, P.A.B., Adams, P., Baxter, P.S and Lee, W.R. (Eds), *Hunter's Diseases of Occupation*, 8th edn, Hodder and Stoughton, London (1992)

Rom, W.N. (Ed.), *Environmental and Occupational Medicine*, Little Brown & Co., Boston (1992)

Seaton, A. et al., *Practical Occupational Medicine*, Edward Arnold, London (1994)

Chapter 3.2

Occupational diseases

Dr A. R. L. Clark

3.2.1 Introduction

Large companies may employ a number of specialists in the field of health, hygiene and safety who co-operate as a team and pool their particular expertise, but in small firms the safety adviser is often the only local source of advice on these matters, other expertise being brought in when considered necessary.

The task of safeguarding the health of persons at work is a formidable one especially during periods of rapid technological and organisational change. Thousands of chemical substances are used by industry and in commerce but only about 800 of those in common use have been recognised as presenting a risk to the health of workpeople.

This chapter provides a brief introduction to some of the more important diseases and materials that cause them. Conveniently, these fall into four major areas covering illnesses and diseases due to:

chemical agents	(sections 3.2.2–3.2.10)
physical agents	(sections 3.2.11–3.2.15)
biological agents	(section 3.2.16)
psycho-social causes	(section 3.2.17).

In addition, section 3.2.18 deals with target organs.

3.2.2 Toxicology

There is no such state as absolute safety in the use of chemicals since all chemicals are toxic to a degree depending on the dose. The toxicity of a substance is its potential to cause harm on contact with body tissues. Toxicology is the scientific study of the medical effects on living beings of poisons[1,2].

To determine these effects, which may be acute or chronic, toxicity testing may be carried out on animals, man (in vivo tests) and in test tubes (in vitro tests). The tests are carried out in three stages.

Stage one – tests to establish acute effects

These tests aim to establish the lethal dose (LD) or lethal concentration (LC) of a substance and are carried out on mice, rats and sometimes larger animals. The degree of toxicity is indicated by the percentage of test animals that are killed by a single dose. Thus LD50 is the dose that will kill 50% of the test animals and is known as the median lethal dose. The values of LD20 and LD90 relate to doses that will kill 20% and 90% respectively. Toxicity units are specified as mg (or g) of poison/kg of body weight.

Typical degrees of LD50 toxicity[3] are:

extremely toxic	1 mg or less
highly toxic	1–5 mg
moderately toxic	50–500 mg
slightly toxic	0.5–5 g
practically non-toxic	5–15 g
relatively harmless	15 g or more

From these measures it is possible to calculate the toxic dose of a man of known weight.

Similarly, the lethal concentrations quoted for LC20, LC50 and LC90 are those that will kill 20%, 50% and 90% of test animals in a certain time (say 48 hours) when exposed in a chamber with a controlled atmosphere of a particular gas, fume, vapour, dust etc. Inhalation tests require the concentrations and exposure times to be recorded to ascertain the uptake by the animal.

In tests for irritant response, the substances can be applied in the eye (Draize test) or on the skin. Skin sensitivity patch tests may be undertaken on man.

Stage two – tests to determine chronic effects

The harmful effects are investigated over 90 days and involve giving regular sub-lethal doses of the substances by the expected human route of exposure, but not exceeding the likely human dose. The tests are carried out on rats and mice. Observations are made of behaviour, growth, food intake, urine, faeces, biochemistry of blood, electrolytes, urea, sugar, fat and the metabolites of the test substances. Teratology studies, using similar techniques, are carried out during the early stages of pregnancy.

Stage three – tests for carcinogenicity

These tests require a large number of test animals with known medical history and of pure strains and involve assessing the rate of tumour induction making allowance for possible spontaneous tumour occurrences.

Other tests include the Ames test in which a strain of bacteria, such as *Salmonella typhimurium*, is mixed with rat liver and the test substance then incubated for two days. The carcinogenicity is indicated by the number of mutants induced. This test is sensitive, quick and cheap. Also tissue cultures tests in which cells from a test animal are cultured in an isotonic fluid medium and the effects of adding the test substance are observed.

Each stage of toxicity testing takes about two years and at the end of it, as with all animal testing, there is difficulty in extrapolating the results to man especially when animals are tested in conditions to which man is not exposed.

3.2.2.1 Portals of entry

Occupational poisons gain entry to the body via the lungs, skin and sometimes the gut. Absorption of a poison depends on its physical state, particle size and solubility. Of the substances entering the lung some may be exhaled, coughed up and swallowed, attacked by scavenger cells and remain in the lung or enter the lymphatics. Soluble particles may be absorbed into the blood stream. The skin is protective unless abraded when soluble substances can penetrate to the dermis, as they may also do via the hair follicles, sweat and sebaceous glands, and then be absorbed into the blood stream.

In pregnancy, harmful substances in the mother's body may cross the placenta to affect the unborn baby.

3.2.2.2 Effects

Effects may be acute, i.e. of rapid onset and short duration; or chronic, i.e. of gradual onset and prolonged. They may be local, occurring at the site of contact only, or general following absorption. Toxic substances may disturb normal cell function, damage cell membranes, interfere with enzyme and immune systems, RNA and DNA activity. Pathological response may be irritant, corrosive, toxic, fibrotic, allergic, asphyxiant, narcotic, anaesthetic and neoplastic.

3.2.2.3 Metabolism

Most substances absorbed will be carried by the blood stream to the liver where they may be rendered less harmful by a change in their chemical composition. However, some may be made more toxic, e.g. naphthylamine which is responsible for bladder cancer and tetra-ethyl lead which is converted into the tri-ethyl form and is toxic to the central nervous system.

3.2.2.4 Excretion

The body eliminates harmful substances in the urine, lungs and less commonly the skin. Some are also excreted in the faeces and milk. The time taken to reduce the concentration of a substance in the blood by 50% is known as its biological half-life. Similarly, the time for a 50% fall in concentration of a substance or its metabolite in urine or breath, after cessation of exposure, is its half-life in that medium. This is important in the design of screening tests[4].

3.2.2.5 Factors influencing toxicity

A number of factors are important when considering the toxic effect of a substance on the body. These include:

1. The inherent potential of a substance to cause harm.
2. Its ease of body contact and entry: work method, particle size and solubility.
3. Dose received (concentration and time of exposure).
4. Metabolism in the body (bio-transformation) and its half-life.
5. Susceptibility of the individual which depends on a number of factors:
   (a) Body weight; the same dose of a substance is more damaging to the smaller person.
   (b) The extremes of age in the working population are more prone to skin damage.
   (c) Fair skinned persons are more liable than the dark skinned to chemically induced dermatitis, and to radiation induced skin cancer.
   (d) Physical and physiological differences between the sexes may cause a variation in toxic response.
   (e) Immunological, nutritional and genetic defects.
   (f) Failure to reach a set health standard for work may expose the individual to greater risk.
   (g) Inadequate level of training, information, supervision and protection.

3.2.2.6 A no-adverse-effect level

In the setting of Occupational Exposure Standards (OES)[5], which are health based exposure concentration limits, it is necessary to establish, with reasonable certainty, the airborne concentrations which will not result in ill-health even if inhaled day after day. This concentration may be derived from the no-adverse-effect level in animal species which is arrived at after careful epidemiological and toxicological tests. This level relates to the average workperson but will not apply to individuals who, under certain circumstances, are susceptible and, therefore, at special risk[6].

3.2.2.7 Epidemiology

Epidemiology may be defined as the study and distribution of disease in human populations. The need for a study may be triggered by suspicions about an individual case, a complaint or the occurrence of a cluster of cases.

An initial *descriptive study* of a cross-section of the affected population is undertaken to determine:

What is the disease	
Who is affected	(sex, age, race, social class and occupation)
Where does it occur	(factory, workshop, room, laboratory, area of site)
When does it occur	(time, day, shift)
How are persons infected	(skin contact, airborne, body fluids)

From this study a hypothesis may be formed which needs to be tested by an analytical investigation involving a case-control study and a cohort study.

A *case-control study* compares persons who have the disease with those without to establish whether the suspect cause occurs more frequently among those with than those without the disease.

A *cohort study* compares those exposed to the suspect cause with those who are not to determine if more persons exposed to the cause develop the disease than those who are not.

An *incidence study* is essential for determining the risk, magnitude and causative factors of a disease:

$$\text{the incidence rate} = \frac{\text{no. of new cases of the disease}}{\text{population at risk}} \text{ over a period of time}$$

3.2.3 Diseases of the skin

3.2.3.1 Non-infective dermatitis

The term 'dermatitis' simply means an inflammation of the skin. When the condition is due to contact with a substance at work it is called 'occupational' or 'industrial' dermatitis. It is a common cause of occupational disease but the number of cases is declining owing to improved work conditions.

The skin has two layers, the outer layer is called the 'epidermis' and the inner the 'dermis'. The epidermis has a protective function. It consists of densely packed flat cells, thicker in some areas, like the palms of the hands, which are more subject to injury. It is covered by a moist film known as an 'acid mantle', made up of secretions from sweat and sebaceous glands, that helps to protect from acids, alkalis, excessive water, heat and friction by preventing the skin from drying out. The natural grease of the skin can be removed by solvents. In the deeper layer of the epidermis are pigment cells which produce the 'tan' following exposure to sunlight and protect the body from ultraviolet radiation.

Some persons are more susceptible to skin damage than others, particularly the young, those with soft, sweaty skin, the fair complexioned and those with poor personal hygiene. Occupational dermatitis can affect any part of the body, but the hands, wrists and forearms are most commonly involved. Damage to the skin may follow exposure to chemical and biological substances as well as physical agents. Dermatitis is of two kinds: irritative and sensitising – the former is four times more common. Chemicals which cause irritative dermatitis include acids, alkalis, cement, solvents, some metals and their salts. Their effect on the skin depends on the concentration and duration of exposure, and will affect most people in contact with them. At first the response may be minor, but it worsens with repeated contact.

Sensitisers, on the other hand, do not cause dermatitis until the individual has first become sensitised by them. This involves an allergic response in the tissues initially, dermatitis follows on subsequent exposure. Once sensitisation has occurred a small dose may be sufficient to cause a rash. Sensitisers include chrome-salts, nickel, cobalt, plastics made of epoxy, formaldehyde, urea or phenolic resins, rubber additives, some woods and plants[7]. Some substances act as both irritant and sensitiser, e.g. chrome, nickel, turpentine and mercury compounds.

3.2.3.1.1 Symptoms

The onset of dermatitis may be unnoticed, especially as it usually clears up when away from work, i.e. at weekends and holidays. On return to work and further exposure the condition recurs, worsening with each subsequent contact. The skin, at first rough and raw, may itch, become cracked and sore, prompting the individual to seek medical advice. The rash may be diffuse, as with eczema, or pimple-like as with acne – the former following exposure to irritative and sensitising agents, and the latter from exposure to mineral oils, pitch and chlorinated hydrocarbons. Patch testing, in which a dilute quantity of chemical is applied to the skin under a plaster and left for several hours to see if a reaction develops, is useful only for determining allergic response to chemicals but requires specialist interpretation.

3.2.3.1.2 Protective measures

Persons with dermatitis or sensitive skin may need to be excluded from certain kinds of work. Good personal hygiene is essential and barrier creams may be helpful. Protective clothing should be considered.

3.2.3.2 Cancer of the scrotum

The first occupational skin cancer was reported by Percivall Pott in 1775, among chimney sweeps. In those days children were apprenticed to master sweeps to climb inside and clean chimneys; their skin became ingrained and their clothes impregnated with soot, and as they seldom washed or changed their clothes the skin was constantly irritated. From puberty onwards a ‘soot wart’ might appear on the scrotum and develop into a cancer. In 1820, Dr Paris wrote of the influence of arsenical fumes affecting those engaged in copper smelting in Cornwall and Wales, giving rise to a cancerous disease of the scrotum similar to that affecting sweeps[8]. From 1870 a number of substances in a variety of industries were found to cause scrotal cancer – shale oil in those engaged in oil refining and cotton mule spinning; pitch and tar in those making briquettes from pitch-containing coal dust; mineral oil used by engineers and gunsmiths, and paraffin in refinery workers. Others at risk include creosote-timber picklers and anthracene chemical workers, and also sheep-dippers using arsenic[9].

Workers' clothes become begrimed with the offending substance, making close contact with the scrotum, the wrinkled skin of which favours the harbouring of the carcinogen.

The cancer begins as a wart, which enlarges and hardens, then breaks down into an ulcer with spread of malignant cells to neighbouring glands and other parts of the body.

Skin cancer is often due to polycyclic aromatic hydrocarbon of the benzpyrene or benzanthracene type. It has also been attributed to sunlight, ionising radiation and arsenic compounds.

3.2.3.2.1 Prevention

The use of non-carcinogen oils: carcinogens can be removed from mineral oil by washing with sulphuric acid or solvents. Workers should be educated to avoid contact as much as possible. The use of splash guards on machinery, protective clothing, avoidance of an oily rag placed in a pocket, which could spread oil through the clothes to the scrotum. To wash the hands before toilet and to have a daily bath. Clothes should be kept reasonably clean and a laundry service provided, so that overalls can be changed once or twice a week as need requires. Workers should not wear their dirty overalls after duty, but be encouraged to change into their home clothes. Workers also should be medically examined prior to employment and periodically to ensure that their skin is clear, and be encouraged to report to the doctor any doubtful 'wart' that might appear.

3.2.3.3 Coal tar and pitch

The destructive distillation of coal yields a variety of products, depending on the temperature at which distillation takes place, e.g.

GAS
↑
Ammoniacal liquid ⟵ COAL ⟶ TAR ⟵
↓ ↓
COKE Residue = pitch

Temp°C	Product
200	light oil
250	carbolic substances
300	creosote
350	anthracene

Distillation at high temperatures results in aromatic polycyclic hydrocarbons retained in the pitch which are harmful to health. Pitch is used in many industries: briquetting of coal, roofing materials, waterproofing of wood, manufacture of electrodes, impermeable paper, optical lenses, dyestuffs and paints.

3.2.3.3.1 Symptoms

Exposure of a worker to pitch dust or vapour may harm the skin by causing irritation, tumour or dermatitis. Irritation is the earliest and commonest reaction, occurring after a few days or weeks of exposure and affecting the face and neck. There is complaint of itching or burning, aggravated by cold, wind or sunlight (Pitch Smarts). Usually it clears up soon after exposure ceases. Benign tumours or warts occur on exposed areas of skin, chiefly the face, eyelids, behind the ears, the neck, arms and, occasionally, on the scrotum and thighs. Their recurrence is related to duration and degree of exposure to pitch[10]. Many regress spontaneously, especially those appearing early, but some undergo malignant change, particularly those appearing in the older age groups. They need to be removed and examined under the microscope, i.e. biopsied, to check for any malignant change. A variety of other skin conditions may occur such as darkening and thickening of the skin, acne, blackheads, cysts and boils, pitch burns and scarring. There is also a risk of damage to the cornea.

3.2.3.3.2 Prevention

Pitch dust and vapour must be avoided by transporting the raw material in a liquid or granular state and enclosing the process as far as possible. Workers require clean protective clothing for head, neck and forearms and eye protection should be worn[11]. Employees ought to be warned of the risk, and advised to report any skin disease which develops. Good personal hygiene is essential, and adequate wash and shower facilities need to be provided. Barrier creams applied before work are helpful. Those susceptible to warts should be excluded from further exposure, and each worker needs to be medically examined regularly to detect possible skin disorders.

3.2.4 Diseases of the respiratory system

3.2.4.1 Pneumoconiosis

The term pneumoconiosis means 'dust in the lung', but medically refers to the reaction of the lung to the presence of dust[12].

3.2.4.1.1 Body defence to inhalation of dust

During inspiration particles of dust in the air larger than 10 μm in diameter are filtered off by the nasal hairs. Others, which enter through the mouth, are deposited in the upper respiratory tract. Particles between 5 μm and 10 μm tend to settle in the mucus covering the bronchi and bronchioles and are then wafted upward by tiny hairs (ciliary escalator) towards the throat. They are then coughed or spat out, though some may be swallowed. Particles less than 5 μm in diameter are more likely to reach the lung tissue. However, fibres (e.g. asbestos) which predispose to disease have a length to diameter ratio of at least 3:1 with a diameter of 3 μm or less; the longer the fibre the more damaging it may be.

3.2.4.1.2 Respirable dust

Respirable dust is that dust in the air which on inhalation may be retained by the lungs. The amount of dust retained depends on the duration of exposure, the concentration of dust in the respired air, the volume of air inhaled per minute and the nature of the breathing. Slow, deep respirations are likely to deposit more dust than rapid, shallow breathing. Dust in the lung causes a tissue reaction, which varies in nature and site according to the type of dust. Coal and silica dust involve the upper lungs whereas asbestos involves the lower lungs.

3.2.4.1.3 Causes of pneumoconiosis

(a) *Benign* The inhalation of some metal dusts, such as iron, tin and barium, results in very little structural change in the lungs and, therefore, few symptoms. The tissue reaction, nevertheless, is detectable on X-ray as a profusion of tiny opacities.
(b) *Symptomatic* The most important causes include coal dust, silica and asbestos. Symptoms of cough and breathlessness develop usually after many years of exposure, but only in the later stages of disease.

Beryllium dust causes acute and chronic symptoms. Early features are breathlessness, cough with bloody sputum and chest pain. Recovery follows removal from exposure, but a chronic state can develop insidiously with cough, breathlessness and loss of weight.

Organic dusts, such as mouldy hay, when inhaled cause a disease known as extrinsic allergic alveolitis with 'flu-like symptoms; cough and difficulty in breathing occur within a few hours of exposure. Repeated exposure leads to further lung damage and chronic breathlessness.

Talc is a white powder consisting of hydrous magnesium silicate. Although some talc presents little risk to health, commercial grades may contain asbestos and quartz and provoke pneumoconiosis and lung cancer.

Cobalt combined with tungsten carbide forms a hard metal used for the cutting tips of machine tools and drills. Inhalation of the dust may give rise to fibrosis of the lungs causing cough, wheezing and shortness of breath.

Man-made mineral fibres irritate the skin, eyes and upper respiratory tract. A maximum exposure limit has been set based on the risk of lung cancer because a 'no-adverse-effect' level cannot be established with reasonable certainty[6].

3.2.4.1.4 Diagnosis of pneumoconiosis

This depends on:

1 A complete occupational history of all jobs.
2 A characteristic appearance on the chest X-ray. There is an international grading system which is used to assess radiologically the extent of the disease.
3 A clinical examination.
4 Lung function tests.
5 In some cases involving organic dust, specific blood tests.

3.2.4.2 Silicosis

Silicosis: the commonest form of pneumoconiosis is due to the inhalation of free silica.

Free silica (SiO_2) or crystalline silica occurs in three common forms in industry: quartz, tridymite and cristobalite. A cryptocrystalline variety occurs in which the 'free silica' is bound to an amorphous silica (non-crystalline). It includes tripolite, flint and chert. Diatomite is the most common form of amorphous silica capable of producing lung disease. Some of these forms can be altered by heat to the more dangerous crystalline varieties, such as tridymite and cristobalite. e.g.

$$\left.\begin{array}{l}\text{Quartz}\\ \text{Cryptocrystalline}\\ \text{Amorphous}\end{array}\right\} \rightarrow \longrightarrow \text{tridymite} \rightarrow \rightarrow \text{cristobalite} \quad (800°C^{+} \longrightarrow)$$

3.2.4.2.1 Lung reaction

Industrial exposure occurs in mining, quarrying, stone cutting, sand blasting, some foundries, boiler scaling, in the manufacture of glass and ceramics and, for diatomite, in the manufacture of fluid filters. Particles of free silica less than 5 μm in diameter when inhaled are likely to enter the lungs and there become engulfed by scavenging cells (macrophages) in the walls of the tiniest bronchioles. The macrophages themselves are destroyed and liberate a fluid causing a localised fibrous nodule which obliterates the air sacs. The nodules are scattered mainly in the upper halves of the lungs. They gradually enlarge to form a compressed mass of nodules. Sometimes a single large mass of tissue may occur, known as progressive massive fibrosis. If much of the lung is affected the remaining healthy tissue is likely to become over-distended during inhalation.

3.2.4.2.2 Symptoms

There are no symptoms in the early stage. Later the initial complaint is of a dry morning cough. Next occurs some breathlessness, at first noticeable on exercise but, as destruction of lung tissue proceeds, breathlessness worsens until it is present at rest. The interval between exposure and the onset of symptoms varies from a few months in some susceptible individuals to, more usually, many years, depending on the concentration of respirable free silica and the exposure time at work. Silicosis is the one form of pneumoconiosis which predisposes to tuberculosis, when additional symptoms of fever, loss of weight and bloody sputum may occur. In the presence of gross lung destruction the blood circulation from the heart to the lung may be embarrassed and result in heart failure.

3.2.4.2.3 Diagnosis

This depends on a history of exposure and, in the early stages, a chest X-ray showing tiny radio opaque nodules and, later, a history of cough and breathlessness and sounds in the chest detectable with a stethoscope. Lung function tests may be helpful, but usually not until the late stages.

3.2.4.2.4 Medical surveillance

Where exposure to free silica is a recognised hazard, a pre-employment medical is advised, which should enquire into previous history of dust exposure, of respiratory symptoms, with examination of the chest, lung function testing and a chest X-ray. The medical should be repeated periodically as circumstances demand.

3.2.4.2.5 Prevention

Reduction of the dust to the lowest level practicable and where necessary by the provision of personal respiratory protective equipment.

3.2.4.3 Asbestosis

There are three important types of asbestos, blue (crocidolite), brown (amosite) and white (chrysotile). Asbestosis is a reaction of the lung to the presence of asbestos fibres which, having reached the bronchioles and air sacs, cause a fibrous thickening in a network distribution, mainly in the lower parts of the lung[13]. There follows a loss of elasticity in the lung tissue (relative to the concentration of fibres inhaled and the duration of exposure), resulting in breathing difficulty.

Among those at risk are persons engaged in milling the ore, the manufacture of asbestos products, lagging, asbestos spraying, building, demolition, and laundering of asbestos workers' overalls.

Symptoms develop slowly after a period of exposure which varies from a few to many years. In some cases exposure may have begun so long ago that it cannot be recalled. Breathlessness occurs first and progresses as the lung loses its elasticity. There may be little or no cough and chest pain seldom occurs. The individual becomes weak and distressed on effort and, eventually, even at rest. Unless periodic medicals are introduced the diagnosis will not be made until symptoms appear. Early diagnosis is essential in order to prevent further exposure and an exacerbation of the condition. Asbestosis predisposes to cancer of the bronchus, a risk increased by cigarette smoking. The chest should be X-rayed every two years and special lung function tests are helpful. Diagnosis depends on history of exposure, chest X-ray, lung function testing, symptoms and physical signs.

3.2.4.4 Mesothelioma

Mesothelioma is a malignant tumour of the lining of the lung (pleura) or abdomen (peritoneum). The abdominal form is less common. The disease is significantly related to exposure to asbestos, especially the blue and brown varieties. However, in some 10–15% of cases there is no such history of exposure[13]. Those at risk are miners, manufacturers of asbestos, builders and demolition workers, and even residents in the neighbourhood of blue asbestos working. While the exposure time may have been minimal, there is no safe threshold of dose below which there is no risk of asbestos-related disease. The onset of the disease is delayed by some 20 to 50 years.

3.2.4.4.1 Symptoms

The lung variety of tumour is more common. Symptoms begin with a gradual onset of breathlessness, particularly noticeable on effort, and due to the growth of tumour and fluid compressing the lung. There may occur pain on one side of the chest, with tenderness, cough and fever. More obvious is a rapid loss of weight and weakness. A chest X-ray reveals an opacity on one side of the chest suggestive of the tumour. The symptoms of the abdominal form also develop slowly, beginning with a swelling, loss of weight, impaired appetite and weakness. Death usually follows within two years of making the diagnosis.

3.2.4.5 Other dust causes of lung cancer

These include: chromate, in the manufacture of chromate from the ore; nickel compounds in the refining of nickel; benzpyrenes in coke-oven work; uranium and radon; and arsenic compounds in mining.

3.2.4.6 Bronchial asthma

Bronchial asthma is defined as breathlessness due to narrowing of the small airways and it is reversible, either spontaneously or as a result of treatment. It may follow inhalation of a respiratory sensitiser or an irritant toxic substance. Symptoms due to sensitisation may be delayed for weeks, months or even years; symptoms due to a toxic substance occur within hours of inhalation, resolve spontaneously but can persist indefinitely. The toxic response is called *reactive airways dysfunction* (RAD) syndrome. Most cases of occupational asthma are due to sensitisation and are listed[14] as prescribed diseases for purposes of statutory compensation. The sensitising substances listed are:

1 Isocyanates.
2 Platinum salts.
3 Epoxy resin curing agents.
4 Colophony fumes.
5 Proteolytic enzymes.
6 Animals and insects in laboratories.
7 Flour and grain dust.
8 Antibiotic manufacture.
9 Cimetidine used in manufacturing cimetidine tablets.
10 Hard wood dusts of cedar, oak and mahogany.
11 Ispaghula used in the manufacture of laxatives.
12 Caster bean dust.
13 Ipecacuanha used in the manufacture of tablets.
14 Azodicarbonamide used in plastics.
15 Glutaraldehyde, a cold disinfectant used in the health service.
16 Persulphate salts or henna used in hair dressing.
17 Crustaceans or fish products used in the food processing industry.
18 Reactive dyes.
19 Soya bean.
20 Tea dust.
21 Green coffee bean dust.
22 Fumes from stainless steel welding.
23 Any other sensitising agent inhaled at work.

Respiratory sensitisers may be referred to as *asthmagens*. In 1989 Surveillance of Work Related Respiratory Disease (SWORD) was started and contains reports by respiratory and occupational physicians.

Other asthma-like diseases are found.

Byssinosis occurs in workers in the cotton processing industry who may develop tightness of the chest on Mondays which decreases as the week progresses. However, with continuing exposure to cotton dust they are affected for more days of the week. Steam treatment of the raw cotton can prevent chest symptoms from this material.

An allergic lung reaction also occurs after exposure to spores on sugar cane (*bagassosis*). The sugar cane spores can be killed by spraying with propionic acid.

3.2.4.7 Extrinsic allergic alveolitis (farmer's lung)

A disorder due to inhalation of organic dust and characterised by chest tightness, fever and the presence of specific antibodies in the blood. Typical examples are:

Disease	*Exposure*	*Allergen*
Farmer's lung	mouldy hay	mould
Malt worker's lung	mouldy barley	mould
Bagassosis	mouldy sugar cane	mould
Bird fancier's lung	bird droppings	protein
Animal handler's lung	rats' urine	protein

3.2.5 Diseases from metals

3.2.5.1 Lead

Lead (Pb) is a relatively common metal, mined chiefly as the sulphide (galena) in many countries – USA, Australia, USSR, Canada and Mexico[15]. In this country we use about 330 000 tonnes of lead annually, much of which comes from recycled scrap.

Lead has a great variety of uses, e.g. (percentages approximated from annual production figures issued by World Bureau of Metal Statistics, London):

Electric batteries	27%
Electric cables	17%
Sheet, pipe, tubes	16%
Anti-knock in petrol	11%
Solder and alloys	9%
Pottery, plastics, glass, paint	4%
Miscellaneous	15%

Lead, as a fume or dust hazard, is therefore met in many industries. The pure metal melts at 327°C and begins to fume at 500°C, but the presence of impurities alters these properties and may form a slag on its surface and thereby reduce fuming, except at higher temperatures. Particle size and solubility are important factors governing the absorption of lead via the lungs. In the gut, however, solubility differences of ingested compounds are of less significance. Among lead miners lead poisoning does not occur due to the insolubility of the sulphide ore.

3.2.5.1.1 Inorganic lead

Inorganic lead can enter the body by inhalation or ingestion[16]. Up to about 50% of that inhaled is absorbed and only about 10% of that ingested. It is then transported in the blood stream and deposited in all tissues, but about 90% of it is stored in the bone. It is a cumulative poison; excretion is slow and occurs mainly in the urine and faeces.

Symptoms Early features are vague and include fatigue, loss of appetite, and metallic taste in the mouth. Constipation is the commonest complaint and is sometimes associated with abdominal pain. This may be so severe as to mimic an acute abdominal emergency. Classically, a blue line appears along the margin between the teeth and gums, but this usually occurs only in the presence of infected teeth and is indicative of lead exposure rather than poisoning.

Lead interferes with the normal formation of haemoglobin, causing anaemia, but the diagnosis of excessive absorption should be made before anaemia appears. The same interfering mechanism causes abnormal products to appear in the urine, e.g. amino laevulinic acid (ALA) which is a useful indicator of excessive lead absorption or poisoning.

Paralysis, though rare nowadays, can occur as wrist or foot drop due to the effect of lead on nerve conduction. It may begin with a weakness in the fingers and wrists, which is a useful early sign.

Lead is transported in the blood and can cross the placental barrier in pregnant women and affect the unborn child. Abortion was common in women employed in lead industries during the nineteenth century and was believed to be due to excessive lead absorption. The brain can also be affected, a condition known as encephalopathy, causing abnormal behaviour, convulsions, coma and death. Children are much more susceptible than adults.

Because of the excretion of lead in the urine, kidney damage is a likely long-term effect.

3.2.5.1.2 Organic lead

Tetra-ethyl and tetra-methyl lead are the most important organic forms used in industry, especially in petrol to improve the octane rating. These substances can be absorbed via the lungs and the skin. In the liver they are changed respectively to tri-ethyl and tri-methyl lead, which are much more toxic. They have a particular predilection for the brain and cause psychiatric disturbance, headache, vomiting, dizziness, mania and coma. Excretion occurs mainly via the urine. The blood is less affected than with inorganic lead.

3.2.5.1.3 Biological monitoring

For lead workers periodic medical examination is a statutory requirement. Blood samples should be taken as required for haemoglobin and lead. Lead level in normal blood is about 20 μg/100 ml but for lead workers can be 40–60 μg/100 ml. The acceptable upper limit of blood lead concentration in adults is 60 μg/100 ml except men who have worked in lead for many years. For young persons is 50 μg/100 ml and for women of child-bearing age the limit is 30 μg/100 ml.

A useful indicator of excessive lead effect is the presence of zinc protoporphyrin (ZPP). It can be measured from a small quantity of blood obtained by finger-prick. For confirmatory evidence of excessive lead absorption or poisoning, urine estimation of amino laevulinic acid is helpful. Inorganic lead is best monitored by blood sampling and organic lead by urine sampling.

3.2.5.2 Mercury

Mercury (Hg) occurs naturally as the sulphide in the ore known as cinnabar, and also in the metallic form quicksilver. It is mined chiefly in Spain, but also in Italy, Russia, USA and elsewhere. The ore is not particularly hazardous to miners, as the sulphide is insoluble. Risk is greater in other industries, such as in the manufacture of sodium hydroxide and chlorine, electrical and scientific instruments, fungicides, explosives, paints and in dentistry.

3.2.5.2.1 Symptoms

Acute mercury poisoning is rare but can occur following the inhalation of quicksilver – it being very volatile at room temperature. There is particular risk should spillage occur in an enclosed space. About 80% of that inhaled can be absorbed[17], and a few hours later there occurs cough, tight chest, breathlessness and fever. Symptoms last a week or so, dependent upon degree of exposure, but its effects are reversible. Acute poisoning may also occur by ingestion of soluble salts, such as mercuric chloride which has a corrosive action on the bowel, causing bloody diarrhoea.

Ingestion of metallic mercury is not generally toxic as it is not absorbed.

Chronic poisoning is the more usual presentation, following absorption by lung or gut of soluble mercury salts. Symptoms develop almost imperceptibly, usually beginning with a metallic taste in the mouth and sore gums. Later tremor of the hands and facial muscles develops; gums may bleed and teeth loosen. Personality changes of shyness and anxiety, inability to concentrate, impaired memory, depression and hallucinations may occur. As excretion is mainly via the urine, the kidney is subject to damage.

Organic mercury can be absorbed via the lung, gut and skin, and also cause chronic poisoning. There are two varieties: aryl and alkyl, and they have different effects on the body. The aryl variety, of which phenyl mercury is an example, has a similar metabolic pathway to inorganic mercury and has a similar clinical effect.

Alkyl mercury is much more dangerous – methyl mercury is an example. It causes irreversible damage to the brain, resulting in a constriction of visual fields, disturbance of speech, deafness and inco-ordination of movement. Most of it (90%) is excreted without change, slowly in the faeces.

All forms of mercury may give rise to dermatitis. Mercury can cross the placental barrier and affect the unborn child of exposed mothers.

3.2.5.2.2 Health surveillance

Those at risk should be medically examined periodically and attention paid to the mouth, tremor of the hands (a writing test is useful), personality, and for those exposed to methyl mercury, vision, hearing and co-ordination. The urine should be checked for protein and mercury excretion. Mercury does not normally occur in urine, but may be detected in some persons with no apparent occupational exposure. In organic exposure, owing to the different metabolic pathway from that of inorganic, the urine concentration does not correlate with body levels. The upper limit which requires further investigation is for inorganic mercury 1000 nmol/litre and, for organic mercury, 150 nmol/litre[18].

3.2.5.3 Metal fume fever

Inhalation of the fume of some metal oxides such as zinc, copper, iron, magnesium and cadmium causes an influenza-like disease. Similar effects may follow the inhalation of polytetrafluoroethylene (ptfe) fumes. Usually there is recovery within one or two days. Zinc fume fever is probably a very common disease, the diagnosis of which is often missed because of the short duration of the illness. Cadmium fume inhalation can be much more serious. It has a half-life of several months (see section 3.2.5.9).

3.2.5.4 Chromium

Chromium (Cr) is a silvery hard metal used in alloys and refractories. Chrome salts are used in dyeing, photography, pigment manufacture and cements. Electroplating tanks contain solutions of chromic acid which forms a mist during the electrolysis process.

Chromates and dichromates used in cement manufacture and chromium plating may cause skin irritation or ulceration and chrome ulcers in the skin of the hands or in the inside of the nose where the ulcer may penetrate the cartilage of the nasal septum.

3.2.5.5 Arsenic (As)

Inorganic arsenic compounds cause irritation of the skin and may produce skin cancer. It is used in alloys to increase hardness of metals, especially with copper and lead.

3.2.5.6 Arsine (arseniuretted hydrogen – AsH_3)

Arsine is a gas which arises accidentally in many metal working industries. It damages the red blood cells, releasing the red pigment haemoglobin from them. This may cause jaundice, anaemia and the urine may appear red due to the presence of haemoglobin pigment. Poisoning by arsine can result in rapid death. Organic arsenic compounds have been used as war gases, and can produce severe and immediate blistering of the skin and severe lung irritation (pulmonary oedema).

3.2.5.7 Manganese (Mn) and compounds

This is used to make manganese alloy steels, dry batteries and potassium permanganate which is an oxidising agent and a disinfectant. Poisoning is rare and follows inhalation of the dust causing acute irritation of the lungs and affects the brain leading to impaired control of the limbs rather like Parkinson's disease.

3.2.5.8 Nickel (Ni) and nickel carbonyl ($Ni(Co)_4$)

Nickel is a hard blue-white metal used in electroplating and in a range of alloys. Nickel salts (green) cause skin sensitivity (nickel itch). Nickel carbonyl (a colourless gas) causes headache, vomiting and later pulmonary oedema.

3.2.5.9 Cadmium (Cd)

This metal is used in alloys, rust prevention, solders and pigments. A fume may be released during smelting, alloy manufacture or when rust-proofed metals are heated, e.g. in welding cadmium-plated metals, which produces irritation of the eyes, nose and throat. With continued exposure tightness of the chest, shortness of breath and coughing may increase and can lead to more severe lung damage which may be fatal.

Long-term damage by smaller quantities of dust or fumes may lead to loss of elasticity of the lungs. Cadmium may cause kidney damage and while it has been suggested that lung cancer may occur after cadmium exposure this has not been proved in man.

3.2.5.10 Vanadium (V)

This material occurs as vanadium ore and is found in petroleum oil. It is also used to make alloy steels and as a catalyst in many chemical reactions. Exposure to the metal occurs when oil-fired boilers are cleaned and manifests itself in eye irritation, shortness of breath, chest pain and cough. The tongue becomes greenish-black in colour. Severe cases may develop broncho-pneumonia. Removal from contact with the dust usually leads to rapid recovery.

3.2.6 Pesticides

3.2.6.1 Insecticides

Various organo-phosphorus compounds are used; two of the commonest are demeton-S-methyl and chlorpyrifos. Poisoning causes headaches, nausea and blurred vision. Further symptoms include muscle twitching, cramps in the belly muscles, severe sweating and respiratory difficulties. Extreme exposure may lead to death. All these effects are due to interference with a chemical enzyme called cholinesterase which is concerned with the passage of nerve impulses. The level of this enzyme in the worker's blood can be measured and if it falls below a certain value the worker must be removed from contact with the chemical until his blood returns to normal. The appropriate protective clothing must be worn at all times when working with these materials.

3.2.6.2 Herbicides

Commonly used as a weedkiller (e.g. paraquat). Ingestion may result in damage to the liver, kidneys and lung. There is no antidote and death occurs in about half the cases.

3.2.7 Solvents

A solvent is a liquid that has the power to dissolve a substance: water is a common example[19]. In industry organic liquids are often used as solvents, and these are mainly hydrocarbons used as degreasing agents and in the manufacture of paints and plastics.

Examples of solvents (classification after Matheson[20])

Hydrocarbons	
(i) Aromatic	Benzene; toluene; styrene
(ii) Aliphatic	Paraffin; white spirit
Aliphatic alcohols	Methyl alcohol; ethyl alcohol
Aliphatic ketones	Methyl-ethyl-ketone
Aliphatic ethers	Diethyl ether
Aliphatic esters	Ethyl acetate
Aliphatic chlorinated	Trichloroethylene; carbon tetrachloride
Non-hydrocarbons	Carbon disulphide

3.2.7.1 General properties

All organic solvents are volatile and have a vapour density greater than one, i.e. their vapours are heavier than air and will therefore settle at floor level; this is important to note when considering ventilation. With the exception of the chlorinated hydrocarbons they tend to be flammable and explosive and in the liquid form most have specific gravities of less than one so will float on water. In the event of a fire, attempt should not be made to extinguish with water, as the solvent will float away and the fire will spread. The chlorinated solvents, being neither flammable nor explosive but heavier than water, have been used as fire extinguishants.

3.2.7.2 Toxic effects

Solvents vary widely in their toxicological properties. In common they cause dermatitis by removing the natural grease from the skin, and narcosis by acting on the central nervous system; additionally some can damage the peripheral nerves, the liver and kidneys and interfere with blood formation and cardiac rhythm. Chlorinated solvents can decompose if exposed to a naked flame to produce acidic fumes (hydrochloric acid and small amounts of phosgene) which are harmful to the lungs. Any harmful effect is related to the amount of solvent absorbed.

Skin penetration varies with the solvent, hence in the list of Occupational Exposure Limits[5] some are designated 'skin', but other factors include surface area exposed and the thickness of the skin, e.g. less may be absorbed via the palms than the forearms while the scrotal area is most absorptive[21].

Absorption is also related to the breathing pattern, activity, obesity and addiction. Because of this individual variation, the amount taken up by the body is a more important estimate of potential harm than the concentration to which the body is exposed. Body uptake correlates well with blood concentration and to a less extent with quantities excreted in the urine[22].

However, periodic urine testing of excreted solvent or its metabolite is a more convenient means of biological monitoring[4]. The biological half-life of solvents is only a few hours. The half-life of some solvents is so short that biological monitoring of urine is not suitable, instead a metabolite must be used, such as mandelic acid for styrene and methyl hippuric acid for xylene.

3.2.7.3 Trichloroethylene

Structural formula:

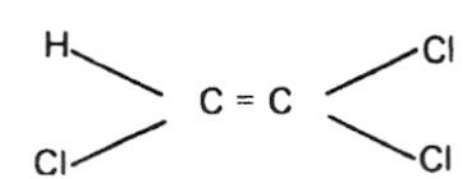

Other names: Tri, Trike, Trilene.
Properties: Non-flammable
Vapour density 4.54
Specific gravity 1.45
Boiling point 87°C
MEL 100 ppm 8 hour TWA (skin)

Exposure to naked flames or red-hot surfaces can cause it to dissociate into hydrogen chloride, possibly with small amounts of phosgene or chlorine.

Use Its main use is as a solvent especially in the degreasing of metals. It has also been used as an anaesthetic. *Figure 3.2.1* shows a single compartment vapour type plant used for cleaning by solvents.

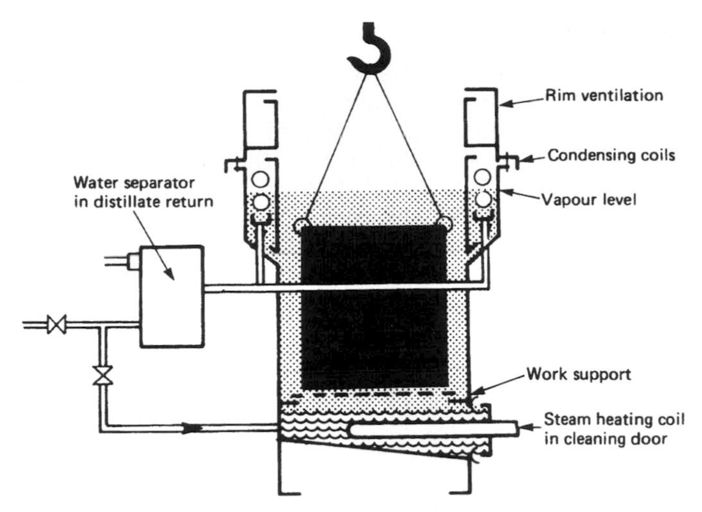

Figure 3.2.1 Cleaning by solvents: single compartment vapour type plant. (Courtesy ICI, PLC, Mond Division)

Metabolism Its main route into the body is via the lungs, where it is rapidly absorbed. Some is excreted into the expired air, while the remainder is converted to trichloroacetic acid and passed in the urine. It is usually cleared quickly from the body, but daily exposure may tend to its cumulation. The estimation of trichloroacetic acid in the urine is a useful test for checking excessive exposure and its concentration should not exceed 100 mg/litre urine, standardised to a specific gravity of 1.016. Samples should be collected at the end of a shift towards the end of a working week.

3.2.7.3.1 Harmful effects

Acute Trichloroethylene is a powerful anaesthetic and can be dangerous in confined spaces. Early features include headache, dizziness, and lack of concentration and eventually unconsciousness. Its vapour may cause irritation of the eye and the skin can be blistered by the liquid.

Chronic The main problem from repeated exposure is a dermatitis of the hands, due to the solvent's action in removing the normal grease of the skin which then becomes rough, red, raw, and cracks – a condition known as eczema. Some people become addicted to trichloroethylene, usually by repeated 'sniffing' of the vapour, or even drinking the fluid, and then display abnormal behaviour known as 'tri-mania'. Rare cases of sudden cardiac arrest have been reported in situations of gross short-term overexposure. After long-term exposure there have been a number of individual case reports of liver damage, and recently, following animal tests in the USA, it has been under suspicion as a carcinogen.

3.2.7.3.2 Prevention

Employees should be made aware of the risks. Local exhaust ventilation around the lips of vapour degreasing tanks is necessary, and in confined spaces good general ventilation is essential. In work areas, atmospheric monitoring is recommended to ensure that exposure is kept to a minimum and certainly below the Maximum Exposure Limit (MEL) of 100 ppm 8 hour TWA. Body absorption can be monitored by a urine sample taken at the end of a shift near the end of a working week and analysed for trichloroacetic acid. Those being tested must refrain from drinking alcohol as it inhibits excretion.

3.2.7.4 Carbon tetrachloride

Structural formula:

```
Cl      Cl
  \    /
    C
  /    \
Cl      Cl
```

Properties: Non-flammable
Vapour density 1.5
Specific gravity 1.6
Boiling point 76.8°C
OES-TWA 2 ppm (skin)

Use Its main use is in the manufacture of chlorofluorocarbons, also aerosols and refrigerants. It has been used in fire extinguishers and grain fumigation. Its use in dry cleaning has declined because of its toxicity.

Metabolism Carbon tetrachloride is absorbed into the blood mainly via the lungs, but also via the skin and gut. Some is excreted in the expired air and the remainder in the urine, but in altered form.

3.2.7.4.1 Harmful effects

Acute In common with other solvents it has a narcotic effect, with features varying from headache and drowsiness to coma and death. If taken by mouth it can cause abdominal pain, diarrhoea and vomiting. Acute over-exposure can result in liver and kidney damage.

Chronic Carbon tetrachloride can also cause damage to the kidneys and liver; in the long term it is more toxic than trichloroethylene. An early sign of kidney damage may be detected by urine examination for protein and cells. Liver damage may be indicated early by special tests or later by the appearance of jaundice. It is also under suspicion as a carcinogen.

3.2.7.5 Other common solvents

3.2.7.5.1 Benzene (C_6H_6)

MEL-TWA 5 ppm. This excellent solvent is seldom used today because of its toxic effects. It may be inhaled or absorbed via the skin and is readily absorbed by fatty tissues. A large proportion of benzene which enters the body is stored in the bone marrow which may be damaged, causing anaemia or more rarely leukaemia. Benzene is altered chemically in the body and then excreted in the urine. For exposures about the MEL, blood benzene is a useful measurement. For lower exposures, breath benzene is suitable. Urinary excretion as a 'phenol' test is no longer recommended.

3.2.7.5.2 Toluene ($C_6H_5CH_3$) (methylbenzene) and xylene ($C_6H_4(CH_3)_2$)

Toluene, OES-TWA 50 ppm and xylene, OES-TWA 50 ppm are frequently used solvents which have toxic effects common to other solvents. They produce narcosis and can damage the liver and kidneys. Blood or breath toluene is suitable for monitoring; for xylene, urine is tested for methyl hippuric acid.

3.2.7.5.3 Tetrachloroethylene ($CCl_2.CCl_2$) (perchloroethylene)

OES-TWA 50 ppm. This solvent is a narcotic and may cause liver damage. Like trichloroethylene it may break down to release phosgene when exposed to naked flames or red-hot surfaces. Monitor using blood sample taken towards the end of the working week.

3.2.7.5.4 Trichloroethane ($CH_3.CCl_3$) (methyl chloroform)

This solvent was once regarded as one of the safest solvents but is being withdrawn to meet requirements of an EU directive[23]. Supplies will not be available after current stocks are used up.

3.2.7.5.5 Carbon disulphide (CS_2)

MEL-8 hour TWA 10 ppm. Carbon disulphide is an inorganic solvent used mainly in the manufacture of viscose rayon fibres. It is absorbed through the lungs and skin and is a multi-system poison affecting the brain, peripheral nerves and the heart. Monitoring is of urinary metabolites.

3.2.8 Gassing

3.2.8.1 Gassing accidents

In the UK those gassing accidents that are reported annually occur in the following approximate order of frequency:

1 Carbon monoxide
2 Chlorine*
3 Hydrochloric acid*
4 Trichloroethylene

5 Sulphur dioxide*
6 Ammonia*
7 Hydrogen sulphide*
8 Phosgene
9 Carbon dioxide
10 Nitrous fumes
11 Phosphorus oxychloride*
12 Carbon tetrachloride

*Highly soluble gases which will irritate the eyes and upper respiratory tract while the less soluble components pass further down the tract to irritate the lung tissue.

Asphyxia caused by gassing falls into two broad categories:

Simple in which oxygen in the lungs is replaced by another gas such as carbon dioxide, nitrogen or methane.

Toxic in which there is a metabolic interference with the oxygen taken up by the body. This occurs with gases such as carbon monoxide, hydrogen sulphide and hydrogen cyanide.

3.2.8.2 Chlorine and hydrochloric acid (HCl)

These highly irritant gases may affect the air passages and lungs causing bronchitis and difficulties in breathing due to fluid in the lungs (pulmonary oedema).

3.2.8.3 Carbon monoxide (CO)

This colourless odourless gas may be found wherever incomplete combustion occurs such as in motor vehicle exhausts, furnaces, steel-works and domestic boilers.

Inhalation results in a rapid rise in CO concentration in the blood within the first hour and a much slower rise thereafter. The gas is more readily absorbed by the blood's red cells to the exclusion of oxygen and so impairs the supply of oxygen to vital organs, particularly the heart.

The effects of the gas are shown in the following table:

Exposure ppm	*Probable concentration of CO in blood after 1 hr exposure (carboxy haemoglobin)*	*Effect*
200	20%	Headache, flushed appearance, breathlessness
400	40%	Dizziness
500	50%	Collapse
600	60%	Unconsciousness

3.2.8.4 Hydrogen sulphide (H_2S)

Occurs in sewers, oil refineries and chemical processes. Its odour of rotten eggs can be detected at concentrations of 0.3 ppm but increasing the concentrations of exposure impairs the sense of smell. Even at low concentrations the gas irritates the eyes. Higher concentrations irritate the lungs causing pulmonary oedema (although the onset may be delayed), headache, dizziness, convulsions and unconsciousness.

3.2.8.5 Carbon dioxide (CO_2)

This occurs in bakeries, breweries etc. and is a result of fermentation. The gas is heavier than air. Low concentrations of CO_2 increase the rate of breathing but higher levels depress respiration causing rapid unconsciousness and even death.

3.2.8.6 Sulphur dioxide (SO_2)

OES-8 hour TWA 2 ppm. A colourless irritant gas with a pungent smell which causes bronchitis and pulmonary oedema.

3.2.8.7 Nitrous fumes (commonest form NO_2)

Pungent brown fumes which cause lung irritation after a delay of a few hours. Occurs in explosions and blasting, silo storage and diesel engine exhaust.

3.2.8.8 Phosgene ($COCl_2$)

Arises from burning chlorinated hydrocarbons, e.g. trichloroethylene. Effects similar to nitrous oxide.

3.2.8.9 Ammonia (NH_3)

OES-8 hour TWA 25 ppm. Ammonia has a corrosive action that will burn the skin, severely irritate or burn the cornea, cause bronchitis and pulmonary oedema.

3.2.9 Oxygen deficiency

Normal respiration requires:

1 An adequate concentration and partial pressure of oxygen in the inspired air.

2 A clear airway to the lungs.
3 Transfer of oxygen in the air sacs to the blood.
4 The transport of oxygen by the red cells to the tissues.

Normal oxygen requirements depend on body size, activity and fitness, and interruption of the supply can occur through failure at any of the above indicated levels. Fresh air contains approximately 21% oxygen, 79% nitrogen, 0.03% carbon dioxide. Although inspired air contains 21% oxygen, that in the air sacs has only 14% which at sea level exerts sufficient partial pressure to cross the lung–blood barrier.

At altitudes above sea level the percentage of oxygen in air is unaltered, but because the barometric pressure is less, the partial pressure of oxygen drops accordingly and makes breathing more difficult. At sea level barometric pressure equals 760 mm Hg, therefore oxygen partial pressure equals $760 \times 21/100 = 160$ mm Hg[24].

In the air sacs, however, there is vapour pressure present. It equals 47 mm Hg irrespective of altitude and diminishes the effective partial pressure which the oxygen would otherwise exert. For example, in the air sacs oxygen partial pressure at sea level equals $(760–47) \times 14/100 =$ 100 mm Hg. In confined spaces the oxygen concentration can fall by several means. It can be displaced by another gas, e.g. a simple asphyxiant such as carbon dioxide. In a disused and ill-ventilated coal mine the oxygen present could be used up in oxidising the coal, resulting in a condition known as 'black damp'. Combustion requires oxygen, so that in a confined space a flame will burn up the oxygen present. Similarly, oxygen can be 'combusted' by ordinary respiration of persons working in the space. Canister type respirators should not be worn in a confined space, because of the danger of a depletion of oxygen in the atmosphere; instead full breathing apparatus should be used.

The presence of disease can also embarrass breathing, as during an attack of bronchial asthma, or in pneumoconiosis, when transfer of oxygen across the lungs is impeded. In anaemia the red cell's capacity for carrying oxygen is diminished, and in heart disease the blood may be inadequately pumped around the body. A similar effect is found with carbon monoxide poisoning, in which the normal uptake of oxygen by the red cells is prevented. Each of these mechanisms results in an inadequate oxygen supply to the tissues, a condition known as anoxia.

3.2.9.1 Oxygen requirement

The 'average man' of 70 kg body weight requires 0.3 litres of oxygen per minute at rest, but considerably more with activity[25].

Degree of work	*Oxygen requirement (l/min)*
Rest	0.3
Light	0.3–1
Moderate	1–1.5
Heavy	1.5–2
Very heavy	2–6

3.2.9.2 Response to oxygen deficiency

At oxygen concentration of 21–18%, the fit body tolerates exercise well. Below 18% the response depends upon the severity of work undertaken. Between 18 and 17% the body will probably not be adversely affected, unless the work undertaken is heavy, when there is likely to develop oxygen insufficiency which could lead to unconsciousness. Between 17 and 16% heavy work is not possible. Light activity will result in an increase in pulse and respiration rate in order to improve oxygen supply to the tissues.

In an environment in which the oxygen concentration is diminished it is the rate of its decline which influences body response. A sudden reduction in which the partial pressure of oxygen is inadequate for it to cross the lung–blood barrier, as might occur when the oxygen supply to an aviator at very high altitude is dramatically cut off, results in convulsions and unconsciousness within a minute and, unless promptly relieved, death. A gradual reduction in oxygen concentration may be unnoticed by the victim, there being at first a feeling of well-being and overconfidence. Then mistakes in thinking and action may occur until, at a level of 10% or lower, unconsciousness follows and, possibly, death. Should the oxygen level be restored and the individual recover, the incident might not be recalled and there could be a repetition of the mistakes as before[26]. Recovery may be complete, or there may be residual headache and weakness for some hours. The most sensitive tissues are the brain, heart and retina, which are liable to sustain damage.

3.2.10 Occupational cancer

Cancer is a disorder of cell growth. It begins as a rapid proliferation of cells to form the primary tumour (neoplasm) which is either benign or malignant. If benign it remains localised, but may produce effects by pressure on neighbouring tissue. A malignant tumour invades and destroys surrounding tissue and spreads via lymph and blood streams to distant body parts (metastasis) such as the lung, liver, bone or kidney (secondary tumours). The patient becomes weak, anaemic and loses weight (cachexia). Pneumonia is the commonest form of death. The incidence of cancer increases with age and is responsible for 24% of all deaths.

Cancer is caused either by the inheritance of an abnormal gene, or exposure to an environmental agent acting either directly or indirectly on the cell genes.

Of all cancers, less than 8% are occupational and due to chemical and physical agents (see *Table 3.2.1*). Occupational cancers tend to occur after a long latent period of some 10–40 years and at an earlier age than spontaneous cancers.

Some carcinogens act together (synergistically); an example is found in asbestos workers who smoke and are much more likely to develop cancer of the bronchus than those who do not.

Table 3.2.1 Table of some causes of occupational cancer in man

Agent	*Body site affected*	*Typical occupation*
Sunlight	Skin	Farmers and seamen
Asbestos	Lung, pleura, peritoneum	Demolition workers, miners
2-naphthylamine	Bladder	Dye manufacture, rubber workers
Polycyclic aromatic hydrocarbons	Skin, lung	Coal gas manufacture, workers exposed to tar
Hard wood dust	Nasal sinuses	Furniture manufacture
Leather dust	Nasal sinuses	Leather workers
Vinyl chloride monomer	Liver	PVC manufacture
Chromium fume	Lung	Chromate manufacture
Ionising radiations	Skin and bone marrow	Radiologists and radiographers

Identification of occupational cancer often depends in the first place on the observation of a cluster of cases, as occurred with cancer of the scrotum in chimney sweeps in 1755, skin cancer in arsenic workers in 1822 and cancer of the liver in PVC manufacture in 1930. Following observation of cases it is necessary to establish the potential link between cause and effect. This requires a descriptive study followed by a cohort or case control study.

Cancer may be suspected where the following are found:

1 Cluster of tumour in particular trades, i.e. chimney sweeps.
2 The chemical substance in use is listed in EH40[5] as 'may cause cancer'.
3 An Ames test proves positive.
4 Among heavy smokers in certain industries involving asbestos, chromate, nickel compounds, coke ovens, chloromethyl ether, uranium, arsenic trioxide etc.
5 The substance in use has a chemical structure that suggests carcinogenicity, e.g. aromatic amine.

Where carcinogenicity is suspected, the epidemiological tests outlined in section 3.2.2 should be carried out.

The classification of carcinogens is based on internationally agreed epidemiological and animal studies[27] and are:

Group 1 Carcinogenic to humans.
Group 2a Probably carcinogenic to humans with sufficient evidence from animal studies.
Group 2b Possibly carcinogenic to humans but absence of sufficient evidence from animal tests.
Group 3 Not classifiable as to its carcinogenicity to humans.
Group 4 No evidence of carcinogenicity in humans or animals.

Many chemical substances have been assigned the risk phrase 'R-45; may cause cancer' in EH40[5].

Although the total number of deaths from cancer in this country is rising there is no evidence that the increase is due to the effect of industrial chemicals. The two most important factors leading to this increase appear to be the ever increasing number of lung cancer deaths due to smoking and fewer deaths from other causes such as infection thus putting more people at risk of developing cancer who otherwise would have died from other causes[28].

3.2.10.1 Angiosarcoma

Angiosarcoma is a rare 'cancer' of the liver, known to be associated with vinyl chloride monomer and, more rarely, with thorium dioxide. Much more commonly, angiosarcoma has occurred without a recognised association with any chemical. Vinyl chloride monomer (VCM) can be polymerised to form polyvinyl chloride (PVC) and was first discovered in Germany in the 1930s[29]. In 1966 VCM was known to cause bone disease, affecting the hands of Belgian autoclave workers employed in the manufacture of PVC. When, in 1971, the chemical was given to animals to reproduce the bone disease, it was found instead to have carcinogenic properties.

3.2.10.2 Vinyl chloride monomer (VCM) ($CH_2 = CH$)

This gas is polymerised when heated under pressure (i.e. molecules of the gas are joined together in long chains) to form polyvinyl chloride (PVC). Although the explosive dangers of the gas have long been recognised, it was not until 1974 that three cases in American factory workers who were making PVC from VCM indicated that it could cause a rare liver tumour, angiosarcoma. Symptoms include abdominal pain, impaired appetite, loss of weight, distention of abdomen, jaundice and death. A Code of Practice[30] gives useful guidance on the control of this substance in the work environment.

3.2.11 Physical agents

In recent years there has been an increasing recognition of the harm that physical agents can do to the health of people at work. Injuries from this source now account for two-thirds of the new successful claims for industrial disease compensation.

3.2.11.1 Hand–arm vibration syndrome (HAVS)

HAVS follows from exposure to vibrations in the range 2–1500 Hz which causes narrowing in the blood vessels of the hand, damage to the nerves

Table 3.2.2 Stockholm scale for the classification of the hand–arm vibration syndrome

Stage	*Grade*	*Description*
1. Vascular component		
1	Mild	Occasional blanching attacks affecting tips of one or more fingers
2	Moderate	Occasional attacks distal and middle phalanges of one or more fingers
3	Severe	Frequent attacks affecting all phalanges of most fingers
4	Very severe	As in 3 with trophic skin changes (tips)
2. Sensorineural component		
0_{SN}	–	Vibration exposed. No symptoms
1_{SN}	–	Intermittent or persistent numbness with or without tingling
2_{SN}	–	As in 1_{SN} with reduced sensory perception
3_{SN}	–	As in 2_{SN} with reduced tactile discrimination and manipulative dexterity

The staging is made separately for each hand.

and muscle fibres and to bones and joint[31] evidenced by pain and stiffness in the joints of the upper arm. The impaired circulation of blood to the fingers leads to a condition known as *vibration white finger* (VWF). The most damaging frequency range is 5–350 Hz.

3.2.11.1.1 Vibration white finger

There is a latent period from first exposure to the onset of blanching which can vary from one to several years depending on the magnitude and frequency of the vibration and the length of exposure. Other symptoms of numbness and tingling, which variably affect the fingers extending from the tips; coldness, pain and loss of sensation may follow. Later, there may be loss of finger dexterity (e.g. picking up objects and fastening buttons) and impairment of grip. Eventually the finger tips become ulcerated and gangrenous. The vascular and nervous effects may develop independently but usually occur concurrently. Disability is graded in accordance with the Stockholm scale (see *Table 3.2.2*).

3.2.12 Ionising radiations

Ionising radiations are so called because they produce 'ions' in irradiated body tissue. They also produce 'free radicals' which are parts of the molecule, electrically neutral but very active.

The biological consequences of radiation depend on several factors:

1 The nature of the radiation – some radiations being more damaging than others. Alpha particles are not harmful until they enter the body

by inhalation, ingestion or via a wound. Beta particles can penetrate the skin to about 1 cm and cause a burn. X-rays, gamma rays and neutrons can pass right through the body and cause damage on the way.
2 The dose and duration of exposure.
3 The sensitivity of the tissue.
4 The extent of the radiation.
5 Whether it is external or internal.

3.2.12.1 Sensitivity of tissue

Tissues vary in their sensitivity to radiation, the most sensitive being the lymphocytes of the blood: they respond to excess radiation by a drop in their number within a couple of days, followed by a fall in other blood cells. Next in sensitivity are the cells of the gonads, the bowel lining, the skin, lung, liver, kidney, muscle and nerves.

3.2.12.2 Extent of radiation

Localised radiation is generally less immediately serious than whole body radiation for the same total dose.

3.2.12.3 Localised external radiation effects

Exposure to a small area of the body may result in redness of the skin, or even a blister, which either heals or ulcerates. The hands are very susceptible to localised radiation, the fingers becoming swollen and tender and, if the blood vessels are affected, gangrene could develop: the nails may become ridged and brittle. Exposure to the eyes in a dose of about 2 sievert may lead to cataract after a lapse of about two years. Exposure to the gonads can cause mutation and loss of fertility.

Injury with a threshold and dose related severity is termed *non-stochastic*; while injury with no threshold and of a random nature, as in neoplasm and DNA damage, is called *stochastic*.

3.2.12.4 Whole body external radiation effects

Dose Sv	*Effect*
Up to 0.25	Probably none. Lymphocyte count might fall in two days. Sperms and chromosomes may be damaged.
0.25–1.00	Damage more likely. Drop in total white cell count.
1.00–2.00	Nausea, vomiting, diarrhoea.
2.00–5.00	Above effects plus increasing mortality.
5.00–10.00	Rapid onset of above symptoms, shock and coma.

3.2.12.5 Acute radiation syndrome

A dose of some 2 sievert or more to the whole body may give rise to an 'acute radiation syndrome'. The response, depending on the intensity of the dose, begins with vomiting and diarrhoea within a few hours. By the second or third day there is an improvement, but the blood count falls. By the fifth day there is a return of symptoms, with fever and infection.

3.2.12.6 Internal radiation

These effects depend upon the nature of the radioactive material, its route of entry and concentration in a particular tissue, and due mainly to α or β particles. Lung cancer has been observed in miners following inhalation of radon, and severe anaemia and bone tumour following ingestion of radium in luminising dial painters.

3.2.12.7 Long-term effects

These may take several years to develop. Cancer of the skin or other organs has a peak incidence about seven years after exposure. The blood can be affected in two ways, either by leukaemia, which is a cancer of the white cells or, less commonly, by a severe anaemia in which the bone marrow fails to produce red cells. Chronic ulceration, loss of hair, cataracts, loss of fingertips, diminished fertility, and mutations may also occur. The maximum permitted doses for persons over 18 years of age are indicated in *Table 3.2.3*. Lower limits apply to trainees under the age of 18 and special limits apply to pregnant women and women of reproductive capacity.

3.2.12.8 Medical examinations

A pre-employment medical is required for employees likely to receive a dose of ionising radiation exceeding three-tenths of the relevant dose limit. The examination will include a test of blood.

Table 3.2.3 Radiation dose limit

Body part	*Dose limit per calendar year mSv*
Whole body	20
Individual organs and tissues	50
Lens of eye	15

A certificate issued by the examining Employment Medical Adviser or factory doctor will be valid for one year.

3.2.12.9 Principles of control

The following simple precautions should be adopted to reduce to a minimum hazards from the use of radioactive materials:

1 Employ the smallest possible source of radiation.
2 Ensure the greatest distance between source and person.
3 Provide adequate shielding between source and person.
4 Reduce exposure time to a minimum.
5 Practise good personal hygiene where there is risk of absorption of radioactive material.
6 Personal sampling by use of (a) film badge and/or (b) thermal luminescent dose meter.
7 A dose of 15 mSv whole body in a year requires investigation of work exposure and control procedures. A cumulative dose of 75 mSv within five years requires further investigation of work, personal circumstances, dose history and advice regarding further exposure to ionising radiations.

3.2.13 Noise-induced hearing loss

3.2.13.1 Mechanism of hearing

What we perceive as sound is a series of compressions and rarefactions transmitted by some vibrating source and propagated in waves through the air[32]. The compressions and rarefactions impinge on the eardrum (tympanic membrane) causing it to vibrate and transfer the movements through three small bones in the middle ear to the fluid of the inner ear. There they are received by rows of hairs (in the organ of corti), which vary in their response to different frequencies of sound, and are then transmitted to the brain and interpreted as sound.

3.2.13.2 Sensitivity of the ear

The ear can interpret frequencies between 20 and 20 000 Hz approximately. Frequencies below (infrasonic) and above (ultrasonic) this range are not heard. The range of frequency for speech is between 400 and 4000 Hz.

3.2.13.3 Definition and effects

Noise is commonly defined as unwanted sound. The definition is dependent on individual interpretation and may or may not include the recognition that some sounds produce harmful effects. Some 'sounds' cause annoyance, fright, or stress; others may interfere with communication. Loud sounds can cause deafness. 'Noise'-induced deafness is of two kinds: temporary and permanent.

3.2.13.4 Temporary deafness

Exposure to noise levels of about 90 dBA for even a few minutes may induce a temporary threshold shift (change of the threshold at which sound can just be heard), lasting from seconds to hours, and which can be detected by audiometry. Temporary threshold shift (TTS) may be accompanied by 'noises' in the ears (tinnitus) and may be a warning sign of susceptibility to permanent threshold shift (PTS) which is an irreversible deafness.

3.2.13.5 Permanent deafness

The onset of permanent deafness may be sudden, as with very loud explosive noises, or it may be gradual. A gradual onset of deafness is more usual in industry and may be imperceptible until familiar sounds are lost, or there is difficulty in comprehending speech. The consonants of speech are the first to be missed: f, p, t, s and k. These are of high frequency compared with the vowel sounds, which are of low frequency. Speech can still be heard, but without the consonants it is unintelligible. There is a risk too that a person exposed to excessive noise may believe himself to be adjusting to it when, in fact, partial deafness has already developed.

3.2.13.6 Limit of noise exposure

As noise effects are cumulative, the noise emission levels should be below 85 dBA. If this is not possible they should be reduced to the lowest level possible and suitable hearing protection provided. Ten years' exposure at 90 dBA ($L_{EP,d}$) can be expected to result in a 50 dB hearing loss in 50% of the exposed population.

If the noise energy is doubled, then it is increased by 3 dBA and requires a halving of the exposure time, e.g.

dBA	*Hours of exposure*
90	8
93	4
96	2
99	1
102	½
105	¼

The above table is helpful provided the noise level remains constant. For variable noise exposure, however, the daily personal noise exposure ($L_{EP.d}$) must be calculated.

Individuals exposed to 85 dBA must be offered hearing protection, but at 90 dBA or more hearing protection must be provided and worn.[32]

3.2.13.7 The audiogram

An audiogram (*Figure 3.2.2*) is a measure, over a range of frequencies, of the threshold of hearing at which sound can just be detected. Early deafness occurs in the frequency range 2–6 kHz and is shown typically as a dip in the audiogram at 4 kHz. The depth of the dip depends on the degree of hearing damage and, as this worsens, so the loss of hearing widens to include neighbouring frequencies. The advantages of an audiogram are that it:

1 provides a base line for future comparison;
2 is helpful in job placement; and
3 can be used to detect early changes in hearing and in the diagnosis of noise-induced deafness.

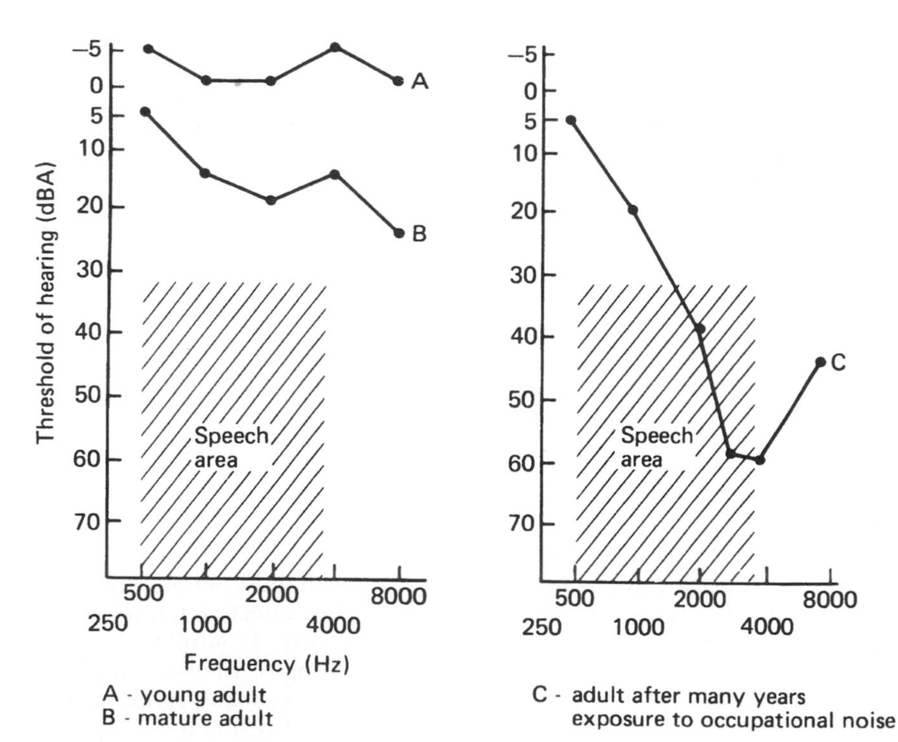

Figure 3.2.2 Audiograms

3.2.13.8 Occupational deafness

Disablement benefit may be awarded if deafness follows from:

1 employment in a prescribed occupation for at least 10 years, or an incident at work, and
2 the hearing loss is 50 dB averaged over the frequencies 1, 2 and 3 kHz in each ear.

3.2.14 Working in heat

Normally the human body maintains its core temperature within the range 36–37.4°C by balancing its heat gains and losses. Maintaining an employee's health in a hot environment requires the control of air temperature and humidity, body activities, type of clothing, exposure time and ability to sweat. To sweat freely the individual must be fit, acclimatised to the heat with sufficient water intake to ensure a urine output of about 2½ pints per day. When the air temperature reaches 35°C plus, the loss of body heat is by sweating only, but this may be difficult when humidity reaches 80% or more.

Body reactions to overheating are:

- An increase in pulse rate. The rate should fall by 10 beats/minute on cessation of exposure.
- Muscle cramp due to insufficient salt intake. Exhaustion with the individual feeling unwell and perhaps confused.
- Fainting and dizziness with pallor and sweating.
- Heat stroke is the most serious with the body temperature very high, the skin dry and flushed.
- Dehydration due to insufficient fluid intake. Prolonged dehydration may lead to the formation of stones in the kidney.

Following first aid care, the patient needs to be referred to a doctor.

3.2.15 Work related upper limb disorders (WRULD)[33]

WRULD covers a number of conditions variously known as telegraphist's, writer's or twister's cramp and tenosynovitis, all of which became prescribed diseases in 1948. Other common conditions include carpal tunnel syndrome, tennis and golfer's elbow.

The condition arises from frequent forceful repetitive arm movement. Early symptoms include aches and pain in the hands, wrists, forearm, elbows and shoulders with tenderness over the affected tendons and muscles. Following rest there is a quick recovery. If similar work is resumed too soon, there is likely to be a worsening to the second phase when, in addition to the symptoms, there appears a redness, swelling and marked limitation of movement. A longer period of rest is then required with possible splinting of the limb and injections of cortisone. If not treated in time the condition can become extremely disabling and may require surgical intervention. In the last few years incidents of WRULD have, numerically, exceeded that of any other group of diseases.

3.2.15.1 Prevention

Identify those jobs involving frequent prolonged rapid forceful movements, forceful gripping and twisting movements of the hand and arm, where the wrist is angled towards the little finger, the arm held above shoulder height or uncomfortably away from the body, and those where repetitive pushing, pulling and lifting are necessary.

Ensure hand tools are designed with good mechanical advantage and have a comfortable grip, are suitable for those who use them and that cutting edges are kept sharp.

Those involved in the work should be warned of the risks and trained in the correct use of the tools. Rest periods and work rotation should be introduced and piece work avoided.

Where the condition is suspected, complaints should be monitored and checks made of first aid records and absence certification. Susceptible persons should be examined by a doctor before further exposure.

3.2.16 Diseases due to micro-organisms

Micro-organisms include a variety of minute organisms such as viruses, bacteria, fungi and protozoa that can only be seen with the aid of a microscope. Micro-organisms gain entry to the body through the lungs, gut or breaks in the skin. If their virulence overcomes the body's defence, disease may result. The term pathogen covers all micro-organisms which cause disease. Diseases of animals transmitted to man are known collectively as zoonoses.

Micro-organisms account for about 10% of successful new occupational disease claims. Typical examples are:

Organism	*Disease*
Viruses:	hepatitis A, B and C, AIDS, orf
bacteria:	anthrax, legionella, leptospirosis, tuberculosis, tetanus, ornithosis, Q fever, dysentery
fungi:	Farmer's lung, ringworm, athlete's foot
protozoa:	malaria, amoebiasis
nematodes:	hookworm
BSE:	new variant C-JD

3.2.16.1 Hepatitis

Hepatitis is an inflammatory condition of the liver. When occurring at work it is usually caused by infections or toxic substances such as alcohol and organic solvents. Most commonly the cause is a virus of which there are three main kinds.

3.2.16.1.1 Infective hepatitis or 'Hepatitis A'

The virus is transmitted from infected stools to the mouth. After 2–6 weeks there occurs fever, nausea, abdominal pain and jaundice. Recovery usually occurs in 1–2 weeks and recurrence is rare.

Precautions to be taken include good personal hygiene, washing hands after the toilet and before handling food. There is a vaccine which gives long-term protection.

3.2.16.1.2 Serum hepatitis or 'Hepatitis B'

The virus is transmitted in infected blood or serum, especially among drug addicts who share needles; there is also a risk in renal dialysis units. The disease manifests itself after 2–6 months with symptoms similar to Hepatitis A but the effects are more prolonged and damaging.

Precautions that should be taken include the screening of donor blood for the presence of antigen and the non-reuse of needles and syringes. People at special risk are those who come into contact with blood or blood products and they should be immunised with Hepatitis B vaccine.

3.2.16.1.3 Hepatitis C

This is transmitted by the same route as Hepatitis B and poses an occupational hazard to a similar group of workers. There is no vaccine available.

3.2.16.1.4 AIDS

AIDS (acquired immune deficiency syndrome) is a breakdown in the body's immune system that can be suffered by both sexes. It can be transmitted from person to person in body fluids during sexual intercourse and infection can occur in transfusion of blood products, donated organs, by mother to child during childbirth or breast feeding and through the use of infected needles used for injections.

Within 12 weeks of infection, antibodies are found in the blood when the individual has become HIV+ve (human immunodeficiency virus positive). The virus can affect a variety of body tissues, particularly a white blood cell known as T4 which plays an important role in providing immunity. Individuals suffer in a variety of ways, from developing painless swellings of the glands in the neck and armpits to acute infections like 'flu from which recovery is usual. Other suffers experience night sweats, loss of weight, diarrhoea and fatigue. A more serious feature is an opportunistic infection whereby, following a failing immunity and fall in T4 count, the body becomes prone to a variety of bacteria, viruses and fungi including tuberculosis and pneumocystis. Diagnosis is difficult since the infecting organisms resist the usual investigative treatments and patients suffer a reduced life span.

3.2.16.2 Leptospirosis

Leptospirosis is caused by bacteria from the urine of infected rats (Weil's disease); dogs (*L. canicola*); cattle (*L. hardjo*) which enter via a break in the skin or into the gut if ingested. A week later a 'flu-like disease occurs sometimes with jaundice. The liver and kidneys may be severely damaged with a 15% mortality. Those most at risk include workers in abattoirs, sewers, mines, tunnels, canals, veterinary workers and those taking part in inland water sports. They should be informed of the risk and issued with a Weil's disease warning card to be presented to their doctor.

3.2.16.3 Legionnaire's disease

In 1976 nearly 200 American Legionnaires attending a convention at a hotel in Philadelphia collapsed with a 'flu-like disease, some with pneumonia, and there were 29 deaths. The disease was later attributed to bacteria, named *Legionella*, of which there are several types differing in pathogenicity.

Legionella bacteria occur in soils, rivers and streams, but the recent dramatic appearance of the disease in hotels, hospitals and industry is related to modern building design which allows water in air-conditioning and water systems to stagnate. At temperatures of 20–50°C bacterial growth is encouraged, which if released as a spray and inhaled, leads to pneumonia. Middle-aged smokers are most vulnerable. The disease has a 15% mortality. Diagnosis is confirmed by the presence of antibodies in the blood.

Water systems and bath shower heads should be cleaned and chlorinated periodically and a record maintained.

Pontiac fever is a less serious non-pneumonic form of the disease.

3.2.16.4 Anthrax

Anthrax is a highly infectious disease of ruminants: goats, cattle, sheep and horses, and is due to a bacillus. Man can be affected by direct contact with the animal, or indirectly by contact with the animal products. The disease is rare among animals in Britain, but it can be introduced into the country by infected materials, such as hides and skin, hair and wool, dried bones and bone-meal, hooves and horn. At risk particularly are those engaged in tanning, wool sorting, manufacture of brushes, bone-meal, fertiliser and glue. Also at risk are dockers and agricultural workers.

3.2.16.4.1 Symptoms

The disease involves the skin in about 95% of cases – the bacillus entering through an abrasion, commonly on the arm. In 2–5 days there appears a red-brown spot or pimple, which becomes a black ulcer surrounded by tiny blisters and inflamed tissue. It is usually painless, but the individual feels unwell with fever, headache, sickness and swollen glands, usually under the arm or in the groin if the leg is infected. Should the bacilli be inhaled, there follows a severe pneumonia with cough and blood-stained sputum. The mortality rate is high. Abdominal infection following ingestion is very rare. Fortunately, the disease responds to an antibiotic like penicillin if given early.

3.2.16.4.2 Prevention

Employees in the risk industries should be informed of the danger and carry with them an HSE card MS(B)3. All cuts must be treated and covered with a dressing while at work. Attention must be given to personal hygiene, washing hands, arms and face before meals and at end of shift. Protection can be provided by immunisation, which requires three injections at three-week intervals, a fourth six months later and then annually. Protective gloves should be worn wherever possible.

3.2.16.5 Humidifier fever

Humidifier fever is a 'flu-like condition which follows the inhalation of a variety of organisms such as amoeba, bacilli and fungi that grow in humidifying system. Symptoms of cough, limb pain and fever occur within a few hours of starting work. The disease is usually short term with recovery by the next day, but symptoms are likely to recur on returning to work after a few days off. It is sometimes known as Monday Morning Fever.

Water systems need to be cleaned and chlorinated periodically and a record of this maintained.

3.2.16.6 Tuberculosis

The incidence of tuberculosis is increasing in some communities owing to resistance to the drugs used in its treatment and to lowered resistance in AIDS patients. Infection is by inhalation of bacteria.

Mainly a disease of the lungs, its symptoms are a persistent cough, bloody sputum, night sweating and loss of weight. Sometimes no symptoms occur and the disease is first discovered on chest X-ray. Those most at risk are medical, veterinary and mortuary staff.

3.2.16.7 Other diseases of micro-organisms

This category has now been widened to include any infection reliably attributable to:

- work with micro-organisms
- provision of treatment for humans

- investigations involving exposure to blood and body fluids
- work with animals or any potentially infected materials so derived.

A list of the relevant conditions is given in schedule 3 of RIDDOR[34].

3.2.17 Psycho-social disorders

This group is probably the largest group of occupational diseases. It stems from the complex interaction of individual, social and work factors and is responsible for a great amount of sickness absence.

3.2.17.1 Stress

Stress is a reaction of the body to external stimuli ranging from the apparently normal to the overtly ill health. It varies with the individual personality but is one of the commonest occupational diseases.

The initial response is physiological and shows as an increase in pulse, blood pressure and respiratory rates. Although the body adjusts, persisting stimuli cause fatigue and the display of signs of 'overstress' with sweating, anxiety, tremors and dry mouth. There is difficulty in relaxing, a loss of concentration, appetite is impaired and sleep disturbed. Some may eventually become depressed, aggressive and try to avoid the cause through absenteeism or the use of alcohol or drugs. Other diseases may appear affecting the skin, peptic ulcer and coronary heart disease.

Studies have identified two personality groups: type A people who are competitive, impatient achievers and who are at greatest risk from the severe effects of stress, whereas type B people are easy-going, patient and less susceptible to pressure. Causes of stress may be considered under a number of headings:

The person – lack of physical and mental fitness to do the job; inadequate training or skill for the particular job; poor reward and prospects; financial difficulties; fear of redundancy; lack of security in the job; home and family problems; long commuting distances.

Work demand – long hours; shift work; too fast or too slow a pace; boring repetitive work; isolation; no scope for initiative or responsibility.

Environment – noise; heat; humidity; fumes; dust; poor ventilation; diminished oxygen; confined space; heights; poor house-keeping; bad ergonomic design.

Organisation – poor industrial relations, welfare services and communications; inconsiderate supervision; remote management.

Common occupational causes of stress are sustained uncertainty, frustration and conflict[35]. Stress has been given a 'social rating scale'[36]

death of a spouse	100
divorce	73
marital separation	65
injury/disease	53
dismissal	47
financial difficulties	38
work responsibilities	29

3.2.18 Target organs

Target organs are those body parts which sustain some adverse effect when exposed to or contaminated by harmful substances or agents.

Many of the body target organs have been referred to in the text and the table below gives a summary of them with causes.

Body part	*Condition*	*Cause*
HANDS/ ARMS	Vibration white finger:	Use of vibratory tools
	Carpal tunnel syndrome:	Use of vibratory tools
	Tenosynovitis:	Repetitive pulling and twisting actions with forceful movements
	Dermatitis:	Exposure to irritants and sensitisers
LUNGS	Pneumoconiosis:	Mineral dust: coal, silica, asbestos, iron, tin, barium; organic dusts
	Extrinsic allergic-alveolitis:	Organic dust
	Asthma:	Proteins and low molecular weight chemicals in toxic dosages
	Irritation/ inflammation:	Nitrous fumes, phosgene, chlorine, hydrogen sulphide, sulphur dioxide, ammonia
	Infection:	Legionella, tuberculosis
	Cancer:	Asbestos, radon, nickel
SKIN	Dermatitis:	Solvents, acids/alkalis, mercury, chrome, nickel, arsenic, mineral oils, wood, plants, resin, heat
	Cancer:	Aromatic polycyclic hydrocarbons, arsenic, UV light, ionising radiations
HEAD: EARS	Deafness:	Noise
EYES	Cataracts:	Ionising radiation, UV light, heat, acids/alkalis, arc flash
	Corneal ulcers:	Ionising radiation, UV light, heat, acids/alkalis, arc flash

NOSE	Ulceration:	Chrome
TEETH	Loosening:	Mercury
	Erosion:	Sulphuric acid
	Mottling:	Fluorides
	Discoloration:	Vanadium, iodine, bromine
BRAIN	Narcosis:	Organic solvents
	Encephalopathy:	Mercury, lead, manganese, carbon disulphide, carbon monoxide
PERIPHERAL NERVES	Neuropathy:	Lead, mercury, carbon disulphide, tetrachloroethane, trichloroethylene, organo-phosphorus compounds, vibration
CARDIOVASCULAR	Anaemia:	Lead, arsine
	White cell count changes:	Benzene, carbon tetrachloride, ionising radiations
LIVER	Hepatitis:	Organic solvents, viruses A,B,C leptospirosis, arsenic, manganese, beryllium
	Cancer:	Hepatitis B and C, vinyl chloride monomer
KIDNEY	Toxicity:	Organic solvents, lead, mercury, cadmium
	Infection:	Micro-organisms
BLADDER	Cancer:	2-naphthylamine
BONE	Osteolysis:	Vinyl chloride monomer, vibration
	Necrosis:	Work in pressurised areas

References

1. William, P.L. and Burson, B.L. (eds) *General Principles of Toxicology, chapter 2: Industrial Toxicology,* Van Nostrand Reinhold, New York (1985)
2. Pascoe, D. *Studies in Biology, No. 149,* Edward Arnold, London (1983)
3. Harrington, J. M. and Gill, F. S., *Occupational Health Pocket Consultant,* Blackwell Scientific Publications, Oxford (1998)
4. Health and Safety Executive, *Biological monitoring in the workplace,* 2nd edn, HSE Books, Sudbury (1997)
5. Health and Safety Executive, *Guidance Note EH40, Occupational Exposure Limits,* HSE Books, Sudbury (latest issue)
6. Health and Safety Executive, Guidance Note, Environmental Health Series No. EH 64 *Occupational exposure limits Summary criteria for 1996,* HSE Books, Sudbury (1999)
7. Fregert, S., *Manual of Contact Dermatitis,* A. Munksguard (1974)
8. Bishop, C. and Kipling, M. D., Dr J. Ayston Paris and Cancer of the Scrotum, 'Honour the Physician with the Honour due unto Him', *J. Soc. Occup. Med.,* **28**, 3–5 (1978)

9. Hunter, D., *The Diseases of Occupations*, 8th edn, Hodder & Stoughton, London (1994)
10. Hodgeson, G. A. and Whiteley, H. J., Distribution of pitch warts – personal susceptibility to pitch, *Brit. J. Ind. Med.*, **27**, 160–166 (1970)
11. Ref. 10, p. 20
12. Parkes, W. R., *Occupational Lung Disorders*, (3rd Edn), Butterworths, London (1990)
13. Ref. 12, p. 231
14. *Social Security (Industrial Injuries) (Prescribed Diseases) Regulations 1985*, The Stationery Office, London (1985)
15. Alexander, W. S. and Street, A., *Metals in the Service of Man*, 6th edn, 30, Penguin Books, London (1994)
16. Waldron, H. A., Health care of people at work – workers exposed to lead, inorganic lead, *J. Soc. Occup. Med.*, **28**, 27–32 (1978)
17. Clarkson, T. W., *Mercury Poisoning Clinical Chemistry and Chemical Toxicology of Metals*, 189–204, Elsevier, Amsterdam (1977)
18. Health and Safety Executive, Mercury – medical guidance notes (Rev), Guidance Notes MS 12, HSE Books, Sudbury (1996)
19. Uvaroy, E. B., Chapman, D. R. and Isaacs, A., *Dictionary of Science*, 7th edn, Penguin Books, London (1993)
20. Matheson, D., *Occupational Health and Safety*, 2085, ILO, Geneva (1983)
21. Bird, M. G., Industrial solvents: some factors affecting their passage into and through the skin, *Annals of Occupational Hygiene*, **24**, No. 2 (1981)
22. Gompertz, D., Solvents: the relationship between biological monitoring stratagem and metabolic handling. A review, *Annals of Occupational Hygiene*, **23**, No. 4 (1980)
23. Commission of the European Communities, *Regulation No. 594/91 on banning the use of methyl chloroform*, EC Publications Department, Luxembourg and The Stationery Office, London (1991)
24. Green, J. H., *An Introduction to Human Physiology*, 78–79, 3rd edn, Oxford University Press (1974)
25. Lamphier, E. H., *The Physiology and Medicine of Diving*, 59–60, Bailliere, Tindall & Cassell, London (1969)
26. Miles, S. and Mackay, D. E., *Underwater Medicine*, 4th edn, 107–108, Adlard Coles Ltd, St. Albans (1976)
27. International Agency for Research on Cancer, *Monograph on the evaluation of carcinogenic risks to humans*, **46**, World Health Organisation, Geneva (1989)
28. Editorial: What proportion of cancers are related to occupation? *Lancet*, 1238, Dec. 9 (1978)
29. Gauvain, S., Vinyl chloride, *Proc. Royal Soc. Med.*, 69 (1976)
30. Health and Safety Executive, Legal guidance booklet no. L67, *Control of vinyl chloride at work (1994 edition) Control of Substances Hazardous to Health Regulations 1994 Approved Code of Practice*, HSE Books, Sudbury (1995)
31. Royal College of Physicians, *Hand Transmitted Vibrations*, 2 vols, Royal College of Physicians, London (1993)
32. Health and Safety Executive, Booklet No. L108, *Reducing noise at work. Guidance on the Noise at Work Regulations 1989*, HSE Books, Sudbury (2002)
33. Health and Safety Executive, Health and Safety Guidance Booklet No. HSG60, *Upper limb disorders in the workplace: a guide to prevention*, HSE Books, Sudbury (2002)
34. Health and Safety Commission, *The Reporting of Injuries, Diseases and Dangerous Occurrences Regulations 1995*, The Stationery Office, London (1995)
35. Gross, R.D., *Psychology, the science of mind and behaviour*, 2nd edn, Hodder and Stoughton Ltd, London (1995)
36. Holmes, T.H. and Rahe, R.H., *Journal of Psychosomatic Research*, 213–218 (1967)

Further reading

National Radiological Protection Board, *Living with Radiation*, 4th edn, National Radiological Protection Board, Didcot (1989)

British Medical Association, *The BMA Guide to Living with Risk*, Penguin Books, London (1990)

Olsen, J., Merletti, F., Snashall, D. and Vuylsteek, K., *Searching for Causes of Work related Diseases*, Oxford Medical Publications, Oxford (1991)

Rose, G. and Barker, D.J.P., *Epidemiology for the Uninitiated*, British Medical Association, London (1979)
James, R.C., Industrial toxicology, Chapter 2 in *General Principles of Toxicology*, William, P.L. and Burson, B.L. eds, Van Nostrand Reinhold, New York (1985)
Pascoe, D., *Studies in Biology No. 149*, Edward Arnold, London (1983).

Chapter 3.3

Occupational hygiene

Dr C. Hartley

Occupational hygiene is defined by the British Occupational Hygiene Society as: 'the applied science concerned with the identification, measurement, appraisal of risk, and control to acceptable standards, of physical, chemical and biological factors arising in or from the workplace which may affect the health or well-being of those at work or in the community'.

It is thus primarily concerned with the identification of health hazards and the assessment of risks with the crucial purpose of preventing or controlling those risks to tolerable levels. This relates both to the people within workplaces and those who might be affected in the surrounding local environment.

Occupational hygiene deals not only with overt threats to health but also in a positive sense with the achievement of optimal 'comfort conditions' for workers, i.e. the reduction of discomfort factors which may cause irritation, loss of concentration, impaired work efficiency and general decreased quality of life.

The American Industrial Hygiene Association in its corresponding definition begins: 'Industrial hygiene is that science and *art* devoted to the recognition, evaluation and control . . .' [author's emphasis] indicating that although much of occupational and industrial hygiene is underpinned by proven scientific theory, a considerable amount relies on 'rule of thumb'; thus in the practical application of occupational hygiene, judgemental and other skills developed by the experienced practitioner are important.

3.3.1 Recognition

People at work encounter four basic classes of health hazard, examples of which are given in *Table 3.3.1*.

Table 3.3.1 Classes of health hazard

Health hazard	*Example*
Chemical	Exposure of worker to dusts, vapours, fumes, gases, mists etc. 100 000 chemicals are believed to be in common use in the UK at present
Physical	Noise, vibration, heat, light, ionising radiation, pressure, ultraviolet light etc.
Biological	Insects, mites, yeasts, hormones, bacteria, viruses, proteolytic enzymes
Ergonomic/ Psychosocial	Personal-task interaction, e.g. body position in relation to use of machine; harmful repetitive work. Exposure to harmful psychological stress at work

3.3.2 Evaluation

When a hazard in the workplace has been identified it is necessary to assess the consequent risk, interpret this against a risk tolerability standard and where appropriate apply further prevention and control measures.

3.3.2.1 Environmental measurement techniques

Some common environmental measurement techniques together with their interpretation as they relate to accepted standards are reviewed.

3.3.2.1.1 Grab sampling

'Grab sampling' is described here in the context of stain detector tubes. This involves taking a sample of air over a relatively short period of time, usually a few minutes, in order to measure the concentration of a contaminant. The results are illustrated in *Figure 3.3.1* where the discrete measured concentrations are plotted.

Stain detector tubes are used in this way to measure airborne concentrations of gases and vapours. Several proprietary types are available which operate on a common principle. A sealed glass tube is packed with a particular chemical which reacts with the air contaminant. The tube seal is broken, a hand pump attached, and a standard volume of contaminated air is drawn through the tube (*Figure 3.3.2*). The packed chemical undergoes a colour change which passes along the tube in the direction of airflow. The tube is calibrated so that the extent of colour change indicates the concentration of contaminant sampled (*Figure 3.3.3*).

The hand pump must be kept in good repair and recalibrated at intervals to check that it is drawing the standard volume of air and care

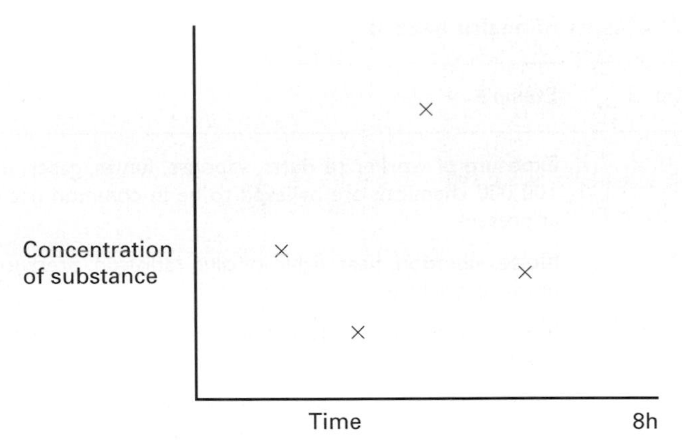

Figure 3.3.1 Results from grab sampling

taken to ensure that a good seal is obtained between the pump and tube.

This method of measurement has several advantages:

1 It is a quick, simple and versatile technique.
2 Stain detector tubes are available for a wide range of chemical contaminants.
3 Measurement results are provided instantaneously.
4 It is a relatively economical method of measurement.

However, it is important to be aware of the limitations of stain tubes:

1 The result obtained relates to the concentration of contaminant at the tubes inlet at the precise moment the air is drawn in. (This can be seen from *Figure 3.3.1.*) Short-term stain tubes do not measure individual worker exposure.
2 Variations in contaminant levels throughout the work period or work cycle are difficult to monitor by this technique.
3 Cross-sensitivity may be a problem since other chemicals will sometimes interfere with a stain tube reaction. For example, the presence of xylene will interfere with stain tubes calibrated for toluene. Manufacturers' handbooks give information on the known cross-sensitivities of their products. This point emphasises the need to consider the whole range of chemicals used in a process rather than just the major ones.
4 Stain tubes are not reusable.
5 Random errors associated with this technique can range up to ±25% depending on tube type.

Using a planned sampling strategy rather than an occasional tube will give a better picture of toxic contaminant levels and manufacturers will

Figure 3.3.2 Stain detector tube and hand pump. (Courtesy Draeger Safety Ltd)

give guidance about this. However, since tubes are not reusable this method could be more costly than some of the long-term sampling techniques.

Some basic practical guidance with regard to toxic substance monitoring is given in an HSE publication[1].

There are other approaches to 'grab sampling' which involve collecting a sample of the workplace atmosphere in a suitable container, for example

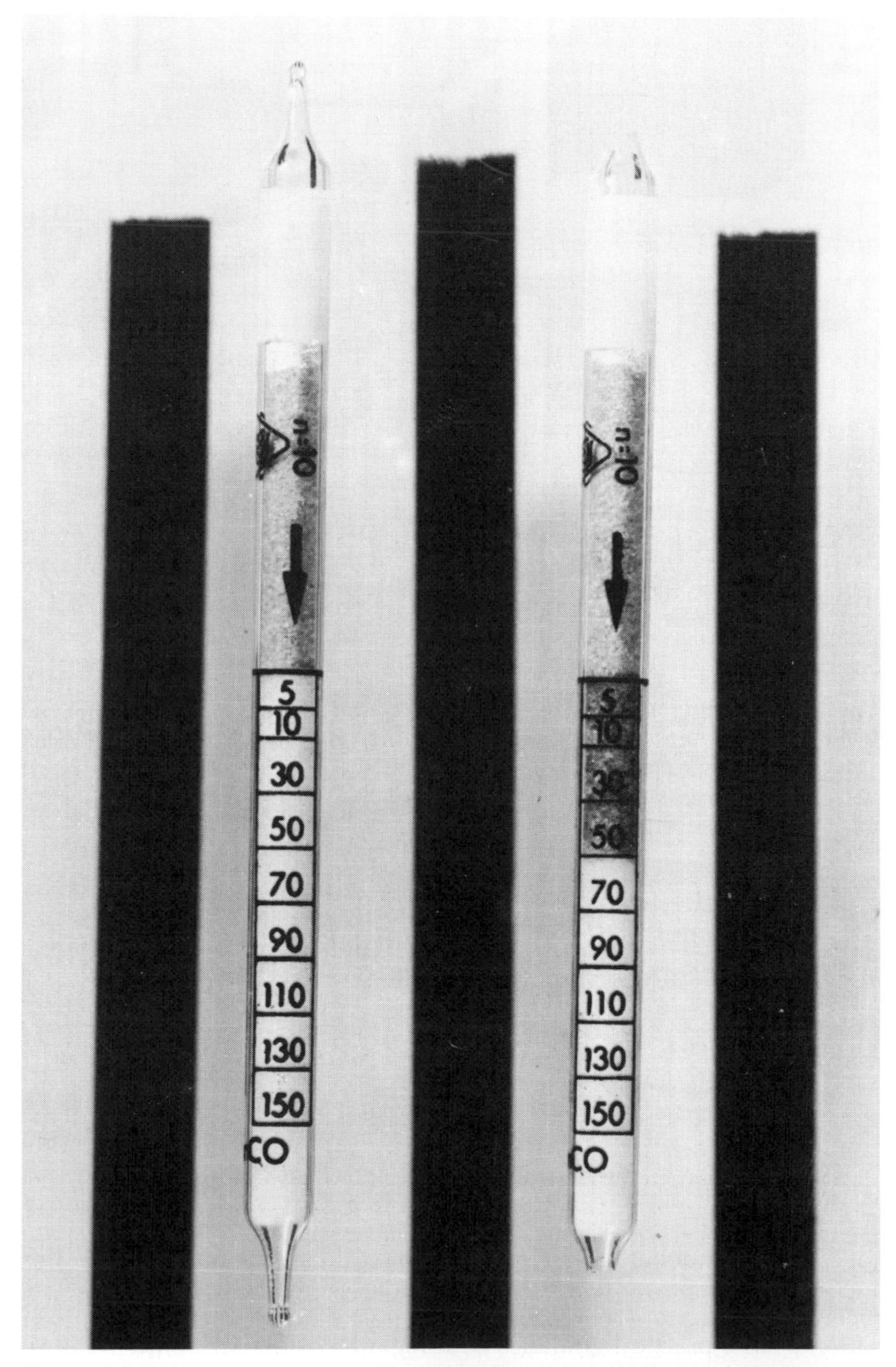

Figure 3.3.3 Stain detector tubes, illustrating the principle of detection. (Courtesy Draeger Safety Ltd)

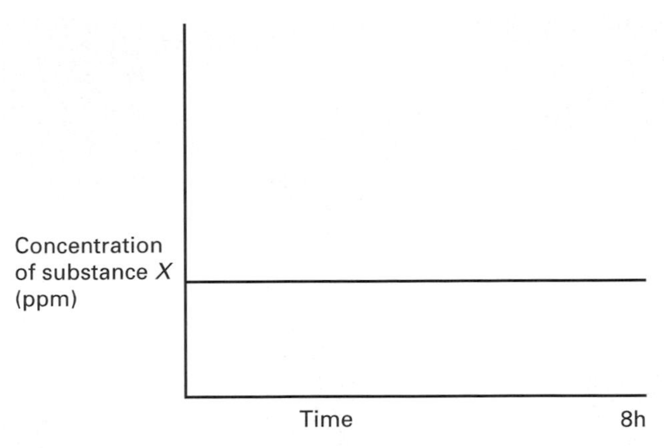

Figure 3.3.4 Results from long-term sampling

a special gas bag, which is taken to a laboratory for detailed chemical analysis.

3.3.2.1.2 Long-term sampling

This involves sampling air for several hours or even the whole work shift. The air sampling may be carried out in the worker's breathing zone (personal sampling) or at a selected point or points in the workplace (static or area sampling). Results give the average levels of contaminant across the sample period. The results are illustrated in *Figure 3.3.4* where the average concentration measured over the (8-hour) sampling period is X ppm.

This long-term sampling approach is discussed below for the monitoring of gas and vapour contaminants and dust and fibre aerosols. Long-term sampling methods are generally reliable, versatile and accurate, being widely used by occupational hygienists in checking compliance with hygiene standards.

3.3.2.1.2.1 Gases and vapours

Long-term stain detector tubes are available for this purpose and are connected to a pump which draws air through the tube at a predetermined constant rate. At the end of the sampling period the tube is examined to measure the extent of staining and hence the amount of contaminant absorbed, from which the average level of contamination can be calculated.

For some substances, direct indicating diffusion tubes are available and these devices may be used for up to 8 hours. In this case the contaminant is collected by diffusion and no sampling pump is required – with obvious cost benefits.

Often more accurate methods are required in the assessment of worker exposure when absorbent (often charcoal) sampling is commonly used.

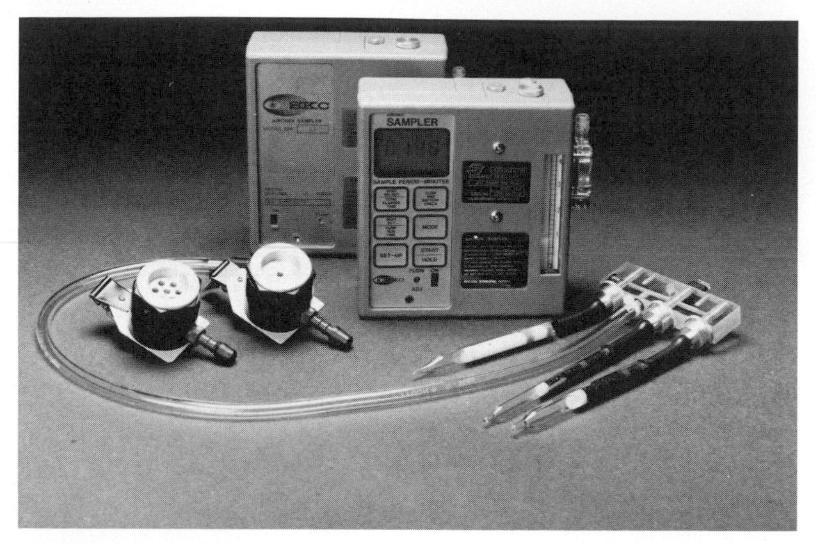

Figure 3.3.5 Sampling pumps suitable for both dust and vapour monitoring with collection heads. (Courtesy SKC Ltd)

Air is drawn by an attached low-flow portable pump through a tube containing an adsorbent (*Figure 3.3.5*). Further, diffusive monitors or badges containing various adsorbents are now becoming much more widely used (*Figure 3.3.6*). These are easy to use since they do not require a sampling pump with the result that the sampling can be carried out by suitably trained non-specialists.

Where charcoal is used, the chemical contaminant is trapped by the adsorbent granules either in the tube or the diffusive monitor. After sampling is completed the tube or diffusive monitor is sealed and taken to the laboratory where the contaminant is removed (desorbed) by chemical or physical means (i.e. heat) followed by quantitative analysis, often using gas chromatography to determine the weight of contaminant collected during sampling.

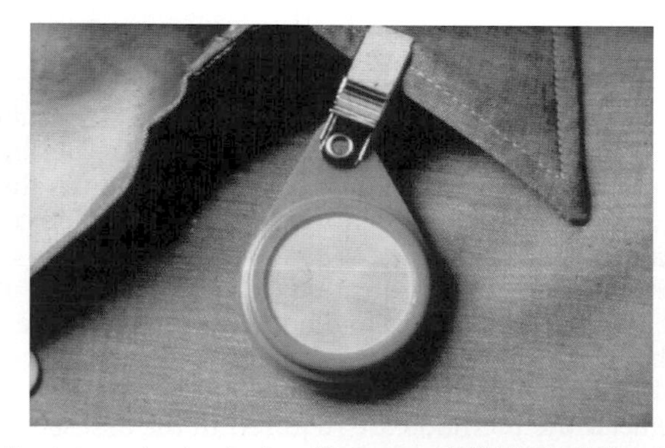

Figure 3.3.6 Passive monitoring badge. (Courtesy of 3M United Kingdom plc)

3.3.2.1.2.1.1 Gas chromatography

Gas chromatography is a common and versatile analytical technique which allows both the identification and quantification of individual components of a mixture of substances. Separation of mixtures occurs after injecting a vaporised sample into a gas stream which then passes through a separating column. Components of the mixture pass through the column at different rates, depending on their individual physical properties such as boiling point and solubility, and emerge sequentially (i.e. at different times) into a detector which allows identification and quantification of each component.

This technique allows determination of the weight of contaminant collected in the tube or on the diffusive monitor. When this weight is related to the known total air volume sampled (from known pump flow rate or calculated flow rate for the passive monitor and sampling time), the average airborne concentration of the substance over the sample period can be calculated.

3.3.2.1.2.2 Dusts and fibres

In the measurement of airborne dusts the collection device is a filter which is positioned in a holder attached to a medium flow pump. After sampling, quantification of the contaminant on the filter may be by gravimetric methods (measuring the weight change of the filter) or by using techniques such as atomic absorption spectrometry which allows the measurement of the quantity of a specific contaminant in the sample (e.g. lead). The precise analytical technique used will depend upon the part of the dust that is of interest. Also, the type of sampling device, which holds the filter, and the filter itself will depend on the nature of the dust and whether the aim is to collect the 'total inhalable' dust fraction or just the 'respirable' dust fraction.

3.3.2.1.2.2.1 Atomic absorption spectrometry

Atomic absorption spectrometry is a commonly used technique for the quantitative determination of metals. Absorption of electromagnetic radiation in the visible and ultraviolet region of the spectrum by atoms results in changes in their electronic structure. This is observed by passing radiation characteristics of a particular element through an atomic vapour of the sample – the sample is vaporised by aspiration into a flame or by contact with an electrically heated surface. The absorbed radiation excites electrons and the degree of absorption is a quantitative measure of the concentration of 'ground-state' atoms in the vapour. With a calibrated instrument it is possible to determine the weight of the metal captured on a filter. By knowing the total volume of air sampled, the average concentration of the metal can be calculated for the sampling period.

3.2.1.2.3 Fibre monitoring

In the measurement of airborne asbestos fibres the collection device is a membrane filter with an imprinted grid. This is positioned in an open-faced filter holder, fitted with an electrically conducting cowl and

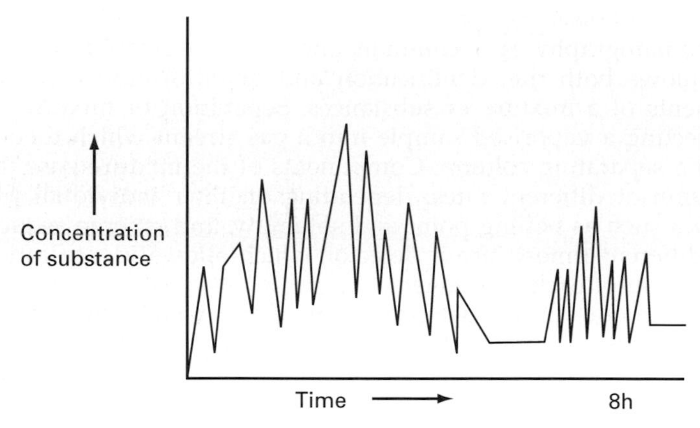

Figure 3.3.7 Results from direct monitoring

attached to a medium-flow pump. After sample collection the filter membrane is cleared, i.e. made transparent, and mounted on a microscope slide.

The airborne concentration of 'countable' fibres can be measured using phase contrast microscopy. 'Countable' fibres are defined as particles with length > 5 μm, width < 3 μm and a length:width ratio >3:1. Fibres having a width of <0.2 μm may not be visible by the phase contrast method so this technique measures only a proportion of the total number of fibres present. However, the specified control limits take this into account.

The method requires that all fibres meeting this size definition be counted. A known proportion of the total exposed filter area is scanned using a phase contrast microscope and a tally made of *all* countable fibres in the examined area of the filter. Fibres are counted using a 'graticule' calibrated eyepiece mounted in the objective lens of the microscope. The graticule defines a known area (field) of the filter in which the fibres are counted. Knowledge of the air sample volume together with the calculated total number of fibres on the whole filter allows calculation of the average fibre concentration over the sample period.

3.3.2.1.3 Direct monitoring instruments

A wide range of instruments are used in the detection of gases, vapours and dusts. These devices make a quantitative analysis giving a real-time display of contaminant level on a meter, chart recorder, data logger or other display equipment. There is a wide variety of commercially available direct monitoring instruments which are based on different physical principles of detection. Many of the physical principles involved and the range of instruments available are considered by Ashton and Gill[2]. A trace from a direct monitoring instrument is illustrated in *Figure 3.3.7*. From such information it is possible to work out the time weighted average concentrations during the sampling period.

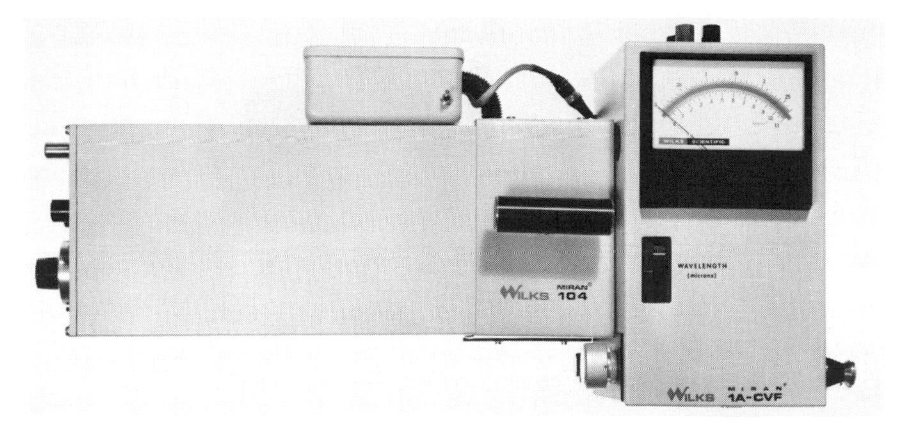

Figure 3.3.8 Portable infrared gas analyser. (Courtesy Foxboro Analytical Ltd)

Direct monitoring is particularly useful where there is a need to have immediate readings of contaminant levels, for example in the case of fast acting chemicals. It is also useful for identifying periods of peak concentration during the work cycle or work shift so that a control strategy can be developed and, for example, the point at which to install a local ventilation system can be determined. This approach is considered below.

3.3.2.1.3.1 Gases and vapours

Many direct reading instruments are available which are specific for particular gases, for example carbon monoxide, nitrogen oxides, hydrogen sulphide and mercury vapour. These can be linked to a chart recorder, data logger or warning device. A portable infrared gas analyser which allows direct monitoring of gases and vapours in the workplace is shown in *Figure 3.3.8*. In infared spectrometry the basic principle utilised is that many gases and vapours will absorb infrared radiation and under standard conditions the amount of absorption is directly proportional to the concentration of the chemical contaminant. This instrument takes a sample of air, detects the extent of interruption, i.e. absorption, of an infrared beam and gives a display of the contaminant concentration. Because so many substances absorb infrared radiation this is an extremely versatile instrument.

The organic vapour analyser featured in *Figure 3.3.9* is a portable, intrinsically safe, direct reading instrument which will monitor total hydrocarbon concentration. It will also operate as a gas chromatograph and will take a sample from the sample stream and carry out identification of mixtures and quantification of component concentrations.

3.3.2.1.3.2 Dusts and fibres

A 'dust lamp' is a very useful and versatile direct reading instrument. Many particles of dust are too small to be seen by the naked eye under

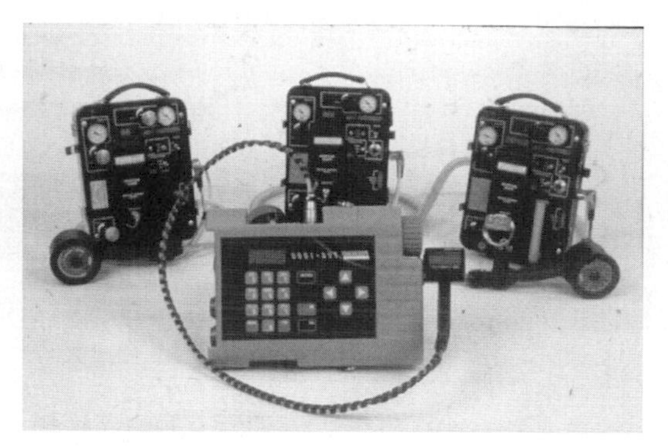

Figure 3.3.9 Organic vapour analyser. (Courtesy Quantitech Ltd)

normal lighting conditions but when a beam of strong light is passed through a cloud of dust the particles reflect the light to the observer – known as forward light scattering – and as a result the particles are readily visible. A natural occurrence of this phenomenon is observed when a shaft of sunlight shines into a dark building highlighting the airborne particles.

Thus, if a portable lamp having a strong parallel beam is set up to shine through a dusty environment, the movement of the particles can be observed. Although this is not a quantitative method, the behaviour of the dust emitted by the work processes can be observed and corrective measures taken. It may be useful to photograph or make a video record of the observation.

Another direct reading dust monitor is the laser nephelometer (*Figure 3.3.10*). Ambient air is drawn continuously through a sensing chamber and illuminated by a laser. Dust particles are detected by the scattering of the laser light. This type of instrument can be used to assess PM_{10} and $PM_{2.5}$ concentrations in ambient air and the thoracic and respirable dust fractions in workplace atmospheres. It can be calibrated for different dusts and will monitor respirable particulate. It is a horizontal elutriator and detects dust particles by light scattering.

Note, a PM_{10} sampling head allows the mass concentration of particles generally less than 10 μm in diameter to be measured, while a $PM_{2.5}$ sampler will allow measurement of particles less than 2.5 μm in diameter.

3.3.2.1.4 Oxygen analysers

Deficiency of oxygen in the atmosphere of confined spaces is often experienced in industry, for example inside large fuel storage tanks when empty. Before such places may be entered to carry out inspections or maintenance work a check must be made on the oxygen content of the atmosphere throughout the vessel. Normal air contains approximately 21% oxygen and when this is reduced to 16% or below people experience dizziness, increased heartbeat and headaches. Such atmospheres should only be entered when wearing air supplied breathing apparatus.

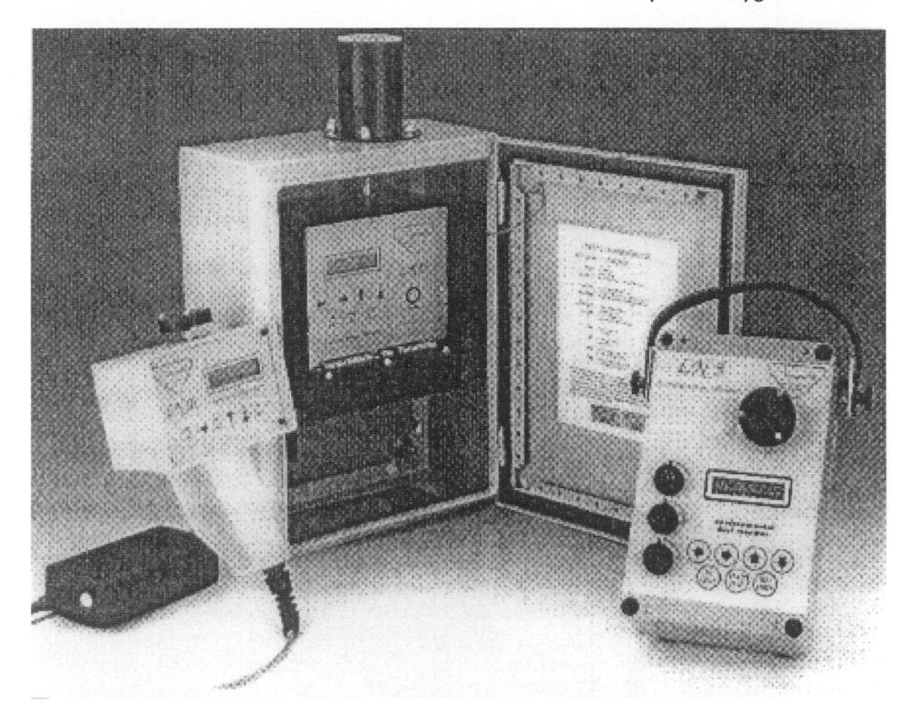

Figure 3.3.10 Laser nephelometer – direct reading dust monitoring instrument. (Courtesy Negretti Automation Ltd)

Portable analysers are available which measure the concentration of oxygen in the air by the depolarisation produced at a sensitive electrode mounted in the instrument (*Figure 3.3.11*). Several different devices are available which vary in sensitivity, reliability and ease of maintenance but they must all be checked and carefully calibrated to the manufacturers' instructions before use. Long extension probes may be attached which allow remote inspection of confined spaces.

3.3.2.1.5 Hygrometry

Hygrometers are instruments used in the measurement of the water vapour content of air, i.e. the humidity. The comfort and efficiency of people depends on their ability to lose heat to the environment so that body temperature may remain constant. In conditions of high temperature and humidity this heat loss cannot occur fast enough. Thus measurement of humidity and its control are important in ensuring the thermal comfort of people at work. There are also some industrial processes, e.g. textile manufacture, whose success depends on a controlled humidity in the mill.

Humidity is generally expressed as relative humidity and quoted as a percentage. It is the ratio at a given atmospheric temperature and

Figure 3.3.11 Portable oxygen analyser. (Courtesy M.S.A. Ltd)

pressure of the mass of water in a given volume of air to the mass if the air had been saturated with water.

3.3.2.1.6 Wet and dry bulb hygrometer

This instrument consists of two normal thermometers, one of which has its bulb exposed to the air while the other has its bulb surrounded by a wick connected to a water reservoir. Evaporation of moisture from the wick to the surrounding air causes the 'wet bulb' thermometer to show a lower reading than the corresponding 'dry bulb' thermometer. The difference between the wet and dry bulb temperature related to the dry bulb temperature defines the hygrometric state of the atmosphere. Tables have been produced, from large numbers of observations, which give the relative humidity corresponding to likely wet and dry bulb temperature combinations.

The Mason hygrometer is mounted in a static position so that readings can be taken whenever required, whilst the 'whirling hygrometer' needs to be rotated rapidly for short periods to obtain a reading.

Instruments are now available which will give a direct reading of relative humidity. In them, air is drawn across two precision matched temperature sensors; one measures dry air (dry bulb) and the other is fitted with a moistened wick (wet bulb). Signals from both sensors are processed electronically to give a direct digital readout of relative humidity.

3.3.2.2 Interpretation of results

3.3.2.2.1 Hygiene standards

When measurements of airborne contamination levels or other parameters have been made it is necessary to interpret results against a *standard*. This interpretation will form the basis for the control strategy. In considering the exposure of workers to chemicals two broad options may be presented:

1 A zero exposure policy.
2 Permit certain tolerable levels of exposure.

To achieve zero exposure to all workplace chemicals is an impossible objective in the light of present day industrial processes. However, this approach has been adopted in some countries for proven human carcinogens.

Since no-exposure as a general policy is not possible, hygiene limits have been introduced in an attempt to quantify 'safe' permissible levels of exposure. These are applied to the workplace environment and attempt to reconcile the industrial use of a wide range of materials with a level of protection of the health of exposed workers.

The setting of a hygiene limit is a two-stage process – it involves firstly the collection and evaluation of scientific data and secondly the decision-making process by a committee which also has to take into account socio-economic and political factors. The HSE claim that its OES type standards are solely based on considerations of health data and that it is only the MEL type of limit that takes account of health and socio-economic data[3].

3.3.2.2.2 Threshold Limit Values

The Threshold Limit Value (TLV) system is that developed and published in the USA by the American Conference of Governmental and Industrial Hygienists (ACGIH), which in spite of its title is a non-governmental organisation. It is a scientific society similar to the British Occupational Hygiene Society.

The preface to 2002 ACGIH List states:

> 'Threshold Limit Values refer to airborne concentrations of substances and represent conditions under which it is believed that nearly all workers may be repeatedly exposed day after day without adverse effect'[4]

Because of individual variation in susceptibility some workers will suffer effects ranging from discomfort to sensitisation to chemicals and occupational disease at exposure levels well below the TLV. The basis for TLVs is intended to be reasonable freedom from irritation, narcosis, nuisance or impairment of health for the majority of workers. Reference to TLVs relates to these US based standards and this system is widely used around the world.

3.3.2.2.3 UK exposure limits

In the past the HSE has incorporated the TLV system in a Guidance Note[5] which was published annually. However, in the mid-1980s the HSE first published a list of British exposure limits to chemical substances[3] which, in their present form, are part of the requirements of the Control of Substances Hazardous to Health Regulations 2002 (COSHH)[6,7]. These standards are used for determining adequate control of exposure (by inhalation) to hazardous substances.

When applied, the limits should not be used as an index of relative hazard and toxicity, nor should they be used as the dividing line between 'safe' and 'dangerous' concentrations. The list is not comprehensive and the absence of a substance does not indicate that it is safe.

The list of exposure limits is divided into two basic parts – namely Maximum Exposure Limits (MELs) and Occupational Exposure Standards (OESs). For substances that have been given MELs, exposure should be reduced as far as reasonably practicable, and in any case, the limit should not be exceeded. With respect to OESs, it will be sufficient to ensure that the level of exposure is reduced to the OES level. This latter requirement reflects the standard demanded for compliance with the COSHH Regulations but it should not obscure the desirable aim of reducing all exposure as far as reasonably practicable.

In order to guide the decision as to which type of standard should be assigned to an individual substance, the HSE has published its *Indicative Criteria*. With respect to OES values these are:

- the ability to identify the concentration (with reasonable certainty) at which there is no indication of injurious effects on people, with repeated daily exposure;
- the OES can reasonably be complied with; and
- reasonably foreseeable over-exposures are unlikely to produce serious short- or long-term effects on health.

An MEL value may be set for a substance which is unable to satisfy the above OES criteria and which may present serious short- and/or long-term risks to man. In some cases where a substance has been assigned an OES value, a numerically higher MEL figure may be assigned because socio-economic factors require that substance's use in certain processes.

With two exceptions, levels embodied in these values relate to personal exposure via the inhalation route, i.e. monitoring is carried out in the person's breathing zone. The EH40 list is reprinted annually, with a list of proposed changes, together with notification of those standards which are priorities for review in the reasonably near future.

The MEL and OES occupational exposure limits are set on the recommendation of the HSC's Advisory Committee on Toxic Substances (ACTS). This follows assessment of the relevant scientific data by another committee known as the Working Group on the Assessment of Toxic Chemicals (WATCH). These committees consider both the type of limit to be set and the precise substance concentration.

The standard setting procedure is briefly outlined in the Guidance Note EH40[3]. Setting an OES is the first option and WATCH comes to a decision based upon consideration of the available information on health effects (often limited) and using the above criteria. If WATCH decides that an MEL is more appropriate, the consideration of the level at which it is to be set is dealt with by ACTS, since this involves balancing health risks against the cost of reducing exposure.

3.3.2.2.3.1 Indicative occupational exposure limit values (IOELVs)

These are concentrations of a hazardous substance in air proposed by the European Commission under the framework of the Chemical Agents Directive[8]. Member States are obliged to introduce an occupational exposure limit for these substances in accordance with national legislation and practice that takes the IOELV into account. IOELVs are based on recommendations by the European Commission's Scientific Committee on Occupational Exposure Limits (SCOEL). In EH40/2002[3], the HSC has implemented these European limits using the UK's domestic system of Occupational Exposure Limits and incorporating them into MEL and OES values.

3.3.2.2.3.2 Maximum exposure limits (MEL)

This is the maximum concentration of an airborne substance, averaged over a reference period, to which employees may be exposed by inhalation under any circumstances. Details of the legal requirements are contained in the COSHH Regulations[6,7], and the relevant substances are listed in the first part of the HSE's Guidance Note EH40. Currently there are around 55 substances listed but this is reviewed each year. A few selected examples are: hardwood dust, rubber fume, cadmium and compounds and trichloroethylene.

3.3.2.2.3.3 Occupational exposure standards (OES)

This is the concentration of an airborne substance, again averaged over a reference period, at which '... according to current knowledge, it is believed that there is no evidence that it is likely to be injurious to employees if they are exposed by inhalation, day after day at that concentration ...'[3]. However, current knowledge of the health effects of some chemicals is 'often limited'[3]. This is particularly the case with respect to long-term health effects on humans.

While control of exposure to the OES level satisfies minimally the requirements of COSHH Regulations, it should not discourage the application of good hygiene principles in reducing the concentration levels still further, especially in view of the limited scientific data available in respect of many of these chemicals. It would be prudent for employers to aim for concentrations of 25–50% of these levels.

3.3.2.2.3.4 Long-term and short-term exposure limits

Two types of exposure limit are listed with the aim of protecting against both short-term effects, such as irritation of the skin, eyes and lungs, narcosis etc., and long-term health effects. Both MEL and OES values are given as time weighted averages (TWA), i.e. the exposure concentrations measured are averaged with time over 8 hours to protect against long-term effects and over 15 minutes for protection against short-term effects.

In both the British and American systems concentrations are given in parts per million (ppm), i.e. parts of vapour or gas by volume per million parts of contaminated air, and also in milligrams of substance per cubic metre of air (mg/m^3).

3.3.2.2.3.5 Time weighted average concentrations (TWA)

The limits refer to the maximum exposure concentration when averaged over a 15-minute period or an 8-hour day. The time weighted average value (Cm) may be obtained from the following formula:

$$Cm = \frac{(C_1 \times t_1) + (C_2 \times t_2) + \ldots (C_n \times t_n)}{t_1 + t_2 + \ldots t_n}$$

where C_1, C_2 = concentrations measured during respective sampling periods;
t_1, t_2 = duration of sampling periods.

A simple example is where the person working an 8-hour day was exposed for 4 hours at 20 ppm vapour and then for 4 hours at 10 ppm.

$$Cm = \frac{(20 \times 4) + (10 \times 4)}{4 + 4}$$

This gives an 8-hour TWA of 15 ppm.

3.3.2.2.3.6 Mixtures

Most of the listed exposure limits refer to single substances or closely related groups, e.g. cadmium and compounds, isocyanates etc. A few exposure limits refer to complex mixtures or compounds, e.g. white spirit, rubber fume. However, exposure in workplaces is often to a mixture of substances and such combinations may, by their nature, increase the hazard. Mixed exposure requires assessment with regard to possible health effects, which should take into account other factors such as the primary target organs of the major contaminants and possible interaction between the latter substances.

General guidance on mixed exposures is given in EH40[3] together with a rule-of-thumb formula which may be used where there is reason to believe that the effects of the constituents of a mixture are *additive*.

$$\text{Exposure ratio} = \frac{C_1}{L_1} + \frac{C_2}{L_2} + \frac{C_3}{L_3} + \ldots$$

where C_1, C_2 = time weighted average concentrations of constituents;
L_1, L_2 = corresponding exposure limits.

The use of this formula is only applicable where the additive substances have been assigned OESs. If the exposure ratio is greater than 1 then the limit for the mixture has been exceeded. If one of the substances has been assigned an MEL then the additive effect should be taken into account when deciding to what extent it is reasonably practicable further to reduce exposure.

Example
If air contains 50 ppm acetaldehyde (OES = 100 ppm) and 150 ppm secbutyl acetate (OES = 200 ppm), applying the formula:

$$\text{Additive ratio} = \frac{50}{100} + \frac{150}{200} = 1.25$$

The threshold limit is therefore exceeded.

This is a relatively crude formula and would not be applicable to a situation where two or more chemicals enhance each other's effects as is the case with synergistic reactions.

3.3.2.3 Physical factors

Physical factors such as heat, ultraviolet light, high humidity, abnormal pressure etc. place added environmental stress on the body and are likely to increase the toxic effect of a substance. Most standards have been set at a level to encompass moderate deviations from the normal environment. However, for gross variations, e.g. heavy manual work where respiration rate is greatly increased, continuous activity at elevated temperatures or excessive overtime, judgement must be exercised in the interpretation of permissible levels.

3.3.2.4 Skin absorption

Some substances have the designation 'Sk' and this refers to the potential contribution to overall exposure of absorption through the skin. In this case airborne contamination alone will not indicate total exposure to the chemical and the 'Sk' designation is intended to draw attention to the need to prevent percutaneous absorption. In the application of the assigned exposure limit it is assumed that additional exposure of the skin is prevented.

3.3.2.5 Sensitisation

Similarly, in the list of exposure standards[3], the designation 'Sen' is assigned to selected substances to indicate that their potential for causing sensitisation reactions has been recognised. Such substances may cause respiratory sensitisation on inhalation, for example allergic asthma, or skin effects where contact occurs, for example allergic contact dermatitis. Although not all exposed persons will become sensitised when exposed to such substances, those that do will develop ill-health effects on subsequent exposure at extremely low concentrations. Once a person has become sensitised to a substance, the occupational exposure limits are not relevant for indicating 'safe' working concentrations.

3.3.2.6 Carcinogens

There are differing views as to whether carcinogens should be assigned exposure limits, with one body of opinion advocating that the only safe level for a substance that can cause cancer is zero. An Approved Code of Practice[7] applies to any carcinogen which is defined in COSHH as:

- a substance classified under CHIP[9] as in the category of danger – Carcinogenic (Category 1) or Carcinogenic (Category 2);
- any substance or preparation listed in Schedule 1 of COSHH.

This definition covers substances and preparations which would require labelling with the risk phrase 'R 45 – may cause cancer' or 'R 49 – may cause cancer by inhalation.'

3.3.2.7 Biological agents

A major change occurred with the consolidation of COSHH in 1994 when 'biological agents' were included in the defined substances. Requirements are contained in Schedule 3 to COSHH and in a supporting Approved Code of Practice[7]. 'Biological agent' is defined in COSHH as:

> Any micro-organisms, cell culture, or human endoparasite, including any which have been genetically modified, which may cause infection, allergy, toxicity or otherwise create a hazard to human health.

This refers to a general class of micro-organisms, cell cultures and human endoparasites, provided that they have one or more of the harmful properties specified in the definition. Most biologial agents are micro-organisms including bacteria, viruses, fungi and parasites. However, DNA is not to be regarded in itself as a biological agent.

Biological agents must be identified as hazards, the risks arising from work involving them assessed, and preventive or adequate control measures applied as for any other defined 'substance hazardous to health'. Biological agents are classified into four hazard groups on the following basis: their ability to cause infection, the severity of the disease that may result, the risk that such infection will spread to the community, and the availability of vaccines and effective treatment.

Key aspects are the potential of the biological agent to cause harm and the nature and degree of worker exposure to it. Risk assessments should reflect the ability they have to infect and replicate, and the possibility that there may be a significant risk to health at low exposures.

Exposure to a biological agent should be either prevented, or where this is not reasonably practicable, adequate control measures applied. Schedule 3 of COSHH is concerned with the special control provisions for biological agents. The selection of control measures should take into account the fact that biological agents do not have any air quality exposure limits and that they are able to infect and replicate at very small doses. An appropriate blend of controls is needed based on perceived levels of risk.

3.3.2.8 Derivation of Threshold Limit Values

Ideally, hygiene standards should be derived from the quantitative relations between the contaminant and its effects, i.e. X ppm of substances causes Y amount of harm. However, such relations are very difficult to establish in humans. The problems involved have been considered in some detail by Atherley[10]. Attempts have been made to relate human disease patterns to industrial experience, but unfortunately sufficient data do not exist. The effects of harmful agents have been studied by various methods. These include chemical analogy, which assumes that similar chemicals have biologically similar effects; and short-term testing, which may involve bacteria, animal exposure experiments and human epidemiology.

The major criteria which have been used to develop the TLV list are the effect of a substance on an organ or organ system (49%), irritation (40%), and thirdly to a lesser extent narcosis (5%) and odour (2%).

An ACGIH publication[11] summarises toxicological information on substances for which TLVs have been adopted and shows that for some substances the hazards are clear, whereas for others there is very little information on human risk. This inherent uncertainty is not reflected in the bland listing of adopted values.

In recent years, the HSE has published summaries[12] of the information used in setting its MEL and OES values. Such information may be useful in assessing the applicability of a standard to a particular workplace situation.

3.3.2.9 Variation in international standards

Hygiene standards vary from country to country depending upon the interpretation of scientific data and the philosophy of the regulations.

International hygiene standards for trichloroethylene illustrate this variation.

	mg/m^3	ppm
+ Australia	535	100
+ UK	535	100
+ USA (ACGIH)	267	50
+ Sweden	105	20
++ Hungary	53	10

+ Time Weighted Average.
++ Maximum Allowable Concentration.

In some countries great emphasis was put on neurophysiological changes in experimental animals as well as behavioural effects in human beings. It should be noted that although low levels may be embodied in national regulations this does not mean they are achieved in practice. In the past this has been acknowledged by former Soviet Union representatives. In the European Community context, there are moves to harmonise exposure limits but there are also simultaneous needs to harmonise compliance strategies.

3.3.2.10 Changes in hygiene limits

With new scientific evidence and changing attitudes, hygiene limits are constantly being revised. A startling example is provided by vinyl chloride monomer (VCM) which occurs in the manufacture of PVC plastics. The acute effects of VCM were identified in the 1930s as being primarily 'narcosis'. To prevent such effects during industrial use a TLV of 500 ppm was set in 1962. After further research VCM was identified as affecting the liver, bones and kidneys and the adopted value (TLV) was lowered to 200 ppm in 1971. In 1974 some American chemical workers died of a rare liver cancer (angiosarcoma) which was traced to exposure to VCM, with the result that in 1978 the adopted value (ACGIH) was dropped to 5 ppm. Hence the adopted TLV for vinyl chloride monomer has been reduced a hundredfold in under 20 years.

A section in the TLV list formally notes chemicals for which a change in the standard is intended.

Similarly, the HSE in its Guidance Note[3] publishes a list of substances where the OES is new, has been changed, or where it is intended to assign an MEL value. There is also a list of substances for which the occupational exposure limits are to be reviewed.

3.3.3 Control measures

When, in a workplace, a hazard has been identified and the risk to health assessed, an appropriate prevention or control strategy is then required. The general control strategy should include consideration of:

Specification.
Substitution.

Segregation.
Local exhaust ventilation.
General dilution ventilation.
Good housekeeping and personal hygiene.
Reduced time exposure.
Personal protection.

These control options should be complemented and underpinned by adequate administrative arrangements which should include the provision for regular reassessment of risks and overall review.

3.3.3.1 Specification

The design of a new plant or process is the ideal stage to incorporate hazard prevention and control features, e.g. limiting the quantities of toxic materials handled, the provision of remote handling facilities, utilising noise control features in the design and layout of new machinery etc. Including safety features at the design stage will be much less costly than having to add them later. This emphasises the need for safety advisers to be involved at the earliest stage of developments.

3.3.3.2 Substitution

This involves the substitution of materials or operations in a process by safer alternatives. A toxic material may be replaced by another less harmful substance or in another context something less flammable. An example is the widespread replacement of carbon tetrachloride by other solvents such as dichloromethane and 1,1,1-trichloroethane. In turn this latter substance has been phased out because of its ozone depleting characteristics and industry has found suitable substitutes.

Care needs to be exercised in the selection of 'safer' substitutes since they may be considered safer simply because there is less information available about their hazards.

Alternatively the process itself may be changed to improve working conditions with a possible benefit of increased efficiency as well.

Arc welding has been widely introduced to replace riveting and subsequently noise levels have been reduced. Again such alterations may introduce new hazards (e.g. welding fumes) and the risks from these must be similarly assessed with the implementation of adequate controls.

3.3.3.3 Segregation

If a substance or process cannot be eliminated, another strategy is to enclose it completely to prevent the spread of contamination. This may be by means of a physical barrier, e.g. an acoustic booth surrounding a noisy machine or handling toxic substances in a glove box. Relocation of a process to an isolated section of the plant is another possibility that reduces the number exposed to the hazard. A particular process may be segregated in time, e.g. operated at night, when fewer people are likely to be exposed. However, in the latter case, such workers are already subjected to the additional stress of night shift working and generally function less efficiently, a point that should be borne in mind by the occupational hygienist when considering alternatives.

3.3.3.4 Local extract ventilation

Where it is not practicable to enclose the process totally, other steps must be taken to contain contaminants. This can be achieved by removing vapours, gases, dusts and fumes etc. by means of a local extract ventilation system. Such a system traps the contaminant close to its source and removes it so that nearby workers are not exposed to harmful concentrations.

Local extract ventilation systems have four major parts:

1. Hoods – collection point for gathering the contaminated air into the system.
2. Ducting – to transport the extracted air to the air purifying device or the outside atmosphere.
3. Air purifying device – charcoal filters are often used to remove organic chemical contaminants.
4. Fan – provides the means for moving air through the system.

There are several different types of local extract ventilation systems and adherence to sound design principles is necessary to achieve effective removal of contaminants. Ventilation systems must also be adequately maintained to ensure that they are operating to design specifications. This subject is dealt with in greater detail in Chapter 3.6.

3.3.3.5 Dilution ventilation

Sometimes it is not possible to extract the contaminant close to its source of origin and dilution ventilation may be used under the following circumstances where there is:

1. Small quantity of contaminant.
2. Uniform evolution.
3. Low toxicity material.

Dilution ventilation utilises natural convention through open doors, windows, roof ventilators etc. or assisted ventilation by roof fans or blowers which draw or blow in fresh air to dilute the contaminant. With both of these systems the problem of providing make-up air at the proper temperature, especially during the winter months, has to be considered.

3.3.3.6 Personal hygiene and good housekeeping

Both have an important role in the protection of the health of people at work. Laid down procedures are necessary for preventing the spread of contamination, for example the immediate clean-up of spillages, safe disposal of waste and the regular cleaning of work stations.

Dust exposures can often be greatly reduced by the application of water or other suitable liquid close to the source of the dust. Thorough wetting of dust on floors before sweeping will also reduce dust levels.

Adequate washing and eating facilities should be provided with instruction for workers on the hygiene measures they should take to prevent the spread of contamination. The use of lead at work is a case where this is particularly important.

Wide ranging regulations[13] and a related guidance booklet[14] dealing with workplace health, safety and welfare require that workplaces are kept 'sufficiently clean' and that waste materials are kept under control. These objectives also apply when considering other control measures.

3.3.3.7 Reduced time exposure

Reducing the time of exposure to an environmental agent is a control strategy which has been used. The dose of contaminant received by a person is generally related to the level of stress and the length of time the person is exposed. A noise standard for maximum exposure of people at work of 90 dB(A) over an 8-hour work day has been used for several years and is now contained in the Noise at Work Regulations 1989[15] as the 'second action level'. Equivalent doses of noise energy are 93 dB(A) for 4 hours, 96 dB(A) for 2 hours etc. (The dB(A) scale is logarithmic.) Such limiting of hours has been used as a control strategy but does not take into account the possibly harmful effect of dose rate, e.g. very high noise levels over a very short time even though followed by a long period of relatively low levels.

3.3.3.8 Personal protection

Making the workplace safe should be the first consideration but if it is not possible to reduce risks sufficiently by the methods outlined above the worker may need to be protected from the environment by the use of personal protective equipment. Where appropriate, the PPE Regulations require the provision of suitable PPE except where other regulations require the provision of specific protective equipment, such as the asbestos, lead and noise regulations. The PPE Regulations are supported by practical guidance[17] on their implementation.

Personal protective equipment may be broadly divided as follows:

1 Hearing protection.
2 Respiratory protection.
3 Eye and face protection.

4 Protective clothing.
5 Skin protection.

Personal protective devices have a serious limitation in that they do nothing to attenuate the hazard at source, so that if they fail and it is not noticed the wearer's protection is reduced and the risk the person faces increases correspondingly.

Making the workplace safe is preferable to relying on personal protection; however, this regard for personal protection as a last line of defence should not obscure the need for the provision of competent people to select equipment and administer the personal protection scheme once the decision to use this control strategy has been taken. Personal protection is not an easy option and it is important that the correct protection is given for a particular hazard, e.g. ear-muffs/plugs prescribed after octave band measurements of the noise source.

Else[18] outlines three key elements of information required for a personal protection scheme:

(i) nature of the hazard,
(ii) performance data of personal protective equipment, and
(iii) standard representing adequate control of the risk.

3.3.3.8.1 Nature of the hazard and risk

The hazards need to be identified and the risks assessed; for example, in the case of air contaminants the nature of the substance(s) present and the estimated exposure concentration, or, with noise, measurement of sound levels and frequency characteristics.

3.3.3.8.2 Performance data on personal protective equipment

Data about the ability of equipment to protect against a particular hazard is provided by manufacturers who carry out tests under controlled conditions which are often specified in national or international standards. Performance requirements for face masks, for example, are contained in two British Standards[19] which specify the performance requirements of full-face and half/quarter masks for respiratory protective equipment. The method used to determine the noise attenuation of hearing protectors at different frequencies (octave bands) throughout the audible range is specified in a European standard[20].

3.3.3.8.3 Standards representing adequate control of the risk

For some risks such as exposure to potent carcinogens or protection of eyes against flying metal splinters the only tolerable level is zero. The informed use of hygiene limits, bearing in mind their limitations, would be pertinent when considering tolerable levels of air contaminants.

A competent person would need these three types of information to decide whether the personal protective equipment could *in theory* provide adequate protection against a particular hazard.

Once theoretically adequate personal protective equipment has been selected the following factors need to be considered:

1 Fit.	Good fit of equipment to the person is required to ensure maximum protection.
2 Period of use.	The maximum degree of protection will not be achieved unless the equipment is worn all the time the wearer is at risk.
3 Comfort.	Equipment that is comfortable is more likely to be worn. If possible the user should be given a choice of alternatives which are compatible with other protective equipment.
4 Maintenance.	To continue providing the optimum level of protection the equipment must be routinely checked, cleaned, and maintained.
5 Training.	Training should be given to all those who use protective equipment and to their supervisors. This should include information about what the equipment will protect against and its limitations.
6 Interference.	Some eye protectors and helmets may interfere with the peripheral visual field. Masks and breathing apparatus interfere with olfactory senses.
7 Management commitment.	This is essential to the success of personal protection schemes.

Appropriate practice should ensure *effective* personal protection schemes are based on the requirements of regulations and codes of practice[16,17].

3.3.3.8.4 Hearing protection

There are two major types of hearing protectors:

1 Ear-plugs – inserted in the ear canal.
2 Ear-muffs – covering the external ear.

Disposable ear-plugs are made from glass down, plastic-coated glass down and polyurethane foam, while reusable ear-plugs are made from semi-rigid plastic or rubber. Reusable ear-plugs need to be washed frequently.

Ear-muffs consist of rigid cups to cover the ears, held in position by a sprung head band. The cups have acoustic seals of polyurethane foam or a liquid-filled annular sac.

Hearing protectors should be chosen to reduce the noise level at the wearer's ear to at least below 85 dB(A) and ideally to around 80 dB(A). With particularly high ambient noise levels this should not be done from simple A-weighted measurements of the noise level, because sound reduction will depend upon its frequency spectrum. Octave band analysis measurements[20] will provide the necessary information to be matched against the overall sound attenuation of different hearing protectors which is claimed by the manufacturers in their test data.

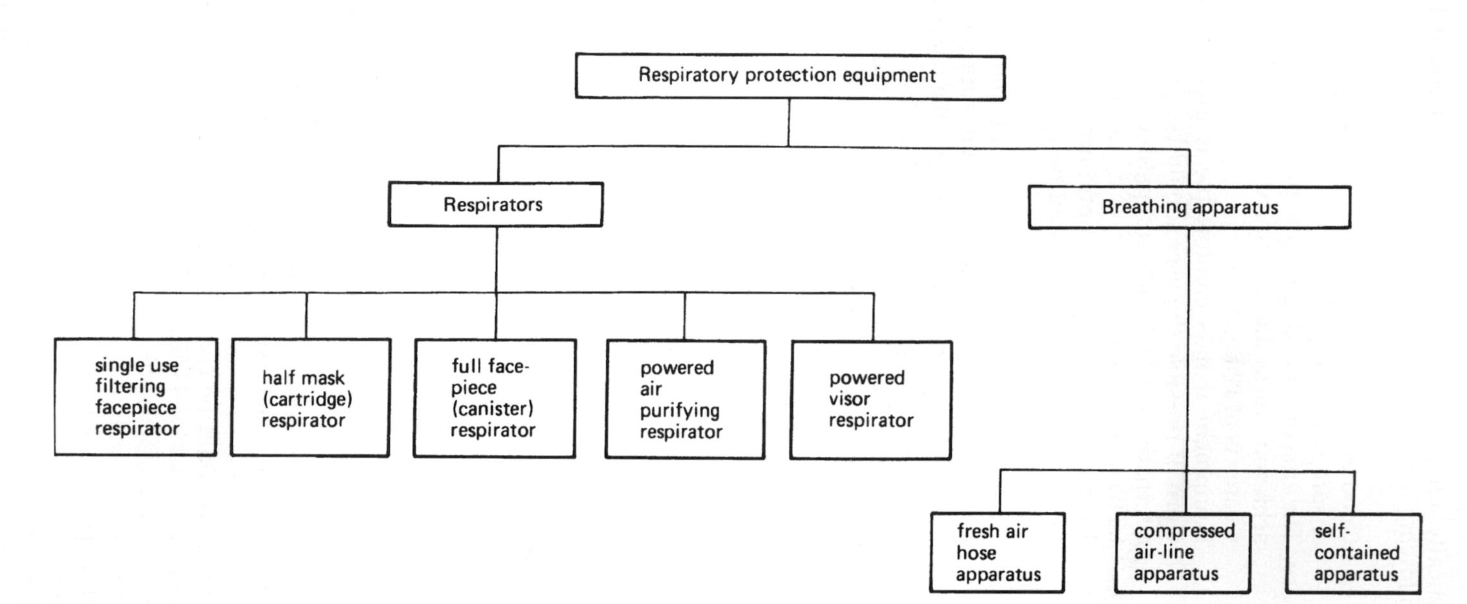

Figure 3.3.12 Types of respiratory protection equipment

3.3.3.8.5 Respiratory protective equipment

This may be broadly divided into two types as shown in *Figure 3.3.12*.

1 Respirators – purify the air by drawing it through a filter which removes most of the contamination.
2 Breathing apparatus – supplies clean air from an uncontaminated source.

3.3.3.8.5.1 Respirators

There are five basic types of respirator:

1 Filtering Facepiece Respirator. The facepiece covers the whole of the nose and mouth and is made of filtering material which removes respirable size particles. (These should not be confused with nuisance dust masks which simply remove larger particles.)
2 Half Mask Respirator. A rubber or plastic facepiece that covers the nose and mouth and has replaceable filter cartridges.
3 Full Face Respirator. A rubber or plastic facepiece that covers the eyes, nose and mouth and has replaceable filter canisters.
4 Powered Air Purifying Respirator. Air is drawn through a filter and then blown into a half mask or full facepiece at a slight positive pressure to prevent inward leakage of contaminated air.
5 Powered Visor Respirator. The fan and filters are mounted in a helmet and the purified air is blown down behind a protective visor past the wearer's face.

Filters are available for protection against harmful dusts and fibres, and also for removing gases and vapours. It is important that respirators are never used in oxygen-deficient atmospheres.

3.3.3.8.5.2 Breathing apparatus

The three main types of breathing apparatus are:

1 Fresh Air Hose Apparatus. Air is brought from an uncontaminated area by the breathing action of the wearer or by a bellows or blower arrangement.
2 Compressed Air Line Apparatus. Air is brought to the wearer through a flexible hose attached to a compressed air line. Filters are mounted in the line to remove nitrogen oxides and it is advisable to use a special compressor with this equipment. The compressor airline is connected via pressure-reducing valves to half-masks, full facepieces or hoods.
3 Self-contained Breathing Apparatus. A cylinder attached to a harness and carried on the wearer's back provides air or oxygen to a special mask. This equipment is commonly used for rescue purposes.

Within these classes there are many different sub-classes of RPE and it is important to choose the correct type of equipment based on a risk

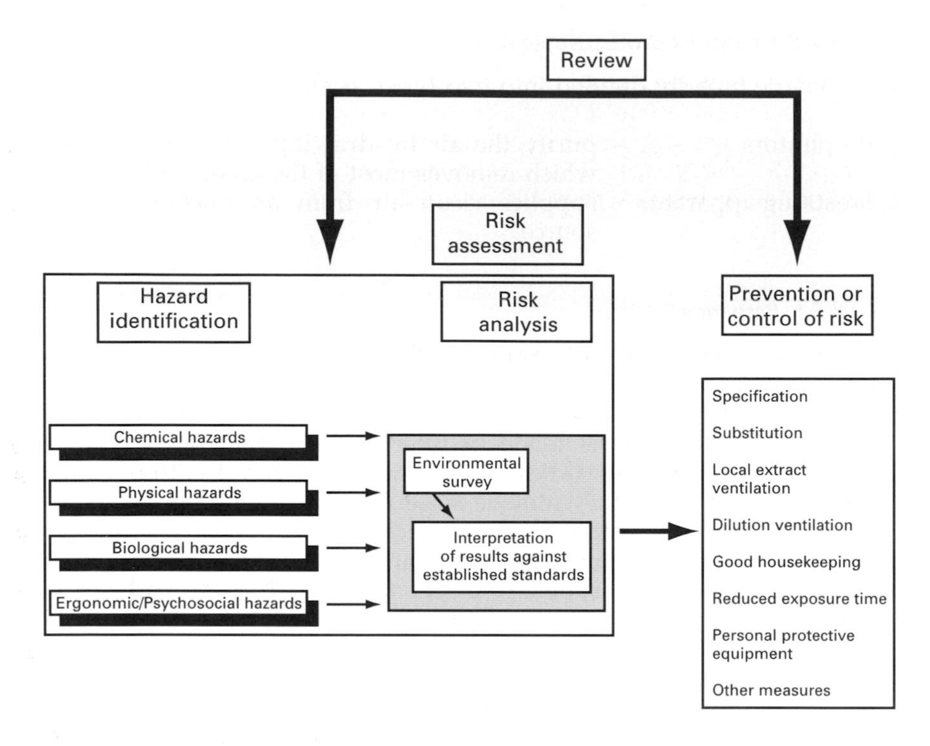

Figure 3.3.13 Diagram of strategy for protection against health risks

assessment. A British Standard[21] gives guidance on the selection, use and maintenance of respiratory protective equipment. From the risk assessment, it is necessary to decide whether to use a respirator or breathing apparatus. The minimum protection required for the situation then needs to be considered:

Minimum protection required (MPR)

$$= \frac{\text{Concentration outside the face piece of the RPE}}{\text{Concentration inside the face piece of the RPE}}$$

The MPR values can then be compared with the Assigned Protection Factors (APFs) listed in the standard[21]. The APFs are intended to be used as a guide and these protection levels may not be achieved where the equipment is not suitable for the environment or the user. The appropriate respirator face piece is combined with a filtering device, for example a cartridge or a canister, to give the desired APF.

Nominal Protection Factors (NPFs) have been used in the past for identifying the capability of different types of RPE. However, this approach has changed because studies have shown that some wearers may not achieve the level of protection indicated by the NPF and this could be misleading.

3.3.3.8.5.3 *Eye protection*

After a survey of eye hazards the most appropriate type of eye protection should be selected. Safety spectacles may be adequate for relatively low energy projectiles, e.g. metal swarf, but for dust, goggles would be more appropriate. For people involved in gas/arc welding or using lasers, special filtering lenses would be required.

3.3.3.8.5.4 *Protective clothing*

Well-designed and properly worn, protective clothing will provide a reasonable barrier against skin irritants. A wide range of gloves, sleeves, impervious aprons, overalls etc. is currently available. The integration and compatibility of the various components of a whole-body personal protection ensemble is particularly important in high risk situations, for example in the case of handling radioactive substances or biological agents.

The factors listed above should be considered when the selection of protective clothing is being made. For example, when selecting gloves for handling solvents, a knowledge of glove material is required:

Neoprene gloves – adequate protection against common oils, aliphatic hydrocarbons; not recommended for aromatic hydrocarbons, ketones, chlorinated hydrocarbons.

Polyvinyl alcohol gloves – protect against aromatic and chlorinated hydrocarbons.

For protective clothing to achieve its objective it needs to be regularly cleaned or laundered and replaced when damaged.

3.3.3.8.5.5 *Skin protection*

Where protective clothing is impracticable, due to the proximity of machinery or unacceptable restriction of the ability to manipulate, a barrier cream may be the preferred alternative. Skin protection preparations can be divided into the following three groups:

1 Water-miscible – protects against organic solvents, mineral oils and greases, but not metal-working oils mixed with water.
2 Water-repellent – protects against aqueous solutions, acids, alkalis, salts, oils and cooling agents that contain water.
3 Special group – cannot be assigned to a group by their composition. Formulated for specific application.

Skin protection creams should be applied before starting work and at suitable intervals during the day.

However, these preparations are only of limited usefulness as they are rapidly removed by rubbing action and care must be taken in their selection since, with some solvents, increased skin penetration can occur. The application of a moisturising cream which replenishes skin oil is beneficial after work.

3.3.4 Summary

The overall strategic approach is summarised in *Figure 3.3.13*. Although this approach to hazard identification, risk assessment and control has long been established in occupational hygiene, detailed supporting legislation has been restricted to selected health hazards, e.g. substances hazardous to health, noise, lead etc. However, new regulations[22] require this same approach to all hazards thus reinforcing hygiene practice.

References

1. Health and Safety Executive, Guidance Booklet No. HSG 173, *Monitoring strategies for toxic substances*, HSE Books, Sudbury (1997)
2. Ashton, I. and Gill, F.S., *Monitoring for health hazards at work*, Blackwell Science, Oxford (2000)
3. Health and Safety Executive, Guidance Note No. EH40, *Occupational exposure limits*, HSE Books, Sudbury (This is updated annually)
4. American Conference of Governmental Industrial Hygienists, *Threshold limit values and biological indices for 2002–2003*, ACGHI, Cincinnati, Ohio (2002)
5. Health and Safety Executive, Guidance Note No. EH15/80, *Threshold limit values*, The Stationery Office, London (1980)
6. *The Control of Substances Hazardous to Health Regulations 2002*, The Stationery Office, London (2002)
7. Health and Safety Executive, Legal Series booklet No. L5, *General COSHH ACOP and Carcinogens ACOP and Biological Agents ACOP* (2002 edn), HSE Books, Sudbury (2002)
8. European Community Council Directive *on the protection of the health and safety of workers from the risks related to chemical agents at work*, Directive no. 98/24/EEC, EU, Luxembourg 1998
9. The Chemical (Hazard Information and Packaging for Supply) Regulations 2002, The Stationery Office, London (2002)
10. Atherley, G.R.C., *Occupational health and safety concepts*, Applied Science, London (1978)
11. American Conference of Governmental Industrial Hygienists, *Documentation of the threshold limit values and biological exposure indices*, 7th edn, ACGIH, Cincinnati, Ohio (2002)
12. Health and Safety Executive, Guidance Note No. EH64, *Summary criteria for occupational exposure limits*, 1996–1999 with supplements for 1999, 2000 and 2001, HSE Books, Sudbury 2001
13. *Workplace (Health, Safety and Welfare) Regulations 1992*, The Stationery Office, London (1992)
14. Health and Safety Commission, *Legislation publication No. L24 Approved Code of Practice: Workplace (Health, Safety and Welfare) Regulations 1992*, HSE Books, Sudbury (1992)
15. *The Noise at Work Regulations 1989*, The Stationery Office, London (1989)
16. *The Personal Protective Equipment at Work Regulations 1992*, The Stationery Office, London (1992)
17. Health and Safety Executive, Legal Series Booklet No. L25, *Personal protective equipment at work, guidance on the Regulations* HSE Books, Sudbury (1992)

18. Else, D., *Occupational Health Practice* (ed. Schilling, R.S.F.), 2nd edn, Ch. 21, Butterworth, London (1981)
19. British Standards Institution, BS EN 136:1998 *Respiratory protective devices. Full face masks. Requirements, testing, marking.* BS EN 140:1999 *Respiratory protective devices. Half masks and quarter masks. Requirements, testing, marking* BSI, London 1999
20. British Standards Institution, BS EN 24869–1:1993 *Acoustics. Hearing protectors. Sound attenuation of hearing protectors. Subjective method of measurement*, BSI, London 1993
21. British Standards Institution, BS 4275:1997 *Guide to implementing an effective respiratory protective device programme*, BSI, London 1997
22. *Management of Health and Safety at Work Regulations 1999*, The Stationery Office Ltd, London (1999)

Chapter 3.4

Radiation

Dr A. D. Wrixon and updated by Peter Shaw and Dr M. Maslanyj

3.4.1 Introduction

Radiation is emitted by a wide variety of sources and appliances used in industry, medicine and research. It is also a natural part of the environment. The purpose of this chapter is to give the reader a broad view of the nature of radiation, its biological effects, and the precautions to be taken against it.

3.4.2 Structure of matter[1–3]

All matter consists of elements, for example hydrogen, oxygen, iron. The basic unit of any element is the atom, which cannot be further subdivided by chemical means. The atom itself is an arrangement of three types of particles.

1 Protons. These have unit mass and carry a positive electrical charge.
2 Neutrons. These also have unit mass but carry no charge.
3 Electrons. These have a mass about 2000 times less than that of protons and neutrons and carry a negative charge.

Protons and neutrons make up the central part of the nucleus of the atom; their internal structure is not relevant here. The electrons take up orbits around the nucleus and, in an electrically neutral atom, the number of electrons equals the number of protons. The element itself is defined by the number of protons in the nucleus. For a given element, however, the number of neutrons can vary to form different isotopes of that element. A particular isotope of an element is referred to as a nuclide. A nuclide is identified by the name of the element and its mass, for example carbon-14. There are 90 naturally occurring elements; additional elements, such as plutonium and americium, have been created by man, for example in nuclear reactors.

If the number of electrons does not equal the number of protons, the atom has a net positive or negative charge and is said to be ionised. Thus if a neutral atom loses an electron, a positively charged ion will result. The process of losing or gaining electrons is called ionisation.

3.4.3 Radioactivity[1–3]

Some nuclides are unstable and spontaneously change into other nuclides, emitting energy in the form of radiation, either particulate (e.g. α and β particles) or electromagnetic (e.g. γ-rays). This property is called radioactivity, and the nuclide showing it is said to be radioactive. Most nuclides occurring in nature are stable, but some are radioactive, for example all the isotopes of uranium and thorium. Many other radioactive nuclides (or radionuclides) have been produced artificially, such as strontium-90, caesium-137 and the isotopes of the man-made elements, plutonium and americium.

3.4.4 Ionising radiation[4]

The radiation emitted during radioactive decay can cause the material through which it passes to become ionised and it is therefore called ionising radiation. X-rays are another type of ionising radiation. Ionisation can result in chemical changes which can lead to alterations in living cells and eventually, perhaps, to manifest biological effects.

The ionising radiations encountered in industry are principally α, β, γ and X-rays, bremsstrahlung and neutrons. Persons can be irradiated by sources outside the body (external irradiation) or from radionuclides deposited within the body (internal irradiation). External irradiation is of interest when the radiation is sufficiently penetrating to reach the basal layer of the epidermis (i.e. the living cells of the skin). Internal irradiation arises following the intake of radioactive material by ingestion, by inhalation or by absorption through the skin or open wounds.

The α particle consists of two protons and two neutrons. It is therefore heavy and doubly charged. Alpha radiation has a very short range and is stopped by a few centimetres of air, a sheet of paper, or the outer dead layer of the skin. Outside the body, it does not, therefore, present a hazard. However, α-emitting radionuclides inside the body are of concern because α particles lose their energy to tissue in very short distances causing relatively intense local ionisation.

The β particle has mass and charge equal in magnitude to an electron. Its range in tissue is strongly dependent on its energy. A β particle with energy below about 0.07 MeV would not penetrate the outer dead layer of the skin, but one with an energy of 2.5 MeV would penetrate soft tissue to a depth of about 1.25 cm. Energy is expressed here in units of electron volt (eV), which is a measure of the energy gained by an electron in passing through a potential difference of one volt. Multiples of the electron volt are commonly used; MeV stands for Mega electron Volts (1 MeV = 1 000 000 eV). As β particles are slowed down in matter, bremsstrahlung (a type of X-radiation) is produced, which will penetrate to greater distances. Thus a β-radiation source outside the body may have more penetrating radiation associated with it than is immediately apparent from the energy of the β radiation. Beta-emitting radionuclides inside the body are also of concern, but the total ionisation caused by β particles is less intense than that caused by α particles.

Gamma-rays, X-rays and bremsstrahlung are all electromagnetic radiations similar in nature to ordinary light except that they are of much higher frequencies and energies. They differ from each other in the way in which they are produced. Gamma-radiation is emitted in radioactive decay. The most widely known source of X-rays is in certain electrical equipment in which electrons are made to bombard a metal target in an evacuated tube. Bremsstrahlung is produced by the slowing down of β particles; its energy depends on the energy of the original β particles. The penetrating power of electromagnetic radiation depends on its energy and the nature of the matter through which it passes; with sufficient energy it can pass right through a human body. Sources of these radiations outside the body can therefore cause harm. With X-ray equipment, the radiation ceases when the machine is switched off. Gamma-ray sources, however, cannot be switched off.

Neutrons are emitted during certain nuclear processes, for example nuclear fission, in which a heavy nucleus splits into two fragments. They are also produced when α particles collide with the nucleus of certain nuclides; this phenomenon is made use of in meters for measuring the moisture content of soil. Neutrons, being uncharged and therefore not affected by the electric fields around atoms, have great penetrating power, and sources of neutrons outside the body can cause harm. Neutrons produce ionisation indirectly. When a high-energy neutron strikes a nucleus in the material through which it passes, some of its energy is transferred to the nucleus which then recoils. Being electrically charged and slow moving the recoiling nucleus creates dense ionisation over a short distance.

3.4.5 Biological effects of ionising radiation[4–8]

Information on the biological effects of ionising radiation comes from animal experiments and from studies of groups of people exposed to relatively high levels of radiation. The best-known groups are the workers in the luminising industry early this century who used to point their brushes with the lips and so ingest radioactivity; the survivors of the atomic bombs dropped on Japan, and patients who have undergone radiotherapy. Evidence of biological effects is also available from studies of certain miners who inhaled elevated levels of the natural radioactive gas radon and its radioactive decay products.

The basic unit of tissue is the cell. Each cell has a nucleus, which may be regarded as its control centre. Deoxyribonucleic acid (DNA) is the essential component of the cell's genetic information and makes up the chromosomes which are contained in the nucleus. Although the ways in which radiation damages cells are not fully understood, many involve changes to DNA. There are two main modes of action. A DNA molecule may become ionised, resulting directly in chemical change, or it may be chemically altered by reaction with agents produced as a result of the ionisation of other cell constituents. The chemical change may ultimately mean that the cell is prevented from further division and can therefore be regarded as dead.

Very high doses of radiation can kill large numbers of cells. If the whole body is exposed, death may occur within a matter of weeks: an instantaneous absorbed dose of 5 gray or more would probably be lethal (the unit gray is defined below). If a small area of the body is briefly exposed to a very high dose, death may not occur, but there may be other early effects: an instantaneous absorbed dose of 5 gray or more to the skin would probably cause erythema (reddening) in a week or so, and a similar dose to the testes or ovaries might cause sterility. If the same doses are received in a protracted fashion, there may be no early signs of injury. The effect of very high doses of radiation delivered acutely is used in radiotherapy to destroy malignant tissue. Effects of radiation that only occur above certain levels (i.e. thresholds) are known as *deterministic*. Above these thresholds, the severity of harm increases with dose.

Low doses or high doses received in a protracted fashion may lead to damage at a later stage. With reproductive cells, the harm is expressed in the irradiated person's offspring (genetic defects), and may vary from unobservable through mildly detrimental to severely disabling. So far, however, no genetic defects directly attributable to radiation exposure have been unequivocally observed in human beings. Cancer induction may result from the exposure of a number of different types of a cell. There is always a delay of some years, or even decades, between irradiation and the appearance of a cancer.

It is assumed that within the range of exposure conditions usually encountered in radiation work, the risks of cancer and hereditary damage increase in direct proportion to the radiation dose. It is also assumed that there is no exposure level that is entirely without risk. Thus, for example, the mortality risk factor for all cancers from uniform radiation of the whole body is now estimated to be 1 in 25 per sievert (see below for definition) for a working population, aged 20 to 64 years, averaged over both sexes[5]. In scientific notation, this is given as 4×10^{-2} per sievert. Effects of radiation, primarily cancer induction, for which there is probably no threshold and the risk is proportional to dose are known as *stochastic*, meaning 'of a random or statistical nature'.

3.4.6 Quantities and units

All new legislation in force after 1986 is required by the Units of Measurement Regulations 1980 to be in SI units. Only the SI system of units is described in full here, although the relationships between the old and new units are given in *Table 3.4.1*.

The *activity* of an amount of a radionuclide is given by the rate at which spontaneous decays occur in it. Activity is expressed in a unit called the becquerel, Bq. A Bq corresponds to one spontaneous decay per second.

Table 3.4.1 Relationship between SI units and old units

Quantity	*New named SI unit*	*In other and symbol*	*Old unit*	*Conversion factor and symbol*
Absorbed dose	gray (Gy)	Jkg^{-1}	rad (rad)	1 Gy = 100 rad
Dose equivalent	sievert (Sv)	Jkg^{-1}	rem (rem)	1 Sv = 100 rem
Activity	becquerel (Bq)	s^{-1}	curie (Ci)	1 Bq = 2.7×10^{-11}Ci

Multiples of the becquerel are frequently used such as the megabecquerel, MBq (a million becquerels).

The *absorbed dose* is the mean energy imparted by ionising radiation to the mass of matter in a volume element. It is expressed in a unit called the gray, Gy. A Gy corresponds to a joule per kilogram.

Biological damage does not depend solely on the absorbed dose. For example, one Gy of α radiation to tissue can be much more harmful than one Gy of β radiation. In radiological protection, it has been found convenient to introduce a further quantity that correlates better with the potential harm that might be caused by radiation exposure. This quantity, called the *equivalent dose*, is the absorbed dose averaged over a tissue or organ multiplied by the relevant radiation weighting factor. The radiation weighting factor for γ radiation, X-rays and β particles is set at 1. For α particles, the factor is 20. Equivalent dose is expressed in a unit called the sievert, Sv. Submultiples of the sievert are frequently used such as the millisievert, mSv (a thousandth of a sievert) and the microsievert, μSv (a millionth of a sievert).

The risks of malignancy, fatal or non-fatal, per sievert are not the same for all body tissues. The risk of hereditary damage only arises through irradiation of the reproductive organs. It is therefore appropriate to define a further quantity, derived from the equivalent dose, to indicate the combination of different doses to several tissues in a way that is likely to correlate with the total detriment due to malignancy and hereditary damage. This quantity, derived for the fractional contribution each tissue makes to the total detriment, is called the *effective dose*. This is defined as the sum of the equivalent doses to the exposed organs and tissues weighted by the appropriate tissue weighting factor. This quantity is also expressed in sieverts.

3.4.7 Basic principles of radiological protection

Throughout the world, protection standards have, in general, been based for many years on the recommendations of the International Commission on Radiological Protection (ICRP). This body was founded in 1928 and, since 1950, has been providing general guidance on the widespread use of radiation sources. The primary aim of radiological protection as expressed by ICRP[5] is to provide an appropriate standard of protection for man without unduly limiting the beneficial practices giving rise to radiation exposure. For this, ICRP has introduced a basic framework for protection that is intended to prevent those effects that occur only above relatively high levels of dose (e.g. erythema) and to ensure that all reasonable steps are taken to reduce the risks of cancer and hereditary damage. The system of radiological protection by ICRP[5] for proposed and continuing practices is based on the following general principles:

(a) No practice involving exposure to radiation should be adopted unless it produces sufficient benefit to the exposed individuals or to society to offset the radiation detriment it causes. (The justification of a practice.)
(b) In relation to any particular source within a practice, the magnitude of individual doses, the number of people exposed, and the likelihood of incurring exposure where these are not certain to be received should be kept as low as is reasonably achievable, economic and social factors being taken into account. This procedure should be constrained by restrictions on the doses to individuals (dose constraints), or risks to individuals in the case of potential exposure (risk constraints), so as to limit the inequity likely to result from the inherent economic and social judgements. (The optimisation of protection.)
(c) The exposure of individuals resulting from the combination of all the relevant practices should be subject to dose limits, or to some control of risk in the case of potential exposures. These are aimed at ensuring that no individual is exposed to radiation risks from these practices that are judged to be unacceptable in any normal circumstances. Not all sources are susceptible to control by action at the source and it is necessary to specify the sources to be included as relevant before selecting a dose limit. (Individual dose and risk limits.)

The ordering of these recommendations is deliberate; the ICRP limits are to be regarded as backstops and not as levels that can be worked up to.

For workers, the effective dose limit recommended by ICRP is 20 mSv per year averaged over defined periods of 5 years with no more than 50 mSv in any single year, the equivalent dose limit for the lens of the eye is 150 mSv in a year and that for the skin, hands and feet is 500 mSv in a year.

For comparison, the principal effective dose limit for members of the public is 1 mSv in a year. However, it is permissible to use a subsidiary dose limit of 5 mSv in a year for some years, provided that the average annual effective dose over 5 years does not exceed 1 mSv per year. The equivalent dose limits for the skin and lens of the eye are 50 mSv and 15 mSv per year respectively.

In the application of the dose limits for both workers and the public, no account should be taken of the exposures received by patients undergoing radiological examination or treatment and those received from normal levels of natural radiation. Guidance on the implementation of the ICRP principles to the protection of workers is given in reference 9.

3.4.7.1 Protection against external radiation[4,6]

Protection against exposure from external radiation is achieved through the application of three principles: shielding, distance or time. In practice judicious use is made of all three. Shielding involves the placing of some material between the source and the person to absorb the radiation partially or completely. Plastics are useful materials for shielding β radiation because they produce very little bremsstrahlung. For γ and X-radiation a large mass of material is required; lead and concrete are commonly used.

Radiation from a point source reduces with the square of the distance and through absorption by the intervening air. Remote handling is one way of putting distance between the source and the person (for example, tweezers may be used when handling β-emitting sources).

3.4.7.2 Protection against internal radiation[4–10]

Protection against exposure from internal radiation is achieved by preventing the intake of radioactive material through ingestion, inhalation and absorption through skin and skin breaks. Eating, drinking, smoking and application of cosmetics should not be carried out in areas where unsealed radioactive materials are used. The degree of containment necessary depends on the quantity and type of material being handled: it may range from simple drip trays through fume cupboards to complete enclosures such as glove boxes. Surgical gloves, laboratory coats and overshoes may need to be worn. A high standard of cleanliness is required to prevent the spread of radioactive contamination and care is necessary in dealing with accidental spills (*Figure 3.4.1*). Anyone working with unsealed radioactive material should wash and monitor his hands on leaving the working area; this is particularly important before meals are taken. Cuts and wounds should be treated immediately and no one should work with unsealed radioactive substances unless breaks in the skin are protected to prevent the entry of radioactive material.

The radiation dose received through the intake of radioactive material depends on the mode of intake, the quantity involved, the organs in which the material becomes deposited, the rate at which it is eliminated (by radioactive decay and excretion) and the radiations emitted.

3.4.7.3 Radiation monitoring

The main objectives of monitoring are to evaluate occupational exposures, to demonstrate compliance with standards and regulatory requirements and to provide data needed for adequate control. For the latter, monitoring can serve the following functions:

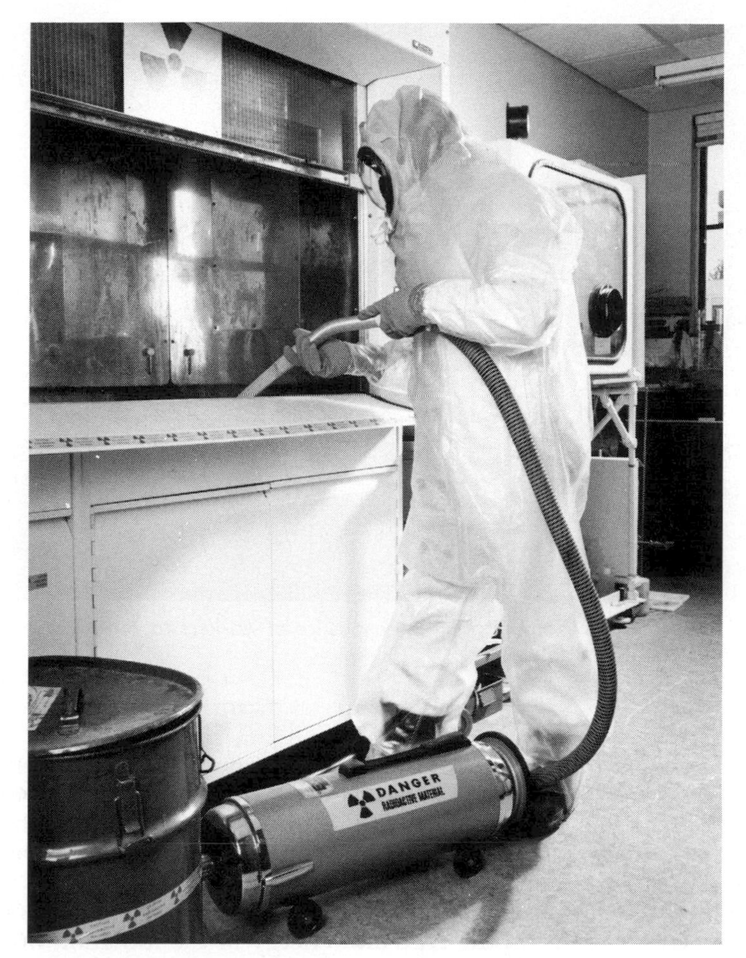

Figure 3.4.1 Decontamination of radioactive area in a laboratory

1 detection and evaluation of the principal sources of exposure,
2 evaluation of the effectiveness of radiation control measures and equipment,
3 detecting of unusual and unexpected situations involving radiation exposures,
4 evaluation of the impact of changes in operational procedures, and
5 provision of data on which the effect of future operations on radiation exposure can be predicted so that the appropriate controls can be devised beforehand and instituted.

The most appropriate means of assessing a worker's exposure to external radiations is through individual monitoring involving the wearing of a 'badge' containing radiation sensitive material, in particular a thermoluminescent chip or powder or a small piece of film (*Figure*

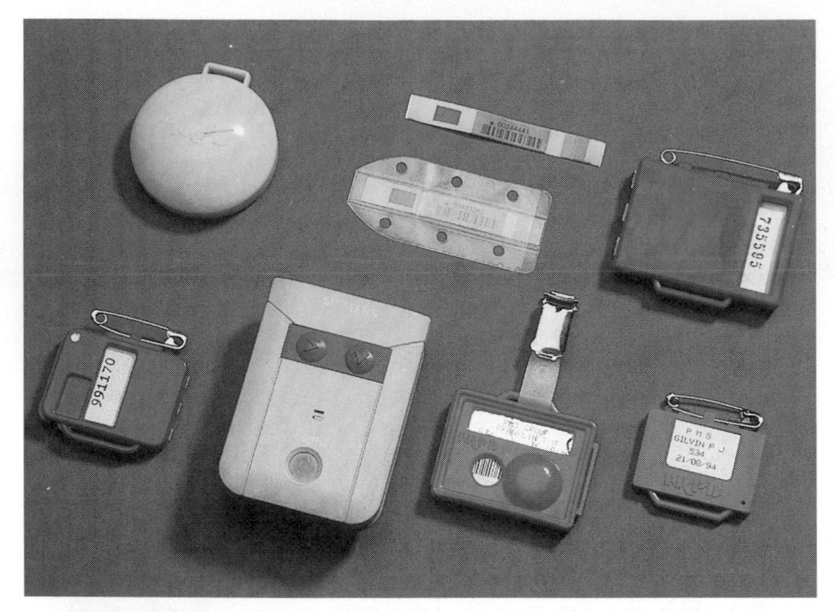

Figure 3.4.2 Devices for monitoring the exposure of workers to various types of radiation. (Courtesy NRPB)

3.4.2). Doses from the intake of airborne contamination can be assessed through the use of air samplers either worn by the person or set up at appropriate points in the workplace. Radioactive material within the body can be determined by excreta or whole body monitoring, depending on the particular radionuclide involved.

The appropriate detector to be used to monitor the workplace environment depends on the type and energy of the radiation involved and whether the hazard arises from external radiation or surface or air contamination. Most survey instruments can be divided into two groups:

(a) Dose rate meters

These measure the radiation in units of dose rate and normally contain an ionisation chamber or Gieger-Müller tube. They are usually used to monitor β, γ and X-radiation fields. Special instruments are used for measuring neutron radiation dose rates.

(b) Contamination monitors

These measure the surface activity of radioactive contamination in counts per unit time. They normally contain a Geiger-Müller, proportional counter tube or scintillation counter. For α contamination, the detector normally employed would be a scintillation counter. The efficiency depends on the particular radionuclide being measured and the instrument should be calibrated for each radionuclide of interest.

The selection and use of monitoring instruments may be complex and should be discussed with a Radiation Protection Adviser (see below) or other suitable expert.

3.4.8 Legal requirements

The principal legislation in the UK affecting the use of ionising radiations in industry is summarised briefly below. However, readers should consult the appropriate documents for full details.

3.4.8.1 The Ionising Radiations Regulations 1999

These regulations, which were made under the Health and Safety at Work etc. Act 1974, came fully into effect on 1 January 2000. They apply to all work with ionising radiation rather than just work in a factory. They took account of the recommendations of ICRP and are in conformity with a Council Directive of the European Communities which lays down basic safety standards for the health protection of the general public and workers against the dangers of ionising radiation[12]. Details of acceptable methods of meeting the requirements of the regulations are given in the supporting Approved Code of Practice[11]. The following is a summary of some of the main requirements of the Regulations.

The Regulations require that employers undertake a *suitable and sufficient* prior risk assessment before commencing activities involving work with ionising radiation. The purpose of this assessment is to identify the measures necessary to restrict the exposure of employees and other persons. The assessment must consider both normal operations and potential radiation accidents.

The dose limits for employees over the age of 18 years are those recommended by ICRP, i.e. the effective dose equivalent limit for employees aged 18 years or over is 20 mSv in a year. Lower limits apply to trainees under the age of 18 years. Special restrictions apply to the rate at which women of reproductive capacity can be exposed and to the exposure of pregnant women during the declared term of pregnancy. The limits for any other person are 1 mSv in a year for the effective dose equivalent and 50 mSv in a year for the dose equivalent to individual organs or tissues other than the lens of the eye for which the value is 15 mSv in a year. The main requirement, however, is for employers to 'take all necessary steps to restrict so far as reasonably practicable the extent to which his employees and other persons are exposed to ionising radiation', in keeping with the emphasis of ICRP. If the effective dose equivalent to an employee exceeds 15 mSv in a year (or a lower level specified by the employer) the employer is required to make an investigation to determine whether it is reasonably practicable to take further steps to reduce exposure.

To facilitate the control of doses to persons, the Regulations specify criteria for designating areas as controlled or supervised areas. The underlying basis of designation is a combination of likely doses and the need for either special work procedures or radiological supervision.

Employers are required to 'designate as classified workers those of his employees who are likely to receive an effective dose in excess of 6 mSv per year or an equivalent dose which exceeds three-tenths of any relevant dose limit'. Only employees aged 18 years or over who have been certified as fit to be designated as a classified person can be so designated. Employees or other persons are only permitted to enter a controlled area if they are classified or enter in accordance with *suitable written arrangements*. In the case of the latter the employer must be able to justify non-classification of the workers involved.

The Radiation Protection Adviser (RPA) is a key figure in the Regulations. His function is to advise the employer 'as to the observance of these Regulations'. He should, for example, be consulted about risk assessments, restricting the exposure of workers, the identification of controlled and supervised areas, dosimetry and monitoring, the drawing up of written systems of work and local rules, the investigation of abnormally high exposures and overexposures and training. By the end of 2004, all RPAs are required to demonstrate their competence, either through accreditation by a competent assessing body or through achieving suitable NVQs.

In relation to employees who are designated as classified persons, the Regulations require employers to ensure that assessments are made of all significant doses. For this purpose, the employer is to make suitable arrangements with an approved dosimetry service (ADS). The employer is also required to make arrangements with the ADS for that service to keep suitable summaries of any appropriate dose records for his employees. The purpose of the approval system is to ensure as far as possible that the doses are assessed on the basis of accepted national standards.

The Regulations also specify requirements for the medical surveillance of employees and the maintenance of individual records of medical findings and assessed doses. The general requirement to keep doses as low as reasonably practicable is strengthened by the inclusion of a basic requirement to control the source of ionising radiation and by subsequent specific requirements to provide appropriate safety devices, warning signals, handling tools etc., to leak test radioactive sources, to provide protective equipment and clothing and test them, to monitor radiation and contamination levels (see *Figure 3.4.3*), to store radioactive substances safely, to design, construct and maintain buildings, fittings and equipment so as to minimise contamination, and to make contingency arrangements for dealing with foreseeable but unintended incidents.

There are also requirements for employers to notify HSE of work with ionising radiation, overexposures and certain accidents and losses of radioactive material. The provision of information on potential hazards and appropriate training are also required. In addition, there are requirements to formulate written local rules and to provide supervision of work involving ionising radiation. Such requirements will necessitate the appointment by management of a radiation protection supervisor (RPS) whose responsibilities should be clearly defined.

Figure 3.4.3 Checking contamination levels after a fire

The RPS should not be confused with the RPA. While the latter may be an outside consultant or body (and this is often the case), the RPS plays a supervising role in assisting the employer to comply with the Regulations and should normally be an employee directly involved with the work with ionising radiations, preferably in a line management position that will allow him to exercise close supervision to ensure that the work is done in accordance with the local rules, though he need not be present all the time. The RPS should therefore be conversant with the Regulations and local rules, command sufficient respect to allow him to exercise his supervisory role and understand the necessary precautions to be taken in the work that is being done.

3.4.8.2 The Radioactive Substances Act 1993[13]

The main purpose of this Act is to regulate the keeping and use of radioactive materials and the disposal and accumulation of radioactive waste. Under the Act those who keep or use radioactive materials on premises used for the purposes of an undertaking (trade, business, profession etc.) are required to register with the Environment Agency (England and Wales), the Scottish Environment Protection Agency or the Northern Ireland Environment and Heritage Service, according to region, unless exempt from registration. Conditions may be attached to registrations and exemptions, and these are made with regard to the amount and character of the radioactive waste likely to arise.

No person may dispose of or accumulate radioactive waste unless he is authorised by the appropriate Agency or Service or is exempt. Whenever possible local disposal of radioactive waste should be used but with many industrial sources, such as those used in gauges and radiography, disposal should be made through a person authorised to do so and advice should be sought from the source supplier, a Radiation Protection Adviser or the appropriate regional Environment Agency or Service.

A number of generally applicable exemption orders have been made under the Act for those situations where control would not be warranted. The orders cover such things as substances of low activity, luminous articles, electronic valves, smoke detectors, some uses of uranium and thorium and various materials containing natural radioactivity. The orders should be consulted for details of the conditions under which exemption is granted. The orders are currently under review.

3.4.8.3 Transport Regulations

Protection of both transport workers and the public is required when radioactive substances are transported outside work premises. The Regulations and conditions governing transport in the UK and internationally follow those specified by the International Atomic Energy Agency. The latest version of the Agency's regulations is listed in reference 14. The particular regulations that apply depend on the means of transport to be used. Those that apply to the transport of radioactive materials by road are given in reference 15. These Regulations came into force on 20 June 1996 and were made under the Radioactive Material (Road Transport) Act 1991. Requirements for sending radioactive materials by post are specified in the Post Office Guide.

A full list of current regulations and guidance concerned with the transport of radioactive materials is obtainable from the Radioactive Materials Transport Division of the Department of the Environment, Transport and the Regions (tel: 020 7271 3870/3868).

3.4.9 National Radiological Protection Board

The National Radiological Protection Board (NRPB) was created by the Radiological Protection Act 1970. The Government's purpose in proposing the legislation was to establish a national point of authoritative reference in radiological protection.

The NRPB's principal duties are to advance the acquisition of knowledge about the protection of mankind from radiation hazards and to provide information and advice to those with responsibilities in radiological protection. Because ICRP is the primary international body to which governments look for guidance on radiation protection criteria, it is important for the UK to be in a position to influence the development of ICRP advice. A number of members of the NRPB staff are therefore actively involved in ICRP work. The NRPB also provides technical services to organisations concerned with radiation hazards, and training in radiological protection. Its headquarters are at Chilton and it has centres at Glasgow, Leeds and Chilton for the provision of advice and services. The services provided relate to both ionising and non-ionising radiations and include: radiation protection adviser (RPA), reviews of design, monitoring of premises, personal monitoring, record keeping, instrument tests, testing of materials and equipment, leakage tests on sealed sources and assistance in the event of incidents and accidents. The Board runs scheduled and custom-designed training courses.

3.4.10 Incidents and emergencies[4,10]

In any radiological incident or emergency, the main aim must be to minimise exposures and the spread of contamination. Pre-planning against possible incidents is essential and suitable first aid facilities should be provided. Where significant quantities of radioactive substances are to be kept, procedures for dealing with fires should be discussed in advance with the local fire service.

Spills should be dealt with immediately and appropriate monitoring of the person and of surfaces should be carried out. Anyone who cuts or wounds himself when working with unsealed radioactive material must obtain first aid treatment and medical advice. This is particularly important as contamination can be readily taken into the bloodstream through cuts. If a radioactive source is lost immediate steps must be taken to locate it and, if it is not accounted for, the appropriate regional environment Agency or Service and the HSE must be notified.

The National Arrangements for Incidents involving Radioactivity (NAIR) enables police to obtain expert advice on dealing with incidents (for example, transport accidents) that may involve radiation exposure of the public and for which no other pre-arranged contingency plans exist or, for some reason, those plans have failed to function. A source of radiological advice and assistance exists in each police administrative area – hospital physicists and health physicists from the nuclear industry, government and similar establishments. The scheme is co-ordinated by the National Radiological Protection Board at Chilton from whom further details are obtainable.

3.4.11 Non-ionising radiation

There are several forms of non-ionising electromagnetic radiation that may be encountered in industry[16,17]. They differ from γ and X-rays in that they are of longer wavelength (lower energy) and do not cause ionisation in matter. They are ultraviolet (a few tens of nanometres (nm) to 400 nm wavelength), visible (400 to 700 nm) and infrared (700 nm to 1 mm) radiations in the optical region, and microwave and radiofrequency radiations and electric and magnetic fields. The ability of radiation within one of these defined regions to produce injury may depend strongly on the wavelength. *Figure 3.4.4* illustrates the monitoring for non-ionising radiation around a mobile phone base station.

Figure 3.4.4 Mobile phone base station signal measurements (photo courtesy NRPB)

3.4.11.1 Optical radiation

Ultraviolet radiation is used for a wide variety of purposes such as killing bacteria, creating fluorescence effects, curing inks and ophthalmic surgery[18]. It is produced in arc welding or plasma torch operations and is emitted by the sun. Short wavelength ultraviolet radiation of wavelength approximately less than 240 nm is strongly absorbed by oxygen in the air to produce ozone which is a chemical hazard. The OES for ozone is 0.1 ppm. Even below this level it may cause smarting of the eyes and discomfort in the nose and throat. It has a characteristic smell.

Ultraviolet radiation does not penetrate beyond the skin and is substantially absorbed in the cornea and lens of the eye. The human organs at risk are therefore the skin and the eyes. The immediate effects are erythema (as in sunburn) and photokeratitis (arc eye, snow blindness). Long-term effects are premature skin ageing and skin cancer, and possibly cataracts. No cases of skin cancer due to occupational exposure to artificial sources of ultraviolet radiation have been identified, but a casual link between skin cancer and exposure to solar ultraviolet radiation is now accepted, particularly for those with white skin[19]. Some chemicals such as coal tar can considerably enhance the ability of ultraviolet radiation to produce damage.

Wherever possible, ultraviolet radiation should be contained[18–21]. If visual observation of any process is required, this should be through special observation ports transparent to light but adequately opaque to ultraviolet radiation. Where the removal of covers could result in accidental injurious exposures, interlocks should be fitted which either cut the power supply or shutter the source. Protection is also achieved by increasing the distance between source and person, covering the skin and protecting the eyes with goggles, spectacles or face shields.

Intense sources of visible light such as arc lamps and electric welding units and, of course, the sun can cause thermal and photochemical damage to the eye; they can also produce burns in the skin. Adequate protection is normally achieved by keeping exposures below discomfort levels.

Infrared radiation is emitted when matter is heated. The principal biological effects of exposure can be felt immediately as heating of the skin and the cornea. Long-term exposure can cause cataracts. Protection is achieved by shielding the source and through the use of personal protective equipment especially eye wear.

The intensity of laser sources in the ultraviolet, visible and infrared regions can be orders of magnitude higher than that of other optical sources. Because of their very low beam divergence some lasers are capable of delivering large high power densities to a distant target. Of particular importance is the injury that can be caused to the eye, such as retinal burns and cornea damage. Protection is achieved by the following hierarchy of controls:

1. by engineering measures through the appropriate design of equipment employing techniques such as enclosure of the device, safety interlocks, shutters, etc.;
2. through administrative means such as adequate training for operators and the provision of suitable warnings both verbal and visual;
3. as a last resort, by the provision of personal protective equipment to protect, in particular, eyes and skin.

It is also necessary to guard against stray reflections.

Lasers are widely used in the workplace for a variety of purposes ranging from cutting and welding to materials analysis and measurement. The types of laser used including their output powers vary depending on the application. The current standard for laser safety[20] provides appropriate advice to both the manufacturer and the user of laser products.

3.4.11.2 Electric and magnetic fields

Time-varying electric and magnetic fields arise from a wide range of sources that use electrical energy at various frequencies. Common sources of exposure include the electricity supply at power frequencies (50Hz in the UK), and radio waves from TV, radio, mobile phones, radar and satellite communications[22].

3.4.11.2.1 Guidelines

In the UK, restrictions on exposure to electric and magnetic fields are covered by NRPB guidelines[23]. The International Commission on Non-ionising Radiation Protection (ICNIRP) has also published international guidelines[24]. Both sets of guidelines aim to provide the rationale and conceptual framework for a system of restrictions on human exposure to electric and magnetic fields and radiation. They are based on the avoidance of adverse consequences of the direct effects of exposure from the consideration of biological responses through extensive reviews of the scientific literature on biological effects, human health and research on dosimetry. Consideration is also given to the avoidance of the indirect effects of exposure such as repeated microshocks (spark discharges), electric shock and radiofrequency burn. Because the basic restrictions are often dosimetric quantities that are not easily measurable the guidelines also incorporate investigation or reference levels used for the purpose of comparing measurable quantities of radiation to establish whether compliance with basic restrictions is achieved. The NRPB guidelines discriminate between occupational and general public exposures and incorporate an additional reduction factor of up to five for the general public.

3.4.11.2.2 Low frequency fields

At extremely low frequencies the guidelines are intended to avoid the effects of induced electric current on functions of the central nervous system and the annoying direct effect of perceptions of electric charge (causing, for example, body hairs to vibrate). At frequencies between 10 Hz and 1 kHz the NRPB guidelines incorporate a basic restriction of 10 mA m^{-2} and this becomes progressively larger at frequencies above and below this range. The NRPB investigation levels for exposure to 50 Hz electric and magnetic fields are 12 kilovolts per metre (kV m^{-1}) and 1600 microtesla (μT) respectively. The corresponding ICNIRP reference levels are 10 kV m^{-1} and 500 μT for occupational exposure, and 5 kV m^{-1} and 100 μT for public exposure. The lower ICNIRP levels for public exposure reflect the lower basic restriction of 2 mA m^{-2} for members of the public.

3.4.11.2.3 Electromagnetic fields

Heating is the major consequence of exposure to RF (including microwave) radiation. Restrictions on exposure in terms of *specific energy absorption rate* (SAR) are intended to prevent responses to increased heat load and elevated body temperature. A whole-body SAR restriction of 0.4 W kg^{-1} in the NRPB guidelines is intended to provide adequate protection against heating. Localised exposure is restricted to 10 W kg^{-1} in the head and trunk and 20 W kg^{-1} in the limbs. The corresponding ICNIRP limits are 0.08 W kg^{-1} for whole body SAR, 2 W kg^{-1} in the head and trunk and 4 W kg^{-1} in the limbs. The guidelines give instructions for time-averaging SAR and averaging of masses of tissue for the partial-body restrictions. Additional restrictions are also incorporated to allow for the interaction of pulsed RF (including microwave) radiation with body tissue.

There is no specific legislation that relates to protection from electromagnetic fields. Nonetheless, there is enabling legislation in the general area of health and safety that places a duty of care on the operators of equipment generating electromagnetic fields. Government departments and agencies such as the HSE have looked to compliance with NRPB guidelines in order to fulfil this responsibility. However, the profile of the ICNIRP guidelines has recently been raised within the UK in relation to the exposure of the general public. The exposure restrictions advised by ICNIRP are incorporated into a European Council Recommendation on public exposure that applies to all Member States including the UK[25].

There is much interest in the possibility of health hazards at levels of electric and magnetic fields much lower than the guidelines. However, a number of recent reviews[26–29] give no clear support for possible health hazards below guideline levels and it has generally been concluded that the scientific evidence is insufficient to require a change in the existing protection levels.

References

Atomic structure and radioactivity

1. Evans, R. D., *The Atomic Nucleus*, McGraw-Hill, New York (1955)
2. Royal Commission on Environmental Pollution, 6th Report, *Nuclear Power and the Environment*, Cmnd. 6618, The Stationery Office, London (1976)
3. Burchman, W. E., *Elements of Nuclear Physics*, Longman, London (1979)

Ionising radiation

4. Bennellick, E. J., Ionising radiation. In *Industrial Safety Handbook*, (Ed. W. Handley), 2nd edn, McGraw-Hill, London (1977)
5. ICRP, 1990 Recommendations of the International Commission on Radiological Protection, ICRP Publication 60, Pergamon Press, Oxford. Ann. ICRP, **21**, No. 1–3 (1991)
6. Hall, G. J., *Radiation and Life*, Pergamon Press, Oxford (1984)
7. Cox, R., Muirhead, C.R., Stather, J.W., et al., *Risk of Radiation Induced Cancer at Low Doses and Low Dose Rates for Radiation Protection Purposes*, Documents of the NRPB, Vol. 7, No. 6, NRPB, Chilton (1995)
8. Edwards, A.A. and Lloyd, D.C., *Risk from Deterministic Effects of Ionising Radiations*, Documents of the NRPB, Vol. 7, No. 3, NRPB, Chilton (1996)
9. ICRP, *General Principles for the Radiation Protection of Workers*, ICRP Publication 75, Pergamon Press, Oxford. *Ann. ICRP*, **27**, No. 1 (1997)
10. Martin, A. and Harbison, S.A., *An Introduction to Radiation Protection*, 4th edn, Chapman and Hall, London (1996)
11. Health and Safety Executive, Publication No. L 121, *Working with ionising radiation*, HSE Books, Sudbury (2000)
12. Council of the European Communities, Council Directive 96/29/Euratom of 13 May 1996 laying down basic safety standards for the protection of the health of workers and the general public against the dangers arising from ionising radiation, *Official Journal of the European Communities*, Vol. 39, L 159, 29 June 1996
13. Department of the Environment, *Radioactive Substances Act 1993*, The Stationery Office, London (1993)
14. International Atomic Energy Authority, IAEA Safety Standards Series publication ST-1, *Regulations for the Safe Transport of Radioactive Material*, 1996 Edn, IAEA, Vienna (1996)
15. Department of Transport, *The Radioactive Material (Road Transport) (Great Britain) Regulations 1996*, The Stationery Office, London (1996)

Non-ionising radiation (general)

16. McHenry, C.R., Evaluation of exposure to non-ionising radiation. In Patty's *Industrial Hygiene and Toxicology, Vol. III, Theory and Rationale of Industrial Hygiene Practice* (Eds L. V. Cralley and L. J. Cralley), John Wiley & Sons, New York (1979)

17. Sliney, D.H., Non-ionising radiation. In *Industrial Environmental Health* (Eds L. V. Cralley et al.), Academic Press, London (1972)

Optical radiation

18. McKinlay, A.F., Harlen, F. and Whillock, M.J., *Hazards in Optical Radiations. A Guide to Sources, Uses and Safety,* Adam Hilger, Bristol (1988)
19. NRPB, *Health Effects from Ultraviolet Radiation,* Documents of the NRPB, Vol. 6, No. 2, NRPB, Chilton (1995)
20. British Standards Institution, BS IEC 60825–1: *Safety of Laser Products, Part 1, Equipment Classification, Requirements and User's Guide*, BSI, London (2001)
21. Health and Safety Executive, *Guidance Notes, Medical Series No. MS 15, Welding,* HSE Books, Subdury, (1978)

Electric and magnetic fields

22. European Union, *Non-ionising Radiation: Sources, Exposure and Health Effects,* EU, Luxembourg (1996)
23. National Radiological Protection Board, *Restrictions on Human Exposure to Static and Time Varying Electromagnetic Fields and Radiation: Scientific Basis and Recommendations for the Implementation of the Board's Statement.* Documents of the NRPB **4** No 5, Chilton (1993)
24. Matthes, R., Bernhardt, J.H. and McKinlay, A.F., International Commission on Non-Ionising Radiation Protection publication ICNIRP 7/99, *Guidelines for Limiting Exposure to Non-ionising Radiation.* Märkl-Druck, München (1999)
25. European Union, Council recommendation no: 1999/519/EC *on the limitation of exposure of the general public to electromagnetic fields (0 Hz to 300 GHz),* Official Journal L 199, EU, Luxembourg (1999)
26. National Radiological Protection Board publication **5** No: 2 *Health Effects Related to the use of VDUs.* NRPB, Chilton (1994)
27. National Radiological Protection Board, publication **12** No: 1 *Electromagnetic Fields and the Risk of Cancer,* NRPB, Chilton (2001)
28. National Radiological Protection Board publication **12** No: 4 *ELF Electromagnetic Fields and Neurodegenerative Disease,* NRPB, Chilton (2001)
29. International Commission on Non-Ionising Radiation Protection publication **109** Supplement 6, *Review of the Epidemiological Literature on EMF and Health. Environmental Health Perspectives.* ICNIRP (2001)

Chapter 3.5

Noise and vibration

R. W. Smith

The first four sections of this chapter explain what noise is, how it is defined and the theory and practice behind the measurement of noise levels. The rest outlines the way the ear works and the damage that can occur to cause noise-induced hearing loss. Some of the problems created by vibrations are considered. Reference is made to the guidelines, recommendations and legislation that exist and which are aimed at limiting the harmful effects of noise in the workplace, and the nuisance effect on the community.

3.5.1 What is sound?

A vibrating plate will cause corresponding vibrations or pressure fluctuations in the surrounding air, which would then be transmitted through to the receiver. For example, when an alternating electrical signal is fed into a loudspeaker, the cone vibrates causing the air in contact with it to vibrate in sympathy, and a sound wave is produced. These pressure waves are transmitted through the air at a finite speed. This is easily

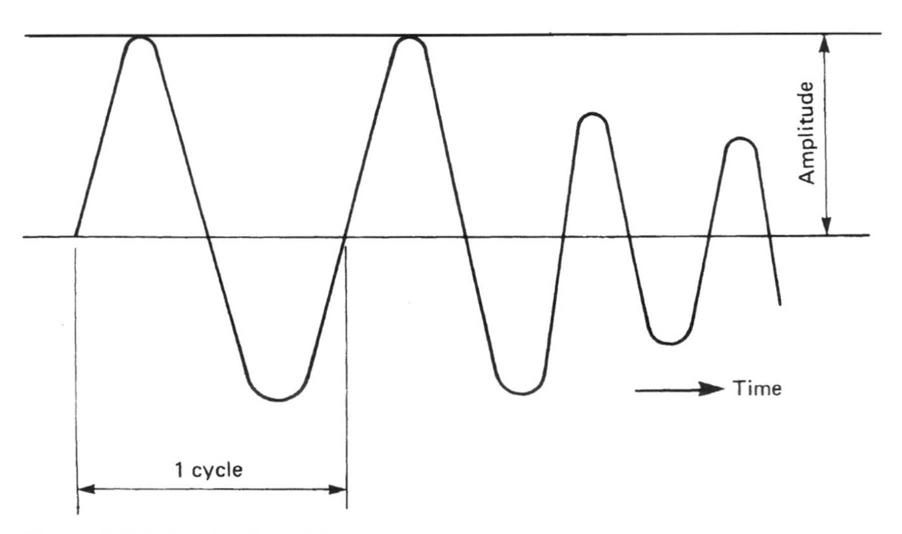

Figure 3.5.1 Amplitude and frequency

demonstrated by observing the time interval between a flash of lightning and hearing a clap of thunder. The velocity of sound in air at normal temperature and pressure is approximately 342 metres per second (1122 feet per minute), at 20°C. Increasing the temperature of air increases the velocity.

These pressure fluctuations or vibrations have two characteristics: firstly the amplitude of the vibration, and secondly the frequency – both are illustrated in *Figure 3.5.1*.

3.5.1.1 Amplitude

The amplitude of a sound wave determines loudness, although the two are not directly related, as will be explained later. Typically these pressure amplitudes are very small. For the average human being the audible range is from the threshold of hearing at 20 μPa up to 200 pascals (Pa) where the pressure becomes painful. This is a ratio of 1 to 10^6. The intensity of noise is proportional to the pressure squared hence the range of intensity covers a ratio of 1 to 10^{12}.

With such a range it becomes more convenient to express the intensity of pressure amplitude on a logarithmic base. The intensity level is proportional to the square of the pressure, thus the sound pressure level (Lp) can be defined as:

$$Lp = 10 \log_{10} (P_1/P_0)^2 \qquad (1)$$

where the sound pressure level (SPL) is expressed in decibels (dB), P_1 equals the pressure amplitude of the sound and P_0 is the reference pressure 20 μPa. All logarithmic calculations are to the base 10. Typical examples of sound pressure levels for a variety of environments are shown in *Figure 3.5.2*.

Note that, since the decibel is based on a logarithmic scale, two noise levels cannot be added arithmetically. Hence, the resultant L_{pr} from adding sources L_{p1}, L_{p2} etc. is obtained thus:

$$L_{Pr} = 10 \log_{10} \left[\left(\frac{P_1}{P_0}\right)^2 + \left(\frac{P_2}{P_0}\right)^2 + \ldots \right] \qquad (2)$$

For two equal sources $L_{p1} = L_{p2}$

$$\therefore L_{pr} = 10 \log_{10} \left[\left(\frac{P_1}{P_0}\right)^2 \times 2 \right] \qquad (3)$$

$$= 10 \log_{10} \left(\frac{P_1}{P_0}\right)^2 + 10 \log_{10} 2 \qquad (4)$$

$$= L_p + 3.1 \text{ (dB)} \qquad (5)$$

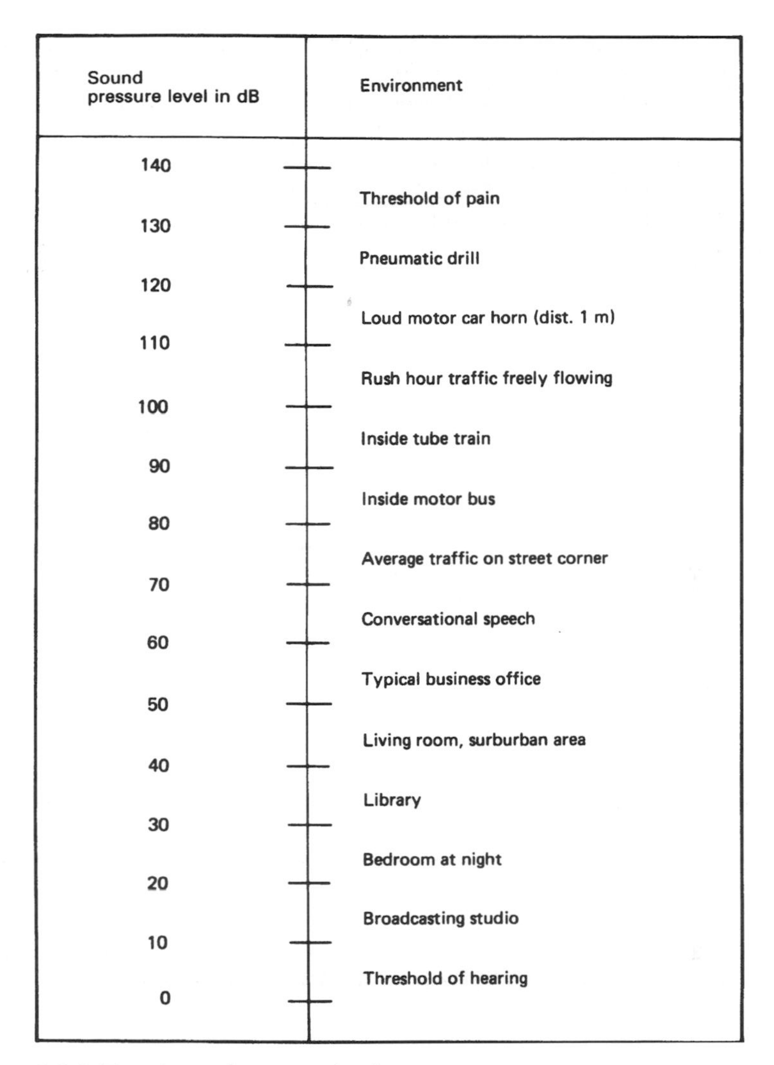

Sound pressure level in dB
Environment
140
Threshold of pain
130
Pneumatic drill
120
Loud motor car horn (dist. 1 m)
110
Rush hour traffic freely flowing
100
Inside tube train
90
Inside motor bus
80
Average traffic on street corner
70
Conversational speech
60
Typical business office
50
Living room, surburban area
40
Library
30
Bedroom at night
20
Broadcasting studio
10
Threshold of hearing
0

Figure 3.5.2 Typical sound pressure levels

Thus for all practical purposes doubling the sound intensity increases the sound pressure level by 3 dB. For example:

$$90\,\text{dB} + 90\,\text{dB} = 93\,\text{dB} \tag{6}$$

Similarly, 103 dB + 90 dB = 103.2 dB, showing that where the effect of a small source is to be added to a large source, the resultant noise level would be relatively unchanged.

Of course, the converse is true, and this is important when considering the control of noise from a number of different sources, since treatment of a minor source may not result in any change in overall levels.

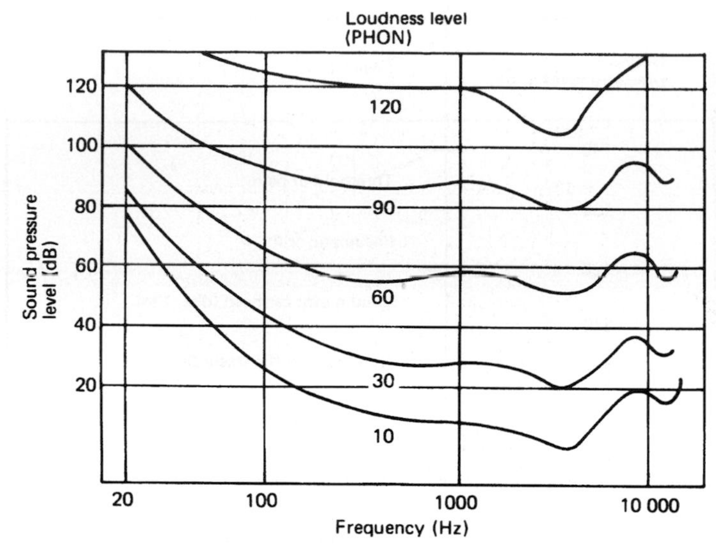

Figure 3.5.3 Equal loudness contours for pure tones

3.5.1.2 Frequency

The rate at which these pressure fluctuations take place is called frequency and is measured in hertz (Hz) or cycles/s. The human ear is normally capable of hearing over a range from 20 Hz to 16 000 Hz (16 kHz) although this range can be considerably reduced at the high frequency end for older people and for those suffering from hearing impairment.

3.5.2 Other terms commonly found in acoustics

3.5.2.1 Loudness

The human ear does not respond equally to all frequencies. To obtain the same subjective loudness at low frequencies as at higher frequencies requires a larger physical amplitude (greater L_p), since the ear is less sensitive at low frequencies. These curves of equal loudness are illustrated in *Figure 3.5.3*. The curves of equal loudness are defined in dB phon, and are obtained experimentally using pure tones to create the same sensation of loudness at different frequencies. From these curves it can be seen that the difference between physical amplitudes at different frequencies required to produce the same loudness curve reduces as the loudness increases.

As an approximation, an increase in sound pressure level of 8–10 dB corresponds to a subjective doubling of loudness. There are a number of different ways of calculating the loudness, and the loudness level in phon should be referenced to the method used.

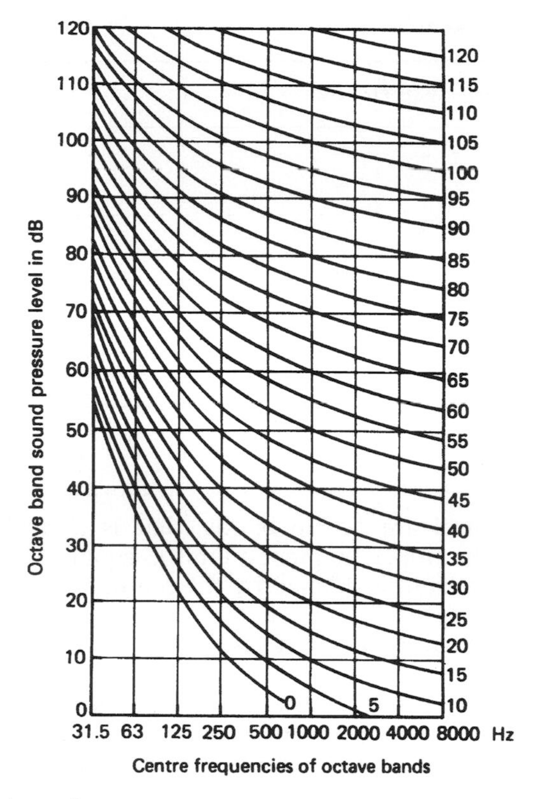

Figure 3.5.4 Noise rating curves

It is apparent that the subjective response to noise is extremely complex and these complexities should always be borne in mind when dealing with individual people. A noise or noise level which is acceptable to one individual may not be to another.

3.5.2.2 Noise rating curves

There are a standard series of Noise Rating (NR) Curves (*Figure 3.5.4* which are stylised forms of the loudness response curves. These NR curves are often used as a criterion for noise control, and as such are internationally accepted. Other criteria may also be encountered such as NC curves (Noise Criteria).

3.5.2.3 Octave bands

The previous two sections have not attempted to define frequency in terms of a bandwidth. The generally accepted bandwidth used within the

Table 3.5.1 Octave bands

Octave band centre frequency (f): Hz	*Octave band range: Hz*
31.5	22–44
63	44–88
125	88–177
250	177–354
500	354–707
1000	707–1414
2000	1414–2828
4000	2828–5657
8000	5657–11814

field of noise control are octave bands, that is a range or band of frequencies with the upper frequency limit f_u equal to twice the lower limit f_l. Each octave band is defined by the centre frequency f_{ob} where

$$f_{ob} = (f_u \times f_l)^{1/2} = (2f_l \times f_l)^{1/2} = 1.414\, f_l \qquad (7)$$

The commonly used octave bands are illustrated in *Table 3.5.1*.

Other bandwidths may be encountered in analysis and noise control work. Typically $\frac{1}{3}$-octave band widths are becoming more frequently used.

3.5.2.4 The decibel

The methods of assessing sound and developing noise criteria are complex and there are two approaches that may be used to obtain a quantitative measure of subjective noise. Either measure the physical characteristics of the sound over the frequency range on a meter and correct for the response of the ear, or alternatively, devise a measurement system which gives an output of loudness rather than amplitude. The first method can be undertaken but is rather long-winded 'by hand'. It can be approximated in instrumentation by feeding the sound pressure level into an electronic weighting network which automatically corrects the octave band levels and adds them together to give an approximation of the subjective level. The unit of measurement used is the decibel (dB).

The shapes of the weighting curves most commonly used are shown in *Figure 3.5.5* and the sound pressure level curve that follows the hearing characteristic of the ear is termed the A weighted curve. This curve gives the most widely used unit (dBA) for quantifying a noise level. Other curves on the weighting network are the B, C and D. Curves B and C are not commonly used and D is sometimes used for aircraft noise.

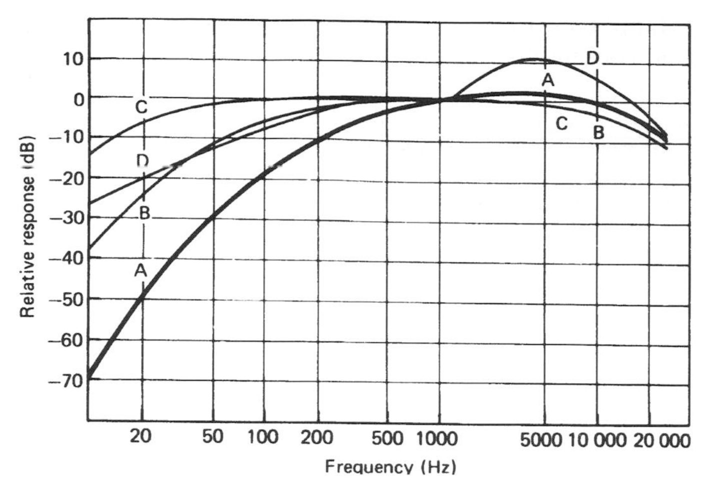

Figure 3.5.5 Weighting curves

3.5.2.5 Sound power level

The sound pressure level (L_p) must always be defined at a specific point in relation to the noise source. It is dependent on the location relative to the source, and the environment around the source and the receiver. The Sound Power Level (L_w) is a measure of the total acoustic energy radiated and is independent of the environment around the source. It is defined as:

$$L_w \text{ (dB)} = 10 \log_{10} \frac{W}{120} \qquad (8)$$

where W is the power in watts. In many cases it is calculated from sound pressure levels by the approximate formula:

$$L_w = L_p + 10 \log_{10} A \qquad (\text{m}^2) \qquad (9)$$

$$\text{or} \quad L_p + 10 \log_{10} A\text{–}10 \qquad (\text{ft}^2) \qquad (10)$$

where A = the surface area of measurement.

The concept of sound power level is best illustrated by the following analogy.

Suppose that the temperature measured at 1 metre from a 1 kW electric fire is the same as that measured 1 metre from a large steam boiler. Although the temperatures are the same, the boiler would be radiating more heat because it is radiating over a larger surface area. Similarly, the sound pressure level 1 metre from a large industrial cooling tower may be the same as that 1 metre from a small pump, but the acoustic power radiated by the cooling tower would be very much greater.

The sound power level is used to calculate the reverberant and pressure level and community noise.

3.5.3 Transmission of sound

The simplest concept in the transmission of noise is the inverse square law. For example, if the distance from the source is doubled, the reduction in noise level would be:

$$10 \log_{10} \left(\frac{2}{1}\right)^2 = 10 \log_{10} 4 = 6\,\text{dB} \qquad (11)$$

Where the sound power level is known the resulting sound pressure level at a distance is given by:

$$L_p = L_w - 10 \log 2\pi r^2\text{, where } r \text{ is in metres} \qquad (12)$$

Where distances exceed 200–300 metres other absorption factors should be taken into account. These include atmospheric attenuation and 'ground' absorption effects. Atmospheric attenuation (molecular absorption) is proportional to distance, but 'ground' absorption is dependent on the intervening terrain between the source and receiver.

3.5.4 The sound level meter

The majority of noise control work undertaken by the safety adviser will involve the measurement and possibly the analysis of noise. It is therefore important that the use and the limitations of sound pressure level measurements and sound level meters are understood. Errors in measurement technique or interpretation could lead to costly mistakes or over-specifying in remedial measures.

3.5.4.1 The instrument

There are many sound level meters on the market, but all work in a similar manner. The basic hand-held set (*Figure 3.5.6*) consists of a microphone, an amplifier with a weighting network and a read-out device in the form of a meter or digital presentation. The microphone converts the fluctuating sound pressure into a voltage which is amplified and weighted (A, B or Linear etc.). The electrical signal then drives a meter or digital read-out. In many instruments an octave or ⅓ octave band filter is incorporated, to enable a frequency analysis to be performed. The signal would then by-pass the weighting network and be fed into the read-out device via the frequency filters.

The difference between the two available grades of sound level meter, precision and industrial, is effectively the degree of accuracy of the

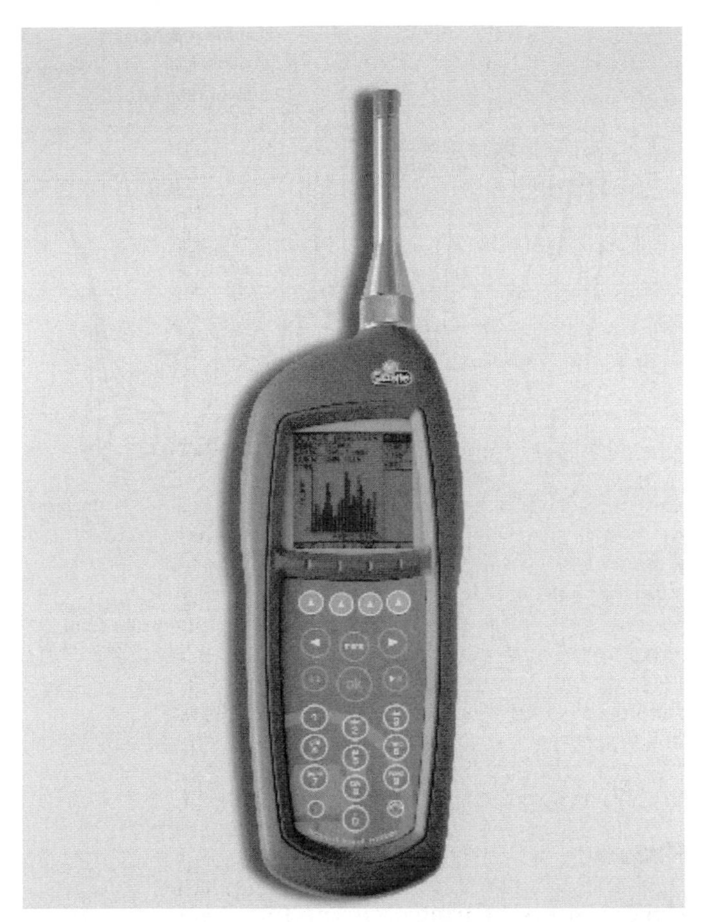

Figure 3.5.6 Industrial hand-held sound level meter. (Courtesy General Acoustics Ltd)

measurements particularly at high and low frequency. Meters should comply with British Standards[1] or with IEC[2] recommendations.

3.5.4.2 Use of a sound level meter

For the operation of sound level meters of different types the safety adviser should refer to the manufacturer's instruction book. The prime requirement for any instrument for noise measurement is that it should not be more sophisticated than necessary and it should be easy to use and calibrate. A typical sound level meter for use by the safety adviser should have the facility for measuring dBA and octave band sound pressure levels. More sophisticated meters have facilities for measuring equivalent noise level (L_{eq}, $L_{EP,d}$) (see section 3.5.6) and undertaking statistical analysis.

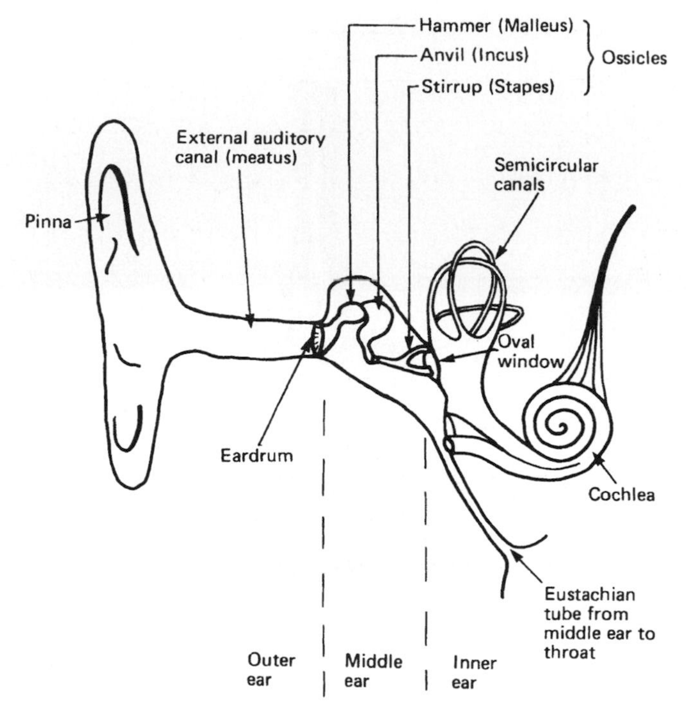

Figure 3.5.7 Diagram of human ear

3.5.5 The ear

3.5.5.1 Mechanism of hearing

Before appreciating how deafness occurs the manner in which the ear works should be understood. Briefly, the sound pressure waves enter the auditory canal (*Figure 3.5.7*) and cause the eardrum (tympanic membrane) to move in sympathy with the pressure fluctuations. Three minute bones called the hammer, anvil and stirrup (the ossicles) transmit the vibrations via the oval window to the fluid in the inner ear and hence to the minute hairs in the cochlea. These hairs are connected to nerve cells which respond to give the sensation of hearing. Damage to, or deterioration of, these hairs, whether by exposure to high noise levels, explosions or ageing will result in loss of hearing.

3.5.5.1.1 Audiogram

The performance of the ear is evaluated by taking an audiogram. Audiograms are normally performed in an Audiology Room which has a very low internal noise level and is vibration isolated. The audiogram is obtained by detecting the levels at which specific tones can be heard and cease to be heard by the person being tested.

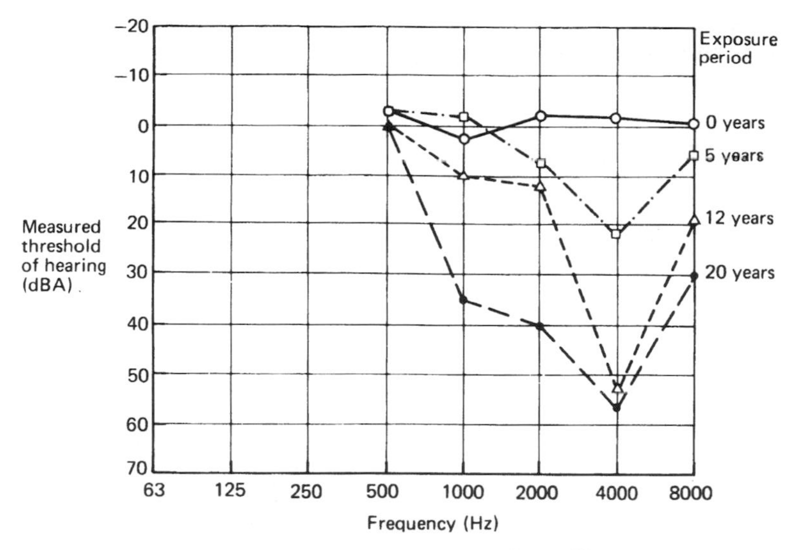

Figure 3.5.8 Typical audiograms showing progressive loss of hearing

3.5.5.1.2 Hearing loss

Noise-induced hearing loss (occupational hearing loss) is caused by over-stimulus of the receptor cells in the cochlea resulting in auditory fatigue. In its early stages it is often shown as an increase in the threshold of hearing – 'temporary threshold shift'. It may be accompanied by a ringing in the ears (tinnitus) which is indicative of temporary hearing damage. If exposure to high noise levels was continued, the result would be a 'permanent threshold shift' or noise-induced deafness. The effects of noise-induced deafness can be seen by the dip in the audiogram at 4 kHz, which deepens and widens with continuing exposure. Eventually the speech frequencies are affected (*Figure 3.5.8*), the sufferer loses the ability to hear consonants, and speech sounds like a series of vowels strung together.

Similar effects occur with age-induced hearing loss, which should always be considered when checking hearing impairment.

3.5.6 The equivalent noise level

In general the noise level in the community or inside a factory will vary with time. The equivalent noise level (L_{eq}) is defined as the notional steady noise level which, over a given period of time, would deliver the same amount of sound energy as the fluctuating level. Thus to maintain the L_{eq} when SPL is doubled, i.e. increased by 3 dB, exposure time must be halved (*Table 3.5.2*). The equivalent noise level concept forms the basis of the exposure criteria used in the Noise at Work Regulations 1989[3] which calls it 'daily personal noise exposure' ($L_{EP,d}$). Where the

Table 3.5.2 Exposure times vs. noise levels for an equivalent noise level of 90 dBA

Noise level (dBA)	*Exposure time (h)*
87	16
90	8
93	4
96	2
99	1
102	½
105	¼

fluctuation is not well defined the calculations can be done electronically using a dosimeter or statistical analyser[4]. Transient noises also require statistical analysis and those measurements often required are:

L_{10} – the noise level exceeded for 10% of the time (average peak).
L_{50} – the noise level exceeded for 50% of the time (mean level).
L_{90} – the noise level exceeded for 90% of the time (average background level).

3.5.7 Community noise levels

Noise generated in a factory may affect not only those employed in the factory but those living in the immediate neighbourhood, especially if machinery is run during night-time. Legislation has been brought into effect to give those affected the right of redress.

3.5.7.1 Control of Pollution Act 1974[5]

The Control of Pollution Act, as amended by EPA90, provides general legislation for limiting community noise as well as other pollutants. No specific limits are set, but the Act empowers the local authority to require a reduction in noise emission and impose conditions for noisy operations, e.g. specify a level of noise emission for a particular operation which must not be exceeded at certain given times. In certain instances they may not only set the limits but specify how they are to be met or how equipment is to be operated, referring to the relevant British Standard[6].

3.5.7.2 Method of rating industrial noise in the neighbourhood

Criteria that can be used to assess the reasonableness of complaints and for setting noise limits for design purposes are contained in a British

Standard[7] which in itself does not recommend specific limits, but predicts the likelihood of complaints. These predictions are based on measured or predicted noise levels, corrections for the noise characteristic and measured or notional background noise levels.

3.5.7.3 Assessing neighbourhood noise

The following procedure is typical for evaluating the existing neighbourhood noise situation:

1 Identify critical areas outside the factory by taking measurements over a reference period, as defined in the standard.
2 Note which equipment in the factory is operating during the measurement period to help evaluate the major sources.
3 Measure the L_{90} or background noise level. If the factory is always operating, the notional background noise level should be assessed using BS 4142[7]. Comparison of the factory noise level with the background level as explained in that British Standard, or with any limit set by the local authority will show whether further action is required.
4 Where there is possibility of complaints, or complaints have been made, a noise control programme should be introduced to reduce the noise from the major sources identified in stage (2).

3.5.8 Work area noise levels

Legislation, in the form of the Noise at Work Regulations 1989[3], has been enacted to limit the risk of hearing damage to workpeople while at work. These regulations give effect to an EC directive[8] on the protection of workers from the risks related to noise at work. Under the regulations, employers are required to take certain actions where noise exposure reaches 'Action Levels'

First action level – $L_{EP,d}$ = 85 dB(A)
Second action level – $L_{EP,d}$ = 90 dB(A)
Peak action level – a peak sound pressure of 200 pascals.

Where employees are likely to be exposed to noise levels at or above the first action or peak action level, the employer must ensure that a noise survey is carried out by a competent person (reg. 4). Records of this assessment must be kept (reg. 5). In some cases the level of noise to which an employee is exposed varies considerably over the duration of a shift making it difficult to assess the exposure from a single or a few spot measurements. The use of a dosimeter will enable a single figure to be derived for a shift. The dosimeter is worn with the microphone close to the ear for the period of the shift and the results either read directly from the instrument or downloaded to a computer. When using the dosimeter it is important that the wearer works to his or her normal pattern and does not spend longer than normal periods in a noisy environment.

Where a noise risk has been identified there is a requirement on the employer to reduce the emission to the lowest reasonably practicable level (regs 6 and 7).

If it is not reasonably practicable to reduce noise exposure levels to below 90 dB(A) $L_{EP.d}$ then the employer is required to provide personal hearing protection (reg. 8). An alternative to personal protection is the provision of hearing havens from which the worker can carry out his duties. Areas where there remains a hearing risk shall be designated 'ear protection zones' and be identified as such (reg. 9).

All hearing protection equipment must be kept in good order (reg. 10) and employees are required to use properly and look after any personal protective equipment issued to them. Considerable emphasis is placed on keeping employees informed (reg. 11) of:

- risks to their hearing,
- steps being taken to reduce the risk,
- procedures for obtaining personal protective equipment, and
- their obligations under the Regulations.

Regulation 12 modifies s. 6 of HSW to impose a duty on manufacturers and suppliers to inform customers when the level of noise emitted by their product exceeds the first action level, i.e. 85 dB(A) $L_{EP.d}$.

3.5.9 Noise control techniques

Before considering methods of noise control, it is important to remember that the noise at any point may be due to more than one source and that additionally it may be aggravated by noise reflected from walls (reverberant noise) as well as the noise radiated directly from the source. With any noise problems there are the three distinct elements shown in *Figure 3.5.9* – source, path and receiver.

Having identified the nature and magnitude of any noise problem, the essential elements of a noise control programme are provided. Where a problem is evident, there are three orders of priority for a solution:

1. Engineer the problem out by buying low noise equipment, altering the process or changing operating procedures.
2. Apply conventional methods of noise control such as enclosures or silencers.
3. Where neither of the above approaches can be used, the last resort of providing personal protection should be considered.

3.5.9.1 Source

Although the control of noise at source is the most obvious solution, the feasibility of this method is often limited by machine design, process or operating methods. While immediate benefits can be obtained, this method should be regarded as a long-term solution.

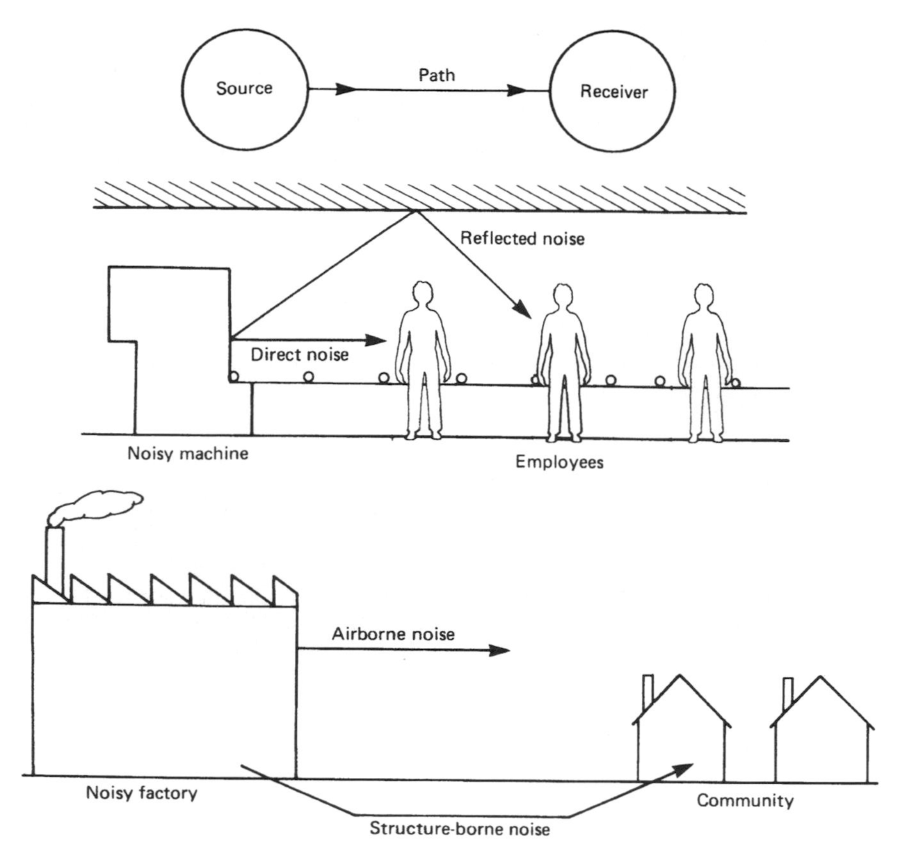

Figure 3.5.9 Noise source, path and receiver

3.5.9.2 Path

(a) Orientation and location

Control may be achieved by moving the source away from the noise sensitive area. In other cases where the machine does not radiate equally in all directions, turning it round can achieve significant reductions.

(b) Enclosure

Enclosures which give an attenuation of between 10 and 30 dBA[9] are the most satisfactory solution since they will control both the direct field and reverberant field noise components. In enclosing any source, the provision of adequate ventilation, access and maintenance facilities must be considered. A typical enclosure construction is shown in *Figure 3.5.10*. The main features are an outer 'heavy' wall with an inner lining of an acoustically absorbent material to minimise reverberant build-up inside the enclosure. An inner mesh or perforate panel may be used to minimise mechanical damage.

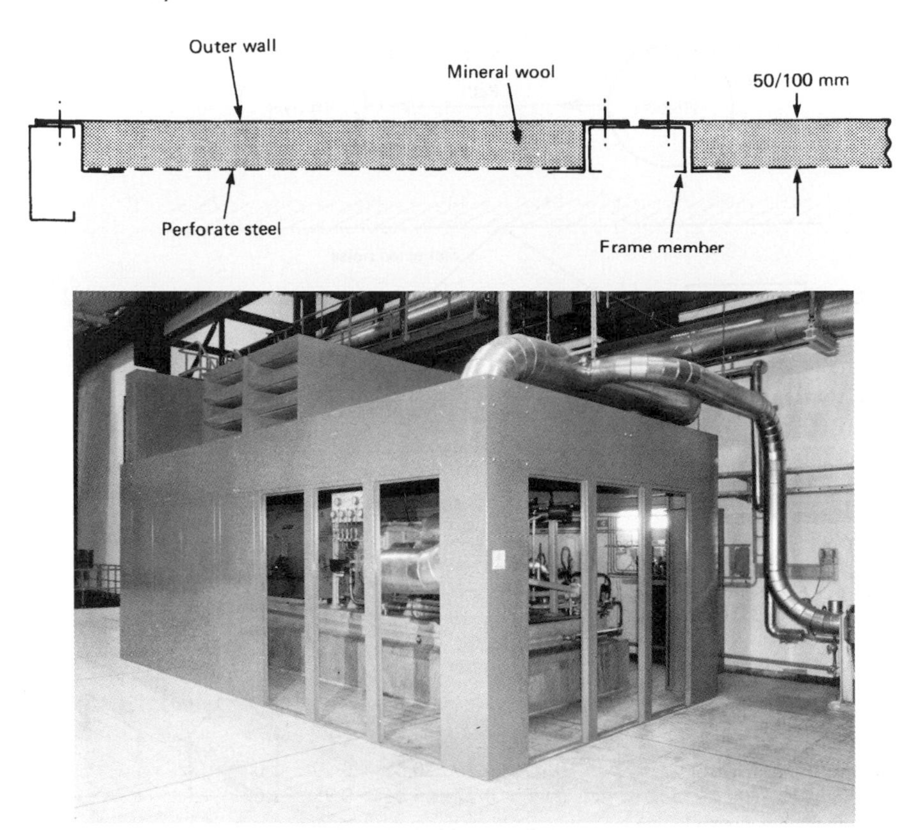

Figure 3.5.10 Noise enclosure. (Above) cross-section through typical noise enclosure wall; (below) noise enclosure with access doors removed. (Courtesy Ecomax Acoustics Ltd)

The sound reduction, attenuation or insertion loss is defined as the difference in sound pressure level or sound power level before and after the enclosure (or any other form of noise control) is installed. The performance of the enclosure will be largely dependent on the sound reduction index (SRI) of the outer wall[10], assuming approximately 50% of the internal surface is covered with mineral wool or other absorption materials[11]. Typical values of sound reduction index for materials used for enclosures are shown in *Table 3.5.3*. Absorption coefficients are shown in *Table 3.5.4*.

When considering an enclosure for any item of equipment, it is essential to consider a number of other aspects as well as the noise reduction required. Ventilation may be required to prevent overheating of the equipment being enclosed. Where ventilation is required each vent should be silenced. Special consideration should be given to the access requirements of maintenance and production, and the designers should ensure that these requirements are considered at an early design stage. In selecting any form of noise control, care should be taken to ensure that the

Table 3.5.3 Sound reduction indices for materials commonly used for acoustic enclosures

Material	*Octave band centre frequency: Hz*							
	63	*125*	*250*	*500*	*1K*	*2K*	*4K*	*8K*
22 g steel	8	12	17	22	25	26	25	29
16 g steel	12	14	21	27	32	37	43	44
Plasterboard	10	15	20	25	28	31	34	38
1/4 in plate glass	12	16	19	21	22	36	31	34
Unplastered brickwork (100 mm thick)	22	31	36	41	45	(50)	(50)	(50)

Bracketed figures refer to practical installations rather than test conditions.

Table 3.5.4 Typical sound absorption coefficients

Material	*Octave band centre frequency: Hz*							
	63	*125*	*250*	*500*	*1K*	*2K*	*4K*	*8K*
25 mm mineral wool	0.05	0.08	0.25	0.50	0.70	0.85	0.85	0.80
50 mm mineral wool	0.10	0.20	0.55	0.90	1.00	1.00	1.00	1.00
50 mm melamine foam	0.07	0.22	0.46	0.95	1.00	1.00	1.00	1.00
100 mm mineral wool	0.25	0.40	0.80	1.00	1.00	1.00	1.00	1.00
Typical perforated ceiling tile with 200 mm air space behind	0.30	0.50	0.80	0.95	1.00	1.00	1.00	1.00
Glass	0.12	0.10	0.07	0.04	0.03	0.02	0.02	0.03
Concrete	0.05	0.02	0.02	0.02	0.04	0.05	0.05	0.06

N.B. Concrete would normally be used for floor of enclosure.

equipment will physically withstand an industrial environment especially if it is particularly hostile. It must be robust and be capable of being dismantled and reassembled.

(c) Silencers[12]

Silencers are used to suppress the noise generated when air, gas or steam flow in pipes or ducts or are exhausted to atmosphere. They fall into two forms:

1 Absorptive, where sound is absorbed by an acoustical absorbent material.
2 Reactive, where noise is reflected by changes in geometrical shape.

Figure 3.5.11(a) shows a typical layout for an absorptive silencer, while *Figure 3.5.11(b)* shows a combination of the two types. The absorptive

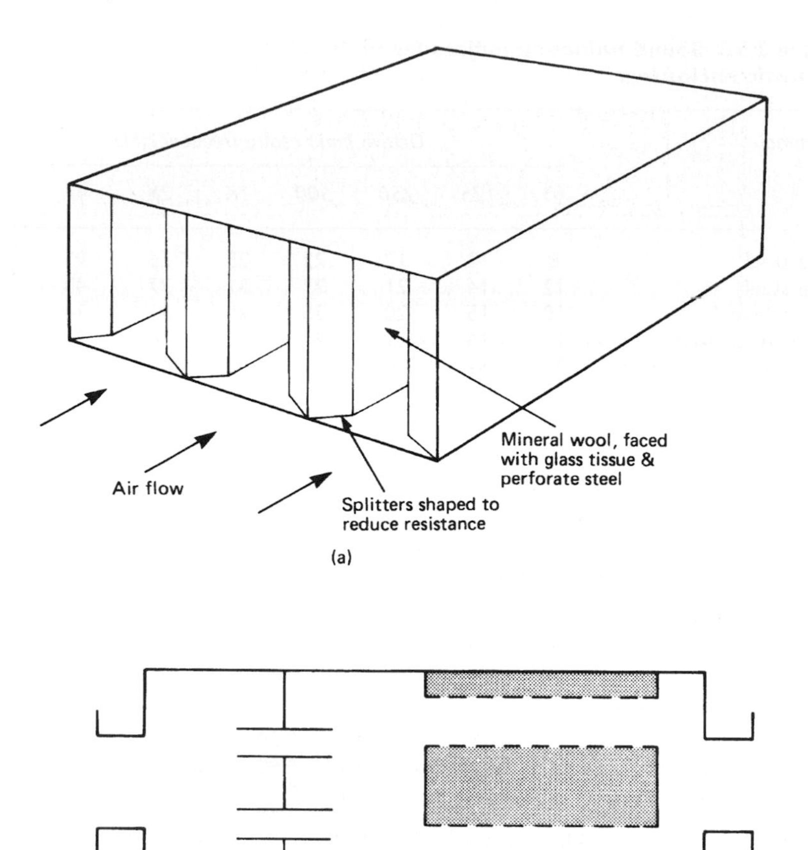

Figure 3.5.11 Typical silencers (a) absorptive splitter silencer; (b) combination reactive/absorptive silencer

silencer normally has the better performance at higher frequencies, whereas the reactive type of silencer is more effective for controlling low frequencies.

The performance of splitter type of silencers is dependent on its physical dimensions. In general:

1 Sound reduction or insertion loss increases with length.
2 Low frequency performance increases with thicker splitters and reduced air gap.

Similarly for cylindrical silencers, the overall performance improves with length and the addition of a central pod. Performance would be limited by the sound reduction achievable by the silencer casing and other flanking paths. Typically no more than 40–50 dB at the middle frequencies could be expected without special precautions.

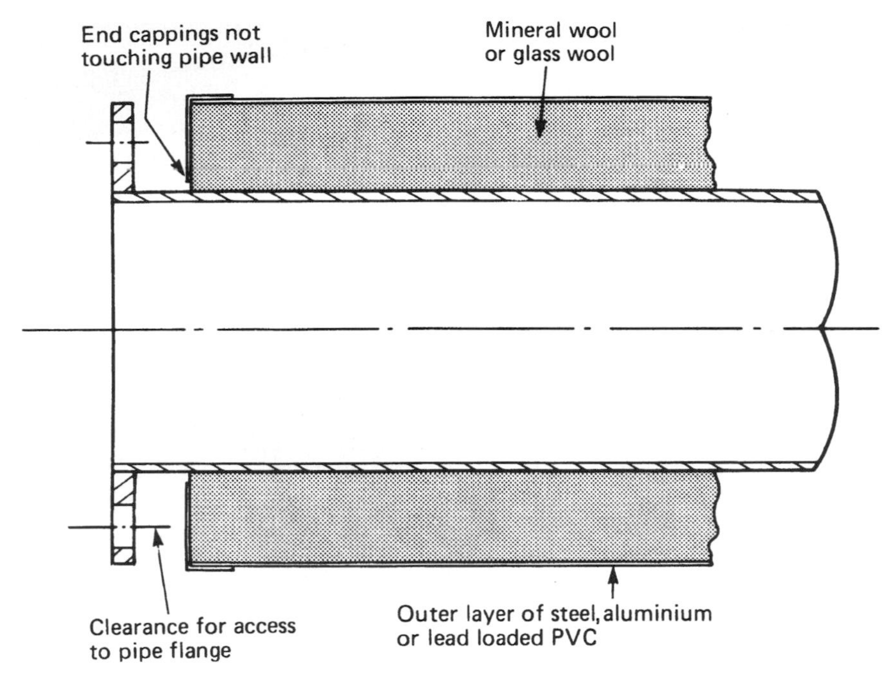

Figure 3.5.12 Pipe lagging

(d) Lagging[13]

On pipes carrying steam or hot fluids thermal lagging can be used as an alternative to enclosure and can achieve attenuations between 10 and 20 dBA, but it is only effective at frequencies above 500 Hz. The cross-section shown in *Figure 3.5.12* illustrates the main features of mineral wool wrapped around the pipes with an outer steel, aluminium or lead loaded vinyl layer. It is important that there is no contact between the outer layer and the pipe wall, otherwise the noise-reducing performance may be severely limited.

(e) Damping[14]

Where large panels are radiating noise a significant reduction can be achieved by fitting proprietary damping pads, fitting stiffening ribs or using a double skin construction.

(f) Screens

Acoustic screens (*Figure 3.5.13*) are effective in reducing the direct field component noise transmission by up to 15 dBA[15]. However, they are of maximum benefit at high frequencies, but of little effect at low frequencies and their effectiveness reduces with distance from the screen.

(g) Absorption treatment

In situations where there is a high degree of reflection of sound waves, i.e. the building is 'acoustically hard', the reverberant component can

Figure 3.5.13 Acoustic screens. (Courtesy Ecomax Acoustics Ltd)

Figure 3.5.14 Acoustic absorption treatment showing suspended panels and wall treatment. (Courtesy Ecomax Acoustics Ltd)

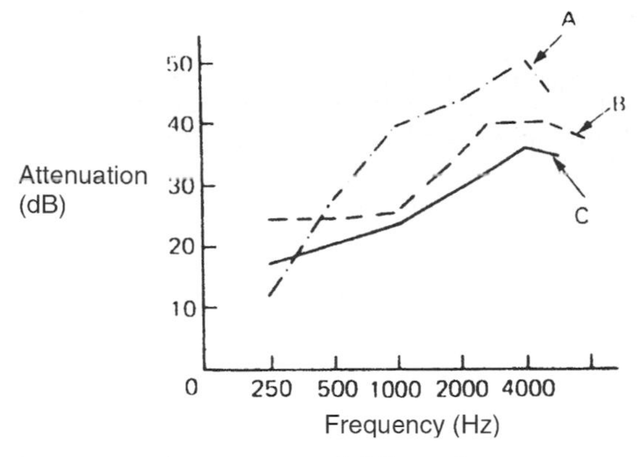

Figure 3.5.15 Attenuation characteristics of different hearing protectors (A: ear-muffs; B: plastic foam ear-plugs; C: acoustic wool ear-plugs)

dominate the noise field over a large part of the work area. The introduction of an acoustically absorbent material in the form of wall treatment and/or functional absorbers at ceiling height as shown in *Figure 3.5.14* will reduce the reverberant component by up to 10 dBA[11], but will not reduce the noise radiated directly by the source. Materials which may be used for absorption include mineral wool, glass fibre or open cell foams. The latter can be supplied as melamine foam which has good fire resistance properties.

3.5.9.3 Personnel protection[16]

The two major methods of personnel protection are the provision of a quiet room or peace haven, and the wearing of ear-muffs or ear-plugs. The peace haven is similar in construction to an acoustic enclosure, and is used to keep the noise out. Ear-muffs or ear-plugs should be regarded only as the final resort to noise control. Their selection should be made with care having regard for the noise source, the environment and comfort of the wearer. Earplugs are only generally effective up to noise levels of 100–105 dBA while ear-muffs can provide protection at higher noise levels to meet a 90 dBA criterion, for noise received by the wearer. Comparative attenuation characteristics for various personal hearing protection devices are shown in *Figure 3.5.15*.

3.5.9.4 Effective noise reduction practices

A number of practical techniques can be used as part of normal day-to-day operational and maintenance procedures that will achieve significant reductions in the noise emitted, will cost nothing or very little to implement and can, additionally, give worthwhile savings in energy. Some of these techniques are listed below:

3.5.9.4.1 On plant

1. Tighten loose guards and panels.
2. Use anti-vibration mounts and flexible couplings.
3. Planned maintenance with programme for regular lubrication for both oil and grease.
4. Eliminate unnecessary compressed air and steam leaks, silence air exhausts.
5. Keep machinery properly adjusted to manufacturer's instructions.
6. Use damped or rubber lined containers for catching components.
7. Switch off plant not in use, especially fans.
8. Use rubber or plastic bushes in linkages, use plastic gears.
9. Resite equipment and design-in noise control.
10. Specify noise emission levels in orders, i.e. 85 dB(A) at 1 metre.
11. Check condition and performance of any installed noise control equipment.

3.5.9.4.2 Community noise

1. Keep doors and windows closed during anti-social hours.
2. If loading is necessary, carry out during day.
3. Vehicle manoeuvring – stop engine once in position.
4. Check condition and performance of any noise control equipment and silencers.
5. Turn exhaust outlets from fans away from nearby houses.
6. Carry out spot checks of noise levels at perimeter fence, both during working day and at other times.

3.5.9.5 Noise cancelling

Recent experiments to cancel the effects of noise electronically by emitting a signal that effectively flattens the noise wave shape have achieved some success in low frequency operation and situations where the receiver position and the source emission are well defined. However, the industrial application of this technique is still in its infancy.

3.5.9.6 Sound masking

Where speech privacy is important, sound masking can be introduced to raise the background noise and hence mask speech or some tonal noises. This principle can only be used successfully in rooms or offices where the conditions are not too reverberant and the background noise level is less than 45 dBA.

3.5.10 Vibration

Vibration can cause problems to the human body, machines and structures, as well as producing high noise levels. Vibration can manifest itself as a particle displacement, velocity or acceleration. It is more commonly defined as an acceleration and may be measured using an accelerometer. There are many types of accelerometer and associated instrumentation available which can give an analogue or digital readout or can be fed into a computerised analysis system. As with sound, the vibration component would be measured at particular frequencies or over a band of frequencies.

3.5.10.1 Effect of vibration on the human body[17]

Generally, it is the lower frequency vibrations that give rise to physical discomfort. Low frequency vibration (3–6 Hz) can cause the diaphragm in the chest region to vibrate in sympathy giving rise to a feeling of nausea. This resonance phenomenon is often noticeable near to large slow speed diesel engines and occasionally ventilation systems. A similar resonance affecting the head, neck and shoulders is noticeable in the 20–30 Hz frequency region while the eyeball has a resonant frequency in the 60–90 Hz range.

The use of vibratory hand tools, such as chipping hammers and drills which operate at higher frequencies, can cause 'vibration induced white finger' (VWF). The vibration causes the blood vessels to contract and restrict the blood supply to the fingers creating an effect similar to the fingers being cold. Currently there are a number of cases where employees have instituted legal proceedings for VWF.

The effects of vibration on the human body will be dependent on the frequency, amplitude and exposure period and hence it is difficult to generalise on what they will be. However, it is worthwhile remembering that in addition to the physiological effects vibration can also have psychological effects such as loss of concentration.

3.5.10.2 Protection of persons from vibration

Where the source of the vibration cannot be removed, protection from whole body vibrations can be provided by placing the persons in a vibration isolated environment. This may be achieved by mounting a control room on vibration isolators in such areas as the steel industry, or simply having isolated seating such as on agricultural machinery.

For segmental vibration such as VWF consideration should be given to alternative methods of doing the job such as different tools.

3.5.10.3 Machinery vibration

For machines that vibrate badly, apart from the increased power used and damage to the machine and its supporting structure, the vibrations can

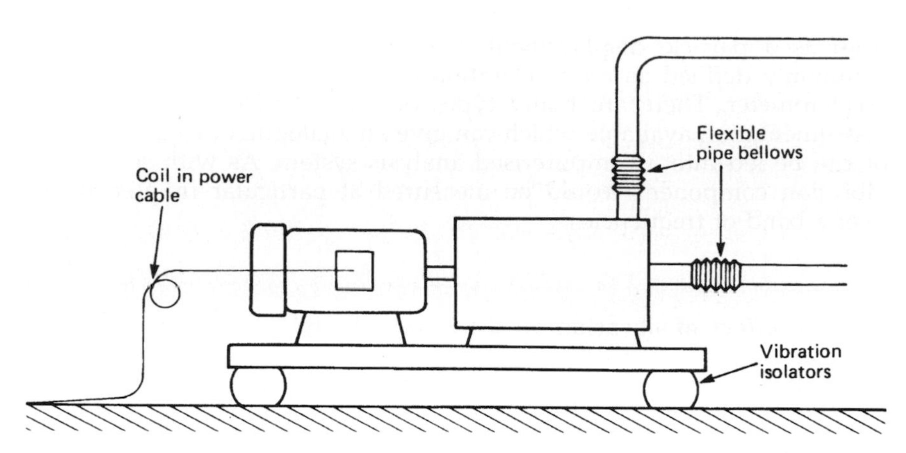

Figure 3.5.16 Vibration isolation

travel through the structure of the building and be radiated as noise at distant points (structure-borne noise).

Where the balance of the moving parts of a machine cannot be improved, vibration transmission can be reduced by a number of methods[18] of which the most commonly used are:

1 Mount the machine on vibration isolators or dampers.
2 Install the machine on an inertia block with a damping sandwich between it and the building foundations.

The method chosen will depend on the size and weight of the machine to be treated, the frequency of the vibration to be controlled and the degree of isolation required. Whichever form of vibration isolation (*Figure 3.5.16*) is selected, care should be taken to ensure that the effect is not nullified by 'bridges'. For example, isolation of a reciprocating compressor set would be drastically reduced by rigid piping connecting it to its air receiver or distribution pipework, or by conduiting the cables to the motor. In severe cases rigid piping would fracture in a very short time.

3.5.11 Summary

The treatment of any noise source or combination of sources may use any of the control techniques individually or in combination. The selection of suitable measures will depend on:

1 The type of noise field – whether dominated by the direct noise radiated from the machine or the reverberant field.
2 The degree of attenuation required.
3 Whether work area limits or community noise limits are to be met.
4 Its cost effectiveness.

References

1. British Standards Institution, BS EN 60651:1994 *Specification for sound level meters*, BSI, London (1994)
2. International Electrotechnical Commission, IEC Standard 651, *Sound Level Meters*, IEC, Geneva (or British Standards Institution, London) (1979)
3. HM Government, *The Noise at Work Regulations 1989*, SI 1989 No. 1790, The Stationery Office, London (1989)
4. Hassall, J. R. and Zaveri, K., *Acoustic Noise Measurements*, Bruel and Kjaer, Naerum, Denmark (1979)
5. *Control of Pollution Act 1974*, part III, The Stationery Office, London (1974)
6. British Standards Institution, BS 5228:1975, *Code of practice for noise control on construction and demolition sites*, BSI, London (1975)
7. British Standards Institution, BS 4142:1997, *Method of rating industrial noise affecting mixed residential and industrial areas*, BSI, London (1997)
8. European Economic Community, Directive No. 86/188/EEC *On the protection of workers from the risks related to exposure to noise at work*, Official Journal No. L137 of 24 May 1986, p. 28, The Stationery Office, London (1986)
9. Warring, R. A. (Ed.), *Handbook of Noise and Vibration Control*, 509, Trade and Technical Press, London (1970)
10. Ref. 9, p. 474
11. Ref. 9, p. 460
12. Ref. 9, p. 543
13. Ref. 9, p. 249
14. Ref. 9, p. 595
15. Ref. 9, p. 504
16. Ref. 9, p. 571
17. Ref. 9, p. 112
18. Ref. 9, p. 586

Further reading

Blitz, J., *Elements of Acoustics*, Butterworth, London (1964)

Beranek, L.L., *Noise and Vibration Control*, McGraw-Hill, New York (1971)

Health and Safety Executive, Health and Safety at Work Series Booklet No. 25, *Noise and the Worker*, HSE Books, Sudbury (1976)

Health and Safety Executive, Report by the Industrial Advisory Subcommittee on Noise, *Framing Noise Legislation*, HSE Books, Sudbury (undated)

Sharland, I., *Wood's Practical Guide to Noise Control*, Woods of Colchester Ltd, England (1972)

Taylor, R., *Noise*, Penguin Books, London (1970)

Webb, J.D. (Ed.), *Noise Control in Industry*, Sound Research Laboratories Ltd, Suffolk (1976)

Burns, W. and Robinson, D., *Hearing and Noise in Industry*, The Stationery Office, London (1970)

Burns, W. and Robinson, D., *Noise Control in Mechanical Services*, Sound Attenuators Ltd, and Sound Research Laboratories Ltd, Colchester (1972)

Health and Safety Executive, *Noise 1990 and You*, HSE Books, Sudbury (1989)

Health and Safety Executive, Legal Series Booklet No. L108, *Guidance on the Noise at Work Regulations 1989*, HSE Books, Sudbury (1998)

Chapter 3.6

Workplace pollution, heat and ventilation

F. S. Gill

The solution of many workplace environmental problems, whether due to the presence of airborne pollutants such as dust, gases or vapours, or due to an uncomfortable or stressful thermal environment, lies in the field of ventilation. Ventilation can be employed in three ways:

1 By using extraction as close to the source of pollution as possible to minimise the escape of the pollutant into the atmosphere. The extraction devices can be either hoods, slots, enclosures or fume cupboards coupled to a system of ducts, fans and air cleaners.
2 By providing sufficient dilution ventilation to reduce the concentration of the pollutants to what is thought to be a safe level.
3 By using air as a vehicle for conveying heat or cooling to a workplace to maintain reasonably comfortable conditions by employing air conditioning or a warm air ventilation system.

A flow of air which may be part of an industrial process can have a substantial effect upon the safety of the workplace by removing – or not – excessive heat, fume or dust. Such processes as ink drying, solvent collection, particulate conveying could fall into this category.

Before embarking upon the design for a ventilation system, it is necessary to assess the extent of the problem, that is, the amount of airborne pollution to be encountered in a workplace, and/or the degree of discomfort or stress expected from a thermal environment. Measurement and analysis techniques need to be devised and criteria and standards applied to the environment under consideration. Where measurement and analysis are concerned, the physics and the chemistry of the properties of the pollutant and its mode of emission need to be studied in such a way that a reliable and accurate assessment of the exposure of a worker can be made. As far as criteria and standards are concerned, medical evidence, biological research and epidemiological methods need to be applied to establish the relationship between the exposure and the long- and short-term effect upon the human body of the worker taking into account the duration of exposure and the work rate. It can be seen, therefore, that many scientific skills require to be involved before a judgement can be made and a method of control devised.

3.6.1 Methods of assessment of workplace air pollution

Airborne pollutants can be divided roughly into three groups:

1 dusts and fibres;
2 gases and vapours;
3 micro-organisms (bacteria, fungi etc.);

although some emissions from workplaces, for example oil mist from machine tools, could contain material from each group.

There is a wide range of techniques available to measure the degree of workplace pollution, some of which are described below. First it is important to decide what information is required as the technique chosen will determine whether the concentration of airborne pollution is measured:

(a) in the general body of the workroom at an instant of time or averaged over the period of work, the latter being known as a time weighted average (TWA); or
(b) in the breathing zone of a worker averaged over the period of work; or
(c) in the case of dusts, as the total or respirable airborne dust (respirable dust is that which reaches the inner part of the lungs).

If a time weighted average is required, it might also be necessary to know whether any dangerous peaks of concentration occurred during the work period.

Owing to air currents in workrooms, pollutants can move about in clouds; thus concentrations vary with place and time and some statistical approach to measurement may be required. Also some workers move about from place to place in and out of polluted atmospheres and, while workplace concentrations might be high, operator exposure levels might be lower.

In addition, workers may be exposed to a variety of different pollutants during the course of a shift, some being more toxic than others, so that the degree of exposure to each might be required to be known. When a mixture of pollutants occurs, the presence of one may interfere with the measurement of another. Therefore, the measurement of the degree of exposure of a worker must be undertaken with care. Occupational (or industrial) hygienists are trained in the necessary skilled techniques and should be called in to carry out the measurements.

3.6.1.1 Airborne dust measurement

The commonest method for measuring airborne dust is the filter method where a known volume of air is drawn through a pre-weighed filter paper (*Figure 3.6.1*) or membrane by means of a pump (*Figure 3.6.2*). The filter can be part of a static sampler located at a suitable place in the

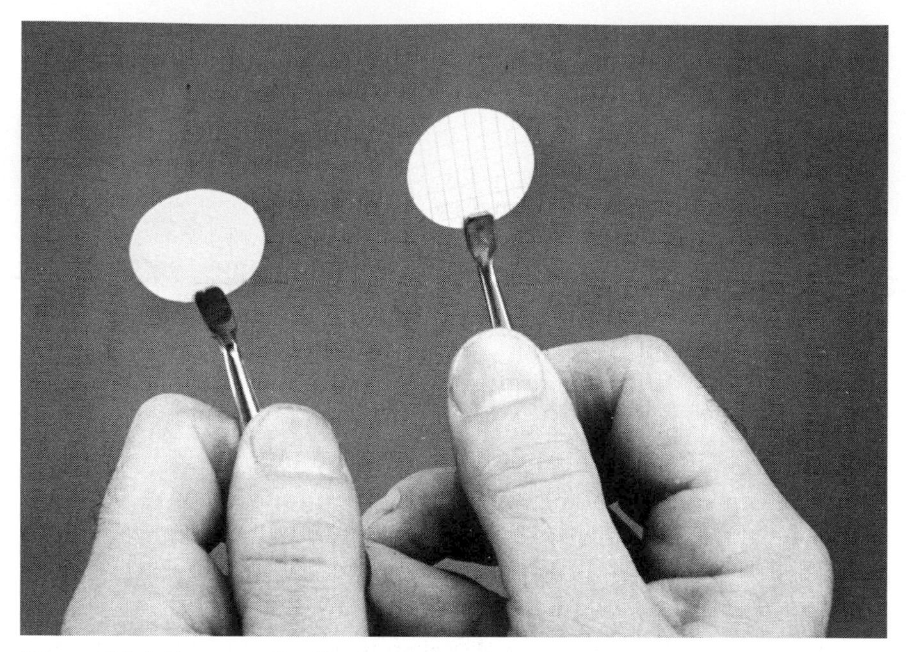

Figure 3.6.1 Dust sampling filters

workroom or it can be contained in a special holder attached to a person as close to the face as possible, usually fixed to the lapel or shoulder strap of a harness and connected by tubing to the pump which is attached to the wearer's belt. At the end of the sampling period the filter is weighed again and the difference in weight represents the weight of dust collected. This, divided by the total volume of air which has passed through the filter, gives the average concentration of dust over the period.

If a membrane-type filter made of cellulose acetate is used, it can be made transparent by the application of a clearing fluid, allowing the dust to be examined under a microscope and the particles or fibres counted if required. This is particularly important if fibrous matter such as asbestos is present.

Other filters can be chemically digested so that the residues can be further examined by a variety of chemical means. Types of filter are available that are suitable for examining the collected dust by X-ray diffraction, X-ray fluorescent techniques or by a scanning electron microscope. Thus, the choice of filter must be related to the type of analysis required. Weighing must be accurate to five decimal places of grams and, as air humidity can affect the weight of some filters, preconditioning may be required.

When on personal samplers the level of respirable dust is required, a device such as a cyclone is used to remove particles above 10 μm in diameter. Static samplers use a parallel plate elutriator for this purpose which allows the larger particles to settle so that only the respirable dust reaches the filter. With all separation devices the airflow rate must be

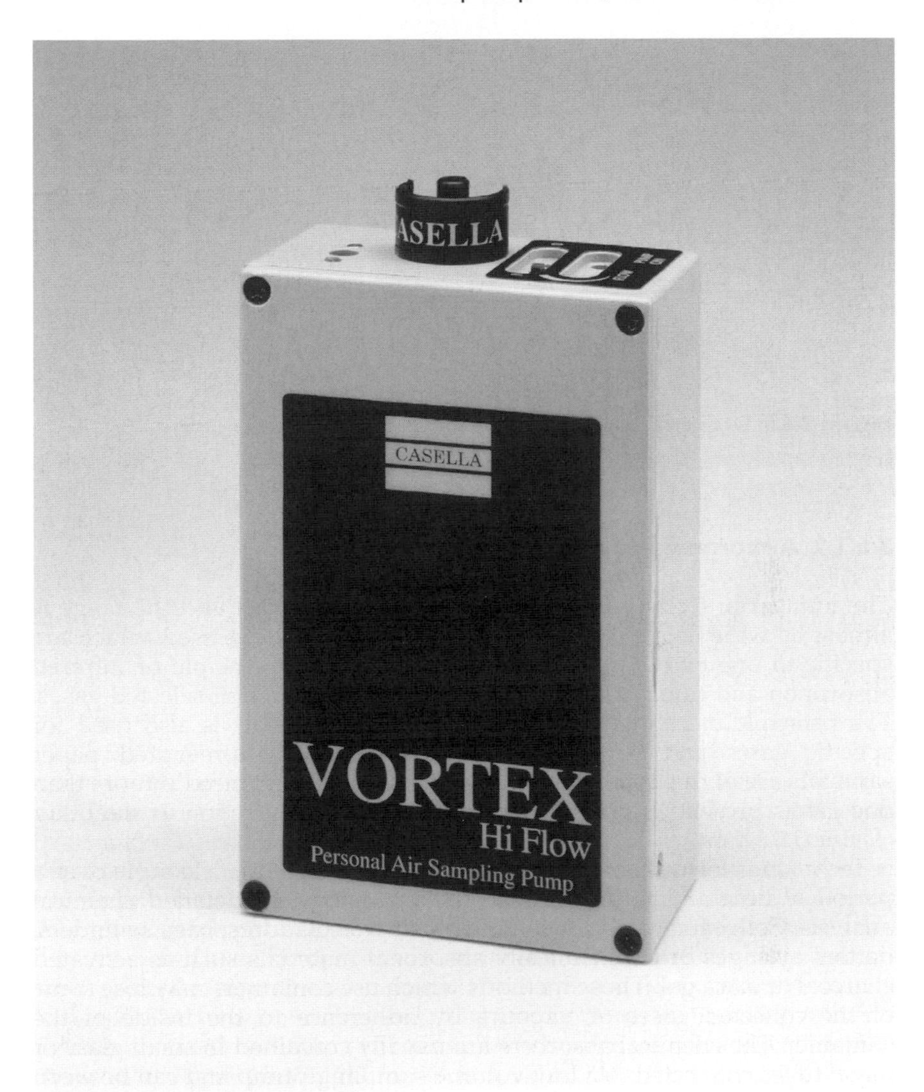

Figure 3.6.2 Personal dust sampling pump. (Courtesy Casella Ltd)

controlled within close limits to ensure that the correct size fraction separation occurs.

Other techniques for static dust measurement include those which use the principle of measuring the amount of light scattered by the dust and one that uses a technique which measures the oscillation of a vibratory sensor which changes in frequency with the amount of dust deposited in it; another uses the principle of absorption of beta-radiation by the amount of dust deposited on a thin polyester film.

Further details of dust sampling and measurement techniques are given in references 1 and 2.

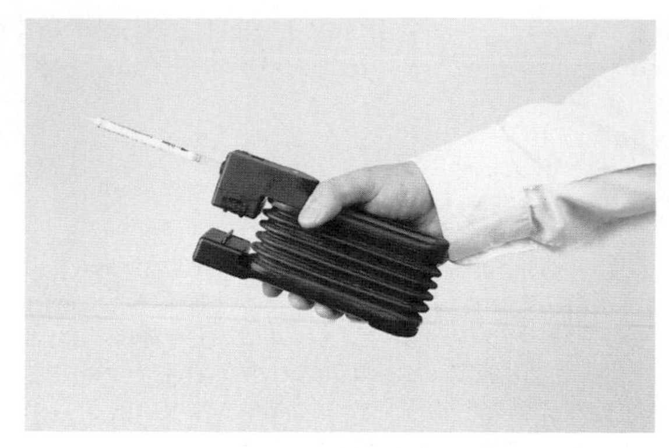

Figure 3.6.3 Draeger bellows pump and tube. (Courtesy Draeger Ltd)

3.6.1.2 Airborne gases and vapours measurement

The number of techniques available in this field is vast and the range is almost as wide as chemistry itself. Instruments can be used which are specific to one or two gases while others use the principle of infrared absorption and can be tuned to be sensitive to a range of selected gases. The principle of change of colour of paper or crystals is also used for specific gases and vapours; detector tube and impregnated paper samplers are of this type, but difficulties can be experienced if more than one gas is present as one may interfere with the detection of the other (*Figure 3.6.3*).

Techniques for unknown pollutants involve collecting a sample over a period of time and returning it to the laboratory for detailed chemical analysis. Collection can be in containers such as bladders, bags, cylinders, bottles, syringes or on chemically absorbent materials such as activated charcoal or silica gel. Those methods which use containers may lose some of the collected gases or vapours by adherence to the inside of the container. The chemical absorbers are usually contained in small glass or metal tubes connected to a low-volume sampling pump and can be worn by a worker in a similar fashion to the personal dust samplers.

All these techniques require a good knowledge of chemistry if reliable results are to be obtained as many problems exist in the collection and analysis of gases and vapours[1,2].

3.6.2 Measurement of the thermal environment

Many working environments are uncomfortable owing to excessive heat or cold in one form or another; some can be so extreme that they lead to heat or cold stress occurring in the workers. When investigating these environments, it is important to take into account the rate of work and the type of clothing of the worker as these affect the amount of heat the body is producing and losing.

To obtain a correct assessment of a thermal environment, four parameters require to be measured together:

1 The air dry bulb temperature.
2 The air wet bulb temperature.
3 The radiant temperature.
4 The air velocity.

If any one of these is omitted, then an incomplete view is obtained. The sling psychrometer (sometimes known as the whirling hygrometer) will measure wet and dry bulb temperatures, a globe thermometer responds to radiant heat, and air velocity can be measured by an airflow meter or a katathermometer. There are several indices which bring together the four measurements and express them as a single value: some also take into account work rate and clothing. A number of these indices are listed below, and their values can be calculated or derived from charts.

1 The Wet Bulb Globe Temperature index (WBGT) can be calculated from the formula:

   For indoor environments
   WBGT = 0.7 × natural wet bulb temperature + 0.3 × globe temperature

   For outdoor environments
   WBGT = 0.7 × natural wet bulb temperature + 0.2 × globe temperature + 0.1 × dry bulb temperature

2 Corrected Effective Temperature index (CET) can be obtained from a chart and can take into account work rate and clothing.
3 Heat Stress Index (HSI) can be calculated or obtained from charts and takes into account clothing and work rate, and from it can be obtained recommended durations of work and rest periods.
4 Predicted four hour sweat rate (P4SR) can be obtained from charts and takes into account work rate and clothing.
5 Wind chill index, as its name suggests, refers to the cold environment and uses only dry bulb temperature and air velocity but takes into account the cooling effect of the wind.

These five indices are considered in detail, including charts and formulae, in references 3 and 4.

3.6.3 Standards for workplace environments

Authorities from several countries publish recommended standards for airborne gases, vapours, dusts, fibre and fume. For many years the main players in the standard setting field have been the United States of America (US) through the American Conference of Governmental

Industrial Hygienists (ACGIH) who publish threshold limit values (TLVs)[6] and the United Kingdom (UK) through the Health and Safety Executive (HSE) who publish Occupational Exposure Limits in their Guidance Note Series EH40[5]. Both these sources update annually. However, recently another important player has emerged and that is the European Union (EU). The Directorate General of the European Commission has set up a Scientific Experts Group whose task is to develop a list of European occupational exposure limits[12] for member states to consider when setting their own standards.

The occupational exposure limits for the UK are published in EH40. Essentially these standards are in two parts: maximum exposure limits (MELs) and occupational exposure standards (OESs), and they have a legal status under regulation 7 of COSHH[7].

3.6.3.1 Maximum Exposure Limits

Regulation 7(6) of COSHH requires that where an MEL is specified the control of exposure shall be such that the level of exposure is reduced as low as reasonably practicable and in any case below the MEL. Where short-term exposure limits are quoted they shall not be exceeded. In the 2000 edition of EH40 there are 64 MELs quoted.

3.6.3.2 Occupational Exposure Standards

Regulation 7(7) of COSHH requires that where an OES is quoted the control of exposure shall be treated as adequate if the OES is not exceeded or, if exceeded, the employer identifies the reasons and takes appropriate action to remedy the situation as soon as is reasonably practicable. There are some 700 OESs published.

3.6.3.3 Units and recommendations

The standards in the above document are quoted in units of parts per million (ppm) and milligrams per cubic metre (mg m^{-3}). They are given for two periods: Long-term exposure which is an 8-hour time weighted average (TWA) value and a short-term exposure which is a 15-minute TWA. Certain substances are marked with a note 'Sk' which indicates that they can also be absorbed into the body through the skin, others have a 'Sen' notation indicating that they may possibly sensitise an exposed person.

It should be remembered that all exposure limits refer to healthy adults working at normal rates over normal shift duration. In practice it is advisable to work well below the recommended value, as low as one-quarter, to provide 'a good margin of safety'.

3.6.4 Ventilation control of a workplace environment

As a result of the COSHH regulations there is a legal duty to control substances that are hazardous to health. The Approved Code of Practice (ACOP)[7] associated with these regulations sets out in order the methods that should be used to achieve adequate control. Extract and dilution ventilation are two of the methods mentioned. These regulations also require the measurement of the performance of any ventilation systems that control substances that are hazardous to health. The places where measurements are required to be taken are listed in para. 61 of the ACOP.

3.6.4.1 Extract ventilation

In the design of extract ventilation it is important to create, at the point of release of the pollutants, an air velocity sufficiently strong to capture and draw the pollutants into the ducting. This is known as the capture velocity and can be as low as 0.25 m/s for pollutants released gently into still air such as the vapour from a degreasing tank or as much as 10 m/s or more for heavy particles released at a high velocity from a device such as a grinding wheel. The capturing device can be a hood, a slot or an enclosure to suit the layout of the workplace and the nature of the work but the more enclosure that is provided and the closer to the point of emission it is placed, the more effective will be the capture.

Difficulty can be experienced with moving sources of pollution such as the particles from hand-held power saws and grinders. In these circumstances high velocity low volume extractors can be fitted to the tools using flexible tubing of 25–50 mm diameter to draw the particle-laden air to a cleaner which contains a high efficiency filter and a strong suction fan (*Figure 3.6.4*).

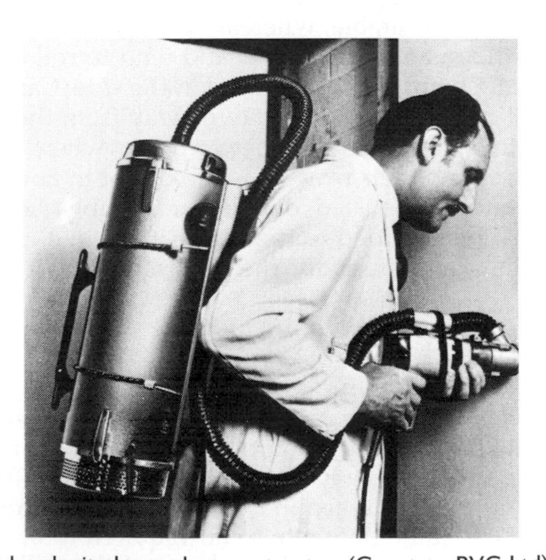

Figure 3.6.4 High velocity low volume extractor. (Courtesy BVC Ltd)

Figure 3.6.5 Portable collecting hood. (Courtesy Myson Marketing Services Ltd)

Hoods attached to larger diameter flexible tubing can be used for extraction from the larger moving sources such as welding over wide areas, but owing to the higher weight of these devices some form of movable support system is required (*Figure 3.6.5*).

When siting a capture hood or slot, advantage should be taken of the natural movement of the pollutants as they are released. For example, hot substances and gases are lighter than air and tend to rise, thus overhead capture might be most suitable, whereas some solvent vapours when in concentrated form are heavier than air and tend to roll along horizontal surfaces, so capture points are best placed at the side. Care must be taken to ensure that all contaminants are drawn away from the breathing zone of the worker – this particularly applies to places where workers have to lean over or get close to their work. It is important to note that whenever extract ventilation is exhausted outside, a suitably heated supply of make-up air must be provided to replace that volume of air discarded.

There are established criteria for the design of extract systems[8].

3.6.4.2 Dilution ventilation

This method of ventilation is suitable for pollutants that are non-toxic and are released gently at low concentrations and should be resorted to only if it is impossible to fit an extractor to the work station. It should not be used if the pollutants are released in a pulsating or intermittent way or if they are toxic. The volume flow rate of air required to be provided must be calculated taking into account the volume of the pollutants released, the concentration permitted in the workplace and a factor of safety which allows for the layout of the room, the airflow patterns created by the ventilation system, the toxicity of the pollutant and the steadiness of its release[9,10].

Hourly air change rates are sometimes quoted to provide a degree of dilution ventilation. The volume flow rate of air in cubic metres per hour is calculated by multiplying the volume of the room in cubic metres by the number of air changes recommended. There are recommended air change rates for a range of situations[11].

3.6.5 Assessment of performance of ventilation systems

In addition to the testing of the airborne concentrations of pollutants, it is necessary, and indeed is a requirement of COSHH, to check airflows and pressures created in a ventilation system to ensure that it is working to its designed performance by measuring:

1 Capture velocity.
2 Air volume flow rates in various places in the system.
3 The pressure losses across filters and other fittings and the pressures developed by fans.

The design value of these items should be specified by the maker of the equipment. Therefore, instruments and devices are required to:

1 Trace and visualise airflow patterns.
2 Measure air velocities in various places.
3 Measure air pressure differences.

Figure 3.6.6 Smoke tube

Figure 3.6.7 Vane anemometer. (Courtesy Air Flow Developments Ltd)

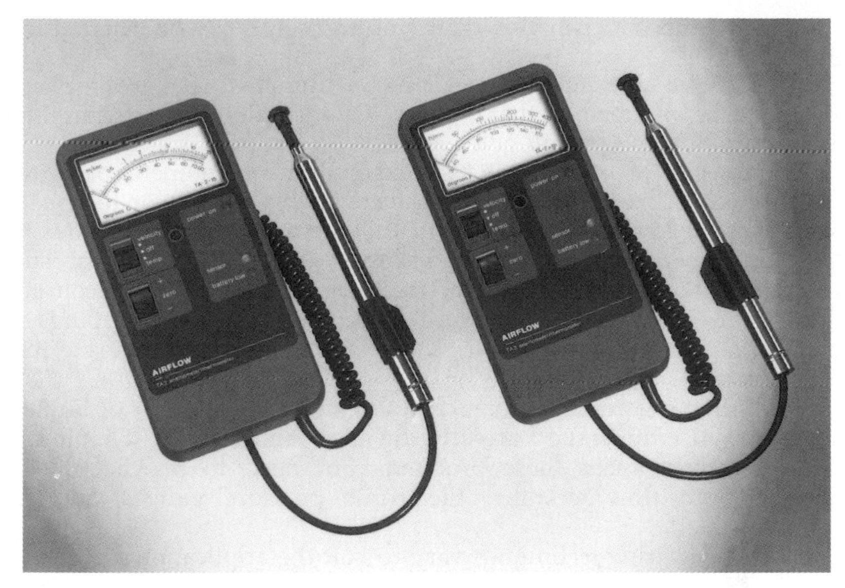

Figure 3.6.8 Heated head air meter. (Courtesy Airflow Developments Ltd)

Air flow patterns can be shown by tracers from 'smoke tubes' which produce a plume of smoke when air is 'puffed' through them (*Figure 3.6.6*). For workplaces where airborne particles are released it is possible to visualise the movement of the particles by use of a dust lamp. This shines a strong parallel beam of light through the dust cloud highlighting the particles in the same way that the sun's rays do in a darkened room.

Air velocities can be measured by a variety of instruments but vane anemometers and heated head (hot wire or thermistor) air meters are the most common. Vane anemometers (*Figure 3.6.7*) have a rotating 'windmill' type head coupled to a meter and are most suitable for use in open areas such as large hoods and tunnels. The heated head type of air meter (*Figure 3.6.8*) is more suitable for inserting into ducting and small slots and is more versatile than the vane anemometers except that it is unsuitable for use in areas where flammable gases and vapours are released. Most air flow measuring instruments require checking and calibration from time to time.

One instrument which requires no calibration but is only effective in measuring velocities above approximately 3 m/s is the pitot-static tube which, in conjunction with a suitable pressure gauge, measures the velocity component of the pressure of the moving air which can be converted to air velocity by means of the simple formula:

$$p_v = \tfrac{1}{2}\rho v^2 \quad \text{or} \quad v = \sqrt{\frac{2p_v}{\rho}}$$

where p_v = velocity pressure (N/m^2 or Pa); ρ = air density (usually taken to be 1.2 kg/m^3 for most ventilation situations); and v = air velocity (m/s).

Pitot-static tubes are small in diameter and can easily be inserted into ducting.

All the above air velocity measuring instruments need to be placed carefully in an airstream so that their axes are parallel to the stream lines; any deviation from this will give errors.

Differences in air pressure can be measured by a manometer or U-tube gauges filled with water or paraffin, placed either vertically or, for greater accuracy, inclined. If the two limbs of the gauge are coupled by flexible plastic or rubber tubing to either side of the place to be measured, such as a fan or a filter, then the difference in height between the two columns of the tube indicates the pressure difference. Pressure tappings in ductings must be at right angles to the air flow to measure what is termed 'static pressure'.

Liquid-filled gauges are prone to spills and the inclusion of bubbles and before use must be carefully levelled and zeroed. Diaphragm pressure gauges avoid these problems but need to be checked for accuracy from time to time. Electronic pressure gauges are also available.

Airflow measuring techniques vary to suit the application[2].

References

1. ACGIH, *Air Sampling Instruments*, 8th edn, American Conference of Governmental Industrial Hygienists, Cincinnati, Ohio (1995)
2. Gill, F.S. and Ashton, I., *Monitoring for Health Hazards at Work*, Chapter 4, 'Ventilation', Blackwell Science, Oxford (2000)
3. Youle, A., 'The thermal environment' chapter in *Occupational Hygiene* (Eds Harrington, J.M. and Gardiner, J., Blackwell Science, Oxford (1995)
4. Harrington, J.M., Gill, F.S., Aw, T.C. and Gardiner, K., *Occupational Health Pocket Consultant*, Blackwell Science, Oxford (1998)
5. Health and Safety Executive, Guidance Note EH40, *Occupational Exposure Limits*, HSE Books, Sudbury, latest issue
6. ACGIH, *Threshold Limit Values for Chemical Substances and Physical Agents in the Workroom Environment*, American Conference of Governmental Industrial Hygienists, Cincinnati, Ohio (2001)
7. Health and Safety Executive, Legal series booklet no. L 5, *General COSHH ACOP (Control of substances hazardous to health), Carcinogens ACOP (Control of carcinogenic substances) and Biological agents (Control of biological agents). Control of Substances Hazardous to Health Regulations 2002. Approved Code of Practice*, HSE Books, Sudbury (2002)
8. British Occupational Hygiene Society, Technical Guide No. 7, *Controlling Airborne Contaminants in the Workplace*, Science Reviews Ltd, Leeds (1987)
9. Gill, F.S., 'Ventilation' chapter in *Occupational Hygiene* (Eds Harrington, J.M. and Gardiner, K), Blackwell Scientific, Oxford (1995)
10. ACGIH, *Industrial Ventilation*, 22nd edn, American Conference of Governmental Industrial Hygienists, Cincinnati, Ohio (1995)
11. Daly, B. B., *Woods Practical Guide to Fan Engineering*, chapter 2, Woods of Colchester Ltd (1978)
12. EEC Council Regulation no. EEC/793/93 *on the evaluation and control of the risks of existing substances*, EC, Luxembourg (1993)

Further reading

Ashton, I. and Gill, F.S., *Monitoring for Health Hazards at Work*, Blackwell Science, Oxford (2000)

Chapter 3.7

Lighting

E. G. Hooper and updated by Jonathan David

3.7.1 Introduction

Lighting plays an important role in health and safety, and lighting requirements are increasingly being included in legislation and standards, albeit that primary legislation tends to specify that lighting shall be 'sufficient and suitable'. Legislation whose content has lighting in its requirements includes that for the workplace[1], work equipment[2], docks[3], the use of electricity[4] and display screen equipment[5]. Most people prefer to work in daylight making the best possible use of natural light, though this may not always be the most energy efficient approach. However, for many working environments natural light is often insufficient for the whole working day, and in deeper spaces may not be adequate at any time. It therefore has to be supplemented or replaced by artificial lighting, usually electric lighting. The quality of the lighting installation can have a significant effect on health, productivity and the pleasantness of interior spaces in addition to its role in safety.

3.7.2 The eye

The front of the eye comprises, in simple terms, a lens to control the focusing point within the eye and an iris to control the light entering the eye. The back of the eye contains the retina which is made up of rod and cone shaped cells which are sensitive to light and are linked by optic nerves to the brain. The lens ensures that the image being viewed is focused on the retina and the iris controls the amount of light. Different cells in the retina are sensitive to different colours, and while the central part of the retina, known as the fovea, is sensitive to colours the peripheral areas are sensitive only to light intensity. A result is that colour vision disappears at low light levels.

3.7.3 Eye conditions

The eye is a very delicate and sensitive structure and is subject to a number of disorders and injuries requiring skilled treatment: some of these disorders are mentioned briefly below.

Conjunctivitis is an inflamed condition of the conjunctiva (the mucous membrane covering the eyeball) caused by exposure to dust and fume and occasionally to micro-organisms.

Eye strain, so called, is caused by subjecting the eye to excessively bright light or glare; the term is also used colloquially to describe the symptoms of uncorrected refractive errors. There is no evidence that the eye can be 'strained' simply by being used normally.

Accommodation is a term for the ability of the eye to alter its refractive powers and to adjust for near or distant vision. As the eye ages the lens loses its elasticity and hence its accommodation, thus affecting the ability to read and requiring corrective spectacles. In addition to this ageing process defects in accommodation can occur early in life, such as by the presence of conditions known as

1 *astigmatism* due to the cornea of the eye being unequally curved and affecting focus;
2 *hypermetropia*, or long sight, in which the eyeball is too short; and
3 *myopia*, or short sight, in which the eyeball is too long.

These defects can usually be corrected by spectacles.

Nystagmus is an involuntary lateral or up and down oscillating and flickering movement of the eyeball, and is a symptom of the nervous system observed in such occupations as mining.

Double vision is the inability of both eyes to focus in a co-ordinated way on an object usually caused by some defect in the eye muscles. It can be due to a specific eye injury, to tiredness or be a symptom of some illness. It may be a momentary phenomenon or may last for longer periods.

Colour blindness is a common disorder where it is difficult to distinguish between certain colours. The most common defect is red/green blindness and may be of a minor character where red merely loses some of its brilliance, or of a more serious kind where bright greens and reds appear as one and the same colour – a dangerous condition in occupations requiring the ability to react to green and red signals or to respond to colour coding of pipework or electrical cables.

Temporary blindness may be due to some illness but it can occur in the following circumstances:

1 Involuntary closure of the eyelids due to glare.
2 Impairment of vision due to exposure to rapid changes in light intensity and to poor dark adaptation or to excessively high light levels.

The act of seeing requires some human effort which is related to the environmental conditions. Even with good eyesight a person will find it difficult to see properly if the illumination (level of lighting) is not adequate for the task involved, e.g. for the reading of small print or working to fine detail. But no standard of lighting, however well planned, can correct defective vision and anyone with suspected visual disability should be encouraged to undergo an eye test and, if advised, wear corrective spectacles. Legislation now requires that employees working with visual display terminals (vdts) be offered free eye tests by their employers if they so request[5].

3.7.4 Definitions[6]

The following terms are used in connection with illumination:

Candela (cd) is the SI unit of luminous intensity, i.e. the measure describing the power of a light source to emit light.

Lumen (lm) is the unit of luminous flux used to describe the quantity of light emitted by a source or received by a surface.

Illuminance (symbol E, unit *lux*) is the luminous flux density of a surface, i.e. the amount of light falling on a unit area of a surface, 1 lux = $1\,lm/m^2$.

Maintained illuminance is the average illuminance over the reference surface at the time maintenance has to be carried out. It is the level below which the illuminance should not drop at any time in the life of the installation.

Luminance (symbol L, unit cd/m^2) is the physical measure of the stimulus which produces the subjective sensation of brightness, measured by the luminous intensity of the light emitted or reflected in a given direction from a surface element divided by the projected area of the element in the same direction.

Luminance = (illuminance $\times$ reflection factor)/π

Brightness is the subjective response to luminance in the field of view dependent on the adaptation of the eye.

Reflectance factor is the ratio of the luminous flux reflected from a surface to the luminous flux incident upon it.

Incandescent lamp is a lamp where the passage of a current through a filament (usually coiled) raises its temperature to white heat (incandescence), giving out light. Oxidisation within the glass bulb is slowed down by the presence of an inert gas or vacuum sealing of the bulb. The most commonly used lamp is the General Service Lamp, but there also exists a wide range of decorative lamps. Higher efficiency incandescent lamps can be created by including in the bulb a small amount of a halogen element such as iodine or bromine. In such lamps, usually known as tungsten-halogen lamps, the halogen combines with the tungsten and is deposited on the inside of the bulb. When this compound approaches the filament it decomposes, owing to the high temperature, and deposits the tungsten back on the filament.

The European Commission has developed a scheme for energy rating of lamps commonly used for domestic purposes. This does not apply to other lamp types or lamps sold to commercial and industrial organisations.

Electric discharge lamp is a lamp where an arc is created between two electrodes within a sealed and partially evacuated transparent tube. Depending on the format of the tube, the remaining gas pressure and the trace elements that are introduced, numerous different types of lamp can be produced:

1 *Low pressure sodium lamp* used chiefly for road lighting which produces a monochromatic yellow light but is highly efficient. However, increased knowledge of the performance of the eye at very low light levels has led to a questioning of whether the low pressure sodium lamp is as effective as previously thought.
2 *Low pressure mercury lamp* – the ubiquitous 'fluorescent tube' in which the ultraviolet radiation from the discharge is converted to visible light by means of a fluorescent coating (phosphor) on the inside of the tube. Fluorescent lamps come in various forms:
   (a) Linear lamps, both full size (600–2400 mm long) and miniature (less than 600 mm long), come in a range of wattages and efficiencies as well as a range of whites and colours. Traditionally, while halophosphate phosphors were used, there was a trade-off between colour quality and efficiency; with modern triphosphor and multi-band lamps this is no longer the case. T12 (38 mm diameter) lamps have largely been superseded by T8 (26 mm) or T5 (15.5 mm) lamps offering higher efficacies and better light control. T5 lamps are offered in two specific ranges: standard and high output. A recent development is T2 (6.5 mm diameter) lamps which offer high efficacy but require dedicated control gear and careful light control. These were originally offered for specialist applications such as under-shelf lighting in retail shops but are finding wider applications.
   (b) Compact lamps, in both retrofit designs intended for existing installations and for newer installations when compatibility with other lamp types does not matter, come in a variety of formats and ratings from 5 W to 55 W.
3 *High pressure mercury lamp* is a largely obsolete type of lamp where light is produced by means of a discharge within an arc tube doped with mercury. The light tends to be bluish in colour and efficiency is lower than other currently used types of discharge lamp. It is still popular in some tropical countries because of its 'cool' light.
4 *High pressure sodium lamp* is similar to a mercury lamp except that the arc tube is doped with sodium giving a yellow light whose colour rendering and whiteness depend on the vapour pressure within the tube.
5 *Metal halide lamp* is similar to the mercury lamp except that the mercury is replaced by a carefully designed cocktail of rare earth elements. Colour rendering can be very good and efficiency is high with additional coloured light being generated by the suitable choice of elements in the cocktail. The small arc tube means that light control can be very good. There can be problems with colour stability over the life of the tube.

Induction lamp in which the lamp itself is simply a glass tube containing an inert gas and coated on the inside with a phosphor to convert the ultraviolet radiation to visible light. The discharge which takes place in the tube is initiated by an electric or microwave field outside the lamp by equipment containing a powerful electromagnet or a magnetron. Different manufacturers have adopted different physical formats. Efficiency is fairly high and, because there are no moving parts in the tube, lamp life can be extremely long making the lamp ideal where maintenance access is difficult.

Luminaire is a general term for all the apparatus necessary to provide a lighting effect. It usually includes all components for the mounting and protection of lamps, controlling the light distribution and connecting them to the power supply, i.e. the whole lighting fitting. Occasionally part of the control gear may be mounted remote from the luminaire.

3.7.5 Types of lighting

The selection of the source of light appropriate to the circumstances depends on several factors. It is important to consider efficiency, ease of installation, costs of installation and running, maintenance, lamp life characteristics, size, robustness and heat and colour output. The efficiency of any lamp (often termed efficacy) can be expressed in terms of light output per unit of electricity used (lumens per watt). Generally speaking, incandescent lamps are less efficient than discharge sources.

Type of lamp	*Lumens per watt*
Incandescent lamps	Up to 15
Tungsten halogen	Up to 22
High pressure sodium	Up to 140
Metal halide	Up to 100
Fluorescent	Up to 100
Compact fluorescent	Up to 85
Induction	Up to 65
Low pressure sodium	Up to 200

Note that smaller ratings are usually less efficient than larger ratings and that the above figures do not include losses within the control gear needed for all but incandescent lamps. Note also that control gear losses can differ markedly between brands. A rating scheme for efficiency of ballasts for fluorescent lamps has been introduced in Europe[7].

In any choice between incandescent and the other types of lamp the total lighting costs must take into account not only running costs but also installation and replacement costs. Incandescent lamps are much cheaper to buy and install, they give out light immediately they are switched on and they can be dimmed easily, but they are more expensive to run and have short lives, thus increasing maintenance costs. High pressure discharge and fluorescent lamps cost more to install but their greater efficiency and longer lives make them more cost effective for general lighting. Linear and compact fluorescent lamps come to full light output

Figure 3.7.1 Factory lighting where fine work is carried out and colour rendering is important, making use of reflector luminaires with tubular fluorescent lamps. (Courtesy Lighting Industry Federation)

reasonably quickly but discharge lamps need some time to strike and then achieve maximum light output, and may need several minutes to cool before they will restrike if accidentally extinguished. Hot restrike is possible for some lamps but is expensive.

In larger places of work the choice is often between discharge and fluorescent lamps. Where colour performance is important the sodium lamp, with its rather warm golden effect, may not be suitable and the choice is usually between the tubular fluorescent lamp and the metal halide lamp. A limitation of the fluorescent lamp is the restricted loading per point (i.e. more lamps are required per unit surface area) and in certain workshops where luminaire positioning at heights is required (in workshops with overhead travelling cranes for example) the high pressure discharge lamp with its higher loading per point (generally up to 1 kW) is often selected. *Figure 3.7.1* shows factory lighting where fine work and colour rendering are important.

3.7.6 Illuminances

The illuminance (lighting level) required depends upon such things as the visual performance necessary for the tasks involved and general comfort and amenity requirements. The average illuminance out of doors in the UK is about 5000 lux on a cloudy day, but may be 10 times that on a sunny day. Inside a workplace, the illuminance from natural light at, say, a desk next to a window, will probably be only about 20% of the value obtaining outdoors. As working areas get further from windows the natural light produces illuminance values of perhaps only 1 to 10% of outdoor values so requires supplementing by artificial lighting. The normal way of expressing the effectiveness with which daylight reaches an interior is termed daylight factor[8].

In normal practice, decisions should be based on the recommendations of the Code for Lighting, produced by the Society of Light and Lighting, part of the Chartered Institution of Building Service Engineers[9] (CIBSE) or the recommendations of a similar national body. Most such recommendations are now based on European standards and/or international recommendations. Typical values of maintained illuminance for certain locations and tasks are given below but for detailed information, for particular industries and tasks, reference should be made to the Society. Guidance and advice can be obtained from an HSE publication[10] and the Lighting Industry Federation[12]. However, HSE requirements deal only with health and safety issues, whereas the Society of Light and Lighting recommendations also take account of cost effectiveness, productivity and amenity.

Although the term maintained illuminance represents levels that are good for general purposes, increases over the figures given may be necessary where tasks of high visual difficulty are undertaken, or low reflection or contrast are present, or where the location is a windowless interior. The Code for Lighting[9] gives criteria on which such adjustments can be based.

Location and task	*Maintained illuminance* (lx)
Storage areas, plant rooms, entrance halls etc.	150–200
Rough machinery and assembling, conference rooms, typing rooms, canteens, control rooms, wood machinery, cold strip mills, weaving and spinning etc.	300–400
Routine office work, medium machinery and assembly etc.	500
Spaces containing vdts used regularly as part of office tasks.	300–500
Demanding work such as in drawing offices, inspection of medium machinery etc.	750
Fine work requiring colour discrimination, textile processing, and fine machinery and assembly etc.	1000
Very fine work, e.g. hand engraving and inspection of fine machinery and assembly	1500

A new requirement, emanating from European standards, is a minimum illuminance of 200 lx for any continuously occupied interior.

For a discussion on average maintained illuminances, minimum measured illuminances and for maximum ratio of illuminances between working and adjacent areas see reference 9.

European standards are being developed for several areas of lighting design. These will normally be taken account of in any revisions of guidance documents such as the Society of Light and Lighting Code for Lighting[9]. However, other than a few mandated standards, European standards are voluntary documents, and there is no compulsion on national lighting societies to adopt them. In addition, there is nothing to stop a national society adopting standards higher than those in a European standard, since documents from professional bodies normally carry no legal status.

3.7.6.1 Maintenance of lighting equipment

Dust, dirt and use will progressively reduce the light output of lamps and luminaires. Attention to good general cleaning and maintenance, and a realistic lamp replacement policy will help maintain the illuminance within recommendations. The expected maintenance regime is an essential factor in calculating the number of luminaires required for an installation. The maintenance regime appropriate to a building will depend on the activities carried out, the amount of dirt and dust carried in from outside and the type of lighting equipment in use. Some modern lamps lose light output much more slowly than older types, though luminaires will soil just as quickly.

3.7.7 Factors affecting the quality of lighting

The eye has the faculty of adjusting itself to various conditions and to discriminating between detail and objects. This visual capacity takes time to adjust to changing conditions as, for example, when leaving a brightly lit workroom for a darkened passage. Sudden changes of illuminance and excessive contrast between bright and dark areas of a workplace should be avoided.

A recent problem, resulting from the introduction of word-processors and other equipment using vdts, is the effect on eye discomfort and general well being of viewing screens for extended periods of time. Problems can be increased if the contrast between the screen and paper task is too great, if there is excessive contrast between the screen and background field of view, and if there are reflections of bright objects (luminaires, windows or even white shirts) in the screen. Lighting installations in such areas must comply with the requirements of the DSE Regulations[5]. The CIBSE has published specific guidance in this area[11] and has recently updated it by means of an addendum to take account of trends in software and VDU screens.

3.7.7.1 Glare

Glare causes discomfort or impairment of vision and is usually divided into three aspects, i.e. disability glare, discomfort glare, and reflected glare.

It is referred to as disability glare if it impairs the ability to see clearly without necessarily causing personal discomfort. The glare caused by the undipped headlamps of an approaching car is an example of this.

Discomfort glare causes visual discomfort without necessarily impairing the ability to see and may occur from unscreened windows in bright sunlight or when over-bright or unshaded lamps in the workplace are significantly brighter than the surfaces against which they are viewed, e.g. the ceiling or walls.

Reflected glare, which can be disability glare or discomfort glare, is the effect of light reflected from a shiny or polished non-matt surface. The visual effect may be reduction of contrast, or distortion, and can be both irritating and, in certain workplaces, dangerous.

3.7.7.2 Glare indices

For many years in the UK, a glare index system has been in use for quantifying the effects of direct glare. It is also in use in certain other countries.

This is now being replaced by the international Unified Glare Rating System (UGR) which has been adopted as standard in Europe. The numerical values will normally be much the same but the derivation formula is different:

$$\mathrm{UGR} = 8 \log [(0.25/L_b) \times \Sigma\ (L^2\omega/P^2)]$$

where L_b = background luminance

L = luminance of the luminous parts of each luminaire in the direction of the observer's eye

ω = solid angle subtended by the luminous parts of each luminaire at the observer's eye

P = Guth position index for each luminaire

A set of tables, based on this formula, has been produced by the Society of Light and Lighting for a range of situations, types of luminaire, etc., and these should be referred to for specific advice[9]. Figures above the recommended levels for a given location may lead to visual discomfort.

Separate advice has been published[11] on reducing glare in premises where VDUs are in use. This includes factories and workshops as well as offices. Draft EU standards for lighting use the Unified Glare Rating system in place of the glare index. These standards are not mandatory except in contracts involving the public sector.

Figure 3.7.2 Factory lighting of correct illuminance, free from shadow and glare, making use of high pressure discharge lamps. (Courtesy Thorn Lighting Ltd)

3.7.7.3 Protection from glare

The most common cause of glare results from looking directly at unscreened lamps from normal viewing angles. Any form of diffuser or louvre fitted over the lamp, or a suitably placed reflector used as a screen will help to reduce the effect of glare from a lamp. The minimum screening angle below the horizontal should be about 20°, though greater angles are specified for areas containing vdts[11]. Reflected glare can only really be eliminated by changing the offending shiny surface for a matt one, or by adjusting the relative positions of light source, reflective surface and viewer.

Glare from sunlight coming through windows can be reduced by using exterior or interior blinds but this reduces the amount of natural lighting. It may be more effective to rearrange the workplace so that the windows are not in the normal direct field of view.

3.7.7.4 Effect of shadow

Shadow will affect the amount of illumination, and its impact on people in working areas will depend on the task being performed, and on the disposition of desks, work benches etc. The remedy is to use physically large luminaires (not necessarily with higher light outputs) or to increase their number. *Figure 3.7.2* illustrates factory lighting where the illuminance is to recommended standards.

3.7.7.5 Stroboscopic effect

The earlier type of tubular fluorescent lamp and discharge lamp were criticised because of the possibility of a stroboscopic effect. The light output from most lamps shows a cyclical variation with the alternating current, although in most circumstances this is not noticeable. However, it can cause a piece of rotating machinery to appear stationary or to be rotating slowly when, in fact, it is rotating at many times a second. This can be extremely dangerous. However, with modern fluorescent lamps and some discharge lamps the problem has been minimised by reducing the flicker effect. Where stroboscopic effects pose a particular danger they can be eliminated since it is possible to operate linear fluorescent and compact fluorescent lamps on electronic control gear at high frequency which both minimises the cyclic variation of light output and changes its frequency so that it is no longer visible as flicker. Alternatively, in most industrial and many commercial buildings it is possible to connect successive luminaires to the three phases of the power supply, which eliminates most flicker and stroboscopic effects.

3.7.7.6 Colour effect

The reflection of light falling on a coloured surface produces a coloured effect in which the amount of colour reflected depends upon the light source and the colour of the surface. For example, a red surface will only appear red if the incident light falling upon it contains red: under the almost monochromatic yellow of sodium street lighting, for example, a red surface will appear brown. The choice of lamp is important if colour effect or 'warm' or 'cool' effect is required and can be as important a consideration as the illuminance itself. Where accurate colour judgements have to be made the illuminance should be not less than 1000 lux and it may be appropriate to use either lamps whose colour rendering index is above 90 (CIE colour rendering group 1) or exceptionally special 'artificial day light' fluorescent lamps – commonly known as DE5 lamps. Forthcoming European standards will require a minimum colour rendering index of 80 for most working interiors, though this may be reduced to 40 for some industrial applications. Fortunately, standard fluorescent lamps now make it easy to achieve this level of colour rendering.

3.7.8 Use of light measuring instruments

The human eye is unreliable as an indicator of how much light is present. For accurate results in the measurement of the illuminance at a surface it is necessary to use a reliable instrument. Light meters are available for this purpose.

A light meter, normally adequate for most locations, is a photocell which responds to light falling on it by generating a small electric current which deflects a pointer on a graduated scale measured in lux or, more commonly nowadays, causes a number to be displayed on a digital display. Most light meters have a correction factor built into their design to allow for using a filter when measuring different types of light (daylight, tubular fluorescent lamps, high pressure sodium lamps etc.). The recommended procedure for taking measurements with a light meter of this type is to:

1 Cover the cell with opaque material and alter the zero adjustment until the pointer reads zero on the scale.
2 Allow a few minutes for the instrument to 'settle down' before taking a reading. A longer period will be required if the light is provided by tubular fluorescent lamps or high pressure discharge lamps which have only just been switched on as they take time to reach full light output.
3 Select the appropriate scale on the instrument, i.e. that which gives the greatest deflection of the pointer or where the reading is closest to the upper end of the range.
4 If readings are to be taken during daylight two readings are necessary:
   (a) with the lights on and with the window blinds drawn back so as to record the combined effect of natural and artificial light, and
   (b) with the same natural light conditions as in (a) but with the artificial lights switched off.

   The result required, i.e. the measure of the artificial light, is the difference between the two readings. If the two readings are large and approximately equal it will be necessary to re-check the artificial light reading after dark.

The measured illuminance should be checked against the maintained illuminance for the location and task, taking account of the requirements, laid down by the CIBSE for the relevant areas[9]. The correct use of a light meter is an important aid to establishing good levels of lighting. However, to ensure accurate readings the instrument should be kept in its case when not in use and away from damp and excessive heat. It is also advisable to have the calibration checked by the manufacturer every year, though this is not cheap and it may be more cost effective to buy a new meter annually.

Do not overestimate the accuracy of the readings you obtain. Few hand-held meters are capable of measuring illuminance more accurately than within 10%, and the position of measurement can affect the measurement considerably. It is possible for measurements to differ from calculations by up to 60% for direct illumination and 20% for calculations involving interreflections. For maximum accuracy, measure at points on a regular grid through the space and average the results. Accuracy will be particularly suspect at low levels even if the meter itself has various ranges.

References

1. *Workplace (Health, Safety and Welfare) Regulations 1992*, regulation 8, The Stationery Office, London (1992)
2. *Provision and Use of Work Equipment Regulations 1992*, regulation 21, The Stationery Office, London (1992)
3. *The Docks Regulations 1988*, regulation 6, The Stationery Office, London (1988)
4. *The Electricity at Work Regulations 1989*, regulation 15, The Stationery Office, London (1989)
5. *Health and Safety (Display Screen Equipment) Regulations 1992*, the schedule, The Stationery Office, London (1992)
6. BS 6100, *Glossary of building and civil engineering terms, Section 3.4 Lighting*, BSI, London (1995), also International Commission on Illumination, publication 17.4, *International lighting vocabulary*, 4th edn, CIE-UK, c/o CIBSE, London (1987)
7. For details contact the Lighting Industry Federation, Swan House, 207 Balham High Road, London SW17 7BQ, tel: 020 8675 5432
8. Building Research Establishment, Digest 309, *Estimating daylight in buildings, Part 1*; Digest 310, *Estimating daylight in buildings, Part 2*, CRC Ltd, London
9. Society of Light and Lighting, *Code for Lighting 2002*, CIBSE, London, 2002
10. Health and Safety Executive, *Lighting at Work*, Health and Safety: Guidance Booklet No. HS(G)38, HSE Books, Sudbury (1989)
11. Chartered Institution of Building Services Engineers, *Lighting Guide 3. The visual environment for display screen equipment*, CIBSE, London (1996 addendum 2001)

Further reading

In addition to the above, numerous booklets and pamphlets on lighting for occupational premises and processes may be obtained from:
Chartered Institution of Building Services Engineers, 222 Balham High Road, London SW12 9BS. Relevant publications on specific types of premises include:
Lighting Guide 2, *Hospitals and health care buildings* (1989)
Lighting Guide 4, *Sports* (1990)
Lighting Guide 5, *The visual environment in lecture, teaching and conference rooms* (1991)
Lighting Guide 7, *Lighting for offices* (1993)
Lighting Guide 8, *Lighting for museum and art galleries* (1994)
Lighting Guide 10, *Daylight and window design* (1999)
Lighting Guide 11, *Surface reflectance and colour. Its specification and measurement for designers* (2001)
Guide to fibre-optic and remote source lighting (joint with the Institution of Lighting Engineers) (2001)
Technical memorandum 12: *Emergency lighting* (1986)
Building Research Establishment, Garston, Watford, Hertfordshire WD25 9XX. Publications available from: CRC Ltd, Bowling Green Lane, London EC1R 0DA
Lighting Industry Federation, Swan House, 207 Balham High Road, London SW17 7BQ

Chapter 3.8

Managing ergonomics

Nick Cook

3.8.1 Introduction

What is ergonomics?

If you visit the aircraft section of London's Science Museum you see an excellent example of what is *not* ergonomics. In a huge hangar sized room on the fourth floor are suspended life sized models of aircraft. One of these is an almost stubby little single seater with swept back wings and a ridiculously small propeller on its dolphin-like nose. By today's standards it has a bolted together look but in 1944 it was far ahead of its time.

In the science museum the Messerschmitt 163B-1 Komet is suspended nearby a Hawker Hurricane and a Supermarine Spitfire. Perhaps it would have been more appropriate to hang it close to a Halifax or a Lancaster for bombers such as these would have been its intended prey.

If the Komet was ahead of its time it had to be. In 1944 Germany was suffering badly. Wave after wave of allied bombers were pounding its cities. So confident were they that they carried out these raids in broad daylight, not even waiting for the cover of darkness.

The Komet was designed to destroy that confidence. It was a daring concept. Its Walter rocket motors provided the thrust for take-off. Once in the air its wheeled undercarriage fell away while the Komet soared to 7600 metres to shoot down the bombers. After ten minutes the liquid fuel in its rockets would be exhausted. At this point the Komet's wings took over. The pilot glided back to the airfield. His landing was cushioned by a retractable sprung skid which descended like a single ski from the fuselage.

At least that was the theory. And it would have worked had more attention been paid to ergonomic considerations.

The first problem was that 250 mph was too high a speed at which to overtake the allied bombers. The Komet was often past them before the pilot had time to aim and fire. The Walter rockets had an unfortunate tendency to explode and even if they didn't the very poor downward view from the cockpit made the Komet very difficult to land. Even if the pilot escaped disintegration or a nose dive into the turf his troubles were not necessarily over. Inadequate springing in the landing skid meant that the impact on the pilot's back was far greater than the impact of the Komet on the allied bombing campaign. Many of those that managed to land the Komet were rewarded with damaged spines.

If there is one thing to be learned from this it is that ergonomics is about people. This is probably the single most important aspect of the subject. It's about different kinds of people; fat people, thin people, tall people, short people, bright people, not so bright people, young people, old people, male people and female people. And increasingly it will include disabled people. It is about taking all these different types of people and assessing their work. It is then about using that assessment to make sure that their tools, their jobs and their work environments do not injure them. It's also about making sure they can do their work as comfortably and as efficiently as possible.

The sheer range of factors to be considered can make the management of ergonomics a daunting prospect. To do it cost-effectively managers and health and safety professionals need a process for identifying and controlling ergonomic risk in the workplace. They need to know when to call in specialists and when to rely on their own in-house resources and common sense.

This chapter aims to give a basic introduction to the subject that will help with the management process. Getting ergonomic management right is important: not only for employee health but also for the health of the business.

3.8.2 Ergonomics defined

Ergonomics is the reason why chairs are made with comfortable, adjustable backrests. It's the reason why VDU screens don't display pink letters on a magenta background and it's the reason why car controls are all in easy reach. And if it isn't, it should be.

A more formal definition was provided by Professor K. F. H. Murrell[1] in 1950. He defined ergonomics as:

> The scientific study of the relationship between man and his environment.

In truth there are probably almost as many definitions of ergonomics as there are practitioners. For example, in 1984 Clarke and Corlett[2] proposed the following definition:

> The study of human abilities and characteristics which affect the design of equipment systems and jobs . . . and its aims are to improve safety and . . . well being.

Other definitions are very detailed indeed in their attempts to capture the essence of this wide ranging and evolving field. Christianson *et al.*[3] in 1988 defined ergonomics as:

> That branch of science and technology that includes what is known and theorised about human behavioural and biological characteristics that can be validly applied to the specification, design, evaluation, operation and maintenance of systems to enhance safe, effective and satisfying use by individuals, groups and organisations.

Although no one could claim this definition is verbally ergonomic, it is certainly comprehensive. It emphasises the fact that the people doing the work and their human attributes (physical and mental) should be considered along with the range of work attributes (the job and the equipment from design to maintenance).

But perhaps the last word on the subject of definition should go to Britain's first Chief Medical Inspector of Factories. In the nineteenth century Sir Thomas Legge[4] proposed the following criteria for assessing work:

> Is the job fit for the worker and is the worker fit for the job?

The field of ergonomics embraces a wide range of disciplines, from psychology to anatomy.

3.8.3 Ancient Egyptians and all that – a brief history of ergonomics

This section aims to put flesh on these definitions by giving some early practical examples of ergonomic issues and a brief outline of the development of the science.

The formal science of ergonomics may be relatively new but ergonomic issues have been around as long as humans. One of the earliest examples dates from over 10 000 years ago. Studies[5] on the female skeletons of Neolithic women who lived in what is now Syria showed specific deformities. These have been attributed to long hours spent kneeling down using a stone shaped rather like a rolling pin to crush corn on another stone. The second stone (because of its shape) is termed a saddle quern. This operation caused damage to the spine, neck, femur, arms and big toe (the injury to the toe was a result of bending it beneath the foot to stabilise the kneeling position adopted for this job).

The recently excavated skeletons of Egyptian pyramid builders tell with grim eloquence of an ergonomic hell. Most of the skeletons show abnormal bony outgrowths (osteophytes) caused by manually dragging the 2.5 tonne blocks used to build the pyramids. Many of their bones also show wear and tear while spines were actually damaged. Some skeletons even had severed limbs or splintered feet. Small wonder that the workers died between the ages of 30 and 35 whereas the nobility lived to 50 and 60[6]. Little was done to improve the lot of these early construction workers. After all Neolithic chieftains and Egyptian Pharaohs had very little incentive to invent ergonomics when they could get away with a 'pass me another worker this one is broken' approach.

It was with the industrial revolution that opportunities for ergonomic improvement really became apparent. Factories and mines in the nineteenth century were death traps. There were few safeguards on machines. Workers, by and large relatively new to an industrial environment, were poorly trained to operate the machinery. In the new factories the emphasis was very much on work rate and long hours of work, both of which made workers susceptible to the hazards inherent in their labour.

Small wonder that people looked back through rose coloured spectacles to pre-industrial times. Even though exploitation undoubtedly existed in cottage industries, at least handloom weavers had a lot more control over how and when they worked. In their own homes they were their own supervisors and could choose when and for how long to take their breaks.

Nor were the mining industries any better. Cornish tin miners were faced with a huge climb to the surface at the end of shifts which were themselves gruelling. Exhausted miners frequently fell from the ladders as they climbed towards the surface. Tragically, these falls tended to occur most often as the miners neared the top of the ladder. Eventually the mines got too deep for ladders, which were replaced by lift cages. But even these were not safe.

If ascending and descending the mine shafts was bad enough life was no more comfortable at the bottom. In the 1930s, George Orwell[7] wrote of the row of 'buttons' down miners' backs. These were the marks left by the always too low roof beams in the tunnels where miners, bent double as they moved along, would scrape their backs.

Industry was clearly crying out for ergonomic help. Ironically the first application of ergonomics was aimed not at diminishing injury and discomfort but increasing profit. F. W. Taylor[8] and F. B. Gilbreth[9] conducted studies with the aim of increasing production efficiency rather than making the job less hazardous for employees. Their mission was to make work more scientific. This involved calculating the most efficient means of working. They took detailed timings of the physical movements made by individuals in the course of their work. Taylor's method focused on breaking down production work into simple functions and allocating each employee one specific task.

Taylor's philosophy became the basis of Henry Ford's success with production lines but even at the time they were controversial enough to attract a congressional investigation. Taylor's attitude, and with it the attitude of this early approach to ergonomics, is perhaps best summed up by his reply[10] to a question concerning those workers unable to meet the demands of the stopwatch:

> Scientific management has no place for a bird that can sing and won't sing.

It is perhaps not surprising that Henry Ford had to pay his workers twice the rate paid by car companies which had not yet adopted the production line. The studies failed to calculate the human cost of the sheer grinding monotony of production line work.

The science of ergonomics gained momentum during the Second World War. The complexity of aircraft, especially when fitted with equipment such as radar, led to confusion and fatigue among aircrew which in turn led to poor performance in an environment where the penalty for poor performance was likely to be very high.

In the early nineteenth century, a Polish scientist, Wojciech Jastrzebowski, first coined a term similar to ergonomics (derived from the Greek ergos meaning laws and nomos meaning work), but the term did not really occur in common use until adopted by Professor K. F. H. Murrell, a founder member of the Ergonomics Society, in the middle of the twentieth century. In the USA the terms *human factors* or *human factors engineering* have been used, although the term ergonomics is being increasingly used. Today ergonomics still has important applications in the armed services and the aerospace industry but is being increasingly applied in the non-military working environment.

3.8.4 Ergonomics – has designs on you

Risk – any risk – is best controlled at source. Unlike issuing personal protective equipment elimination at source protects everybody. The risk to hearing from a noisy motor is best controlled by replacing the motor with a quieter one rather than supplying people with ear-muffs. And the risk of silicosis to workers doing grinding operations was reduced by making grinding wheels from carborundum rather than sandstone. In ergonomics the same principle applies. Whether considering a job, the tools or the equipment needed to do it, the aim should always be to control the risk at the design stage. Ergonomics has many concepts and techniques to help achieve this goal. Some of the main ones are discussed below.

3.8.5 Ergonomic concepts

3.8.5.1 Usability

Usability is the capability of a system to be used safely and efficiently. The fact that all humans are different must be taken into account when assessing usability. For example, shorter stockier pilots are better able to deal with the G-forces experienced when executing tight turns in a fighter plane. Their hearts don't have to work so hard to get the blood to the head. If it is not possible to design planes or flying suits to eliminate the effects of G-forces then it may be necessary to select short stocky people to become fighter pilots. This is a shame. This is fitting the person to the job. In general it is more desirable to fit the job to the person. An example of this latter approach is the development of voice controlled word processing software for workers handicapped by repetitive strain injury (RSI).

In addition to the diversity of individuals likely to operate the system, the specific range of physical and environmental conditions must be specified. For example, are controls easily accessible? Is the room temperature and humidity satisfactory? The specific social and organisational structure should also be taken into account.

3.8.5.2 The human–machine interface

The human–machine interface is an imaginary boundary between the individual and the machine or equipment. When humans operate tools or machinery, information and energy have to cross this boundary. Consider a helicopter pilot. Information passes across the interface from the machine to the pilot via the control panel display. In response to this information energy then passes from the pilot to the machine via the controls. This example is the basic model for the interaction between humans and machines. It has been described as a closed-loop system. The human receives the information from the machine, processes the information and responds by operating controls as appropriate. The machine responds to the controls and then sends information to the human via a display.

The ergonomic design of the interface (e.g. the controls and panel display) is very important. It has to fit the individual's physical and mental capabilities. Getting it wrong can be fatal. For example, the pilot of a British Airways helicopter that crashed into the sea off the Isles of Scilly claimed that he didn't see the warning light on the altimeter[11]. For people of his particular height the joystick obscured the view. Clearly human variability had not been taken into account for this particular human–machine interface. It was a costly oversight. Out of 26 people aboard the helicopter 20 died.

The following sections consider displays and controls, the two fundamental elements in the human–machine interface, in more detail.

3.8.5.2.1 Displays

The type of display must meet the needs of the human operating the machine or equipment and the display itself must be as clear and as easy to read as possible. It should not overload the operator with too much data but must take into account the information needed and how quickly it needs to be assimilated. The importance of getting this right is underlined by the fact that poor display design was a contributory factor to the nuclear power station incident at Three Mile Island.

The type of display should be appropriate to the data displayed. For example, analogue displays are better for showing rates of change. A needle on a dial or even a column of mercury in a thermometer gives a human operator a very clear picture of the rate of change of temperature. This will be much better than a digital display which will simply show a changing number. This can be very confusing especially if the rate of change is rapid. Analogue displays are also very good at indicating whether the temperature remains within a desired range, especially if that range is marked on the gauge. A digital display on the other hand is very good where more precise readings are required. For example, provided the temperature is not varying too quickly, it is much easier to get an accurate reading from a digital display than from an analogue display.

A visual display may not be the most important way to present data. Using the example above, there may be serious consequences if the temperature strays from a pre-defined range. In this case an audible alarm might be needed to draw immediate attention to the divergent condition.

Much has been written about the relative merits of different types of displays and how they should be fitted to the needs of the operator. Grether and Baker[12] consider the preferred display by information type. Wilson and Rajan[13] in a comprehensive review of systems control consider the relative merits of panel and VDU displays. As far as the nuts and bolts of display criteria is concerned, one of the best introductions is the very practical account given by Cushman *et al.*[14]

Ergonomists use the term 'coding' when referring to the specific way the information is represented. Letters or numbers can be used to represent elements within a system. The size of a symbol on a screen can be used to represent magnitude. Brightness could be used to represent temperature and colour could be used to help classify data. However, colour should only be used as 'redundant code'. In other words colour should not be the only means of displaying the information. Other coding, e.g. letters or pictograms should be the primary code. Relying on colour alone could introduce the possibility of error, for example under certain lighting conditions or cases of colour blindness. Before changes resulting from an EU directive, UK fire extinguishers were colour coded according to type. However, reliance was not placed on the colour coding alone and the type of extinguisher was also indicated by text on the extinguisher.

3.8.5.2.2 Text clarity

A large amount of time is spent reading (or in some cases deciphering) text. It is surprising that writing style has been relatively neglected as an ergonomic issue. Text is a very important interface, not only between operators and machines and equipment (e.g. operating manuals) but also between humans and the organisation in which they work (e.g. work procedures, conditions of employment, policies etc.). At a more fundamental level, text is the interface between workers and the subject knowledge necessary for their jobs. But this relatively neglected area is beginning to find its rightful place in texts on ergonomics. A recent textbook on the methodology of ergonomics[15] has devoted a whole chapter to this important topic where one of the critical concepts covered is readability.

A possible tool for evaluating readability is the *Gunning Fog Index* which works on the premise that long sentences and long words make text difficult to understand. In order to work out the Gunning Fog Index the following equation is used:

Gunning Fog Index = (average sentence length + number of long words) × 0.4

The average sentence length can be calculated by counting the number of sentences in 100 words of text and dividing 100 by that number. Long words in that sample of 100 words are any words that contain three or more syllables (excluding proper names).

The Gunning Fog Index is a number which gives a rough guide to the readability of the sample of text. A guide to the significance of specific Fog Indices is shown in the table below:

Index	Interpretation
10-	Would be readily understandable by the average 15-year-old secondary school pupil.
14–16-	Would be readily understandable by university students.
>18-	the text is now becoming too difficult to understand without serious study.

Many academic texts and corporate documents have Fog Indices which greatly exceed 18. It is almost as if the writers deliberately make their texts complicated to make them look more authoritative. There is no excuse for this. Text should be as user friendly as possible and follow the KISS principle – Keep It Short and Simple.

There are other indices of readability (e.g. the Flesch Reading Ease Index). Many word processing packages will not only calculate a readability index on input text but also give guidance on other aspects of style.

Applying the following guidelines will help keep writing clear and focused:

- keep the average sentence length to about 16;
- do not have any sentences longer than about 26 words;
- punctuate longer sentences with shorter ones;
- keep paragraphs short. This will help break up otherwise indigestible looking blocks of text. Remember, a paragraph should contain a single idea.

I must emphasise that these are guidelines rather than rules. Many writers produce good text without obeying them. Having said that, following the guidelines will in general, help you to produce clear, readable and even entertaining, prose.

While readability is an important factor, the text layout and type and range of font sizes should be chosen carefully to ensure that nothing in the appearance of the written text creates a mental barrier to the assimilation of the information.

3.8.5.2.3 Controls

Controls – e.g. levers, buttons, switches, foot pedals – represent the other half of the man–machine interface. To optimise a design it is necessary to take into account factors such as:

(a) Speed: where a fast response is required the control should be designed so that it can be operated by the finger or the hand. The reason for this is that the hand and finger give the greatest speed and dexterity with the least effort.
(b) Accuracy: using a mouse to select icons on a screen is an example of an operation where accuracy may be required. For such operations it is important to get the control/display ratio (C/D ratio) right. This is the ratio between the amount of movement of the control device to the degree of effect. In the case of a mouse it would be the ratio of the distance required to move the mouse to the resulting distance moved by the cursor on the screen. Where the C/D ratio is low a large movement of the mouse is required to achieve a relatively small movement of the cursor on the screen. This results in a slower but more accurate operation.
(c) Force: it should not be necessary to have to use excessive force in order to operate controls. On the other hand, a certain amount of force may be necessary to prevent a control being accidentally tripped.
(d) Population stereotypes[16]: this term refers to the expectations that controls work in certain ways. For example, we expect to have to turn the steering wheel of a car clockwise to go right and anticlockwise to go left. We expect the effect exercised by vertical levers to increase as we pull them towards the body and to decrease when pushed away. These stereotypes are the expectations of most of the population.

In some cases, however, certain stereotypes will differ from country to country. For example, to switch a light on in the UK the switch has to be moved to the down position. In the USA this convention is reversed. When a dial is used to alter a display the direction of movement on the display conventionally moves in the direction of the nearest point of the dial to the display. Dials are normally located below the display to leave the display in view when the dial is being operated. Therefore clockwise movement of a dial would be expected to move a pointer on a display to the right.

Conforming to the prevailing conventions dictated by stereotyping is desirable. There is always the danger that controls which do not conform may inadvertently be turned the wrong way. Population stereotypes also exist for displays. For example, red signifies danger while green signifies that it's safe to go.

3.8.5.3 Allocation of function

What proportion of a job should be done by the human operator and what proportion should be done by the machine? Hand tools are, inevitably, controlled by the user but for many machine tools automation is increasingly being employed.

To allocate function effectively, ergonomists need to consider the differences in capability between humans and machines. In an early attempt to allocate function at the design stage Fitts[17] listed the relative abilities of humans and machines. This list, subsequently modified by Singleton[18], included observations such as the fact that machines were much faster and more consistent than humans but that humans were a lot easier to reprogram and were by and large better at dealing with the unpredicted and the unpredictable. However, many have warned against the rigid application of such lists. The pace of technological change means that such lists will always be obsolete. Computer technology has made it possible to be more flexible on this issue. The allocation of function can be variable according to the type of person and operation[19].

3.8.5.4 Anthropometry

Anthropometry is the science of measurement of physical aspects of the human body. These include size, shape and body composition. Using data obtained from anthropometry, tools, equipment and workstations can be designed in such a way as to ensure the maximum comfort, safety and efficiency for individuals using them. Some anthropometric data is given in EN standards[20].

3.8.5.5 Error

How do we deal with human error in the workplace? Some employers take the attitude that as long as the equipment is working properly the blame for any errors must lie with the worker, possibly stemming from a lack of motivation, skill, training, or ability. This human error can be tackled by implementing a zero defects programme involving publicity posters, refresher training, job transfer and ultimately disciplinary action such as verbal warnings, written warnings, demotion and finally dismissal.

A more realistic approach is to assume that even the best people are prone to make mistakes. Jobs and equipment should be designed to work, not only from a functional point of view but also taking into account the fact that people can and do make mistakes. Equipment and methods of work should be designed to minimise the possibility and consequences of mistakes. Carried to extremes, the elimination of errors could lead to complete automation of the job dispensing with the operator altogether. Where this is not possible, reliance must be placed on interlocking and self-checking systems so that a wrong action or sequence of actions cannot be performed.

There is, thus, a need to assess human error at the design stage and one such approach is the technique of Human Reliability Assessment[21]. The aims of this technique can be summarised as:

(a) identify what can go wrong;
(b) quantify how often human error is likely to occur; and
(c) control the risk from human error either by preventing it in the first place or reducing its impact when it does occur.

Essentially, in health and safety terms this approach involves the assessment and control of risk where the steps to take include:

1 Define the problem: basically this means looking at what work has to be done and identifying the ways in which mistakes by human beings may interfere or prevent this happening.
2 Analyse the task: what precise actions do humans have to carry out in order to do the job?
3 Identify the human errors and specify the ways in which they can be recovered. At each stage of the process under analysis, one technique[22] is to ask simple questions such as:
   (a) what if a required act is omitted?
   (b) what if required actions are carried out
      (i) incorrectly?
      (ii) in the wrong order?
      (iii) too early?
      (iv) too late?
      (v) too much?
      (vi) too little?
   (c) what if a wrong act is carried out?
4 Assess the probability of errors occurring and their likely seriousness.
5 Control: identify and implement steps to reduce the risk of human errors occurring. This could be done by improving human performance through training and/or improving the design of the work equipment and the environment.
6 Review and audit: check that the control measures are effective. A quality assurance programme (such as ISO 9001) should be implemented to ensure the continuation of the error reduction programme.

For more detail on human reliability assessment see Kirwan[21].

Focusing on people is very important. Their actions (or inactions) are responsible for up to 80% of workplace accidents[22a]. Recent disasters resulting from human error include: the release of radiation from the Three Mile Island nuclear power station in the USA in 1979; the Clapham Junction Rail disaster of 1988; the Union Carbide release of methyl cyanate in Bhopal, India in 1984 (responsible for a death toll of 2500) and the Piper Alpha explosion in 1988.

Two types of human failure have been identified. These are:

- Active failures; which usually have immediate consequences and are made by workers 'at the coal face'. Examples include drivers, machine operators and of course coal miners. Active failures have an immediate consequence.

- Latent failures can be made by people further away from the 'coal face'. These people tend to be designers, planners and managers.

Very often an accident can result from a combination of these two failure types. For example the Three Mile Island release occurred when operators failed to diagnose a stuck open valve. This was the active failure. A latent failure (poor design of the control panel used by the operators) also contributed to the accident.

Lack of sleep has also been implicated in human error[22b]. Dr Stanley Coren of the University of British Columbia believes that before 1913 it was more common for people to average nine hours sleep a night. In fact evolution had programmed us to require this amount of sleep. What happened in 1913? The invention of the light bulb! This, combined with the rise of automated factories robbed people of their sleep. Dr Coren traces blames lack of sleep for many major accidents. He quotes the Challenger space shuttle disaster to illustrate his argument, pointing out that NASA personnel were forced to work up to 14 hours a shift for as many as 26 days before the accident. And although the captain of the ship-wrecked Exxon Cadiz was drunk in his cabin at the time of the accident, it was the third mate falling asleep over the wheel that caused the oil tanker to run aground.

Lack of sleep makes us stupid, concludes Dr Coren, and this stupidity leads to accidents. His theory is supported by a recent article[22c]. This puts the case that sleep is an important part of our learning process. It gives the long-term memory the opportunity to 'discuss' and make sense the events of the day with the short-term memory. This learning process takes about 8 hours.

3.8.6 Managing ergonomic issues in the workplace

This chapter has so far considered some of the steps that ergonomic practitioners might take during the design stage of processes and equipment, i.e. anticipating ergonomic problems and designing them out before they are inflicted on the workforce. Managers and health and safety professionals need to be aware of these issues when selecting equipment and machinery or managing the implementation of new processes.

It is equally important to have a procedure for managing ergonomics in an existing workplace. Very often managers and health and safety professionals inherit a workplace where, at the design stage, little if any thought had been given to ergonomic issues. Even where consideration has been given at the design stage improvements can still be made. Getting ergonomics right is an iterative process since often it is only after a process or piece of equipment has been used for some time that problems become apparent. All of which makes a good ergonomics programme that much more important.

The basic approach is one of risk assessment followed by action to eliminate any unacceptable risks and a maintenance and review process to ensure that the agreed work methods are followed. Workers should be informed of the reasons for, and given adequate training in, any necessary control measures. One of the keys to success is to involve the operators who carry out the work. Their involvement can result in risk assessments which are more realistic and controls which are practical and acceptable. Involvement also results in greater awareness of the health and safety issues by both managers and operators.

3.8.7 Work-related upper limb disorders (WRULD)

3.8.7.1 Background

In October 1993 High Court Judge John Prosser dismissed a claim for industrial injury. Reuters journalist Rafiq Mughal[23] claimed to have contracted repetitive strain injury (RSI) while manning the Reuters Equities Desk. In dismissing the case, Judge Prosser said that 'RSI has no place in the medical textbooks'. He went on to refer to 'eggshell personalities who needed to get a grip on themselves'. Which was an unfortunate remark because if you have RSI it is extremely difficult to get a grip on anything!

However, there is still controversy over the exact meaning of the term RSI. In many cases patients suffer symptoms (in the case of Rafiq Mughal it was a tingling and numbness in his hands which eventually spread up his forearm and a painful right shoulder). But the precise causes are often not identifiable. Some doctors maintain that the term RSI should be reserved solely for this situation. Others, however, may use the term in the broader sense.

RSI as a term is a media rather than a medical invention. In the UK, work-related upper limb disorder (WRULD) is used as a blanket term for a collection of disorders, most of which, unlike RSI, have symptoms which can be attributed to identifiable causes. In the USA the preferred term is cumulative trauma disorder (CTD).

3.8.7.2 Physiology of WRULD

WRULDs arise because of the physiology of the upper limbs. When performing an action with the hands it is the arm muscles that do the work. The force is transmitted to the bones in the fingers by a network of cables known as tendons. These tendons reach the fingers via the carpal tunnel, a bony arch located where the wrist joins the hand. In the course of moving a finger a tendon can travel as much as five centimetres. Tendons were well designed by evolution. To prevent wear and tear from friction each tendon is surrounded by synovial fluid which is contained in the synovial sheaf which covers the tendon in much the same way as electrical insulation covers a wire.

Superb as this system is it was never designed to cope with the stresses of modern work activities such as keyboard use. Mill[24] describes how such conditions can lead to injury, the main factor being overuse.

Keyboard work, for example, requires an enormous amount of very quick, repetitive finger actions. These can lead to strain and injury particularly if carried out in awkward positions and for too long. Overuse can lead to quite specific injuries such as:

- Tenosynovitis: inflammation of the synovial sheath.
- Tendinitis: inflammation of a tendon.
- Epicondylitis: an inflammation of the tendons which attach muscle to the elbow. Depending on its exact nature, this condition is commonly called either tennis elbow or golfers elbow.
- Carpal tunnel syndrome: tendons and nerves entering the hand have to squeeze through the carpal tunnel. This is an archway where the roof is made of cartilage and the sides and floor are composed of bones. Swelling of the synovial sheath surrounding the tendons puts pressure on the nerve and restricts the flow of blood into the hand causing swelling of the wrist with pain, numbness and tingling which are typical symptoms.

Other specific conditions include supraspinatus tendinitis which affects the shoulder, de Quervain's disease which affects the tendon sheath in the thumb, and ganglions, small cysts which appear on the wrists. These specific conditions are often associated with modern keyboard use. Advice on assessing the risks of upper limb disorders is given in an HSE leaflet[25].

Recent research has shown the use of the mouse to be a significant contributory factor in the development of symptoms in the hand, wrist and fore-arm. In many cases these symptoms can be alleviated by the use of alternative pointing devices. Pointing devices using an electronic 'pen' in conjunction with a tablet have proved particularly effective.

The effects of overuse can be made much worse by two contributory factors. The first of these is static posture. Problems for certain limbs may be caused by repetitive movements but for other parts of the body it is the absence of movement. While the fingers are pounding away on the keyboard or operating the mouse, other parts of the body, e.g. arms and shoulders are often held rigidly still. Muscles are continuously contracted, continuously under tension. This can restrict blood flow and causes nerves to become trapped and compressed.

The other contributory factor is stress. The human reaction to stress comes straight out of our hunter-gatherer past, i.e. the fight or flight syndrome. Hormones to stimulate physical action are poured into the blood and while this was very necessary for a stone age man facing a sabre tooth tiger, it is not so good for today's humans sitting rigidly in front of a word processor. This lack of physical activity to disperse the hormones round the body can have long-term medical effects.

3.8.7.3 Managing the risks from display screen equipment (DSE)

The first step in managing the risks from DSE is to develop a strategy and identify the actions that need to be taken. This can be achieved through writing a Company Display Screen Equipment Standard which should contain the following sections:

(a) Management intent: this should state the company policy on the use of workstations and that they must not pose a risk to the health of users. Also, that users will be given information, instruction and training so that they know the reason for control measures and how to implement them.
(b) Performance expectations: these concern what needs to be done and cover items such as risk assessment, training, provision of work breaks and eyesight testing. They will specify that new equipment and accessories comply with the relevant national or international Standards and they will require consideration of the suitability of the software from the point of view of the health, safety and ability of the user. Guidance on this topic is available from the HSE[26,27].
(c) Guidance: this section should give any additional guidance deemed necessary to meet the performance expectations.
(d) Appendices: these should contain additional information, e.g. diagrams showing the correct posture to adopt at workstations. They should also contain examples of any forms used, e.g. the risk assessment checklist form.

A well-written standard provides staff with a ready source of reference and helps to ensure a consistent approach across the organisation. The very act of preparation helps to clarify what needs to be done. But note that the key words are well written (see the section on text clarity above). A badly written standard which confuses the reader can result in a reluctance on the part of employees to get involved in a DSE management programme.

Employee involvement is vital. It starts with the Purchasing Department ensuring that all equipment purchased conforms to the appropriate Standards. It then continues with the installer ensuring that all the equipment operates satisfactorily. Finally the employees can be involved by carrying out a risk assessment using a simple checklist to assess his or her workstation. The checklist should cover all the elements of the workstation, i.e. the chair, the desk, the keyboard, the pointing device (e.g. the mouse), the display screen itself and the general environment.

Any risks deemed to be high should be eliminated. This could be as simple as giving an employee a larger mouse, supplying a wrist rest for the keyboard, or the provision of a suitable footrest. The office politics of workstations should also be taken into account. A recent article by Steemson[28] argues for allocating workstations on the basis of the amount of space the work requires rather than on the basis of status. The biggest workstation should not necessarily go to the most senior person in the section.

The training of DSE users should emphasise the importance of following the recommended method for operating the equipment. The showing of a suitable video followed by a discussion with hands-on experience has proved effective. Suitable videos can be obtained from many commercial organisations.

Finally the effectiveness of the measures taken should be reviewed and, if found wanting, should be modified to achieve the required standard. Routine workplace inspections should identify any workstations that do not comply. Similarly, the written standard and assessment pro-formas should be reviewed periodically, particularly whenever any changes of workplace equipment or software occur.

It should be stressed that WRULD is not specifically a keyboard disease. Anyone carrying out work requiring repetitive use of their upper limbs is potentially at risk. Other occupations at risk include chicken process workers, telephonists, mail sorters, signers for the deaf, fabric sewers and cutters and musicians. However, the procedure outlined above is applicable to all occupations where WRULD is a risk.

3.8.8 Back issues

3.8.8.1 Background

In 1994–1995 116 million working days were lost as certified sickness through back pain[29]. This is the biggest single cause of lost working time in the UK. The annual cost to industry is estimated at over £5 billion.

When managing the risks of damage to the lower back, which is the site of most back injuries, it is important to realise that the causes are not limited to heavy lifting or heavy repetitive work. They are often the result of something we do more and more of in our working lives – sitting down. Poor sitting posture can damage the discs, the joints between the vertebrae, and the ligaments which connect them. Sitting exerts about 40% more pressure on the discs in the lower back (the lumbar region) than standing. Leaning forward exacerbates this situation where the extra pressure on the discs increases from 40 to 90%. The pressure can rise even more when leaning forward to pick up something like a book or a heavy file.

It is much better to have seating arranged so that the angle between the back and the thighs is greater than 90°. This will help decrease the pressure on the discs and vertebrae in the lumbar region.

The insidious thing about bad posture is that its effects manifest themselves gradually. Because it takes less effort, slumping can often seem more comfortable than sitting properly, but all that extra pressure on the lower back has a cumulative effect. The pressure can cause the annulus fibrosis (the elastic tissue which makes up the outer ring on the disc) to become worn and tattered. It fails to contain the nucleus pulposus (the fluid 'plug' inside the disc) which then presses on the nerve. Even when the annulus fibrosis heals the fluid plug will remain out of position. This is what is termed a prolapsed disc. It is also called a slipped disc but, in fact, this is a misnomer since these discs cannot actually slip.

The second major cause of back pain arises from wear and tear on the joints between the vertebrae (the facet joints). These joints are lubricated by synovial fluid held in place by a cartilage sheath. Over the years wear and tear can roughen these joints, causing them to lose their flexibility. Proper exercise can help keep them lubricated thus prolonging their useful life.

Ligaments can become permanently stretched by poor posture which can result in extra pressure on joints and discs and a permanently distorted posture. The damage accumulates until a single action triggers a long lasting and painful condition. This action can be as simple as picking up a child, twisting to get a case off the back seat of a car or even pulling up a weed. Ironically it can be the single, simple action that gets the blame rather than the years of abuse which caused the cumulative damage.

Part of a good ergonomics management programme should be not only to look at the job but to encourage individuals to adopt the correct posture and to develop the muscles around the spine. An employee awareness programme could usefully include recommendations for a regular (and of course properly supervised) programme of exercise in a gymnasium.

Training is an important element in ergonomic management and increasingly equipment manufacturers are providing health and safety training packages specific to their own products. One such example is Vauxhall Motors Ltd which has produced a training package for the owners of its cars. It consists of a video and booklet[30] and emphasises the importance of adjusting the seat in order to provide adequate support for the lower back. For salesmen who spend much of their time driving company cars, getting posture right is very important. Review and refresher training may also be necessary because the correct posture does not, at first, feel as comfortable as sitting incorrectly.

More general guidance can be obtained from the HSE[31,31a].

3.8.9 Managing the ergonomics of disability

Managing the ergonomic requirements of disabled workers poses a unique but rewarding challenge. The rewards include the resources saved by not having to dismiss disabled workers and recruit and retrain their replacements, and the retention of the experience and training already invested in the individual. An added incentive for the effective ergonomic management of disability comes from the obligations placed on employers by the Disability Discrimination Act 1995 (see section 3.8.10 below). This Act has been criticised as 'toothless' because it does not have an enforcing authority, relying instead on civil action brought by the claimant. Nevertheless it does provide a clear outline of employers' duties.

One potential barrier to the management of the ergonomics of disability is the general lack of awareness about the disability itself. In the absence of an in-house occupational health department this can pose a serious problem. Most disabilities have their own associated charities who are a ready source of often well-produced informative material. Additionally, many of these charities have experts who focus on the implications of employing people with specific disabilities. Another source of information is the Employers Forum on Disability (Tel. 0207 492 8460).

Disability is an issue which, increasingly, is attracting the attention of the health and safety press. An example of such coverage is the Working Wounded series of articles published in RoSPA's journal, *Occupational Safety & Health*. These gave examples of ergonomic adjustments which employers can make to help people with specific disabilities including:

- Arthritis[32]: ramps for wheelchair users, stair lifts, toilets with wide doors. Special consideration should be given to emergency escape from buildings. Advice given with the Fire Precautions (Workplace) Regulations 1997 emphasises the need to consider the disabled when reviewing means of escape arrangements.
- RSI[33]: voice recognition software makes possible hand-free operation of software packages such as Word for Windows. However, if keyboards are used, preventive steps such as the provision of ergonomic keyboards (e.g. the keyboard manufactured by PCD Maltron) could be considered.
- Occupational deafness[34]: loop systems which transmit speech straight to a hearing aid enabling speech to be heard above the extraneous noise. 'Minicoms' enable deaf people to type text messages to each other's screens using ordinary telephones.
- Visual impairment[35]: speech synthesisers on computers which, for example, will say the words as the user moves the cursor round a Word for Windows document. Electronic braille: pins rise and fall under a strip on the keyboard to create a Braille impression of the document on the screen. Magnification programs: to make screen text larger for those with visual impairment.

Financial help may be available to pay for special measures such as those outlined above from government agencies.

3.8.10 Legal requirements

The main reasons for implementing an effective ergonomics management system should always be the health of the workforce and the wealth of the business and not merely legal compliance. Nevertheless an awareness of the legal requirements will provide useful guidance in formulating a management strategy. The relevant requirements are summarised below.

3.8.10.1 The Workplace (Health, Safety and Welfare) Regulations 1992

- adequate lighting (reg. 8)
- adequate space (reg. 10)
- suitable workstations, including suitable seating (reg. 11).

3.8.10.2 The Provision and Use of Work Equipment Regulations 1998

- equipment to be suitable for the intended purpose (reg. 4)
- suitable guards and protective devices (reg. 11). Ergonomic principles are inherent in the advice given about reg. 11(2) and contained in HSE Guidance[36], namely the requirement for:
  (a) fixed enclosing guards
  (b) other guards or protecting devices
  (c) protection appliances (jigs, holders, push sticks etc.)
  (d) the provision of information, instruction training and supervision.
  The guidance makes it quite clear that in considering which combination of the above measures to adopt, they should be considered in strict hierarchical order from (a) to (c). In other words the best ergonomic solution (fixed guarding so that workers do not have to do anything themselves in order to be protected) must always be the first to be considered
- controls for work equipment must be clearly visible and identifiable (reg. 17)
- suitable and sufficient lighting (reg. 21).

3.8.10.3 The Personal Protective Equipment at Work Regulations 1992

- that PPE should only be worn as a last resort (reg. 4(2)). This requirement takes into account that wearing PPE is never as ergonomically desirable as not having to wear it at all because the hazard has been removed
- that PPE should take account of the ergonomic requirements and the state of health of people who wear it (reg. 4(3)(b)). The Guidance[37] accompanying these Regulations states that when selecting PPE, not only the nature of the job and the demands it places on the worker should be taken into account, but also the physical dimensions of the worker.

3.8.10.4 The Manual Handling Operations Regulations 1992

- employers, if possible, to avoid giving employees work which could result in injury (reg. 4(1)(a))
- employers to do a risk assessment on all work which involves carrying and which could represent a risk of injury to employees (reg. 4(1)(b))
- reduce the risk of injury from jobs where manual handling cannot be avoided and which could cause injury (reg. 4(1)(b)(ii)).

The Guidance[38] accompanying these Regulations contains an enormous amount of illustrative advice: it is virtually a practical ergonomics textbook in its own right.

3.8.10.5 The Health and Safety (Display Screen Equipment) Regulations 1992

- employers to assess and control the risks to users of display screen equipment workstations (reg. 2)
- the Guidance[26] and its schedule outline the physical considerations necessary to meet the Regulations
- the schedule also takes into account the mental stress which may result from using the software and in Schedule 4 stipulates that:
  - software must be easy to use and, where appropriate, adaptable to the level of knowledge or experience of the operators or user;
  - no quantitative or qualitative checking facility may be used without the knowledge of the operators or users;
  - the principles of software ergonomics should be applied in particular to human data processing;
  - systems should display information in a format and at a pace which are adapted to operators or users.

3.8.10.6 The Supply of Machinery (Safety) Regulations 1992

- From an ergonomic point of view the requirements of these Regulations eliminate at source a great many of the risks from machinery by requiring manufacturers to comply with the appropriate safety-of-machinery standard and confirm the fact in a declaration of conformity they issue with the equipment which they supply (see 4.3.1.3).

3.8.10.7 The Management of Health and Safety at Work Regulations 1999

- require that employers take account of the capabilities of employees when allocating work (reg. 13(1)).

3.8.10.8 The Reporting of Injuries, Diseases and Dangerous Occurrences Regulations 1995

- require an employer to report incidents of RSI resulting from work activity. This includes conditions such as cramp, tenosynovitis or carpal tunnel syndrome.

3.8.10.9 The Disability Discrimination Act 1995

- makes it unlawful for an employer to discriminate against a disabled employee (s. 4(1))
- requires an employer to make reasonable adjustments in order that a disabled person is not placed at a disadvantage compared to people who are not disabled (s. 6). These adjustments could be ergonomic as discussed above.

3.8.11 Conclusion

A chapter such as this cannot hope to provide examples of all factors which need to be considered in order to manage ergonomics. The subject is too vast. It encompasses psychology on the one hand, and occupational hygiene issues such as heat, noise and lighting on the other. Managing these issues effectively is as much a question of managing specialist consultants who may have to be called in to supplement 'in-house' expertise. Whether bought-in expertise is used will depend upon the complexity of thc issuc and the depth of knowledge required to deal with it.

Effective ergonomics can be both operationally and cost effective making sense both from a health and safety and from a business point of view.

Acknowledgement

The author wishes to thank Mrs J. Steemson of the Royal Society for the Prevention of Accidents (RoSPA) for invaluable help and guidance with the research and writing of this chapter.

References

1. Murrell, K.F.H., *Ergonomics, Man and his Working Environment*, Chapman and Hall, London (1965)
2. Clark, T.S. and Corlett, E.N., *The Ergonomics of Workspaces and Machines: A Design Manual*, Taylor and Francis, London (1984)
3. Christiansen, J.M., Topmiller, D.A. and Gill, R.T., Human factors definitions revisited, *Human Factors Society Bulletin*, 31, 7–8 (1988)
4. Legge, Sir Thomas
5. Molleson, Theya, The eloquent bones of Abu Hureyra, *Scientific American*, 60 (1994)
6. *New Scientist*, p. 8, 20 January 1996
7. Orwell, G., *The Road to Wigan Pier*, Penguin Books, 24, ISBN 0 14 018238 1
8. Taylor, F.W., *What is Scientific Management?* Classics in Management, rev. edn, American Management Association, New York (1970)
9. Gilbreth, F.B., *Science in Management for the One Best Way to do Work*, Classics in Management, rev. edn, American Management Association, New York (1970)
10. Beynon, Huw, *Working for Ford*, 2nd edn, Penguin Books, 147 (1984) ISBN 0 14 022590 0
11. *The Guardian*, 24 February 1984
12. Grether, W.F. and Baker, C.A., *Visual presentation of information*, Ch. 3, van Cott and Kinkade (1972)
13. Wilson, R. and Rajan, J.A., *Human–machine interfaces for systems control, Ch. 13, Evaluation of Human Work*, eds Wilson, J.R. and Corlett, E.N., Taylor and Francis, 383 et seq. (1995)
14. Cushman, W.H. et al., *Equipment Design, Ch. 3, Ergonomic Design for People at Work*, eds Eggleton, E.M. and Rodgers, S.H., Eastman Kodak Company, New York, 91 (1983)
15. Hartley, J., *Methods for evaluating text, Ch. 11, Evaluation of Human Work*, eds Wilson, J.R. and Corlett, E.N., Taylor and Francis, 286 et seq. (1995)
16. Cushman, W.H. et al., *Equipment Design, Ch. 3, Ergonomic Design for People at Work*, eds Eggleton, E.M. and Rodgers, S.H., Eastman Kodak Company, New York, 107 (1983)
17. Fitts, P.M., *Handbook of Experimental Psychology*, chapter on Engineering psychology and equipment design, John Wiley, London (1951)
18. Singleton, W.T., *Man-machine Systems*, Penguin Books, Hammondsworth (1974)

19. Clegg, C., Ravden, S., Corbett, M. and Johnson, G., Allocating functions in computer integrated manufacturing: a review and a new method, *Behaviour and Information Technology*, 8/3, 175–190 (1989)
20. British Standards Institution, BS EN 294, *Safety of machinery – Safety distances to prevent danger zones being reached by the upper limbs* (1992) BS EN 349, *Safety of machinery – Minimum gaps to avoid crushing of parts of the human body* (1993) BS EN 614–1, *Safety of machinery – Ergonomic design principles – Part 1: Terminology and general principles* (1995) BS EN 999, *Safety of machinery – The positioning of protective equipment in respect of approach speeds of parts of the human body* (to be published) Available from BSI, London
21. Kirwan, B., *Human reliability assessment, Ch. 31, Evaluation of Human Work*, eds Wilson, J.R. and Corlett, E.N., Taylor and Francis, 921 (1995)
22. Swain and Guttman, *A Handbook of Human Reliability Analysis with Emphasis on Nuclear Power Plant Applications*. Nureg/CR–1278, USNRC, Washington, DC (1983)
22a. Health and Safety Executive, Guidance booklet no. HSG 48, *Reducing Error and Influencing Behaviour 2nd edn.*, HSE Books, Sudbury (1999)
22b. Coren, S., *Sleep Thieves*
22c. Phillips, H., Perchance to dream, *New Scientist*, 25 September 1999, p. 26
23. Rafdiq Mughal v. Reuters
24. Mill, W.C., *RSI*, Thorsons, London (1994) ISBN 0 7225 2919 8
25. Health and Safety Executive, Leaflet Pack no. INDG 171, *Work related upper limb disorders: Assessing the risk*, HSE Books, Sudbury (1995)
26. Health and Safety Executive, Legal Series booklet no. L26, *Display Screen Equipment at Work. The Health and Safety (Display Screen Equipment) Regulations 1992. Guidance on the Regulations*, HSE Books, Sudbury (1999)
27. Health and Safety Executive, Leaflet Pack no. IND 36, *Working with VDUs*, HSE Books, Sudbury (1998)
28. Steemson, J., Space craft, *Occupational Safety and Health*, October (1997)
29. Department of Social Security, Statistical Unit. *Statistics of certified incapacity*
30. Vauxhall Motors Limited, *Are you Sitting Comfortably?*, video and booklet, Vauxhall Motors Limited, Luton (1997)
31. Health and Safety Executive, Leaflet Pack no. INDG 242, *In the driving seat*, HSE Books, Sudbury (1997)
31a. Health and Safety Executive, leaflet no. IND 333, *Back in work. Managing back pain in the workplace, a leaflet for employers and workers in small businesses*, HSE Books, Sudbury
32. Cook, N., The working wounded: arthritis and rheumatism, *Occupational Safety and Health*, April, 17 (1995)
33. Cook, N., The working wounded: RSI, *Occupational Safety and Health*, August, 36 (1996)
34. Cook, N., The working wounded: hearing impairment, *Occupational Safety and Health*, January, 28 (1996)
35. Cook, N., The working wounded: visual impairment, *Occupational Safety and Health*, June, 43 (1995)
36. Health and Safety Executive, Legal Series booklet no. L22, *Work equipment. Provision and Use of Work Equipment Regulations 1998 – Guidance on the Regulations*, HSE Books, Sudbury (1998)
37. Health and Safety Executive, Legal Series booklet no. L25, *Personal protective equipment at work. Personal Protective Equipment at Work Regulations 1992 – Guidance on the Regulations*, HSE Books, Sudbury (1992)
38. Health and Safety Executive, Legal Series booklet no. L23, *Manual handling. Manual Handling Operations Regulations 1992 – Guidance on the Regulations*, Second edition, HSE Books, Sudbury (1998)

General reading

Dul, J. and Weerdmeester, B., *Ergonomics for Beginners: A Quick Reference Guide*, Taylor and Francis. Full of clear, practical advice on doing risk assessments.

Chalmers Mill, W., *Repetitive Strain Injury*, Thorsons Health. A very clear introduction to RSI and its prevention.

Health and Safety Executive, Guidance booklet no. HSG 121, *A pain in your workplace? Ergonomic problems and solutions*, HSE Books, Sudbury (1994)

Wilson, J.R. and Corlett, E.N. (eds), *Evaluation of Human Work*, 2nd edition, Taylor and Francis (1995). An extremely comprehensive survey of the methodology of ergonomics. Very academic and possibly of more value to specialists than industrial health and safety professionals.

Rogers, S.H. and Eggleton, E.M. (eds), *Ergonomic Design for People at Work* (2 volumes), The Human Factors Section, Eastman Kodak, published by van Rostrand Reinhold Company, New York (1983) Despite its age this is an excellent, clearly written and comprehensive survey of the subject. Written with practical health and safety professionals in mind.

Office Health and Safety: a guide to risk prevention, Unison, 1 Mabledon Place, London WC1H 9AJ. A clear and well-written brief guide for union members and safety representatives

Useful contacts:

The Ergonomics Society, Devonshire House, Devonshire Square, Loughborough, Leicestershire LE11 3DW. Tel: 01509 234904. Publishes a register of consultancy firms in their membership.

Ergonomics Information Analysis Centre, School of Manufacturing and Mechanical Engineering, University of Birmingham, Birmingham B15 2TT. Tel: 0121 414 4239

The Computability Centre: a national charity which aims to help workers disabled by RSI return to work. PO Box 94, Warwick CV34 5WS.

Chapter 3.9

Applied ergonomics

J. R. Ridley

3.9.1 Introduction

The body is an amazing organism. It will stand an enormous amount of misuse and still continue working. However, it does eventually reach a stage when it says 'enough is enough' and either ceases working altogether or makes continuation of work so painful as effectively to prevent further work. While the main aim of ergonomics must be to optimise the working conditions for both worker comfort and maximising output, one of its features is to establish the limits of what the body will stand and ensure that working conditions do not exceed them. This applies to both the physical and mental aspects of work activities as well as to the environment in which the work has to be carried out. When applying ergonomic principles, account needs to be taken of the physique and behavioural characteristics of the operators involved.

The word 'ergonomics' is occurring with increasing frequency in standards and legislation[1] and is being referred to by those concerned with safety in the workplace. But what does it mean? Its literal meaning taken from its Greek derivation is 'work rules' but in the modern concept it has come to refer to the relationship between a person and the conditions in which his activities, whether work operations or leisure activities, place him. More commonly, it is being applied to the interface between the individual and work equipment and takes account of the surroundings in which the equipment is used. However, the design and layout of equipment is not the only influence acting on workpeople that affects their attitudes towards work. Work colleagues are also an influence and the psychological aspects of relationships between employees also needs to be borne in mind.

A commonly heard expression is 'matching the machine to the individual' but there are many more factors that affect the well-being of a person at work and at play than the equipment or implement being used. Ergonomics requires an understanding of how and within what limits various limbs and organs of the body can be employed, the effects of the climate and environment in which work has to be carried out and the stresses resulting from constraints or pressures imposed in the name of productivity.

Ergonomics is basically a process of adjusting the circumstances and conditions under which a person has to work to make the experience more pleasurable, rewarding and safer. Many of the current regulatory health and safety measures have an ergonomic content.

This chapter is aimed, not at the professional ergonomist and designer, but at the manager, the safety practitioner and student who want an overview of the subject so they can understand the implications of the layout of the equipment with respect to the demands put on the operator, the need for the right working climate and some of the limitations to which the human body is prone. These various aspects are considered in the following sections of this chapter which also indicates convenient and comfortable movement limits for various limbs. These limits are purely indicative and may vary with individuals. For detailed statistical information on movement limits reference should be made to the many ergonomic books covering this subject, some of which are listed at the end of this chapter. Similarly, for detailed anthropological data the many published books on the subject should be consulted.

3.9.2 Physiology

3.9.2.1 The human body

The human body is a marvellous and adaptable machine but it does have limitations. Different limbs have developed to fulfil particular functions which they do very effectively. Since ergonomics is the study of fitting the machine to the man it would seem sensible that priority of consideration should start with what a human body can do before deciding how the machine should fit it.

All movements of any part of the human body are controlled by the brain but the actual execution of movement is carried out by muscles acting on a particular limb. Individual limbs have a considerable range of movement but are most effective over the inner 50–60% of that range. Beyond that point the ability of the motor muscles to do work is reduced, the effort can become more stressful and, eventually, painful.

3.9.2.2 Muscles

The skeleton is, in effect, a series of linked struts or bones held together by ligaments that allow relative movement between the bones they link. The joints between bones sit on cartilage cushions that facilitate movement. The cartilage does not have a blood supply and relies on the compression and decompression due to the movement of the joint for its supply of nourishment from the joint fluid. Movement of the limbs is by voluntary motor muscles (section 3.1.8.6) that are flexible and contract and extend to provide movement. The ends of these motor muscles are attached to the bones of the limbs by bundles of tissue fibres called tendons.

Energy needed for the muscles to contract is stored in the muscles in the form of phosphate chemical compounds. High energy phosphates break down into low energy compounds releasing energy and allowing the muscles to do work. Regeneration of the low energy phosphate compounds requires glucose from the blood stream in the presence of oxygen. A side product of the breaking down of the high energy phosphates to generate energy is the creation of waste matter (lactic acid) which is carried away by the blood flow. If the blood flow is low, the waste matter accumulates giving rise to fatigue and eventually pain. Thus for muscles to work effectively and consistently they require a good supply of blood.

The amount of blood flowing is influenced by the type of activity undertaken. Static effort tends to restrict the flow of blood through the muscles, whereas the movement of the muscles in dynamic effort tends to assist the circulation of the blood. The supply of blood is provided by the action of the heart pumping the blood through the arteries which have muscles and can assist in the circulation. The flow of blood back to the heart is through veins which have poor musculation but do contain valves that ensure the blood flows only towards the heart. Flow in the veins is assisted by compression of the muscles during movement. Where there is lack of movement, in jobs such as hairdressers that require long periods of standing with little movement, the pressure of the blood can cause the vein to dilate when the valves cease to work and the veins become subject to additional pressures from the heart resulting in the condition of varicose veins.

Static effort, i.e. effort without muscle movement, can only be maintained for a limited time because the restricted blood flow causes a reduction in the regeneration of the muscle's energy store. Conversely, with dynamic effort there is an increased supply of blood to regenerate the energy store and the effort can be maintained indefinitely. After static efforts greater rest periods are necessary to restore the high energy resource than after dynamic effort.

Similarly, the position of the muscle above or below the heart is a factor that can decide the quantity of blood that flows to a limb. The higher above the heart the lower the flow and the lower below the heart the greater the flow and the pressure. Thus, the region for the hands and arms to be the most effective, i.e. where they get the highest blood flow and hence energy regeneration, is between waist and shoulder level.

In other limbs, such as the legs that by virtue of their position below the level of the heart, are subject to high blood pressures, normal walking and use requires the expansion and contraction of the muscles that assist the veins in returning blood flow to the heart.

3.9.2.3 Blood circulation

Blood is circulated round the body by the pumping action of the heart (*Figure 3.1.4*). The returning blood is received by the right auricle which passes it through a non-return valve to the right ventricle which, in turn, pumps the blood through the lungs where it picks up oxygen. The blood then passes through the left auricle into the left ventricle that pumps it into the arteries and hence the various organs and muscles.

Essential elements in the circulation system, if the muscles and organs are to perform effectively, are a sound heart and lungs whose oxygen transfer mechanism is unimpaired by diseases such as tuberculosis, asbestosis, pneumoconiosis, asthma, etc. The flow of oxygen-carrying blood to the brain must be maintained. A break of four minutes or more can result in irreparable brain damage. A good blood flow is needed through the muscles to encourage regeneration of the stored energy and to remove waste matter created during the chemical reaction when the phosphates stored in the muscles degenerate to create energy.

3.9.2.4 Spinal column

In the process of evolution, the bony spinal column developed in animals (quadrupeds) for a horizontal alignment. Since primates adopted the vertical posture there has not been sufficient time for the human spinal column to adjust its anatomy to accommodate a vertical alignment. The present shape of the human spinal column is inappropriate for a vertical alignment resulting in considerable disadvantages in human activities.

The spinal column consists of a series of linked bones or vertebrae that give structure to and support the torso. Between each vertebra is a pad of cartilage which acts as a cushion and gives the spine flexibility of movement to accommodate the different positions the body needs to adopt. The spinal column comprises a total of 33 vertebrae, 7 at the neck (*thoracic* or *cervical* vertebrae), 12 in the region of the chest (*dorsal* vertebrae), 5 in the loins (*lumbar* vertebrae), 5 in the pelvic region that have fused together to form the *sacrum* and 4 below the sacrum that have also fused together as the *coccyx*. These are shown in *Figure 3.9.1*.

At its upper end the spinal column supports the skull and at its lower end the sacrum forms part of the pelvic girdle. The whole weight of the torso and skull is supported by the sacrum which, as an integral part of the pelvis, is effectively built-in or *encastré* and rotates with the pelvis, as shown in *Figure 3.9.2*, changing the shape of the spine. The spinal column has four curvature sections in its length corresponding to the upper four groups of vertebrae that, when standing upright, lie about a line drawn from the skull to the pelvis, so that, effectively, the weights of the various parts of the body pass straight down the vertebrae and the discs without putting any load on the back muscles.

Similarly, the reaction to any load that is carried in the arms must pass through the spinal column but, because the load is external to the body, not only does it put an additional vertical load on the spine, it also imposes a bending load which in turn results in a stress being put on the back muscles. To understand the effect of this it is necessary to appreciate the structural mechanics involved. *Figure 3.9.3* shows diagrammatically an individual carrying a weight W at a distance of Y from the spine. Distance Y can depend on a number of factors including load shape, position of its centre of gravity, the anatomical characteristics of the carrier, etc. This load imposes a bending moment on the spine of $M = W \times Y$.

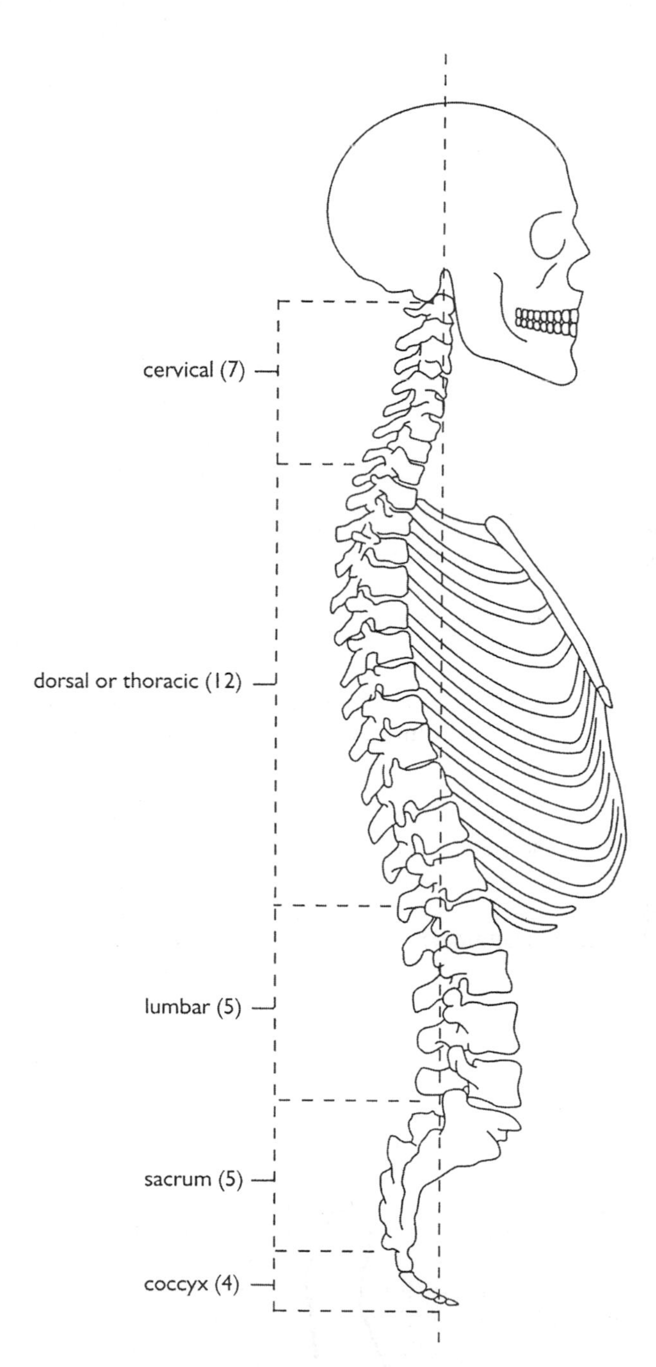

Figure 3.9.1 Structure of the spine

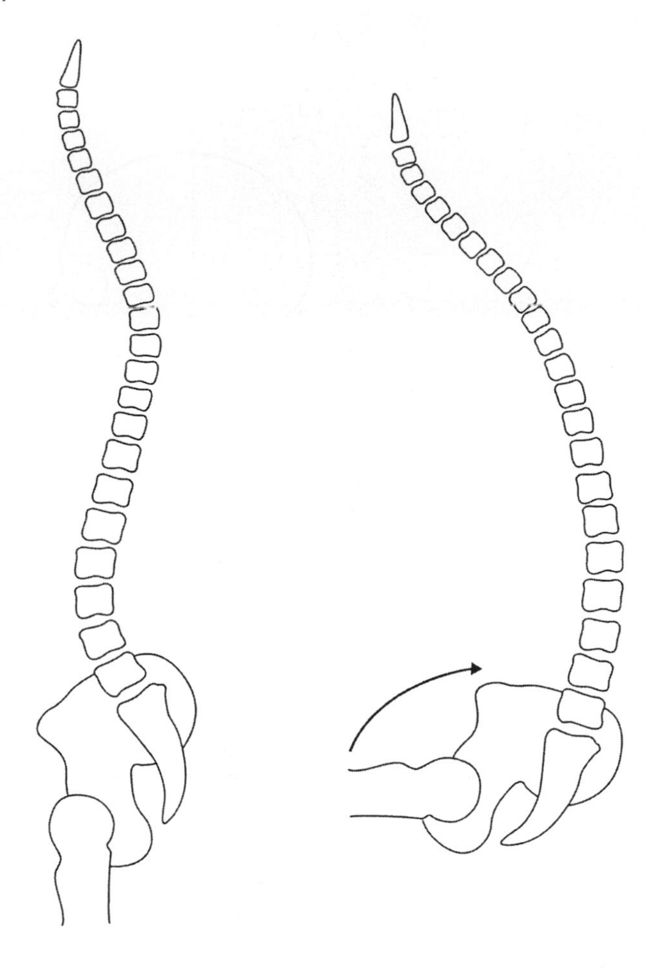

Figure 3.9.2 Effect of rotation of the pelvis on the spine

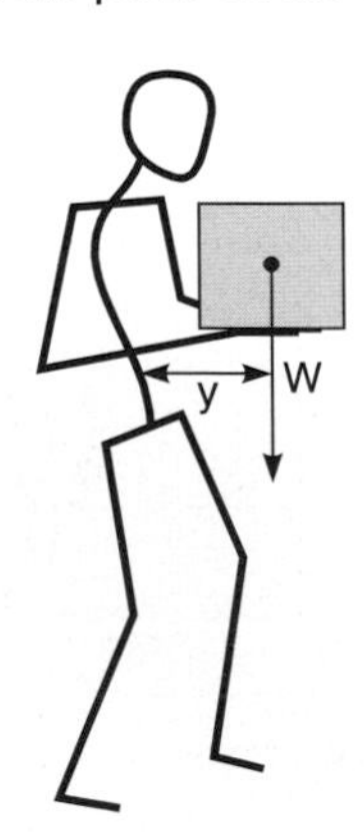

Figure 3.9.3 Body mechanics when lifting

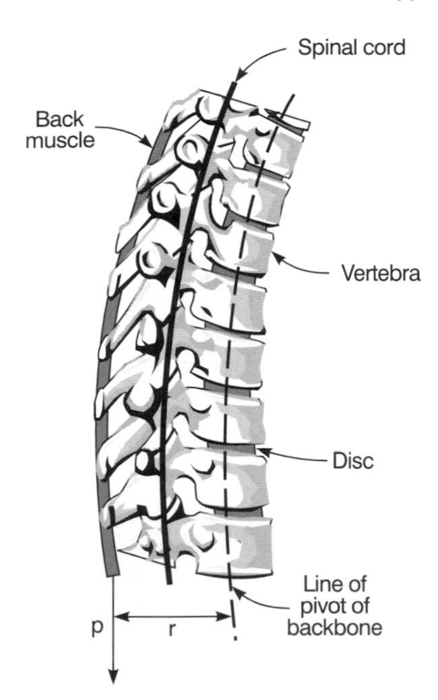

Figure 3.9.4 Structure of the spine

Figure 3.9.4 shows the construction of the spine with the positioning of the discs, spinal cord and the back muscles. The bending moment created by the weight W pivots about the discs and is resisted by the back muscles. Thus:

$$M = p \times r$$

hence for stability

$$p \times r = W \times Y$$

and the load on the back muscle

$$p = (W \times Y) \div r$$

If r is small compared to Y the load put on the back muscles will be many times greater than the actual load being lifted. Not only does this put a considerable load on the back muscles, but because the disc acts as a fulcrum it must carry the combined weight plus the muscle reaction – in addition to the normal body weight. This is shown diagrammatically in *Figure 3.9.5*.

$$\text{Disc load} = 11\ W + \text{Body weight}$$

Carrying a load on one shoulder means that the weight is carried close to the spine but to one side. In most cases, the spine can adjust its shape so that the load does, in fact, pass straight down the spine.

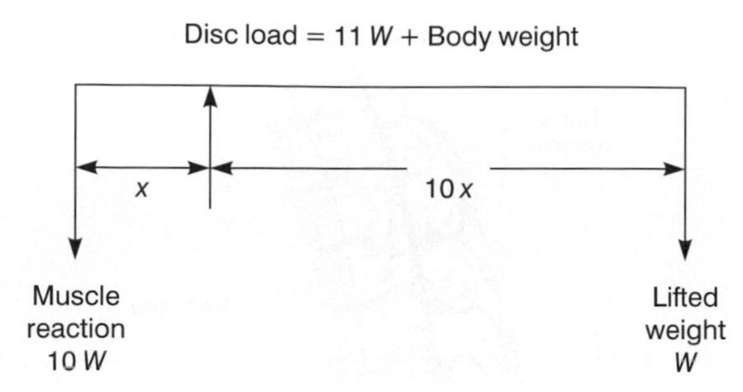

Figure 3.9.5 Diagram of loads on a vertebrate

3.9.2.5 Arms

The arms are very flexible limbs since they can transcribe three-dimensional movement. While it is possible to raise the arms to the upright position, any position above the horizontal can be very tiring with, because of their height above the heart, a reduced supply of blood and hence a reduced ability to do work. The arms can also be lowered to the vertically downward position but work with the arms below waist height is less efficient since it is difficult to see what they are doing. An exception to this is in the control of machine tools where the operator is watching the effect at the machining point of the movements performed by his hands which may be at a lower part of the machine. Generally, the arms are most effective in the region between waist and shoulder height as shown in *Figure 3.9.6*.

In the horizontal plane, the arms are capable of being moved to a position behind the shoulder line but this puts a considerable strain on the pectoral muscles. For continuous and efficient working the sweeping range of arm movement should be restricted to between the centre of the body and approximately 45° from the shoulder. Any sweeping movement beyond this may involve whole body twisting. In reaching to pick items, the whole arm length can be used, but for grasping the reach should be restricted to about one and a half times forearm length (*Figure 3.9.7*).

Where loads have to be lifted this can most effectively be carried out when standing provided that the correct lifting techniques are used (section 3.9.4). However, it should be noted that the size and weight of an item that can safely be handled will depend on the strength and physique of the individual. The ability to handle loads when seated is much less than when standing. In these cases, the whole of the lifting load is taken by the back muscles in the lumbar region.

3.9.2.6 Hands and wrists

The hands are the grasping and gripping limbs and have flexibility of movement in the vertical and horizontal planes plus the ability to rotate. The wrists allow the hands to be bent away from the palm (extension)

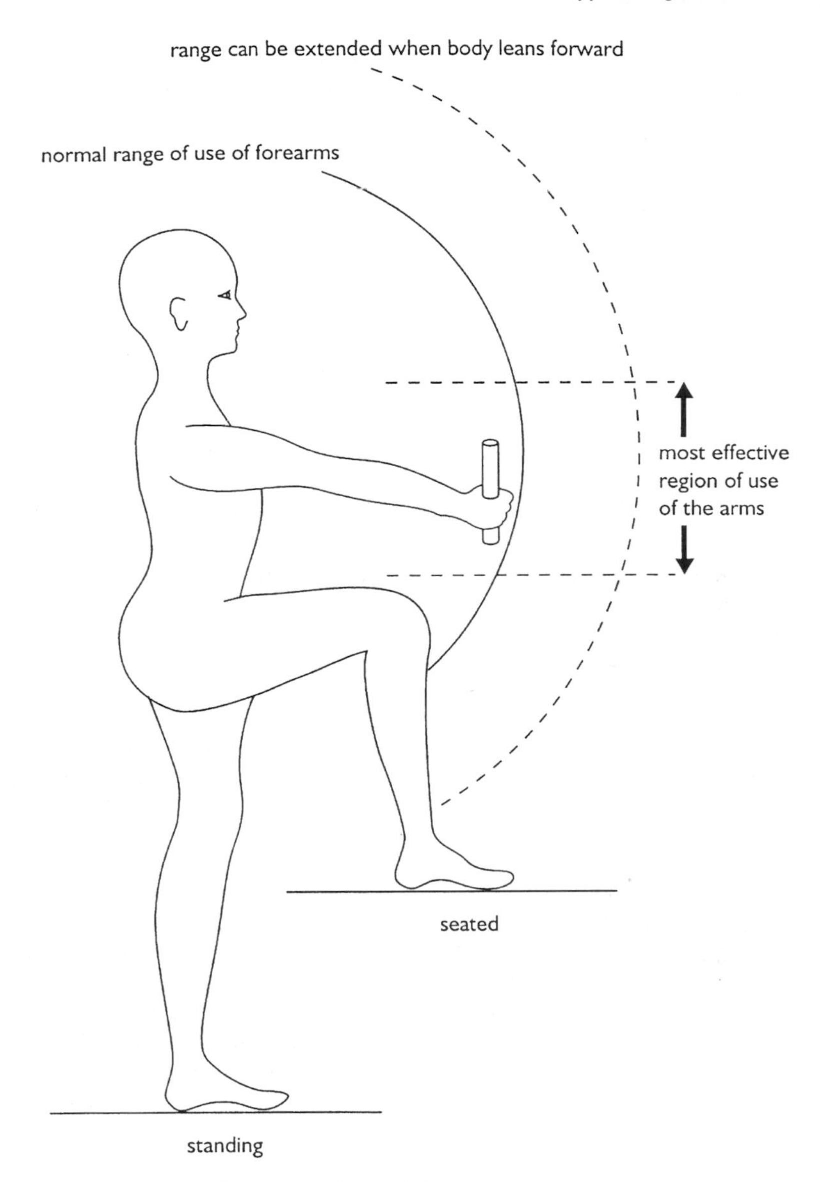

Figure 3.9.6 Comfortable range of use of the arms

and towards the palm (flexion) (*Figure 3.9.8*). In addition, the wrist allows the hand to be swivelled towards the thumb (radial movement) and away from it (ulnar movement) (*Figure 3.9.9*).

Hands also possess the ability to rotate about the axis of the forearm, the amount depending on whether the arm is bent or straight. In the latter case, the upper arm provides additional rotation particularly in the inward direction (palm downwards or pronation) (*Figure 3.9.10*).

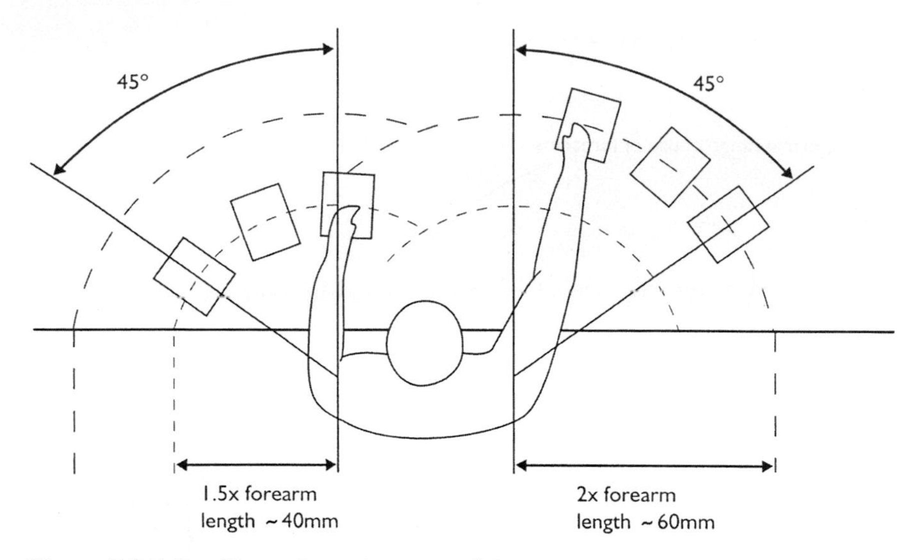

Figure 3.9.7 Reaching and grasping range of the arms

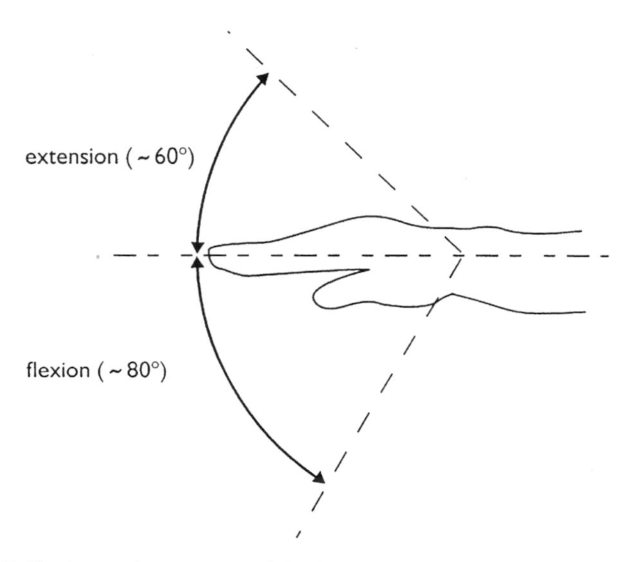

Figure 3.9.8 Flexion and extension of the hand

When using tools that are gripped in the hand, it should be noted that in the relaxed and natural position the line of grip, which follows the line of the knuckles, lies at an angle of approximately 75° to the line of the lower arm (*Figure 3.9.11*). Any tilting of the wrist necessary to allow tools to be used can reduce the power of the grip and, if used repeatedly, increases the risk of wrist and lower arm disorders.

By utilising the independent movement of the fingers and thumb, the hands are able to grip and to manipulate objects and controls. The size of

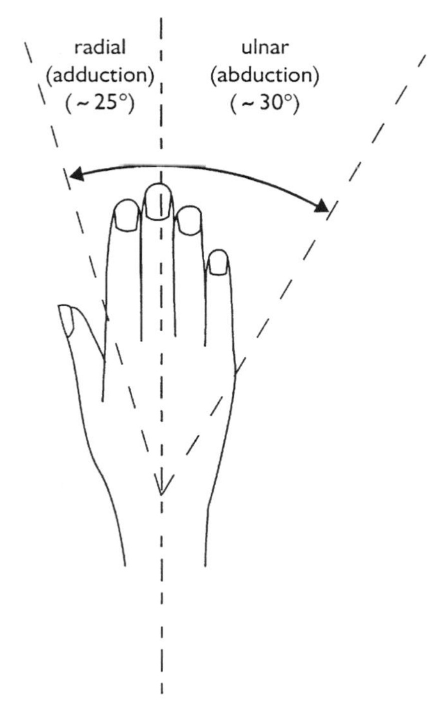

Figure 3.9.9 Swivelling of the hand

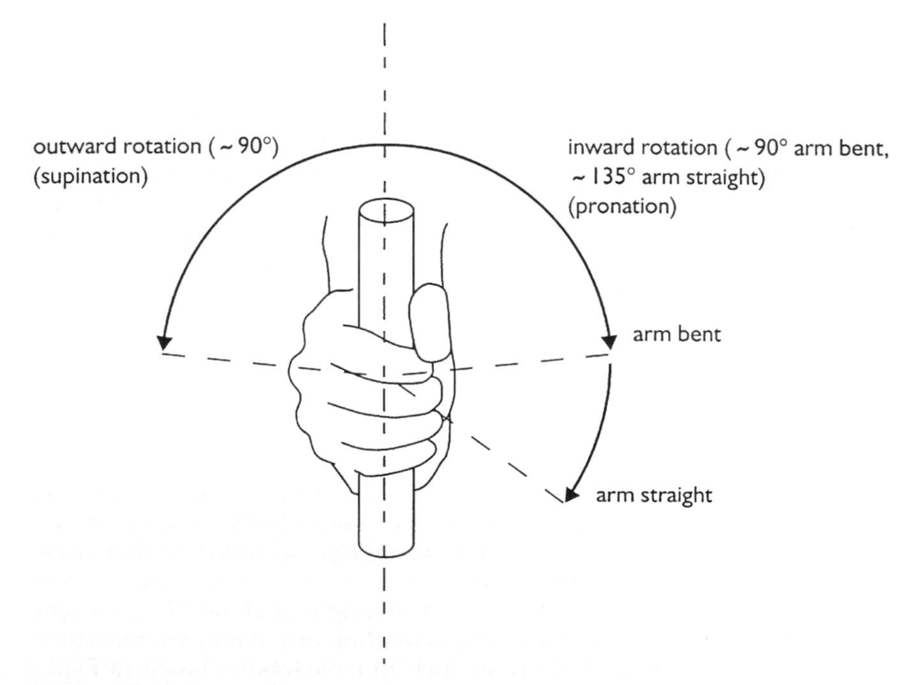

Figure 3.9.10 Rotation of the hand and wrist

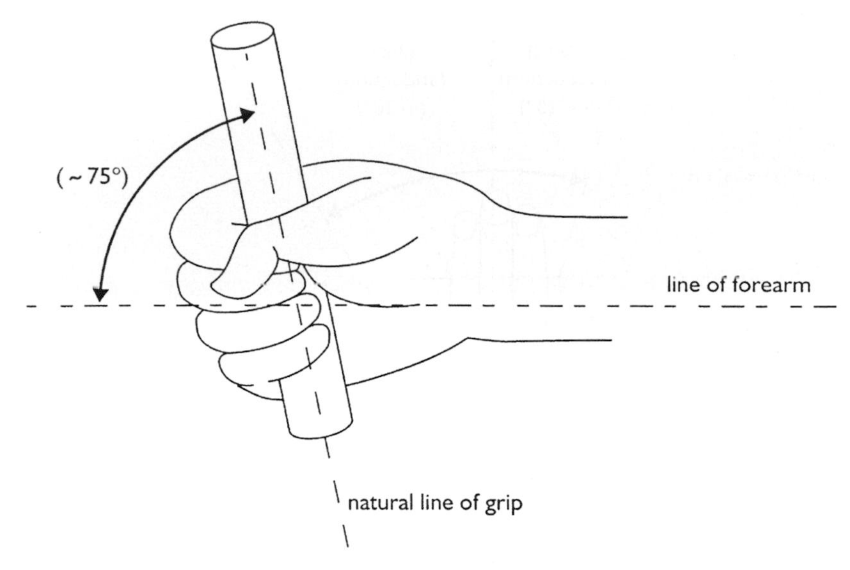

Figure 3.9.11 Line of grip of the hand

the object that can be gripped, the ability to manipulate it and the degree of dexterity exhibited varies considerably between individuals.

3.9.2.7 Legs and feet

The main movement of the leg is a forward and backward swing from the hip with a small amount of lateral movement. The greater the flexing at the hip, the greater the strain put on the muscles. Crouching down puts a very great static strain on the leg and knee muscles. This interferes with the flow of blood leading to a build-up of waste matter in the muscles that can quickly become painful. Such working positions should be avoided. Kneeling is better but if it is necessary to work at low levels, a small stool should be provided and the work carried out from the seated position.

If it is necessary to raise the leg this should be done with the knee bent. To keep the knee straight stretches the muscles at the back of the thigh and restricts the upward movement of the whole leg. By bending the knee, the thigh can be raised to a higher level.

The need to stand on one foot for any length of time to allow the other foot to operate a pedal or control should be avoided since this can cause early fatigue in the leg that carries the whole body weight. Where operations by foot are necessary, they should be arranged so that either foot can be used.

In its natural position, the foot lies at an angle of about 75–80° to the line of the lower leg. The flexing and extending of the foot are measured from this position. Typical range of such movements is shown in *Figure 3.9.12*.

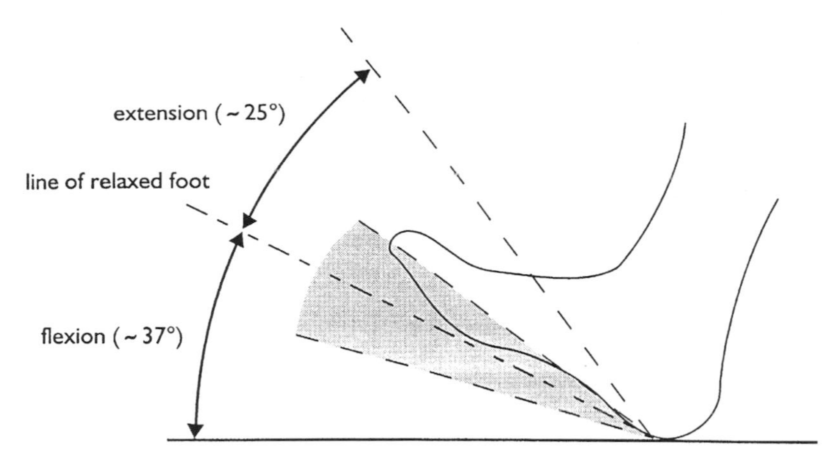

Figure 3.9.12 Flexion and extension of the foot

Where the foot is used to control a machine, the heel should rest on a support and in operating the control the foot should only need to be raised (extended) a minimum amount and the control actuated by flexing the foot to give a fine degree of control. Foot controls that require the whole foot to be raised allow only very coarse control and are best restricted to operations that require only a simple either/or position of the control.

3.9.2.8 Eyes

The eye is a completely flexible organ able to move freely within the constraints of its containing sclera (*Figure 3.1.7*). For this reason it is difficult to determine a datum line of sight that can be used to determine the positioning of work equipment. However, Kroemer and Grandjean[2] suggest that the 'Ear-Eye Line' (EE line) from the ear hole through the meeting point of the eyelids gives a convenient datum (*Figure 3.9.13*). For an erect head, this datum would be about 15° above the horizontal and would be the normal position of the head for distance sight, i.e. the normal line of sight is about 15° below the EE line. The nearer the object being viewed, the smaller the angle between the EE line and the horizontal. For DSE work the angle could be about 5° while for reading papers on a desk the EE line may be as much as 45° below the horizontal.

Movement of the eye can occur without discomfort within an included cone of 30° of the nominal line of sight (*Figure 3.9.14*), and although the eye can move beyond this cone some degree of discomfort may be experienced. To give good viewing, workpieces should be positioned so that the head does not need to be lowered more than 60° below the erect head position. By the same token, instruments that have to be read should be within the 30° included cone angle of sight and, ideally, so that the head does not have to be raised above the erect position.

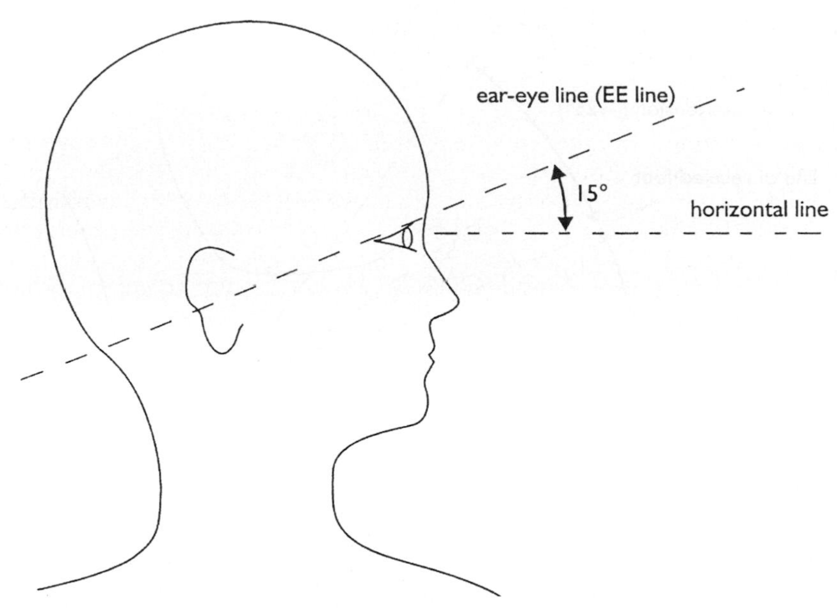

Figure 3.9.13 Datum line of sight using the 'EE line' (after Kroemer and Grandjean)

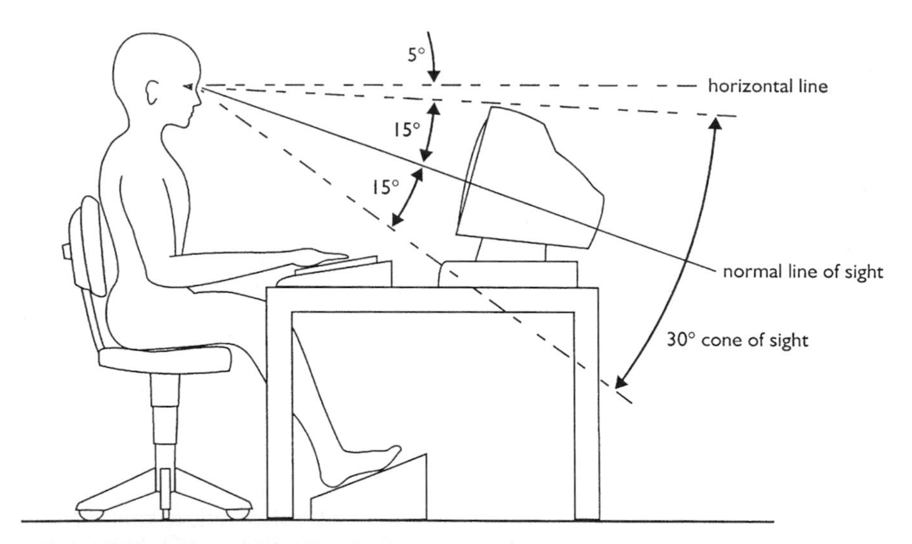

Figure 3.9.14 Cone of sight

The eye acts like a photo-electric cell and focuses on the bright parts of the image before it and tends not to see the darker parts. Therefore the lighting of workstations should be arranged so that it is uniform across the whole of the workpiece with no areas of dark contrast. The eye is capable of accommodating a wide range of light intensities but for comfort, and in some cases, accuracy of work, the intensity levels should be adjusted to suit the work within the range of maintained illuminance of 150–200 lux for storage and pedestrian areas to 1500 lux needed for very close work such as hand engraving, fine machinery inspection, etc. (section 3.7.6).

The eye is a very delicate organ that is susceptible to aggressive and hazardous agents in the atmosphere and where these are likely to be met suitable eye protection should be provided. Many occupations requiring a high level of visual concentration can cause eye strain, which is manifest by headaches, stinging eyes, fatigue, etc. This is a particular problem in the use of VDUs. Whenever eye strain is experienced the individual should be referred to an optician for an eye test.

3.9.2.9 Ears

In our everyday life we are bombarded by a vast array of sounds, some wanted some unwanted. Certain sounds are essential components of our day-to-day life and are needed for communication, warning and balance. Other sounds that are not wanted are referred to as *noise*. The ear (*Figure 3.5.7*), like the eye, is a very sensitive organ that feeds signals to the brain. These signals are interpreted as sounds by the brain, which is capable of discriminating between various types of sound and of ignoring sounds that occur regularly or are not relevant to the current situation.

Sound is a series of pressure waves whose frequency determines the 'pitch' and whose amplitude determines the loudness of the sound. The ear is capable of reacting to sound at frequencies from about 150 to 10 000 cycles per second (hertz) and pressure variations from 0.0002 μbar (0 dB(A)) up to 1 mbar (140 dB(A)), a range of 1 to 10 million. Sound levels related to hearing protection are measured on a decibel scale that is weighted to match the hearing characteristics of the human ear, written as dB(A).

Sound levels of 85 dB can cause hearing loss in a small percentage of the population and above 90 dB are likely to affect all those exposed to it. Exposure to sound levels at or above 104 dB are likely to give rise to pain and possible disorientation.

3.9.3 Working environment

The term *working environment* is a general phrase that is given to any or all of the media, agents and conditions likely to occur in the workplace atmosphere which can impinge on the effectiveness of an operator. It ranges from the physical conditions provided by the employer, such as cleanliness, lighting, decoration, etc., to the social conditions reflected in the interaction within working groups and personal relationships between individuals. The psychological aspects of individual's conceptions, ambitions, status and outside social commitments also have an influence.

Examples of these various aspects are considered in the following sections.

3.9.3.1 Cleanliness

The degree of cleanliness will depend upon the type of work being carried out, with the strictest standards being in the food and catering, medicine manufacture, nuclear, etc. industries. Generally high standards of cleanliness engender high standards of, and a pride in, workmanship. The conditions of the toilets is generally a good indicator of the attitude of a workforce to cleanliness.

The layout of plant and the design of individual machines should allow adequate access for cleaning. Where high levels of cleanliness are essential, as in the food and pharmaceutical industries, equipment should not have acute internal corners or crannies that can harbour dirt. If washing down is necessary, the electrical equipment should be either encapsulated or protected to the appropriate IP[3] level.

Particular attention should be paid to safe access where cleaning is carried out when machinery is running. The type of flooring provided will depend on the industry but it must be suitable to allow regular cleaning whether by washing, manual or mechanical sweeping. If floors are washed down, adequate drainage will be required. There should be effective arrangements for the removal and disposal of rubbish and waste. Gangways should be kept clear and not be encroached upon by product or waste.

3.9.3.2 Colour

Careful use of colour can be a factor in safe working.

In general, the colours of notices, posters and safety items in the workplace should follow the international protocol:

red – danger, emergency, stop
green – safety, normal, go
yellow – warning, caution, abnormal
blue – mandatory, obey

The use of colours on controls should comply with the international standard[4]. Contrasting colours can be used to draw attention, for example red emergency buttons on a yellow background or to create an association between related controls or parts. Conversely different colours can identify different functions such as different coloured copies of a permit-to-work. They can also be used to highlight hazards particularly in identifying when machine guards have been removed. If the guard itself is painted the same colour as the machine and the area covered by the guard is painted in a warning colour (red or yellow/black diagonal stripes) a missing guard becomes immediately obvious.

The effects of colour can also be illusory:

Colour	Illusory effects		
Blue and green	Cold	Restful	Makes objects look more distant
Orange and yellow	Warm	Stimulating	Makes items look nearer
Brown and buff	Neutral	Restful	Makes items appear much closer

Light colours create the impression of spaciousness whereas dark colours tend to be oppressive. In some areas of labour intensive work, the operators may have strong preferences for the colour of the general decoration. Meeting their wishes can have a beneficial effect on productivity.

3.9.3.3 Space

The minimum space per employee recommended in the Approved Code of Practice[5] is 11 m^3 per person. The actual space can be arrived at by dividing the workplace volume below a height of 3 m by the number of persons working in it. If there is much machinery or furniture the volume figure may need to be adjusted to allow for it.

It is also important to recognise the psychological behaviour of individual employees who, having been given a job and a space in which to do it, becomes possessive of that space and resent intrusions into it – similar to animal territorialism. This attitude can give rise to considerable stress in the individual when another person is given a job to carry out in the same area. It can result in resentment, loss of or interference with concentration, reduction in productivity, loss of interest in the job and even open aggression. The situation is different if, in the first instance, a number of workers are allocated to work in the space when an equitable working relationship will be arrived at between them. It is only where a 'stranger' is thrust into an established 'territory' without consultation, even though prior notice may have been given, that resentment surfaces and stresses develop.

3.9.3.4 Temperature

Although there are no legislative requirements concerning the working temperatures, only that they should be reasonable, a Code of Practice[5] suggests a minimum workplace temperature of 16°C except for strenuous physical work where a minimum temperature of 13°C may be acceptable. There are no requirements regarding maximum temperature or for outdoor working.

The normal temperature of the major organs of the body is 37°C (98.4° F). Other organs, such as the muscles and skin, can accommodate a range of temperatures above and below this level depending on the activity being undertaken. Heat is carried round the body by the blood, from areas of high to areas of low temperature. The skin is the main agent for transferring heat to or from the body, i.e. receiving warmth from external sources or getting rid of surplus heat by sweating.

At excessive (high) temperatures the blood flows to the surface of the skin making it red and resulting in sweating as a means of reducing the body temperature. Conversely, at low temperatures the blood withdraws from the skin to prevent further heat loss, leaving it white and the adjacent muscles short of blood. This can reduce the effectiveness of muscular efforts and can result in a reduced ability to hold or grip items. Shivering is the body's attempt to generate heat to encourage the flow of blood.

Workplace temperatures should be such that they cause neither sweating nor shivering in the workpeople. For indoor working the temperature can normally be controlled and should be adjusted to suit the type of work and clothing normally worn. In cold areas, such as cold stores and working outside in the winter months where the temperature is dictated by service or natural conditions, additional warm clothing may need to be provided. Working in high temperatures, such as in bakeries, potteries, etc., will require the provision of special lightweight clothing, or in extreme cases, such as in steel making, the provision of protective heat barriers supplemented by directed cooling air supply. Working for periods in areas of high temperature may cause heat stroke but the susceptibility varies between individuals.

Table 3.9.1 lists temperature ranges for various working circumstances, but these must be recognised as indicative only since individual comfort temperatures vary according to sex, hereditary background, ethnic origin, etc. It is likely that the workpeople will give a vocal indication of their preferred temperature and while a commonly acceptable figure may be arrived at, it is unlikely to satisfy everyone.

In working areas, it is important that the temperature at floor level should be approximately the same as that at body level. Where ventilation is provided care should be taken to ensure that there is no temperature stratification and that there are no large temperature gradients from floor to ceiling.

Table 3.9.1 Temperatures in the working environment

Type of work	*Temperature range, °C*
Sedentary and office work	20–24
Sedentary production involving some body movement	18–23
Standing production involving regular body movement	17–21
Manual production work involving some physical effort	16–20
Heavy manual work	15–18

3.9.3.5 Humidity

The degree of moisture in the air needs to be controlled within certain limits. Excessive levels of moisture (high humidity) can seriously interfere with the body's ability to sweat and can cause considerable discomfort. Where the production process requires high humidity, such as in papermaking, exposure times should be kept to a minimum. A dry atmosphere (low humidity) can cause dryness of the throat and de-hydration. Normal comfort levels of humidity lie between 40% and 50% relative humidity but may vary slightly between different types of work. Extended exposures to a relative humidity below 30% can give rise to adverse pulmonary health effects.

In considering optimum temperatures and humidity account should be taken of the clothing normally worn, whether personal choice or company issue, the physical nature of the work, exposure to sources of heat (from the process or naturally from sunlight) and the amount of ventilation provided.

The measurement of the thermal environment is discussed in section 3.6.2.

3.9.3.6 Lighting

To be able to carry out any work effectively and accurately proper and appropriate lighting is essential. The eye reacts to strong or bright light such that areas of shadow or darkness are not seen in as much detail if at all. With all work situations suitable and sufficient lighting that enables the eye to see all the facets of the work and the surrounding area is necessary. Recommended levels of illuminance for various locations and tasks are give in section 3.7.6.

While ensuring an adequate level of illumination, care must be taken to avoid positioning illuminaires where they can interfere with the clarity of vision. Typical situations to avoid include:

(a) Glare and dazzle from a source of light positioned behind the object to be viewed effectively prevents the object from being seen. This can occur with low level lighting on access ways (*Figure 3.9.15*) or high level lights in areas of lifting operations. Similarly, viewing is interfered with if the emissions from a source of light shine directly on the eye.
(b) Areas of sharp contrast since the eye reacts to the bright areas with the result that the darker areas will either not be seen or be seen only with difficulty by straining the eyes (*Figure 3.9.16*). Deep shadows and fluctuating levels of light have the same effect.
(c) Reflections of a light source on the object being viewed whether paper, metal, desk top or monitor screens.
(d) Flicker, which is a cyclic variation of light intensity that is more noticeable at frequencies below 50Hz. It is particularly noticeable at the edge of the visual field and can be distracting, cause fatigue and, in some cases, epileptic seizures.

Figure 3.9.15 Disability glare from a light fitting (Courtesy The Stationery Office)

Figure 3.9.16 Sharp contrast between exterior light and interior shadow (Courtesy the Stationery Office)

(e) Stroboscopic effect occurs when the flicker from fluorescent lamps coincides with the speed of rotating objects making them appear stationary. This can be avoided by utilising twin tube fittings wired 90° out of phase.

3.9.3.6.1 Types of illuminaires

Sources of artificial light split broadly into two types:

(i) point sources such as the tungsten filament lamp where a glass envelope containing either a vacuum or a filling of halogen, mercury or sodium vapour at pressure. Since the filament is heated to white heat to provide the illumination the surrounding glass envelope can get hot. Adequate arrangements for cooling are needed and the lamp should not be located near flammable materials. Because this type of illuminaire is a point source of light it is important that it does not:

- create areas of bright light and deep shadows,
- reflect on work surfaces and
- mask information on VDU monitors.

The gas used to fill the bulb – mercury, sodium or halogen – creates a colour bias in the light emitted. This must be allowed for in processes where colour recognition is important, such as electrical wiring, paint colour matching, etc.

(ii) fluorescent strip light fittings in which the light emanates from the fluorescent coating on the inside of the tube. Although giving a much more even spread of light than tungsten lamps they can still cause reflections on surfaces. This effect can be reduced to a minimum by the use of diffusers and louvre fittings. Problems that are met with this type of illuminaire include:

- flicker and
- stroboscopic effect.

The positioning of illuminaires is important to ensure they do not create interference with viewing. Where interference does occur, the object being viewed should be moved or the illuminaire repositioned. Advice on the type and positioning of luminaires is given in a guidance note[5].

3.9.3.7 Ventilation

The presence of contaminants in the atmosphere is a potential source of distraction and annoyance. They may be there as a result of fumes or dust leaking from the process or of someone's personal habits such as smoking. With contaminants there is also an associated potential health risk (from hazardous fumes and dusts, tobacco smoke, etc.). Legislation[7] requires the supply of *a sufficient quantity of fresh or purified air*. It does not specify quantities but guidance[5] suggests a minimum of 5–8 l/s per occupant (18–29 m^3/h). However, this does not allow for the effects of the production process nor the type of work so these quantities may need to be increased. Fresh air is that drawn from outside but care needs to be taken to ensure the intake point is clear of exhaust outlets or other sources that might contaminate the air or be hazardous or evil smelling. Purified air refers to recirculated air that has been 'conditioned' but the Code of Practice[5] recommends that *some* fresh air should be added to it although again no quantity is specified.

In many situations, an adequate supply of fresh air can be obtained from an open window but in the larger open plan offices and workshops some form of forced air ventilation may be required. The outlets from ventilation systems should be arranged so that they do not play on an individual since this can be a source of annoyance and also interfere with sweat rates to become a health hazard. Outlet velocities and directions of flow should ensure that the air velocity at any one workstation is not so high as to be unpleasant or uncomfortable. In general, the more active the work an airflow as high as 0.5 m/s can be tolerated. However, for sedentary work the flow should be less than 0.1 m/s while in jobs requiring deep concentration even that level of air movement can be distracting.

3.9.4 Manual handling

Ideally, if objects have to be moved it should be done mechanically. Where this is neither technically feasible nor economically viable manual handling will have to be employed. Manual handling is a known and well documented source of occupational injury and in spite of publicity and training accident attributed to manual handling remains the highest cause of absences. Legislation[8] sets out the actions to be taken to reduce the hazards with advice on ways to achieve them given in a Code of Practice[9]. The ability to handle loads varies with the position of the load with respect to the body and *Figure 3.9.17* indicates a suggested range of maximum

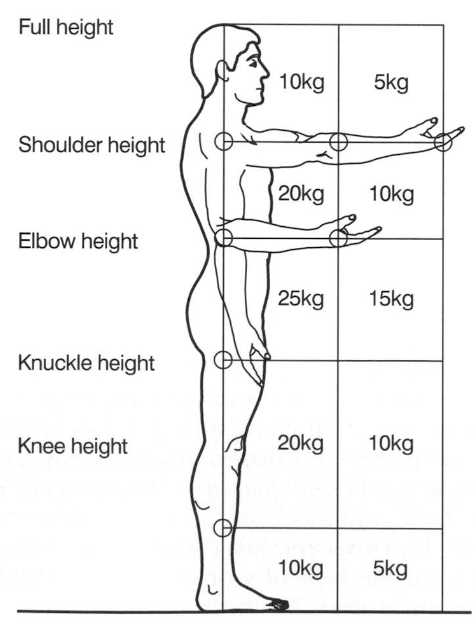

Figure 3.9.17 Suggested maximum loads at various distances from the body

weights that can be lifted and carried. The values given are typical and will need to be adjusted to suit the physique and ability of the operator. It is important to remember that it is not only what is picked up but how.

For manual handling, work should be arranged so that:

- the load to be lifted is the smallest technically feasible and economically viable;
- loads that cannot be broken down to safe weights are handled by mechanical means such as sack barrow, special purpose handling equipment, lift trucks, etc.;
- the level from which the load is lifted should be as high as possible up to waist level;
- ideal height for picking up a load is waist level;
- if necessary an intermediate resting platform is provided;
- close body approach is possible to the delivery platform to prevent the need to lean with the load;
- the final delivery level is not above shoulder height;
- for placing loads at higher than shoulder level a lift truck or suitable step ladder is used.

Where loads have to be carried manually, the floor surface should be level, smooth and in good condition.

Where manual handling has to be carried out from the sitting position, the ability to lift may need to be reduced to as little as 20% of the equivalent load when standing.

3.9.5 Repetitive actions

Actions that involve putting repeated loads on particular muscles, especially on the arms and the wrist, can cause a number of muscular conditions variously referred to as repetitive strain injury (RSI), tenosynovitis, carpal tunnel syndrome, work-related upper limb disorder, etc. Symptoms exhibited include soreness in the muscles that initially disappears when ceasing work but rapidly returns when work is recommenced. If the same work is continued, the condition can become very painful and have long lasting effects.

On jobs where this condition is a known or suspected risk arrangements should be made to:

- eliminate the type of work that causes the condition and replace it by alternative work methods;
- restrict the time engaged on the suspect activity;
- rotate jobs during the shift so that operators carry out a number of different functions using different muscles;
- ensure tools and equipment used on suspect operations are, and are maintained, in good condition and do not require excessive force for their proper use;
- build into the work programme adequate rest periods;
- instruct supervisors and operators in the symptoms and the action to be taken if they occur, i.e. move to alternative work and seek medical advice.

3.9.6 Plant design

Layout of plant should ensure that any movements the operators need to make are direct, free and unimpeded by other parts or equipment. Operator work areas should be clear, clean, well lit with a good floor surface. If the work platform is at a raised level, it should have a safety rail and be provided with access steps if the height warrants. The treads of steps should be wide enough to accommodate the full length of a normal shoe. Steps at an angle greater than 45° should be avoided, but if space limitations dictate steeper steps, proper permanent ladders with hand rails should be provided. Any step up (or down) should not be greater than 25 cms (10 ins). Steps higher than this can greatly increase the strain on the knee and hip muscles with consequent increased fatigue.

Adequate space should be left around each machine to permit free and easy movement for operating it and to allow for maintenance activities. Walkways should be identified by suitable lining and not allowed to be used for storage purposes. Services, such as air, water, electrical power, necessary for the work being carried out should be conveniently situated for the operators' use. Machines in sequential operations should be positioned to require the minimum amount of handling of product. Wherever possible that handling should be automated or by mechanical means.

The emission of noise and fumes by machinery which can affect the operator and those on adjacent machines should be reduced to a minimum.

3.9.7 Controls and indicators

Controls and instruments are the main interface between the operator and the machine or plant. In the design and layout of them:

- The movement of all controls must be consistent with the natural movement of the limb operating it.
- Movement of a control in a clockwise direction, to the right or towards the operator should cause an increase in the machine function – the exception to this is a tap or valve where clockwise movement results in a decrease in output, i.e. the valve is shut.
- Coarse adjustment and adjustments that require some force should utilise the full arm, leg or hand movement.
- Where foot pedals are used, if actuation is by movement of the whole leg the pedals should be arranged so they can be operated by either foot. If movement of the foot only is required it should be by pivoting on the heel. In both cases the arrangement should ensure that the operator is not required to stand on one leg for long periods.
- Quick, precise or fine adjustments that require little physical effort should be by the fingers.
- Hand operated controls should be located at a height between waist and shoulder level and be in clear view.

- Controls that have to be actuated frequently should be positioned adjacent to or within easy reach of the operator's hands. Other controls should be within easy arms reach.
- Adjacent hand or finger operated controls such as push buttons, toggle switches and rotating knobs should be spaced at least 25 mm (1 in) apart to prevent inadvertent operation.
- In the layout and shape of control buttons:
  - start buttons should be recessed into the control panel, shrouded, or gated to prevent inadvertent operation;
  - stop buttons should be positioned adjacent to the start control, stand proud above the panel surface and be red in colour;
  - emergency stop buttons should be red, of the mushroom headed type and lock in the open circuit condition when actuated.
- The function of all control actuators should be clearly indicated either by words or symbols.
- Where the condition of the control is important and may need to be known without looking at it, a datum mark such as a small pin or notch should be made in the mounting panel and a matching pin or notch made in the control handle. The two should line up at either neutral or normal operating position so any deviation from it can easily be sensed.
- Control handles for separate operations should have a unique tactile identity[10].
- Instruments that are important should be in clear view of the operator, ideally at eye level or within 20° of the normal eye line but must not interfere with the operator's view of the machine or plant.
- The movement of the condition indicator of an instrument should be consistent with the change in condition, i.e. increase in the condition shows as a clockwise movement or, in linear gauges, to the right or upwards.
- Instruments that measure associated parameters should be positioned together and arranged so that the pointers or condition indicators all lie in the same orientation for normal operation allowing any deviant reading to be seen easily.
- Where controls have to be actuated over periods of time with little body movement, seating should be provided for the operator and the positioning of the controls and instruments arranged accordingly.

Controls that are operated by the feet fall into two categories, those in which the whole leg is moved giving only a very coarse degree of control and those using the foot only, when a fine degree of control can be achieved. In the former case, movement of the foot is from the hip allowing only a basic ON/OFF type of control without any intermediate positioning, such are used for the initiation of a press stroke or the foot pedals of an organ.

Where a fine degree of control over the operating range is necessary this can be achieved by pivoting the heel of the foot on the floor or suitable rest. An even finer degree of control, such as the accelerator pedal in a car, can be achieved by providing a support at the outer side of the foot about which the foot can pivot. If a foot control is operated from a standing position, the arrangement should ensure that part of the body weight can be taken by the operating foot, or if this is not possible, the control should allow operation by alternate feet to prevent the excessive strain imposed when one leg takes the full body weight.

3.9.8 Noise and vibrations

In all walks of life sound is a necessity, for communication, for warning and leisure enjoyment (music and the theatre). Unfortunately there are differing views about what sound is useful and what is an adequate amount of sound. Any unwanted sound is regarded as noise and as such should be eliminated or reduced to the lowest level possible. In general sound that interferes with people's enjoyment of their private lives and pursuits becomes a nuisance and has been legislated against[11]. But excess sound can also interfere with concentration at work and become a potential hazard as well as reducing the operating performance of those subject to it. Examples of typical noise levels are shown in *Figure 3.5.2*.

Noise in an area is likely to be a hazard if it is necessary, when standing 1 metre apart, to have to shout to carry on a conversation. Where there appears to be a noise problem, sound level readings should be taken to establish the extent of the problem.

The presence of noise has long been recognised as one of the factors that reduces the quality of working life. While the human brain can 'tune out' consistent and/or irrelevant noises it can only do this up to a point. As noise levels rise so they become more insistent and invasive. Similarly unexpected changes in even quite low levels of noise can stimulate a subconscious response and, in some cases, break completely the current train of thought. The problems of noise from machinery and advice on the measures to combat it are well documented in HSE publications[12, 13] and it is not proposed to iterate them here.

Vibrations on the other hand, where there is a finite movement of the plant, equipment or a pulsing of the air, are much more invasive and can interfere with certain body organs ultimately causing ill health.

Measures that can be taken to reduce the distracting effects of noise include:

- elimination of sources of noise;
- if that is not possible then:
  - enclose the source of noise in a sound proof room but ensure adequate cooling and ventilation is provided;
  - provide sound havens or soundproof operating rooms ensuring there is adequate ventilation;
  - use sound absorbing screens and barriers;
  - separate work areas from noise sources;
  - position potential noise sources away from work areas – the frequency hum from a transformer can be very invasive;
- directing the outlet ducts from ventilating systems, dust extraction systems, etc. away from affected areas. This can include private house bedrooms where fan exhausts can become a nuisance and subject to abatement orders;

- in offices, replacing noisy matrix and daisywheel printers by inkjet or laser printers;
- installing floor covering that deadens the sound of footsteps particularly the clacking of heels on a hard floor;
- ban the use of personal radios in the workplace – they can interfere with the reception of warning signals;
- in open plan offices, the segregation of those with penetrating telephone voices;
- arrange for operations that generate noise, such as use of pneumatic drills, etc., to be carried out in 'non-working' hours. This includes work on part of the structure of reinforced concrete framed buildings since noise travels through the concrete;
- as a last resort, provide suitable personal protective equipment.

Mechanical vibrations generated by the movement of parts of the plant and machinery can travel through the machine and be transmitted to the building and those working in it. Air vibrations occur as a pulsing of the air and can be generated at the outlet of fans and blowers and from the exhaust of slow running engines. Both mechanical and air vibrations can be a health hazard since they can induce sympathetic vibrations in certain human organs resulting in damage to that organ.

The transmission of mechanical vibrations can be reduced by:

- mounting the equipment on anti-vibration mounts;
- providing flexible connections between the vibrating plant and other equipment.

Air vibrations can be reduced by:

- changing the speed of the fan or blower;
- installing diffusers;
- changing the flow resistance of the air circuit;
- ensuring the intake to the fan is not obstructed.

3.9.9 Stress

Stress has many causes including an inability to do what ought to be done or failure to meet the targets set. The cause may be within the individual or it may be imposed from outside. Internally caused stress can only be resolved by the individual himself but imposed stress causes can be reduced or eliminated by following ergonomic principles. The build up of stress in an individual will make him less efficient in his work and may even make him a safety hazard. To optimise an individual's performance the stress suffered should be reduced to a minimum.

Typical stress situations, with possible ways to resolve them, include:

- working at a machine led rate which is either faster or slower than the individual's natural work rate. *Wherever possible suitable adjustments should be made to the machine speed;*

- being required to undertake work which is either well below or well above his inherent ability. *This may require a re-assessment of the operator and moving to other more appropriate work;*
- being given inadequate or excessively complex instructions about his job. *Instructions should be realistic and comprehensive and in terms and language that the operator can understand;*
- being prevented from working at his own natural rate. *Some means should be provided to adjust the demanded rate of work;*
- having to do a job in a less efficient manner than he knows it can be done. *Listen to the operator's suggestions and act on them or explain why not;*
- being uncertain of his position in the organisation and not knowing who his bosses are. *Provide training in the role and position within the company covering areas of responsibility, extent of authority, subordinates and superiors, etc.;*
- having to wait for materials or data. *Improve planning and expediting;*
- being unable to understand and follow work methods. *Further training and the provision of back-up information;*
- working in software in which he has not been properly trained and without back-up. *Ensure adequate training and provide competent back-up to resolve queries;*
- at loggerheads with his supervisor. *A personal matter to be resolved by the individuals or by separating them;*
- being pressurised by his peer group;
- under a threat of redundancy without having any details. *Ensure kept informed of the latest position;*
- family affairs;
- frustration with lack of progress on agreed action affecting his work and working conditions. *Initiate suitable action or explain why it has not been possible;*
- lack of recognition for ideas put forward. *Improve human relations in the company;*
- irritating noises. *Investigate and eliminate;*
- boredom from repetitive uninteresting work. *Re-assess ability and move to more demanding work.*

3.9.10 Display screen equipment (DSE)

The ergonomic aspects of the use of DSEs has been well documented in the HSE's guidance publication[14] particularly those aspects concerned with the physical comfort of the users and operators such as:

- chair with adjustments for seat height and back rest;
- suitable foot rest;
- adequate leg room below work table;
- adjustable screen both rotating and tilting;
- document holder to reduce amount of eye movement;
- limit on time of continuous operation;
- training in the use of the software with back-up immediately available in case of queries;

- screen should have adjustments to ensure stable picture, enable change of polarity of characters and control over contrast;
- work surface to be large enough to accommodate keyboard, all papers/documents and any peripherals such as the mouse, printers, disc, imager, etc.

DSEs can make the atmosphere very dry and cause discomfort. Sources of moisture, such as house plants, should be installed to improve the humidity.

3.9.11 Signs and signals

Signs and signals are a vital means of passing information where verbal contact is not possible or reasonable. The signs, generally, in the form of posters, warnings, etc., are passive while signals, usually by hand or light, are dynamic. It is important that those who need to read signs know their correct meaning. In the passing of operational information by hand signals, such as in the use of cranes, it is important that both the signaller and the receiver (crane driver, etc.) use the same codes[15] and that both are fully conversant with the full range of hand signals. Familiar hand signs have different meanings in different countries and care must be exercised when selecting hand signals to ensure they are not in common usage in workers' mother countries where they may have a totally different, and sometimes insulting, meaning.

With the number of migrating workers travelling to work in countries foreign to them, meeting obligations to provide information presents difficulties of language. This can largely be overcome by the use of pictograms. Standards[16], incorporating the requirements of a directive, specify a range of pictogram safety signs with the aim of their being understood regardless of the language of the viewer. The standard signs are intended to be stand-alone but text may be added where necessary to provide additional information.

The positioning of signs is important. They must be placed where they will be clearly visible by those at whom they are aimed. Emergency signs, such as fire exits, fire points, etc., should be clearly visible from all places to which employees, visitors and others may have recourse as a normal part of their activities. The height at which signs are located should be considered. A fire emergency exit sign at chest height is of no use if, in an emergency, the rush of people from the area completely cover it. Emergency safety signs should be positioned above head height so they are clearly visible from all parts of the area they serve. Conversely, signs placed at high levels are often overlooked because the general trend is to look downwards rather than upwards. Signs should be mounted within a sight line of 20° above horizontal when viewed from all the areas served.

Care must be exercised when selecting audible signals to ensure, first, that they do not add to the general noise to the extent of raising it above the accepted safe levels. Second, they must be clearly distinguishable from all other audible signals and from other normal sounds in the area.

Where audible warnings are used, such as fire alarms, reversing vehicles, etc., the signal must be audible to all those likely to be in a position of risk from the danger warned against. Audible warnings should not be used with such frequency that they become a part of the general background noise, also that their use does not become an irritant to others working nearby. Where audible warnings are employed all those in areas covered by the warning should be familiar with the sound and with the action to be taken.

Public places such as cinemas, theatres, stores, supermarkets, etc., present a particular problem in an emergency. Staff should be trained in advising the public what to do should an alarm be sounded. The use of broadcast verbal warnings or safety instructions should be avoided since the message may be inaudible in some areas, can be misunderstood and give rise to confusion.

3.9.12 Coda

The application of ergonomic principles to work activities can make life safer and more pleasant for employees. Many of the ergonomic techniques are being incorporated into regulatory requirements and into standards but there are still many techniques that the employer can adopt that will further improve not only the safety and quality of working life but productivity.

References

1. British Standards Institution, BS EN 614–1 *Safety of Machinery – Ergonomic design principles – Part 1: Terminology and general principles*, BSI, London (2000)
2. Kroemer, K.H.E. and Grandjean, E., *Fitting the Task to the Human*, 5th edn, Taylor & Francis, London (1999)
3. British Standards Institution, BS IEC 60529 *Degrees of protection provided by enclosures* (IP code), BSI, London (1991)
4. British Standards Institution, BS IEC 60204 *Safety of machinery – Electrical equipment of machines – Part 1: General requirements*, Clause 10.2, *Push buttons*, BSI, London (1997)
5. Health and Safety Executive, Legal series publication L24 *Workplace health, safety and welfare. Workplace (Health, Safety and Welfare) Regulations 1992, Approved Code of Practice and Guidance*. HSE Books, Sudbury (1992)
   See also health and safety guidance series publication HSG 202 *General ventilation in the workplace*, HSE Books, Sudbury (2000)
6. Health and Safety Executive, Health and safety guidance series publication HSG 38 *Lighting at work*, HSE Books, Sudbury (1998)
7. *Workplace (Health, Safety and Welfare) Regulations 1992*, Regulation 6, *Ventilation*, The Stationery Office, London (1992)
8. *Manual Handling Operations Regulations 1992*, The Stationery Office, London (1992)
9. Health and Safety Executive, Legal series publication L23 *Manual Handling, Manual Handling Operations Regulations 1992, Guidance on the Regulations*, HSE Books, Sudbury (1992)
   See also Health and safety guidance series publication HSG 115, *Manual handling solutions you can handle*, HSE Books, Sudbury (1994)
10. British Standards Institution, BS IEC 61310, *Safety of machinery – Indication, marking and actuation – Part 1: Requirements for visual, auditory and tactile signals*, BSI, London (1995)

11. *Environmental Protection Act 1990*, The Stationery Office, London (1990)
12. Health and Safety Executive. Legal series publication L108 *Guidance on the Noise at Work Regulations 1989*, HSE Books, Sudbury (1998)
13. Health and Safety Executive, Health and safety guidance series publication L138 *Sound solutions, techniques to reduce noise at work*, HSE Books, Sudbury (1995)
14. Health and Safety Executive, Legal series publication L26 *Display screen equipment work – Health and Safety (Display Screen Equipment) Regulations 1992, Guidance on the Regulations*, HSE Books, Sudbury (1992)
15. British Standards Institution, BS 7121 *Code of Practice for the safe use of cranes*, BSI, London
16. British Standards Institution, BS 5378 *Safety signs and colours* and BS 5499 *Fire safety signs, notices and graphic symbols*, BSI, London

Suggested reading

Kroemer, K.H.E. and Grandjean, E., *Fitting the Task to the Human*, 5th edn, Taylor & Francis, London (1999)
Bridger, R.S., *Introduction to Ergonomics*, McGraw Hill, Singapore (1995)
Pheasant, S., *Ergonomics, Work and Health*, Macmillan Press, London (1991)
Helander, M., *A Guide to the Ergonomics of Manufacturing*, Taylor & Francis, London (1995)
Chartered Institution of Building Services Engineers, *Code for Interior Lighting*, CIBSE, London (1994)
McKeown, C. and Twiss, M., *Workplace Ergonomics: a Practical Guide*, IOSH Publishing Services Ltd., Leicester (2001)

PART IV

Workplace safety

Much of the work undertaken by safety advisers requires an understanding of technical industrial processes. Even in a single factory unit the safety adviser may be called upon to advise on avoiding the hazards from a chemical reaction, guarding particular types of machinery, the standards of safe working to be expected of a building contractor, the precautions to be taken to prevent fire and the fire fighting equipment that should be provided, etc.

To carry out his duties effectively, the safety adviser should have an understanding of basic physics and chemistry and of the current safety techniques for reducing the risks associated with the more commonly met industrial processes. This Part considers some of these processes and the basic sciences from which they stem.

Chapter 4.1

Science in engineering safety

J. R. Ridley

4.1.1 Introduction

In the construction of machines, plant and products, materials are selected because they have particular physical and chemical properties. Wood, metals, concrete, plastics and other substances all have their uses but there are limitations as to what they can do and how long they can do it. Properties may change with use, temperature, operating atmosphere, contamination by surrounding chemicals and for many other reasons. It is necessary to know the properties of the materials and how and why they have been used so that an assessment can be made of whether likely changes in the properties may give rise to hazards.

These properties stem from the chemical and physical characteristics of the different materials and substances used and their behaviour under certain conditions can determine the safety or otherwise of a process or operation. This chapter looks at some of the characteristics and properties of materials in common use, their application, circumstances of use and possible causes of hazards.

4.1.2 Structure of matter

Everything that we use in our work and daily life is made up of chemical substances, by themselves or in combination of one sort or another. Each substance consists of elements which are the smallest part of matter that can exist by itself. In its free state, an element comprises one or more atoms. When atoms combine together they form molecules of the element or, if different atoms combine, of compounds. The ratio in which atoms combine is determined by their combining power or valency.

Atoms are made up of three particles:

- protons which have a unit mass and carry a positive charge,
- neutrons which have a unit mass but carry no charge, and
- electrons which have negligible mass (i.e. 1/2000 proton) but carry a negative charge.

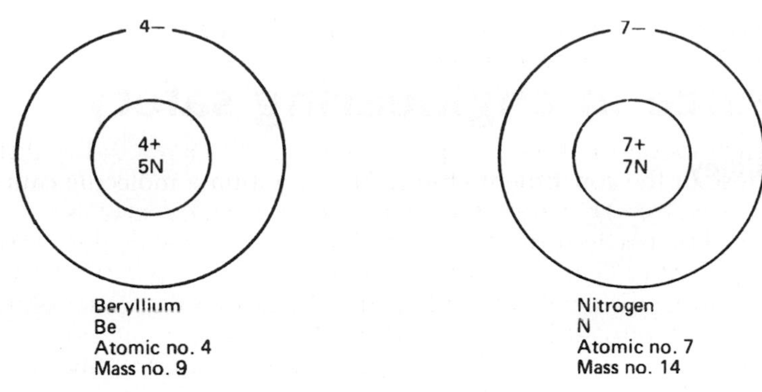

Figure 4.1.1 Atomic structures

The core or nucleus of the atom consists of protons and neutrons with electrons travelling in orbits around the nucleus (*Figure 4.1.1*). Elements normally have no overall charge since the number of protons is matched by an equal number of electrons. However, it is possible to upset this balance by removing either a proton or an electron resulting in the atom carrying a charge when it is said to be ionised.

In chemistry, atoms are given 'atomic numbers' which equal the number of protons or electrons in the atom. They are also given 'mass numbers' which equal the sum of the number of protons plus neutrons. The mass number is always equal to or greater than 2 × the atomic number except in the case of hydrogen. Some elements can occur in conditions where they have the same atomic number, and hence the same name, but different mass numbers; they are then known as isotopes and are generally referred to as nuclides. Very large heavy atoms, such as uranium, can be unstable and easily break down to produce smaller atoms with the production of particles or energy. These atoms are radioactive and provide the source of energy in nuclear reactors.

Approximately 100 different atoms have been identified and each has been given a name and a coded symbol which usually is the first one or two letters of its name: carbon – C, lithium – Li, titanium – Ti, etc. Exceptions in this coding system arise because when it was evolved in the early 1800s some chemicals were still known by their Latin names, such as copper (cuprum – Cu) and tin (stannum – Sn).

When atoms join together their molecular formulae are written as groups of atomic symbols to indicate the number of those atoms present to form a stable molecule.

Molecular formulae

Br_2	bromine
O_2	oxygen
H_2O	water
NaOH	sodium hydroxide (caustic soda)

$CBrF_3$	bromotrifluoromethane (BTM or Halon 1301)
HCHO	formaldehyde
$C_2H_5NO_3$	ethyl nitrate

Each compound has its own properties which may be vastly different from those of the constituent atoms. Atoms within a molecule cannot be separated unless the compound undergoes a chemical reaction.

A chemical reaction occurs when the atoms in molecules rearrange either by decomposing into smaller molecules or by joining with other atoms to form different molecules; in both cases the atoms reorganise themselves to form different structures. Endothermic reactions require the input of heat to make them happen whereas exothermic reactions occur with the evolution of heat.

$$2Na + 2H_2O = 2NaOH + H_2 + \text{heat}$$
$$2H_2 + O_2 = 2H_2O + \text{heat}$$

These chemical equations show in chemical shorthand, using the chemical codes, the rearrangement of atoms which occurs in these reactions, a balance of the number of atoms being maintained during the reaction. The molecular mass of molecules can be obtained by adding together the mass numbers of the constituent atoms.

Compounds which contain atoms of elements other than carbon, but including carbon dioxide (CO_2), carbon monoxide (CO) and the carbonates (e.g. calcium carbonate $CaCO_3$) are called inorganic chemicals. All other compounds which contain carbon atoms are known as organic chemicals.

Carbon is an unusual element; not only is it able to form simple compounds where one or two carbon atoms are joined to atoms of other elements, but carbon atoms can link together to form chains or rings of atoms. Almost all other atoms can be joined into these chains and rings to create millions of different organic compounds, from the comparatively simple ones consisting of one carbon with one other type of atom to the highly complex molecules with hundreds of linked carbon atoms joined with other different atoms. Organic chemicals include most of the solvents, plastics, drugs, explosives, pesticides and many other industrial chemical substances.

4.1.3 Properties of chemicals

Properties of chemicals are to a large extent determined by how the atoms are bonded.

4.1.3.1 Metals

Metals are different in structure from both types of compounds described below, existing in the solid state as an ordered array of atoms held together by their electrons which circulate freely between them. Applica-

Table 4.1.1 Properties of typical elements

Element	*Symbol*	*Properties*
Reactive metals		
Aluminium	Al	m.p. 660°C, good conductor, surface oxide formation resists attack by air or water
Barium	Ba	m.p. 850°C, soft, spontaneously flammable in air, reacts with water
Lithium	Li	m.p. 186°C, soft, burns vigorously in air, reacts with water
Less reactive metals		
Cobalt	Co	m.p. 1490°C, hard, not attacked by air or water
Iron	Fe	m.p. 1525°C, burns in oxygen, reacts slowly with water
Mercury	Hg	liquid, slowly attacked by oxygen, no reaction with water
Silver	Ag	m.p. 961°C, ductile, not attacked by oxygen or water
Non-metals		
Bromine	Br	dark red liquid, b.p. 59°C, very reactive, not flammable
Phosphorus	P	red form, m.p. 600°C or white form, m.p. 43°C, burns readily to P_2O_5, insoluble in water
Sulphur	S	yellow or white, m.p. 115°C, burns to SO_2, insoluble in water

tion of an electric potential across a metal allows the electrons to undergo a directional flow between the atoms making metals good electrical conductors. *Table 4.1.1* lists the properties of some typical metals and other elements.

4.1.3.2 Inorganic compounds

In some compounds one or more of the bonds joining the atoms are the result of an unequal sharing of electrons between the two atoms and these produce ionic compounds which are crystalline solids, usually with a high melting point. They are often soluble in water giving a solution which conducts electricity.

Many other compounds have bonds based on an equal sharing of electrons and so do not ionise. These compounds can be solids having low melting points, or liquids or gases. Usually they are not soluble in water unless they react with it. There are also many more compounds with types of bond intermediate between the two described and which exhibit properties that relate to both types. *Table 4.1.2* lists some of the properties of a selection of inorganic compounds.

With the exception of sulphur, stannic chloride and potassium chloride, all the elements and compounds listed in *Tables 4.1.1* and *4.1.2* present hazards to health.

Table 4.1.2 Properties of a selection of inorganic compounds

Compound	*Formula*	*Properties*
Ammonia	NH_3	gas, b.p. −33°C, dissolves readily in water giving a basic solution
Carbon monoxide	CO	gas, b.p. −190°C, odourless, almost insoluble in water
Hydrogen chloride	HCl	gas, b.p. −85°C, dissolves readily in water giving hydrochloric acid
Hydrogen sulphide	H_2S	gas, b.p. −61°C, strong odour, burns in air
Hydrogen peroxide	H_2O_2	liquid, decomposes violently on heating, powerful oxidising agent
Stannic chloride	$SnCl_4$	liquid, b.p. 114°C, fumes, reacts rapidly with water
Sulphuric acid	H_2SO_4	liquid, decomposes at 290°C giving SO_3, strong acid, reacts violently with water
Aluminium silicate	$Al_2Si_2O_7$	solid, infusible, unreactive, clay silicate
Phosphoric acid	H_3PO_4	solid, m.p. 39°C or syrupy liquid, strong acid
Potassium chloride	KCl	solid, m.p. 770°C, very soluble in water, unreactive
Sodium hydroxide	NaOH	solid, m.p. 318°C, deliquescent, strong alkali

4.1.3.3 Organic compounds

As most organic compounds contain a relatively large percentage of carbon and hydrogen atoms they are flammable and many are toxic. All living matter is constructed of complex interdependent organic chemicals and it is because organic compounds interfere with the normal functioning of living matter that they constitute fundamental health and hygiene hazards.

Although there are very many organic compounds they can be grouped into a small number of classes according to their reactive properties. These broad groups are listed in *Table 4.1.3* which gives examples of compounds in each group.

4.1.3.4 Acids and bases

Acids are compounds which dissolve in water to give hydrated hydrogen ions:

$$HCl + H_2O = H_3O^+ + Cl^-$$
$$H_2SO_4 + H_2O = H_3O^+ + HSO_4^-$$

Strong acids completely dissociate into ions in solution; weak acids only partially dissociate. A concentrated acid is one which is not diluted with water, and the terms *strong* and *concentrated* should not be confused.

Acids are corrosive in that they react with both metals and with body proteins. Acids are dangerous not just because of their acidity but they can be oxidising agents (HNO_3, $HClO_4$), violently reactive with water

Table 4.1.3 Examples of the main groups of organic compounds

Group	*Example*	*Formula*	*Use*
Aliphatic hydrocarbons	Methane Butane	CH_4 C_4H_{10}	natural gas petroleum gas
Aromatic hydrocarbons	Benzene Toluene	C_6H_6 $C_6H_5CH_3$	toxic solvent solvent
Halocarbons	Bromomethane Trichloroethane	CH_3Br CH_3CCl_3	fumigant solvent
Alcohols	Ethanol Glycerol	C_2H_5OH $C_3H_5(OH)_3$	'alcohol' glycerine
Carbonyl compounds	Formaldehyde (methanal)	$HCHO$	fumigant
	Benzaldehyde Acetone	C_6H_5CHO CH_3COCH_3	manfacturing solvent
Ethers	Ethyl ether Dioxan	$C_2H_5OC_2H_5$ $C_4H_8O_2$	anaesthetic solvent
Amines	Methylamine Aniline	CH_3NH_2 $C_6H_5NH_2$	manufacturing manufacturing
Acids	Ethanoic acid Phthalic acid	CH_3CO_2H $C_6H_4(CO_2H)_2$	acetic acid manufacturing
Esters	Ethyl acetate	$CH_3CO_2C_2H_3$	solvent
Amides	Acetamide Urea	CH_3CONH_2 $CO(NH_2)_2$	manufacturing by-product

(H_2SO_4) and many are toxic. Phenol (C_6H_5OH) is one of the most dangerous acidic organic compounds.

Bases are of two types, solid alkalis such as metal hydroxides which dissolve in water to give hydroxide ions, and gases and liquids such as ammonia and the amines, which liberate hydroxide ions on reaction with water:

$$NaOH + H_2O = Na^+ + OH^- + H_2O$$
$$NH_3 + H_2O = NH_4^+ + OH^-$$

Some of the bases are toxic, many react exothermally with water and all are highly corrosive or caustic towards proteins. Alkalis spilled on the skin penetrate much more rapidly than acids and should be leached out with copious water and not sealed in by attempting neutralisation.

The reaction between an acid and a base is a vigorous, exothermic neutralisation forming a salt. The strength of acids and bases can be measured in terms of hydrogen ion concentration by the use of either meters or test papers, and it is expressed as a pH value on a scale from 0 (acid) to 14 (base). Pure water has a neutral pH of 7.

4.1.3.5 Air and water

Air and water deserve to be considered separately since they are ever present and are necessary for the operation of many processes and responsible for the degradation of many materials.

Air is a physical mixture of gases containing approximately 78% nitrogen, 21% oxygen and 1% argon. These proportions do not vary greatly anywhere on the earth but there can be additional gases as a result of the local environment: carbon dioxide and pollutants near industrial towns, sulphur fumes near volcanoes, water vapour and salts near the sea etc.

Air can be liquefied and its constituent gases distilled off; liquid nitrogen (b.p. −196°C) has many uses as an inert coolant, liquid oxygen (b.p. −183°C) is used industrially in gas-burning equipment and in hospitals, and argon (b.p. −185.7°C) is used as an inert gas in certain welding processes. Liquid oxygen is highly hazardous as all combustible materials will burn with extreme intensity or even explode in its presence. Combustion is a simple exothermic reaction in which the air provides the oxygen needed for oxidation. If the concentration of oxygen is increased the reaction will accelerate. This effect was experienced in the fire on HMS Glasgow[1].

Water is a compound of hydrogen and oxygen that will not oxidise further and is the most common fire extinguishant. However, caution must be exercised in its use on chemical fires since a number of oxides and metals react energetically with it, in some cases forming hazardous daughter products and in others producing heat and hydrogen which further exacerbate the fire.

4.1.4 Physical properties

All matter, whether solid, liquid or gas, exhibits properties that follow patterns that have been determined experimentally and are well established and proven. This section looks at some of the factors that influence the state of matter in its various forms.

4.1.4.1 Temperature

Temperature is a measure of the hotness of matter determined in relation to fixed hotness points of melting ice and boiling water. Two scales are universally accepted, the Celsius (or Centigrade) scale which is based on a scale of 100 divisions and the Fahrenheit scale of 180 divisions between these two hotness points. Because Fahrenheit had recorded temperatures lower than that of melting ice he gave that hotness point a value of 32 degrees. Converting from one scale to the other:

$$(°F - 32) \times 5/9 = °C$$
$$(°C \times 9/5) + 32 = °F$$

Man has long been intrigued by the theory of an absolute minimum temperature. This has never been reached but has been determined as being –273°C. The Kelvin or absolute temperature scale uses this as its zero, O K; thus on the absolute scale ice melts at +273 K.

Devices for measuring temperature include the common mercury in glass thermometer, thermocouples, electrical resistance and optical techniques.

4.1.4.2 Pressure

Pressure is the measure of force exerted by a fluid (i.e. air, water, oil etc.) on an area and is recorded as newtons per square metre (N/m^2). With solids the term stress is used instead of pressure. Datum pressure is normally taken as that existing at the earth's surface and is shown as zero by pressure gauges which indicate 'gauge pressure' (i.e. the pressure above atmospheric). However, at the earth's surface the weight of the air of the atmosphere exerts a pressure of 1 N/m^2 or 1 bar. Beyond the earth's atmosphere there is no pressure and this is taken as the base for the measurement of pressure in absolute terms. Thus:

$$\text{gauge pressure} = \text{absolute pressure} - 1\ N/m^2$$
$$\text{or absolute pressure} = \text{gauge pressure} + 1\ N/m^2$$

The pressure at the top of a mercury barometer, where the force due to the weight of the atmospheric air outside the tube is balanced by the force exerted by the weight of the column of mercury inside, is normally taken as zero (0 N/m^2 or absolute vacuum), although scientifically there is a small vapour pressure from the mercury.

Pressure can be measured by means of manometers which show the pressure in terms of the different levels of a liquid in a U-tube, by mechanical pressure gauges which record the differential effect of pressure forces on the inside and outside surfaces of a coiled tube or of a diaphragm, and electronic devices which measure the change of electrical characteristic of an element with pressure.

4.1.4.3 Volume

Volume is the space taken up by the substance. With solids which retain their shape, their volume can be measured with comparative ease. Liquid volume can be measured from the size of the containing vessel and the liquid level. Gases, on the other hand, will fill any space into which they are introduced, so to obtain a measure of their volume they must be restrained within a sealed container.

Each type of material reacts to changes of temperature and, to a lesser extent with solids and liquids, to changes of pressure, by increases or decreases in their volume and this fact can be made use of, or has to be allowed for, in many industrial processes and plant.

4.1.4.4 Changes of state of matter

At ordinary temperatures, matter exists as solid, liquid or gas but many substances change their state as temperatures change – for example, ice melts to form water at 0°C and then changes into steam at 100°C. The stages at which these changes of state occur are also influenced by the pressure under which they occur.

4.1.4.4.1 Gases

In gases the binding forces between the individual molecules are small compared with their kinetic energy so they tend to move freely in the space in which they exist. When heated, i.e. additional kinetic energy is given to them, they move much more rapidly and if restrained in a fixed volume impinge more energetically on the walls of the containing vessel, a condition that is measured as an increase in pressure. The relationship between temperature, pressure and volume of gases is defined by the general Gas Law:

$$\frac{PV}{T}\text{ (initial)} = \frac{PV}{T}\text{ (final)}$$

where P = absolute pressure, V = volume and T = absolute temperature.

Thus in a reaction vessel which has a fixed volume, if the temperature is increased, so the pressure will increase. If the reaction is exothermic and the temperature increase is not controlled there is a risk that the pressure in the vessel could rise above the safe operating level with consequent risk of vessel failure, a situation that may be met in chemical processes that use autoclaves and reactor vessels.

This general law applies with variation when gases are compressed in that the temperature of the gas rises. In air compressors where there is likely to be oil present the temperature of the compressed air must be kept below a certain level to prevent ignition of the contained oil. Conversely, when the pressure of a gas is decreased, the temperature drops, a condition that can be seen with bottles of LPG where a frost rime forms and where in cold weather there is a danger of the temperature of the gas dropping so low that the control valve freezes up.

Some gases can be compressed at normal temperature until they become liquids (e.g. carbon dioxide, chlorine, etc.), and can conveniently be stored in that state, while others, called permanent gases, cannot be liquefied in this way but are stored either as compressed gases (e.g. hydrogen, air etc.) or under pressure in an absorbent substance (e.g. acetylene).

Air, carbon dioxide and a number of other gases which dissolve in water become more soluble as the pressure increases or the temperature decreases. Increase in temperature or decrease in pressure causes the dissolved gases to come out of solution, e.g. tonic water or fizzy lemonade. This is why hydraulic systems need venting. A similar condition arises with divers who surface too quickly and release dissolved gases from their blood causing the 'bends'.

The specific gravity (sp. gr.) of a gas is determined by comparing its weight with that of air and this has important implications at work. The charging of the batteries of fork lift and other trucks results in the generation of hydrogen (which has a sp. gr. of 0.07) which will either become trapped under any covers or lids left over the battery creating a serious explosion risk or will rise into the ceiling space where good ventilation is necessary to prevent a fire risk. At the other end of the scale carbon dioxide (CO_2, sp. gr. – 1.98) and hydrogen sulphide (H_2S, sp. gr. = 1.19) being heavier than air tend to sink to the floor and will fill the lower part of pits and storage vessels making entry hazardous. Similarly, the vapours of most flammable liquids are heavier than air and tend to collect in low places in the floor and will flow like water to the lowest point creating fire hazards possibly remote from the site of the leakage.

4.1.4.4.2 Liquids

In liquids, the kinetic energy of the molecules is sufficient to allow them to slide over each other but not sufficient for them to move at random in space since they are subject to the binding force of cohesion and remain a coherent mass with a definite volume. Thus the substance will flow to take up the shape of the container but cannot be compressed. When heated the kinetic energy of the molecules increases and the liquid expands until boiling point is reached when the cohesive forces can no longer hold adjacent molecules together, the liquid evaporates and behaves as a gas. If the temperature falls, i.e. the kinetic energy of the vapour is reduced, the vapour will revert to its liquid state by

Table 4.1.4 Properties of some flammable liquids and gases

Substance	*Sp. gr. of liquid*	*Sp. gr. of vapour*	*Boiling point, °C*	*Sp. heat of liquid at 20°C*	*Vapour pressure at 20°C in mmHg*	*OES long-term ppm*
Acetaldehyde	0.78	1.52	21	–	760	20
Acetone	0.79	2.00	56.5	0.53	185	500
Acetylene	–	0.91	–84	–	–	*
Ammonia	–	0.60	–33.4	–	–	25
Carbon disulphide	1.3	2.64	46	0.24	298	10 (MEL)
Carbon dioxide	–	1.98	–79	0.82	–	5000
Diethyl ether	0.71	2.56	35	0.54	442	100
Hydrogen	–	0.07	–253	–	–	*
Commercial propane	0.50	1.4–1.55	–45	1.53	9	*
Toluene	0.87	3.14	111	0.39	23	50

OES = occupational exposure standard.
MEL = maximum exposure limit.
* = asphyxiant, requires monitoring for oxygen content of atmosphere.

condensing. However, at temperatures well below the boiling point, some surface molecules gain sufficient energy to escape from the surface. This means that the liquid is constantly losing some of its surface molecules by vaporisation and the extent to which this is occurring is measured as vapour pressure (*Table 4.1.4*). Vapour pressure will increase as temperature rises.

In passing from the liquid to the vapour or gaseous state a considerable additional input of energy is required without raising the temperature. This energy is the 'latent heat of vaporisation' (H_v in *Table 4.1.5*) of the liquid and can be quite substantial. This characteristic is made use of in fire fighting where a fine spray of water absorbs the heat of the flames to become steam and in so doing reduces the gas temperature to below that required to maintain combustion.

Table 4.1.5 Heats of vaporisation and fusion

Substance	*M.P., °C*	*B.P., °C*	*Hv, kJ/kg*	*H_f, kJ/kg*
Aluminium	660	2450	10500	397
Copper	1083	2595	4810	205
Gold	1063	2660	1580	64
Water	0	100	2260	335
Ethanol	−114	78	854	105
Oxygen	−219	−183	210	13.8

H_v = heat of vaporisation.
H_f = heat of fusion.

As liquids are heated they expand volumetrically and this effect has to be taken into account in the storage of liquids in closed vessels. It is normal to leave an air space above the liquid, the ullage, to allow for this expansion, and to fit pressure relief valves to release any excess pressure generated. An example of what can happen if these precautions are not taken was seen at Los Alfaques Camping Site at San Carlos de la Rapita in Spain in July 1978 when a road tanker carrying 23.5 tonnes of propylene exploded killing 215 people. The circumstances of the incident were that the unlagged tanker was filled completely with propylene during the cool of the early morning, leaving no ullage. The tank was not fitted with any form of pressure relief. As the tanker was driven in the heat of the sun, the tank and its contents heated up to a temperature where the vapour pressure of the contents exceeded atmospheric pressure and the propylene (an incompressible liquid), expanding at a greater rate than the tank, caused the tank to rupture as it was passing a holiday camp at 14.30 hours creating a BLEVE (boiling liquid expanding vapour explosion) that caused devastation over an area of five hectares.

The incompressibility of liquids is utilised in many ways in industry particularly in hydraulic power transmission. Two common applications are as power sources for machines and vehicles and in hydraulic cylinders. Normally the hydraulic medium is an oil and hazards can arise where leaks occur either as oil pools on a floor or from high-pressure lines as a fine mist which is highly flammable.

4.1.4.4.3 Solids

In a solid at room temperature the molecules are tightly packed together, have little kinetic energy and vibrate relative to each other. A solid has form and a rigid surface as a result of the strength of the cohesive forces between the molecules. As a solid is heated the vibrations of the molecules increase and the solid expands. Eventually a temperature is reached where the kinetic energy allows the molecules to slide over one another and the solid becomes a liquid.

Solids, and particularly metals, have a number of characteristics that are of great value in almost everything to do with modern standards of living. They have stability of form over a wide range of temperatures which enables them to be formed and machined into shapes, a measurable strength over a wide range of stresses and temperatures, and permanent physical characteristics such as rates of expansion and contraction with temperature.

The change of state from solid to liquid requires the input of a considerable quantity of heat – the latent heat of fusion (H_f in *Table 4.1.5*) – without raising its temperature. This can easily be seen with melting ice where the temperature of the water remains at 0°C until all the ice has melted. At room temperatures some solids, such as solid carbon dioxide, do not pass through the liquid stage but change directly from a solid to a gas, a process known as 'sublimation', the liquid state only occurring at low temperature and/or high pressure. Organic substances such as wood also have no liquid stage but decompose on heating, rather than melting, reducing their very large molecules into smaller ones some of which are given off as vapours.

As solids are heated they expand at rates determined by their 'coefficients of expansion' that vary with the different substances. The heat needed to raise the temperature of different solids varies and can be determined from its 'specific heat'. Problems can arise where dissimilar metals are welded together that have different specific heats and different coefficients of expansion. Allowance has to be made for differential expansions where different metals are moving in contact and have different coefficients of expansion, such as in bearings. However, use is made of these different expansion rates in bimetal strips for temperature measurements and in thermostats.

Physical properties of some solids are shown in *Table 4.1.6*.

Solids, and particularly metals, have great strength and an ability to withstand high levels of stress and strain. By alloying metals they can be made to exhibit particular characteristics to suit particular applications. Characteristics of metals can vary from very ductile lead and gold through high-strength high-tensile steels to brittle cast iron, each having its particular use. One exception is the metal mercury which is a liquid at room temperature.

Table 4.1.6 Physical properties of some solids

Material	*Density kg/m³*	*Coefficient of linear expansion μm/mK*	*Specific heat kJ/kgK*
Aluminium	2700	24	900
Brass	8500	21	380
Cast iron	7300	12	460
Concrete (dry)	2000	10	0.92
Copper	8930	16	390
Glass	2500	3–8	0.84
Polythene	930	180	2.20
Steel	7800	12	480

4.1.5 Energy and work

There are two types of energy, kinetic and potential. Kinetic energy is involved in movement and may be available to do work; it can be present in a number of forms such as heat, light, sound, electricity and mechanical movement. Potential energy is stored energy and usually requires an agent to release it, as, for instance, the potential energy of a loose brick lying on a scaffold which needs to be kicked off to produce kinetic energy.

Heat is a form of energy where the degree of hotness of a material is related to the rate of movement of the atoms and molecules which make it up. Heat is transferred from one material to another by conduction, convection or radiation. Conduction occurs within and between materials that are in contact through the physical agitation of molecules by more energetically vibrating neighbours. Some materials are better heat conductors than others depending on the ease with which molecules can be made to vibrate. Convection is the conveyance of heat in gases and liquids by the heated fluid rising and heating any surface with which it comes into contact. Initial spread of fire within a building is most likely to be by convection and a method of control is by venting the hot gases to outside the building so preventing the lateral spread of fire. Radiation of heat from a heat source occurs as infrared emissions which pass through the atmosphere by wave propagation.

If a force is exerted on an object, no energy is expended until the body moves when the force is converted into kinetic energy which is equal to the work done on the body by the force. Mechanical work is done whenever a body moves when a force is applied to it.

$$\text{Work done (joules)} = \text{force (newtons)} \times \text{distance moved (metres)}$$

In all mechanical devices or machines the work done is never as great as the energy expended, as some of the energy is lost in overcoming friction. The ratio of work done to energy expended is the mechanical efficiency of the machine and is always less than one.

Power is the rate of doing work and is measured in joules per second:

$$1 \text{ joule per second} = 1 \text{ watt of power}$$
$$1\,\text{kWh} = 3.6 \times 10^6 \text{ joules}$$
$$746 \text{ watts} = 1 \text{ horsepower}$$

The rate at which machines work is given in either horsepower or kilowatts.

Pressure is unreleased potential energy since when contained in a pressure vessel the fluid exerts equal and opposite forces on the vessel walls but no movement takes place. Pressure is independent of the shape of the vessel and is exerted at right angles to the containing surfaces. With gases the pressure is virtually the same throughout the containing vessel but with a liquid the pressure varies according to the depth, i.e. the weight of liquid above the point of measurement. Pressure can be converted into kinetic energy through the movement of a piston in a pneumatic or hydraulic cylinder.

4.1.6 Mechanics

Mechanics is that part of science that deals with the action of forces on bodies. Much of the theory is based on Newton's three laws of Motion:

1 Every body continues in its state of rest or of uniform motion in a straight line except in so far as it is compelled by external impressed force to change that state.
2 Rate of change of momentum is proportional to the force applied and takes place in the direction in which the force acts.
3 To every action there is always an equal and contrary reaction.

If a force is applied to a body resting on a plane, initially the body will not move because of the friction between itself and the plane. The force is resisted by an equal and opposite force due to friction. Once the applied force exceeds the friction force the body will move in the direction of the applied force. However, if the same force continues to act on the body its speed will accelerate because sliding friction is less than limiting (i.e. static) friction. Hence:

$$\text{Before movement the force } F = W\mu_L$$

where μ_L is the coefficient of static or limiting friction, and W is the reaction between the body and the plane (i.e. its weight = mass × gravity).

$$\text{After movement the excess force } F_e = F - W\mu_s$$

where μ_s is the coefficient of sliding friction.

This excess force overcomes the *inertia* of the body and causes it to accelerate at a rate f given in the formula

$$F_e = mf$$

This is why a sticking object will shoot away once it has been freed. Friction has an ambivalent role – it is necessary to enable us to walk, for cars to move and to enable us to control motion through the use of brakes; on the other hand in machines friction absorbs energy and makes the machine less efficient.

Bodies can possess potential energy as a result of the height at which they are located above a datum and that potential energy can be converted into kinetic energy by the object falling. Thus a mass m falling through a height h does work equal to

$$\begin{aligned} W &= m \times g \times h \\ &= \tfrac{1}{2} \times m\, v^2 \text{ its kinetic energy} \end{aligned}$$

where v is its velocity.

For a body to move in the direction of the force applied to it, the line of that force must pass through the centre of gravity of the body. If the line of the force does not pass through the body's centre of gravity, the force will apply a turning moment to the body. Two equal and opposite forces acting on a body and not passing through its centre of gravity will apply a turning couple.

In machines, forces act on component parts in overcoming friction, accelerating or decelerating or in transmitting power or loads. These forces can be tensile, compressive, shear, bending or torsional.

4.1.7 Strength of materials

For machines, plant and buildings to work and give an economic life, their component parts must be capable of resisting the various forces to which they are subject during normal working. Any load applied to a component induces stresses and strains in it. The type of stress depends on the manner in which the load is applied. Stress is force per unit area (N/m^2). Strain is the proportional distortion when subject to stress ($\delta l/l$) and is often quoted as a percentage. The ratio of stress to strain is constant and known as Hooke's Laws after its discoverer, thus:

$$\frac{\text{stress}}{\text{strain}} = \text{constant} = E$$

E is called Young's modulus or modulus of elasticity and its value depends on the material and the type of stress to which it is subjected.

Forces acting on a component can be tensile, compressive, shearing, bending or torsional. The effect of these is shown diagrammatically in *Figure 4.1.2*.

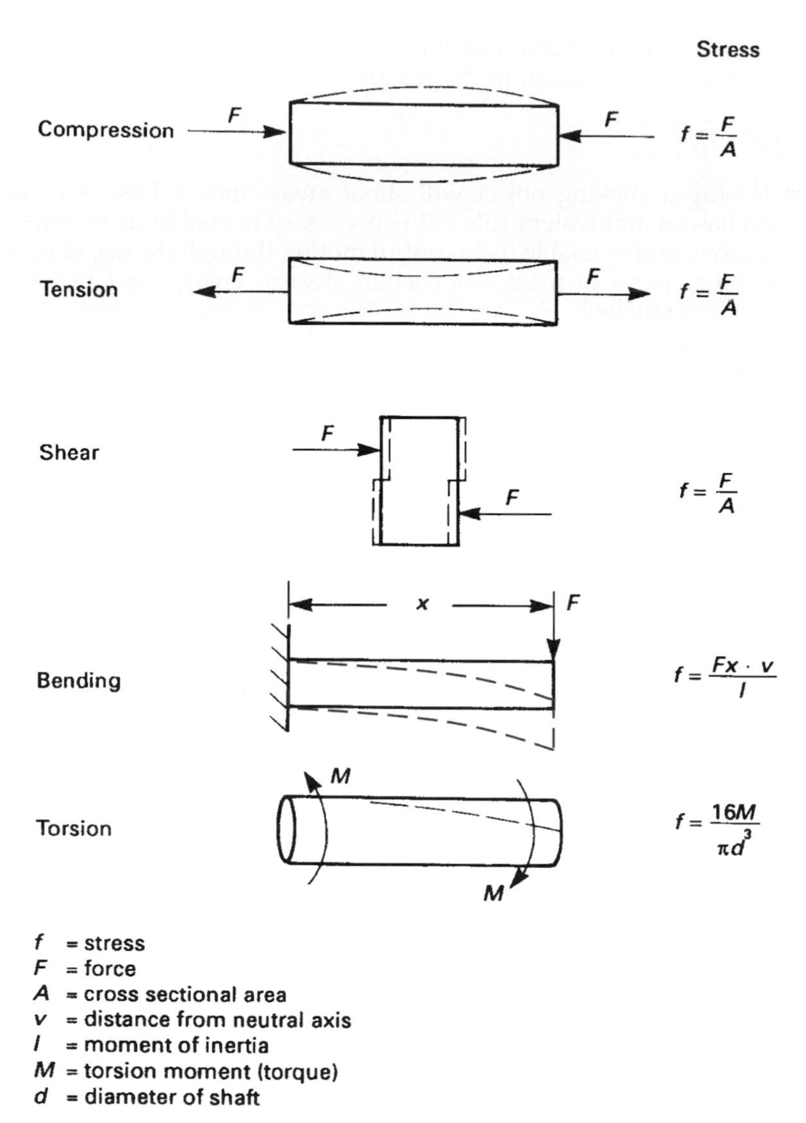

Figure 4.1.2 Showing the different deformations produced in a material by the different forces acting on it

The calculation of tensile, compressive and shear stresses is relatively easy, but to obtain bending stresses it is necessary to know the moment of inertia of the section of the beam and in the case of torsion, although it is a shear stress, because it is spread along the length of the shaft its calculation is complex. With bending moments, a tensile stress is induced in the outer surface of the beam and an equal and opposite compressive stress occurs in the inner surface.

While each material exhibits particular physical characteristics, they all follow a broad pattern of behaviour as shown in *Figure 4.1.3*. Where an

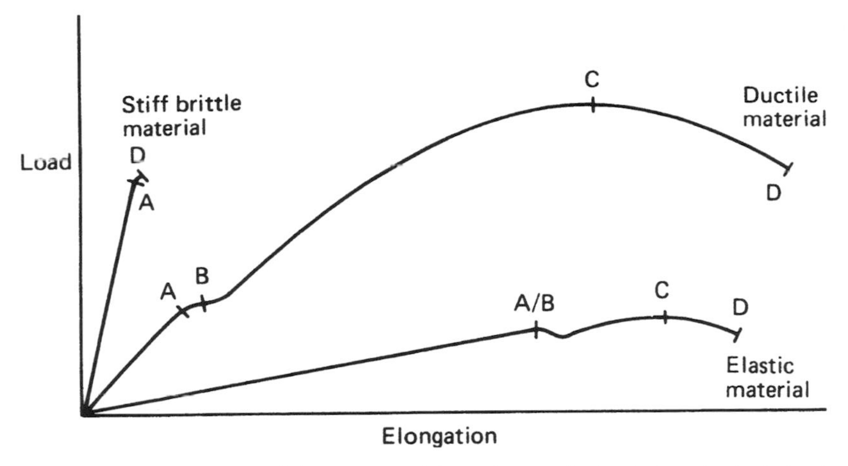

Figure 4.1.3 Behaviour of different materials under tensile load. A = elastic limit; B = yield point; C = ultimate tensile strength; D = breaking point

item has a load applied to it, provided that the strain induced does not go beyond the elastic limit, then when the load is removed the item will return to its original size, i.e. there will have been no permanent deformation. However, if the elastic limit is exceeded permanent deformation occurs and the characteristic of the material will have changed. Typical stiff brittle materials are cast iron, glass and ceramics which have virtually no ductility and will fracture at the elastic limit. Ductile materials include mild steel and copper, and elastic materials include plastics.

4.1.8 Modes of failure

Even in the best of designed machinery and plant failures can occur. When they do, the cause is normally investigated to determine the mode of failure so that repetition can be prevented. Modes of failure relate to the type of stress to which the component has been subjected and the characteristic features of failures due to tension, compression, shear and torsion are well known. Sometimes the failure is related more to the operating process than to the stress, particularly when there is repeated stress cycling of the part when it can suffer fatigue failure.

In processes using certain chemicals, stresses have been shown to make the parts prone to corrosive attack which can reduce their strength. Similarly failures have occurred where a chemical in contact with the part has affected its ability to carry stresses through causing embrittlement such as zinc embrittlement of stainless steel[2] and hydrogen embrittlement of grade T chain[3].

4.1.9 Testing

While there are common legislative requirements for the testing of finished products where, in normal work, the equipment is stressed, such as pressure vessels, cranes, lifting equipment etc., a great deal of testing takes place long before the product is manufactured. It starts at the material producer who tests his product for quality and to ensure that it meets specification. Such materials testing is usually on specially shaped test pieces that are tested to destruction and by chemical analysis to ensure that the constituent materials are present in the correct amounts. The supplier will issue the appropriate test certificates.

In the manufacture of plant and machinery, and following subsequent periodic inspections, should they become necessary testing is normally non-destructive and a number of specialised techniques have been developed. The aim is to check the condition of the material of which the plant is constructed and to identify faults that cannot be seen by eye. Special detection techniques used to highlight the weaknesses or faults in the material include the use of magnetic particles, penetrant dyes, X-ray and gamma-ray sources, ultrasonic vibrations, microwave and infrared rays. Equipment is now available to enable visual inspections to be carried out of inaccessible places using fibre optics and remote-controlled television.

4.1.10 Hydraulics

This side of engineering is concerned with fluids, both liquids such as water, oil etc., and gases such as air and other gaseous elements and compounds. Steam can be classified as a gas although it has characteristics that make it more hazardous than normal compressed gases. These two types of fluid exhibit disparate characteristics in that gases can be compressed while liquids are incompressible.

4.1.10.1 Compressible fluids

When gases are compressed, work is done on them and they acquire considerable potential energy as pressure. They also become hot and it is sometimes necessary to provide cooling to ensure that critical temperatures are not exceeded. Conversely when gas pressure is released, the gas does work or gives up energy and heat and becomes cold. This latter effect is the basis by which refrigerators work and it can be seen, even in the middle of summer, as rime on the outside of LPG cylinders.

Compression of gases and air to achieve high pressures with relatively low volumes is by positive displacement compressors of either the piston or rotary type, while for large volumes at lower pressures centrifugal compressors are used as for example in gas turbines. Gas flows can be measured by means of a vane or hot wire (heated head) anemometer, by measuring the pressure difference across an orifice[4], or through the use of pitot tubes.

Compressed gases have a large amount of potential energy which can be utilised for actuating thrust cylinders. Once pressurised gas has been admitted to a cylinder, even if the supply is then cut off, the cylinder will continue to operate until all the energy of the gas has been absorbed in work done. A dramatic demonstration of the energy of compressed gases can be seen when a containing vessel ruptures with heavy sections being projected very considerable distances or the jet effect when a valve breaks off a gas cylinder. Because of the energy it contains, compressed air is not used for pressure testing of vessels except in certain very specific cases where the use of water is not practicable, e.g. in nuclear reactor vessels.

4.1.10.2 Incompressible fluids

Normally referred to as liquids, incompressible fluids have many advantages when used for power transmission in that, even at very high pressures, they do not in themselves contain a great deal of energy so that should a failure of a containing vessel or pipework occur the result is not catastrophic. These fluids also have the advantage that their flow and pressure can be controlled to a very fine degree which makes them ideal for use in applications where control is critical. They also have the advantage that they can be pumped at very high pressures and so permit the application of very large forces through hydraulic jacks and high torques through hydraulic motors. The power of a fluid jet has been harnessed in cutting lances where very high-pressure liquid, normally water, is emitted from a very small orifice with sufficient force to cut through a variety of materials including steel. These lances have a number of applications but can be very dangerous so their use requires strict control.

The pumps used in handling these high-pressure liquids can suffer considerable damage from cavitation. Incompressible liquids will not compress, nor will they withstand tension; thus if the suction inlet to a pump is restricted the fluid will release any contained air to form cavities. This condition seriously affects the performance of the pump, can cause damage to its rotor and generates a great deal of noise. Gas or air entrained in a hydraulic fluid is detrimental to its effectiveness as a power transmission medium, an effect that may be experienced in the braking system of a vehicle.

4.1.11 Summary

Although much of occupational safety is now recognised as being behavioural, engineering with its foundation in science still has a major role to play. Many of the scientific principles learnt in the lecture room and scientific laboratory have application across a spectrum of work activities and have important connotations as far as occupational safety is concerned. This chapter has reviewed some of those applications – there are many more for the student to discover.

References

1. Health and Safety Executive, *Investigation report: Fire on HMS Glasgow, 23 September 1976*, HSE Books, Sudbury (1978)
2. Health and Safety Executive, *Guidance Note, Plant and Machinery series No. PM 13: Zinc embrittlement of austenitic stainless steel*, HSE Books, Sudbury (1977)
3. Health and Safety Executive, *Guidance Note, Plant and Machinery series No. PM 39: Hydrogen cracking of grade T (8) chain and components*, HSE Books, Sudbury (1998)
4. British Standards Institution, BS 1042: *Measurement of fluid flows in closed conduits, Part 1: Pressure differential devices and Part 2: Orifice plates, nozzles and venturi tubes inserted in circular cross section conduits running full*, BSI, London.

Reading list

Basic Engineering Science

Titherington, D. and Rimmer, J.G., *Engineering Science*, Vol. 1 (1980), and Vol. 2 (1982), McGraw-Hill, Maidenhead

Deeson, E., *Technician Physics*, Longman, Harlow (1984)

Harrison, H.R. and Nettleton, T., *Principles of Engineering Mechanics*, 2nd edn, Edward Arnold, London (1993)

Hannah, J. and Hillier, M.J., *Applied Mechanics*, 3rd edn, Pitman, London (1995)

Houpt, G.L., *Science for Mechanical Engineering Technicians*, McGraw-Hill, Maidenhead (1970)

Berry, J., Bryden, P., Graham, T. and Porkess, R., *MEI Structured Mathematics–Mechanics vols 1–3*, 2nd edn, Hodder & Stoughton, London (2000)

Properties of Materials

Peapell, P.N. and Belk, J.A., *Basic Materials Studies*, Butterworth, London (1985)

Timings, R.L., *Engineering Materials*, Vol. 1, Longman, Harlow (1998)

Bolton, W., *Engineering Materials Technology*, 3rd edn, Butterworth-Heinemann, Oxford (1998)

Crane, F.A.A. and Charles, J.A. *Selection and Use of Engineering Materials*, 2nd edn, Butterworth, London (1989)

Hanley, D.P., *Introduction to the Selection of Engineering Materials*, van Nostrand, Wokingham (1980)

Young, W.C., *Roark's Formulas for Stress and Strain*, 6th edn, McGraw-Hill, New York (1989)

Higgins, R.A., *Properties of Engineering Materials*, 2nd edn, Hodder and Stoughton, Sevenoaks (1994)

Gordon, J.E., *The New Science of Strong Materials*, 2nd edn, Pitman, London (1991)

Hydraulics

Turnbull, D E., *Fluid Power Engineering*, Butterworth, London (1976)

Pinches, M.J. and Ashby, J.G., *Power Hydraulics*, Prentice-Hall, Hemel Hempstead (1996)

Massey, B.S., *Mechanics of Fluids*, 6th edn, van Nostrand Reinhold, London (1989)

Basic Introductory Chemistry

Lewis, M. and Waller, G., *Thinking Chemistry*, Oxford University Press, Oxford (1986)

Jones, M., Johnson, D., Netterville, J. and Wood, J., *Chemistry, Man and Society*, Holt, Rinehart and Winston, Eastbourne (1980)

Hazardous Chemicals

Schieler, L. and Pauze, D., *Hazardous Materials*, van Nostrand Reinhold, Wokingham (1976)

Hazardous Chemicals, A manual for schools and colleges, Oliver and Boyd, Edinburgh (1979)

Chapter 4.2

Fire precautions

Ray Chalklen

4.2.1 Introduction

Fire is one of the most destructive, disruptive and costly causes of damage to any building yet fires don't just happen they are caused. While many fires start because of a momentary act of carelessness, ignorance or failure to take account of fairly obvious hazards, an increasing number are started deliberately. Fire has been a great friend of man through the ages enabling him to cook, improve his range of tools and lengthen his day by giving him a source of light. Fire has also been a constant enemy and its destructive powers have been both worshipped and feared.

Most fires can be prevented by a few simple precautions and those that do start can usually be held in check or quickly controlled by fire safety measures. These measures can be incorporated into buildings either during construction or renovation work but in addition, well trained staff can play an equally important role in preventing and tackling fires. Fire can be responsible for the loss of jobs, loss of businesses and loss of life as well as serious damage to the environment. Many companies go out of business following a serious fire.

In recent years the introduction of the Fire Precautions (Workplace) Regulations 1997[1] (as amended) has introduced the need for a formal fire risk assessment for most places of work. This requires the owners and occupiers of buildings to play a much more proactive role in identifying the most likely risks of a fire starting and of the likely risks to staff. It then requires them to take measures to reduce these risks.

4.2.2 Basic fire technology

Fire, or combustion, is a chemical reaction in which a substance reacts with the oxygen in the air and emits heat and light. It is a particular example of the more general process known as oxidation.

4.2.2.1 The fire triangle

For combustion to occur, three things need to be present – oxygen (usually from the air); a fuel which can either be a liquid, a gas or a solid;

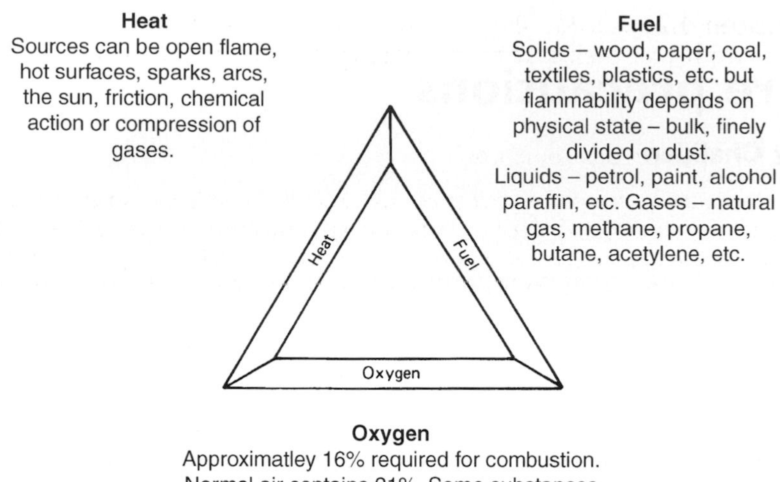

Heat
Sources can be open flame, hot surfaces, sparks, arcs, the sun, friction, chemical action or compression of gases.
Fuel
Solids – wood, paper, coal, textiles, plastics, etc. but flammability depends on physical state – bulk, finely divided or dust.
Liquids – petrol, paint, alcohol paraffin, etc. Gases – natural gas, methane, propane, butane, acetylene, etc.
Heat
Fuel
Oxygen
Oxygen
Approximatley 16% required for combustion. Normal air contains 21%. Some substances contain sufficient oxygen to support combustion.

Figure 4.2.1 The fire triangle

and heat from some external source. These three elements are usually represented graphically as the fire triangle (*Figure 4.2.1*).

4.2.2.2 Fuel

The flammability of solids depends on their physical state. Usually the more finely divided they are the more rapidly they will burn, e.g. sawdust will burn quicker than a tree trunk and coal dust more quickly than large lumps of coal. Solids and liquids do not burn but when heated give off a gaseous substance or vapour and it is this substance that burns. For example coal when heated gives off methane, benzene and other hydrocarbon gases and it is these gases which burn rather than the coal itself. Flammable liquids always have a gaseous layer above their surface and it is this layer that burns and not the actual liquid.

4.2.2.3 Oxygen

The main source of oxygen is from the air which contains 21% oxygen. The level of oxygen must be reduced to below 16% for combustion not to occur or be suppressed. Some substances, such as chlorates and organic peroxides, contain sufficient oxygen in their chemical makeup to sustain combustion without any additional oxygen being available.

4.2.2.4 Heat

The most usual source of heat which provides sufficient energy for combustion to commence is an external source such as a match, heat from friction or an electrical short-circuit. Other forms of ignition include a spark from static electricity or making and breaking of an electrical contact.

4.2.2.4.1 Spontaneous combustion

Spontaneous combustion occurs in materials where some internal chemical or biological action causes a rise in temperature sufficient to ignite the material. Other materials can react with oxygen at normal temperatures to generate heat. If the heat is not able to dissipate quickly enough then the temperature of the material will continue to increase until eventually it reaches its ignition point when it will ignite. When stored in bulk, most organic materials such as coal, hay, straw, wood or paper, are prone to spontaneous combustion particularly if damp when initially stored. The risk of spontaneous combustion of such materials can be reduced by ensuring they are dry when stored, by providing sufficient circulation of air through the storage area and by careful monitoring of the temperature build-up during storage. Some materials when stored in bulk, such as chlorates, are prone to explosive ignition with disastrous results.

4.2.2.4.2 Spontaneous ignition temperature

This is the temperature at which a material will ignite spontaneously. Some materials have so low an ignition temperature that if exposed to normal room temperatures the substance will ignite spontaneously without the need for an external source of heat or flame. This is referred to as auto-ignition and the temperature at which it occurs the auto-ignition temperature. Solids that exhibit this characteristic include white phosphorus which must be kept immersed in a liquid to exclude it from contact with the air.

4.2.2.4.3 Smouldering

Smouldering is the very slow combustion in air and can exist for long periods without any noticeable flame. It can occur in porous materials such as paper, sawdust and latex rubber and produces large volumes of flammable smoke which accumulates until it reaches its lower flammability limit when it will ignite. Smouldering can be transformed into a flaming fire if the supply of oxygen increases. An example of smouldering combustion occurs in upholstered furniture, ignited by a cigarette, which can lie dormant for a considerable time before bursting into flame.

4.2.2.5 Fire spread

Once a fire has started and there is sufficient fuel and oxygen to sustain it, there are three ways in which it can spread through a building: conduction, convection or radiation.

4.2.2.5.1 Conduction

Conduction is the carrying of heat through or along a material and can occur in solids, liquids or gases although it is most evident in solids. Heat energy is passed from molecule to molecule and flows away from the source of heat towards areas of lower temperature. The ability to conduct heat varies between materials. In general, good conductors of heat are also good conductors of electricity, i.e. metals. In a fire, a metal girder passing through a fire compartment wall may conduct enough heat to ignite materials in the neighbouring compartment. A steel door with no insulation will conduct heat much better than a wooden door although initially it may resist a fire better.

4.2.2.5.2 Convection

Convection is the carrying of heat by the internal movement of molecules within a material and only occurs in liquids and gases. It occurs when water in a saucepan is heated. As the water at the bottom of the saucepan heats up it becomes less dense and rises with the colder denser water taking its place at the bottom of the pan. A similar effect occurs in a fire with the hot air or smoke rising and cooler air being drawn in at the base of the fire. The rising heated smoke forms a plume until it reaches a horizontal surface such as a ceiling where it spreads out. As the smoke gets further from the fire it cools and drops towards the floor forming a 'mushroom effect'. As this process continues the temperature of the plume will gradually rise until it reaches a temperature at which it will ignite any combustible materials with which it comes into contact.

4.2.2.5.3 Radiation

Radiation is the emission of rays that transfer heat energy through the atmosphere. It will heat solids and liquids but not gases. It does not involve any physical contact between the source and the target material. For example heat energy from the sun travels through empty space and the earth's atmosphere to warm the surface of the earth. Heat is radiated as infra-red radiation and its transfer can be likened to the transfer of light. Radiated heat can pass through some materials such as glass, and ignite combustibles on the other side.

Other radiations occur as electromagnetic radiations where the transmitter emits electromagnetic waves that act directly on the molecules of a solid or liquid increasing their energy and hence raising the temperature of the material. Microwave ovens and radio frequency welding equipment emit radiations of this sort.

4.2.3 Fire hazards and their control

There are many causes of fires but the most common have been identified through the collation of statistics over many years. Control of smoking in the workplace has led to a decline in the number of fires, however arson remains one of the major causes of fires. This section examines briefly some of the most common causes of fire and some of the basic precautions that can be taken to prevent them.

4.2.3.1 Arson

Arson is the single largest cause of fires in the workplace. Those most likely to commit arson are:

1. people with a grudge against the company or individuals within it;
2. intruders with the intention of destroying evidence of another crime;
3. staff to cover a fraud from which they have gained financially;
4. members of action groups campaigning against particular products or practices;
5. vandals, usually children or youths who are opportunists rather than deliberate arsonists;
6. arsonists who enjoy the spectacle of a fire.

A number of measures can be taken to prevent arson and these include:

1. security arrangements to control visitors entering a premises;
2. securing all windows and doors at the end of the working day;
3. good perimeter fencing and external lighting with CCTV coverage;
4. good housekeeping to prevent the build-up of combustible materials and rubbish both inside the building and, most importantly, outside the building. Skips and other rubbish containers should be located well away from buildings;
5. careful selection of new employees and pre-employment checks including the following up of references from previous employers.

In addition, should arson occur, the extent of damage can be minimised if an early warning of a fire can be given. This can be achieved by a sprinkler system, which will restrict the spread of the fire, coupled with an automatic fire detection installation linked to a permanently manned control centre who can call out the fire brigade.

4.2.3.2 Combustible dusts

Combustible dusts, such as sawdust, flour, corn starch, etc., in bulk will only smoulder if ignited. However, if the dust occurs as a cloud in the atmosphere, i.e. presents a large oxidising surface area, it can ignite with explosive force and cause extensive and devastating damage to a building. An explosive concentration of dust would be a health hazard and as such would not be acceptable in a working atmosphere. The normal sequence of events is that a minor explosion occurs in a part of the plant and dislodges dust that has settled on joists, roof trusses and other parts of the building. This disturbed dust forms the explosive cloud that causes a secondary devastating explosion. Such dust explosions can be prevented by:

1 enclosing the process plant to prevent the escape of dusts;
2 installing an effective dust extraction system;
3 maintaining high standards of housekeeping particularly on surfaces at high level, i.e. roof trusses and girders;
4 excluding sources of ignition, such as friction, static electricity, naked flames and the effective containment of electrical switchgear.

4.2.3.3 Electricity

Approximately 25% of fires in industrial premises are caused by electricity. A major cause is overloading of the conductor resulting in its overheating and causing a breakdown of its insulating sheath. This can lead to a short circuit which, in turn, can ignite flammable materials or vapours. Electrical fire from supply cables and wiring can be prevented by ensuring that:

1 the supply cables, connections and connectors are in good condition;
2 the supply cable has the capacity to meet the electrical demands;
3 the rating of fuses or circuit breakers that protect individual circuits are appropriate for the current-carrying capacity of the cable;
4 all electrical appliances and equipment are inspected and tested at regular intervals;
5 the condition of supply cables and wiring, especially of portable leads, is inspected and tested at regular intervals;
6 all appliances, except double insulated, are effectively earthed;
7 appliances and equipment are protected by an RCD or similar earth leakage protection device;
8 the provision of an adequate number of socket outlets or multi-point extension leads to eliminate the need for socket adapters.

Other electrical causes of fires include:

9 arcing when contactors make or break a circuit. This risk can be reduced by ensuring the contactor loading is within manufacturer's limits;
10 electrostatic discharge. Electrostatic charges can acquire sufficient energy to ignite materials. Static build-up can be prevented or reduced by earthing (grounding) or the use of static eliminators.

4.2.3.4 Smoking

The number of fires caused by cigarette ends has decreased in recent years, largely through much tighter controls on smoking in the workplace, either a complete ban or through the provision of allocated smoking areas that have been provided with fireproof receptacles for cigarette ends. However, there is a need for constant monitoring to ensure that smoking does not occur in private places such as toilets.

4.2.3.5 Hot work

Hot work, involving gas cutting and burning, gas and electric arc welding and burning, is a high fire risk operation and should only take place when suitable fire prevention measures are in place. These should include the provision of local portable extinguishers, suitable fire/spark resistant screening and, in areas of high fire risk, the issue of a permit-to-work. A trained fire fighter should remain in attendance until at least half an hour after the work is completed, and the area should be checked for at least a further hour.

4.2.3.6 Heating systems

Hot pipes and ductwork carrying high temperature and high pressure substances can pose a high fire risk. They should be lagged particularly where they pass close to combustible materials. Combustible materials should not be stacked or stored against hot pipes but a suitable space to allow circulation of cooling air should be left. Where portable heaters are used in the workplace they should be tested to ensure that they are in a safe working condition. Portable heaters should be securely mounted or fixed in position since loose heaters can be knocked over with the risk of igniting adjacent materials. They should not be placed near combustible materials when switched on. Heating appliances, whether portable or fixed, that use gas as the heating medium should be isolated at the main supply valve and not rely on the valve on the appliance itself – the flexible supply pipes can become porous and leak gas causing a potentially explosive atmosphere. The siting of portable heaters is important and they should not have product stored in front of them nor should clothing be put on top of them to dry.

4.2.3.7 Housekeeping

The fuel required for a fire can be provided by rubbish left in the workplace. High standards of housekeeping play an important role in fire prevention. Any rubbish should be cleared up and disposed of in a proper receptacle. If the rubbish is contaminated by a flammable liquid or substance, it should be stored in a fireproof container.

4.2.3.8 Lighting

Modern fluorescent luminaires do not generate a great deal of heat. However, tungsten lamps, by their nature rely on very high temperatures in the element to create light. Heat from the element can be transmitted to the bulb casing which can reach temperatures high enough to ignite combustible materials placed close to them. Where tungsten lighting is used, combustible materials should be stowed well clear of them.

4.2.4 Fire alarms and detectors

Should a fire occur, the priority response must be the saving of life. To this end, detection of the outbreak of a fire and the sounding of an alarm play a crucial role. The means of detection and the type of alarm employed will depend on the type of organisation, the operations it carries out and the number of employees at risk. A characteristic of an alarm is that it must be capable of being heard in every area to which an employee may need to have access, including inner offices, stores, toilets, etc.

4.2.4.1 Manually operated fire alarms

The simplest form of fire alarms are manually operated devices such as gongs or bells. These are suitable only for the smallest premises where one of the devices can be heard throughout the whole building.

For larger premises an electrical fire alarm system will be required incorporating break glass call points and sounders (audible warning) such as bells, sirens or hooters. In a medium sized building the alarm system could employ a single circuit but for more complex buildings and plants a multi-zone alarm system may be required that divides the building or site into discrete fire warning areas. This can give a local area evacuation warning and also give a general alert warning to other areas.

Audible messages can also be part of the alarm system particularly where members of the public are present. These will alert the occupants to the fact of a fire and can be used to direct them to the nearest fire exits by the safest routes. Tests have shown that the public react more quickly to an audible message than they do to a bell or other sound. The standards required for voice alarms are contained in BS 5839: Part 8[2] and EN 60849[3].

4.2.4.2 Automatic fire alarms

Automatic fire alarms have the advantage of being able to raise the alarm in the event of a fire in an unoccupied or unmanned area. They operate by detecting particular changes in the environment. Different types of detectors have been developed to react to the different stages in the development of a fire. The different characteristics that may be used in detecting a fire include:

1 variations in the strength of transmitted beams or rays between transmitter and receiver, whether visible light, infra-red or radioactive, caused by the rising hot products of combustion;

2 the visual interference (obscuration) of a light beam caused by smoke from burning materials;
3 light from flames impinging on a photo-electric cell, but this may be distorted by the level of local illumination;
4 changes in temperature levels either to above a preset level or by a rate of temperature increase above an ecological norm;
5 changes in the reception of existing remote controlled visual security monitoring systems.

The stage at which fire is detected depends on the type of detector used. The three most common types are:

(a) smoke detectors;
(b) heat detectors;
(c) flame detectors.

4.2.4.2.1 Smoke detectors

Smoke is a complex mixture of gases, liquids and solid particles depending on the material that is burning and on the conditions of combustion. Each of the constituents of smoke displays particular optical and physical properties which are exploited in the two main types of detectors – optical and ionisation. Aspirating smoke detectors use either optical or ionisation detectors. The general requirements for fire alarm smoke detectors are contained in BS EN 54[4] with detailed requirements for smoke detectors covered in Part 7.

Recent research has resulted in the development of carbon monoxide detectors. They are most effective in detecting slow smouldering fires where high levels of carbon monoxide are produced before there is any smoke. They are not intended as replacements for smoke detectors but as an addition to the range.

4.2.4.2.2 Ionisation detectors

Ionisation detectors contain a radioactive source which ionises the air within a containing chamber resulting in a small current flow between two electrodes. Particles of smoke entering the chamber interfere with the ion transport and lead to ion–electron recombination, thus reducing the current flow. This reduction in current is sensed and triggers the alarm (*Figure 4.2.2*). Ionisation detectors respond more quickly to smoke containing small particles but have a less rapid response to smouldering fires involving polymers (plastics etc.).

4.2.4.2.3 Optical detectors

Optical detectors contain a light emitting diode and a receiver. Smoke particles entering the detector either obscure or scatter the light beam causing the output signal from the receiver to vary. This in turn triggers the alarm. Optical detectors respond more effectively to dense, heavy particulate smoke such as that generated by oil and plastic fires. Optical detectors are of two main types:

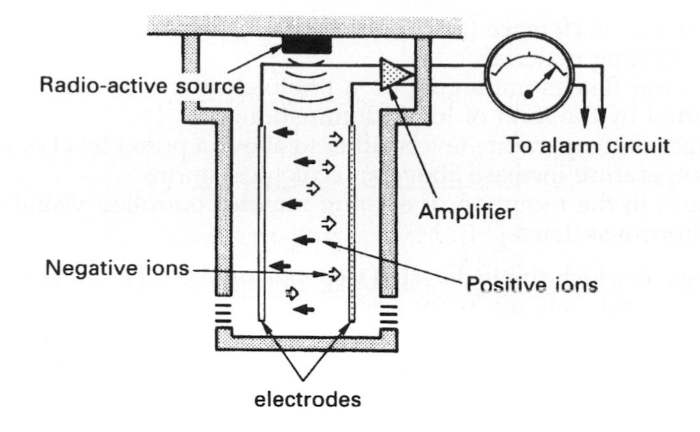

(i) 'non-fire' condition

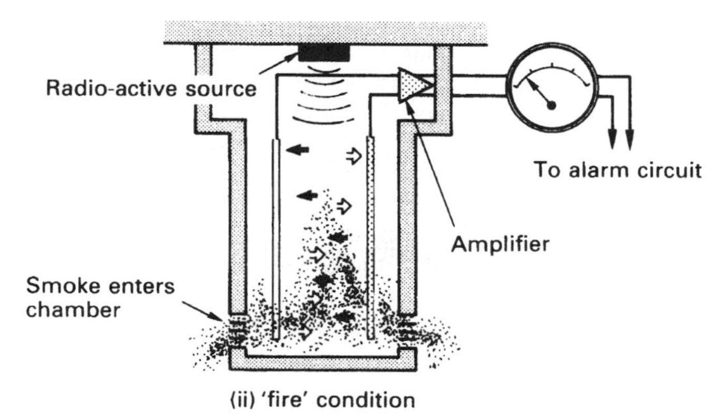

(ii) 'fire' condition

Figure 4.2.2 Diagram of smoke detector – ionising type (Manual of Firemanship. Courtesy Controller SO)

(a) *Light scatter type* where smoke entering the detector reflects the light from a light source onto a photo-electric cell. The small electrical charge produced by this is amplified and actuates the alarm relay (*Figure 4.2.3*).
(b) *Obscuration type* detectors are usually installed to span large areas. A light source and lens is positioned at one end of the area and a receiving photo-electric cell is located at the other end. Rising smoke from a fire passes through the beam deflecting the beam or obscuring the light. This results in a reduction in the intensity of light falling onto the photo-electric cell and causes the alarm signal to be triggered (*Figure 4.2.4*).

4.2.4.2.4 Aspirating smoke detectors

Aspirating smoke detectors comprise probe tubes which run from a control unit into the fire risk zone. A monitoring unit contains the actual smoke detector which may be of either the optical or ionisation type

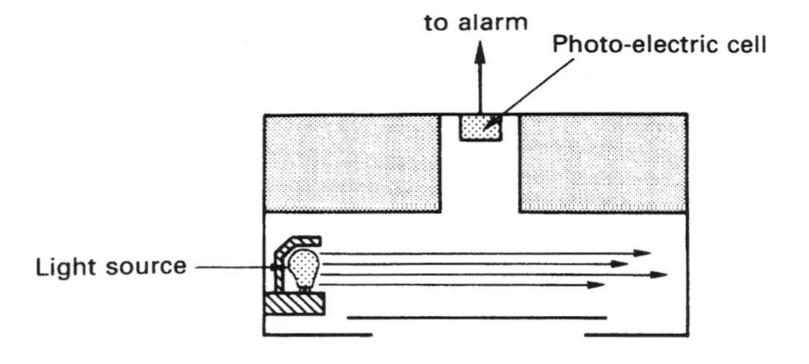

(i) 'non-fire' condition

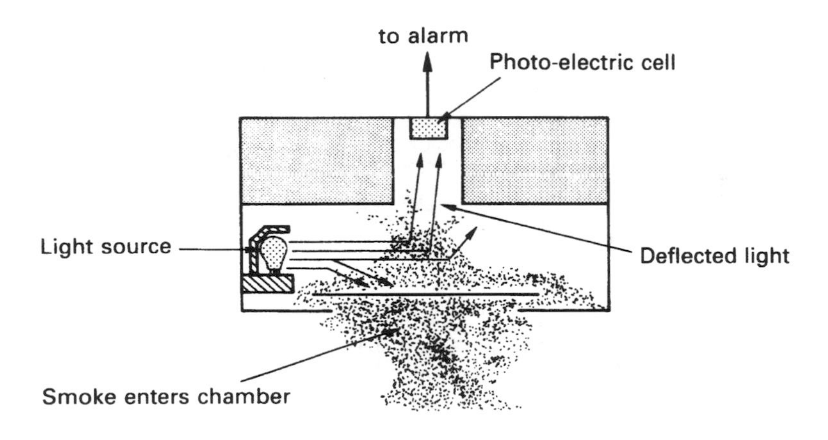

(ii) 'fire' condition

Figure 4.2.3 Diagram of smoke detector – optical light scatter type (Manual of Firemanship. Courtesy Controller SO)

depending on the likely type of fire. This monitoring unit continuously samples the atmosphere in the protected zone by drawing air in through the small holes in the probe tube. This air is passed through a detection chamber which compares it with the normal ambient atmosphere and is set to trip the alarm when combustion products reach a certain level of concentration. The detector sensitivity can be set to give an alert warning at a pre-alarm level when the system indicates that conditions exist where ignition might take place. This type of detection system is very effective in the detection of diffused smoke in challenging environments such as those with high air flows.

4.2.4.2.5 Video smoke detectors

A recent development in the range of smoke detectors is a system which utilises existing video security equipment to detect smoke at the same

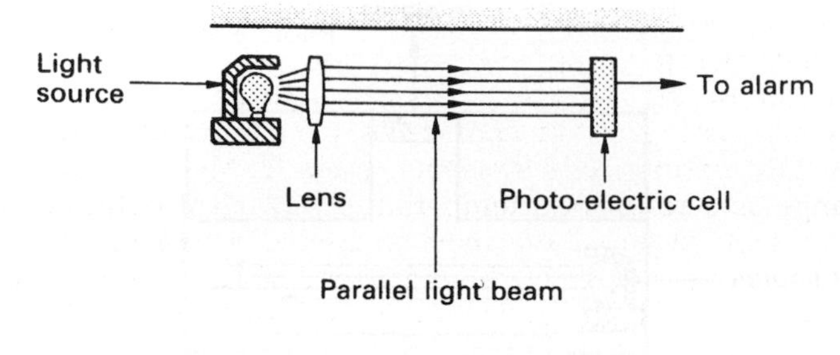

(a) 'Non-fire' condition

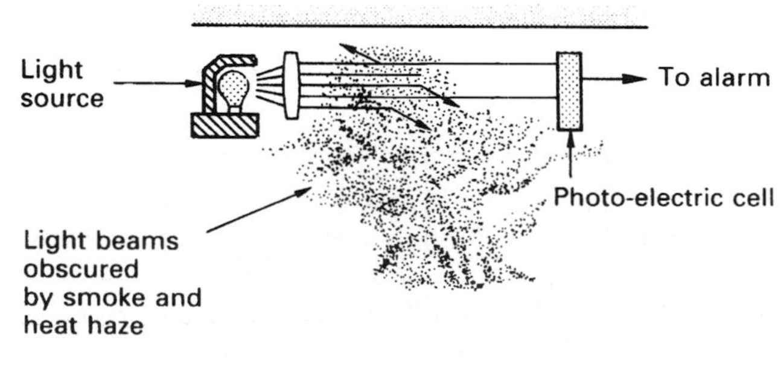

(b) 'Fire' condition

Figure 4.2.4 Optical smoke detector – obscuration type (Manual of Firemanship. Courtesy: The Controller, the Stationery Office)

time as performing its security function. The system works by looking for small areas of change within the image at the picture digitisation stage and passing these pixel changes to the main processor where a series of filters seek particular characteristics associated with smoke. The system checks that these amount to a confident prediction of the presence of smoke. The system can be programmed to sense varying amounts of smoke and the time the smoke has been present to allow for situations where smoke exists as a normal part of the production process. Alarm signals are displayed on a video, which may be located in a central control room, providing the operator with an instantaneous and highlighted view of the cause of the alarm, thus enabling a visual assessment to be made of the situation before actuating the fire alarm.

4.2.4.3 Heat detectors

Heat detectors can be of either the fixed temperature, rate of temperature rise or linear type. Fixed temperature detectors are activated when the temperature in the area reaches a predetermined level and operate in a similar way to a thermostat. Rate of temperature rise detectors are activated when there is an abnormally rapid increase in the temperature at the detector above that experienced with heating systems or sunlight. They can incorporate an upper temperature setting to provide warning where the rate of temperature rise is below the detectable level. Linear heat detectors comprise a detection cable that runs through the area to be protected. The alarm is raised when the electrical characteristics of the cable change as a result of a temperature rise. Heat detectors are most effective in areas where there may be smoke or steam under normal conditions i.e. boiler rooms, kitchens or areas where, in the event of a fire, it can be expected to be a flaming fire with little or no smoke.

4.2.4.4 Radiation detectors

As well as producing hot gases, fire releases radiant energy in the form of:

1 visible light
2 infra-red radiation
3 ultra-violet radiation.

These forms of energy radiate in waves from their point of origin and the detectors are designed to respond to measurements of this radiation.

The use of the visible light band detector has a number of disadvantages associated with the fact that it is not able to differentiate between the various legitimate sources of visible light and those created by a fire. In practice, radiation detectors are designed to respond to either infra-red or ultra-violet radiations. Both of these types of detectors 'look' at the flame and memorise it before having a second 'look' after a short delay to confirm that the flame is still there. If it is, then the alarm signal is triggered.

Infra-red detectors have the ability to scan large areas and identify the wave length and pattern of radiation given out by a flame. They can also be used as spark detectors to protect dust extraction systems.

Ultra-violet detectors are used in specialised areas such as aircraft engine compartments.

Radiation detectors generally complement heat and smoke detectors, especially in tall, unobstructed compartments and are effective in special applications such as flammable liquid storage areas.

4.2.4.5 Carbon monoxide detectors

Normal heat and smoke detectors rely on the presence of convection currents caused by the fire to carry either the products of combustion or heat past the detector which is usually mounted at high level. If there is little heat from the fire then there can be a significant delay before the detector receives enough information from the fire to actuate the alarm. Carbon monoxide detectors overcome this problem since the carbon monoxide produced by the fire dissipates into the atmosphere of the protected area and can reach the detector without the need for convection currents. This type of detector must not be confused with the carbon monoxide gas sampler used to protect people from the presence of this gas. The two instruments perform quite separate functions with widely different trip settings.

4.2.4.6 Laser detectors

Laser detectors operate in a similar way to optical beam detectors. The laser beam is directed across the area to be protected and is deflected or obscured by either the heat, the flame or the smoke rising from a fire, thus changing the intensity of light at the receiver. This triggers the alarm.

4.2.4.7 Wiring

Whatever type of fire alarm system is installed the integrity of the control circuit is crucial to its effectiveness and the condition of the wiring plays a vital part in ensuring that the system maintains its reliability. The cable used must be of a type that is not susceptible to mechanical damage or is suitably protected in metal conduit. The circuit should be monitored by the control equipment so that any fault, deterioration or damage in it is detected and the alarm is tripped.

4.2.4.8 Radio fire alarm systems

Radio alarm systems are especially useful where alarm systems need to be installed into existing buildings that have preservation orders or are listed buildings where any visible disfiguration, such as that caused by wiring, is not permitted. These systems are also suitable for protecting temporary buildings as the system can be installed quickly and is easily removed for use elsewhere when the building is no longer required. The system comprises independent call points, detectors, sounders and control panels each of which has its own power source. They are linked, not by hardwiring but by radio transmissions. Care must be taken to ensure there is no local electromagnetic contamination that may interfere with the system's integrity and also that the system's radio signal does not interfere with local, especially emergency service, radio systems.

4.2.4.9 Control and indicating equipment

The 'heart and brains' of any fire alarm installation, the control and indicating equipment provides the power as well as monitoring the system and indicating the location of any detected fire. The control equipment should be in an easily accessible location so that it can be seen quickly and easily in the event of an alarm and give immediate information on the location of the fire. More advanced systems (known as programmable or intelligent systems) can identify exactly which device has actuated and also self monitor to give early warning of a fault or failure of a component of the system.

4.2.4.10 False alarms

Automatic fire detection systems are notorious for giving false alarms. However, many of the so-called false alarms, while not detecting a fire, have in fact detected an abnormal condition within the building. These need investigating since they may be an early indication of a condition that could build up to a fire. However, false alarms do occur and common causes include:

1. cooking fumes being detected in an area adjacent to the kitchen. This can be due to inadequate or faulty extraction equipment causing the staff to prop open the kitchen door to clear the fumes, allowing the fumes to reach a detector;
2. optical smoke detectors being activated by steam leaking from the process plant due to poor maintenance;
3. contamination of detectors by insects, pollen, dust, etc., causing a change in the sensitivity of the detector. In optical smoke detectors it can cause a false activation;
4. incorrect type of detector. A circumstance that can result from the change in the use of a room or work area;
5. dust from contractors working on site affecting optical smoke detectors or whose equipment causes electric interference with the alarm control system. Detectors in the area likely to be affected in this way should be temporarily isolated from the system, but this should be cleared with the fire insurers before disconnection is made;
6. failure to inform the alarm monitoring station that an alarm system is being tested or maintained;
7. unsatisfactory maintenance or an inadequate testing programme;
8. the wrong type of detector installed or the right type installed in an unsuitable location.

4.2.4.11 Standards for fire alarm systems

The manufacture and installation of fire detection and alarm equipment are subject to a growing number of standards. A current list is given at the end of this chapter.

4.2.5 Classification of fires

Five different categories of fire are listed in a standard[5] where each category is related to the type of substance that forms the fuel. These

Table 4.2.1. Classes of fires and suitability of types of extinguishers

	Water	*Carbon dioxide*	*Dry powder*	*Foam*	*Wet chemical*	*Fire Blanket*
Class A Paper, wood	✓	X	?	?	?	?
Class B Flammable liquids	X	?	✓	✓	?	✓
Class C Flammable gases	X	X	✓	X	X	X
Class D Metals	X	X	Special powders only	X	X	X
Class F Deep fat fryers	X	X	X	X	✓	X
Electrical	X	✓	?	X	X	X

✓ Suitable extinguishant
? Can be used but not ideal
X Unsuitable and should not be used

categories are listed in *Table 4.2.1* which also includes reference to the means for extinguishing the fire. The identification of the type of fire is important to ensure selection of the correct type of fire extinguisher.

4.2.5.1 Class A fires

Class A fires include those involving solid materials normally of an organic nature such as wood, paper, natural fibres, etc., in which combustion occurs with the formation of glowing embers. Water is the most effective extinguishing agent and acts by reducing the temperature available to ignite further material.

4.2.5.2 Class B fires

Class B includes fires involving liquids such as petrol, oil, paints, and liquefiable solids such as fats, waxes, greases, etc. The most effective extinguishing agents are foam and dry powder which blanket the burning material and exclude oxygen. Cooking oils and fats are excluded from this class and are now designated as class F.

4.2.5.3 Class C fires

Class C includes fires involving gases such as butane and propane. Extinguishing these fires is by shutting a valve in the supply line, i.e. removing the fuel. If this is not immediately possible special techniques that require expert knowledge need to be used and should be left to the specialists. Until this can be effected, gas fires should be left burning and the main preventative action should be protecting buildings and property in the surrounding area. If the flames of a gas fire are extinguished before the supply can be isolated there is a danger of the build up of an explosive gas mixture.

4.2.5.4 Class D fires

Class D fires involve metals such as aluminium, magnesium and sodium. Most metal fires are difficult to extinguish as they burn at very high temperatures and react violently with oxygen, either in the air or in the extinguishing medium. In the reaction, the oxygen is removed from the water releasing hydrogen which then ignites violently. Metal fires will normally only respond to the correct extinguishing medium for the type of fire. Wherever it is safe to do so, metal fires should be allowed to burn themselves out with the surrounding area being protected from fire spread. The use of water or allowing the fire to burn out will not be possible where metal fires occur within a building. Suitable extinguishing substances, such as powdered graphite, powdered talc, soda ash, limestone and dry sand, should be used. These must be applied gently to form a coating and hence smother the fire. Premises where a risk assessment has identified metal fires as one of the hazards should have suitable extinguishing medium available and staff specially trained in the techniques of dealing with this type of fire.

4.2.5.5 Class F fires

Class F includes fires involving commercial deep fat and oil fryers. These fires have been given a separate classification because the high temperatures involved make the inclusion of these substances in class B inappropriate. Most of the extinguishers suitable for class B fires would not extinguish a fire involving cooking oil. A special wet chemical extinguisher has been developed which cools and smothers the fire by emulsifying the cooking oil and sealing its surface with a non-combustible crust to prevent re-ignition.

4.2.5.6 Electrical fires

There is no separate classification for electrical fires since electricity, while being a cause of fires, is not a flammable material itself. It is important in any fire involving live electrical equipment that the electrical supply is isolated before attempting to tackle the blaze. If that is not possible, carbon dioxide or dry powder extinguishers only should be used. Once the supply has been isolated the extinguisher appropriate to the burning material can be used.

Figure 4.2.5 Clear unobstructed fire point (Courtesy Reed Medway Sacks Ltd)

4.2.6 Portable fire-fighting equipment

Portable fire extinguishers are intended to be used by individuals in the early stages of a fire. They are restricted in weight to be manageable and should be mounted so as to be easily accessible. Portable extinguishers should comply with BS EN 3[6], which stipulates that the body of all extinguishers will be red. However BS 7863[7] qualifies BS EN 3 and allows a coloured panel to be fixed to an extinguisher using the UK colour coding system to indicate the contents. This panel must be on or above the operating instruction label and must not cover more than 5% of the extinguisher's body area. Existing extinguishers which have been manufactured to BS 5423[8] and have their bodies painted in the indicative colours, do not need to be replaced until they reach the end of their useful life.

A new standard BS 7937:2000[9] specifies requirements for wet chemical extinguishers for use on deep fat fires. Portable extinguishers should be inspected at least once a year and certified as in good operational condition.

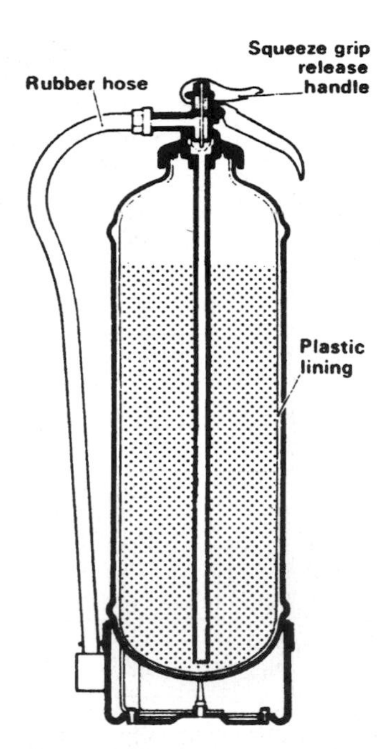

Figure 4.2.6 Section through a stored pressure type of water extinguisher (Manual of Firemanship. Courtesy Controller SO)

4.2.6.1 Quantities, siting and types of portable extinguishers

A standard BS 5306[10] contains details of the number and type of extinguishers required for a particular risk or premises. These requirements may be added to by the local fire prevention authority to meet particular risks found when they inspect a premises.

4.2.6.1.1 Siting of portable extinguishers

Extinguishers should where possible be sited:

1 on escape routes and adjacent to exit doorways (*Figure 4.2.5*);
2 in multi-storey buildings, at the same location on each floor level;
3 away from extremes of temperature;
4 wherever possible, in groups to form identified fire points (*Figure 4.2.5*).
5 specific extinguishers should be located as close as possible to the site of any special risks.

Extinguishers should be hung on wall brackets with the top of the extinguisher about 1 metre from floor level and should be clearly visible to anybody using an escape route.

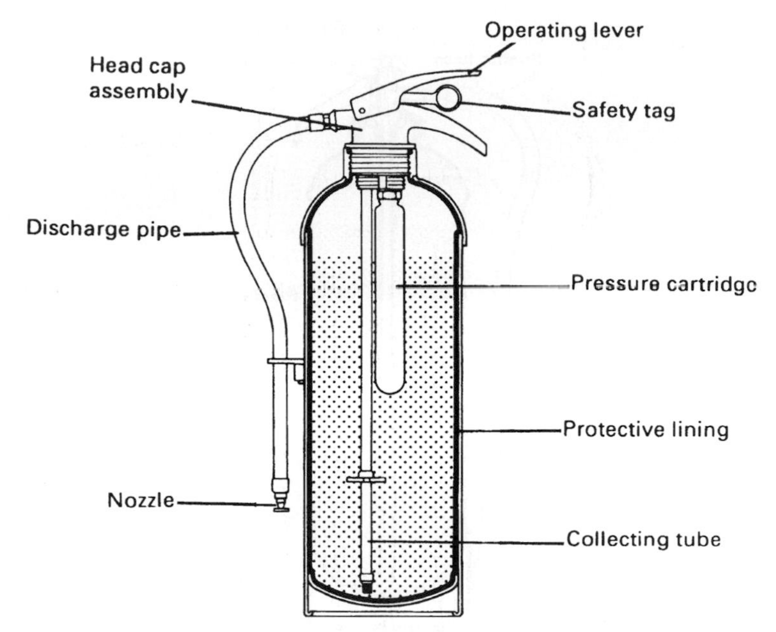

Figure 4.2.7 Section through a water extinguisher (Courtesy Thorn Security Ltd.)

4.2.6.1.2 Water extinguishers

Water extinguishers extinguish a fire by cooling the burning material which not only extinguishes the fire but also prevents re-ignition. They should not be used on any live electrical equipment or on flammable liquid fires. The water content of the extinguisher is propelled either by stored pressure in the extinguisher itself (*Figure 4.2.6*) or by gas released from a high pressure cartridge (*Figure 4.2.7*).

4.2.6.1.3 Foam extinguishers

Foam extinguishers, identified by a pale cream panel, contain either AFFF (aqueous film forming foam), FFFP (film forming fluoroprotein foam) or FPF (fluoroprotein foam). These types of extinguishers all form a foam blanket over the fire which excludes the oxygen but their contents react differently when subjected to heat to form different fire suppressant daughter products. Foam extinguishers are most suitable for fires involving flammable liquids such as petrol but are not suitable for use on fires involving live electrical equipment or deep fat fryers.

4.2.6.1.4 Wet chemical extinguishers

Wet chemical extinguishers have a canary yellow panel to identify them. They are for use on fires involving cooking oils and fats in deep fat fryers which do not ignite until heated to above their auto-ignition temperature

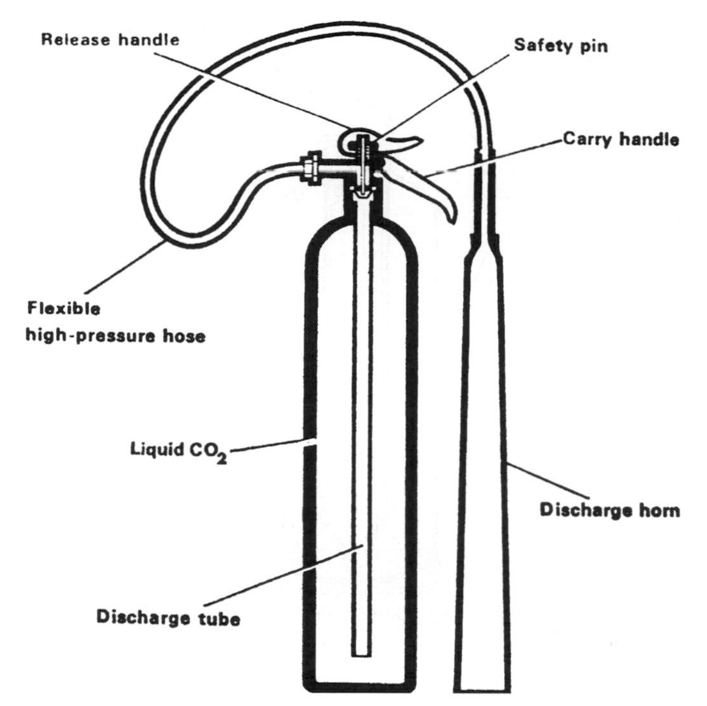

Figure 4.2.8 Section through a carbon dioxide extinguisher showing the piercing mechanism, control valve and discharge horn (Manual of Firemanship. Courtesy Controller SO)

which is between 285° and 385°C. Once ignited, the auto-ignition temperature of oils and fats is reduced by 30°C which means the whole content of the fryer must be reduced to below 255°C to ensure there is no re-ignition. Wet chemical extinguishers work by saponifying the liquid oils and fats to form a soapy scum on the surface. This soapy scum traps the flammable vapours, generates steam, extinguishes the flames and allows the oil to cool to below its lower auto-ignition temperature. Wet chemical extinguishers achieve saponification through the use of an alkaline liquid solution of up to 20% potassium acetate, potassium citrate and potassium carbonate which is applied to the fire in the form of a fine mist. The method of operation is similar to that of pressurised water extinguishers.

4.2.6.1.5 Carbon dioxide extinguishers

Carbon dioxide extinguishers are identified by a black panel and in operation blanket the fire with an inert gas. This extinguishes the fire by reducing the oxygen content of the air to below a level that will support combustion. They are most effective in enclosed spaces but can present an oxygen-depletion hazard to any operators in the area. They are particularly suitable for fires of electrical origin as the gas can penetrate

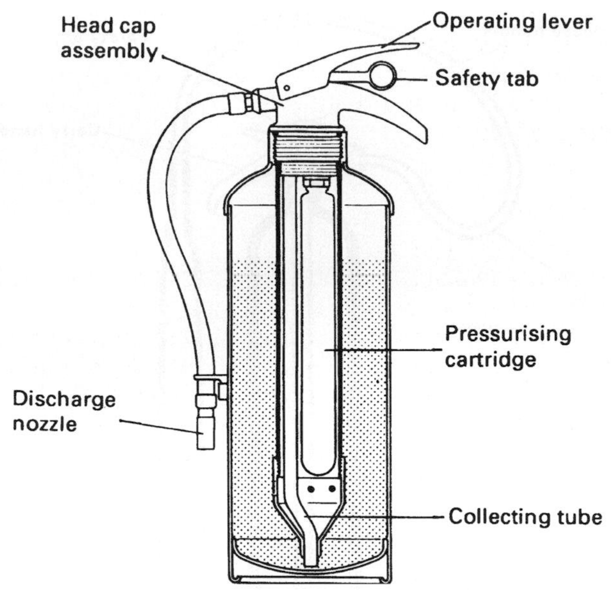

Figure 4.2.9 Section through a dry powder extinguisher (Courtesy Thorn Security Ltd)

into spaces where access is difficult and they can also be used while the equipment is still live. They provide little or no cooling so there is a risk of the fire re-igniting once the supply of carbon dioxide is exhausted. Carbon dioxide extinguishers should not be used on computer systems since it can react with component parts.

4.2.6.1.6 Dry powder extinguishers

Dry powder extinguishers (*Figure 4.2.9*)have an identifying panel of blue colour. They extinguish fires by spreading the heat over a large surface area effectively reducing the temperature and suppressing the flames. Some multi-purpose powders also form a skin across the top of the fire and smother it. Operators should be careful not to inhale the powder. Generally dry powder has little or no cooling effect on the bulk of the fire so there is a risk of re-ignition if the supply of powder runs out. These extinguishers can be used on live electrical equipment.

4.2.6.1.7 Halon extinguishers

Halon extinguishers are identified by emerald green panels and extinguish fire by chemically inhibiting combustion. They can produce hazardous daughter products when heated and the free gas contributes to the 'greenhouse' effect of global warming. The production of halon has been banned since 1 January 1994 and its sale must cease from 31 December 2002. The use of all existing halon is required to cease by 31 December 2003. Exemptions to the ban have been granted for military and aviation purposes and also for some marine use. All halon extinguishers in use should be withdrawn and returned to an approved manufacturer for recycling or disposal. Alternatives to halon have been developed for fixed installations but these are not yet available for use in portable extinguishers.

4.2.6.1.8 Fire blankets

Fire blankets extinguish fires by excluding the oxygen and smothering the flames. They are particularly useful for fires in containers such as chip pans but can also be used to extinguish burning clothing. By holding the blanket up in front of themselves, operators can protect themselves from the heat of the flames before placing the blanket over the fire. Once the fire has been extinguished the blanket should be left in place until the burning material has cooled completely, removing the blanket too quickly can allow the fire to re-ignite. Fire blankets are normally stored in red coloured cylindrical containers with an open end at the bottom allowing speedy removal of the blanket when needed. BS 7499[11] and EN 1869[12] lay down requirements for fire blankets for industrial and domestic use respectively.

4.2.6.1.9 Installed hose reel

Installed hose reels are wall mounted and comprise a length of hose wound on a drum. The feed pipe to the reel should be charged at all times. In some installations a valve is incorporated into the spindle of the reel drum that automatically opens when the reel is pulled out. The flow of water is controlled by a valve at the free end which allows adjustment from a water jet to a fine spray. The siting of the installations and the length of the hoses should be such that no part of the area to be protected is more than 6 m from any nozzle. Hose reels should be checked regularly for leaks and fully run out once a year so the condition of the full length of the hose can be checked.

4.2.6.2 Extinguisher ratings

To enable comparisons to be made of the extinguishing capabilities of different extinguishers, BS 5306 part 8[10] specifies conditions for tests by which designated ratings can be assessed. The tests cover the ability to extinguish fires of solid materials (Class A) and liquids (Class B).

Class A ratings indicate the length of the longest wooden crib, 0.5 m wide × 0.56 m high, that the extinguisher could extinguish. The rating figure given is the length in decimetres of the crib. Thus an extinguisher which extinguishes a fire in a crib 1.3 m long is given a rating of 13A. This is the usual rating achieved by 9 litre water extinguishers.

Class B ratings are based on the time taken to extinguish a fire involving varying quantities of flammable liquids. For example, an extinguisher that will extinguish a fire of 55 litres of flammable liquid is given the rating 55B. This is the usual rating achieved by a 3 kg dry powder extinguisher. Some extinguishers that can be used on both solid and liquid fires are given an A

Table 4.2.2 Typical portable extinguisher ratings

Extinguishant Type	Capacity litres	Weight kg	Discharge		Fire rating	
			Range (m)	Time (s)	A	B
Water	6	9–11	6	45	8	–
	9	12–15	6	70	13	–
Foam	6	10–11	4	27	8	144
	9	14–16	4	45	13	183
	kg					
Powder	2	3–4	4	10	13	55
	4	7.5–8.5	5	10	21	144
	6	9–14	5	12	34	183
	9	14–16	6	15	43	233
Carbon dioxide	2	4.5–8	2	14	–	34
	5	11–18	2	24	–	55
	7	17–23	3	26	–	55

and B rating which indicates that they have achieved a satisfactory extinguishing performance with both types of fires (*Table 4.2.2*).

4.2.6.3 Maintenance of portable fire extinguishers

The value of a fire extinguisher lies in the fact that it is immediately available and ready for use at all times. This can best be assured by regular maintenance which should include a brief monthly check with a more detailed examination by a competent person at least once a year. This latter examination should include a check on the contents and of the pressurising arrangements, whether by cartridge or pressurisation of the extinguisher itself. A written record, giving the date and the result of the examination, should be kept. It is usual for this to be in the form of a signed label securely attached on the side of the extinguisher which can be checked at any time.

The monthly or intermediate checks should ensure that the extinguishers:

1 are still in their correct locations and are properly mounted;
2 are not obstructed by other equipment or product;
3 still have the tamper-proof seals intact;
4 have no obvious defects such as leaks or damage to the extinguisher body.

Should any fault be found, the extinguisher should be removed for repair or servicing and replaced by an operational one.

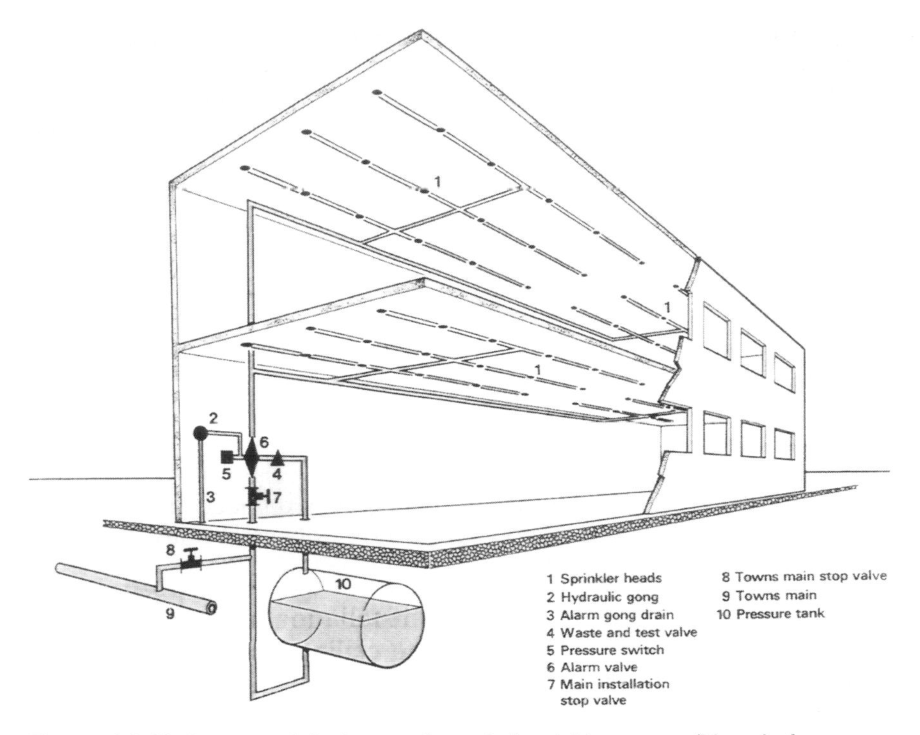

Figure 4.2.10 Diagram of the layout of a typical sprinkler system (Manual of Firemanship. Courtesy the Controller, Stationery Office)

4.2.7 Fixed fire-fighting equipment

Fixed fire-fighting equipment is permanently fixed in a building and is designed to work automatically in the event of a fire. Essentially it comprises a system of fixed pipes with discharge heads or points for the extinguishing medium at predetermined positions. The positioning of the discharge heads is determined by the type of building, its contents, the fire risk and the type of extinguishant and is normally to standards developed by fire insurers. Once installed and commissioned fixed fire-fighting equipment must be kept in a state of instant readiness. Sprinkler systems have the advantage that only the discharge heads in the immediate vicinity of the fire rupture, thus limiting the extent of water damage to areas in the immediate vicinity of the outbreak. In older wet systems that drew their water supply from ponds and rivers, the system was kept pressurised by air pressure above the water in a storage tank. This provided an immediate supply of water to a fractured head which continued until the main pumps cut in. The flow of water actuated a hydraulic alarm gong.

In modern sprinkler systems (shown diagrammatically in *Figure 4.2.10*) water is supplied at pressure from the town mains which charges a pressure tank. When a sprinkler head ruptures, the loss of water pressure in the supply pipe triggers the alarm valve. The alarm is actuated by the alarm valve either by diverting part of the water flow to a hydraulic gong or by actuating an electrical alarm. The pressure tank acts as a buffer to absorb any water hammer effects and to provide a temporary water supply if the mains flow is interrupted.

Gas systems contain an extinguishant gas at high pressure supplied from gas cylinders.

Fixed fire fighting systems have the advantages that they can:

1 detect the fire;
2 raise the alarm;
3 attack the fire; and
4 prevent the fire from spreading;

even when the building is not occupied. The most common type of fixed installation uses water as the extinguishant. However, in situations where water could present a hazard or do unacceptable damage an alternative extinguishant can be used. The different types of systems are described below.

4.2.7.1 Sprinkler systems

Water sprinkler systems were first developed in the mid-19th century and the original designs used a discharge head sealed by a fusible soldered strut. These have been superceded by glass bulb sprinkler heads which, by selecting particular bulbs, can be made to rupture at a range of temperatures.

The main type of sprinkler systems currently in use are:

wet – where all pipes are permanently filled with water;
dry – where the pipes above the main control valve is filled with compressed air that holds back a pressurised water supply. On the actuation of a sprinkler head the compressed air is released allowing water to enter the pipes and discharge through the broken head onto the fire;
alternate – are used in areas where the installation is liable to freeze. The system is normally wet but is temporarily drained during winter and converted to a dry system;
pre-action – systems are normally dry but are linked to an automatic fire detection system. On actuation of the detection system water enters the supply pipes and the system becomes a wet one, but water does not discharge until a sprinkler head breaks.

In the UK, sprinkler systems are installed in accordance with the Loss Prevention Council (LPC) Rules[13] which incorporate Part 2 of BS 5306[10]. Further requirements are contained in Part 1 of BS EN 12259[14]. During 2002 the Fire Protection Association (FPA) produced a new set of

Sprinkler Rules based on the contents of the new BS EN standard. The LPC Rules together with the Comité Européen des Assurances (CEA) Rules[15] are now being considered with a view to producing a European standard for sprinkler installations. This may become a requirement of the Construction Products Directive, *Interpretative Document for Safety in Case of Fire*[16].

In the USA the National Fire Protection Association (NFPA) has revised its sprinkler rules and produced the new NFPA 13 Sprinkler Code[17].

4.2.7.1.1 Sprinklers in the protection of buildings

Fire prevention arrangements in buildings include a requirement to 'compartmentise' the building with at least one hour fire separation between compartments. In buildings fitted with sprinklers an Approved Document B[18] allows some relaxations of this requirement. For example, office, shop, commercial, assembly, recreation, industrial, storage and other non-residential premises a 30 minute reduction in the fire separation is allowed. Other requirements of this document include the fitting of sprinklers in all buildings over 30 metres high and sprinkler protection in all new single storey buildings having a retail floor area of 2000 m^2 or more.

4.2.7.1.2 Sprinklers for life safety

Traditionally sprinkler systems were installed to protect buildings and contents rather than save life. However, in 1979 a fire occurred in a retail store in Manchester in which 10 people died. One of the recommendations from the investigation into the cause of the fire was that if sprinklers had been installed the life loss may have been reduced or eliminated. At that time, while sprinklers had been well proven in the protection of building structures, there was no evidence that their installation had any effect on life safety. Research, subsequently undertaken into the role of sprinklers in saving life, has resulted in official recognition of a fire engineering approach to fire prevention in buildings. This allows a sprinkler system to be part of the overall fire safety measures in buildings to which the public have access as well as in workplaces.

4.2.7.2 Drencher or deluge system

These systems are usually employed to protect high risk installations such as above ground storage tanks for highly flammable liquids or chemicals. A common arrangement, shown in *Figure 4.2.11*, is for the system to comprise a series of open drencher discharge heads arranged in a pattern to ensure that all parts of the installation are drenched. A second system of conventional dry sprinkler heads is interspersed with the drencher heads. The conventional system is pressurised by air which also retains the main drencher water valve in the closed position. In the event of a fire, a conventional sprinkler head will rupture releasing the air pressure and allowing the main drencher valve to open resulting in the installation being drenched.

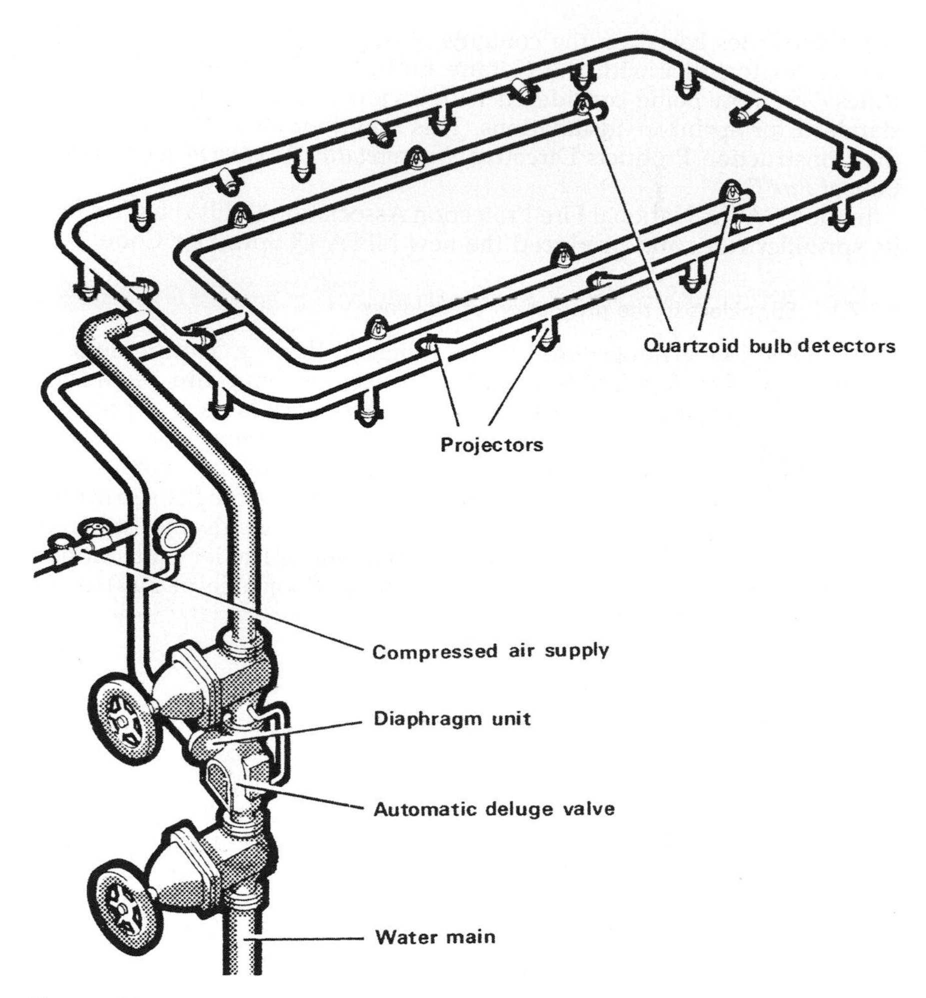

Figure 4.2.11 Diagram of a typical drencher (deluge) system (Manual of Firemanship. Courtesy the Controller, Stationery Office)

4.2.7.3 Water mist systems

A development to replace halon systems has been water mist systems in which a system of open discharge heads is used. The system is actuated by smoke detectors located within the area to be protected. Operation of one detector raises the alarm and it requires a second detector to be actuated before the water is released into the discharge pipework. Water for the system is stored in cylinders at high pressure and the discharge heads are designed to produce a fine water mist. Water mist systems use significantly less water than a sprinkler system and extinguish fires by:

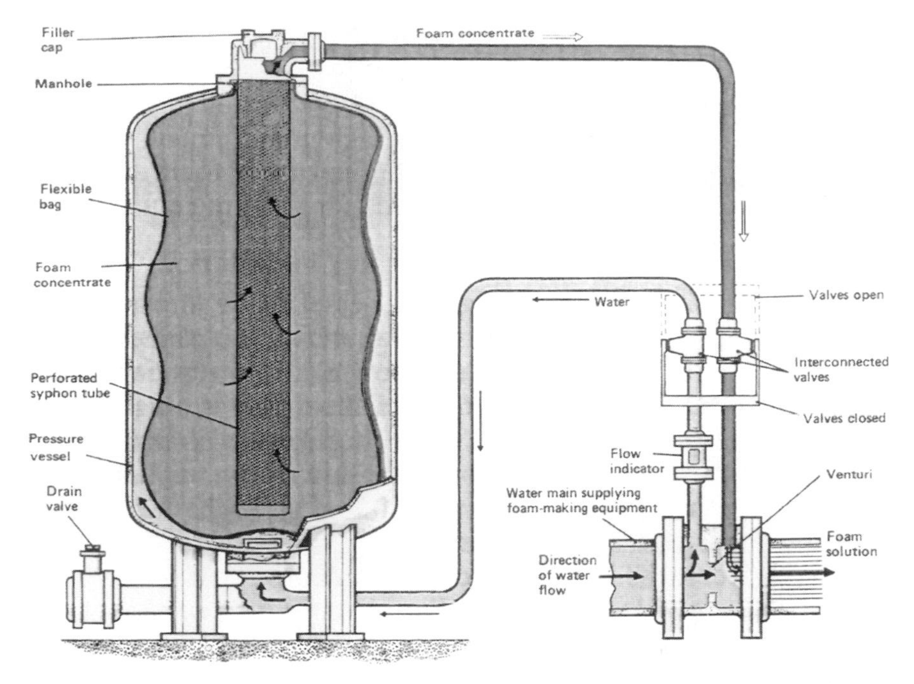

Figure 4.2.12 Diagram of a proportional-tank mechanical foam installation (Manual of Firemanship. Courtesy the Controller, Stationery Office)

1 cooling – fine droplets of water create large cooling surface area. Air entrained in the fire plume assists in the penetration of water droplets into the fire;
2 radiant heat suppression – high density of water mist and water vaporisation create a condition for high heat absorption and provide a screen to reduce radiant heat transfer;
3 oxygen level reduction – when water droplets vaporise, they expand about 1700 times and displace the air with its oxygen, thus inhibiting growth of the fire at the flame base.

4.2.7.4 Foam systems

There are several different types of foam systems which can either use a traditional sprinkler pipework system employing a proportional-tank mechanical foam installation, shown in *Figure 4.2.12*, in the supply line to the sprinkler heads or have their own alarm system and discharge pipework incorporating a pre-mix foam installation, shown in *Figure 4.2.13*.

Installations using high expansion foam are used to protect large areas such as aircraft hangars. In these installations a number of foam generators, each producing foam with expansion ratios of between 200 and 1200 to 1, can be connected together. For areas containing oil or

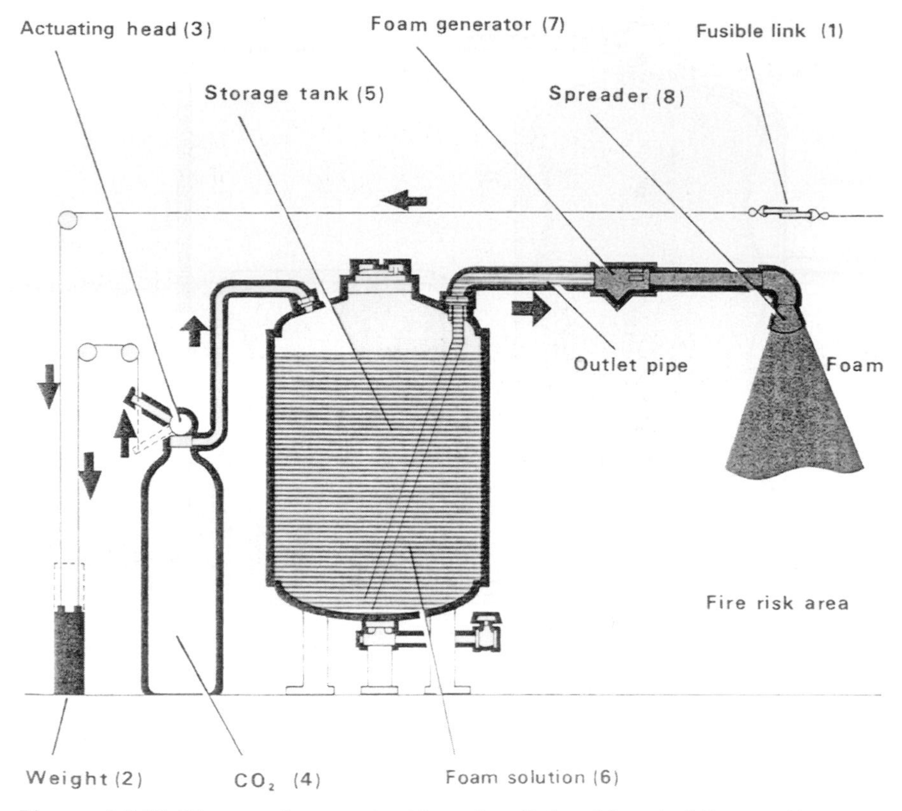

Figure 4.2.13 Diagram of a pre-mixed foam installation (Manual of Firemanship. Courtesy the Controller, Stationery Office)

flammable liquids, foam inlets are provided on the exterior of the building to which the fire brigade can connect their foam-making branch pipe. Fixed piping transports the foam to the area being protected.

4.2.7.5 Dry powder systems

A dry powder system comprises a container in which the powder is stored and which is connected to pipework leading to discharge nozzles in the protected area. On actuation of the system the powder is fluidised in the expellant gas (an inert gas such as nitrogen or carbon dioxide) and is conveyed to the nozzles. These systems are suitable for fires involving flammable liquids, electrical equipment or where water damage must be kept to a minimum. They are not suitable in situations where there is likely to be a heat sink such that re-ignition could occur. Standards which cover dry powder systems are contained in BS EN 12416–2:2001[19].

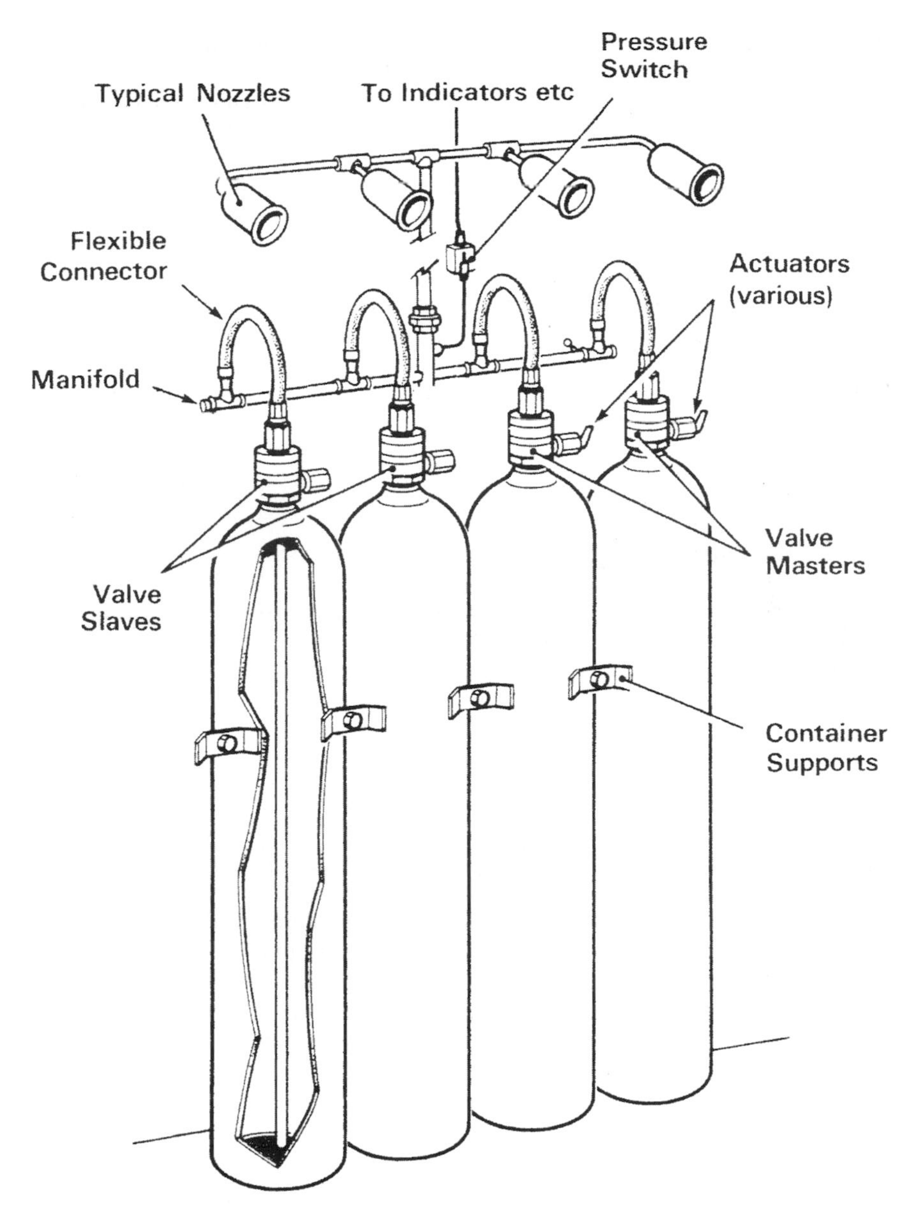

Figure 4.2.14 Diagram of battery of CO_2 cylinders supplying a small gas installation (Manual of Firemanship. Courtesy the Controller, Stationery Office)

4.2.7.6 Gas systems

Gas systems are used to protect sensitive equipment such as computer suites and electrical equipment and also areas where water damage is not acceptable, such as archive stores. The gases used depend on the application but fall into three main categories:

1 inert gases such as argon, carbon dioxide and nitrogen;
2 chemically produced gases;
3 halon gases which are now strictly controlled and the only permitted use is in specialist military and aerospace applications.

The gas, at high pressure, is stored in cylinders (*Figure 4.2.14*) which are usually located outside the area being protected. Actuation of the system is by automatic fire detectors. Because the gases present a health hazard when installed in a workplace area, discharge is initiated in two stages. A pre-discharge alarm is sounded when the first detector operates allowing any staff present to evacuate the area. Gas discharge occurs only when the second detector is actuated.

BS ISO 14520[20] series of standards covers gas fire-extinguishing systems with each part dealing with a different extinguishant. General requirements can be found in part 1 of this series.

4.2.8 Fire safety signs

The posting of fire safety signs has been a legislative requirement for many years. Their primary purpose is to ensure that the occupants and users of buildings are aware, not only, of escape routes and exits but also of any hazards and dangers that may be present.

4.2.8.1 Current legislation

The EU identified the need to have a unified system of signs and symbols that could be recognised throughout the Union and introduced the Safety Signs Directive[21] that has been implemented in the UK in Regulations[22] which require the provision and maintenance of appropriate signs to:

1 warn workers of any risk to their health and safety;
2 indicate safe exit routes;
3 identify the location of fire-fighting equipment.

The signs specified in these Regulations also satisfy the requirements for fire safety signs required by other legislation (such as the Fire Precautions Act 1971, The Fire Precautions (Workplace) Regulations 1997 (as amended)).

The signs required by these Regulations are pictograms of specified shapes, colours and patterns that give the appropriate information. The pictograms may be supported by a written caption but written notices on their own are not acceptable as fire safety signs. The object of using pictograms is to make them understandable to all regardless of knowledge of the local language. Thus, the words FIRE EXIT on their own are not permitted but the words FIRE EXIT used in conjunction with a pictogram is allowed. To prevent confusion with other safety signs, all safety signs relating to fire alarm and fire-fighting equipment are square or rectangular in shape with a white pictogram on a red background. The signs used on fire escape routes are the same as emergency exit signs, i.e. square or rectangular with a white pictogram on a green background. Typical examples are shown on the inside covers of this book.

The appropriate fire safety signs should be used to indicate the location of fire-fighting equipment and fire alarm call points which should be on fire escape routes. If equipment and call points are located together they can be mounted on a panel to form a Fire Point. In crowded and irregular shaped work areas high level signs can make the location of fire exit routes and fire-fighting equipment easier to identify.

4.2.9 Means of escape in case of fire

The principle aim in establishing a means of escape in case of fire is to provide a safe escape route that everyone who may be on the premises can follow by their unaided efforts regardless of where the fire occurs. When designing escape routes consideration must be given to the abilities and needs not only of the fit but also of the old, the young, the infirm and the disabled.

The factors to be taken into consideration when assessing means of escape include:

1 an assessment of the fire risk;
2 the construction and surface finish of walls, floors and ceilings in workplaces and on escape routes;
3 the likely maximum number of occupants in the building;
4 the ability of the occupants to respond to the fire alarm and make their way to safety;
5 the distance of travel to a place of safety;
6 the degree of fire protection provided by escape corridors and staircases;
7 the provision of information and guidance such as exit, directional signs, emergency lighting, etc.

Certain types of equipment that are not officially recognised as means of escape include:

1 lifts (except for a specially designed lift for evacuation of disabled persons);
2 portable ladders and throw-out ladders;
3 manipulative apparatus and appliances such as fold-down ladders, chutes, self-rescue and lowering devices.

4.2.9.1 Assessment of the fire risk

When assessing the risk to the occupants of a premises from fire and the safety of the means of escape, account must be taken of the type and materials of construction of the building, the type and condition of the contents and particularly its flammability, the likely rapidity of spread of smoke and flames and the use to which the building is being put. Fire risk assessments are dealt with in detail in section 4.2.13.

4.2.9.2 Construction and surface finish of walls, floors and ceilings

The construction and surface finish of escape routes are important factors in the degree of protection provided in the event of a fire. In general, the fire resistance of escape routes should be at least 30 minutes but in high risk buildings a higher standard of protection may be necessary. Regulations[23] require that circulation spaces and escape routes should be Class O (the highest standard), rooms (except small rooms) Class 1 and small rooms and small areas other than rooms Class 3 (see also *Table 4.2.4*). These latter two classifications are determined by reference to criteria given in a standard[24].

These documents also cover the fire resistance of walls, floors and doors. Additional information is given in a guide[25].

4.2.9.3 Occupancy of a building

The rated occupancy of a building is the maximum number of persons, including staff, visitors or customers, that can safely be accommodated in a building at any one time. Occupancy rates can be calculated using floor space factors detailed in the documents referred to above and range from $0.3\,m^2$ per person for standing spectator areas to $30\,m^2$ per person for warehouse and storage areas.

4.2.9.4 Ability of occupants to respond

Account must be taken of the types of occupants likely to be in an area or premises and of the proportion who may experience difficulties in responding to the sounding of an alarm. People in wheelchairs or who walk with sticks or crutches have an obvious disability and may well need physical assistance to evacuate the building, while people with poor sight or those suffering from hearing loss have a less obvious disability but may still need assistance in an emergency. Elderly people or young children will also need special consideration. In hospitals and other care premises the patients who are bedridden will be totally reliant on assistance to escape.

Guidance on the means of escape provisions for disabled people is given in Part M of the Building Regulations[23] and also in a Code of Practice[26]. The concept of refuges and the use of evacuations lifts are covered but the most important aspect is the need for the effective management of any evacuation. All premises that are likely to contain people who are not able to respond immediately or quickly to a fire evacuation should have an emergency plan that makes special provisions

Table 4.2.3 Travel distances for various types of premises

Type of premises or part of premises	*Maximum travel distance*[1]	
	In one direction (m)	*in more than one direction (m)*
Institutions[2]	9	18
Residential		
a) to exit door of bedroom	9	18
b) from bedroom door to exit from floor	9	35
c) elsewhere in accommodation	18	35
Offices	18	45
Shops and commercial premises[3]	18	45
Assembly and recreational		
a) buildings used primarily by the disabled	9	18
b) schools	18	45
c) places with seating in rows	15	32
d) all other assembly or recreational premises	18	45
Industrial		
a) normal risk	25	45
b) high risk	12	25
Storage and other non-residential	25	45
Premises with a special fire hazard	9	18
Plant rooms, in the building and on the roof top		
a) within the plant room	9	35
b) escape route not in open air (overall distance)	18	45
c) escape route in open air (overall distance)	60	100

Notes:

[1] Where the layout of furniture, equipment and fixtures is not known, the travel distance should be assumed to be 1.5 times the direct distance of the remotest position a person is likely to be from the fire exit.

[2] In hospitals and health care premises the recommendations of the Department of Health's 'Firecode' should be followed.

[3] Where a covered shopping complex has only one exit the more restrictive conditions of BS 5588: Part 10[26] apply.

to ensure that these people are catered for. The special provision could include reduced distances of travel or upgraded means of escape.

4.2.9.5 The distance of travel

Guidance on acceptable distances of travel is contained in the various documents dealing with fire safety where the recommendations are based on the risk categorisation of the building (see section 4.2.13.2) and data on the speed at which escaping persons can travel. These recommendations are summarised in *Table 4.2.3*. In designing escape routes for new buildings, because the final layout of furniture and other obstructions is not known, the travel distance is calculated at 1.5 times the direct distance from the remotest location of a person to the nearest fire exit.

4.2.9.6 Protection of escape routes

Escape routes from work areas must be constructed to provide protection from the fire – both smoke and flames – and arranged so that smoke seepage is kept to a minimum. All doors on escape routes should open in the direction of travel and be self-closing. They should be provided with a viewing window of Georgean wired or similar protective glass. Smoke doors should be fitted with seals to reduce the seepage of smoke.

4.2.9.7 Fire escape signage

All fire exit doors should be identified as such by a standard sign and additional signs, that can be seen from all parts of the occupied area, should be posted to identify emergency escape routes and doors. Directional signs should be used over the full length of the escape route. The final door to the place of safety should be identified on both sides and the area beyond it kept clear of all obstructions to allow a free and rapid egress.

4.2.10 Fire engineering

Fire engineering is the logical application of proven principles based on the current state of knowledge of materials, structures, behaviour and the assessment of fire risks to the prevention of loss of life and property due to fire.

4.2.10.1 Fire engineering in building design

In line with other legislation, that concerning fires has moved from the prescriptive to the proscriptive using a risk based assessment to determine the precautionary action needed. This change reflects advances in technology and building methods and allows the provision of a total fire safety package to be based on a fire engineered solution. This is particularly suitable for large and complex buildings since it allows flexibility in the application of fire safety principles while maintaining the required level of fire safety. For an engineered solution to be acceptable the following aspects need to be assessed:

1 the risk of a fire occurring in the first place;
2 the likely severity of a fire based on the fire load of the building and its contents;
3 the structural resistance of the building to the spread of fire and smoke;
4 the potential risk to people in and around the building if a fire occurred.

To achieve a satisfactory standard in respect of the above aspects, the following factors need to be considered in the building design:

(a) adequate arrangements for preventing fires;
(b) quick and effective warning of the outbreak of a fire;
(c) use of active fire extinguishing systems (e.g. sprinklers);
(d) effective control over smoke movement;
(e) facilities to assist the brigade in fighting a fire (e.g. fire-fighting staircases, rising mains);
(f) training of staff in fire safety procedures and fire evacuation;
(g) continuing control of the building under other fire safety legislation.

In order to achieve and prove a satisfactory engineered solution, features such as zone modelling, virtual reality simulations, computational fluid dynamics, risk analysis, fire threat factors and statistical analysis need to be employed. These are specialist techniques and guidance on them is given in two standards[27,28].

A new British Standard[29] is being prepared to reflect recent changes in fire safety practice and procedures, in particular fire risk assessment. This standard has been produced to assist architects and designers who do not want to adhere to the prescriptive guidance found in the Approved Documents to the Building Regulations, Scottish Technical Standards and most Home Office Guidance, nor get involved with the complexities of a fire engineered solution.

4.2.10.2 Fire loading

The fire loading of a building is an empirical value given to indicate the potential gross heat output of the contents if they were completely burnt. The fire loading is becoming an important factor in the calculations for the fire engineered solution needed to ensure protection of a building from the effects of a fire. There is an inverse relationship between the fire loading density (gross heat output per unit area of floor) and structural failure in a building, i.e. the higher the fire loading density the quicker will structural failure occur.

The concept of fire loading has its main application in the design of those buildings that come outside the scope of the Building Regulations and, particularly, in those buildings and premises subject to the COMAH Regulations.

4.2.11 Fire protection measures

Fire protection measures in buildings are those features in the design and use that play a role in controlling the spread of fire and allowing the occupants to escape. They can be considered to operate in two separate ways – active and passive.

4.2.11.1 Active fire protection measures

Active measures are those that react to the presence of fire or the products of combustion and initiate actions aimed at extinguishing the fire and ensuring the escape of the building's occupants. These measures may have been installed as a result of the fire risk assessment carried out under the 1997 Regulations or they may have been a mandatory requirement of the fire authority as a condition for the issue of a fire certificate. Typical active fire protection measures are considered below.

4.2.11.1.1 Fire detection and alarm systems

Automatic fire detection and alarm systems can range from a simple single smoke detector in a small office or home to a sophisticated zoned system for a large office or works complex. Such sophisticated systems, based on a central control board, can identify the location of a fire, give differential warnings to adjacent areas and be linked automatically to the local fire brigade. They have the advantage that they give protection even when the premises are unoccupied.

Between these two extremes are the manually operated system, whether by break glass electric alarm points or mechanically operated, which rely on someone detecting the outbreak and taking the appropriate action to raise the alarm. Whichever system is used it is essential that the alarm can be heard in all areas to which the occupants have access, including toilets, remote stores, etc.

4.2.11.1.2 Emergency lighting

In buildings occupied during the hours of darkness or where the fire escape route passes through areas that do not have natural lighting, emergency lighting should be provided. It must be additional to the normal artificial lighting and come into operation in the event of the failure of the mains supply. It is usually operated from batteries either within the light fitting or at a central location. The level of illumination provided must be sufficient to allow the persons escaping to see their way safely along the whole of the escape routes to the exit doors and a place of safety. It should illuminate fire safety signs and may also provide lighting at fire extinguishers and fire alarm call points.

4.2.11.1.3 Smoke extraction systems and fire dampers

To prevent the build-up of smoke in larger buildings, separate smoke extraction systems may be installed. They work independently of the normal ventilation system and are connected direct with the outside of the building. Their most common application is in below-ground areas. Operation of the smoke extraction system is triggered by the fire alarm which would also shut down the normal ventilating system to reduce the chance of smoke being distributed around the building. In normal ventilating as well as smoke extraction systems the air is carried in ducts and where these pass through fire break walls between fire compartments they should be fitted with fire dampers which can be actuated either from a central control or by a fusible link arrangement.

4.2.11.1.4 Automatic fire extinguishing systems

Automatic fire extinguishing systems such as sprinklers, drenchers or gas systems operate when the heat of the fire in the region of a detector head reaches predetermined levels and cause the seal in the detector head to fracture releasing the extinguishant. These systems can incorporate a facility to raise the alarm.

4.2.11.1.5 Active fire protection and smoke stop doors

Pedestrian and vehicular movement between fire compartments should be through fire resisting doors which, ideally, should be kept closed except when allowing passage of people or vehicles. If the volume of traffic is high the fire doors can be held open during normal working hours provided they are arranged to close automatically in the event of a fire, either by magnetic catches linked to the alarm system such that the doors are released when the alarm is actuated or by fusible links. Fire doors form part of the compartmentation of a building.

4.2.11.1.6 Staff training and emergency planning

The aim of a fire alarm system is to give warning of a fire and allow the building's occupants to escape. To obtain the greatest benefit from this warning, employees should be trained and practised in evacuation procedures including assembly and roll call. The training of staff is doubly important in buildings to which the public have access. Particular emphasis in the training needs to be put on directing members of the public to the nearest escape route and ensuring the various areas have been completely evacuated. An emergency plan should be drawn up detailing actions and procedures in the event of a fire and the need to evacuate the building. Fire drills should be carried out regularly to ensure the evacuation procedures are understood and remembered.

4.2.11.1.7 Housekeeping

Piles of rubbish and debris in the workplace are one of the most common causes of fires and high standards of housekeeping are an important feature of fire prevention. Horizontal surfaces in overhead areas, such as structural beams and trusses and attic spaces, allow large amounts of dust to collect. Dust shaken down from roof trusses is a major factor in the severity of dust explosions in factories, while flames running across the surface of dust can result in a very rapid spread of fire to unexpected areas. This latter feature was one of the causes of the spread of fire in St George's Chapel at Windsor Castle in 1992.

4.2.11.2 Passive fire protection measures

Passive measures relate to the ability of a building to withstand the effects of fire and also to prevent the spread of fire. These measures are usually part of the structure of the building and can result from a requirement of the Building Regulations or they may be a mandatory requirement of the fire authority before they will issue a fire certificate.

4.2.11.2.1 Fire compartments

The spread of fires can be effectively prevented if the building is divided into discrete isolatable areas or fire compartments. This effectively divides the building into 'boxes' of fire resisting construction with the aim of preventing the spread of fire into other compartments and containing the heat and smoke within the compartment of origin. The size and fire resistance of each compartment should comply with the recommendations of the Building Regulations. These take account of the type and quantities of materials and their flammability characteristics and of any fixed fire fighting measures such as sprinklers which are installed. Some openings between compartments, such as doors, stairways etc., are necessary for the movement of people and vehicles but each opening must be protected by a door of fire resisting construction that closes automatically in the event of a fire to maintain the integrity of the compartment.

4.2.11.2.2 Passive fire protection and smoke stop doors

Fire and smoke stop doors in corridors are installed to prevent the spread of smoke and fire throughout a building. They should be fitted with smoke seals or intumescent strips at their edges to fill the small gaps that are necessary to allow the doors to open properly. Fire and smoke doors should be self-closing. If fire doors are held open to facilitate the flow of traffic, the retaining device must be such that it automatically releases the door in the event of a fire. Fire and smoke stop doors must not be wedged open since this will annul their effectiveness and greatly increases the risk of fire spreading throughout the building. It is important that the automatic closing equipment on fire doors is maintained in good working condition.

Table 4.2.4 The surface spread of flame for wall and ceiling materials and linings

Class	*Protection area*	*Typical materials*
O	Circulating spaces and escape routes	Brickwork, blockwork, concrete ceramic tiles, plaster finishes, woodwool slabs, thin vinyl and paper coverings
1	Larger rooms and places of assembly but not escape routes	Timber, hardboard, blockboard, chipboard, flock wallpaper, flame retardant thermosetting plastics
3	Small rooms (i.e. $< 30\,m^2$)	Timber, hardboard, blockboard, chipboard, heavy flock wallpapers, thermosetting plastics, expanded polystyrene

4.2.11.2.3 Wall linings

Fire can spread very rapidly along the linings of walls and it is important that the materials of the fire escape route wall and ceiling linings prevent this. Escape routes must be constructed of materials that have the lowest rates of surface spread of flame. The permitted materials for lining walls and ceilings in different work and access areas are determined by the requirements of the Building Regulations, examples are given in *Table 4.2.4*.

4.2.11.2.4 Building structural stability

The stability of a building is determined by the architect and the structural engineer at the design stage when a number of features may be incorporated into the structure to ensure it is not subjected to high temperatures in the event of a fire. These features can include lining structural parts with a suitable fire resistant covering. Where used, it is essential that the lining is kept intact and any damage repaired as soon as possible. While steel is an immensely strong material, it rapidly loses its strength when subjected to high temperatures. For this reason the steel structure of a building should be encased in a suitable insulating material.

The incident on 11 September 2001 which led to the collapse of the twin towers of the World Trade Center in New York is an example of structural steel losing its strength when subjected to high temperatures, on this occasion caused by the ignition of aviation fuel.

4.2.12 Legal requirements

For many years, premises of all types have been subject to legislative requirements to have adequate fire precautions. These requirements were often enforced through the Fire Authority's powers to issue fire certificates. However, recent legislation has shifted the emphasis of the responsibility for compliance from the Fire Authority to the employer/ occupier. The legislation recognises that the fire precautions should be risk based and places a duty on the employer to ensure that the precautions he implements match the fire risk faced as determined from a fire risk assessment. However, ultimately the Fire Authority still have a responsibility for checking that the precautions taken are adequate.

4.2.12.1 The Fire Services Act 1947

This Act[30] is concerned with the setting up of suitable fire-fighting organisations in local authorities' areas. It places a number of obligations on fire authorities including the requirements to:

1 establish a fire brigade and equip it;
2 train staff in fire-fighting and rescue;
3 make arrangements for dealing with emergency calls and for summoning members of the brigade;
4 obtain information about buildings in their area;
5 take effective steps to mitigate damage resulting from fire-fighting;
6 make arrangements for giving advice on fire prevention, restricting the spread of fire and means of escape in case of fire;
7 ensure adequate supplies of water will be available for use in case of fire.

The Act allows fire brigades to:

8 use any convenient and suitable supply of water (but they may be liable to pay compensation for any damage done);
9 pay water undertakings to upgrade water supplies and to provide, fit and mark fire hydrants.

The Act also gives the Fire Authority considerable powers to:

10 enter or, if necessary, break into any premises or place in which a fire has or is reasonably believed to have broken out, or any premises or place it is necessary to enter for the purpose of extinguishing a fire or of protecting the premises from the effects of a fire;
11 to prosecute any person who obstructs or interferes with fire fighting activities;
12 the senior fire brigade officer who has sole charge of fire fighting operations at the scene of a fire can close streets or stop traffic if no police are present;
13 prosecute anybody who gives a false alarm of a fire where the penalties can be a fine or imprisonment for up to three months.

4.2.12.2 The Fire Precautions Act 1971

The specific requirements contained in the Fire Precautions Act 1971[31] (FPA) have been considerably modified and extended by the Fire Safety and Safety of Places of Sport Act 1987[32] and it is necessary to refer to the two Acts when considering fire safety requirements.

No specific requirements in respect of fire, fire prevention, or fire precautions are contained in the Health and Safety at Work etc. Act although under s. 78, FPA was extended to include places of work.

4.2.12.2.1 Fire certificates

Section 1 of the FPA authorises the designation of those premises that are required to have a Fire Certificate. Only two Designating Orders have been made in respect of:

1 hotels and boarding houses;
2 factories, offices, shops and railway premises.

In the case of hotels and boarding houses, a fire certificate is required if sleeping accommodation for more than six people, being staff or guests, is provided above the first floor or below the ground floor. For factories, offices, shops and railway premises a fire certificate is required if more than 20 persons are at work in the building at any one time or more than 10 elsewhere than on the ground floor. A fire certificate is also required for factories storing or using highly flammable or explosive materials regardless of how many people are working in the premises.

Fire Authorities have the power to exempt premises from the requirement to have a fire certificate if they consider the fire risk to be low and there are adequate fire safety arrangements. Before issuing a certificate the fire authority may inspect the building, ask for more information including plans or require remedial work to be carried out.

An application for a fire certificate must be made to the fire authority on form FP1(Rev). For single occupancy buildings, the occupier is responsible for making the application while in multi-occupancy buildings it is the owner's responsibility. The responsibility for ensuring compliance with the conditions of a Fire Certificate rests with the person making the application.

Once the application for a fire certificate has been made and prior to an inspection or the issue of a fire certificate the occupier must:

1 ensure that the existing means of escape are kept clear and useable at all times;
2 maintain all existing fire fighting equipment in good working order;
3 train all staff in the actions to be taken in the event of a fire.

When the fire authority are satisfied that the fire safety arrangements are of an adequate standard in a building, they will issue a fire certificate which will detail:

(a) the use or uses to which the premises may be put;
(b) the means of escape in case of fire, usually indicated on a plan of the building;
(c) the measures for ensuring that the means of escape can be safely and effectively used at all times, covering fire doors, emergency lighting and fire exit signs;
(d) the means for fighting fires for use by persons in the building (fire extinguishers and hose reels);
(e) the means for giving warning in case of fire;
(f) in the case of a factory, particulars of the quantities of any highly flammable or explosive materials that may be stored or used in or under the premises.

The fire certificate may also impose requirements to:

(i) maintain the means of escape and keep the escape route free from obstruction;
(ii) maintain in effective working order the fire fighting equipment and the fire alarm system;
(iii) train staff in what to do in the event of a fire;
(iv) limit the number of people who may be in the premises or a part of the premises at any one time;
(v) keep records of the maintenance of the means of escape, the fire-fighting equipment and the training of staff;
(vi) take any other precautions considered necessary.

Once a fire certificate has been issued it must be kept on the premises to which it relates. For buildings which are occupied by more than one company then the owner of the building should have a copy of the fire certificate for the whole building and each occupier should have a copy of the certificate relating to their part of the premises.

Any proposed alterations to the building that could affect any item covered by the fire certificate must be notified to the fire authority and their approval obtained before the work commences.

4.2.12.2.2 Appeals and offences

If the owner or occupier of a building considers any requirement imposed by the fire authority is unreasonable including the time given to carry out any work, they can appeal to a Magistrate's or Sheriff's court within 21 days of being made aware of that requirement. If an occupier or owner contravenes any requirement of a fire certificate or puts a designated premises to use without a fire certificate, he will be guilty of an offence.

4.2.12.2.3 Powers of the fire authority

The powers of the Fire Authority were considerably extended by the Fire Safety and Safety in Places of Sport Act 1987 to allow them to issue statutory notices. For premises that are not subject to a Fire Certificate, if the occupier has not provided reasonable means of escape and fire-fighting equipment, the Fire Authority can issue an Improvement Notice requiring the matters to be put right. In the case of designated premises, if the fire risk gives rise to immediate danger the Fire Authority can issue a Prohibition Notice. These Notices have the same status as those issued by other enforcing agencies but appeal against them is to a Magistrate's or Sheriff's court.

The powers of the Fire Authority are exercised by Fire Prevention Officers or Inspecting Officers who are authorised to:

1 enter and inspect at any reasonable time, any premises to which the Act applies or appears to apply;
2 make such enquiries as is necessary to ensure compliance with the Act;
3 require the production of and inspect the Fire Certificate;
4 require the owner, occupier or other responsible person to render any assistance needed in the carrying out of an investigation or enquiries.

The Fire Authority may make a charge for the issue or amendment of a Fire Certificate. The charge may relate only to the amount of work involved in the preparation of the certificate and not any additional costs incurred if it is decided to carry out an inspection of the premises.

4.2.12.3 The Fire Certificate (Special Premises) Regulations 1996

These Regulations[33] apply to industrial premises where there is a high risk from a fire involving the materials being processed, such as nuclear sites, explosives factories, etc. In such cases, the Regulations transfer the responsibility for issuing fire certificates from the Fire Authority to the Health and Safety Executive.

4.2.12.4 The Fire Precautions (Workplace) Regulations 1997

The Regulations[34] were made to bring into UK law fire precaution requirements from the EC Framework Directive[35] and the Workplace Directive[36]. In doing so these Regulations were extended to include vehicles, offshore installations, tents or moveable structures as workplace premises, but do not include those premises already covered by or that have applied for a Fire Certificate.

They place obligations on the employer or persons who have control of premises to meet the requirements of the regulations in respect of fire-fighting and detection equipment and alarms, training of employees in fire escape procedure and to make contact with the local emergency services. Fire escape routes and exits are to be kept clear at all times and provided with adequate signage. Fire fighting equipment and facilities provided must be properly maintained and kept in effective working order. The Regulations also amend MHSW Regulations to require the carrying out of fire risk assessments. Enforcement of these Regulations is by the Fire Authority.

4.2.12.5 Environment and Safety Information Act 1988

This Act[37] requires that Prohibition Notices issued by Fire Authorities under FPA or by the Health and Safety Executive under HSW shall be kept in a register and be available for inspection by the public on request.

4.2.12.6 Building Regulations

The Regulations[23] are concerned with the structure and construction of buildings and are enforced by local authorities who have responsibilities to ensure that the fire safety arrangements in new and altered buildings conform to the appropriate standard. For buildings which will be put to a designated use under the FPA, the local authority are required to consult the Fire Authority before approving plans and are required to issue a completion certificate when they are satisfied that the finished work complies with the Regulations.

Where a building has been built or altered in accordance with the Buildings Regulations a *Statutory Bar* applies to the Fire Authority which prevents them from requiring additional structural or other work relating to means of escape prior to issuing a fire certificate.

The equivalent building controls in Scotland are the Building Standards (Scotland) Regulations 1990[38] which apply to building construction, demolition and change of use. They are administered by the Regional or District Council who grant a warrant if they are satisfied that the building operations will conform with the Building Operations (Scotland) Regulations 1975[39] and that the completed building will comply with the Regulations. They require that the building operations do not cause a hazard to passers-by. The warrant is issued on completion of the work.

4.2.13 Fire risk assessment

The process of carrying out a fire risk assessment is similar to that of a general risk assessment except that the identification of hazards is restricted to fire matters. It is the findings of a fire risk assessment that will determine the type and number of fire extinguishers to be provided. Under the Fire Precautions (Workplace) Regulations 1997 the findings of a fire risk assessment must be recorded. The fire risk assessment should cover a defined area of the workplace. In small premises the defined area could be the whole premises whereas for larger places of work the premises should be divided into discrete sub-areas and each made the subject of a separate fire risk assessment.

A fire risk assessment involves a number of discrete stages that are listed below and commence with the identification of fire hazards. Typical fire hazards can include the flammable nature of the material being processed, the condition of the machinery, the state of housekeeping, the presence of rubbish, oil leaks, temporary or faulty wiring, evidence of smoking, hot spots and hot surfaces, building structures or equipment layout that may impede escape, likely rapidity of spread of smoke and flames, etc. Where found, each of these should be recorded with notes of the action proposed or taken to eliminate the hazard. If the fire hazard cannot be eliminated, action should be taken to reduce its effects to a minimum. Thereafter the risk from any remaining hazard should be assessed. The fire risk is the likelihood that a fire will occur and the consequences of that fire for staff in the building.

A simple strategy involving a predetermined sequence of steps will assist in ensuring that a fire risk assessment is effective. Typical steps that can be followed are:

1 identify the fire hazards;
2 remove or eliminate the hazards where possible;
3 for each of the remaining hazards, identify the people who would be at risk if a fire started;
4 give each remaining hazard a risk category (low, medium or high);
5 decide on the additional fire safety measures necessary to protect those at risk;
6 implement the additional measures and check that they are effective;
7 monitor the effectiveness of the control measures and review the risk assessment at regular intervals.

The records made of each fire risk assessment should be used as the basis for monitoring the effectiveness of the measures taken.

4.2.13.1 Fire risk categories

For ease of application it is convenient to categorise the fire risk in each defined area as either high, medium or low. The criteria used for determining the category of each area should be consistent and *Table 4.2.5* lists typical factors that can be used in arriving at a category rating.

4.2.13.2 Emergency plan

To support the arrangements made for dealing with a fire, a plan should be prepared for dealing with unexpected emergencies. This should outline the action to be taken to cope with an emergency should one arise and include the nomination and training of individuals to undertake key roles. When developing an emergency plan the worst case scenario should be considered. Typical features to be included in the plan should be:

1 action in the event of an emergency – means of escape, assembly point, importance of a roll call to account for everybody on the premises;

Table 4.2.5 Summary of factors in fire risk categorisation

Features of building	*Factors in the categories of risk*		
	High	*Medium*	*Low*
Use of premises	Sleeping and accommodation	Office, large shopping centre, printing works	Engineering works, paper mill
Flammability of materials in use	Easily ignitable, flammable liquids & gases, solvents, foam plastic, wood shavings	Takes time to ignite – paper, furniture, plastics	Non-flammable materials
Rapidity of spread of smoke and flames	Large roof voids, lack of compartmentation	Some compartmentation	Large volume, high roof carries fire & smoke upwards
Construction features of building affecting escape	Large complex of separate work or occupied areas, restricted egress, multi-storey, below ground, low ceilings	Open office with high density occupancy, escape routes internal to building	Open work areas with high ceilings, short escape routes to safe place, steel & concrete building
Areas of special risk	Flammable stores, foam plastic store, electrical substation, below ground rooms	Storage of loose papers, mixed storage warehouses	None
Number of occupants	Low number if sleeping or working in upper storeys or below ground level	Medium to high number if escape routes are short	Low density occupancy with large floor area per person
Occupants with special difficulties	Many floors with escape routes with stairs above and below ground	Disabled restricted to ground floor	No disabled present
General comments	High rate of fire spread, difficult egress, widely spaced occupants in separate rooms, high density open offices, presence of public	Fire spread likely to be slow, reasonable egress, all accommodation above ground, low rise offices	Low rate of fire spread, good access to safe place, high ceilings to carry smoke away

2 ensuring staff know the sound and pattern of audible alarms;
3 nominate person or persons to call emergency services;
4 nominate a person to act as guide for the emergency services when they arrive on site;
5 appoint an emergency co-ordinator who should be the point of contact with the emergency services and who should make themselves known to the relevant services before an emergency arises;
6 appoint area wardens whose duties will include ensuring their area of responsibility has been evacuated;
7 make arrangements to ensure the safe evacuation of any staff who need special assistance, e.g. disabled person;
8 train all employees in the emergency plan procedures.

The emergency plan should be in writing to prevent any ambiguity and so it can be referred to when required. When making appointments for specific duties under the plan, account must be taken of the possibility of absences due to holidays, illnesses or other reasons.

4.2.14 Access and facilities for the fire brigade

For the fire brigade to be able to deal effectively with a fire and protect life, it is essential that they have adequate access to all parts of the premises and that suitable facilities, such as a supply of water, are available to them.

A section of the Building Regulations[23] requires that the building design and construction incorporate facilities to assist fire-fighters in the protection of life and include:

1 sufficient means of external access to enable fire appliances to be brought near to the building for effective use;
2 sufficient means of access into, and within, the building for fire-fighting personnel to effect rescue and fight fire;
3 the building to be provided with sufficient internal fire mains and other facilities to assist fire-fighters in their tasks;
4 the building to be provided with adequate means for venting heat and smoke from a fire in the basement.

4.2.14.1 Access for fire appliances

Depending on the height of a building access may be required for pumping appliances and turntable ladders/hydraulic platforms. For small buildings (of less than 2000 m^2 floor area or a top storey no higher than 11 m above ground level) access for pumping appliances is required to within 45 m of any point on the 'footprint' or plan area of the building or 15% of the perimeter whichever is less onerous. For the largest of buildings (with a floor area of more than 24 000 m^2 with a top storey more than 11 m above ground level) access is required to 100% of the perimeter for a pumping and a high reach vehicle. These distances can be reduced if the building is fitted with internal fire mains. The access roadway needs to withstand 12.5 tonnes for pumping appliances and 17 tonnes for high reach appliances. Roads and gates should be wide enough, there should be adequate turning circles and the height clearance of overhead pipelines, wires and bridges should be sufficient to accommodate the largest fire appliance.

4.2.14.2 Fire-fighting access shafts

Larger buildings with floors more than 18 m above or more than 10 m below ground level should be provided with a fire-fighting shaft that contains fire-fighting lifts. Fire-fighting shafts should be separated from the accommodation areas by a fire resistant lobby at each floor level. A fire main outlet should be located in the fire resistant lobby at each floor level. The fire-fighting shaft provides, first, a protected means of escape for persons in the building and, second, gives fire-fighters a safe route by which to reach the upper floors without the need to wear breathing apparatus. A normal lift shaft and surrounding stairs serves this purpose provided it and the lobby at each floor have separation from the main floor area by a fire resistant construction.

4.2.14.3 Fire-fighting information about the premises

Under the Fire Services Act fire authorities are required to obtain information about buildings in their area. This information is designed to assist them in preparing fire-fighting plans for the building and to ensure that they are familiar with access points, water supplies and any special hazards that the building contains. The Fire Precautions (Workplace) Regulations place reciprocal obligations on occupiers to liaise with the local fire brigade and enable them to become familiar with the premises. Also to ensure that when the fire brigade attend an incident they are met by somebody who can guide them to the source of the fire, report on the evacuation of occupants and be a point of contact for any queries.

4.2.15 Fire terminology

An increasing wide range of fire terminology is now being used and it is important to have a basic understanding of the most frequently used terms. This is important in discussions with Fire Prevention Officers to ensure there is no misunderstanding. Some of the more commonly used terms are listed below.

Accommodation stairway is a stairway provided for the convenience of occupants and is separate from those required for means of escape purposes.

Active fire protection are measures to contain the spread of fire which require some form of mechanical actuation, i.e. operation of smoke vents or release of fire shutters.

Alternative escape routes are escape routes located so that should a fire occur in any part of the building it will not affect both routes at the same time.

Auto-ignition temperature is the temperature at which a material will decompose and ignite without the application of an external source of heat.

Backdraught occurs when the air supply to a fire is restricted and the oxygen in the air is used up more quickly than it can be replaced making the fire appear to die down but remain above the auto-ignition temperature. As soon as a fresh supply of oxygen becomes available, such as by opening a door, it will form an explosive mixture which will ignite with explosive force. This 'explosion' is called a backdraught.

Cavity barrier is the arrangement provided in a cavity, such as a loft space, or a concealed space to prevent the penetration or movement of smoke or flame within such a space.

Dead end is an area from which escape is possible in one direction only.

Distance of travel is the actual distance a person needs to travel from any point in a building to the nearest storey exit having regard to the layout of walls, partitions, furniture and plant.

Dry rising main is a rising water main that is normally kept empty but with facilities at ground level to enable the fire brigade to connect their pumps.

Dry sprinkler system consists of sprinkler pipework pressurised with air. When a sprinkler head ruptures water enters the system.

Emergency escape lighting is that part of the emergency lighting that ensures that the means of escape can be used at all times.

Escape route forms the means of escape from any point in a building to the final exit.

Final exit is the termination of an escape route from a building that gives direct access to a place of safety. The final exit must be sited to ensure the rapid dispersal of persons from the vicinity of a building so that they are no longer in danger from smoke or fire.

Fire door is a door or shutter which together with its frame and furniture is intended to resist the passage of fire or smoke.

Fire engineering is an approach to fire prevention that takes into account the total fire safety package and sets a range of fire safety features against an assessment of the fire hazard and the fire risk for the particular premises.

Fire load is the amount of fuel and combustible materials within a room or area which will burn to generate heat and so feed the fire.

Fire point is the lowest temperature at which a liquid gives off sufficient flammable vapours to produce a sustainable flame when ignited. The fire point is normally a few degrees higher than the flash point. (Also refers to the panels on which fire alarms and extinguishers are mounted.)

Fire risk is a combination of the probability of ignition and the consequent life and property loss.

Flammable liquid is a liquid with a flash point between 32°C and 55°C.

Flashover is the very quick acceleration of a fire when the temperature of the combustible material in a compartment reaches a level at which it all ignites simultaneously. This requires a plentiful supply of fresh air.

Flash point is the lowest temperature at which a liquid produces sufficient flammable vapours to cause a momentary flame when ignited. It should be noted that as soon as the source of ignition is removed the flame extinguishes.

Highly flammable liquid is a liquid with a flashpoint below 32°C.

Ignition temperature is the temperature to which a material has to be heated for sustained combustion to take place once the material has been ignited. Ignition temperature applies to all substances whereas fire point is only applicable to liquids.

Intumescent materials are special materials that expand and form an insulating or sealing layer when heated to a predetermined temperature.

Lower explosive limit is the lowest concentration of a flammable gas or vapour mixture in air which is capable of ignition and subsequent flame propagation. Below this concentration ignition cannot take place.

Means of escape refers to the safe route by which persons in a building can travel to a place of safety.

Refuge is a place of temporary safety within a building.

Storey exit is the final exit from a storey or floor.

Travel distance represents the actual distance a person must travel to reach the nearest storey exit having regard to the layout of furniture, equipment, etc.

Upper explosive limit is the highest level of concentration of a flammable gas or vapour mixture in air which is capable of ignition and subsequent flame propagation. Above this concentration level ignition cannot take place.

References

1. *The Fire Precautions (Workplace) Regulations 1997 as amended by the Fire Precautions (Workplace) (Amendment) Regulations 1999*, The Stationery Office, London 1999.
2. BS 5839, *Fire detection and alarm systems for buildings, (6 parts)*, BSI, London.
3. BS EN 60849, *Sound systems for emergency purposes*, British Standards Institution, London
4. BS EN 54:2001, *Fire detection and alarm systems, (7 parts)*, BSI, London.
5. BS EN 2, *Classification of Fires*, BSI, London (1992)
6. BS EN 3, *Portable fire extinguishers, (6 parts)*, BSI, London
7. BS 7863, *Recommendations for colour coding to indicate the extinguishing media contained in portable fire extinguishers*, BSI, London (1996)
8. BS 5423 (withdrawn – see BS EN 3; BS 7863:1996; BS 7867:1997)
9. BS 7937:2000, *Specification for portable extinguishers for use on cooking oil fires (Class F)*, British Standards Institution, London 2000
10. BS 5306, *Fire extinguishing installations and equipment on premises, (7 parts)*, BSI, London
11. BS 7944, *Type 1 heavy duty fire blankets* and *Type 2 heavy duty heat protection blankets*, see also BS EN 1869 BSI, London
12. BS EN 1869: 1997, *Fire blankets*, BSI, London (1997)
13. LPC Rules for Automatic Sprinklers Installations, The Loss Prevention Council, London
14. BS EN 12259: *Fixed fire fighting systems (4 parts)*, BSI, London
15. CEA rules, Commité Européen des Assurances, Paris.
16. European Council Directive, *Interpretative Document for safety in case of fire (Construction Products Directive)*, EU, Luxembourg, in preparation
17. *Sprinkler Code*, Code 13, National Fire Protection Association, Quincy, USA
18. The Building Regulations 1991, Approved Documents. Approved Document B: Fire Safety 2000, The Stationery Office, London (2000)
19. BS EN 12416: *Fixed fire fighting systems. Powder systems*
    Part 1 – Requirements and test methods for components,
    Part 2 – Design, construction and maintenance, BSI, London (2001)
20. BS ISO 14520 – Part 1: 2000, *Gaseous fire fighting systems: Physical properties and systems design*, British Standards Institution, London, 2000
21. European Council Directive 92/58/EEC, *Safety Signs Directive*, EU, Luxembourg, 1992
22. *The Health and Safety (Safety Signs and Signals) Regulations 1996*, The Stationery Office, London (1996)
23. The Building Regulations 1991, The Stationery Office, London (1991)
24. BS 476: *Fire tests on building materials and structures: (17 parts)*, BSI, London
25. Guide to Fire Precautions in Existing Places of Work that Require a Fire Certificate (The Blue Guide), The Stationery Office, London.
26. BS 5588: *Fire precautions in the design, construction and use of buildings*,
    Part 8, Code of Practice for means of escape for disabled people,
    Part 10, Code of practice for shopping complexes, BSI, London
27. BS DD 240, *Fire safety engineering in buildings*,
    Part 1, Guide to the application of fire safety engineering principles,
    Part 2, Commentary on the equation given in part I, BSI, London (1997)
28. BS 7974:2001, *Application of fire safety engineering to the design of buildings – Code of practice*, BSI, London (2001)

29. BS 5588, *Fire precautions in the design, construction and use of buildings*
30. *The Fire Services Act 1947*, The Stationery Office, London (1947)
31. *The Fire Precautions Act 1971*, The Stationery Office, London (1971)
32. *The Fire Safety and Safety in Places of Sport Act 1987*, The Stationery Office, London (1987)
33. *The Fire Certificate (Special Premises) Regulations 1996*, The Stationery Office, London (1996)
34. *The Fire Precautions (Workplace) Regulations 1997*, The Stationery Office, London (1997)
35. European Council Directive no. 89/391/EEC, *On the introduction of measures to encourage improvements in the safety and health of workers at work (the Framework Directive)*, EU, Luxembourg, (1989)
36. European Council Directive no. 89/654/EEC, *concerning the minimum safety and health requirements for the workplace (the Workplace Directive)*, EU, Luxembourg (1998)
37. *The Environment and Safety Information Act 1988*, The Stationery Office, London (1988)
38. *The Building Standards (Scotland) Regulations 1990*, The Stationery Office, London (1990)
39. *The Building Operations (Scotland) Regulations 1975*, The Stationery Office, London (1975)

Statutes, Publications and Guidance available from various sources

The following listed documents can be obtained from:

The Stationery Office, HMSO Publications Centre, London SW8 5DT. Tel: 020 7873 9090/0870 6005522. Website: www.tso-online.co.uk

The Building Regulations 1991
  Approved Document B: *Fire Safety* (2000)
  Approved Document M: *Access and facilities for disabled* (1998)
Building Regulations and fire safety procedural guidance
Code of Practice for fire precautions in factories, offices, shops and railway premises not required to have a fire certificate (1994)
Guide to fire precautions in existing places of entertainment and like premises (1990)
Fire Precautions Act 1971: *Guide to fire precautions in existing places of work that require a fire certificate: factories, offices, shops and railway premises* (1993)
Fire Precautions Act 1971: *Guide to fire precautions in premises used as hotels and boarding houses which require a fire certificate* (1991)
Fire Safety: An employer's guide.
Technical standards for compliance with the Building Standards (Scotland) Regulations 1990, as Amended by the Building Standards (Scotland), (1999)
Fire fighting. Halon phase out: advice on alternatives and guidelines for users (1995)

HSE Books, PO Box 1999, Sudbury, Suffolk CO10 6FS, Tel: 01787 881165. Website: www.hsebooks.co.uk

HSE 8 *Oxygen: Fire and explosion hazards in the use and misuse of oxygen* (1998)
FIS2 *Dust explosions in the food industry* (1993)
L21 *Management of health and safety at work – Management of Health and Safety at Work Regulations 1999 Approved Code of Practice and Guidance* (2000)
L54 *Managing construction health and safety – The Construction (Design and Management) Regulations 1994, Approved Code of Practice* (1995)
INDG98 *Permit to work system* (1997)
INDG227 *Safe working with flammable substances* (1996)
INDG236 *Maintaining portable electrical equipment in offices and other low-risk environments* (1996)
INDG237 *Maintaining portable electrical equipment in hotels and tourist accommodation* (1996)
INDG314 *Hot work on small tanks and drums* (2000)
HSG51 *The storage of flammable liquids in containers* (1998)
HSG71 *Chemical warehousing: Storage of packaged dangerous substances* (1998)
HSG118 *Electrical safety in arc welding* (1994)
HSG140 *Safe use and handling of flammable liquids* (1996)

HSG168 *Fire safety in construction work* (1997)
HSG176 *The storage of flammable liquids in tanks* (1998)
Fire precautions in the clothing and textile industries (2000)
Fire safety in the paper and board industry (1995)
Guide to general fire precautions in explosives factories and magazines. Fire Certificate (Special Premises) Regulations (1976 revised 1990)

Fire Protection Association: Bastille Court, 2 Paris Garden, London SE1 8ND. 020 7902 5303. www.thefpa.co.uk
LPC Library of Fire Safety:
FSB8 Vol 1 *Fire protection yearbook*
FSCO2 Vol 2 *Fire and hazardous substances*, C-D ROM edition (2002)
FSB36 Vol 4 *Guide to fire safety signs* (1997)
FSB22 Vol 5 *Fire risk management in the workplace: A guide for employers* (2000)
FSB40 Vol 6 *The Prevention and Control of Arson* (1999)
FSB41 Vol 7 *Fire Protection Equipment and Systems: A guide* (in development)
FSB42 Vol 8 *Fire Safety for Electrical Equipment* (2002)
FSB6 *Heritage under fire (A guide to the protection of historic buildings)* (1995)
FSB71 *Fire safety management in hotels and boarding houses* (1996)
Essential Fire Safety:
FSB15 *Fire safety in retail premises* (1994)
FSB20 *Fire safety in offices* (1996)
FSB32 *Fire safety in village halls and community centres* (1996)

Building Research Establishment: Fire Research Station: Bucknalls Lane, Garston, Watford WD2 7JR. 01923 664000. www.bre.co.uk.
BR225 *Aspects of fire precautions in buildings* (1993)
Digest 320 *Fire doors*
IP 13/92 *False alarms from automatic fire alarm detection systems*
IP17/89 *Photo illuminescent markings for escape routes*
Digest 288 *Dust explosions*
FN7 *Dust explosions: Flame and pressure effects outside vents: Guidance for industry*

British, European and International Standards

British Standards Institution, 389 Chiswick High Road, London W4 4AL. 020 8996 9001. www.bsi.org.uk

British Standards
BS 476 *Fire Tests on building materials and structures (Parts 20–24)*
BS 1635:1990 *Recommendations for graphic symbols and abbreviations for fire protection drawings*
BS 3169:1986 *Specification for first aid reel hoses for fire fighting purposes*
BS 3251:1976(1993) *Specification. Indicator plates for fire hydrants and emergency water supplies.*
BS 4422: *Glossary of terms associated with fire (8 parts)*
BS 5266: *Emergency lighting (2 parts)*
BS 5268: *Structural use of timber*
BS 5306: *Fire extinguishing installations and equipment on premises (parts 0–7)*
BS 5445: *Components of automatic fire detection systems (parts 5, 7–9)*
BS 5499: *Fire safety signs, notices and graphic symbols (parts 1–3)*
BS 5588: *Fire precautions in the design, construction and use of buildings (Parts 1, 4–11)*
BS 5810: 1979 *Code of practice for access for the disabled to buildings*
BS 5839: *Fire detection and alarm systems for buildings (parts 1–6, 8)*
BS 5908: 1990 *Code of practice for fire precautions in the chemical and allied industries*
BS 6575: *Specification for fire blankets*
BS 6643: *Recharging fire extinguishers (parts 1 & 2)*
BS 7863: 1996 *Recommendations for colour coding to indicate the extinguishing media contained in portable fire extinguishers.*

BS 7937: *Specification for portable fire extinguishers for use on cooking oil fires (class F)*
BS 7939:1999 *Smoke security devices. Code of practice for manufacture, installation and maintenance*
BS 7974: *Application of fire safety engineering principles to the design of buildings. Code of practice*
DD 240: *Fire Safety Engineering in Buildings*
PD 6520: 1998 *Guide to the Fire Test Methods for Building Materials and Elements of Construction*

European Standards
BS EN 2: 1992 *Classification of fires*
BS EN 3: *Portable fire extinguishers (parts 1–6)*
BS EN 54: *Fire detection and alarm systems (7 parts)*
BS EN 469:1995 *Protective clothing for fire fighters. Requirements and test methods for protective clothing for fire fighting.*
BS EN 615: 1995 *Fire Protection. Fire Extinguishing Media. Specification for Powders (other than Class D powders)*
BS EN 671: *Fixed fire fighting systems. Hose systems. Part 1, Hose reels with semi-rigid hose*
Part 2, Hose systems with lay-flat hoses
Part 3, Maintenance of hose reels with semi-rigid hose and hose systems with lay flat hose
BS EN 1838:1999 *Lighting applications. Emergency Light*ing. (same as BS 5266 – 7: 1999)
BS EN 25923:1994 *Fire extinguishing media. Carbon dioxide*

Chapter 4.3

Safe use of machinery

J. R. Ridley

4.3.1 Introduction

Machinery and equipment have been evolved to meet a need whether for producing, changing or moving materials and components. Initially they were essentially functional with scant regard for the health and safety of those using them. But attitudes have changed and all work equipment must now be designed and built so that it does not put the user at risk of damage to health or injury. However, equipment does not work by itself – it needs someone to work it, drive it or, in the case of robots, tell it what to do. It is at this interface between equipment and operator that the risks to health and injury arise.

Many techniques have been developed to reduce these risks to a minimum and this chapter looks at some of those techniques that are available to the designer and user of modern equipment. What is not dealt with is the other vital element in the interface – the operator – and the training necessary to ensure his/her safety and how it matches the equipment, the culture and the working methods of the particular organisation. Essentially the term *work equipment* encompasses any equipment used in the course of work. However, in this chapter work equipment will be considered in four major functional areas: machinery, power trucks, cranes and lifts, and pressure systems.

4.3.1.1 Legislative arrangements

With the gradual demise of the FA and the growing influence of the EU, UK health and safety legislation concerning work equipment has polarised into that dealing with the facilities provided with 'new' equipment where the emphasis is on the provision of safeguards, and that dealing with the use of all work equipment which extends to include the provision of safeguards for 'existing' (pre-1993) equipment. In both these cases the relevant UK legislation stems from EU directives which rely on *harmonised EN standards*, or where they do not exist, extant national standards, to specify the conditions to be met for conformity with the directive.

The relevant legislation is the Machinery Directive[1] and its UK manifestation – the Supply of Machinery (Safety) Regulations 1992 (SMSR) – and the Work Equipment Directive[2] with its UK counterpart, the Provision and Use of Work Equipment Regulations 1998 (PUWER 2) respectively.

4.3.1.2 Machinery directive

When the first Machinery Directive was adopted, it excluded a number of special purpose machines because of pressure from specialist sectors on the grounds that they had special needs that could not be covered in the general safety requirements that applied to run-of-the-mill machines. To prevent delaying adoption of the directive these exclusions were allowed. However, since 1993 the European Commission has studied these particular cases and amended the directive to include additional conditions having specific application so that these particular cases could be brought within its compass.

The Machinery Directive was drawn up using the *new approach to legislative harmonisation* whereby the main body of the directive itself lays down broad objectives to be achieved, lists in annexes the areas to which safety attention should be directed and relies on European harmonised (EN) standards to specify the conditions to give conformity.

4.3.1.3 The Supply of Machinery (Safety) Regulations 1992

These Regulations, which incorporate the contents of the Machinery Directive, lay down the requirements to be met by the manufacturers of new machinery, i.e. machinery which is currently being put on the market, and indicate what a purchaser of a new machine, whether manufactured in the UK or elsewhere in the EU, can expect by way of safeguards on the machine and information about it from the supporting documentation. Their aim is to ensure that the machinery meets the standards necessary to ensure safety in use. Where machinery conforms with the requirements of the Regulations (and hence the directive) it will have open access to the whole of the EU market. Compliance with these conditions is also required of machinery imported into the EU. However, the requirements do not apply to machinery that is to be exported to non-EU countries.

The Regulations cover a wide range of machinery within the definition (reg. 4):

(a) an assembly of linked parts or components, at least one of which moves including . . . the appropriate actuators, control and power circuits, joined together for a specific application, in particular for the processing, treating, moving or packaging of a material
(b) an assembly of machines, that is to say, an assembly of items of machinery which . . . in order to achieve the same end are arranged and controlled so that they function as an integral whole . . ., or

(c) interchangeable equipment modifying the function of the machine . . .

but reg. 5 refers to a list of exclusions which includes, *inter alia*, machinery whose risks are covered by other directives plus lifting equipment for raising and/or moving persons. However, requirements for construction of lifts, both for goods and people, are now covered by the Lifts Regulations 1997.

General duties are put on suppliers of machinery to conform with these Regulations, whether the machine is manufactured in the UK or imported from a non-EU country (reg. 11). Documentary evidence, in the form of a technical file and certificates, is required to prove that the machinery conforms to the Regulations. The procedures for preparing the documentation are laid down and vary according to whether the machine is:

- constructed to EN or equivalent standards
- constructed to other equally effective safety standards or
- machines posing special hazards and which are listed in Schedule 4.

These procedures are shown diagrammatically in *Figure 4.3.1*.

An essential common element in the conformity assessment procedure is the preparation of the technical file which should contain:

- drawings of the machine
- a list of:
  - the relevant essential health and safety requirements
  - transposed, harmonised or other standards complied with
  - technical specifications
- a description of the safety devices incorporated
- copies of any test reports
- operating instructions.

Where a number of machines of the same type are to be made, only one technical file need be prepared but the manufacturer must provide documentary evidence of the procedure he will follow to ensure that all machines are manufactured to the same standard.

For all machines, except those posing high risks, where the machinery has been designed and manufactured to comply with EN or transposed national standards, the manufacturer completes a Certificate of Conformity and attaches the CE mark to the machine (reg. 13) (*Figure 4.3.2*).

In the case of the high risk machines that are listed in Schedule 4 and which comply with an EN or transposed standard (reg. 14), the manufacturer can either:

- send a copy of the technical file to an approved body for its retention or
- submit the technical file to an approved body requesting:
  - verification that the standards have been correctly applied and
  - the issue of a Certificate of Adequacy or

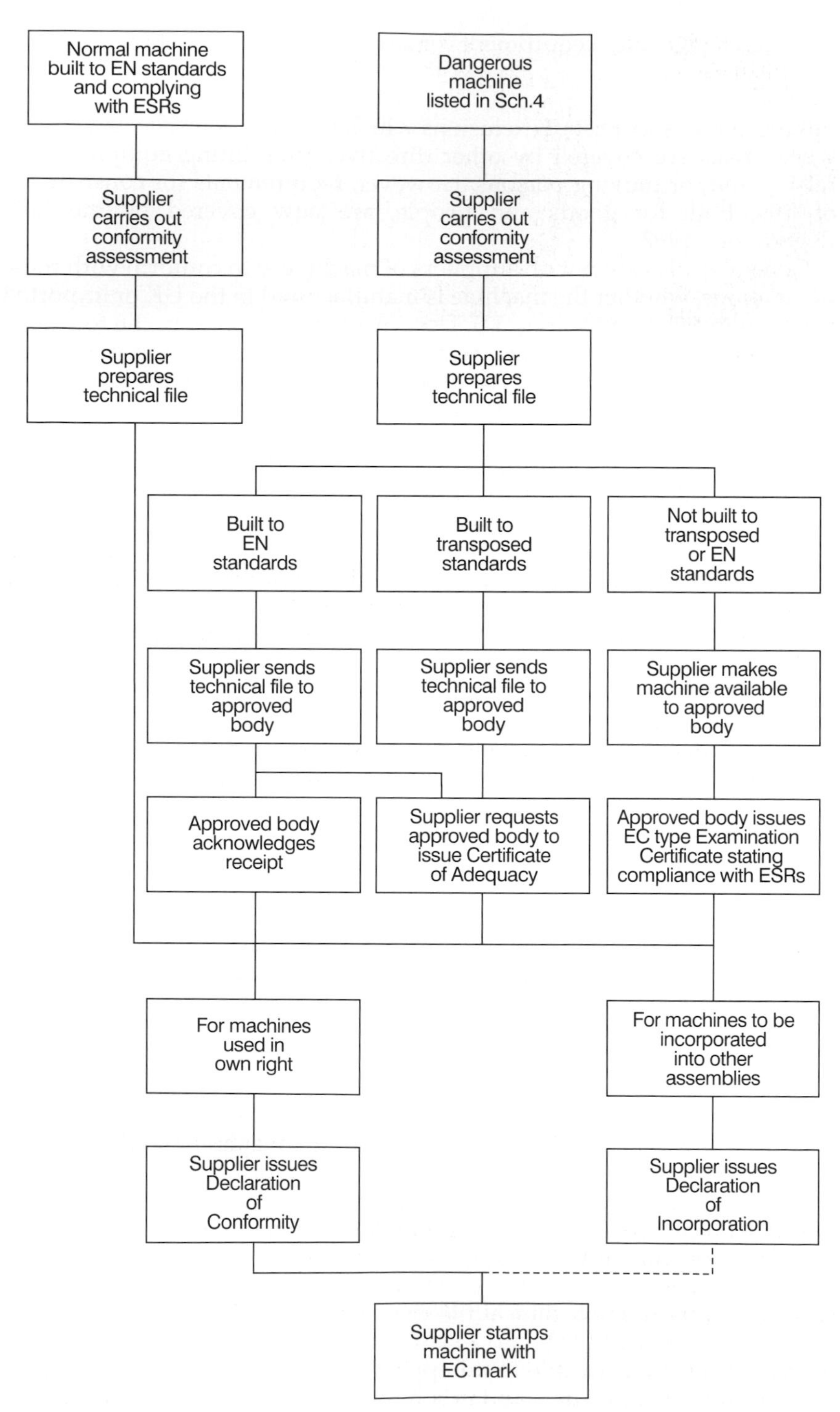
Normal machine built to EN standards and complying with ESRs
Dangerous machine listed in Sch.4
Supplier carries out conformity assessment
Supplier carries out conformity assessment
Supplier prepares technical file
Supplier prepares technical file
Built to EN standards
Built to transposed standards
Not built to transposed or EN standards
Supplier sends technical file to approved body
Supplier sends technical file to approved body
Supplier makes machine available to approved body
Approved body acknowledges receipt
Supplier requests approved body to issue Certificate of Adequacy
Approved body issues EC type Examination Certificate stating compliance with ESRs
For machines used in own right
For machines to be incorporated into other assemblies
Supplier issues Declaration of Conformity
Supplier issues Declaration of Incorporation
Supplier stamps machine with EC mark

Figure 4.3.1

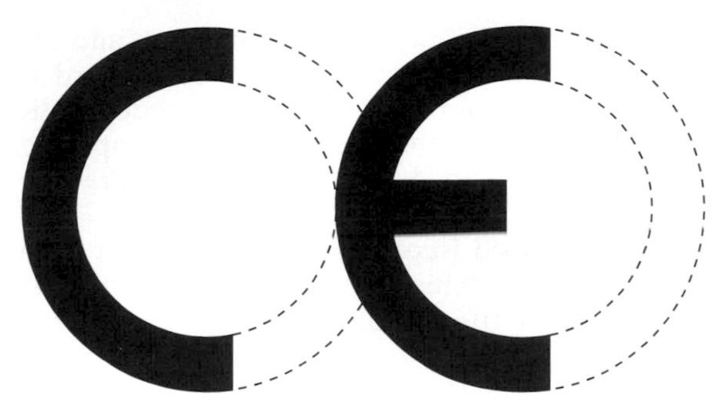

Figure 4.3.2

- submit the technical file to an approved body, arrange for an example of the machine to be available for EC-type testing and request an EC Type-examination Certificate.

Where a high risk machine is not manufactured to harmonised standards or no harmonised standards exist, an example of the machine must be made available to the approved body who must satisfy themselves that the machine complies with the relevant essential health and safety requirements and can be used with safety. They will then issue an EC Type-examination Certificate. The details to be included in the various certificates are laid down in the Regulations.

An 'approved body' is appointed by a Member State (in the UK by the Secretary of State) and notified to the European Commission as an organisation which has the qualifications and necessary resources to undertake the examination and certification work. Approved bodies may charge the manufacturer fees for the certification work it undertakes on his behalf (reg. 19).

While the body of the regulation lays down the procedures to be followed for ensuring conformity with the required standards, it relies on a list of ESRs given in Schedule 3, to detail specific aspects requiring attention. Evidence of conformity with any of the ESRs is through compliance with the appropriate harmonised EN standard.

4.3.1.4 The Provision and Use of Work Equipment Regulations 1998 (PUWER 98)

These Regulations, which are supported by an Approved Code of Practice[3] (ACoP), are concerned with safety in the use of all work equipment and extend to lay down safeguarding requirements for existing equipment, i.e. that which had been purchased for use at work before 31 December 1992, and is not covered by SMSR. Although these Regulations have been made under HSW certain of the detail Regulations demand higher levels of protection than *so far as is reasonably practicable*. They expect *foreseeable* hazards to be anticipated while in other cases the test of *practicability* may need to be applied, i.e. costs do not come into the consideration. However, in general the level of protection demanded is no greater than that required by earlier, but now superseded, UK legislation, i.e. FA and associated regulations. Guidance on complying with the requirements is contained in an Approved Code of Practice (ACoP)[3].

These Regulations have been extended to cover work equipment such as power presses and woodworking machinery that present particular hazards. Advice on compliance with the supplementary requirements relating to protecting against the hazards from these equipments is contained in separate ACoPs[29,30].

The Regulations cover work equipment (reg. 1) which the ACoP describes as anything provided for use at work, from a scalpel to scaffolding, a ruler to a reactor but excludes livestock, chemical substances, building structures and private cars. They cover all work situations except shipboard activities on a sea-going ship (reg. 3). The responsibility for compliance is placed on employers (reg. 4) and extends to the self-employed, anyone who has control of non-domestic premises used for work and to the occupier of a premises used for work. Employers must ensure that all work equipment they provide is suitable for the intended work (reg. 5) and is used only for that purpose. In providing the equipment, employers must carry out an assessment to identify reasonably foreseeable risks to operators taking into account the equipment and the working conditions. If the equipment is adapted to a further purpose, the employer must ensure that it is suitable for its new use, can be used safely and that no new risks are introduced.

All work equipment must be kept well maintained (reg. 6) and in the case of machines a maintenance log must be kept. This could form the basis for a system of planned maintenance, covering routine maintenance such as lubrication and cleaning as well as examinations and overhauls, thus ensuring that machines are maintained before they fail or become dangerous. Where the operation, servicing or maintenance of any work equipment involves specific hazards (reg. 7) the employer must ensure that the work is carried out safely and that only properly trained and competent operators are allowed to carry out the particular tasks.

Suitable information and instruction must be given to those who use the work equipment but the obligation has been extended to cover foreseeable abnormal use of the equipment (reg. 8). Any such information and instruction must be comprehensible to the recipients making allowance for different assimilation abilities and also those whose first language is not English. Similarly, operators must be trained (reg. 9) to use the equipment safely and any supervision provided must be familiar with and competent in its safe use.

Where the protective measures demanded for work equipment under other relevant EU directives differ from that demanded by regs 11 to 24, the higher standard applies (reg. 10). Regulation 11 lays down a hierarchy of measures to be taken to protect against dangerous parts of machinery. First, it places *absolute* requirements on employers to prevent access to rotating stock bars and to provide means to ensure that dangerous parts of machinery have stopped before contact can be had with them. The hierarchy of measures demand the highest level of protection that it is *practicable* to achieve in the following priority order:

1 fixed guards to EN standards[4],
2 other guards or devices that prevent contact with dangerous parts. Many parts are only dangerous when moving, so measures, such as interlocking guards that stop the machine before contact can be made, are acceptable. EN standards[4,5] cover suitable safeguarding means as do two other publications[6,31],
3 the use of jigs, holders, push sticks etc. that keeps the operators hands away from the dangerous parts. These are particularly relevant in the use of woodworking machinery,
4 systems of work requiring the provision of information, training, instruction and supervision. In the UK, systems of work have never been accepted as primary means of protection against dangerous parts of machinery but only as back-up for other safeguards.

The *practicablity* condition does not allow consideration of cost but expects practices to match the state-of-the-art knowledge. Wherever a decision is made to use a lesser standard of protection, the reasons for selecting it should be recorded since the decision may later be called into question and have to be justified, possibly before a court.

Any guards provided shall:

(a) be suitable for their intended purpose
(b) be of good construction, sound material and adequate strength
(c) be kept in good repair and effective working order
(d) not introduce hazards
(e) not be easily defeated
(f) be an adequate distance from the dangerous parts[7,8,9]
(g) not interfere with the operation of the equipment
(h) allow maintenance to be carried out safely.

The intent of earlier law often required interpretation in the courts especially where a legal action following an accident revolved round compliance with a particular section or phrase[10]. In reg. 12 the opportunity has been taken to bring the substance of some of those decisions into statute law covering such matters as:

(a) articles or substances falling or being ejected from the machine
(b) rupture or disintegration of the work equipment[11]
(c) equipment overheating or catching fire
(d) the unintended or premature discharge from the equipment of articles or substances
(e) explosion of the equipment or anything in it.

The measures to be taken, which exclude the use of personal protective equipment and systems of work, are aimed at preventing or minimising the effects of any hazards likely to be met in the workplace. Also excluded from these requirements are processes covered by other specific legislation, such as that dealing with lead, asbestos, radiations, noise and hazardous chemicals.

Employers are required to provide protection against high and low temperatures (reg. 13), and ensure machinery has adequate controls:

- for starting or changing the state of the equipment (reg. 14)
- to stop the equipment (reg. 15)
- for emergency stopping (reg. 16)
- located so operator can see all of the machine or, if not, incorporate an audible warning with time delay start (reg. 17)
- that do not cause dangers even under fault conditions (reg. 18).

All powered work equipment should be capable of being isolated from its source of power (reg. 19). This is a requirement of the Electricity at Work Regulations 1989 but PUWER 2 extends it to include hydraulic, pneumatic and any other source of power. When used, work equipment must be stable either inherently, by clamping or by using stabilising devices such as the outriggers on mobile cranes and on access towers (reg. 20), and be sufficiently well lit to enable the work to be carried out without risk to health or safety (reg. 21).

Maintenance work should be carried out with the work equipment shut down and locked off. However, if maintenance work must be done with the machine running, all necessary precautions must be taken to protect those carrying out the work (reg. 22). Where work equipment presents a danger to health and safety it should be clearly identified as such (reg. 23) and, if appropriate, be fitted with warning devices (reg. 24).

Guidance on the techniques and practices for achieving compliance with the requirements of these Regulations can be found in the publications[3,29,30,31].

4.3.2 Strategy for selecting safeguards

At the design stage of new work equipment, when considering the purchase of additional equipment or the alteration of existing equipment allowance must be made for the safeguards necessary to ensure safety, or continuing safety, of the equipment and any consequent effect on the working methods to be followed to ensure its safe use. To this end, a strategy should be developed that includes the following stages:

1 Identify the hazards.
2 Eliminate the hazards or reduce to a minimum.
3 Carry out a risk assessment of the residual hazards.
4 Design/select safeguards.
5 Develop safe operating methods.

6 Inform operators of hazards and train in safe operating methods.
7 Install safeguards.
8 Monitor effectiveness and acceptability of safeguards and modify as necessary.

These stages are considered below.

4.3.2.1 Identifying the hazards

The foreseeable hazards at all stages of the equipment's use need to be taken into account and include:

(a) its physical dimensions
(b) method of drive and power requirements
(c) parameters of speed, pressure, temperature, size of cut, mobility etc.
(d) materials to be processed or handled and method of feed
(e) operator position and controls
(f) access for setting, adjustments and maintenance
(g) environmental factors such as dust, fumes, noise, temperature, humidity etc.
(h) operating requirements including what the operator needs to do.

Thus an overall picture can be obtained of any limitations or constraints on the design of suitable safeguards and give a pointer to providing the most effective measures for the safe operation of the equipment.

Typical hazards that can be met are discussed by Ridley and Pearce[31] and include:

- crushing
- shearing
- cutting or severing
- entanglement
- drawing-in or trapping
- impact
- stabbing or puncture
- friction or abrasion
- high pressure fluid ejection
- electrical shock
- noise and vibration
- contact with extremes of temperature.

Identification of these hazards can be from personal knowledge of the equipment, from the descriptions given in EN 292–1[12] or using one of the techniques described in annex B of EN 1050[14]. Reports of incidents involving other similar types of equipment can also prove a useful source of information.

4.3.2.2 Eliminating or reducing hazards to a minimum

Examples of the elimination or reduction of hazards include:

- changing process material for something less hazardous
- modifying the process
- reducing the operating limits of speed, pressure, temperature, power etc.
- automating the production or handling process
- controlling the process from a remote safe position.

4.3.2.3 Assessment of residual risks

A risk assessment is defined[12] as:

> A comprehensive estimation of the probability and the degree of possible injury or damage to health in a hazardous situation in order to select appropriate safety measures:

Evaluation of the risk requires an estimation of the likely severity of any injury or damage and of the probability of it happening. A procedure for obtaining information needed to make the assessment of the risk is outlined in EN 1050[14].

The assessment of lesser risks tends to be qualitative with the assessor making a subjective evaluation. However, for more complex risks or where there are a number of parallel risks some form of quantitative assessment may be necessary to determine priorities. Typical techniques that can be employed are described in annex B of EN 1050 and include:

- Preliminary hazard analysis (PHA)
- 'What-if' method
- Failure mode and effects analysis (FMEA)
- Fault simulation for control systems
- Method organised for a systematic analysis of risk (MOSAR)
- Fault tree analysis (FTA)
- DELPHI technique.

There is also a growing body of reliability data for component items of work equipment which can be used in a quantitative assessment of the risk.

4.3.2.4 Design and selection of safeguarding measures

Where the hazard cannot be eliminated it will be necessary to provide safeguarding measures which for machinery could be any of the techniques described in section 4.3.1.4. Selection of the type of safe-guard will depend on a number of factors including:

- the operating methods and systems of work for the equipment
- proximity of the operator to hazardous areas or points
- need for access for cleaning, setting, adjustment, maintenance etc.
- stopping time of the machine
- the severity of the potential injury or ill-health from the residual hazard.

4.3.2.5 Operator training

Where working methods change because of the introduction of new equipment or as a result of modifications to existing equipment, the operators should be trained in any new techniques involved. Ideally the operators should be involved in the process of making changes, whether to new equipment or of the addition of guards to existing equipment, since this can facilitate its introduction. Once a decision has been taken concerning the new measures to be adopted, a programme of training and instruction can be developed and implemented so that when the changes are put into effect, the operators are familiar with the new operating methods and procedures.

4.3.2.6 Monitoring the effectiveness of the safeguards

Once the safeguarding measures have been installed the operations should be monitored to ensure that:

- the operators are following the new work method
- the safeguards do not interfere with the control of the process
- the safeguards provide an adequate degree of protection
- the safeguards are robust enough and are not likely to fail.

4.3.3 Safeguarding techniques

In new machines, guards should be designed in as an inherent part of the machine while in existing machines any added guards should be designed to provide the necessary protection while allowing the machine to be operated with the minimum of disruption. There is a wide range of types of guards and guarding techniques on which a designer can draw and these are outlined in EN standards[12,13] and in Ridley and Pearce[31] with design requirements specified in BS EN 953[4]. The use of the word 'guard' implies a physical barrier whereas 'safety device' means other non-physical measures for providing the desired level of protection, such as interlocks, pressure sensitive trips, electro-sensitive protective devices (photo-electric curtains) etc. Safeguard is a general term that refers to the means provided to protect against access to dangerous parts and can be either a guard or a safety device or a combination of both.

4.3.3.1 Fixed guards

A fixed guard is defined[12] as a guard that is kept in place permanently by welding or by fasteners that can only be released by the use of a tool and when in position it should not be capable of being displaced casually. Where fixed guards need to be removed periodically for maintenance or clearing a jam-up they could be a hinged door or a slot-in panel secured by a bolt or other suitable device. The simpler the device the more likely is it to be replaced after removal especially if it is backed by commonsense, training and good supervision.

4.3.3.2 Distance guard

A distance guard is simply a barrier sited at an appropriate distance from the danger. The degree of risk being faced will determine whether a fixed rail or fence is necessary; in the latter case, the distance from the dangerous part will determine the opening or mesh sizes, or vice versa. Guidance on safe distances is given in two EN standards[7,15].

4.3.3.3 Adjustable guards

Adjustable guards comprise a fixed guard with adjustable elements that the setter or operator has to position to suit the job being worked on. They are widely used for woodworking and toolroom machines. Where adjustable guards are used, the operators should be fully trained in how to adjust them so that full protective benefit can be obtained.

4.3.3.4 Tunnel guard

Where there is an automatic feed to or delivery from a dangerous part of a machine, operator safety can be provided through the use of tunnel guards which should be of a cross-section to permit the free movement of the product and long enough (at least 1 m) to prevent the operator reaching the dangerous part. If metal components are being delivered from a machine and the chute is lined with acoustic material an additional benefit will be obtained from the reduction in noise generated.

4.3.3.5 Fixed enclosing guard

The provision of a fixed guard enclosing the whole of a dangerous machine achieves a high standard of guarding. Work can be fed to the machine through an opening in the guard by a manual or automatic device arranged so that the operator cannot reach the dangerous parts.

4.3.3.6 Interlocked guards

Where guards need to be moved or opened frequently and it is inconvenient to fix them, they can be interlocked mechanically, electrically, pneumatically, etc., to the machine controls. Two basic criteria must be observed: until the guard is closed the machine should not be capable of being started; and the machine should be brought to rest as soon as the guard is opened. Where there is a run down time, the guard may need to be fitted with a delay release mechanism. The interlock system can provide either control interlocking which acts through the machine controls, power interlocking that operates by interrupting the primary power supply or by mechanical disconnection of the machine from its power source. Different arrangements of interlocking systems are reviewed in BS EN 1088[5].

An essential feature of an interlock or safety circuit is that it must be completed before the machine can start and that any break in it trips the controls and brings the machine to rest. With hydraulic and pneumatic safety circuits, the circuit must be pressurised in the safe condition, any loss of pressure causing the system to trip.

Interlocked guards can be hinged, sliding or removable but the integrity of the design of the interlocking mechanisms is crucial. The mechanisms must be reliable, capable of resisting interference and the system should not fail to danger.

Interlocking guards allow ready access while ensuring the safety of the operator. However, there are circumstances that may require the machine to be moved when the guard is open, i.e. for setting, cleaning, removing jams, etc. Such movement is only permitted under the following circumstances:

1 As part of a 'permit-to-work' system.
2 On true inch control.
3 On limited inch control with each movement of the producing part not exceeding 75 mm (3 in) at a predetermined minimum speed.
4 If for technical reasons the machine or process cannot accept intermittent movement, a continuous movement is permitted provided it is

only at a predetermined minimum speed and is controlled by a hold-on switch, release of which causes the machine to stop immediately. Also, there should be only one such control operable on a machine at any one time and it should override all other controls except the emergency stop switch.

The choice of interlocking method will depend on power supply and drive arrangement to the machine, the degree of the risk being protected against and the consequences of failure of the safety device. The system chosen should be as direct and as simple as possible. Complex systems can be potentially unreliable, have unforeseen fail-to-danger elements and are often difficult to understand, inspect and maintain, and can have low operator acceptability.

4.3.3.6.1 Types of interlocks

Interlocks can be actuated in a number of ways.

(a) Direct manual switch or valve interlocks (*Figure 4.3.3*) where the switch or valve controlling the power source cannot be operated until the guard is closed, and the guard cannot be opened at any time the switch is in the run position.
(b) Mechanical interlocks provide a direct mechanical linkage from the guard to the power transmission shaft. The most common application is on power presses (*Figure 4.3.4*).
(c) Cam-operated limit switch interlocks are versatile, effective and difficult to defeat. They can be rotary (*Figure 4.3.5a*) or linear (*Figure 4.3.5b*) and in each case the critical feature is that in the safe operating position the switch is relaxed, i.e. the switch plunger is not depressed. Any movement of the guard from the safe position causes the switch plunger to be depressed, breaking the safety circuit and stopping the machine. This is the 'positive mode' of operation (*Figure 4.3.6(2)*) and must be used whenever there is only one interlock switch. 'Negative

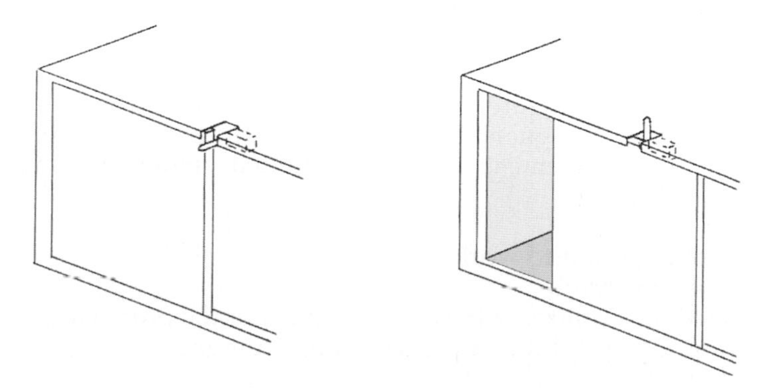

Figure 4.3.3 Direct manual switch interlock

Figure 4.3.4 Interlocking guard, mechanically linked to a crankshaft arrestor fitted to a power press. (Courtesy J. P. Udal Company)

mode' of operation (*Figure 4.3.6(1)*) occurs when the switch plunger is depressed as the guard moves to the safe position and is not acceptable for single switch applications. However, a combination of the two in series is used on high risk machines such as injection moulding machines. This arrangement can incorporate a switch condition monitoring circuit as shown in *Figure 4.3.7*. The type of electrical switch used in interlocking is important. They must be of

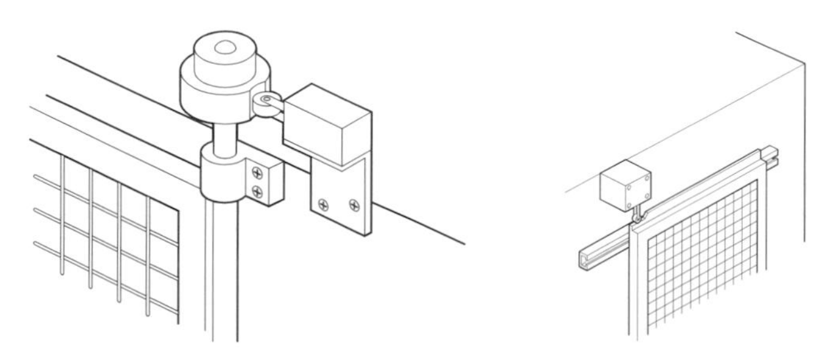

(a) rotary cam operated switch (b) linear cam operated switch

Figure 4.3.5 Positive action cam operated interlocking switches

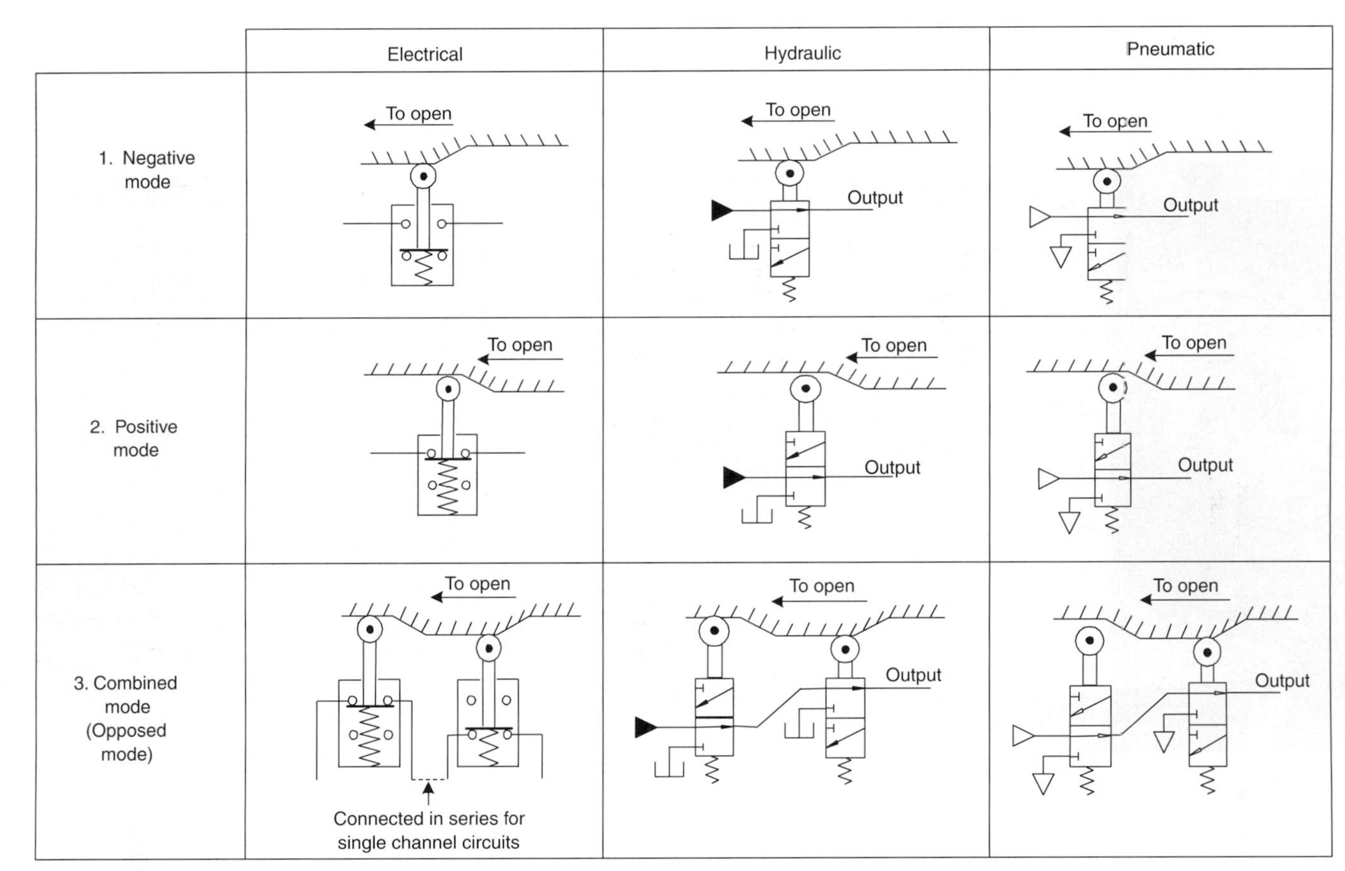

Figure 4.3.6 Modes of operation of cam operated interlocking switches and valves (Ridley and Pearce[31])

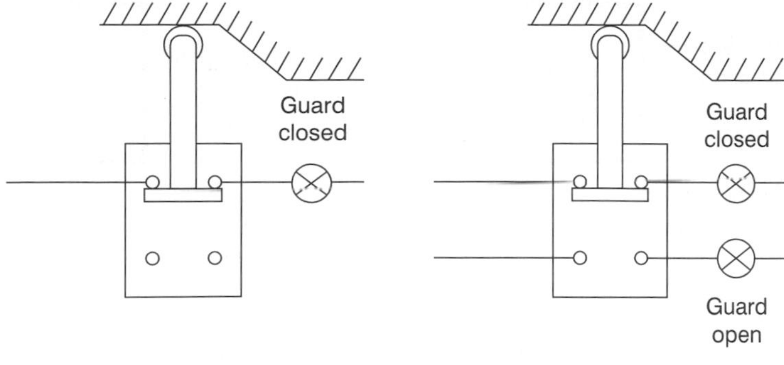

Figure 4.3.7 Interlocking switch condition indication lights. Note: (a) Use of a single indicator light follows the safety circuit principle that a completed safety circuit (when the light is lit) is safe, interruption of the circuit (when the light is out) shows danger. (b) Dual lights can be used to show the state of the interlocking switch or valve

Figure 4.3.8 Limit switch with cover removed showing the principle of fail-safe operation. (Courtesy Dewhurst & Partner plc)

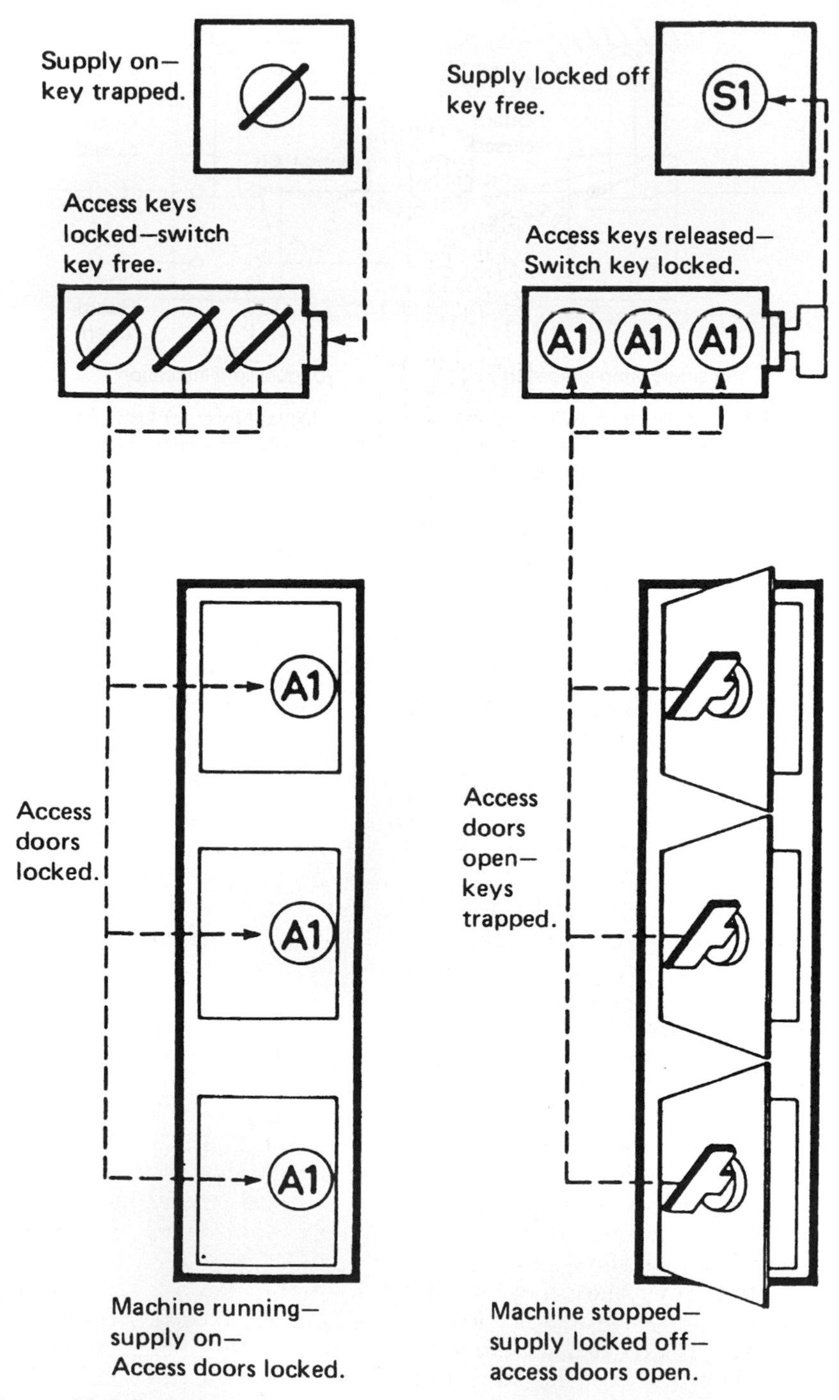

Figure 4.3.9 Diagrammatic arrangement of a key exchange system applied to a machine with a number of access doors required to be open at the same time. (Courtesy Castell Safety International Ltd)

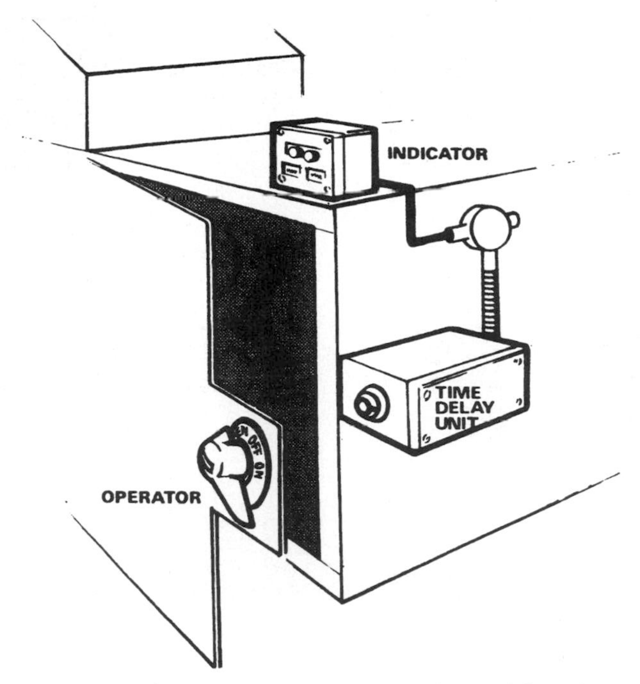

Figure 4.3.10 Captive key safety switch with electrical time delay release device. (Courtesy Unimax Switch Ltd)

the positive make-and-break type (known as 'limit' switches) that fail to safety and have contacts capable of carrying the maximum current in the circuit. The principle of operation of such switches is shown in *Figure 4.3.8*. Micro-switches relying on leaf spring deflection for contact breaking are not acceptable as safety interlocking switches.

(d) Trapped key interlocks (key exchange system) work on the principle that the master key, which controls the power supply to the machine through a switch at the master key box, has to be turned OFF before the keys for individual guards can be released. The master switch cannot be turned to ON until all the individual keys are replaced in the master box. Each individual key will enable its particular guard to be opened, releasing the guard key which the operator should take with him when he enters the machine. The individual key is trapped in the guard lock until the guard is replaced and locked by the guard key. A diagram of a typical installation is shown in *Figure 4.3.9*.

(e) Captive key interlocking involves a combination of an electrical switch and a mechanical lock in a single assembly where usually the key is attached to the movable part of the guard. When the guard is closed, the key locates on the switch spindle. First movement of the key mechanically locks the guard shut and further movement actuates the electrical switch to complete the safety circuit (*Figure 4.3.10*).

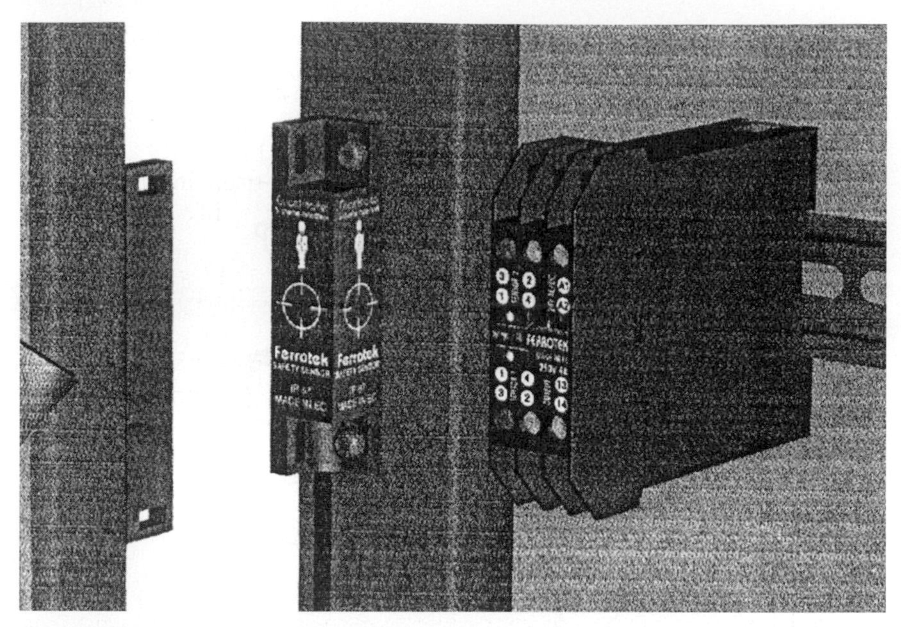

Figure 4.3.11 Magnetic switch with coded magnet actuator, sensor and control box (Courtesy Guardmaster Ltd)

(f) The type of magnetic switch shown in *Figure 4.3.11* which uses a number of magnets uniquely configured to match components of the switch part, provides a high degree of protection. It also has the advantage, since it is encapsulated, of withstanding washing, a benefit in the food industry. Other non-contact switches work through inductive circuits between an actuator and the switch.

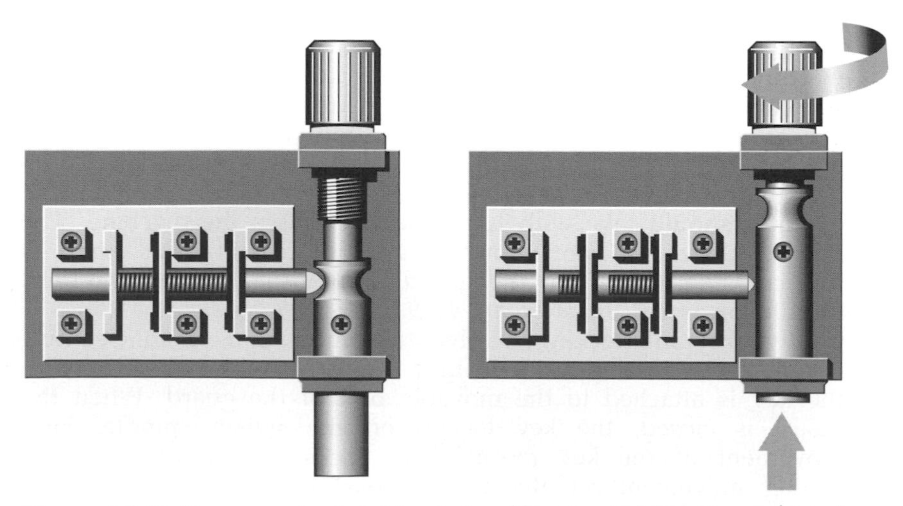

Figure 4.3.12 Electro-mechanical delay device. (Courtesy EJA Ltd, Guard master)

(g) Time delay arrangements are necessary when the machine being guarded has a large inertia and, consequently, a long rundown time on stopping. An electromechanical device is shown in *Figure 4.3.12* where the first movement of the bolt trips the machine, but the bolt has to be unscrewed a considerable distance before the guard is released. A solenoid-operated bolt can also be used in conjunction with a time delay circuit that is actuated from either the machine controls or the trip circuit.

(h) Mechanical scotches are required on certain types of presses to protect the operator when reaching between the platens. These scotches can be linked to the guard operation so that they are automatically positioned each time the guard is opened. Similar devices are needed to restrain the raised platform of a tipper lorry when work is done on the chassis. However, with scissor lifts the scotch must be inserted between the bottom rollers and the base frame.

4.3.3.7 Automatic guard

This type of guard closes automatically when the machine cycle is initiated and is arranged so that the machine will not move until the guard is in the safe position. The machine then commences its movement automatically restraining the guard in the closed position. Where trapping points occur as the parts of the guard come together, trip devices should be fitted. A version of this type of guard moves across the danger zone as the machine operation is initiated, removing any part of the operator that is in that area. This type is sometimes known as a 'sweep-away' guard but is no longer acceptable on modern machinery such as paper cutting guillotines.

4.3.3.8 Trip devices

A trip device is any device which, as the operator approaches the dangerous part, automatically trips the safety circuit. A simple flap trip is shown in *Figure 4.3.13* but trip devices can include: trip bars and wires,

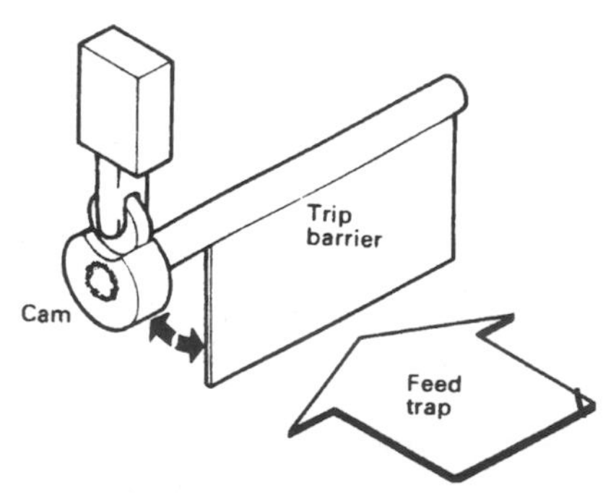

Figure 4.3.13 Hinged barrier with cam-operated limit switch. (Courtesy Engineering)

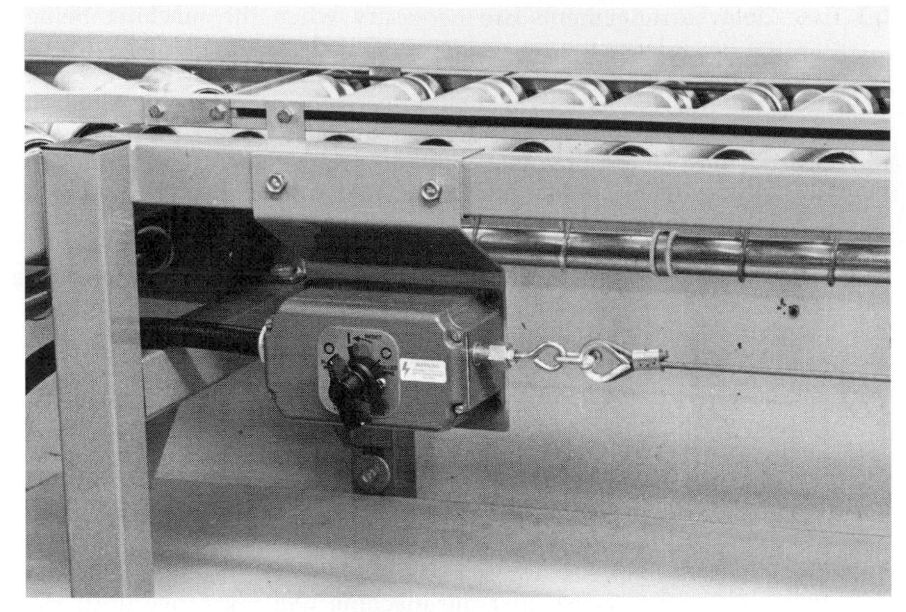

Figure 4.3.14 Roller conveyor with emergency grab wire switch. (Courtesy Craig & Derricott Ltd)

sensitive probes (on drilling machines), photoelectric devices, pressure sensitive strips and mats, and emergency stop switches. *Figure 4.3.14* shows a pull trip switch on a conveyor. It is important that trip devices are properly adjusted and that the machine's brake is in good working order.

The safe distance of trip barriers and electrosensitive screens from the dangerous part depend on the speed at which a person can move into the danger area and the rapidity with which the dangerous parts can be brought to rest. An EN standard[9] gives data on the approach speeds of parts of the body.

4.3.3.9 Two-hand control devices

Two-hand controls are used on machines having a single cycle operation where the work is placed in the machine and the machine struck on. They are applicable only to machines with a single operator. It is essential that the control buttons are positioned more than a hand-span apart, the control circuit arranged so that both controls must be activated simultaneously to start the cycle and that both controls must be released after each cycle before the next cycle can be initiated. Release of either button during the dangerous part of the cycle must stop or reverse the machine movement. Details are specified in an EN standard[16].

4.3.3.10 Control guards

On some small machines, such as forming presses and gold blocking machines, in which the guard has to be moved out of the way to place and remove the work, the guard can be arranged so that as it is moved to the safe position it actuates the machine cycle, and all the time the machine is moving the guard is locked in position. This type of guard also has application on some large panel presses.

4.3.3.11 Guard material

To ensure a high standard of protection is afforded, guard material must have sufficient strength and durability and not interfere with the functioning of the machine. Typical materials include:

- sheet steel – first choice of material especially where fluids such as lubricating oil need to be contained
- weld mesh or expanded metal – useful where ventilation is required
- polycarbonate – gives clear view of work, used extensively in the food industry since it can easily be cleaned by washing. It is extremely tough but has the disadvantage of scratching easily.

4.3.3.12 Openings in guard materials

Where a guard material has openings or gaps in it, care must be exercised to ensure that the guard is positioned such that it is not possible for any parts of the human body to pass through and reach the dangerous parts. The size and shape of the opening are important in determining what parts of the body can pass through. Information on the relationship between opening size and shape and distance from the danger area is contained in publications[6,31] and BS EN standard[17].

4.3.3.13 Reaching over guards

When employing a distance guard or fence it is important that the height of the fence and its distance from the dangerous parts are sufficient to ensure the dangerous parts cannot be reached over the top of the guard. Information on the safe distance and heights is given in two BS EN standards[7,15].

4.3.4 Powered trucks

Powered trucks, which come within the equipment encompassed by PUWER 98, cover a vast range of equipment from small pedestrian operated units through the various sizes of counterbalance and reach fork-lift trucks, rough terrain trucks to the enormous straddle vehicles used for lifting and transporting 40 tonne containers. Each of these types has its own operating techniques but there is one common aspect applicable to them all – training of the operators. In general, these trucks are powerful and expensive. In inexperienced hands a great deal of costly damage can be done to and by the trucks.

4.3.4.1 Training truck operators

Training should be to a formalised programme and the trainers should be skilled in the operation of the particular type of truck, well versed in teaching skills and the training organisation should have the necessary resources. Advice on training is given in an Approved Code of Practice[18].

It is equally important that the candidates selected for training should be suitable and in particular they should:

(i) be over 18 years of age and under 60
(ii) be medically examined and passed as fit to operate the particular type of truck
(iii) have stereoscopic vision (with glasses if necessary). Monocular vision (one eye only) can cause difficulties in estimating distances
(iv) not be colour blind particularly where stacking and storage is by colour coding
(v) not be known illegal drug users.

Three major aspects of training that should be covered are:

1 Theory of operation of trucks, particularly the effects of rear wheel steering.
2 Familiarisation with the truck and its controls.
3 Operating in the working environment.

At the end of the training, the candidates should be examined on their knowledge and skills and, if found to have reached a satisfactory standard, be given a certification of competence or 'licence'. The licence should give details of the truck or trucks which the operator is competent to drive.

4.3.4.2 Conditions for the safe operation of powered trucks

Although rider operated reach and counterbalance trucks are the most common in use, the operating conditions that apply to them also largely apply to other types of truck. Legislative requirements for the safe operation of mobile work equipment are contained in part III of PUWER 98 and include:

1 Controls and structure of the truck should give good all-round visibility.

Figure 4.3.15 High lift fork truck with load rest, overhead and rear guards. (Courtesy Hyster Europe Ltd)

2 The truck should be fitted with overhead and back guards as shown in *Figure 4.3.15*.
3 State of the truck, its cleanliness and mechanical condition.
4 Floor surface – level and in good condition without potholes, capable of withstanding the point loading of truck wheels and well drained. Gullies should be covered by substantial well fitting grills or bridging plates.
5 Lighting must be adequate, whether outside or within buildings with particular attention being paid to the level of lighting in racking aisles. Care should be taken to avoid glare and areas of high light and shadow contrast. When operating away from lit areas, the truck should have its own lights. Guidance on lighting is given in publications by CIBSE[19] and HSE[20].

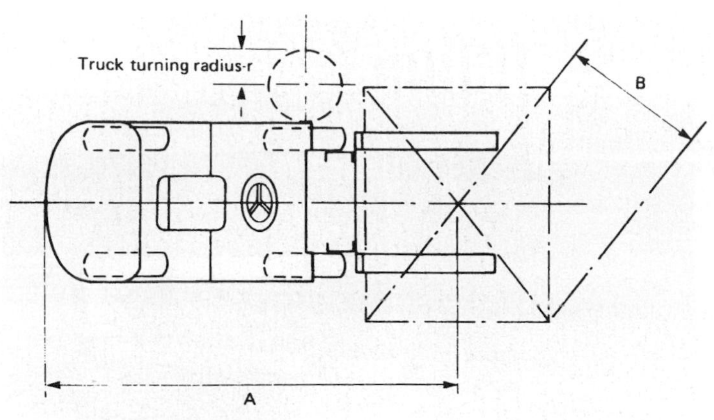

Aisle width $W = A + B + 10\%$

Example: if $A = 1800$ mm
$B = 800$ mm
aisle width $W = (1800 + 800) + 10\%$
$= 2600 + 10\% = 2860$ mm

Where trucks have a large turning circle (e.g. i/c engined trucks) the inner turning radius 'r' must be added, thus aisle width $W_1 = (A + B + r) + 10\%$

Figure 4.3.16 Width of stacking aisle for fork trucks

6 There must be adequate space to allow trucks to manoeuvre between stacking positions for positioning and recovery of loads and between vehicles being loaded or unloaded. *Figure 4.3.16* shows an empirical method for calculating the manoeuvring space a truck requires for placing loads.
7 Where ramps or slopes have to be negotiated the trucks should always move straight up or down the incline. Turning on or moving across a slope can result in the truck becoming unstable. When travelling up or down a slope, the load must be above the driver, i.e. towards the top of the slope.
8 Loads must be lowered as far as possible before being transported and never transported in the raised position.
9 Pedestrians should not be allowed in areas where ride-on trucks are operating. Truck and pedestrian routes should be separated and suitably marked, as shown in *Figure 4.3.17*.
10 However, in areas where pedestrians and trucks cannot be separated such as stores and picking areas, priority right of way should be given to pedestrians.
11 The maker's specified maximum loads should never be exceeded.
12 Ignition keys or start cards should not be left in unattended trucks to prevent use by untrained employees. However, some companies train all employees in an area as truck operators and provide trucks as a facility to be used by all.
13 Multi-item loads should be tied or bonded to prevent load movement during travel, as shown in *Figure 4.3.18*.

Figure 4.3.17 Separation of truck and pedestrian gangways

Figure 4.3.18 Bonded and wrapped pallet load

14 Passengers must not be carried unless the truck is fitted with a purpose made seat.
15 Adequate high level ventilation should be provided in the charging areas of battery electric trucks.
16 Where powered trucks are used for lifting people, special working platforms should be used and the advice given in HSE guidance note[21] followed.

Guidance on the safe operation of lift trucks is given in an HSE Guidance Booklet[22]. The main responsibility for the safe operation of powered trucks lies with the operator.

4.3.5 Lifting equipment

The term lifting equipment covers any equipment used in the process of lifting loads or people and includes lifts, cranes, hoists and lifting accessories that join the load to the crane. The legislative requirements for the construction of new lifting equipment and for its day-to-day use are contained in two separate sets of regulations. The construction of new lifting equipment is subject to SMSR supplemented by the Lifts Regulations 1997 (LR) which impose additional requirements that are specific to lifts. In use, lifting equipment is subject to PUWER 98 supplemented by the Lifting Operations and Lifting Equipment Regulations 1998 (LOLER).

4.3.5.1 Definitions of lifting equipment

Cranes are any lifting machine and associated parts where the movement of the load is not restricted by guides or rails. The description includes permanent fixed installations in buildings such as overhead travelling cranes, temporary installations typically found on construction and building sites, self-contained mobile cranes and hand-operated chain pulley blocks.

Lifting accessories are defined in SMSR as:

> '. . . components or equipment not attached to the machine and placed between the machinery and the load or on the load in order to attach it;'

and *separate lifting accessories* as:

> '. . . accessories which help to make up or use a slinging device, such as eyehooks, shackles, rings, eyebolts, etc.'

Lifts are defined in LR as:

> '. . . an appliance serving specific levels, having a car moving
> (a) along guides which are rigid, and

Table 4.3.1 Table of test coefficients for lifting equipment

Lifting equipment	Test coefficients	
	Static	*Dynamic*
Powered equipment	1.25	1.1
Manual equipment	1.5	1.1
Lifting ropes	5	
Lifting chains	4	
Separate accessories:		
Metallic rope eyes	5	
Welded link chains	4	
Textile ropes and slings	7	
Metallic components of slings	4	

(b) along a fixed course even where it does not move along guides which are rigid (for example, a scissor lift),

and inclined at an angle of more than 15 degrees to the horizontal and intended for the transport of:

- persons
- persons and goods
- goods alone if the car is accessible, that is to say, a person may enter it without difficulty, and fitted with controls situated inside the car or within reach of a person inside.'

4.3.5.2 Construction of lifting equipment

4.3.5.2.1 Cranes

Cranes and their accessories are work equipment and as such must be designed and manufactured to conform with SMSR with supporting documentation as evidence of conformity. In addition, before they are put into service they must be subjected to the tests summarised in *Table 4.3.1*. The supplier should issue a Test Certificate on completion of the test.

4.3.5.2.2 Lifts

The manufacture of lifts follows the same procedural requirements as other work equipment but with the added requirements contained in the Lifts Regulations (LR) which recognises the value of quality assurance schemes and also the fact that many of the components may be supplied by specialist manufacturers. Similar obligations are placed on both the lift manufacturer (reg. 8) and the component manufacturer (reg. 9) to ensure their products meet the required standard. These obligations include:

- lifting equipment and components must satisfy the appropriate ESRs. Evidence of this is through compliance with a harmonised (EN) standard
- carrying out a conformity assessment
- drawing up a Declaration of Conformity
- affixing the CE mark to the inside of the lift car or the component
- ensuring it is, in fact, safe.

The Lifts Regulations also recognise the important role quality assurance schemes play in ensuring high standard of product, and consequently safety, and use it as a core requirement in the conformity assessment procedure. The conformity assessment (reg. 13) is undertaken by 'notified bodies' who:

- may carry out unannounced inspections during manufacture
- examine and check details of the quality assurance scheme under which the lift or component was manufactured
- carry out a final inspection.

In an alternate certification procedure, the lift maker can request the notified body to carry out a 'unit verification' on his product to confirm that it conforms to the requirements of the Regulations.

Where a quality assurance scheme has been part of the manufacturing process of a lift but the design has not been to harmonised standards, the manufacturer can request the notified body to check that the design complies with the requirements of the Lifts Directive[23].

Notified bodies (reg. 16) are bodies or organisations with suitable technical and administrative resources to carry out inspections and conformity assessments. They are appointed by the Secretary of State who notifies the European Commission and their appointment is published in the Official Journal of the EU. When a lift is being installed, the builder and the installer are responsible for ensuring that the lift shaft contains no pipework or cabling other than that necessary for the operation of the lift (reg. 11).

In addition to the ESRs contained in SMSR, lifts must meet the ESRs listed in the Lifts Regulations which include:

- Take precautions to prevent the car falling, such as double suspension ropes or chains, the incorporation of an arrester device and means to support the car in the event of a power or control failure.
- Ensure the functions of the controls are clearly indicated and that they can be reached easily, especially by disabled persons.
- The doors of the car and at the landings must be interlocked to prevent movement of the car when any of the doors are open or prevent any doors being opened except when a car is at the landing.
- Access to the lift shaft must not be possible except for maintenance or in an emergency and there must be arrangements at the ends of travel to prevent crushing.
- The car must be provided with:
  - suitable lighting

- means to enable trapped persons to be rescued
- a two-way communication system to contact emergency services
- adequate ventilation for the maximum allowed number of passengers
- a notice stating the maximum number of passengers to be carried.

For information on the detailed requirements, the Regulations and appropriate EN standards should be consulted.

4.3.5.3 Safe use of lifting equipment

The requirements to be met for the safe use of lifting equipment are contained in PUWER 2 supplemented by the Lifting Operations and Lifting Equipment Regulations 1998 (LOLER) and a supporting Approved Code of Practice[24]. These Regulations cover all work equipment for lifting loads including accessories that connect the load to the crane and they revoke the Hoists Exemption Order 1962. A load is defined to include persons (reg. 2). These Regulations are proscriptive and risk based and require the carrying out of risk assessments of lifting operations.

The obligations imposed on the employer (reg. 3) have been extended to the self-employed, to anyone who has control of lifting equipment and to anyone who controls the way lifting equipment is used.

Lifting equipment must be suitable for its purpose (reg. 4) and constructed of materials of adequate strength with a suitable factor of safety taking account of any hostile working environment. It should be stable when used for its intended purpose and this is particularly pertinent for mobile lifting equipment which should be provided with outriggers. Access to operating positions and, where necessary, other parts should be safe and precautions should be taken to prevent slips, trips and falls whether on the equipment itself or when moving in the work area during a lifting operation. Protection must be provided for the operator especially where he is likely to be exposed to adverse weather. Instruments should be provided to detect dangerous weather conditions such as high winds so precautions can be taken and, if necessary, the equipment taken out of use.

Additional measures have to be taken for lifts that carry people (reg. 5) including enhanced strength of lifting ropes, means to prevent crushing or trapping, falling from a carrier and to allow escape from a carrier in an emergency. The lifting equipment should be positioned to minimise the risk of equipment or load striking someone (reg. 6), loads should not be carried over people and hooks should have safety catches. Where carriers pass through shafts or openings in floors, the openings should be fenced to prevent anyone falling through.

The safe working load or maximum number of passengers, as appropriate, should be marked on all lifting equipment (reg. 7). All lifting operations should be properly planned and supervised (reg. 8) and measures taken to ensure that no loads pass over places where people are working and that people do not work under suspended loads. The operator should have a clear view of the load or be directed by a banksman using signs or signals clearly understood by himself and the operator. Lifting equipment should not be used for operations likely to cause it to overturn, for dragging loads or used in excess of its safe working load. Lifting accessories should be used within their safe working loads and stored where they will not deteriorate or be damaged.

All lifting equipment must be regularly inspected (reg. 9) to a programme laid down either as a result of an assessment of its use or based on past experience. The inspection should be carried out by someone competent and knowledgeable in the equipment – such as an insurance surveyor – and a report containing the prescribed particulars prepared for the employer. Any faults affecting the safe operation must be reported to the enforcing authority (reg. 10). Reports of inspections and documents accompanying new equipment must be kept available for inspection (reg. 11).

The requirements of these Regulations are more flexible in implementation than earlier prescriptive requirements and allow realistic duties to be developed to match the actual conditions of use.

4.3.5.3.1 Safe use of cranes

Perhaps the most commonly used piece of handling equipment is the crane, which over the years has been developed to meet highly specialised applications, with the result that there is now a great range of types and sizes in use in industry, the docks and on construction sites. As a result of accidents in the past, a body of legislation has grown up which covers the construction and use of cranes. This body of legislation has been consolidated into SMSR for the design and manufacture and LOLER for the safe use and periodic inspections of cranes. There are a number of common techniques and safety devices that contribute to the safe operation of cranes and some of these are summarised below:

Overtravel switches

To prevent the hook or sheave block from being raised right up to the cable drum, a robust limit switch should be fitted to the crab or upper sheave block. Checks of this limit switch should be included in routine inspections.

Protection of bare conductors

Where bare pick-up conductors are used to carry the power supply they must be shielded from accidental contact particularly if near cabin access. Suitably worded notices, e.g. WARNING – BARE LIVE WIRES, should be posted on the walls or building structure. The power supply isolating switch should be provided with means for locking-off during maintenance work.

Controls

The controls of cranes, whether cabin, pendant or radio, should be clearly identified to prevent inadvertent operation. On overhead electric travelling (OET) cranes with electric pendant controls the directions of travel should be unambiguously marked. Controls should be of the 'dead-man' type.

Load indicators
Load indicators are required to be fitted to jib cranes and can be used with benefit on all cranes.

Safety catches
Crane hooks should be fitted with safety catches to prevent slings, chains, ropes etc. from 'jumping' off the hook.

Emergency escape
Where, on travelling cranes, access to the cab is not an integral part of the crane, suitable escape equipment should be provided to enable the driver to reach the ground quickly and safely in an emergency.

Access
Safe means of access should be provided to enable:

1 the driver to reach his operating position;
2 the necessary inspections and maintenance work to be carried out safely.

Operating position
The arrangement of the driver's cab should ensure:

1 a clear view of the operating area and loads;
2 all controls are easily reached by the driver without the need for excessive movement of arms or legs;
3 all controls are clearly marked as to their function and method of operation.

Passengers
No one, other than the driver, should be allowed on the crane when it is operating unless there is a special reason for being there and it has been authorised. 'Riding the hook' is prohibited but should it be necessary to carry persons, the properly designed and approved chair or cradle should be used.

Safe working load
All cranes should be marked with their safe working load which must never be exceeded except for test purposes. If there is any doubt of the weight to be lifted, advice should be sought.

Controlling crane lifts
With many cranes including overhead electric travelling, mobile jib and construction tower cranes, the safe moving of loads relies on team effort involving the driver, slinger and sometimes a separate signaller (or banksman). Only one person, the signaller or if there is no signaller the slinger, should give signals to the driver and these should be clearly

understood by both. The basic signals[25] shown in *Figure 4.3.19* are similar to those given in an EU Directive[37] and an HSE publication[38].

Slingers, signallers and drivers should be properly trained, medically fit and of a steady disposition. Detailed advice on the safe use of cranes, lifting accessories and mobile cranes is given by Dickie, Short and Hudson[26,27,28].

4.3.6 Pressure systems

Pressure systems refer to any system of pipes, vessels, valves or other equipment for containing or transferring gases and liquids at high pressure.

However, as a result of moves to comply with EU directives, new legislation in respect of pressure systems has polarised into two discrete areas, manufacture of systems and their use. The earlier Pressure Systems and Transportable Gas Containers Regulations 1989 have been revoked and requirements concerning transportable gas containers have been incorporated into the Carriage of Dangerous Goods (Classification, Packaging and Labelling) and Use of Transportable Pressure Receptacles Regulations 1996[32].

4.3.6.1 Pressure equipment

The legislation on pressure equipment, the Pressure Equipment Regulations 1999[33] (PER), is concerned with the quality of the equipment that is manufactured and supplied and incorporates the requirements of the Pressure Equipment Directive[34]. This Directive is aimed at reducing the barriers to trade in respect of pressure equipment.

The Regulations define pressure equipment as:

> Vessels, piping, safety accessories and pressure accessories; where applicable, pressure equipment includes elements attached to pressurised parts, such as flanges, nozzles, couplings, supports, lifting lugs, and similar;

and fluid as:

> Gases, liquids and vapours in pure phase as well as mixtures thereof; a fluid may contain a suspension of solids.

It divides fluids into two groups, Group 1 are those fluids which are in themselves hazardous to health, i.e. explosive, flammable, toxic or oxidising. All other fluids are in Group 2.

The Regulations apply to all pressure equipment where the contained pressure exceeds 0.5 bar above atmospheric pressure (7.25 psig). No pressure equipment may be put on the market unless it complies with these Regulations.

NOTE The signaller should stand in a secure position where HE CAN SEE THE LOAD AND CAN BE SEEN CLEARLY by the driver. Face the driver if possible. Each signal should be distinct and clear.

Figure 4.3.19 Crane signals (BS 7121)

A number of pressure equipment and assemblies, listed in schedule 1 of the Regulations, are excluded from these requirements.

Regulation 7 qualifies the different vessels and systems covered, using a measure of either bar-litre (bar-L) (pressure in bars × volume in litres) or a maximum allowable pressure (PS) as the criteria and includes all:

(a) Unfired vessels handling fluids in Group 1 which must comply where bar-L > 25 or PS > 200 bar and those handling fluids in Group 2 where the criteria are bar-L 50 > or PS > 1000 bar. All fire extinguishers and breathing apparatus air bottles are included.
(b) Fired and heated vessels for the generation of steam or super- heated water, where there is a risk of overheating, operating at more than 110°C and having a volume > 2L. This includes pressure cookers.
(c) Piping handling Group 1 fluids having a nominal bore (ND) >25 mm and Group 2 fluids having an ND > 32 mm and a product of ND× PS > 1000 bar. For piping containing liquids whose vapour pressure at the maximum allowable temperature (TS) < 0.5 bar handling Group 1 fluids where ND > 25 mm and ND × PS > 2000 bar and Group 2 fluids where ND > 200 mm and ND × PS > 5000 bar.

Under reg. 8 all pressure equipment and systems that come within the scope of the Regulations must:

i. Satisfy the relevant *essential safety requirements* (ESRs) listed in schedule 2 of the Regulations. Conformity with a pertinent harmonised standard presumes compliance with the ESRs.
ii. Have been subject to the appropriate conformity assessment procedure which is outlined in schedule 3 of the Regulations.
iii. Carry the CE mark
iv. In fact, be safe.

Pressure systems and assemblies used for experimental purposes are excluded from these requirements.

Any pressure equipment must carry the CE mark (reg. 9) and:

(a) be designed and manufactured in accordance with sound engineering practice in order to ensure it is safe;
(b) be accompanied by adequate instructions for its safe use;
(c) have adequate identification marks; and
(d) be safe.

These requirements do not apply to pressure equipment for use outside the EU, whether the equipment was manufactured in or just handled by a Member State.

Pressure equipment is divided into four categories determined by the type of vessel or system as described in reg. 7 and by its bar-L relationship according to tables in schedule 3. For each category the relevant conformity assessment procedure is listed in reg. 13(3) according to a series of modules described in schedule 4. Conformity assessments may only be carried out by *notified bodies* who have been either approved by the Secretary of State or notified to the EU commission.

Regulation 23 recognises that any pressure equipment or assembly that carries the CE mark and is accompanied by a *declaration of conformity* complies with these Regulations. Non-compliance is an offence that, on conviction, carries a custodial sentence, a fine or both. However a defence of *due diligence* is allowed.

4.3.6.2 Pressure systems safety

Once pressure equipment and assemblies have been installed and put to work, it is essential that they are used and maintained in a manner that ensures they remain safe throughout their operating life. Criteria, procedures and requirements for ensuring this are contained in the Pressure Systems Safety Regulations 2000 (PSSR)[35] which refers not only to newly purchased and commissioned plant but also to pressure equipment and systems that have been in service for a number of years. The definitions contained in PER and PSSR are complementary but with important differences in the definition of *fluid*. PER is concerned with the pressure element of contained fluids and with the safety integrity of the containing vessels and pipework in preventing failures and leaks. PSSR, on the other hand, is concerned with protecting the operator and others from the harmful effects of escaping fluids, particularly steam at any pressure at or above atmospheric with its potential to cause harm, such as scalds and burns, resulting from its high level of latent heat. Conversely, PSSR is not concerned with the chemical and biological hazards of the contained fluids since these are the subject of other statutory provisions. PSSR is supported by an ACoP[36]. Because of the widely differing operating circumstances of pressure systems, the regulations recognise the need for a flexible approach to ensuring safe operation and acknowledge the value of an operating system based on an assessment of the possible risks should the system fail.

The Regulations apply to all who design, manufacture, import, supply or use any pressure system or vessel for work purposes, whether for profit or not, but individual responsibilities extend only to matter under a person's direct control. There are a number of exclusions listed in Schedule 1 which largely refer to pressure systems that are a necessary ancillary part of other equipment or processes.

Pressure systems must (reg. 4):

i. be properly designed and constructed to prevent danger over the whole of its expected operational life with allowance made for the characteristics of the fluid contained;
ii. allow any examination necessary for ensuring the safe operation of the system to be carried out;
iii. ensure that any access into vessels can be made without danger;
iv. be provided with suitable safety devices which, if they release the contents, do so safely.

Where a pressure system is designed, supplied or modified, the person carrying out the work must provide all the information necessary for the safe operation and maintenance of the system (reg. 5). If vessels are involved, they must be marked with:

- manufacturer's name;
- identifying number;
- date of manufacture;
- standard to which the vessel was built;
- maximum (or minimum) allowable pressure; and
- the design temperature.

Any imported vessel must carry the same information.

When a pressure system is being installed (reg. 6), the installer should ensure:

- only competent workmen and supervision are employed;
- components have adequate foundations and supports;
- suitable lifting equipment is available;
- the component parts are in good order and are protected from damage;
- access for operating and carrying out examinations is not obstructed; and
- the system is cleaned before being put into operation.

The user is responsible for ensuring the pressure system is operated within specified safe limits and that the design conditions are not exceeded. A written scheme of examinations (reg. 8) must be prepared by a competent person before the system is put into operation and should include details of:

- the operating conditions on which they are based;
- the nature and frequency of examinations;
- the preparations necessary for carrying out of examinations; and
- the initial examination before the system is put to work.

Provisions should be made for the recording and storing the results of these examinations (reg. 14). If operating conditions change or the system is modified, the written scheme should be reviewed and adjusted accordingly. Examinations are to be carried out in accordance with the written scheme (reg. 9). A written copy of the report of the examination has to be sent to the user within 28 days or if the user carries out the examination, the report must be completed within 28 days. The report should:

- list the parts examined;
- detail any repairs necessary to maintain the safety of the system and the date by which those repairs must be completed;
- nominate the date for the next examination; and
- comment on the adequacy of the written scheme.

Where repairs are necessary, the system must not be run until the repairs have been completed. If it is necessary to operate the system beyond the date of the next due examination without it being examined, this should only be with the agreement of the examiner and after notifying the enforcing authority. On mobile systems, the date of the next examination must be clearly and indelibly marked.

Preparations necessary to ensure that examination can be carried out safely should include:

- ensuring the system is cool;
- dispersing any toxic or harmful gases or fumes;
- if the lagging contains asbestos, warning the examiner and ensuring appropriate precautions are taken;
- providing suitable means of access, including staging if necessary;
- isolating the system from others that may still be pressurised; and
- removing components and safety devices as appropriate.

If the examiner is of the opinion that the system is likely to cause imminent danger unless repairs are carried out (reg. 10), he must notify the user in writing immediately specifying the necessary repairs. A copy of the report must be sent to the enforcing authority within 14 days. The user must ensure that the system is not used until the required repairs have been carried out. It is the responsibility of those carrying out the work to ensure that any modification or repairs do not reduce the safety of operation of the system (reg. 13).

The user of a pressure system must ensure that the system operators are fully instructed in the safe operating techniques (reg. 11) including maximum operating limits and the action to be taken in an emergency. Operating instructions should include start-up and shut-down procedures, the functions and use of controls and dealing with potentially hazardous situations. Any vessel intended to be operated at atmospheric pressure (reg. 15) should be provided with a secure, unrestricted, connection direct to atmosphere.

Pressure systems must be kept properly maintained (reg. 12) to ensure that they can continue operating with safety. The extent of the maintenance will be determined by consideration of:

(a) the age of the system;
(b) the materials contained and operating conditions;
(c) the working environment;
(d) the maker's maintenance recommendations;
(e) the maintenance history and modifications made;
(f) recommendations from periodic examinations; and
(g) the results of a risk assessment of the likely effects of a system failure.

Records must be kept (reg. 14) of:

- the technical and operating instructions provided by the supplier;
- all reports of periodic examinations;

- details of any modifications or repairs; and
- the maintenance and operating logs of the system.

The records should be kept at either the operating site of the system or the office from which the system is controlled. They can be written (hard copy) or in electronic form provided they cannot be interfered with, i.e. read-only format. Whichever method is used it must be such that a hard copy can easily be produced if requested by the enforcing authority.

In any proceedings for an offence under these Regulations, a defence can be pleaded that:

i. the offence was due to the act or default of another person; or
ii. all due diligence had been exercised.

In either case, the defendant must justify his defence plea.

4.3.7 Coda

The use of machines and machinery is an essential fact of industrial and commercial life. But there is no reason why the use of this equipment should cause damage or injury. Over the years, manufacturing standards and operating techniques have been developed that ensure safe and efficient operation over the economic life of the equipment. It is up to the maker and user to ensure that those standards and practices are adhered to so that any risks from the use of the equipment are kept to a minimum.

References

1. European Union, *Council Directive on the approximation of the laws of Member States relating to machinery*, No. 89/392/EEC as amended by Directive No 91/368/EEC, EU, Luxembourg (1991) and consolidated in Directive No. 98/37/EC
2. European Union, *Council Directive concerning the minimum health and safety requirements for the use of work equipment by workers at work*, as amended by Directive No. 95/63/EC, EU, Luxembourg (1995)
3. Health and Safety Executive, Legal Series booklet No. L22, *Safe use of work equipment. Provision and Use of Work Equipment Regulations 1998. Approved Code of Practice.* HSE Books, Sudbury (1998)
4. British Standards Institution, BS EN 953, *Safety of machinery – Guards – General requirements for the design and construction of fixed and movable guards*, BSI, London (1998)
5. British Standards Institution, BS EN 1088, *Safety of machinery – Interlocking devices associated with guards – Principles for design and selection*, BSI, London (1995)
6. British Standards Institution, Published Document, PD 5304:2000, *Safe use of machinery*, BSI, London (2000)
7. British Standards Institution, BS EN 294, *Safety of machinery – Safety distances to prevent danger zones being reached by upper limbs*, BSI, London (1992)
8. British Standards Institution, BS EN 349, *Safety of machinery – Minimum gaps to avoid crushing of parts of the human body*, BSI, London (1993)
9. British Standards Institution, BS EN 999, *Safety of machinery – The positioning of protective equipment in respect of approach speeds of parts of the human body*, BSI, London
10. Uddin *v.* Associated Portland Cement Manufacturers Ltd [1965] 2 All ER 213

11. Close *v.* Steel Company of Wales [1962] AC 367; [1961] 2 All ER 953, HL
12. British Standards Institution, BS EN 292–1, *Safety of Machinery – Basic concepts and general principles for design – Part 2: Basic terminology and methodology*, BSI, London (1991)
13. British Standards Institution, BS EN 292–2, *Safety of machinery – Basic concepts and general principles for design – Part 2: Technical principles and specifications*, BSI, London (1991) and Part 2/Al (1995)
14. British Standards Institution, BS EN 1050, *Safety of machinery – Principles for risk assessment*, BSI, London (1997)
15. British Standards Institution, BS EN 811, *Safety of machinery – Safety distances to prevent danger zones being reached by lower limbs*, BSI, London (1997)
16. British Standards Institution, BS EN 574, *Safety of machinery – Two-hand control devices – Functional aspects – Principles for design*, BSI, London (1996)
17. British Standards Institution, BS EN 547, *Safety of machinery – Human body dimensions (7 parts)*, BSI, London (parts 1 & 2, 1996)
18. Health and Safety Executive, Publication L117, *Rider operated lift trucks: Operator training*, Approved Code of Practice and guidance, HSE Books, Sudbury (1999)
19. Society of Light and Lighting, *Code for Lighting 2000*, CIBSE, London (2000)
20. Health and Safety Executive, Guidance booklet No. HSG 38, *Lighting at work*, HSE Books, Sudbury (1998)
21. Health and Safety Executive, Guidance Note No. PM28, *Working platforms on fork lift trucks*, HSE Books, Sudbury (2000)
22. Health and Safety Executive, Guidance booklet No. HSG 6, *Safety in working with lift trucks*, HSE Books, Sudbury (2000)
23. European Union, *Council Directive on the approximation of the laws of Member States relating to lifts*, No. 95/16/EC, EU, Luxembourg (1995)
24. Health and Safety Executive, Publication L113, *Safe use of lifting equipment, Lifting Operations and Lifting Equipment Regulations 1998. Approved Code of Practice*, HSE Books, Sudbury (1998)
25. British Standards Institution, BS 7121, *Code of practice for the safe use of cranes, Part 1 – General; Part 2 – Inspection, testing and examination*, BSI, London
26. Dickie, D.E., *Lifting Tackle Manual* (Ed. Douglas Short), Butterworths, London (1981)
27. Dickie, D.E., *Crane Handbook*; (Ed. Douglas Short), Butterworth, London (1981)
28. Dickie, D.E., *Mobile Crane Manual* (Ed. Hudson, R.W.), Butterworth, London (1985)
29. Health and Safety Executive, Publication L112, *Safe use of power presses*, HSE Books, Sudbury (1998)
30. Health and Safety Executive, Publication L114, *Safe use of woodworking machinery*, HSE Books, Sudbury (1998)
31. Ridley, J. and Pearce, R., *Safety with machinery*, Butterworth-Heinemann, Oxford (2002)
32. *The Carriage of Dangerous Goods (Classification, Packaging and Labelling) and Use of Transportable Pressure Containers Regulations 1996*, The Stationery Office, London (1996)
33. *The Pressure Equipment Regulations 1999*, The Stationery Office, London (1999)
34. European Union, *Council Directive on the approximation of the laws of Member States concerning pressure equipment*, No. 97/23/EC, EU, Luxembourg (1997)
35. *The Pressure Systems Safety Regulations 2000*, The Stationery Office, London (2000)
36. Health and Safety Executive, Publication L122, *Safety of pressure systems. Pressure Systems Safety Regulations 2000. Approved Code of Practice*, HSE Books, Sudbury (2000)
37. European Union, *Directive on the minimum requirements for the provision of safety signs at work*, No. 92/58/EEC, EU, Luxembourg (1992)
38. Health and Safety Executive, Publication L64, *Safety signs and signals. The Health and Safety (Safety Signs and Signals) Regulations 1996*, HSE Books, Sudbury (2000)

Further reading

King, R. W. and Magid, J., *Industrial Hazard and Safety Handbook*, pp. 567–603, Butterworth-Heinemann, Oxford (1979)

Ridley, J. and Pearce, R., *Safety with machinery*, Butterworth Heinemann, Oxford (2002)

Health and Safety Executive, the following publications which are available from HSE Books, Sudbury:
Guidance booklets:
HSG 39 *Compressed air safety* (1998)
HSG 54 *The maintenance, examination and testing of local exhaust ventilation* (1998)
HSG 87 *Safety in the remote diagnosis of manufacturing plant and equipment* (1995)
HSG 89 *Safeguarding agricultural machinery. Advice for designers, manufacturers, suppliers and users* (1998)
HSG 93 *The assessment of pressure vessels operating at low temperatures* (1993)
HSG 113 *Lift trucks in potentially flammable atmospheres* (1996)
HSG 129 *Health and safety in engineering workshops* (1999)
HSG 136 *Workplace transport safety: Guidance for employers* (1995)
HSG 172 *Health and safety in sawmilling. A run-of-the-mill business?* (1997)
HSG 180 *Application of electro-sensitive protective equipment using light curtains and light beam devices to machinery* (1999)
Guidance notes:
PM24 *Safety in rack and pinion hoists* (1981)
PM28 *Working platforms on lift trucks* (2000)
PM55 *Safe working with overhead travelling cranes* (1985)
PM63 *Inclined hoists used in building and construction work* (1987)
PM65 *Worker protection at crocodile (alligator) shears* (1986)
PM66 *Scrap baling machines* (1986)
PM73 *Safety at autoclaves* (1998)
PM79 *Power presses. Thorough examination and testing* (1995)
PM83 *Drilling machines. Guarding of spindles and attachments* (1998)

Chapter 4.4

Electricity

E. G. Hooper and revised by Chris Buck

4.4.1 Alternating and direct currents

4.4.1.1 Alternating current

An alternating current (ac) is induced in a conductor rotating in a magnetic field. The value of the current and its direction of flow in the conductor depends upon the relative position of the conductor to the magnetic flux. During one revolution of the conductor the induced current will increase from zero to maximum value (positive), back to zero, then to maximum value in the opposite direction (negative) and, finally, back to zero again having completed one cycle. A graph plotted to show the variation of this current with time follows a standard sine wave. The number of cycles completed per second, each comprising one positive and one negative half cycle, is referred to as the frequency of the supply, measured in hertz (Hz). Mains electricity is supplied in the UK as ac at a nominal frequency of 50 Hz (50 cycles per second).

4.4.1.2 Direct current

Direct current (dc) has a constant positive value above zero and flows in one direction only, unlike ac. A simple example of direct current is that produced by a standard dry battery.

DC is really ac which has its positive or negative surges rectified to provide the uni-directional flow. In a dc generator the natural ac produced is rectified to dc by the commutator.

DC will be found in industry in the form of battery supplies for electrically powered works plant, such as fork lift trucks, together with associated battery charging equipment. Otherwise, dc, obtained by rectification of the mains ac supply, will be encountered only for specialist applications, e.g. electroplating.

Danger from electricity may arise irrespective of whether it is ac or dc. Where dc is derived from ac supply the process of rectification will result in some superimposed ripple from the original ac waveform. Where this exceeds 10%, the electrical shock hazard must be considered to be the same as for an ac supply of equivalent voltage. Additionally, both ac and dc can cause injury as a result of short circuit flashover. The dangers of electricity are discussed in more detail later in this chapter.

4.4.2 Electricity supply

Alternating current electricity is generated in thermal power stations (coal-, gas- and oil-fired), in nuclear power stations, as well as using natural resources (e.g. wind farms). It is then transmitted by way of overhead lines at 400 kV (i.e. at 400 000 volts) (*Figure 4.4.1*), 275 kV or 132 kV to distribution substations where it is transformed down to 33 kV or 11 kV for distribution to large factories, or further transformed down to 230 V for use in domestic and commercial premises and smaller factories.

Following privatisation, many changes have and are still taking place in the electricity industry. The demand for electricity varies considerably from day to day as well as throughout each day. Generation must be matched to meet this continually varying demand, ensuring that there is always sufficient generating capacity available at minimum cost. This is achieved through arrangements operated by the electricity pool. There is now an open market for all power users to shop around suppliers to achieve the best price.

Electricity is received by most industrial and commercial consumers as a 'three-phase four-wire' supply at a nominal voltage of 400/230 V. The three phases are distinguished by the standard colours red, yellow and blue, the fourth wire of the supply serving as a common neutral conductor. There are proposals to adopt European standard phase colours at a future date which will result in a change from the present colours. The individual phase voltages (230 V) are equally displaced timewise (phase displacement of 120°) and because of this the voltage across phases is higher. Three-phase supplies can be used to supply both single-phase equipment (with the loads balanced as equally as possible between the three phases) or three-phase equipment such as to large motors.

Consumers with large energy requirements often find it more economical and convenient to receive electricity at 11 kV and to transform it down to the lower values as required. In all cases, however, it is essential that the consumer's electrical installation and equipment, both fixed and portable, meet good standards of design, construction and protection, are adequately maintained and correctly used.

The British Standards Institution[1] (BSI) issues standards and codes giving guidance on electrical safety matters. One such standard is BS 7671, otherwise known as the IEE Wiring Regulations[2]. These non-statutory Regulations specify requirements for low voltage electrical installations, i.e. those operating at voltages up to 1000 V. BS 7671 takes account of the technical matters contained in a number of European Standard Harmonisation Documents published by the European Committee for Electrotechnical Standardisation (CENELEC). However, it is important to appreciate the legal obligations relating to the safe use of electricity and electrical machinery at places of work.

Figures 4.4.1 400 kV suspension towers on the National Grid's Sizewell-Sundon 400 kV transmission line. (Courtesy National Grid)

The Health and Safety at Work etc. Act 1974 (HSW) is an enabling Act providing a legal framework for the promotion of health and safety at all places of work. Although the Act says nothing specific about electricity it does require, among other things, the provision of safety systems and methods of work; safe means of access, egress and safe places of employment; and adequate instruction and supervision. These requirements have wide application but they are, in general terms, also relevant to the safe use of electricity. But for specific advice on the electrical legal requirements we must turn to a set of Regulations[3] made under the HSW Act.

4.4.3 Statutory requirements

4.4.3.1 The Electricity at Work Regulations 1989

The Electricity at Work Regulations 1989 (EAW) is the primary piece of legislation dealing specifically with electricity and came into force on 1 April 1990. Since these Regulations were made under the umbrella of the HSW they apply in all cases where the parent Act applies. They are thus work activity, rather than premises, related and are therefore of wide application. The Regulations do not implement a corresponding EU Directive, as is the case with other health and safety legislation, and therefore the requirements are enforceable only in Great Britain. Some of the individual regulations are relevant to all industries while others apply only to mines. Separate, but virtually identical, Regulations have been made for Northern Ireland – the Electricity at Work Regulations (Northern Ireland) 1991. The requirements of the Regulations now also apply to offshore installations by virtue of the Offshore (Electricity and Noise) Regulations 1997.

In line with modern health and safety legislation, the EAW are 'goal setting' aimed at specifying, albeit in general terms, the fundamental requirements for achieving electrical safety. Thus they provide flexibility to accommodate future electrical developments. They specify the ends to be achieved rather than the means for achieving them. With regard to the latter, guidance is provided in a number of booklets published by the HSE as well as in BSI and other authoritative guidance. The main supporting documents are:

1 a Memorandum of Guidance[4], and
2 two Approved Codes of Practice[5,6] dealing respectively with the use of electricity in mines and in quarries.

The Memorandum of Guidance referred to above gives technical and legal guidance on the Regulations and provides a source of practical help. Similarly, the two Codes of Practice provide essential advice for mines and quarries.

The Electricity at Work Regulations comprise 33 individual regulations which place firm responsibilities on employers, the self-employed, managers of mines and of quarries and employees to comply as far as they relate to matters within their control. Additionally, employees have a duty to co-operate with their employers so far as is necessary for the employers to comply with the Regulations.

Topics covered by particular regulations include:

(a) Construction and maintenance of systems, work activities and protective equipment (Reg. 4).
(b) Strength and capability of electrical equipment (Reg. 5).
(c) Adverse and hazardous environments (Reg. 6).
(d) Insulation, protection and placing of conductors (Reg. 7).
(e) Earthing and other suitable precautions (Reg. 8).
(f) Integrity of 'referenced' conductors (Reg. 9).
(g) Connections (Reg. 10).
(h) Means for protecting from excess current, for cutting off supply and isolation (Regs. 11 and 12).
(i) Precautions for work on dead equipment and on or near live conductors (Regs. 13 and 14).
(j) Working space, lighting and access (Reg. 15).
(k) Persons to be competent to prevent danger and injury (Reg. 16).
(l) Regulations applicable to mines only (Regs 17–28).

4.4.3.2 Status of regulations

Certain of the individual regulations are subject to the qualification 'so far as is reasonably practicable'. This means that any action contemplated should be based on a judgement balancing the perceived risk against the cost of eliminating it, or at least reducing it to an acceptable level; in other words a risk assessment.

The remaining regulations are of an 'absolute' nature, which means that their requirements must be met regardless of cost. Nevertheless, in the event of a criminal prosecution for an alleged breach of statutory duty under one of these regulations, regulation 29 allows a defence to be pleaded that all reasonable steps were taken and all due diligence was exercised to avoid the commission of the offence.

4.4.4 Voltage levels

Unlike their predecessors, the 1989 Regulations apply equally to all systems and equipment irrespective of the voltage level. The duty is to avoid danger and prevent injury from electricity. Voltage is but one factor determining the presence of danger and, therefore, the risk of injury; examples of other matters requiring consideration when evaluating the electrical risk are the equipment type, its standard of construction and the nature of the work environment.

4.4.5 Electrical accidents

Electricity is a safe and efficient form of energy and its benefits to mankind as a convenient source of lighting, heating and power are obvious. But, if electricity is misused, it can be dangerous – a statement of the obvious, but one which must be made so as to keep the matter in proper perspective.

In the UK, every year, up to 20 people may be killed, at work, as a result of an electrical accident. In addition, around 750 or so are injured. These figures, considering the widespread use of electricity in industry and when compared with the numbers killed and injured as a result of other types of accident, are relatively small. Nevertheless, a knowledge of electrical safety is important because, by comparison with the proportion of serious injury resulting from accidents arising from all causes, an electrical accident is more likely to lead to serious injury. There is the potential also for expensive damage to plant and property due to fires of electrical origin, e.g the overloading of cables.

4.4.6 The basic electrical circuit

For an electrical current to do its job of providing lighting, heating and power, it must move safely from its source, through the conducting path and back from whence it came. In short, electric current requires a suitable circuit to assist its flow without danger. The circuit must be of suitable conducting material, e.g. copper, covered with a suitable insulating material (to stop the current 'leaking' out) such as PVC or rubber.

For an electrical current measured in amperes to flow in a circuit it requires pressure (voltage), measured in volts. As it flows it encounters resistance from the circuit and apparatus and this characteristic is measured in ohms.

This relationship between volts, amps and ohms is brought together in the famous Ohm's law known to most schoolboys. Thus, to put it simply, the current in a circuit is proportional to the voltage driving it and inversely proportional to the resistance it has to overcome:

$$\text{amps} = \frac{\text{volts}}{\text{ohms}} \quad (1)$$

Alternatively, this may be written as:

$$\text{ohms} = \frac{\text{volts}}{\text{amps}} \quad (2)$$

or

$$\text{volts} = \text{amps} \times \text{ohms} \quad (3)$$

A further useful relationship is that between power (measured in watts) and the voltage and current. Thus:

$$\text{watts} = \text{volts} \times \text{amps} \quad (4)$$

From Ohm's law, this may be expressed also as:

$$\text{watts} = \text{amps}^2 \times \text{ohms} \tag{5}$$

$$\text{or watts} = \frac{\text{volts}^2}{\text{ohms}} \tag{6}$$

4.4.6.1 Impedance

As an alternating current passes around a circuit under the action of an applied voltage it is impeded in its flow. This may be due to the presence in the circuit of resistance, inductance or capacitance, the combined effect of which is called the impedance and is measured in ohms.

In a pure resistance circuit the applied voltage has to overcome the ohmic value of the resistance as is the case for direct current (see equations (1) to (6) above).

If, however, the circuit contains inductance, such as the presence of a coil, the alternating magnetic field set up by the alternating current will induce a voltage in the coil which will oppose the applied voltage and cause the current to lag vectorially behind the voltage (up to 90° where the circuit contains pure inductance only). This property is called reactance and is measured in ohms. Sometimes a circuit may contain a capacitor: the applied voltage 'charges' the capacitor and the effect is such as to cause the current vectorially to lead the voltage. This property is also called reactance. Now most circuits contain resistance, inductance and capacitance in various quantities, and the effect of impedance is found as follows:

$$\text{impedance}^2 = \text{resistance}^2 + \text{reactance}^2$$

Strictly speaking this is a vectorial calculation. It is beyond the scope of this section to go further into these relationships and readers are referred to a standard textbook on electricity[7] and to BS 4727[8].

4.4.7 Dangers from electricity

It has been said that properly used electricity is not dangerous but out of control it can cause harm, if it passes through a human body, by producing electric shock and/or burns. Electricity's heating effect can also cause fire but we will first deal with the electric shock phenomenon.

4.4.7.1 Electric shock

If a person is in contact with earthed metalwork or is inadequately insulated from earth then, because the human body and the earth itself are good conductors of electricity, they can form part of a circuit (albeit an abnormal one) through which electricity, under fault conditions, can flow. How can such fault conditions occur?

If, for any reason, there was a breakdown of insulation in a part of an electric circuit or in any apparatus such as, say, a hand-held metal-cased electric drill, it is conceivable that current would flow external to this supply circuit, if a path were available. For example, the metalwork of the drill may be in contact with a live internal conductor at the point of insulation breakdown (an example of indirect contact). Or, take the example of someone working at a switch or socket outlet from which the cover had been removed before the electricity supply had been isolated. In such cases the person concerned could touch live metal or a live terminal and, if the conditions were right, would thereby cause an electric current to flow through his body to earth (an example of direct contact). If the total resistance of the earth fault path were of a sufficiently low value, the current could kill or maim.

Electric shock is a term that relates to the consequences of current flow through the body's nerves, muscles and organs and thereby causing disturbance to normal function. Owing to a current's heating effect the body tissue could also be damaged by burns. A particular danger with electric shock from alternating current is that it so often causes the person concerned to maintain an involuntary grip on the live metal or conductor (particularly hand-held electric tools) and this prolongs current flow. The passage of shock current through the body may interfere with the correct functioning of the lungs and heart. Death could occur when the rhythm of the heart is disturbed such as to affect blood flow and hence the supply of oxygen to the brain, a condition that is known as ventricular fibrillation. Unless prompt medical attention is given, ventricular fibrillation can be irreversible. However, it is still fortunately the case that most electric shock victims recover without permanent disability or lasting effect.

Although the effect of a direct current shock is generally not as dangerous as with ac (there is no dangerous involuntary grip phenomenon for example), it is recommended that similar precautions against shock be taken. In any case it will be recalled from section 4.1.2 that the dc electrical shock hazard could be similar to that of an equivalent ac voltage as a result of the amount of superimposed ripple.

The severity of an electrical shock depends on a number of factors, the most significant of which are the combination of the magnitude and the duration of the flow of shock current through the body.

Personal sensitivity to electric shock varies somewhat with age, sex, heart condition etc., but for an average person the relationship between shock current, and time for which the body can accommodate it, is given by a formula of the following kind:

$$\text{current} = \frac{116}{\sqrt{\text{time}}}$$

where the current is measured in milliamps (mA) and the time is measured in seconds (s). Above the duration of one heart beat a lower current threshold is recommended.

Thus a 50 mA shock current (i.e. 0.05 A) could probably flow through a body, without much danger, for up to 4 seconds; whereas a 500 mA current (0.5 A) flowing for only 50 ms (0.05 s) could be fatal.

The maximum safe 'let go' current is less than 10 mA, whereas 20 mA to 40 mA directly across the chest could arrest respiration or restrict breathing; currents above 500 mA flowing for as little as 50 ms can be fatal. However, even 'safe' currents at the level of about 5 mA to 10 mA could still cause a minor shock sensation and cause someone to fall if working at a height.

From all this it will be concluded that at normal mains voltage of 230 V, and given the average value of resistance of a human body at 1500 Ω, the current flowing through the body would, from equation (1), be a maximum of

$$\frac{230}{1500} = 0.15\,\text{A approximately (150 mA)}$$

– a dangerously high value. Under normal circumstances there will be additional resistances (or impedances as they are more correctly called) such as, for example, the resistance of the circuit, the earth electrode, and any footwear worn. It must also be remembered that body resistance varies from person to person depending upon biological, environmental and climatic conditions. But even so, given the very small value of current that could cause harm, all possible sources of contact with live electric parts must be avoided. Live work presents a high risk since a hand-to-hand shock path may be established where one hand comes into contact with an exposed live part while the other is simultaneously touching the earthed metal equipment case. The precautions to be taken are discussed in later sections of this chapter.

4.4.7.2 Burns

Burn injuries may be associated with shock and can be seen as burn marks on the body at the points of current entry and exit or may also occur in the burning of internal tissue. However, severe burn injuries are more likely to arise as a consequence of short circuit flashover. In fact the number of fatalities arising from this latter cause is similar to that resulting directly from electric shock.

Short circuit flashovers caused during the course of live work are likely to result in serious injury for the simple reason that the worker is in close proximity to and probably directly facing the equipment that has been inadvertently short circuited. The extent of the flashover will depend on the amount of electrical energy available to flow into the fault. This will be determined by the fault level (the amount of current that the incoming electrical supply is capable of feeding into the fault) and the speed of operation of the electrical protection, e.g. a fuse or circuit breaker, to interrupt the flow of fault current.

In the case of a factory installation the fault level is likely to be of the order of several thousand amperes. This large current will be capable of generating severe arcing during the short period required for the electrical protective devices to see the fault and safely disconnect the supply. Many will be aware of the flashover capability from even low voltage dc supplies, such as a 12 V car battery, if the terminals are accidentally shorted by dropping a metal tool across them.

4.4.7.3 Fires

Fires may occur due to a variety of electrical problems, in particular as a result of the overheating of cables or equipment, arcing due to loose connections or the use of unsuitable electrical equipment in a flammable atmosphere. Such problems often arise due to deficiencies in the design or construction of the electrical installation or incorrect equipment specification. The EAW address all these issues by specifying fundamental requirements to ensure that the design and construction of installations is such as to prevent, so far as is reasonably practicable, danger.

4.4.8 Protective means

4.4.8.1 Earthing and other suitable precautions

To prevent danger where it is 'reasonably foreseeable' that a conductor (other than a circuit conductor) may become charged with electricity, earthing or other suitable precautions need to be taken. In the case of earthing, all metalwork forming part of the electrical installation (metal conduit and trunking housing cables) or apparatus (metal equipment casings of switchgear, transformers, motors etc.) should be adequately and solidly connected to earth. Such earthing is provided by means of 'protective conductors' which may comprise a separate conductor, as in the 'twin and earth' cable or, where appropriate, the cable armouring or metal conduit or trunking. However, flexible or pliable conduit is not acceptable for this purpose.

It is important to ensure that the resistance of the earth return path, comprising the protective conductor and connection with earth, is as low as possible. This is to ensure that, in the event of an earth fault, there will be sufficient current to 'blow' the fuse or operate any other form of device protecting the circuit in question. The IEE Wiring Regulations (BS7671)[2] specify maximum permitted disconnection times for different types of installation. There is also a BS Code of Practice on the subject of earthing[9].

4.4.8.2 Work precautions

When work is to be carried out on a part of a circuit or piece of electrical equipment, certain precautions need to be taken to protect the worker concerned from electrical danger. The electricity supply should first of all be switched out, locked off and warning notices posted. This ensures that the circuit or apparatus being worked on is effectively electrically isolated and cannot become live.

Using a suitable voltage proving device, that part of the circuit to be worked on should be checked to ensure that it is dead before work is allowed to commence. Correct operation of the proving device should be confirmed immediately before and after use.

In some circumstances further precautions will need to be taken, such as earthing, to counter the effects of any stored or induced electrical charge. A permit to work system (PTW), explained in more detail in section 4.4.10, may also be used. Although EAW regulation 14 permits live working this must first be properly justified and then suitable precautions must be taken to prevent injury (an absolute duty!). Thus dead working is the norm and the preferred choice.

The HSE have published a number of guidance documents concerning those work activities where previous accident history has shown a need for more understanding to ensure electrical safety[10,11,12].

4.4.8.3 Insulation

Mention has already been made of the need to ensure that electrical conductors etc. are adequately insulated. Insulating material has extremely high resistance values to prevent electric current flowing through it. The principle of insulation is used when work has necessarily to be carried out at or near uninsulated live parts. Such parts should always be made dead if at all possible. If this cannot be done then properly trained people, competent to do the work, can make use of protective equipment (insulated tools, gloves, mats and screening materials) to prevent electrical shock and short circuit flashover. The provision and use of such equipment must meet the requirements of the Personal Protective Equipment at Work Regulations 1992 as well as regulation 4(4) of EAW. It is important that all protective equipment provided is suitable for the intended use (i.e. designed and constructed to an appropriate specification such as a British Standard), adequately maintained and properly used. A number of BSs cover the specification of such equipment[13,14,15,16].

4.4.8.4 Fuses

A fuse is essentially a thin wire, placed in a circuit, of such size as would melt at a predetermined value of current flow and therefore cut off the current to that circuit. Obviously a *properly rated fuse* is a most useful precaution because, in the event of abnormal conditions such as a fault, when excess current flows, the fuse would 'blow' and protect the circuit or apparatus from further damage. A fuse needs to be capable of responding to the following types of abnormal circuit conditions:

- overload
- short circuit (phase to neutral or phase to phase)
- earth fault (phase to earth).

To operate effectively and safely the fuse should be placed in the phase (live) conductor and never in the neutral conductor, otherwise even with the fuse blown or removed, parts of the circuit, such as switches or terminals, could still be live. Fuses come in various sizes with different construction characteristics and degrees of protection. Good practice advises that every fuse must be so constructed, guarded and placed as to prevent danger from such things as overheating and the scattering of hot metal when it blows. Modern cartridge fuses, the simplest variant of which is contained in the standard 13 amp fused plug, are well constructed to meet these requirements but it is difficult to tell at a glance if they have 'blown'. Simple battery continuity tests are available for easy checking and for the larger industrial sizes of cartridge fuse an automatic indication can be provided.

Overfusing, that is to use a fuse rating higher than that of the circuit it is meant to protect, is dangerous because in the event of a fault a current may flow to earth without blowing the fuse. This could endanger workpeople and the circuit or apparatus concerned. In addition it could result in the cable carrying an excessive current leading to considerable overheating with the risk of fire.

4.4.8.5 Circuit breakers

A circuit breaker, although more expensive than a fuse, has several advantages for excess current circuit protection. The principle of operation is that excess current flow is detected electromagnetically and the mechanism of the breaker automatically trips and cuts off electricity supply to the circuit it protects. A 'blown' fuse must be replaced with one with the correct current rating whereas a circuit breaker simply needs to be reset once the fault condition has been cleared. Miniature circuit breakers (mcbs) are designed for fitting into distribution boards in place of fuses. Residual current operated circuit breakers are also available to detect earth leakage current and, indeed, units are available that detect both over-current and earth leakage currents and thereby give very good circuit protection.

The majority of electric shock injuries occur when the body acts as conductor between line and earth. A general level of protection against such shocks is provided by the inclusion of a current sensitive earth leakage circuit breaker in the supply line. A typical example is shown in *Figure 4.4.2*. Residual current devices (RCDs) are discussed further in section 4.4.15.

4.4.8.6 Work near overhead lines and underground cables

Work near overhead electricity lines and underground electricity cables has caused many serious and fatal accidents over the years. The precautions to be taken are dealt with in two HSE guidance notes[10,11].

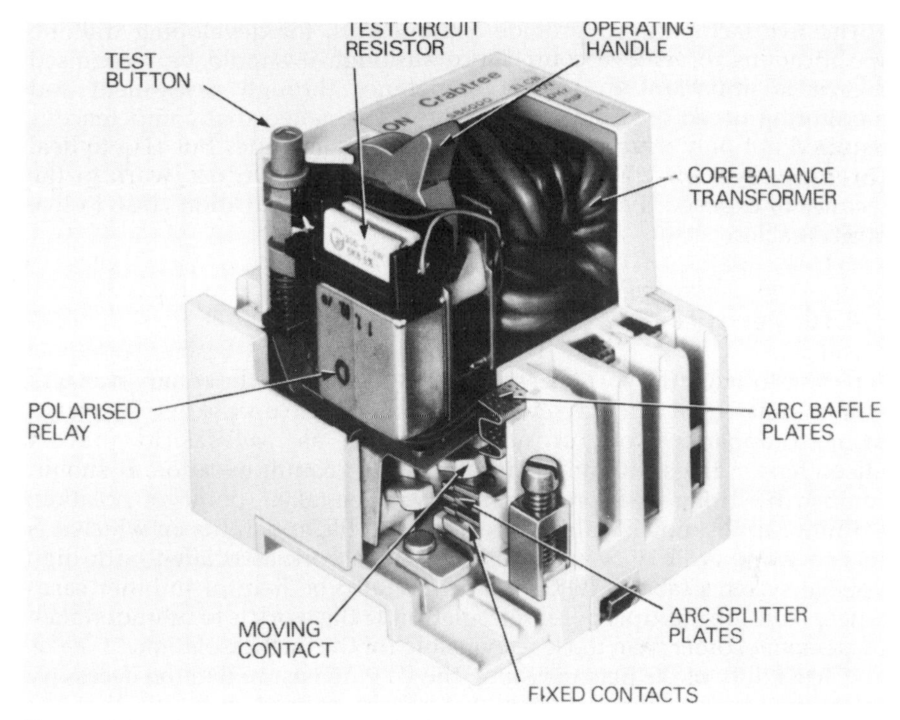

Figure 4.4.2 Cut-away illustration of a 30 mA current operated earth leakage circuit breaker. (Courtesy Crabtree Electrical Industries Ltd)

4.4.9 Competence

Regulation 16 of the EAW requires that no person shall be engaged in any work activity where technical knowledge or experience is necessary to prevent electrical danger or injury unless that person has the appropriate knowledge or experience having regard to the nature of the work. The Memorandum of Guidance[4] lists five factors to be considered when evaluation the scope of 'technical knowledge or experience'. These are:

- adequate knowledge of electricity
- adequate experience of electrical work
- adequate understanding of the equipment to be worked on
- understanding of the hazards that may arise during the work
- ability to recognise whether it is safe for work to continue.

Where technical knowledge or experience may be lacking then regulation 16 requires that the person concerned shall be under an appropriate level of supervision. The legal duty allows flexibility in that competence is required in relation to the task to be performed and the need to prevent danger and/or injury from electricity. Thus competence is not expected, nor would it be realistic to expect it, across the complete spectrum of work – only in relation to the activities in which the individual will be involved.

The five factors listed provide a framework for developing training specifications to achieve competence although it should be recognised that it is important to verify competence through assessment and monitoring on an ongoing basis. It should be noted that competence is required not only in respect of electrical work activities but also to deal with any situation where electrical danger may arise, e.g. work in the vicinity of exposed live overhead conductors or excavation close to live buried cables.

4.4.10 Permits-to-work

A permit-to-work (PTW) is an essential prerequisite to the commencement of certain classes of work involving special danger to people. A PTW serves to hold apparatus out of normal service as well as to prevent misunderstandings through a lack of or poor communication. It should confirm in writing what precautions have been taken (points of isolation, earthing, application of safety locks etc.) and the apparatus on which it is safe to work. A PTW is invariably used for work associated with high voltage systems (above 1000 V) and may also be helpful in other cases where there are multiple points of isolation or the work is to be undertaken by personnel other than those responsible for the initial isolation.

It is the duty of the person issuing the PTW to ensure that the necessary safety precautions detailed in it have been carried out, and that the person receiving the permit is fully conversant with the nature and extent of the work to be done. Proper arrangements for the issue, receipt, clearance and cancellation of PTWs are essential.

It is important to recognise that a PTW alone does not constitute the safe system of work but serves as a further precaution by confirming that the safeguards necessary as part of the system of work have been implemented. The effectiveness of a PTW system is dependent upon and is only as good as the safety culture existing in the company. A model electrical PTW is shown on pages 783 and 784.

Some companies use a multi-purpose PTW covering a range of hazards, of which electricity may be one. In other cases, PTWs are used simply as safety documents to warn of the presence of hazards such as overhead lines or underground cables. It is important, therefore, to understand the circumstances in which an electrical PTW is to be used. The generally accepted application, as covered by the model PTW, is that it is intended to be issued only in respect of apparatus that has been isolated from all supply sources and made safe for the required work. Different forms of safety document are best adopted for other types of work, e.g. a 'sanction-for-test' for live testing and a 'limitation-of-access' for work in close proximity to exposed live conductors.

4.4.11 Static electricity

When two dissimilar bodies or substances meet, electrons pass from one to the other at the surface contact area. When the bodies separate,

MODEL FORM OF PERMIT-TO-WORK – FRONT

NAME OF FIRM

PERMIT-TO-WORK

1. ISSUE NO........................

To ....

I hereby declare that it is safe to work on the following Apparatus, which is dead, isolated from all live conductors and is connected to earth:-

ALL OTHER APPARATUS IS DANGEROUS

Points at which system is isolated ....

....

....

Warning Notices posted at ....

....

....

The apparatus is efficiently connected to earth at the following points

....

....

Other precautions ....

....

....

The following work is to be carried out ....

....

....

Signed ....
being an Authorised Person possessing authority to issue a Permit-to-Work.

Time................................ Date................................

MODEL FORM OF PERMIT-TO-WORK – BACK

2. RECEIPT

I hereby declare that I accept responsibility for carrying out the work on the apparatus detailed on this Permit-to-Work and that no attempt will be made by me, or by persons under my control, to carry out work on any other apparatus.

Signed ....

Time .... Date ....

Note: After signature for the work to proceed this Receipt must be signed by and the Permit-to-Work be retained by the person in charge of the work until the work is suspended or completed and the Clearance section has been signed.

---

3. CLEARANCE

I hereby declare that the work for which this Permit-to-Work was issued is now *suspended/completed, and that all persons under my charge have been withdrawn and warned that it is no longer safe to work on the apparatus specified on this Permit-to-Work, and that gear, tools and additional earthing connections are all clear.

Signed ....

Time .... Date ....

*Delete word not applicable.

4. CANCELLATION

This Permit-to-Work is hereby cancelled.

Signed ....

being an Authorised Person possessing authority to cancel a Permit-to-Work.

particularly if they are of an insulating material, a difference of potential occurs across the separating medium which manifests itself as static electrical charge. Such an effect can, for example, occur when there is dispersion of liquid from a nozzle, powder from a tray or paper from a reel. In the movement of flammable liquid or a powder or dust with fine particles, static electricity can be generated that can give rise to sparking of sufficient energy to ignite the vapour or dust. The precautions to be taken to prevent fire and explosion as a result of static electricity depend upon the nature of the materials and process[17]. Examples of the approach to the more common problems are as follows:

1 All pipework and containers used for conveying flammable liquids should be effectively bonded together and earthed so that any static electricity produced is immediately discharged to earth before it builds up to a dangerous energy level.
2 Workshop atmospheres where flammable solvents are used for spreading as in, say, material proofing processing, can be artificially humidified. Where practicable, specialised radioactive static eliminators or earthed metal combs near the charged material etc. can also be used. It is also a wise precaution to ensure adequate ventilation such as to keep the gas/air mixture well below the lower explosive limit (LEL) of the flammable solvent concerned.
3 High speed rotating flat belts and pulleys are known to produce dangerous static charges and should not be used at or near flammable solvents. Where there are difficulties, special conducting belts, belt dressings and earthing of drive and pulley shafting can help to eliminate the build-up of static electricity.
4 Certain processes, such as electrostatic paint spraying, make use of the characteristics of static electricity and special precautions against solvent ignition are required[18].

4.4.12 Use of electricity in adverse or hazardous environments

The safe use of electricity can pose particular problems in adverse or hazardous work environments. Such environments may degrade equipment to the extent that it becomes unsafe or even develops a fault thus increasing the electric shock risk. Flammable and explosive atmospheres present special risks.

The importance of ensuring that the specification and selection of electrical equipment is appropriate to the environment and conditions of use is recognised in the EAW. Regulation 6 specifically mentions various kinds of adverse or hazardous environments, namely the effects of the weather, natural hazards, temperature, pressure, wet, dusty and corrosive conditions as well as exposure to flammable or explosive dusts, vapours or gases. The requirement is that electrical equipment which may 'reasonably foreseeably' be exposed to such conditions shall be of such construction, or as necessary protected, as to prevent, so far as is reasonably practicable, danger arising from such exposure.

Work on construction sites provides a good example of circumstances where an adverse environment may be encountered. The temporary nature of, and frequent changes to, such electrical installation may also encourage wiring and equipment to be installed in an unsuitable manner without adequate protection. Because of the special risks associated with the use of electricity on construction sites the HSE has issued guidance[10,11,19]. A number of BSs and BS ENs are concerned with the specification of electrical equipment for use in adverse environments. BS EN 60529[20], known as the 'IP' code, provides a specification for degrees of protection provided by enclosures, such as for electrical switchgear, against the ingress of solid objects, dusts and water. BS 7375[21] provides a code of practice for the distribution of electricity on construction and building sites.

4.4.13 Electrical equipment in flammable atmospheres

4.4.13.1 Explosive and flammable atmospheres

The techniques to be adopted to prevent danger when using electrical equipment in the vicinity of potentially explosive or flammable atmospheres have changed over the years and present legal requirements are contained in regulation 6 of the EAW.

4.4.13.2 Construction of equipment for use in flammable or explosive atmospheres

The construction of electrical equipment to be used where a flammable or explosive atmosphere is likely to occur must be such as to prevent ignition of that atmosphere. The selection and installation of such equipment are detailed and specialised matters requiring expert knowledge.

The relevant standards have been affected by the standard harmonisation process within the EU through the European Committee for Electrotechnical Standardisation (CENELEC) and through the International Electrotechnical Commission (IEC). Individual parts of BS EN 60079 deal with different aspects concerning electrical apparatus for potentially explosive atmospheres, namely: classification of hazardous areas[22], electrical installations in hazardous areas (other than mines)[23] and inspection and maintenance of electrical installations in hazardous areas (other than mines)[24]. BS 50014[25] covers general requirements.

The first step in the selection of appropriate equipment will be the classification of the hazardous area from the viewpoint of the likelihood of a flammable atmosphere being present. Other considerations will include temperature class or ignition temperature of the gas or vapour involved and external influences. A number of different safeguards may be employed in the design and construction of the equipment to minimise the risk of ignition. Equipment is normally certified for use in a particular situation and marked accordingly.

Currently certification in Great Britain is through the British Approval Service for Electrical Equipment in Flammable Atmospheres (BASEEFA). As a further step in the harmonisation process a new Directive, known as the ATEX Directive, will be implemented in July 2003. Thereafter, all equipment and components intended for use in potentially explosive atmospheres (gases, vapours, mists or dusts) will need to comply with the requirements of that Directive and certification will be undertaken throughout the EU by 'Notified Bodies' following the same rules and procedures.

4.4.13.3 Classification of hazardous areas

In industry, with the exception of mining, areas that are hazardous, so far as flammable gases and vapours are concerned, are classified according to the probability of occurrence of explosive concentrations of gas or vapour. These classifications, called zones, are as follows:

Zone 0 is a zone in which a flammable atmosphere is continuously present or for long periods.
Zone 1 is a zone in which a flammable atmosphere is likely to occur in normal working (10–1000 hours per annum).
Zone 2 is a zone in which a flammable atmosphere is unlikely to occur except under abnormal conditions and then only for a short time (i.e. less than 5–10 hours per annum).

The particular zone determines the types of protection required for electrical equipment in use in that zone. Ideally, the prime method of protection should be to exclude electrical apparatus from any hazardous area. However, where this is not practical or economic, the next consideration should be whether the electrical apparatus can be segregated by fire-resistant impermeable barriers. Where the installation of electrical apparatus in hazardous areas is unavoidable, the following types of protection may be used according to the circumstances:

4.4.13.4 Type 'N' equipment

Type 'N' equipment is so constructed that in normal operation it is not capable of igniting a surrounding explosive atmosphere. Such equipment is designed for use in Zone 2 areas.

4.4.13.5 Type 'e' equipment

Type 'e' equipment, also known as 'increased safety' equipment, employs a protection method to electrical apparatus that does not, in normal operation, produce sparks, arcing or excessive temperatures. Examples are transformers and squirrel cage motors. Some type 'e' equipment may be suitable for use in Zone 1 as well as Zone 2 areas.

4.4.13.6 Pressurising and purging (type 'p')

Pressurising is a method used whereby clean air, drawn from outside, is blown at a pressure slightly above atmospheric into the room or enclosure to maintain the atmosphere at a pressure sufficiently high to prevent ingress of the surrounding potentially flammable atmosphere. *Purging* is a method whereby a flow of air or inert gas of sufficient quantity is maintained in a room or enclosure to reduce or prevent the flammable atmosphere occurring.

4.4.13.7 Electrical fittings

Electrical installations used in flammable atmospheres require special consideration as regards design, construction and installation of the metallic cable sheathing, cable armouring or conduit, junction boxes, cable gland sealing, wiring etc. Particular attention should be given to the earthing arrangements. This is a specialised subject for which expert opinion should be sought.

4.4.13.8 Intrinsically safe systems (type 'i')

In a circuit where the amount of electrical energy available to cause a spark is below that necessary for igniting flammable vapour or gas, the equipment is considered to be intrinsically safe. It is particularly suitable for use in instrumentation, remote control etc.

There are two categories of intrinsically safe equipment recognised in Europe. Type 'ia' must be incapable of causing ignition with two independent faults while type 'ib' must be incapable of causing ignition with a single fault. The former may be used in all Zones while the latter is suitable for use only in Zones 1 and 2.

4.4.13.9 Flameproof equipment (type 'd')

Flameproof equipment is regarded as safe for use when exposed to the risk of explosive atmosphere for which certification has been given. Electrical apparatus, defined as flameproof, has an enclosure that will withstand an internal 'explosion' of the flammable vapour or gas in question which may enter the enclosure. The joints of the enclosure which are designed with clearance gaps to prevent a build-up of internal pressure also prevent any internal 'explosion' igniting vapour or gas surrounding the equipment. The surface temperature of the enclosure must be below the ignition temperature of the vapour or gas in question. Flameproof enclosures are primarily intended for use in Zone 1 or Zone 2 classification but *not* in Zone 0. A characteristic of such equipment is its robust construction. Older equipment, certified in accordance with earlier standards may be marked 'FLP'.

4.4.14 Portable tools

Portable electric tools are a convenient aid to many occupational activities. However, the necessity to use flexible cables to supply electricity to the tool introduces hazards. For example, such cables are often misused and abused resulting in damaged insulation and broken or exposed conductors. The tool itself could also become unsafe if, say, its metalwork became charged with electricity due to a fault. Constant care and adequate maintenance and storage are essential to safe use.

Of the hand-held power operated electrical tools, the most common are the electric drill and portable grinding wheel. The major safety requirements for portable tools are:

1 The supply cable should be of the flexible type (i.e. with stranded conductors) with its connections correctly made, being electrically and mechanically sound at both tool and point of supply. Plugs and sockets should be of a type appropriate to the work environment and conform to the relevant BS or BS EN. The cable sheath should be clamped securely at plug and tool entries.
2 Metal-cased (class I) tools should be efficiently earthed, normally achieved by a connection from the case to the protective (earth) conductor in the supply cable.
3 For preference, tools should be powered from a low voltage supply, i.e. below mains voltage. A common arrangement in the case of a single-phase supply is to provide this via a portable 230/110 V CTE step down transformer, where the output winding has a centre tap connection to earth (CTE). While the supply to the tool is 110 V any shock voltage to earth will be restricted to 55 V because of the centre point to earth connection. In the case of 110 V three-phase supplies the equivalent shock voltage will be 64 V approximately.
4 Suitable means should be provided for cutting off the supply as well as for isolation purposes. In the case of portable tools, the plug and socket connection will achieve this; otherwise an isolating switch or switch-fuse will need to be installed in a readily operable position.

It is important to ensure that all portable electrical equipment is regularly inspected and adequately maintained to minimise the risk of danger to the user. Equipment of the double-insulated or all-insulated types (class II), to the relevant BS[26], has no provision for earthing and is not earthed. Such equipment should be marked with the symbol ⧈. There is no symbol for class I equipment.

4.4.15 Residual current devices

Additional back-up protection can be provided by residual current devices that ensure that in the event of an earth fault the current is cut off before a fatal shock is received. This form of protection works on the principle of monitoring any differential between (i) the current entering a circuit to supply power to the portable apparatus and (ii) the current returning to the supply point. For normal safe operation this current differential is zero but if there is a fault, such as leakage to earth, a differential current occurs which the device rapidly senses, tripping to cut off the supply to the apparatus.

Thus an RCD will not prevent electric shock because shock current must flow through the body to cause sufficient 'out-of-balance' between the conductors to be detected by the device. However, their sensitivity and speed of operation will limit the current flow to a few fractions of a second, thus making the shock more survivable. A typical current level that will result in operation of the device (tripping current) is 30 mA (0.03 A), with operation taking place within 40 ms (0.04 s). The design and construction parameters for RCDs are specified in BSs[27,28,29].

4.4.16 Maintenance

It is a requirement of the EAW that all electrical apparatus and conductors shall, among other things, be adequately maintained. Maintenance does not simply mean general care and cleanliness but implies a system for regular inspection and, where appropriate, testing to ensure serviceability. For maintenance to be cost effective, its extent and frequency should be determined on the basis of an assessment of the risks, i.e. taking account of factors such as the age and type of construction of the equipment and the nature and environment of its use. Maintenance will require a thorough examination to check for signs of damage or defects. This will need to be undertaken by someone with the necessary competence, who knows what to look for, is able to recognise damage or defects that may be significant and who knows the appropriate actions that need to be taken. Electrical testing may be necessary to confirm aspects concerning the condition of the equipment that might otherwise not be possible to verify by visual examination alone.

In the case of portable equipment, testing is often carried out using a proprietary portable appliance tester (PAT). One useful test is the measurement of the insulation resistance to confirm that it is sufficiently high to prevent undue leakage. Additionally, for class I equipment (that must be earthed) it is important to verify that the connection to earth is sound, i.e. its electrical resistance is low, and that the conductor is capable of carrying the sort of high current that may occur under fault conditions. Both the HSE and the IEE have produced guidance on electrical equipment maintenance[30,31].

4.4.17 Conclusion

The EAW cover everything in the life of an electrical system, from initial concept in terms of its specification and design, through the construction phase to eventual commissioning for use, its use over many years allowing for possible alteration of or addition to it during its life, to eventual dismantling at the end of its useful life. However, the legal duties are expressed only in goal setting terms or as objectives to be achieved. The means for achieving these objectives are many and varied and it is for dutyholders under the Regulations, who in general are the employer, manager of a mine or quarry, the self-employed and the individual employee, to determine and put in place measures to ensure satisfactory standards of electrical safety.

Dutyholders' responsibilities are covered in regulation 3 and require the dutyholder to take appropriate action, but only in respect of matters within his or her control. In practice, employees will be governed by, and need to work within, the framework established by the employer. This should be clearly set down in company policy with the organisation and arrangements for implementation, such as work procedures and safety rules.

The HSE and other bodies, such as the BSI and IEE, have published guidance covering many aspects of the EAW which serve as useful information concerning the means for achieving compliance. Many of these publications are referred to in the list of references.

References

1. British Standards Institution, *British Standards Catalogue*, BSI, London (latest edn)
2. Institution of Electrical Engineers, *Requirements for Electrical Installations*, 16th (or latest) edn, IEE, London (BS 7671) (2001)
3. H. M. Government, *The Electricity at Work Regulations 1989*, SI 1989 No. 635 as amended, HMSO, London (1989)
4. Health and Safety Executive, *Memorandum of Guidance on the Electricity at Work Regulations 1989*, Health and Safety Series Booklet HSR25, HSE Books, Sudbury (1989)
5. Health and Safety Executive, *Approved Code of Practice No. COP 34, The Use of Electricity in Mines*, HSE Books, Sudbury (1989)
6. Health and Safety Executive, *Approved Code of Practice No. COP 35, The Use of Electricity in Quarries*, HSE Books, Sudbury (1989)
7. Hughes, E., *Electrical Technology*, Longmans, London (1978)
8. British Standards Institution, BS 4727, *Glossary of electrotechnical, power, telecommunications, and electronics, BSI, London*
   Part 1 – Terms common to power, telecommunications and electronics
   Part 2 – Terms particular to power engineering
   Part 3 – Terms particular to telecommunications and electronics
9. British Standards Institution, *BS 7430, Code of practice for earthing, BSI, London (1998)*
10. Health and Safety Executive, *Avoidance of Danger from Overhead Electricity Lines*, Guidance Note GS 6, HSE Books, Sudbury (1997)
11. Health and Safety Executive, *Avoiding Danger from Underground Services*, Health and Safety Guidance Booklet No. HSG47, HSE Books, Sudbury (2000)
12. Health and Safety Executive, *Electricity at work; safe working practices*, Booklet HSG 85, HSE Books, Sudbury (1993)
13. British Standards Institution, *BS EN 60900, Hand tools for live working up to 1000 V ac and 1500 V dc*, BSI, London (1994)
14. British Standards Institution, *BS EN 60903, Gloves and mitts of insulating material for live working*, BSI, London (1993)
15. British Standards Institution, *BS EN 50321, Electrically insulating footwear for working on low voltage installations*, BSI, London (2000)
16. British Standards Institution, *BS IEC 61111, Matting of insulating material for electrical purposes*, BSI, London (2001)
17. British Standards Institution, *BS 5958, Code of Practice for control of undesirable static electricity*, BSI, London (1991)
18. British Standards Institution, *BS 6742, Electrostatic painting and finishing equipment using flammable materials*, BSI, London (1987/1990)

19. Health and Safety Executive, *Electrical safety on construction sites*, Booklet HSG 141, HSE Books, Sudbury (1995)
20. British Standards Institution, *BS EN 60529, Specification for degrees of protection provided by enclosures (IP code)*, BSI, London (1992)
21. British Standards Institution, *BS 7375, Code of practice for distribution of electricity on construction and building sites*, BSI, London (1996)
22. British Standards Institution, *BS EN 60079–10, Electrical apparatus for potentially explosive gas atmospheres. Part 10, Classification of hazardous areas*, BSI, London (1996)
23. British Standards Institution, *BS EN 60079–14, Electrical apparatus for potentially explosive gas atmospheres, Part 14, Electrical installations in hazardous areas (other than mines)*, BSI, London (1997)
24. British Standards Institution, *BS EN 60079–17, Electrical apparatus for potentially explosive gas atmospheres. Part 17, Inspection and maintenance of electrical installations in hazardous areas (other than mines)*, BSI, London (1997)
25. British Standards Institution, *BS EN 50014, Electrical apparatus for potentially explosive atmospheres, General requirements*, BSI, London (1998)
26. British Standards Institution, *BS 2769, Hand held electric motor-operated tools*, BSI, London
27. British Standards Institution, *BS EN 61008–1, Residual current operated circuit-breakers without integral overcurrent protection for household and similar uses (RCCBs). Part 1, General rules*. BSI, London (1995)
28. British Standards Institution, *BS EN 61009–1, Residual current operated circuit-breakers with integral overcurrent protection for household and similar uses (RCBOs). Part 1, General rules*. BSI, London (1995)
29. British Standards Institution, *BS 7071, Specification for portable residual current devices*, BSI, London (1998)
30. Health and Safety Executive, *Maintaining portable and transportable electrical equipment*, Booklet HSG 107, HSE Books, Sudbury (1994)
31. Institution of Electrical Engineers, *Code of practice for in-service inspection and testing of electrical equipment*, IEE, London (2001)

Chapter 4.5

Statutory examination of plant and equipment

J. McMullen and updated by J. E. Caddick

4.5.1 Introduction

The industrial revolution began in the late 18th century with the mechanisation of the textile industry and subsequent major developments in mining, transport and industrial production, based upon Britain's rich mineral resources such as coal and iron ore, and the use of steam power.

The great industrial towns such as Manchester began to expand with steam power providing the impetus, and this great human exploit was subsequently to make Manchester the world's first industrial city. The industrial conurbation within a 10 mile radius of central Manchester contained in excess of 50 000 boilers – the largest concentration of steam boilers in the world – supplying power to textiles, engineering and to other industries of the period.

The demand by industry for ever-higher boiler operating pressures was outstripping the ability of engineers to meet it safely. As a consequence there was great public concern over boiler explosions which were occurring around Greater Manchester at an alarming rate. The boilers were literally blowing to pieces, causing multiple deaths and injuries.

From 1851 the famous Manchester engineer, Sir William Fairbairn, began arguing for periodic inspection of boiler plant. An advocate of high pressure, in the interests of economy, he decided that the explosions arose from avoidable mechanical causes which could be located in time by carrying out periodic inspections. So in 1854 he enlisted the help of two other eminent Manchester men – Henry Houldsworth, master cotton spinner, and the celebrated engineer, Sir Joseph Whitworth – to form an Association for boiler inspections. This became known as the Manchester Steam Users' Association.

The Boiler Explosions Act 1882 made the reporting of all boiler explosions compulsory and required enquiries to be conducted by Board of Trade surveyors as to the causes and circumstances of the explosions.

At this stage there was still no statutory requirement to have boilers inspected, but, as the 20th century approached, the evidence of over 1000 enquiries held under the Boiler Explosions Act clearly demonstrated the value of regular thorough inspections by competent engineers. The result, included in the Factories and Workshops Act of 1901, was a requirement that steam boilers be subjected to periodic thorough internal examinations by competent persons, and so became the first piece of legislation requiring an item of engineering plant to be inspected, and laid the foundations for the extensive and varied provisions of present day UK law. Electrical equipment inspections and insurance became a prominent part of the insurance portfolio towards the end of the 19th century, with lifts and cranes from early in the 20th century. Dust extraction plant (now referred to as local exhaust ventilation) and power presses followed much later.

4.5.2 Legislation

The legislation requiring the statutory examination of plant and machinery is undergoing significant changes and is taking a more flexible approach to periodic examinations than earlier prescriptive laws. However, many of the requirements of the latter have been retained in the text in the following sections since they offer guidance on sound and proven inspection procedures. A summary of the principal inspection requirements under existing legislation for pressure systems, lifting and handling equipment, power presses, press brakes, local exhaust ventilation equipment and electrical equipment and installations is contained in *Table 4.5.1*.

4.5.3 Pressure systems

Under the Pressure Systems Safety Regulations 2000 the requirements for examination have become much less prescriptive in that statutory reporting forms no longer need be used and the phasing of the examinations can be related more closely to operating circumstances. Its main requirements are summarised below.

Many of the techniques and practices that developed under the older legislation provide sound safety and engineering guidance and it is sensible to continue using them where they do not clash with the latest requirements. Some of them are described below.

4.5.3.1 Requirements of the Pressure Systems Safety Regulations 2000

These Regulations have been made under HSW and deal with broad objectives to be achieved in the operation of complete pressure systems at all places of work.

Table 4.5.1 Summary of principal statutory inspection requirements

Statute	*Class of plant*	*Period between examinations – months*
The Lifting Operations and Lifting Equipment Regulations 1998	Lifting equipment	6–12 or as scheme of examination
Pressure Systems Safety Regulations 2000, reg. 9	All pressure plant	As scheme of examination
Control of Substances Hazardous to Health Regulations 2002, reg. 9	Local exhaust ventilation plant and dust/fume extraction plant	1–14
Electricity at Work Regulations 1989 (IEE Regs.)	Electrical installations	3–60 depending on the application
Lifts Regulations 1997	Applies to construction of new lifts	
The Provision and Use of Work Equipment Regulations 1998, reg. 32	Power press	6–12 depending on guard type

Pressure systems are defined as:

(a) a system comprising one or more pressure vessels of rigid construction, any associated pipework and protective devices;
(b) the pipework with its protective devices to which a transportable gas container is, or is intended to be, connected; or
(c) a pipeline and its protective devices which contain, or are liable to contain, a relevant fluid, but do not include a transportable gas container or a transportable pressure vessel.

Relevant fluid is defined as:

(a) steam;
(b) any fluid or mixture of fluids which is at a pressure greater than 0.5 bar above atmospheric, and which fluid or mixture of fluids is:
   (i) gas, or
   (ii) a liquid which would have a vapour pressure greater than 0.5 bar above atmospheric pressure when in equilibrium with its vapour at either the actual temperature of the liquid or 17.5 degrees Celsius; or
(c) a gas dissolved under pressure in a solvent contained in a porous substance at ambient temperature and which could be released from the solvent without the application of heat.

The Regulations place obligations on anyone who manufactures or constructs a new pressure system, and anyone who repairs or modifies a new or existing pressure system or part of it, to ensure that no danger will arise when it is operated within the safe operating limits specified for that plant.

The other main requirements of the Regulations are:

- the user, or owner in the case of a mobile system, must establish the safe operating limits of the system
- the user must have a written scheme of examination for the system
- the user must maintain the system
- the user must have operating instructions and ensure that the system is only operated in accordance with those instructions.

The written scheme of examination is one of the most important features of the Regulations and is required to be compiled before a pressure system can be operated. All protective devices, as well as those pressure vessels and pipework where a defect could give rise to danger, must be included in the written scheme. In addition, it must specify the nature and frequency of examinations and the measures necessary to prepare the system for safe examination over and above the precautions that the user or owner of the system would reasonably be expected to take.

The onus for ensuring that these requirements are met lies with the user of an installed pressure system or the owner of a mobile system.

A report of the periodic examination by the competent person must be given to the user or owner of the system within 28 days. However, if there is imminent danger from the continued operation of the system a copy should be provided to the enforcing authority within 14 days. The Regulations also require that certain records are retained. These include:

- the last report of examination
- any previous reports that required changes to the safe operating limits or repairs
- any documentation provided by the manufacturers.

These Regulations are supported by an Approved Code of Practice[1]. The Approved Code of Practice outlines, in relative terms, the qualifications to be held by a competent person to enable them to compile a written scheme for a major, intermediate or minor system. Guidance on the frequency of examination of pressure plant can be found in these ACOPs.

4.5.3.2 Fired pressure vessels and associated pipework

Any reference to a boiler also includes all fitments and attachments. All existing boilers should have a safe operating limit which for a new boiler is the pressure specified by the manufacturer, but for existing boilers is the pressure specified on the last report of thorough examination. This is the lift-off pressure of the safety valve although it is normal to run the boilers at 90–95% of this to minimise continual leakage from the safety valves.

All boilers should have a safety valve so adjusted as to prevent the boiler being worked at a pressure greater than the safe operating limit, a suitable stop valve, steam pressure gauge, at least one water level gauge, means for attaching a test pressure gauge and be provided with a low water alarm device. The nature and frequency of examination of boilers will be specified in the written scheme of examination.

The examination of a steam boiler is normally in two parts:

1 A thorough examination, both internally and externally by a competent person, when it is cold, after the interior and exterior have been properly prepared in accordance with one or more of the following provisions:
   (a) Opening up and cleaning out and scaling of the fire and water sides including the removal of all access doors from manholes, mudholes and handholes.
   (b) Opening out for cleaning and inspection of fittings including pressure parts of automatic controls, safety valves and water gauges, and blow down valve.
   (c) In the case of water-tube boilers, the removal of drum internal fittings as required by the competent person.
   (d) In the case of shell-type boilers, the dismantling of firebridges if made of brick, and furnace protective brickwork at specified periods.
   (e) All brickwork, baffles and coverings must be removed for the purpose of the thorough examination, to the extent required by the competent person.
2 The boiler (including economiser and superheater if fitted) should be thoroughly examined by a competent person when it is under normal steam pressure. This examination should be made on the first occasion when the steam is raised after the examination when the boiler is cold, and must consist of ensuring that the safety valve is adjusted so as to prevent the boiler being worked at a pressure greater than the safe operating limit. It should then be locked off to prevent unauthorised tampering with the setting. The operation of the pressure gauges, water gauges, controls, alarms and cut-outs must also be checked.

Whilst the examination of the boiler, like any other item of statutory plant, is based upon a visual inspection, it may be supplemented by: material thickness measurement, proving a clear waterway through the tubes, withdrawal of sample tubes for evidence of deterioration of wall thickness, non-destructive testing for cracking/flaws by ultrasonic, radiographic, magnetic particle or dye penetrant testing, or by hydraulic testing. Also, as part of the examination the peaking (deviation from circular shape) at the longitudinal seam should be measured. The periods between ultrasonic examination of the seam can be determined from information in the SAFed guidance on the examination of longitudinal seams[2]. For certain boilers with flat end plates it is normal to undertake ultrasonic examination of the end plate welds to detect any cracking from the water side of the weld. Guidance on the nature and frequency of these inspections can be found in SAFed publications[3,4].

Entry into any boiler, which is one of a range of two or more, is forbidden unless all inlet pipes through which steam or hot water could enter the boiler have been disconnected, or the valves controlling entry of steam or hot water have been closed and locked. Particular attention should be paid to blow-down valves which discharge into a common line or sump, where special safety procedures must be followed[5]. Suitable precautions should also be taken to ensure that the boiler is free from dangerous fumes and has adequate ventilation.

The purpose of the examination is to ascertain the material condition of the boiler with particular reference to any defects which could affect the continued safe working of the boiler at its current safe operating limits. During the examination when cold the competent person will be checking for the following types of typical defects:

1 *External* – Wastage from corrosion due to leakage at joints between fittings and shell, tubes and tubeplates, or at seams in riveted boilers; wastage of manhole, mudhole and handhole joint seatings; general physical damage.
2 *Fire side* – Wastage in combustion chamber plates due to leaking tubes, stay tubes or stays; furnace flame impingement damage; erosion of material by fast moving gases and/or entrained particles – particularly where gases are hottest and changing direction (i.e. combustion chamber ends of tubes in horizontal shell-type boilers); overheating damage due to scale or sludge build-up, or water shortage; furnace bulges and combustion chamber crown bulges.
3 *Water side* – Chemical damage causing wastage, pitting, thinning and necking of material due to oxygen or other chemical corrosion; mechanical damage caused by expansion and contraction, and grooving at plate flange bends, evidence of 'peaking' at the longitudinal weld caused during the manufacturing process, cracking at the tube/tubeplate joints, and other slowly developing fatigue cracks at a variety of locations.

Examples of typical defects are shown in *Figure 4.5.1(a)* and *(b)*.

Reports which should cover both parts of the examination should be issued by the competent person. Electronic storage of data is acceptable subject to suitable security arrangements for access to it and the facility to produce a hard copy if required. A list of the items that should be included in the report is given in Approved Codes of Practice[1].

When the examination reveals defects which affect the safe working of the boiler at the current safe working limit, the report must make recommendations on conditions to be met before the boiler can be returned to operation. The conditions may include certain defined repairs, or the reduction of the safe working limit.

Where a defect has been found in the boiler and subsequent repairs carried out, the competent person should satisfy himself that the repair

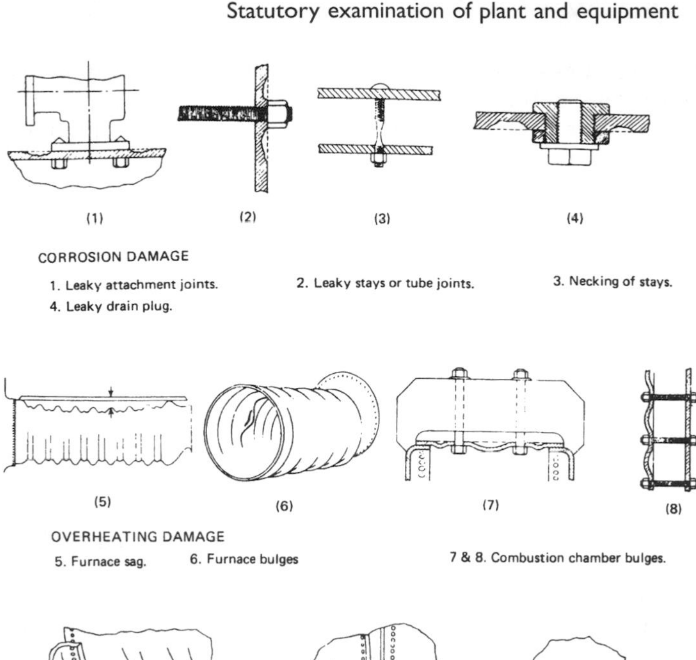

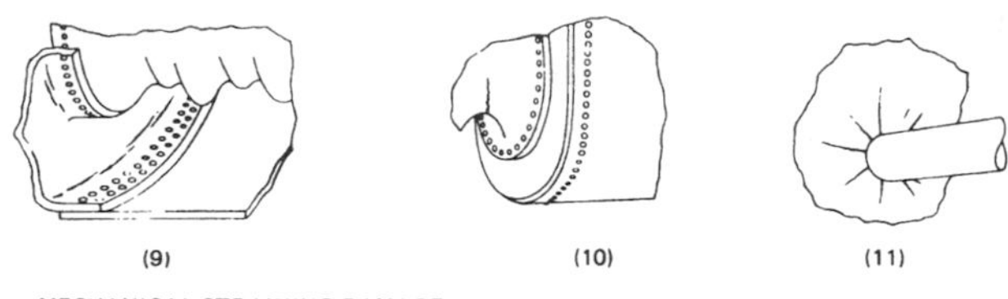

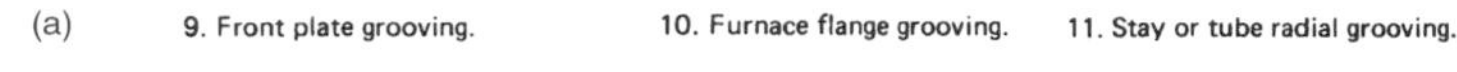

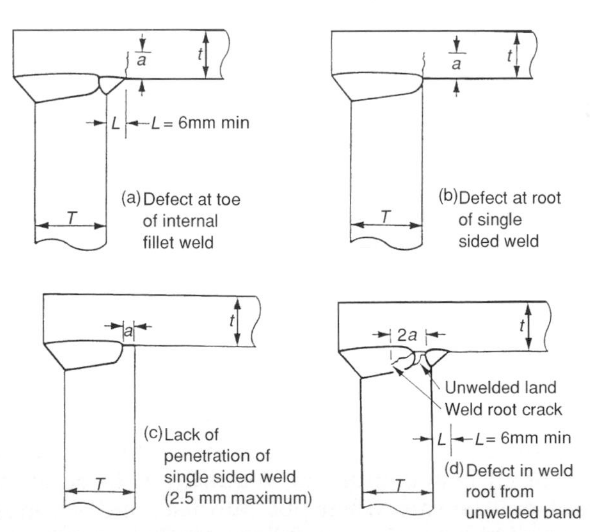

Figure 4.5.1 (a) Some typical examples of different types of boiler defects. (Courtesy Milton and Leach[6]). (b) Examples of surface and buried defects in boiler and plates. T = endplate/tube plate thickness (mm); t = furnace or shell plate thickness (mm); a = depth of surface defect; L = fillet weld leg size

work has in fact been carried out satisfactorily before the boiler is released for service. This will normally include review of material certificates for new components, review of weld procedures, welders qualification and NDT reports. Additionally the competent person will examine the boiler during repair and witness a hydraulic test before it is returned to service.

4.5.3.3 Unfired pressure vessels and associated pipework

The safe operating limit is a legal term which applies to pressure vessels. In the case of a new vessel, it means the pressure specified by the manufacturer, and in the case of a vessel that has been examined in accordance with the written scheme of examination, that specified by the competent person.

Where the vessel is not suitable for the pressure of the source of supply, a suitable reducing valve, pressure gauge and safety valve should be fitted in the supply line. The vessel should be provided with a suitable manhole, handhole or other means which will allow the interior to be thoroughly cleaned.

The frequency of examination of this equipment will be specified in the written scheme. If a vessel is so constructed that the internal surface cannot be thoroughly examined, then either non-destructive testing or a hydraulic test may be carried out.

As with boilers the scheme of examination for these vessels should be drawn up by a competent person before they are put into service in a system. A report of the examination should be sent to the user or owner within 28 days.

Where a defect is found which could give rise to imminent danger, a copy of the report should be sent to the enforcing authority within 14 days.

4.5.4 Lifting equipment

The range of what has generically become known as lifting and handling equipment has expanded over the years in both scope and sophistication, with many innovative designs now being used not only in industry but in a wide spectrum of applications across the whole field of employment. Because of the very nature of this type of plant, together with its versatility, legislative controls aimed at ensuring its safe use have been developed in a fragmented manner. However, the legislative requirements regarding lifting equipment have been rationalised, largely as a result of developing directives in the EU. The pattern of the legislation is in two stages, first that dealing with the design and manufacture of all new equipment, requiring it to be to standards that will ensure it will be safe when put into use, and second that dealing with the use of the equipment where further requirements are laid down in supplementary legislation covering operating procedures aimed at ensuring the equipment is properly maintained and is used in a safe manner.

4.5.4.1 Lifting equipment manufacture

All new work equipment of whatever type must be manufactured to conform with the Machinery Directive (Council Directive 89/392/EEC[7]) which has been embodied in UK law as the Supply of Machinery (Safety) Regulations 1992[8]. These incorporate the essential health and safety requirements (ESRs) that have to be met by all machinery including lifting machinery, lifting equipment and lifting accessories. The regulations also specify the overload test conditions for the different types and parts of lifting equipment. However, schedule 5 of these regulations specifically excludes most lifts for carrying persons.

This loophole is filled by the Lifts Regulations 1997 with which all new lifts put on the market after 1 July 1997, whether for carrying goods or passengers, must comply. The Lifts Regulations are not stand-alone regulations but complement the Machinery Regulations by imposing additional requirements, specific to lifts, aimed at protecting against the risks that are peculiar to lifts. Under these regulations lifts must:

- satisfy the relevant ESRs
- be subject to the conformity assessment procedure
- have a CE mark attached
- have a declaration of conformity drawn up, and
- be safe to operate.

However, there is no statutory requirement for lifts to be tested before being supplied but the manufacturer and supplier are required to include in the documentation accompanying the lift recommendations in respect of inspections and periodic checks and provide a log book in which the periodic checks can be recorded.

Certain types of lifts having specific functions are excluded from the requirements of these regulations. In these regulations the word 'lift'

> means an appliance serving specific levels, having a car moving–
>
> (a) along guides that are rigid; or
> (b) along a fixed course even where it does not move along guides which are rigid (for example, a scissor lift)

and inclined at an angle of more than 15 degrees to the horizontal and intended for the transport of:

- persons
- persons and goods
- goods alone if the car is accessible, that is to say, a person may enter it without difficulty, and fitted with controls situated inside the car or within reach of a person inside.

4.5.4.2 Safe use of lifting equipment

The position changes once a lift, or any other lifting equipment, has been delivered and installed ready for putting into operation when it becomes subject to the Provision and Use of Work Equipment Regulations 1998[9] (PUWER 98). PUWER 98 covers all work equipment in use and while it has a requirement, in regulation 6, for all equipment to be inspected where the safe use of the equipment depends on the conditions of installation and where it is exposed to conditions causing deterioration which could result in danger, it excludes lifting equipment from this specific regulation.

The above Regulations, which specify general requirements for the safe use of all work equipment, are complemented by the Lifting Operations and Lifting Equipment Regulations 1998[10] (LOLER), which detail the conditions necessary to ensure the safety in use of all lifting equipment. LOLER provides a single set of modern goal-setting regulations for the safe use of lifting equipment and has removed many of the anomalies of the earlier sectoral specific legislation. These regulations are supported by a Code of Practice[11].

LOLER are aimed primarily at the type of equipment that was covered by previous lifting legislation, i.e. cranes, lifts, hoists and components including chains, ropes, slings, hooks, shackles and eyebolts but extends its application to all employments including those not previously covered, such as agriculture, forestry, hospitals, schools, places of entertainment and offshore oil and gas installations.

LOLER defines '*lifting equipment*' as '*work equipment for lifting or lowering loads and includes its attachments used for anchoring, fixing or supporting it*'. Included are any *lifting accessories* which are the various pieces of equipment for attaching the load to the lifting machine. The *lifting machine* is the equipment or machine that raises or lowers the load.

Typical of the range of equipment now covered are:

- passenger lifts in office blocks;
- ropes and gin wheels for raising buckets of cement on a building site;
- dumb waiters in restaurants or hotels;
- the vacuum lifting attachment to a crane;
- vehicle inspection hoists; and
- scissor lifts.

4.5.4.3 Need for a thorough examination

Obligations are put on employers to ensure that before lifting equipment is put into use it is thoroughly examined for any defects unless the equipment is new, i.e. has not been used before, and it is supported by a valid EC declaration of conformity. With lifting equipment that has to be installed before it can work, the employer must ensure that the safe operation of the equipment is not prejudiced by the manner in which it is installed. This applies particularly to lifting equipment that is moved from one site to another, such as tower cranes.

LOLER requires that thorough examinations of lifting equipment be carried out. Where operating conditions may adversely affect the safe working of lifting equipment, the Regulations require that:

- for equipment and any associated accessory used for lifting persons they should be thoroughly examined every 6 months; and
- for other lifting equipment, at least once in every 12 months; or
- where a risk based scheme of examination has been drawn up by a competent person, examinations should be in accordance with that scheme;
- a thorough examination should also be made after any exceptional circumstances that might prejudice the safe operation of the equipment, such as very high winds, heavy snow, overload, mechanical damage, etc.;
- LOLER also recommends that, in adverse operating conditions, an intermediate inspection should be carried out by a competent person between thorough examinations.

These regulations, which are goal-setting rather than prescriptive, allow the employer to select between following a system of thorough examinations prescribed by the regulations or a scheme of examinations drawn up by a competent person based on the findings of a risk assessment.

4.5.4.4 Hired lifting equipment

Where lifting equipment is hired, either for semi-permanent use or for one particular lifting operation, the person from whom the equipment is hired has a duty under LOLER to ensure that the equipment is covered by a current certificate of thorough examination. A copy of this certificate should accompany the equipment and the user should check the certificate and its validity. With installed equipment, such as tower cranes, that must be thoroughly examined before being used, the installer, who has the necessary expertise, may carry out the thorough examination, or it may be undertaken by an independent body such as a specialist inspection company or insurance company engineer surveyor.

Where the lifting equipment is on long-term hire, it is the responsibility of the user to ensure it is thoroughly examined at the appropriate times. These examinations could be carried out by either the user's insurers or by the hiring company. Typically for fork lift trucks on long hire, the obligation to carry out the thorough examinations could be linked in with a maintenance contract placed with the hiring company.

4.5.4.5 Resale of lifting equipment

Where lifting equipment is disposed of to another person and it has the potential for further use, the disposer must ensure that the equipment is accompanied by evidence of a current examination, and the acquirer must ensure the equipment is so covered. Lifting equipment that is moved from one site to another owned by the same organisation is not considered to have been disposed of and does not need to be accompanied by a copy of the examination certificate issued to the original site, providing always that a copy of the certificate is readily available at the company's offices should it be required.

4.5.4.6 Thorough examination

While the regulations require the carrying out of a thorough examination, they do not specify any details of what is required, but rely on the competency of the person carrying out the examination to ensure that the examination is sufficiently thorough taking into account the operating circumstances and experiences. The competent person may be an employee, but if so his employment must be sufficiently protected to enable him to act independently. However, it is normal for the insurer who provides insurance cover for the equipment to provide a competent person, an engineer surveyor, to carry out the examinations.

All lifting equipment may deteriorate in use and should be thoroughly examined periodically to detect any possible deterioration in sufficient time to allow remedial action to be taken before failure occurs. Deterioration is more rapid the more hostile the operation environment, such as in wet, abrasive or corrosive atmospheres, and the examiner will need to determine what greater frequency of examination is necessary. Certain types of thorough examination may demand detailed examination of specific parts such as the operating mechanisms or control gear. The aim of the examinations is to identify and remedy faults that could reduce the level of safety during lifting operations.

The extent of a thorough examination will be determined by the operating circumstances of the lifting equipment and on information on operating experiences and conditions provided to the competent person by the user. This information should be sufficiently detailed to enable the examiner to identify areas of potential faults that could adversely affect safety of operations before the next examination. The examination may involve the testing of parts or of the whole equipment.

Different items or specific parts of the lifting equipment that may be known to be subject to severe operating conditions or may be identified as presenting a high risk if they fail could be made subject to additional separate thorough examinations between the specified periodic examinations.

4.5.4.7 Testing

Testing of lifting equipment or a particular part of it can be required when the competent person believes it is necessary for the continuing safe use of the equipment. The testing can involve the removal of a part for non-destructive testing, either on site or in a laboratory. In other cases the competent person may require the carrying out of a load test on the equipment to check the overall adequacy of its lifting ability and stability.

4.5.4.8 Inspections

In addition to periodic thorough examinations, the regulations recommend, where operating conditions are severe, the carrying out of *inspections* between the examinations. These inspections would be by a visual check of the equipment to identify damage or component deterioration that may affect safe operations. They could be carried out by a responsible person or someone familiar with the equipment. For example, in operations employing slingers, it would be prudent for them to inspect the lifting accessories every time before they use them.

In other cases, routine inspections and checking of equipment should be implemented. This should be by the operators or supervisors and should be part of the operating procedures for the equipment. Typical of items checked could be load limiters and indicators on cranes, tyre pressures on mobile lifting equipment, hoist limit switches on the hoisting mechanism of overhead travelling cranes, the adjustment of brakes on hoists, etc. The inspection may identify other defects which could include:

- textile slings – damaged, cut, abraded or stretched;
- chains – deformed, stretched links, cracks in links, severe corrosion;
- wire ropes – broken wires, needles, kinks;
- fibre ropes and slings – knots, cuts, splices, unravelled lay;
- in lifting machines – loose or missing bolts, damaged structural members.

Any equipment found to be defective on an inspection should be taken out of service and the employer informed. Defective lifting accessories should be either repaired or scrapped, but defects to lifting machines should be reported to the competent person who may wish to carry out a special examination.

Recommendations on daily checks and weekly inspections for cranes are made in BS 7121[12].

4.5.4.9 Identification of lifting equipment

It is essential that equipment to be examined is properly identified and it is incumbent upon the user to ensure that any identification system is clear and does not allow confusion between different pieces of lifting equipment. Ideally a register should be kept of all lifting equipment which should be made available to the examining engineer for him to use.

4.5.4.10 Records of examinations

A record should be kept of every examination. It can be by entry in a plant register, the completion of a standard form, a written report or in electronic format provided the record cannot be interfered with, is immediately available for inspection and can produce a hard copy when required.

LOLER lays down the data that should be contained in a report of a thorough examination. Such data includes:

(a) name and address of the employer for whom the examination was made;
(b) address at which the examination was made;
(c) sufficient information to identify the equipment examined;
(d) the date of the last thorough examination;
(e) the safe working load of the equipment or the safe working load for each configuration of which it is capable;
(f) the circumstances of the examination, i.e. after installation;
(g) details of any defects found and remedial action taken or recommended including details of any testing;
(h) name, address and qualification of the person making the report;
(i) date of the report;
(j) date of the next thorough examination.

Where a defect is found that could affect the safe operation of the equipment, a copy of the examination report should be given to the user and, if the equipment is on hire, to the hirer. If the defect is such as to give rise to the imminent risk of injury the user should be notified forthwith and a copy of the report must be sent to the appropriate enforcing authority.

4.5.5 Power presses and press brakes

Power presses and press brakes are complex pieces of equipment that have the potential to cause serious injury. For this reason it is important to ensure safe systems of work are in place and that power presses and their associated guarding arrangements are subjected to regular examinations and testing to ensure that all systems are operating properly and have not suffered damage likely to reduce the required level of safety. The Power Press Regulations 1965 provided for such examinations to be carried out at prescribed intervals by competent persons.

The Regulations defined a power press as being a press or press brake which in either case is used wholly or partly for the working of cold metal by means of tools or for the purpose of die proving is power driven and embodies a flywheel and clutch. Hydraulic and pneumatic presses were beyond the scope of these Regulations.

Following the introduction of the Regulations in 1965 a number of certificates were issued by HM Chief Inspector of Factories which exempted from the requirements certain classes of power presses including those used for compacting metal powders, presses with a stroke not exceeding 6 mm, turret punch presses and other machinery such as combination machines used solely for punching, shearing and cropping. Other machines beyond the scope of these Regulations include guillotines, riveting machines and special purpose machines such as those used for the manufacture of zip fasteners, eyelets, and similar components.

The Provision and Use of Work Equipment Regulations 1998 (PUWER 98) revokes the Power Press Regulations but contains specific requirements for power presses in part IV of the Regulations. PUWER 98 is supported by the HSC Code of Practice and Guidance *Safe use of power presses*[13].

The scope of PUWER 98 has been extended beyond that of the Power Press Regulations 1965 which applied only to power presses in factories. The PUWER 98 requirements apply to all power presses used at work. This includes, for example, power presses used in research and educational establishments. The exemptions for certain classes of power presses issued under the 1965 Regulations have been carried forward to PUWER 98 and are listed in Schedule 2 to the regulations.

The main thrust of the regulations is towards the thorough examination and testing of machines and guarding systems before the machine is first taken into use and thereafter at periodic intervals depending on the nature of the tool guarding arrangements. In addition, a thorough examination is required each time that exceptional circumstances have occurred which are liable to jeopardise the safety of the power press, its guards or protective devices.

Where the dangerous tool trapping area is protected exclusively by fixed guards the interval between thorough examinations is 12 months, but in all other cases including where interlock guards, automatic guards or photo-electric safety systems are used, the interval is 6 months. The examination requires close scrutiny of the safety critical parts such as the clutch and brake mechanisms and will, at some stage, require dismantling of the flywheel assembly in order to expose the condition of the inner components. The need for dismantling is confirmed in HSE guidance[14].

The results of the thorough examination and test must be entered on a report of the thorough examination which must be kept for at least 2 years. Any defects found that may affect safety at the tools must be notified immediately in writing to the user and a copy of the report of the thorough examination sent to the HSE. Otherwise the user must be provided with a copy of the report within 14 days of the examination.

The setting, resetting, adjusting or trying out of tools on a power press and the proving of the guarding arrangements may only be carried out by someone who:

- has been suitably trained
- is competent
- has been appointed by the employer in writing.

He would normally be the tool setter.

In addition, the tool guards and safety devices must be inspected and tested during the first 4 hours of a shift by the setter who must sign a register or certificate, which should be kept on or near the press, to verify that the guards are in position and in good working order. Such registers must be kept until 6 months after the date of the last entry.

Although hydraulic press brakes are not included in the Regulations, in their publication on press brakes[15] the HSE recommend that the presses and their associated guarding systems should be treated the same as power presses and subjected to a thorough examination every 6 months in addition to the checks carried out on each shift.

4.5.6 Local exhaust ventilation

Local exhaust ventilation (LEV) equipment is intended to control mechanically the emission of contaminants such as dust and fumes that are given off during a manufacturing process or in a chemical laboratory. Normally this is done as close to the point of emission as possible using a stream of air to remove the airborne particulate matter and transport it to where it can be safely collected for ultimate disposal. The physical layout and setting of LEV equipment is critical for it to work effectively, and comparatively minor alterations can affect its performance. It is, therefore, important that LEV equipment[16] should be properly designed, manufactured, installed, operated and maintained.

The main elements of the equipment would normally comprise:

1 captor hood,
2 exhaust ducting,
3 extraction fan, and
4 filter/collecting bags.

In addition, the following items would be regarded as LEV equipment:

5 parts of machinery such as integral machine casing or guarding which has a dual purpose of controlling and venting emissions from the process to atmosphere;
6 vacuum cleaners when permanently connected to an exhaust system and fitted to portable tools;
7 flues from a furnace/oven when the plant is producing hazardous or toxic fumes – but not when the flue only serves to create a draught for a combustion process;
8 fume cupboards in chemical laboratories; and
9 low volume/high velocity (lv/hv) extraction for cutting processes.

The major legislative requirements concerning the examination and testing of LEV plant are contained in the Control of Substances Hazardous to Health Regulations 2002 (COSHH) and its supporting Approved Codes of Practice[17,18].

Where an assessment has identified an exposure level above the occupational exposure standard (OES) or maximum exposure limit (MEL) for the substance, then under a hierarchy of preferred controls, LEV equipment must be installed wherever practicable, as opposed to providing personal protective equipment, as the means for reducing the employee's exposure. Such equipment must be properly used by the operator and visually inspected for obvious defects. Further, the employer must ensure that the equipment is maintained and the statutory inspections are carried out. As a part of this inspection it may be necessary to monitor the working environment by air sampling to ensure that the plant is continuing to operate effectively.

COSHH lists, in schedule 4, the frequency of examination of LEV as:

1 All LEVs other than those specific processes mentioned below – every 14 months.
2 Processes in which blasting is carried out in, or incidental to, the cleaning of metal castings, in connection with their manufacture – monthly.
 (It must be noted that these monthly inspections refer to castings only. In the case of blasting other metallic items the inspection frequency is 6 months.)
3 Processes, other than wet processes, in which metal articles (other than gold, platinum, or iridium) are ground, abraded or polished using mechanical power, in any room for more than 12 hours in any week – every 6 months.
4 Processes giving off dust or fumes in which non-ferrous metal castings are produced – every 6 months.
5 Jute cloth manufacture – every month.

Where plant is old, the frequency of inspections may need to be increased.

COSHH does not extend to include asbestos or lead which are covered by extant regulations.

The Control of Asbestos at Work Regulations 2002 requires, *inter alia*, that every employer shall take control measures necessary to protect the health of employees from inhaling asbestos. Details of the pressure and air velocity tests and inspection procedures to be used are given in an Approved Code of Practice[19] which provides for three levels of inspection, i.e. weekly inspections by a responsible person, an examination of new and substantially modified plant (referred to as Part 1 examination), and six monthly inspections by a competent person (referred to as Part II examination).

The competent person may enlist the assistance of specialists to carry out certain tests, but he must make a report within 14 days of the examination. The report of Part I tests should include technical details of the plant.

The Control of Lead at Work Regulations 2002 requires employers to provide such control measures, other than by the use of respiratory equipment or protective clothing, as will prevent the exposure of his employees to lead. An Approved Code of Practice[20] outlines the weekly inspections to be carried out by a responsible person and requires that the annual examination and test covers the condition of the LEV plant, static pressures and air velocities at various points and a check that the lead dust or fume is being effectively controlled. A competent person should thoroughly examine and test any exhaust ventilation equipment at least once every 14 months.

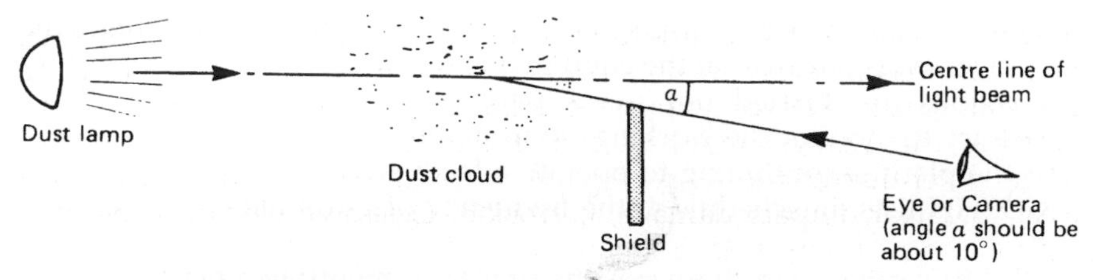

Figure 4.5.2 Use of dust lamp to see or photograph dust

Apart from the visual inspection of the LEV equipment, a statutory thorough examination under COSHH would include some of the following tests:

1 Tyndall (dust) lamp (*Figure 4.5.2*) and/or fuming sulphuric acid test;
2 static pressure behind the captor hood. This is perhaps the most important test as this will determine whether the performance of the LEV equipment has altered;
3 air velocities across the face of the captor hood;
4 centreline velocity of the small duct between the captor hood and the main exhaust ducting;
5 velocity and static pressure at the main exhaust ducting;
6 static and total pressures at the inlet and outlet of the exhaust fan; and
7 static pressure at the inlet and outlet of the dust collector and filters to obtain the differential pressure – particularly required for systems that recirculate the filtered air back into the workplace.

4.5.7 Electrical equipment and installations

The periodic inspection and testing by a competent person of electrical equipment and installations as a statutory requirement arises only under the Cinematograph (Safety) Regulations 1955 which require inspection and testing by a competent person every 12 months.

However, enforcing authorities may impose licensing conditions requiring petrol dispensing and storage (under the Petroleum (Consolidation) Act 1928), launderettes, sports grounds and buildings used for public entertainment, to be inspected, tested and certified periodically (see also below). Enforcing authority licensing conditions specify annual inspections, testing and certification that the installations comply with the current (16th) edition of the Institution of Electrical Engineers (IEE) Regulations[21], now incorporated into BS 7671[22].

In Scotland compliance with the IEE Regulations is a requirement of the Building (Scotland) Act 1959.

The Electricity at Work Regulations 1989 apply to electrical systems in all work situations. They refer to design parameters such as the circuit strength and capability, insulation, earthing, protective devices and the precautions to be taken when working on electrical equipment. All systems must be maintained, as far as is reasonably practicable, so as to prevent danger. A technique to ensure proper maintenance is through programmed preventive maintenance of which inspection and testing are essential parts. The frequency of maintenance is not specified, but practical experience should be a useful indicator. Factors to consider when determining the frequency of maintenance include mechanical wear and tear, impact damage, corrosion, excessive electrical loading, ageing and environmental conditions.

The Inspection and Testing Guidance Notes to the IEE Regulations suggest the following periods between inspections:

domestic premises	10 years
commercial premises	5 years
educational premises	5 years
hospitals	5 years
industrial works	3 years
public buildings	1 year
(i.e. cinemas, leisure complexes, restaurants, hotels etc.)	
special installations	1 year
(i.e. fire alarms, launderettes, petrol filling stations etc.)	

Two Approved Codes of Practice written under the Electricity at Work Regulations deal with electrical systems in mines[23] and quarries[24]. Both these codes require periodic inspection and test of electrical equipment.

In the case of testing portable electrical equipment reference should be made to HSE Guidance Note HSG 107. Portable electrical tools should be inspected each time they are returned to stores and tested annually, although under certain hostile operating conditions testing should be more frequent.

The testing of electrical installations and equipment should include:

- polarity,
- earth fault loop impedance/earth continuity,
- insulation resistance,
- operation of devices for isolation and switching,
- operation of residual current devices,
- verification of overcurrent protective devices
- relay and starter contact erosion, and
- wiring integrity.

Portable equipment should be tagged to identify it and indicate its inspection status. This can provide the data for a register which can be used to maintain a record of inspection dates. Similar records should be kept of inspections and examinations of fixed installations.

4.5.8 Other considerations

A number of common issues arise from the various legislative requirements for examination and inspection of plant and equipment.

4.5.8.1 Competent person

In general where the law requires particular items of plant to be thoroughly examined and tested it requires that the work be carried out by a 'competent person'.

There is no precise guidance in the statute or in subsequent case law as to what constitutes 'competence'. However, the competency of a person to carry out particular examinations or tests is a matter of fact on which the occupier or owner of the statutory equipment must be satisfied. In the event of legal proceedings it will require to be demonstrated in court that the person chosen was indeed competent to carry out the statutory surveys.

An often quoted definition is that 'the competent person should have such practical and theoretical knowledge and actual experience of the type of machinery or plant which he has to examine, as will enable him to detect defects or weaknesses which it is the purpose of the examination to discover, and to assess their importance in relation to the strength of the machinery or plant in relation to its function'. It is not sufficient for the person making the examination to be able to detect faults, he must also, from his knowledge and experience, be able to assess their seriousness.

The competent person need not necessarily be from an independent authority, but can be an internal appointment of a suitably qualified person by the company concerned, but in exercising his responsibilities for carrying out examinations he must be separated from any other functions that might cause a clash of interests. The competent person, whether employee or from a specialist organisation, must be allowed to act objectively and in a professional manner.

For complex plant and equipment it is doubtful whether any one individual would have sufficient knowledge and expertise to carry out the full examination on his own. In such circumstances, the support of a team of suitably qualified specialists may be needed.

The Code of Practice on Pressure Systems[1] considers that competence should be related to the size, complexity and hazard associated with the plant concerned and suggests different degrees of qualification for minor, intermediate and major pressure systems. The largest employers of competent persons, the engineer-surveyors, are the specialist engineering inspecting authorities associated with insurance companies, but there is an increasing number of small inspection companies providing 'competent person' inspection services who have no interest in insurance. While there is no requirement in law to have any item of plant, whether subject to statutory inspection or not, insured, it is prudent to do so. Some large companies – particularly in the steel and petrochemical industries – have their own in-house inspection departments.

The onus of responsibility for compliance with the statutory inspection obligations has been placed firmly with the company or organisation that owns the plant, and the enforcing authorities have always taken a consistent and firm line on this point. Thus, if a specialist inspection organisation is engaged to carry out all statutory examinations and it fails to do so, either properly or within the statutory period of time, the inspecting organisation may be in breach of contract, but it is the company that owns the plant that is held responsible for the breach of the particular legislation.

However, Regulations recognise the fact that a person, other than the owner and not employed by him, may, by his actions or failure to act, cause a breach resulting in an offence and offer the owner a defence if he identifies the person who caused the breach and can prove that he – the owner – exercised all due diligence to avoid the breach. Nevertheless, under s. 36 of HSW the owner may still be liable.

Under certain regulations there is a requirement for a 'responsible person' to carry out routine daily or weekly inspections of plant. That responsible person could be the employee who regularly operates the plant or equipment provided he has been trained in the checks to be carried out.

In recent years a national voluntary accreditation scheme – based on BS EN 45004[26] – has been established for inspection bodies under the aegis of the United Kingdom Accreditation Service (UKAS). Inspection bodies operating within this scheme are required to establish and maintain a quality system within their scope of activities which is subject to ongoing assessment by UKAS. The scheme is supported by HSE.

4.5.8.2 Thorough examination

A thorough examination is a detailed visual examination, both stationary and under working conditions, carried out as carefully as the conditions permit in order to arrive at a reliable conclusion as to the safety of the plant or machinery and hence an assurance that the operation of the plant will be safe until the next statutory inspection.

Where the competent person needs to satisfy himself as to the condition of internal component parts he can require the plant to be dismantled. In addition, whenever considered necessary the visual examination can be supplemented by non-destructive testing of components to determine their internal or surface condition without causing any detrimental change in the material.

4.5.8.3 Reporting

The completion of a report of an examination is a statutory requirement in respect of certain types of plant. Reports of examinations can be in a form to suit the existing record keeping system although they should contain as a minimum the particulars listed in the relevant regulations or supporting ACOPs. The reports of an examination usually indicate the condition of the plant, e.g. 'in good order', list the defects requiring correction or place restrictions on the use of the plant because of its condition. Where defects are found which may give rise to imminent danger to persons, some regulations require a copy of the report to be sent to the enforcing authority within a prescribed time limit. Defects can affect the safe working of the plant or machinery and where identified must receive the attention specified within the time limit given in the report.

It is common practice for the engineer-surveyor to make observations in his report to draw attention to other matters of a less serious nature. These observations are advisory and do not affect the continuing safe operation of the plant and machinery.

The reports of the statutory inspections must be kept readily available for inspection by the enforcing authorities.

4.5.9 Conclusion

This chapter has endeavoured to cover the principal statutory inspection requirements in the UK that are likely to be of concern to occupational safety advisers. Certain areas have not been covered, such as gasholders nor the slightly differing requirements in Eire, Northern Ireland, the Channel Islands and the Isle of Man.

It should not be assumed that because there is no statutory requirement to periodically inspect a particular type of plant or machine, that it need not be so inspected. Section 2 of HSW places a duty on employers to provide and maintain plant in a safe condition and to provide a safe place to work generally. To meet this general obligation it is prudent to carry out regular inspections and tests on a range of plant and machines and to record the results. This is underlined by the increasing number of industry-produced Codes of Practice, such as for injection moulding machines, die-casting machines and concrete pumping booms, which recommend inspections and tests as being the best practical means of ensuring continuing compliance with HSW.

Modern legislation permits inspection and test requirements to be determined following a risk assessment conducted by the user/owner. This gets away from the prescriptive requirements of earlier legislation which had proved inappropriate in a number of circumstances. The current goal setting approach ensures the user/owner has much greater flexibility in selecting appropriate inspection regimes but at the same time imposes a responsibility on him to select suitable and effective regimes which ensure the continuing safety of work equipment.

References

1. Health and Safety Executive, Approved Code of Practice No. L122, *Safety of pressure systems. Pressure Systems Safety Regulations 2000*, HSE Books, Sudbury (2000)
2. Safety Assessment Federation, *Shell boilers: guidelines for the examination of longitudinal seams of Shell boilers*, SAFed, London (1998)
3. Safety Assessment Federation, *Guidelines on periodicity of examinations*, SAFed, London (1997)
4. Safety Assessment Federation, *Guidelines on the examination of boiler shell to end plate and furnace to end plate welded joints*, SAFed, London (1997)
5. Health and Safety Executive, Guidance Note No. PM 60, *Steam boiler blowdown systems*, HSE Books, Sudbury (1998)

6. Milton, J.H. and Leach, R.M., *Marine Steam Boilers*, 4th edn, Butterworth-Heinemann, Oxford (1980)
7. European Union, *Council Directive on the laws of Member States relating to machinery*, No. 89/392/EEC as amended by Directive No. 91/368/EEC and consolidated in Directive No. 98/37/EC, Luxembourg (1998)
8. Health and Safety Executive, *The Supply of Machinery (Safety) Regulations 1992*, SI 1992 No. 3073, The Stationery Office, London (1992)
9. Health and Safety Executive, *The Provision and use of Work Equipment Regulations 1998*, SI 1998 No. 2306, The Stationery Office, London (1998)
10. Health and Safety Executive, *The Lifting Operations and Lifting Equipment Regulations 1998*, The Stationery Office, London (1998)
11. Health and Safety Executive, publication L113, *Safe use of lifting equipment; Approved Code of Practice and Guidance*, HSE Books, Sudbury (1998)
12. British Standards Institution, *BS 7121* – Part 1:1989, *Code of Practice for the safe use of cranes. General*; Part 2:1991, *Code of Practice for the safe use of cranes – inspection, testing and examination*; Part 3:2000, *Code of Practice for the safe use of cranes – Mobile cranes* BSI, London
13. Health and Safety Executive, publication L112, *Safe use of power presses; Approved Code of Practice and Guidance*, HSE Books, Sudbury (1998)
14. Health and Safety Executive, Guidance Note No. PM 79, *Power presses: Thorough examination and testing*, HSE Books, Sudbury (1995)
15. Health and Safety Executive, *Press brakes*, HSE Books, Sudbury (1984)
16. Health and Safety Executive, Guidance Booklets Nos: (a) HSG 37, *Introduction to local exhaust ventilation* (1993) (b) HSG 54, *The maintenance, examination and testing of local exhaust ventilation* (1998). HSE Books, Sudbury
17. Health and Safety Executive, Legislation Booklet No. L 5, *General COSHH ACOP and Carcinogens ACOP and Biological Agents ACOP (2002)*, HSE Books, Sudbury (2002)
18. Health and Safety Executive, Legislation booklet L86, *Control of substances hazardous to health in fumigation operations*, HSE Books, Sudbury (1996)
19. Health and Safety Executive, Legislation Booklet No. L27, *The control of asbestos at work*, HSE Books, Sudbury (1999)
20. Health and Safety Executive, Approved Code of Practice No. COP 2, *Control of lead at work*, HSE Books, Sudbury (1998)
21. The Institution of Electrical Engineers, *Requirements for Electrical Installations*, 16th edn, IEE, London (2001)
22. British Standards Institution, *BS 7671:2001, Requirements for electrical installations, IEE wiring regulations, 16th edition*, BSI, London (2001)
23. Health and Safety Executive, Approved Code of Practice No. COP 34, *The use of electricity in mines*, HSE Books, Sudbury (1989)
24. Heath and Safety Executive, Approved Code of Practice No. COP 35, *The use of electricity in quarries*, HSE Books, Sudbury (1989)
25. Health and Safety Executive, publication HSG 107, *Maintaining portable and transportable electrical equipment*, HSE Books, Sudbury (1994)
26. British Standards Institution, *BS EN 45004 General criteria for the operation of various types of bodies performing inspections*, BSI, London

Further reading and references

General

Sinclair, T. Craig, *A Cost-Effective Approach to Industrial Safety*, HMSO, London (1972)

Legal

Fife, I. and Machin, E.A., *Redgrave Fife and Machin; Health and Safety*, Butterworth-Heinemann, Oxford (1998)

Munkman, J., *Employer's Liability at Common Law*, 11th edn, Butterworth, London (1990)

Pressure vessels

Jackson, J., *Steam Boiler Operation: Principles and Practices*, 2nd edn, Prentice-Hall, London (1987)

Robertson, W.S., *Boiler Efficiency and Safety*, MacMillan Press, London

Brown, Nickels and Warwick, *Periodic Inspection of Pressure Vessels*, (A.O.T.C.) I. Mech. E. Conference, London (1972)
British Standards Institution, London:
BS 470:1984 Specification for inspection, access and entry openings for pressure vessels
BS 709:1983 Methods of destructive testing fusion welded joints and weld metal in steel. This is gradually being replaced by a series of BS EN standards
BS 759 (Pt 1):1984 Specification for valves, gauges and other safety fittings for application to boilers, and to piping installations for and in connection with boilers
BS 1113:1999 Specification for design and manufacture of water tube steam generating plant (including superheaters, reheaters and steel tube economisers)
BS 1123–1:1987 Safety valves, gauges and fusible plugs for compressed air or inert gas installations. Code of practice
BS 2790:1992 Specification for the design and manufacture of shell boilers of welded construction
BS 5169:1992 Specification for fusion welded steel air receivers
BS PD 5500:2000 Unfired fusion welded pressure vessels
BS 6244:1982 Code of Practice for stationary air compressors
BS EN 12952–4:2000 Water-tube boilers and auxiliary installations. In-service boiler life expectancy calculations
BS EN 1435:1997 Non-destructive testing of welds. Radiographic examination of welded joints.

ANSI/ASME Boiler and pressure vessel code
Sec. I – Rules for the construction of power boilers
Sec. VII – Recommended guidelines for the care of power boilers

HSE Guidance Notes, HSE Books, Sudbury
GS 4 Safety in pressure testing (1992)
PM 5 Automatically controlled steam and hot water boilers (1989)
HS (G) 29 Locomotive boilers (1986)
PM 60 Steam boiler blowdown systems (1998)
L 101 Safe working in confined spaces, Confined Spaces Regulations 1997 Approved Code of Practice
L122 Safety of pressure system, Pressure Systems Safety Regulations 2000 Approved Code of Practice

Lifting equipment
Phillips, R.S., *Electric Lifts*, Pitman, London (1973)
Dickie, D.E., *Lifting Tackle Manual*, (Ed. Douglas Short), Butterworth, London (1981)
Dickie, D.E., *Crane Handbook*, (Ed. Douglas Short), Butterworth, London (1981)
Dickie, D.E., *Rigging Manual*, Construction Safety Association of Ontario (1975)
Associated Offices Technical Committee (A.O.T.C.) *Guide to the testing of cranes and other lifting machines*, 2nd edn, A.O.T.C., Manchester (1983)
British Standards Institution, London:
BS 466:1984 Specification for power driven overhead travelling cranes, semi-goliath and goliath cranes for general use
BS 1757:1986 Specification for power driven mobile cranes
BS 2452:1954 Specification for electrically driven jib cranes mounted on a high pedestal or portal carriage (high pedestal or portal jib cranes)
BS 2573 (Pt 1): 1983 Specification for the classification, stress calculations and design criteria for structures
BS 2573 (Pt 2): 1980 Specification for the classification, stress calculations and design of mechanisms
BS 2853: 1957 Specification for the design and testing of steel overhead runway beams
BS 4465: 1989 Specification for design and construction of electric hoists for both passengers and materials
BS 5655 (10 parts) covering safety of electric and hydraulic lifts, dimensions, selection and installation, control devices and indicators, suspension eyebolts, guides, and the testing and inspection

Parts have been superseded by:
BS EN 81–1:1998 Safety rules for the construction and installation of lifts. Electric lifts
BS EN 81–2:1998 Safety rules for the construction and installation of lifts. Hydraulic lifts
BS EB 81–3:2000 Safety rules for the construction and installation of lifts. Electric and hydraulic service lifts.
BS 7172–1:1989 Code of practice for the safe use of cranes. General
BS 7121–2:1991 Code of practice for the safe use of cranes. Inspection, testing and examination
BS 7121–3:2000 Code of practice for the safe use of cranes. Mobile cranes
ISO 4309:1990 Cranes – Wire ropes – Code of Practice for examination and discard
ISO 4310 Cranes – test code and procedures

HSE Guidance Notes, HSE Books, Sudbury
PM 3 Erection and dismantling of tower cranes (1976)
PM 8 Passenger carrying paternosters (1987)
PM 9 Access to tower cranes (1979)
PM 24 Safety at rack and pinion hoists (1981)
PM 27 Construction hoists (1981)
PM 34 Safety in the use of escalators (1983)
PM 43 Scotch derrick cranes (1984)
PM 45 Escalators: periodic thorough examination (1984)
PM 54 Lifting gear standards (1985)
PM 55 Safe working with overhead travelling cranes (1985)
PM 63 Inclined hoists used in building and construction work (1987)
HSG 150 Health and safety in construction

Power presses
Joint Standing Committee on Safety in the Use of Power Presses:
Safety in the use of power presses, HSE Books, Sudbury (1979)
Power press safety; Safety in material feeding and component ejection systems, HSE Books, Sudbury (1984)
British Standards Institution, London:
BS 4656 (Pt 34): 1985 Specification for power presses, mechanical, open front
BS EN 61496–1:1998 Safety of machinery. Electro-sensitive protective equipment. General requirements and tests
BS IEC 61496–2:1997 Safety of machinery. Electro-sensitive protective equipment. Particular requirements for equipment using active opto-electronic protective devices (AOPDs)

HSE Guidance Notes, HSE Books, Sudbury
HSG 180 Application of electro-sensitive protective equipment using light curtains and light beam devices to machinery

Local exhaust ventilation plant
Industrial Ventilation, American Conference of Government Industrial Hygienists, Cincinatti, Ohio

Principles of Local Exhaust Ventilation, Report of the Dust and Fume Sub Committee of the Joint Standing Committee on Health, Safety and Welfare in Foundries, HSE Books, Sudbury (1975)
Relevant Standard: BS EN 779:1993 Particulate filters for general ventilation. Requirements, testing, marking

HSE Guidance Notes, HSE Books, Sudbury
MS 13 Asbestos (1999)
EH 10 Asbestos – Exposure limits and measurements of airborne dust concentrations (1995)
EH 25 Cotton dust sampling (1980)
HSG 37 Introduction to local exhaust ventilation (1993)
COP 2 Control of lead at work (1998)

Electrical installations
Institution of Electrical Engineers, *Requirements for Electrical Installations*, 16th edn, London (2001). See also BS 7671:2001
British Standards Institution, London:
BS 2754:1976 Construction of electrical equipment for protection against electric shock
BS EN 60204, Safety of machinery – Electrical equipment of machines, part 1 Specification for general requirements
BS 4444:1989 Guide to electrical earth monitoring
BS 5958 (Pt 1):1991 Code of Practice for control of undesirable static electricity – general considerations
BS 5958 (Pt 2):1991 Code of Practice for control of undesirable static electricity – recommendations for particular industrial situations
BS 6233:1982 Methods of test for volume resistivity and surface resistivity of solid electrical insulating materials

HSE Guidance Notes, HSE Books, Sudbury
GS 6 Avoidance of danger from overhead electrical lines (1997)
PM 29 Electrical hazards from steam/water pressure cleaners etc. (1995)
PM 38 Selection and use of electric hand lamps (1992)

Chapter 4.6

Safety on construction sites

R. Hudson

The construction industry has always been plagued with an abundance of reportable accidents coming to an all time high of over 45 000 accidents in 1966 with, over the previous decade, an average of 250 persons killed each year. These appalling figures occurred in spite of a considerable volume of safety legislation aimed at improving safe working in the construction industry.

Many of the causes of these accidents are reflected in the detailed requirements of the relevant Regulations[1] which lay down the preventive measures to be taken. This chapter looks at the safety legislation for the construction industry and some of the techniques for meeting the required safety standards.

It should be remembered, however, that since the Health and Safety at Work etc. Act 1974 (HSW) came into effect all subordinate legislation, such as Regulations, made under it apply to all employer/employee relationships and though the title may not include 'construction' this does not mean they do not apply to construction works.

4.6.1 Construction accidents

An indication of the size and seriousness of the problem can be obtained by considering the annual HSE report[2] containing data on fatal and major accidents and respective incidence rates.

While overall until 1996–7 there has been a reduction in fatal accidents this may be accounted for by a smaller workforce or the change in, rather than improved, standards. If the incidence rate for construction projects, covering the range from large civil and high rise building to refurbishment and low rise structures, is compared with manufacturing it is, year on year, consistently six times more dangerous. As some 70% of the accidents investigated could have been prevented by management action[3] this continued to be an unacceptable situation.

Further analysis of both fatal and major accidents gives a good indication of the problem areas. While the numbers vary from year to year the pattern remains fairly constant with 'falls from height' accounting for some 40% of major injuries and 50% of fatalities.

4.6.2 Safe working in the industry

In considering safe working and accident prevention in the construction industry, this chapter will follow broadly the progression of a construction operation. All stages should be adequately planned making allowance for the incorporation of safe systems of work.

Planning has been the province of the main contractor but with the coming into effect of the Construction (Design and Management) Regulations 1994[4] (CDM) this responsibility has been clarified. Under these Regulations the client has an obligation to appoint a competent planning supervisor for the project. In many instances this role will be filled by professional advisers such as architects or engineers who act on behalf of the client.

The planning supervisor is required to:

- ensure the designers have fulfilled their responsibilities under the regulations and the design includes adequate information about the design and materials to be used where they might affect the health and safety of those carrying out the construction work;
- prepare a health and safety plan, to be included with the tender documentation, which details the risks to health and safety of any person carrying out the construction work so far as is known to the planning supervisor or are reasonably foreseeable, and any other relevant information to enable the contractor to manage the works;
- prepare and deliver to the client a health and safety file on the as-built structure which the client retains for reference during subsequent construction works on the structure.

The client is required to appoint a competent principal contractor for the project.

The principal contractor must for his part:

- adopt and develop the health and safety plan and provide information for the health and safety file;
- ensure the health and safety plan is followed by all persons on the site; and
- co-ordinate the activities of others on the site and ensure that all co-operate in complying with the relevant statutory provisions that affect the works.

For these purposes the principal contractor can give directions or establish rules for the management of the construction works as part of the health and safety plan. Such rules must be in writing and be brought to the attention of all affected persons.

Finally, one of the main provisions of these Regulations defines the responsibility which designers, such as architects, have for health and safety during the construction stages. Designers have to ensure, so far as is reasonably practicable and provided the structure conforms to their design, that persons building, maintaining, repairing, repainting, redecorating or cleaning the structure are not exposed to risks to their health and safety. In addition, the designer must ensure that included in the design documentation is adequate information about the design and materials used, particularly where they may affect the health and safety of persons working on the structure.

Basically, these requirements place designers of buildings and structures, such as architects, under similar obligations to those who design articles and substances, whose obligations are contained in s.6 of HSW.

As work gets under way, the principal contractor, who has responsibility for the construction phase of the project, has to ensure that all those employed are properly trained for their jobs. Under HSW, now amplified by the Management of Health and Safety at Work Regulations 1999[1] (MHSWR), the employer is required to provide training in specific circumstances, i.e. on joining an employer, when work situations change and at regular intervals.

In addition, specific job training is prescribed in numerous statutory provisions such as the Construction (Health, Safety and Welfare) Regulations 1996 (construction activity where training is necessary to reduce risks) and the Provision and Use of Work Equipment Regulations 1998[1] (PUWER) (adequate training in the use of work equipment).

In an industry increasingly reliant upon the use of subcontractors, the main contractor retains the onus for health and safety on site. This onus can extend to training employees of subcontractors where their activities may affect the health and safety of the employees of the main contractor and of the subcontractor himself (ss. 2 and 3 of HSW). This is more clearly defined by CDM which requires the principal contractor to ensure that other employers on the construction work provide their employees with appropriate health and safety training when they are exposed to new or additional risks due to:

- changes of responsibilities, i.e. promotion,
- use of new or changed work equipment,
- new technology, or
- new or changed systems of work.

The provision of information is also an essential contribution to reducing risks to health and safety. As with training, ss. 2 and 3 of HSW apply to both main and subcontract employers in relation to informing each other of risks, within their knowledge, arising out of their work. The decision in *Regina* v. *Swan Hunter Shipbuilders Ltd*[5] clarified this in respect of 'special risks'. In this case, a number of fatalities resulted from a fire in a poorly ventilated space in a ship which had become enriched with oxygen due to an oxygen supply valve being left open by a subcontractor's employee. An intense fire developed when another contractor struck an electric arc to do some welding. Swan Hunter were well aware of the fire risk associated with oxygen enrichment and provided detailed information for their own, but not subcontractors' employees. Swan Hunter were prosecuted under ss. 2 and 3 of HSW and convicted for failing to ensure the health and safety of their own employees by not informing the employees of subcontractors of special risks which were within its, Swan Hunter's, knowledge, i.e. from fires in oxygen enriched atmospheres. This decision has been overtaken by MHSWR which requires employers who share a workplace to take all reasonable steps to inform other employers of any risks arising from their work. Further, CDM places a duty on the principal contractor to inform other contractors of the risks arising out of or in connection with the works and ensure that those subcontractors inform their employees of:

- risks identified by the contractor's own general risk assessment,
- the preventive and precautionary measures that have to be implemented,
- any serious or imminently dangerous procedures and the identity of any persons nominated to implement those procedures, and
- details of the risks notified to him by the principal or another contractor.

Apart from the overall obligations placed on both the main and sub-contractor employers by the Health and Safety at Work Act, more extensive requirements specific to the building and construction industry are contained in Regulations dealing with particular aspects of safety in building and construction work.

4.6.2.1 Notification of construction work

The CDM and Construction (Health, Safety and Welfare) Regulations have redefined the work that has to be notified extending it from 'building operation or work of engineering construction' to a broader term 'construction work'. This latter term is defined in the Regulations as including every aspect of the carrying out of the work from beginning to end of a project. It includes site clearance and site investigation, the assembly and disassembly of fabricated units (site huts), the demolition and removal of spoil and the installation, commissioning, maintenance and repair of services such as telephones, electricity, compressed air, gas etc. and on small projects such as extensions to the engineering work involved in the installation, maintenance and dismantling of major process plants.

Responsibility for making the notification lies with the planning supervisor who must provide to the HSE the information listed in schedule 1 of the Regulations before any work starts on the site. Official form F10 (rev), which calls for the necessary information, can be used but is not mandatory. Notification of construction work must be made where the work being undertaken is expected to last more than 30 days or where it involves a total of more than 500 person-days. A working day is any day on which work of any sort is carried out on the site and includes weekends and all other times outside the 'normal' working week.

However, whether there is a need to notify or not, full compliance with all the requirements of the relevant health and safety legislation is necessary.

4.6.2.2 The Construction (Health Safety and Welfare) Regulations 1996

4.6.2.2.1 Responsibilities

The responsibility for complying with the requirements of these Regulations is placed on employers, the self-employed person, the person controlling construction works, employees and every person at work.

The requirements of the Regulations cover several subject areas which are dealt with in greater depth below, by including practical advice on the separate subjects to give a greater understanding of how compliance with the Regulations can be achieved.

4.6.2.2.2 Safety in excavations

In any excavation, earth work, trench, well, shaft, tunnel or underground working where there is a risk of material collapsing or falling, proper support must be used as early as practicable in the course of the work to prevent any danger from an earth fall or collapse. Suitable and sufficient material should be available for this purpose or alternative methods used such as:

1 *Battering the sides*, i.e. cutting the sides of the excavation back from the vertical to such a degree that fall of earth is prevented.
2 *Benching the sides*. The sides of the excavation are stepped to restrict the fall of earth to small amounts. Maximum step depth 1.2 m (4 ft).

Figure 4.6.1 shows typical examples of these trenching techniques.

Inspection of any excavation which is supported must be made:

(a) at the start of every shift before any person carries out any work;
(b) after any event likely to have affected the strength or stability of the excavation or any part of it;
(c) after any accidental fall of rock, earth or other material.

A report of the inspection containing the prescribed particulars (no form or official register is necessary) shall be made within 24 hours of the inspection and retained until 3 months after the work has been completed. Only one report needs to be made every 7 days in respect of item (a) above for excavations and items (a) and (b) for coffer-dams and caissons. Reports of inspections following the other incidents listed above must be made before the end of the working period.

All material used for support should be inspected before use and material found defective must not be used. Supports must only be erected, altered or dismantled under competent supervision and whenever practicable by experienced operatives. All support must be properly constructed and maintained in good order. Struts and braces must be fixed so that they cannot be accidentally dislodged. In addition, in the case of a coffer-dam or caisson, all materials must be examined and only if found suitable should they be used.

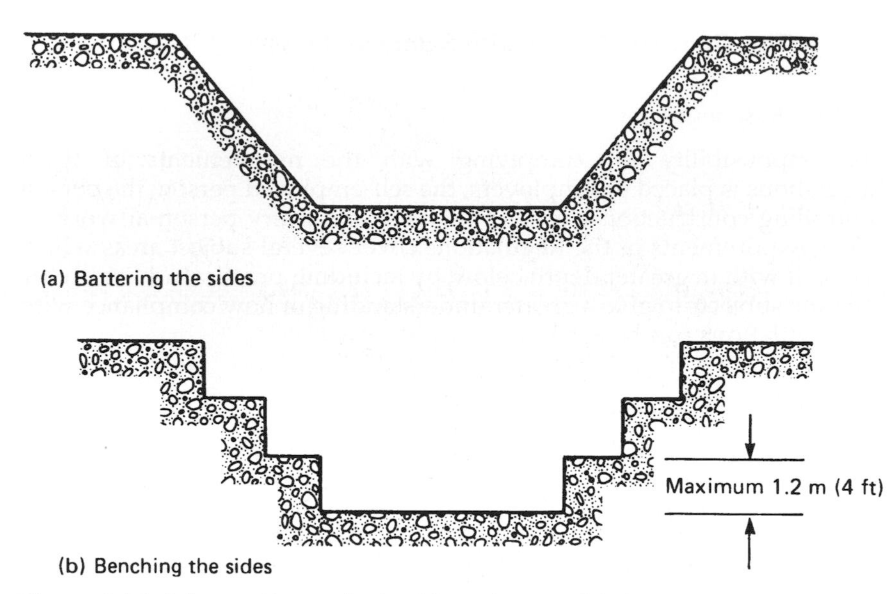

Figure 4.6.1 Safe trenching methods without the use of timber

If there is risk of flooding, ladders or other means of escape must be provided.

When excavating in close proximity to existing buildings or structures, be they permanent or temporary, there is a requirement to give full consideration to their continued stability. This is intended to protect persons employed on site. However, under the Health and Safety at Work Act this responsibility is extended to the safety of the public, i.e. those not employed on the site, and may relate to private dwelling houses, public buildings or public rights of way. It is particularly important when excavating near scaffolding.

Where any existing building or structure is likely to be affected by excavation work in the vicinity, shoring or other support must be provided to prevent collapse of the building or structure. Examples of trench shoring are given in *Figures 4.6.2, 4.6.3* and *4.6.4.*

Excavations more than 2 m deep near which men work or pass, must be protected at the edge by guardrails or barriers or must be securely covered. Guardrails, barriers or covers may be temporarily moved for access or for movement of plant or materials but must be replaced as quickly as possible.

Where the excavation is not in an enclosed site and is accessible to the public the standard for protection is more onerous. Even the most shallow depressions should be fenced so that members of the public are not exposed to risks to their health and safety.

Materials, plant, machinery etc. must be kept away from the edge of all excavations to avoid collapse of the sides and the risk of men falling in, or material falling on men.

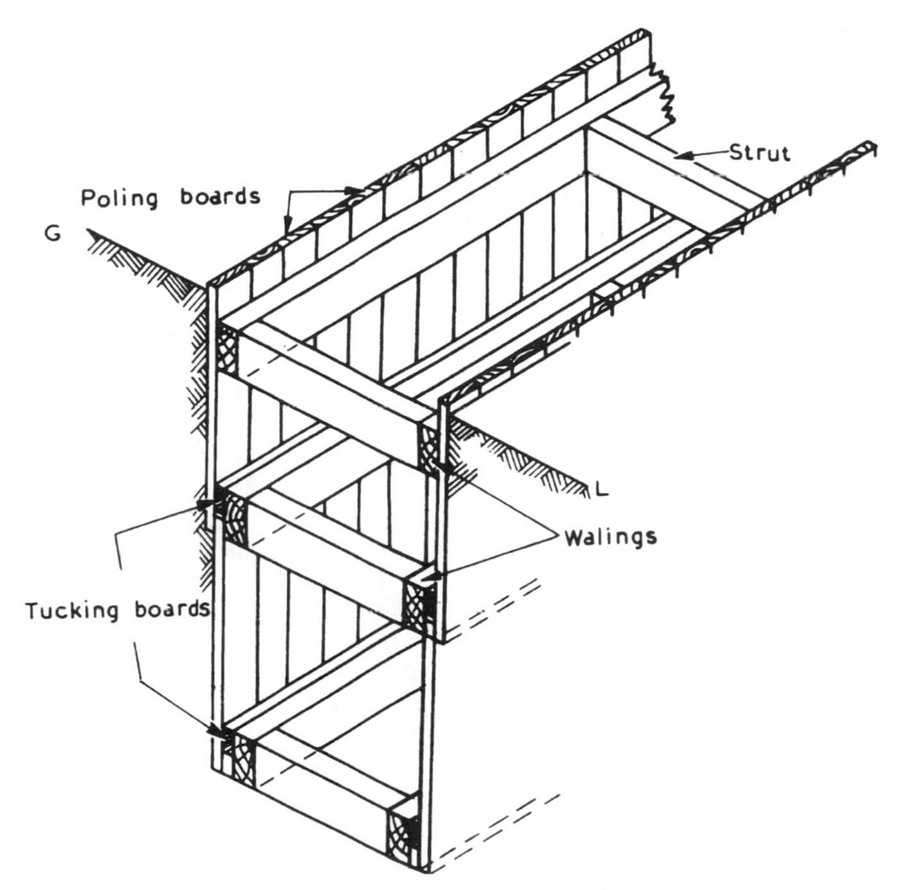

Figure 4.6.2 Close poling with tucking frames. (BS 6031)[6]

Severe weather conditions such as heavy rain, or where timber has become wet then followed by a hot dry spell, could so affect timbering etc. as to cause it to become dangerous. In these circumstances, where the strength or stability of an excavation could be adversely affected, an inspection would be required together with a report. Guidance on the construction of trenches, pits and shafts is given in the British Standard CP 6031 Code of Practice for Earthworks[6].

On sites where mobile machinery such as tippers, diggers, rough terrain fork lift trucks etc. are used, special care should be taken to ensure that operators are fully aware of the stability of their machines and of the maximum slope on which they can be safely used. Particular attention should be paid to the condition of the ground and whether it is capable of bearing the vehicle weight. Information on safe ground conditions and angles of tilt can be obtained from the machine manufacturers.

Where overhead cables cross the line of excavations, particular care must be taken in the selection of the type of plant to be used and

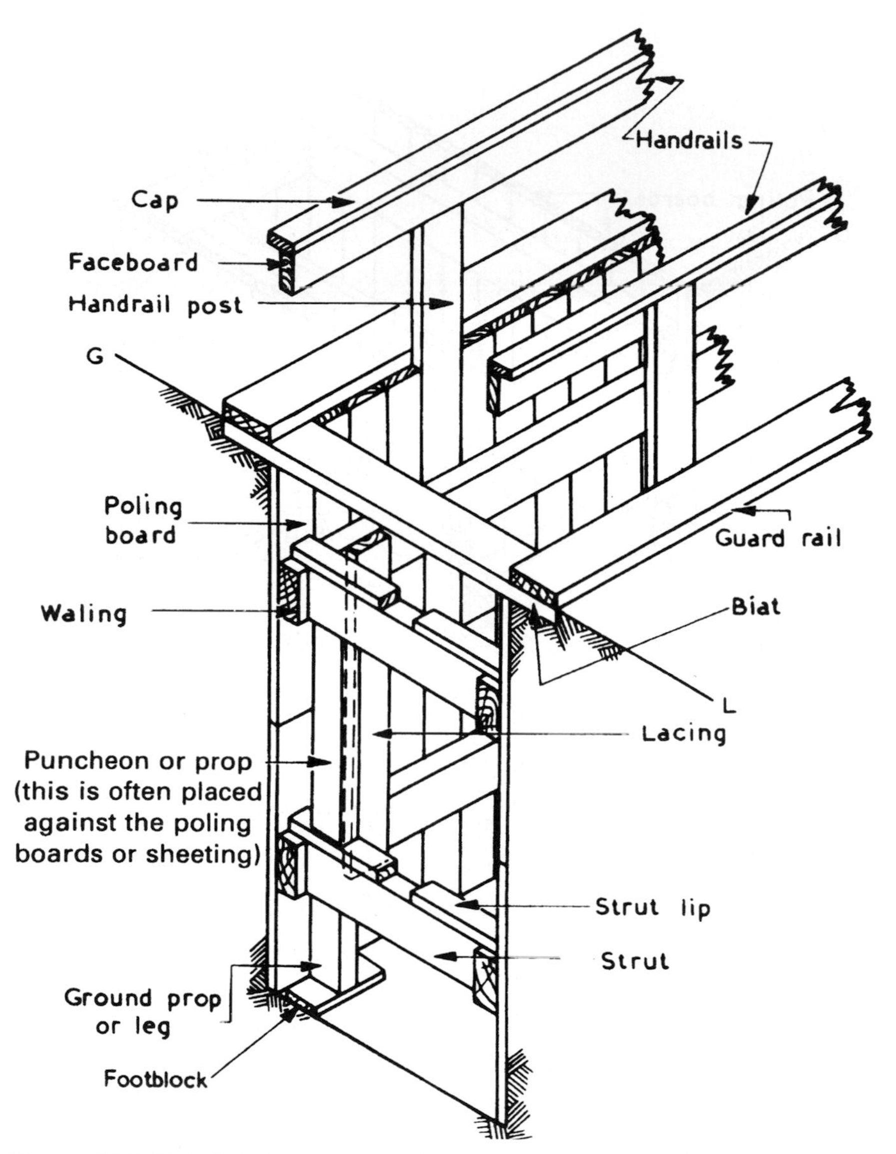

Figure 4.6.3 Typical single or centre waling poling frame. (BS 6031[6])

precautions taken to ensure that the equipment does not or cannot touch live high voltage conductors. Underground cables, be they high or low voltage, telephone or television links, together with gas piping, present a more difficult problem which, in the main, rests with the excavating contractor. Advice is given in an HSE publication[7] but the contractor should approach each of the service authorities asking for accurate information on the actual location, run and depth of their services. This should be in writing, and the information supplied by the authority

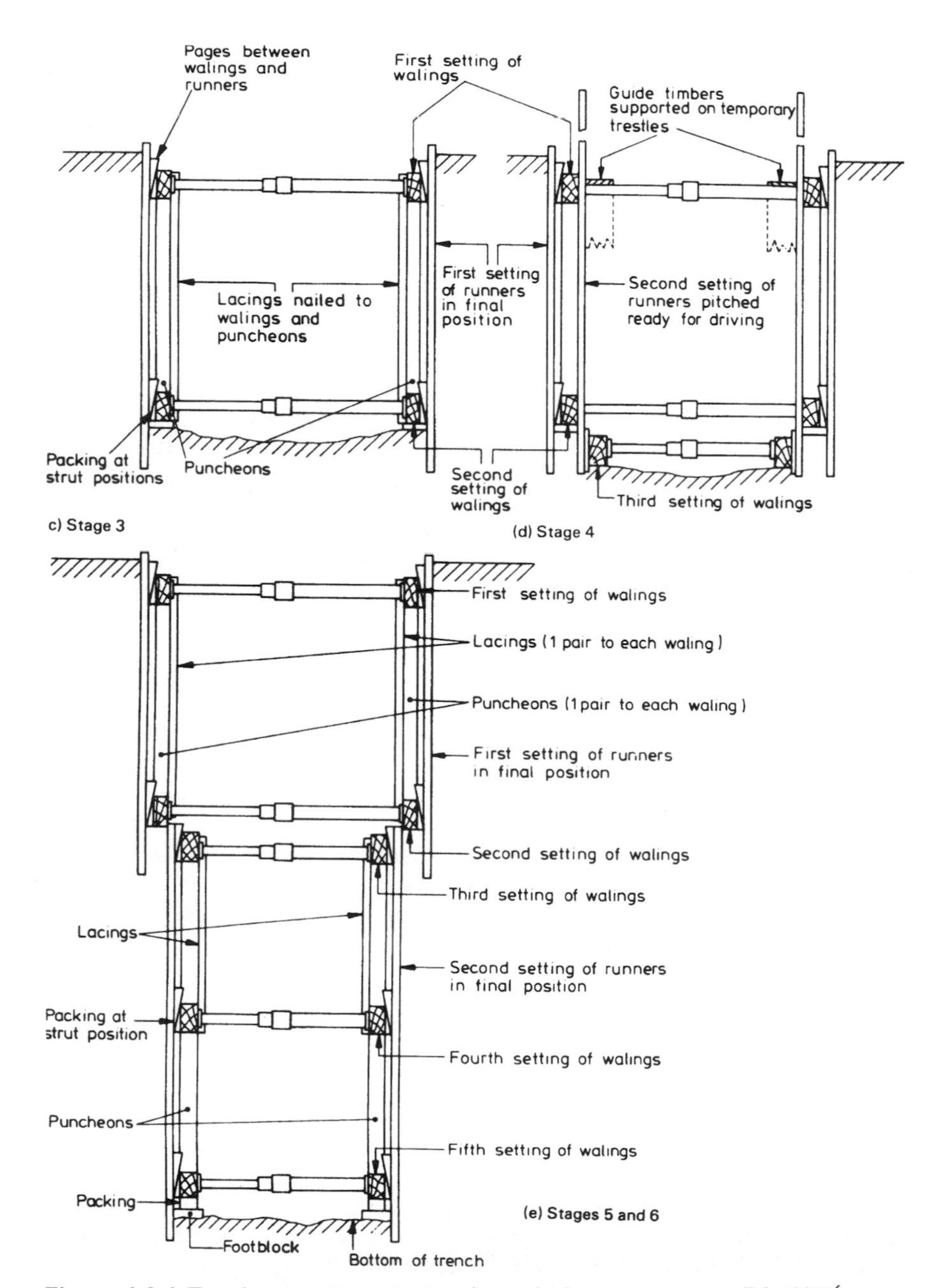

Figure 4.6.4 Trench excavation using steel trench sheets as runners. (BS 6031)[6]. Note Stages 1 and 2 are shown in the British Standard

should preferably be marked on a drawing; ideally the authority should authorise the drawing as correct. The location should be confirmed using devices for locating cables and other services and the route of the service marked on the surface. Services should be carefully exposed by hand-dig methods to verify their precise location and depth before mechanical means are employed. Mechanical equipment such as excavators should not be used within 0.5 m of the suspected cable until its route has been specifically located.

Knowledge of the whereabouts of underground services is also necessary where heavy plant or vehicles are used since many such services are at shallow depth and can be damaged by the sheer weight of equipment. Although injury is not likely to result, considerable cost to the contractor could be involved.

Careful planning, including the selection of the correct plant and equipment is essential, for both safety and economic reasons, when carrying out excavation work. For example, the correct size of excavator can act as a crane eliminating the need to bring extra specialised plant onto site. However, when using an excavator either for digging or lifting, the machine must comply with all the requirements of the Lifting Operations and Lifting Equipment Regulations 1998.

Before work is started on a construction site there are a number of matters that should be checked, from both a prudent and statutory point of view:

1 Provide site security – particularly to stop children getting in.[8]
2 Investigate the nature of the ground before excavations begin and decide the form that the support work will need to take and ensure that adequate supplies of sufficiently strong materials are available. Special precautions may be needed where trenches pass near adjacent roads or buildings.
3 Locate all public services, water, gas, electricity, telephone, sewers etc., and avoid if possible; if not, take necessary precautions.
4 Provide material for barriers and authorise traffic notices.
5 Provide adequate lighting.
6 Position spoil heap at a distance not less than the depth from the edge of excavation. Tip if possible on blind side of excavator to ensure operator has visibility when swinging back to trench excavation.
7 Provide personal protective equipment.
8 Provide sufficient ladders of suitable length, strength and type.
9 Query necessity for bridges and gangways.
10 Take note of all overhead services, the arrangements made for their protection and the safety of all working in their vicinity.

4.6.2.2.3 Mechanical plant and portable tools

All machinery for use at work is now subject to the Provision and Use of Work Equipment Regulations 1998[1] (PUWER) with its requirements for guarding dangerous parts, controls and associated matters.

The general principles of guarding are contained in BS 5304[9] but this standard is being replaced by a number of harmonised European (EN) standards. Other publications[10] are available that give advice on the safety and care of site plant and equipment.

Where guards are removed to enable maintenance work to be carried out, they must be replaced before the machinery is returned to normal work.

Portable tools are used extensively on construction sites and commonly suffer damage. Arrangement should be made for regular checks on the condition of portable tools, paying particular attention to the integrity of electrical insulation on the tool itself and on the lead and to damage to rotating parts remembering that the Electricity at Work Regulations 1989 require such equipment to be properly maintained, involving regular inspection and testing. In the case of compressed air equipment, the Pressure Systems Safety Regulations 2000 also have effect.

Under PUWER, not only the operators of any plant and machinery, being work equipment, but their supervisors and managers must be trained:

- in the correct method of use,
- on the risks such use may involve, and
- on the precautions to be taken.

Operators of plant and machinery on a construction site inevitably will be over 18 years old unless being trained and under the direct supervision of a competent person as the risks associated with persons under 18 years of age will generally preclude them and they will not have sufficient experience to operate such equipment.

4.6.3 Site hazards

The extensive use of temporary or semi-permanent wiring on construction sites, the rough usage that equipment gets, the hostile conditions under which it is used and, often, the lack of knowledge of those using the equipment contribute to the high risk potential of the use of electricity. Compliance with the Electricity at Work Regulations[11] will reduce the hazards, which broadly can be divided into three categories:

1 Electrocution.
2 Fire.
3 Glare.

4.6.3.1 Electrocution

There are three operations that carry the highest risk of electrocution: use of portable tools, striking a buried cable, and cranes and excavators making contact with overhead power lines.

Portable tools are used extensively on sites and maintaining them and their connecting cables in good repair is a critical factor of their safe use. Electrocution occurs when the body acts as the conductor between a power line and earth, often because the earth connection on the tool has broken or, less commonly, as a link between differently charged conductors. All portable tools must be securely earthed or be of double or all insulated construction and the plug on the lead must be correctly fused. Unfortunately it is frequently difficult to keep track of every item, so reliance has to be placed on the person using them.

Protection for nominal 240 V supplies can be obtained by the use of residual current devices (RCDs) either in the supply circuit or on the connection to the particular equipment. In addition, on construction sites nominal 240 V supplies should be carried by armoured, metal sheathed or other suitably protected cables.

The greatest protection from electrical shock on construction sites is through the use of a reduced voltage system. Where portable hand-held tools are used this is essential and the system recommended is 110 V ac with the centre point of the secondary winding tapped to earth, so that the maximum voltage of the supply above earth will be 55 V which normally is reckoned to be non-fatal. A further alternative is to use a low voltage supply at 24 V, but in this case, because of the low voltage, the equipment tends to be heavier than with higher voltages. The risk of electrocution from power tools is increased if they are used in wet conditions.

Electrocution from striking an underground cable can be spectacular when it occurs and the most effective precaution is to obtain clearance from the local Electricity Board or the location's electrical engineer that the ground is clear of cables. If doubt exists, locating devices are available that enable underground cables to be traced.

All too often overhead power cables cross construction sites and are a potential hazard for cranes and mechanical equipment. If the supply cannot be cut off, suspended warning barriers should be positioned on each side and below the level of the cable and drivers warned that they may only pass under with lowered jib.

4.6.3.2 Fire

Usually caused through overloading a circuit, frequently because of wrong fusing. Repeated rupturing of the fuse should be investigated to find the cause rather than replacing the blown fuse by a larger one in the hope that it will not blow. A second cause can be through water getting into contact with live apparatus and causing a short circuit which results in overheating of one part of the system. Where electrical heaters are used on sites, they should be of the non-radiant type, i.e. tubular, fan or convector heaters. Multi-bank tubular heaters used for drying clothes should be protected to prevent clothing, paper etc. being placed directly on the tubes. This protection can be achieved by enclosing the heaters in a timber frame covered with wire mesh.

4.6.3.3 Glare

Not usually recognised as a hazard, glare can prevent a crane driver from seeing clearly what is happening to his load and it can cause patches of darkness in accessways that prevent operators from seeing the floor or obstacles. Electric arc welding flash can cause a painful condition known as 'arc-eyes', so welding operations should be shielded by suitable flame-resistant screens. Floodlights are designed to operate at a height of 6 m or more and must never be taken down to use as local lighting as the glare from such misuse could create areas of black shadow and may even cause eye injury. Floodlights should not be directed upwards since they can dazzle tower crane drivers.

4.6.3.4 Dangerous and unhealthy atmospheres

Conditions under which work is carried out on construction sites is largely dictated by the weather, ranging from soaking wet to hot, dry and dusty and suitable protection for the health of the operators has to be provided. However, there is also a considerable range of substances[12] and working techniques now in use that have created their own hazards. A number of these are considered below.

4.6.3.4.1 Cold and wet

Cold is most damaging to health when it is associated with wet, as it is then very difficult to maintain normal body temperature. Being cold and wet frequently and for substantial periods may increase the likelihood of bronchitis and arthritis and other degenerating ailments. The effects of cold and wet on the employee's health and welfare can be mitigated by three factors: food, clothing and shelter. Where practicable, shelter from the worst of the wind and wet should be provided by sheeting or screens. The accommodation which has to be provided 'during interruption of work owing to bad weather' could also be used for warming-up and drying-out breaks whenever men have become cold, wet and uncomfortable.

4.6.3.4.2 Heat

Excessive heat has tended to be discounted as a problem on construction sites in the UK, and cases of heat exhaustion which do occur during heat-waves are often attributed to some other quite irrelevant cause. Common forms of heat stress produce such symptoms as lassitude, headache, giddiness, fainting and muscular cramp. Sweating results in loss of fluids and salt from the body, and danger arises when this is not compensated for by increased intake of salt and fluids. If the body becomes seriously depleted it can lead to severe muscular cramps.

4.6.3.4.3 Dust and fumes

Despite the general outdoor nature of the work, construction workers are not immune from the hazards of airborne contaminants. Although natural wind movement will dilute dust and fumes throughout the site, operatives engaged on particular processes may have a dangerous concentration in their immediate breathing zones unless suitable extraction is provided. This is particularly relevant for work in shafts, tunnels and other confined spaces where forced draught ventilation may have to be provided.

Certain processes commonly met on construction sites create hazardous dusts and fumes. Typical are:

Cadmium poisoning from dust and fumes arising from welding, brazing, soldering or heating cadmium plated steel.
Lead poisoning resulting from inhalation of lead fumes when cutting or burning structures or timber that has been protected by lead paint.
Silicosis due to inhaling siliceous dust generated in the cleaning of stone structures, polishing and grinding granite or terrazo.
Carbon monoxide poisoning caused by incomplete combustion in a confined space or from the exhausts of diesel and petrol engines.
Metal fume fever from breathing zinc fumes when welding or burning galvanised steel.

Each of these hazards would be eliminated by the provision of suitable and adequate exhaust ventilation or, in the case of silicosis, by the provision of suitable breathing masks. In each case food should not be consumed in the area, and the medical conditions can be exacerbated through habitual smoking. A good standard of personal hygiene is also an important factor in maintaining good health on site.

4.6.3.4.4 Industrial dermatitis

The use of an increasing range of chemical based products on sites poses a potential health risk to those who handle them unless suitable precautions are taken. The complaint is neither infectious nor contagious, but once it develops the sufferer can become sensitised (allergic) to the particular chemical and will react to even the smallest exposure. All chemical substances supplied to sites should carry instructions for use on the label and if the precautions recommended by the maker are followed little ill-effect should be experienced.

Barrier creams may be helpful but suffer the disadvantage of wearing off with rough usage or being washed off by water. Effective protection is provided by the use of industrial gloves and, where necessary, aprons, face masks etc. Again good personal hygiene is important and the use of skin conditioning creams after washing is beneficial.

4.6.3.4.5 Sewers

Sewers, manholes and soakaways are all confined spaces and before any work is carried out in them an assessment of the risks to health and safety from the work to be done must be made to determine the control measures necessary to avoid those risks as required by the Confined Spaces Regulations 1997. Some precautions that may need to be taken include the testing of the atmosphere for toxic and flammable gases and lack of oxygen. Where the atmosphere is foul, respirators or breathing apparatus, as appropriate, should be worn. Due consideration must be given to preventing the onset of 'Weil's Disease', a 'flu-like disease which if untreated can have a serious or fatal outcome. It is transmitted in rat's urine and enters the body through breaks in the skin or, more rarely, by ingestion of contaminated food.

4.6.3.5 Vibration-induced white finger

The vibrations from portable pneumatic drills and hammers can produce a condition known as white finger or Raynaud's phenomenon in which the tips of the fingers go white and feel numb as if the hand was cold. Anyone showing these symptoms should be taken off work involving the use of these drills or hammers and found alternative employment.

4.6.3.6 Ionising radiations

There are two main uses for radioactive substances that give off ionising radiations on construction sites. Firstly, tracing water flows and sewers where a low powered radioactive substance is added to the flow and its route followed using special instruments. Only authorised specialists should be allowed to handle the radioactive substance before it is added to the water. Once it is added, it mixes rapidly with the water and becomes so diluted as not to present a hazard.

The second application is in the non-destructive testing of welds where a very powerful gamma (γ) source is used. Because of its penetrating powers and the effects its rays have on human organs, very strict controls must be exercised in its use. The relevant precautions are detailed in Regulations[13] whose requirements must be complied with.

4.6.3.7 Lasers

Lasers are beams of intense light, they are radiations but do not ionise surrounding matter. Hazards stem from the intensity of the light which can burn the skin, and, if looked into, can cause permanent damage to eyesight. Ideally, lasers of classes 1 or 2 should be used as these present little hazard potential. Class 3A lasers give rise to eye hazards and should only be used in special cases under the supervision of a laser safety adviser. Class 3B and above generally should not be used on construction work, but if the necessity arises only adequately trained persons should operate them. When eye protection is assessed as being necessary, the type supplied must be certified as providing the required attenuation for the laser being used[14].

4.6.3.8 Compressed air work

The health hazards of work in compressed air and diving are decompression sickness ('the bends') and aseptic bone necrosis ('bone rot'). Both these illnesses can have long-term effects varying from slight impairment of mobility to severe disablement. The protective measures, including decompression procedures, are laid down in the Work in Compressed Air Special Regulations[15] and where diving work is involved, the Diving at Work Regulations 1997[1] apply.

4.6.4 Access

4.6.4.1 General access equipment

Although there is a trend in the construction industry towards specialised plant to meet a particular need, the most common material at present employed to provide access scaffolding is scaffold tube and couplers. Large-scale or difficult projects are best carried out by experts but there is a very large amount of scaffold erection of the smaller type in short-term use which can be quickly and safely erected by craftsmen who are to work on them, provided they have been trained in the basic techniques and requirements of the British Standard Code of Practice[16].

As with all structures, a sound foundation is essential. Scaffolds must not be erected on an unprepared foundation. If soil is the base it should be well rammed and levelled and timber soleplates at least 225 mm (9 in) wide and 40 mm (1½ in) thick laid on it so that there is no air space between timber and ground.

The standards should be pitched on baseplates 150 mm × 150 mm (6 in × 6 in) and any joints in the standards should occur just above the ledger. These joints should be staggered in adjacent standards so that they do not occur in the same lift. Ledgers should be horizontal, placed inside the standards and clamped to them with right-angle couplers. Joints should be staggered on adjacent ledgers so that they do not occur in the same bay.

Decking will generally be 225 mm × 40 mm (9 in × 1½ in) boards and each board should have at least three supports but this is dependent upon the grade of timber used for the boards. The British Standard[17] recommends that they do not exceed 1.2 m (4 ft). Boards are normally butt jointed but may be lapped if bevel pieces are fitted or other measures taken to prevent tripping. A 40 mm (1½ in) board should extend beyond its end support by between 50 mm and 150 mm (2 in and 6 in).

Guardrails must be fitted at the edges of all working platforms at a height of at least 910 mm with an intermediate guardrail so there is no opening greater than 470 mm between any guardrail or toe board. An alternative to an intermediate guardrail is the use, between the top guardrail and the decking, of in-fill material which should be of sufficient strength to prevent a person from falling through the gap.

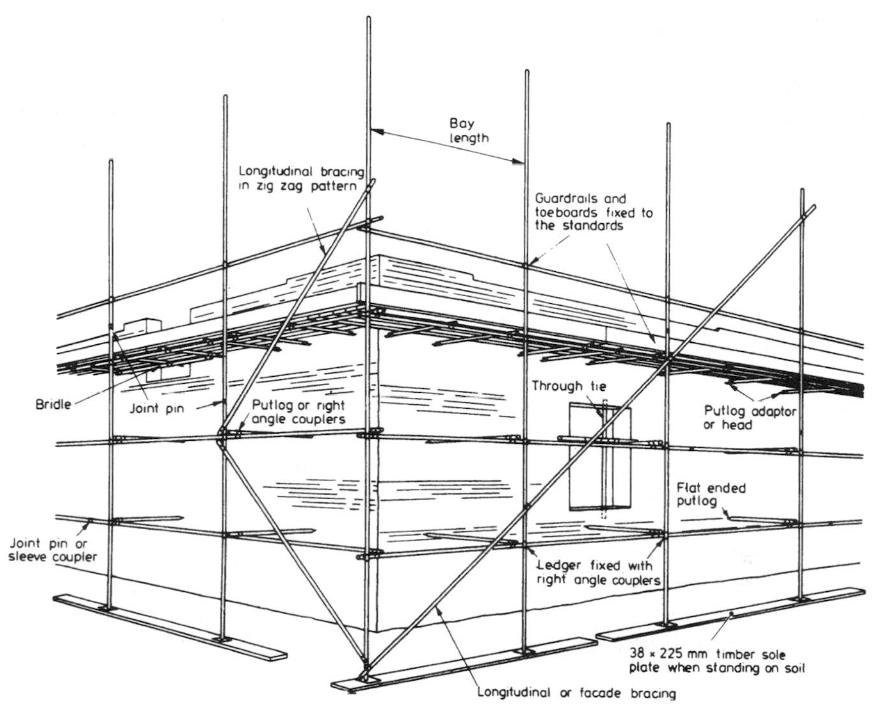

Figure 4.6.5 Typical putlog scaffold. (BS 5973)[16]

Note: The Construction (Health, Safety and Welfare) Regulations 1996 require the provision of an intermediate guardrail at working platforms on scaffolding. BS 5973 is currently under review to incorporate this requirement.

Ladders must stand on a firm level base and must be secured at the top and bottom so that they cannot move. All ladders must extend at least 1.07 m (3 ft 6 in) beyond the landing level. To preserve them they may be treated with a clear preservative or be varnished but must not be painted. All rungs must be sound and properly secured to the stile. No ladder, or run of ladders, shall rise a vertical distance exceeding 9 m unless suitable and sufficient landings or rest areas are provided.

Unless properly designed to stand on their own, all scaffolds must be sufficiently and effectively anchored to the building or structure by ties which are essential to ensure stability of the scaffold. Before using a scaffold, the employer has a duty to arrange for it to be inspected by a competent person, then ensure that it is inspected every seven days and a record maintained of the inspection. All scaffold material must be kept in good condition and free from patent defect. Damaged equipment should be stored separately and identified as 'damaged' or destroyed. Metal scaffold tubes and fittings and timber scaffold boards should comply with the appropriate British Standard[17,18]. *Figure 4.6.5* shows a simple scaffolding structure where only an outside row of standards are used to support the platforms, with putlogs fixed into brickwork joints.

4.6.4.2 Mobile towers

A mobile access tower[19] (*Figure 4.6.6*) is a tower formed with scaffold tube and mounted on wheels. It has a single working platform and is provided with handrails and toeboards. It can be constructed of prefabricated tubular frames and is designed to support a distributed load of 30 lb/ft^2. The height of the working platform must not exceed three times the smaller base dimension and no tower shall have a base dimension less than 4 ft. Rigidity of the tower is obtained by the use of diagonal bracing on all four elevations and on plan. Castors used with the tower should be fixed at the extreme corners of the tower in such a manner that they cannot fall out when the tower is moved and shall be fitted with an effective wheel brake. When moving mobile towers great care is essential. All persons, equipment and materials must be removed from the platform and the tower moved by pushing or pulling at the base level. Under no circumstances may mobile towers be moved by persons on the platform propelling the tower along.

4.6.5 The Lifting Operations and Lifting Equipment Regulations 1998

These Regulations apply to all equipment used for lifting or lowering loads (including persons) on construction sites and include fixed, mobile and travelling cranes, hoists used for both goods and passengers and also the ropes, chains, slings etc. that support the load being lifted.

There are certain requirements that are common to all this equipment in that they must be of adequate strength and stability. When erected the equipment must be properly supported and secured and that ground conditions are such as to ensure stability. Erection must be under competent control.

All lifting equipment must be thoroughly examined regularly with safe means of access provided for those carrying out these examinations. The safe working load must be clearly indicated and jib cranes must have an automatic safe load indicator which must be tested. The specified safe working load must not be exceeded.

The positioning of travelling or slewing cranes should be such that a clear passageway, 0.6 m (2 ft), is ensured at all times. Where drivers or banksmen require platforms that are more than 2 m (6 ft 6 in) above an adjacent level, suitable guardrails must be provided and a cab with safe access should be provided for drivers exposed to the weather. Any communication between banksman and driver must be clear. *Figure 4.3.19* (p. 761) shows the visual signals in common use[20].

Lifting tackle such as chains, rings, hooks, shackles etc. must not be modified by welding unless by a competent person and followed by a test. Hooks should have a safety clip and slings must not be used in such a way that is likely to damage them.

Records must be kept of the examinations of lifting equipment.

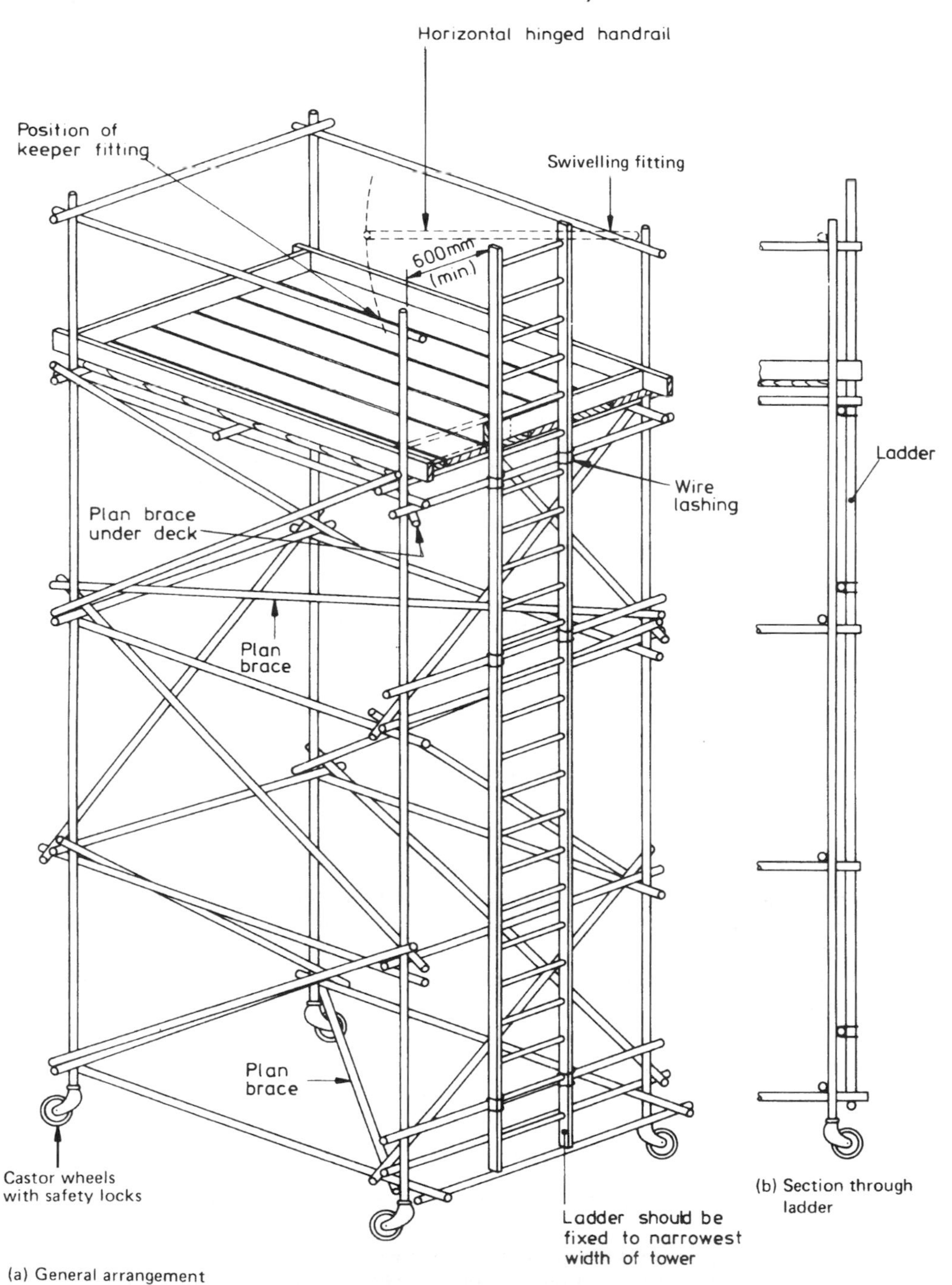

Figure 4.6.6 Mobile access tower. (BS 5973)[16]
Note: The Construction (Health, Safety and Welfare) Regulations 1996 require the provision of an intermediate guardrail at working platforms on scaffolding. BS 5973 is currently under review to incorporate this requirement.

To facilitate compliance with these requirements, checklists can be used for the different items of lifting gear and tackle and the following are typical lists.

4.6.5.1 Checklists

4.6.5.1.1 Mobile cranes

Prior to work commencing ensure a competent 'lifting co-ordinator' has been appointed:

When was the crane selected, and what information was available/used at the time?
Has the selected crane been supplied?
Check that the ground is capable of taking the loads (outriggers/crane/load/wind). If in doubt get ADVICE from specialist departments/firms.
Ensure that the approach and working area are as level as possible.
Ensure that the area is kept free of obstructions – minimum 600 mm (2 ft) clearance.
Ensure that the weights of the loads are known, and that the correct lifting gear is ordered/available.
Ensure that there is a competent, trained banksman available.
Check that there are no restrictions on access, i.e. check size(s) of vehicles etc.
Ensure that the work areas are adequately lighted.
Check that the Plant Department/Hirer has provided information re the cranes etc.

Whilst work is in progress:

Check that there is an up-to-date thorough examination certificate.
Check that the inspections are being carried out and a written record maintained.
Ensure that the crane is operating from planned/approved positions only.
Ensure that the banksman is working in the correct manner.
Ensure that the correct lifting gear is being used.
Ensure that outriggers are being used, and are adequately supported.
Check that the safe load/radius indicator is in working order.
Check that the tyres/tracks are at the correct pressure and in good, clean condition.
Check that the crane is kept at a safe, predetermined distance from open excavations etc.
Check that, when travelling, the load is carried as near to the ground as possible and that hand lines are being used.
Check that when travelling on sloping ground the driver changes the radius to accommodate the moving of the load.
Check that loads are not being slewed over persons and that persons are not standing or walking under the load.

4.6.5.1.2 Automatic safe load indicator

Automatic safe load indicators must be fitted to all cranes and it is the responsibility of the operator to:

(a) determine the type of indicator fitted;
(b) determine how the adjustments are made;
(c) ensure that it is correctly adjusted for the various lifting duties;
(d) ensure that the electrical circuit is tested for serviceability;
(e) take immediate action when an overload is indicated.

The signals given by the indicator take the form of coloured lights, a dial indicator or both and a bell.

Green/white – Indicator adjusted for 'free' duties
Blue – Indicator adjusted for 'blocked' duties
Amber – Maximum safe load being approached
Red – Overload condition reached.
The red light will be supported by a bell to give an audible warning of overload.

4.6.5.1.3 Goods hoists, static and mobile – safe working checklist

(a) Erect the hoist in a suitable position.
(b) Make the hoistway as compact as possible.
(c) Hoistway to be efficiently protected by a substantial enclosure at least 2 m (6 ft 6 in) high.
(d) Hoist gates – guards to be at least 2 m (6 ft 6 in) high.
(e) Engine or motor must also be enclosed to a height of 2 m (6 ft 6 in) where practicable.
(f) Make sure that no one can come into contact with any moving part of the hoist.
(g) Enclosures at the top may be less than 2 m (6 ft 6 in) but in no case less than 0.9 m (3 ft) providing that no one can fall down the hoistway and that there is no possibility of anyone coming into contact with any moving part.
(h) All intermediate gates will be 2 m (6 ft 6 in) unless this is impractical, i.e. confined space etc.
(i) The construction of the hoist shall be that it can only be operated from one position at any one time.
(j) It shall not be operated from inside the cage (unless designed for the purpose).
(k) The person operating the hoist must have a clear and unrestricted view of the platform throughout.
(l) The safe working load shall be plainly marked on every hoist platform and this load must not be exceeded.
(m) No person shall ride on the hoist (unless so designed), and a notice to this effect must be exhibited on the hoist so that it can be seen at all levels.

(n) Every hoist must be fitted with an efficient automatic device to ensure that the platform does not overrun the highest point for which it is intended to travel.
(o) Every hoist must be fitted with an efficient device which will support the platform and load in the event of the failure of the ropes or lifting gear.
(p) All movable equipment or plant must be scotched, to prevent its displacement while in motion.
(q) All materials will be so placed as to prevent displacement.
(r) Gates should be kept closed on all landing stages. Every person using hoists must close landing place gates immediately after use. (This is a statutory duty imposed upon the person actually using the hoist but the employer also has the duty of seeing that the regulation is obeyed, and the employer's representative on site is the General Foreman or Site Agent.)
(s) Landing stages should be kept free from materials and plant.
(t) No person under the age of 18 must be allowed to operate or give signals to operator.
(u) Only a competent person should operate the hoist.
(v) Signals to be of distinct character; easily seen or heard by person to whom they are given.
(w) Every hoist must be inspected and a written record should be made.
(x) Every hoist should be thoroughly examined every six months by a competent person and certificated.

4.6.5.1.4 Chains, rope slings and lifting gear – safe working checklist

Prior to commencing work:

(a) Examine the slings provided and check that the 'thorough examination' has been carried out and recorded.
(b) Determine and clearly mark the Safe Working Loads for all slings.
(c) Ensure that the correct and up-to-date copies of the Sling Chart and Safe Working Load Tables are available, when using multi-leg slings.
(d) Ensure that a copy of the correct crane signals is available.
(e) Ensure that a suitable rack is available for storing slings, etc. not in use. N.B. Wire ropes should be stored in a dry atmosphere.
(f) Ensure that the weights of loads to be lifted are known in advance, and that load weights are clearly marked.
(g) Find out the type of eye bolt fitted to the load, in advance, to ensure that the correct equipment, shackles/hooks/lifting beams, is available on site.

Whilst work is in progress:

(h) Ensure that the 'right' techniques are being used.
(i) Ensure that the copies of the Sling Chart and the Safe Working Loads Tables are being used, where necessary.

(j) Ensure that the correct crane signals are being used, and that signals are given only by 'approved' banksmen.
(k) Ensure that regular inspections of the equipment are being carried out.
(l) Ensure that unfit slings are destroyed, or at least removed from site.
(m) Stop persons 'hooking back' onto the legs of slings.
(n) Ensure that slingers understand that 'doubling up' the sling does NOT 'double up' the Safe Working Load: avoid this practice if possible.
(o) Limit the use of endless wire rope slings.
(p) Ensure that wire rope slings are protected from sharp corners of the loads, by suitable packings.
(q) Prevent/stop slings/ropes from being dragged along the ground.
(r) Ensure that the hooks used for lifting are NOT also carrying unused slings.
(s) Ensure that the crane hook is positioned above the load's centre of gravity.
(t) Ensure that the load is free before lifting and that all legs have a direct load.
(u) Ensure that the load is landed onto battens to prevent damage to slings.
(v) Ensure that a sling is NOT passed through more than one eye bolt.
(w) Ensure that 'snatch' loading does NOT take place.
(x) Ensure that NO ONE rides on a load that is being slung.

Additional guidance on the standards to be achieved is given in British Standards[20–22] and in HSE Guidance Notes[23].

4.6.6 Welfare facilities

Under the Construction (Health, Safety and Welfare) Regulations 1996 every contractor or employer has a duty to provide, or ensure there is provided, certain health and welfare facilities for his own employees who must have proper access to them. Because a number of contractors may be working on the site some of the facilities may be shared or alternatively arrangements may be made by the contractor to use the facilities offered by adjacent premises. Such arrangements should be agreed in writing. When such agreements are terminated, to prevent confusion it is advisable to give notice in writing.

If facilities are shared with another employer or contractor on the site, then the one who provides the facilities or equipment must:

1 in deciding what facilities to provide, assume that he employs the total number of men who are to use the facilities: e.g. say own employees = 50, other employees = 40 – therefore, for the purposes of providing facilities, assume that he employs 90;
2 Keep a record showing the facilities to be shared and the names of the firms sharing them.

Contractors or employers on the same site can jointly appoint the same man to take charge of first aid and ambulance arrangements.

4.6.6.1 Facilities to be provided on site

Clearly marked 'FIRST AID' boxes must be provided and put in the charge of a responsible person whose name must be displayed near the box. After assessing the level of risk, the availability of emergency services and other matters detailed in the Regulations[24] and the Approved Code of Practice[25], it may be necessary for the responsible person to be a trained and certificated first aider.

Where there is a large workforce on a site a suitably staffed and equipped first aid room should be provided. However, where a large workforce is divided into several dispersed working groups or the location of the site makes access to places of treatment outside it difficult, the needs of such a site may be better met by the provision of first aid equipment and trained first aiders at different parts of the site.

Regardless of the number of employees there must be at least one first aid box on site, and provision should be made for every employee to have reasonably rapid access to first aid.

Construction workers are frequently exposed to the weather and facilities must be provided to store the clothes they do not wear while working, to warm themselves, and to dry their clothing when not in use. In addition, a supply of drinking water must be available and suitable arrangements for warming and eating food.

Suitable washing facilities must be provided and toilets, accessible from all workplaces on the site, must be under cover, partitioned from each other, have a door with fastening, be ventilated and provided with lighting. They must not open directly into workrooms or messrooms and must be kept clean. Separate conveniences must be provided for men and women unless each convenience is in a separate room, the door of which can be secured from the inside.

4.6.7 Other relevant legislation

4.6.7.1 Personal protective equipment

The requirements to be met in the application and use of this equipment are laid down in the Personal Protective Equipment at Work Regulations 1992[1] (PPER).

The general assessment required by MHSWR should identify the hazards, the extent of the risks faced and enable the necessary preventive and precautionary measures to be decided. If personal protective equipment is considered appropriate, PPER sets out the steps to be taken in the process of selecting suitable and effective equipment which the employer has to provide to his employees.

These Regulations revoke early regulations made under FA 1961, such as the Protection of Eyes Regulations 1974, and have marginally modified subsequent regulations that include provisions for personal protective equipment, such as the Noise at Work Regulations 1989, the Construction (Head Protection) Regulations 1989 etc., which continue to apply.

4.6.7.2 The Construction (Head Protection) Regulations 1989[1]

Under these Regulations employers are required to provide, maintain and replace, as necessary, suitable head protection for their employees and others working in the areas over which they have control. They must ensure that the head protection is worn unless there is no risk of a head injury occurring other than by falling over. To be 'suitable' head protection must be:

- designed, so far as is reasonably practicable, to provide protection against foreseeable risks of injury to the head to which the wearer may be exposed,
- adjustable so that it can be made to fit the wearer, and
- suitable for the circumstances in which it is to be used.

Persons who have control over a site, such as management contractors, may make written rules concerning the wearing of head protection by anyone, employees and others (with the exception of Sikhs wearing turbans), working on that site, and should make arrangements for enforcing those rules. Employees are required to take care of equipment and to report cases of damage, defect or loss.

4.6.7.3 Fire Certificates (Special Premises) Regulations 1976

(a) These Regulations were introduced as a result of the operation of the Health and Safety at Work Act with effect from 1 January 1977. Most temporary site buildings used as offices or workshops at building operations and works of engineering construction were previously subject to certain fire precautions provisions, under the Offices, Shops and Railway Premises Act, or occasionally under the Factories Act. Health and Safety Executive Inspectors already inspect building sites for general inspection purposes, and by including these premises into the new Regulations the Health and Safety Executive now becomes responsible for fire precautions at them. It is not intended to issue a fire certificate for every temporary site building, however, and the Regulations are designed to exclude small site buildings conditionally from the certification procedure.

Where a site building, used as an office or workshop, does not require a fire certificate, the Construction (Health, Safety and Welfare) Regulations require certain fire detection and firefighting measures to be implemented similar to those in buildings where construction work is carried out. Steps must be taken to:

1. provide suitable and sufficient firefighting equipment
2. provide suitable and sufficient fire detectors and alarms

3 maintain, inspect, examine and test any equipment provided
4 ensure non-automatic equipment is readily accessible
5 provide training for the workforce.

Some large site buildings, or those with special risks, will require a certificate.

(b) A fire certificate will be required for any building or part of a building which is 'constructed for temporary occupation for the purposes of building operations or works of engineering construction, or, which is in existence at the commencement there of any further such operations'. But a fire certificate will not be required for these buildings if all of the following conditions are complied with:

(i) Not more than 20 persons are employed at any one time in the building or part of the building.
(ii) Not more than 10 persons are employed at any one time elsewhere than on the ground floor.
(iii) No explosive or highly flammable material is stored or used in or under the building.
(iv) The building is provided with reasonable means of escape in case of fire for the persons employed there.
(v) Appropriate means of fighting fire are provided and maintained and so placed as to be readily available for use in the building.
(vi) While anyone is inside, no exit doors may be locked or fastened so that they cannot be easily opened from inside.
(vii) If more than 10 people are employed in the building, any doors opening on to any staircase or corridor from any room in the building must be constructed to open outwards unless they are sliding doors.
(viii) Every exit opening must be marked by a suitable Notice.
(ix) The contents of every occupied room in the building must be arranged so that there is a free passageway for everyone employed in the room to a means of escape in case of fire.

(c) To obtain a fire certificate application must be made to the HSE (Construction) office for the area in which the site is located, using form F2003 to give the required particulars, which are listed below. The HSE will then carry out an inspection of the site buildings. In practice, all final exit doors and any door opening on to a staircase or corridor from any room in the building should open outwards. Fire extinguishers to be hung on wall brackets adjacent to the final exit, with the top of the extinguisher 1.07 m (3 ft 6 in) from the floor. A fire blanket should be hung on wall bracket adjacent to the cooker in the canteen with the top of the blanket 1.52 m (5 ft 0 in) from the floor. LPG cylinders should be sited outside huts and have an isolating valve at the cylinder and an ON/OFF control valve as near as practicable to the heater.

Particulars required on form F2003 when applying for a fire certificate include:

(a) Address of premises.
(b) Description of premises selected from those listed in Schedule 1 to the Regulations.
(c) Nature of the processes carried on, or to be carried on, on the premises.
(d) Nature and approximate quantities of any explosive or highly flammable substance kept, or to be kept, on the premises.
(e) Maximum number of persons likely to be on the premises at any one time.
(f) Maximum number of persons likely to be in any building of which the premises form part at any one time.
(g) Name and address of any other person who has control of the premises.
(h) Name and address of the occupier of the premises.
(j) If the premises consist of part of the building, the name and postal address of the person or persons having control of the building or other part of it.

4.6.7.4 Food Safety (General Food Hygiene) Regulations 1995

The Food Safety Act 1990 is now one of the main pieces of legislation concerning food hygiene. It is an enabling Act allowing specific subordinate legislation (Regulations) to be made as necessary. An example is the Food Premises (Registration) (Amendment) Regulations 1993 which require all places where food is served, including site canteens, to be registered with the local authority. The 1995 Regulations establish a set of rules within which businesses handling and preparing food must operate. All food premises, particularly food rooms, must be arranged so that surfaces can be cleaned properly, waste disposed of hygienically and a high standard of personal hygiene maintained by the food handlers. The proprietor must have a system which identifies the points in the food handling process where contamination could occur and have in place control measures to avoid the risk. He can achieve this by following the principles of the Hazard Analysis and Critical Control Points (HACOP) outlined in Schedule 2 to the Regulations.

If a person handling food suffers a specified infection of the digestive system, the Medical Officer of Health must be notified immediately. An adequate supply of hot and cold water and hand wash basins must be supplied for the use of food handlers, and separate sinks with hot and cold water for preparing vegetables and for washing equipment. No toilet may connect direct with a room used for preparing or eating food. Food rooms must be adequately lit and ventilated and walls, floors, doors, windows, ceilings, woodwork etc., must be kept clean and in good repair. Waste must not be allowed to accumulate.

4.6.7.5 Petroleum Consolidation Act 1928
Petroleum Spirit (Motor Vehicles etc.) Regulations 1929
Petroleum Mixtures Order 1929

Broadly, the Act and Regulations deal with petroleum spirit and mixtures whether liquid, viscous or solid. Petroleum means substances giving off a flammable vapour at a temperature of less than 23°C (73°F). It will be appreciated that this is less than body heat so there is a need for strict control on their uses. By prior arrangement with the local authority, up to 60 gallons may be kept on site without licence but this is subject to a review annually. Where quantities in excess of 60 gallons are to be kept, a petroleum licence must be obtained from the Petroleum Officer. These requirements are subject to discussion on their repeal.

4.6.7.6 Quarries Regulations 1999

The requirements of these Regulations apply to sites, or parts of sites, that fall within the description of a quarry. The most common example being a borrow pit used in motorway construction, although it can also apply where old quarries are being in-filled with spoil. Enforcement of these Regulations is by the HSE who cover building on the surface as well as work within the quarry itself.

Notification of the commencement and termination of work in a quarry must be made to the HSE. The owner of the quarry is required to appoint a manager and a deputy manager.

4.6.7.7 Asbestos

Asbestos is a strong, durable, non-combustible fibre, and these physical properties make it ideal as a reinforcing agent in cement, vinyl and other building materials, e.g. vinyl floor tiles, bath panels, cold water cisterns, roof felts, corrugated or flat roof sheets, cladding sheets, soffit strips, gutters and distribution pipes. It is also a good insulant and has been used for protecting structures from the effects of fire.

Asbestos has been used extensively in the past and may be found in many forms in existing buildings including: industrial wall and roof linings; internal partitions; duct and pipe covers; suspended ceilings; fire doors and soffits to porch and canopy linings. It also occurs in plant rooms and boiler houses and as asbestos coatings and insulating lagging on structures and pipework.

Work with asbestos products is governed in the UK by specific legislation:

The Asbestos (Licensing) Regulations 1983 and 1998
The Asbestos (Prohibitions) Regulations 1992 and 1999

The Control of Asbestos at Work (Amendment) Regulations 1992 and 1998
The Control of Asbestos in the Air Regulations 1990

The main purpose of the 1983 Regulations is to control the manner in which companies and the self-employed carry out work which involves the disturbance of asbestos insulations or coatings. All persons carrying out such work must be in possession of an HSE licence and receive regular medical examinations. Wherever asbestos products are encountered the Regulations should be complied with, using advice given in HSE guidance notes[26]. Useful information is also available from the Asbestos Information Centre Ltd.

Control limits for asbestos dust in the working environment are set out in an Approved Code of Practice[27]. These control limits are not intended to represent safe levels of airborne dust but the upper limits of permitted exposure. There is still a statutory duty to reduce exposure to the lowest level that is reasonably practicable.

Where asbestos containing materials have to be stripped, its disposal is governed by special regulations[28].

References

1. Statutory Instruments relevant to construction work:
   The Lifting Operations and Lifting Equipment Regulations 1998
   The Electricity at Work Regulations 1989
   The Reporting of Injuries, Diseases and Dangerous Occurrences Regulations 1995
   The Highly Flammable Liquids and Liquefied Petroleum Gases Regulations 1972 (SI 1972 No. 917)
   The Diving at Work Regulations 1997
   Work in Compressed Air Regulations 1996
   Noise at Work Regulations 1989 (SI 1989 No. 1790)
   Construction (Head Protection) Regulations 1989 (SI 1989 No. 2209)
   The Management of Health and Safety at Work Regulations 1999 (SI 1999 No. 3242)
   The Provision and Use of Work Equipment Regulations 1998 (SI 1998 No. 2306)
   The Manual Handling Operations Regulations 1992 (SI 1992 No. 2793)
   The Personal Protective Equipment at Work Regulations 1992 (SI 1992 No. 2966)
   The Workplace (Health, Safety and Welfare) Regulations 1992
   Confined Spaces Regulations 1997 (SI 1997 No. 1713)
   All published by The Stationery Office, London.
2. Health and Safety Executive, Manufacturing Services and Industries – Annual Report (published each year), HSE Books, Sudbury
3. Health and Safety Executive, *Blackspot Construction, A study of five years fatal accidents in the building and civil engineering industries*, HSE Books, Sudbury (1988)
4. *The Construction (Design and Management) Regulations 1994*, The Stationery Office, London (1994)
5. Regina *v.* Swan Hunter Shipbuilders Ltd and Telemeter Installations Ltd [1981] IRLR 403
6. British Standards Institution, BS 6031, *Code of Practice for earthworks*, BSI, London (1981)
7. Health and Safety Executive:
   Health and Safety Guidance Booklet No. HSG47, *Avoiding danger from underground services*

Health and Safety Guidance Note No. GS6, *Avoiding of danger from overhead electrical lines*
HSE Books, Sudbury
8. Health and Safety Executive, *Health and Safety Guidance Note No. GS7, Accidents to children on construction sites*, HSE Books, Sudbury
9. British Standards Institution, BS EN 953, *Safety of Machinery – Guards – General requirements for the design and construction of fixed and movable guards*, BSI, London
10. Health and Safety Executive, Guidance Notes:
No. MS 15 Welding
No. PM 17 Pneumatic Nailing and Stapling Tools
No. PM 14 Safety in the Use of Cartridge Operated Tools
No. PM 5 Automatically Controlled Steam and Hot Water Boilers
No. PM 1 Guarding of Portable Pipe Threading Machines
11. *The Electricity at Work Regulations 1989*, The Stationery Office, London
12. *The Control of Substances Hazardous to Health Regulations 2002*, The Stationery Office, London (1994)
13. *The Ionising Radiations Regulations 1999*, The Stationery Office, London (1999)
14. British Standards Institution, BS EN 60825, *Radiation, Safety of laser products*, BSI, London
15. *The Work in Compressed Air Regulations 1996*, The Stationery Office, London (1996)
16. BS 5973, *Code of Practice for access and working scaffolds and special scaffold structures in steel*, British Standards Institution, London
17. BS 2482, *Specification for timber scaffold boards*, British Standards Institution, London
18. BS 1139 – Metal scaffolding
Part 1: *Specification for tubes for use in scaffolding*
Part 2: *Specification for couplers and fittings for use in tubular scaffolding*
Part 4: *Specification for prefabricated steel splitheads and trestles* British Standards Institution, London
19. Health and Safety Executive, Health and Safety Guidance Notes No. GS 42, Tower Scaffolds, HSE Books, Sudbury
20. BS 5744, *Code of Practice for safe use of cranes (overhead/underhung travelling and goliath cranes, high pedestal and portal jib dockside cranes, manually operated and light cranes, container handling cranes and rail mounted low carriage cranes)*, British Standards Institution, London
21. CP 3010, *Safe uses of cranes (mobile cranes, tower cranes and derrick cranes)*, British Standards Institution, London (1972) (see also ref. 22)
22. British Standards Institution, BS 7121, *Code of Practice for the safe use of cranes*, and BS EN 12077, *Cranes – Safety – Requirements for health and safety*, BSI, London
23. Health and Safety Executive, Guidance Notes:
PM 3 Erection and Dismantling of Tower Cranes
PM 9 Access to Tower Cranes
PM 8 Passenger Carrying Paternosters
PM 16 Eyebolts
PM 27 Construction Hoists
GS39 Training of Cranes Drivers and Slingers,
HSE Books, Sudbury
24. *The Health and Safety (First Aid) Regulations 1981*, The Stationery Office, London (1981)
25. Health and Safety Executive, Legal series booklet No. L74, First aid at work.
Health and Safety (First Aid) Regulations 1981. Approved Code of Practice and Guidance, HSE Books, Sudbury (1997)
26. Health and Safety Executive, Guidance Notes:
EH35 Probable asbestos dust concentrations at construction processes
EH36 Work with asbestos cement
EH37 Work with asbestos insulating board
EH40 Occupational exposure limits,
HSE Books, Sudbury
27. Health and Safety Executive, Legislation booklet L27, The control of asbestos at work. The Control of Asbestos at Work Regulations 2002, Approved Code of Practice, HSE Books, Sudbury (2002)
28. *The Environmental Protection Act 1990*,
The Special Waste Regulations 1996,
The Stationery Office, London

Further reading

Dickie, D.E., *Crane Handbook,* (Ed.: Douglas Short), Butterworth, London (1981)
Dickie, D.E., *Lifting Tackle Manual,* (Ed.: Douglas Short), Butterworth, London (1981) *Construction Safety Manual,* Construction Safety, Crawley, Sussex
Dickie, D.E., Ed. Hudson, R., *Mobile Crane Manual,* Butterworth, London (1985)
King, R. and Hudson, R., *Construction Hazard and Safety Handbook,* Butterworth, London (1985)

Chapter 4.7

Managing chemicals safely

John Adamson

4.7.1 Introduction

In 1974 a flammable vapour cloud destroyed the Nypro cyclohexane oxidisation plant at Flixborough[15] in North Lincolnshire. Twenty-eight people were killed. All chemicals can be handled safely, but get it wrong and the situation can result in a disaster.

A chemical incident has the potential to affect many people, the operators, other employees on site, members of the public and, in the wider sense, it can cause widespread damage to the ecology. The results of some incidents such as at Seveso and Bhopal have been of epic proportions and were truly catastrophies. Such events carry an massive price tag in terms of human suffering, loss of life, business interruption and damage to the environment. All incidents, regardless of their scale, damage the reputation of the organisation and of the industry as a whole.

Management needs to provide resources to furnish a safe working environment, to allow systems of control to be identified and reviewed, and equipment installed and maintained. Prevention of incidents requires company employees to know the nature and hazardous properties of the materials so that suitable precautionary arrangements can be made for their safe storage, handling and use. Employees need to comply with company policy, procedures and help identify areas of concern and unsafe practices.

4.7.2 Chemical data

Two keys to the safe use and handling of hazardous materials are, first, to know what substances are and, second, to know the characteristics of those substances. The Control of Substances Hazardous to Health Regulations 2002[1] (COSHH) require that data on each of the hazardous substances on the site should be kept and be available to operators at all times. A supporting Approved Code of Practice[2] recommends that a list of all hazardous materials is also kept and made available. Hazardous substances are defined in COSHH as:

'Those substances specified as very toxic, toxic, corrosive or irritant within the meaning of the Chemicals (Hazard Information and Packing for Supply) Regulations 2002[3] (CHIP 3).'

The details are given in an associated Approved Supply List[4]. In essence, the substances with any of the following properties are hazardous to some degree:

(a) explosive
(b) oxidising
(c) flammable
(d) highly flammable
(e) extremely flammable
(f) toxic
(g) very toxic
(h) harmful
(i) corrosive
(j) irritant
(k) carcinogenic (causes cancer)
(l) mutagenic (causes inherited changes)
(m) teratogenic (causes harm to the unborn)
(n) micro-organisms that create a hazard to health
(o) substantial concentrations of dust
(p) radioactive materials
(q) any substance not mentioned above which creates a comparable hazard.

4.7.3 Source of information

Under s. 6 of HSW, manufacturers, importers, designers and suppliers must ensure that articles and substances supplied for use at work are safe and without risk to health. They also have a duty to provide adequate safety information about the substances they produce, which they usually do in the form of chemical safety data sheets. Some employers use these data sheets as they stand, others prefer to generate their own in a format that matches their in-company documentation. Guidance on the information a chemical safety data sheet should contain is given in an HSE publication no. L62[5]. For many of the commonly used substances comprehensive information can be obtained from reference books[6–8]. Where chemicals used in a process are modified during the process, information should also be available concerning the intermediates produced.

Where no hazard data are available for a new substance and it is not possible to estimate its characteristics from substances of similar molecular structure, that substance must be treated as toxic until proved otherwise. Before new substances are put on the market, they must be notified under the Notification of New Substances Regulations 1993 (NONS) with its supporting guide[9] and some may need to be notified under the Chemical Weapons Act 1996[10]. The purpose of NONS is to ensure that adequate technical information is available on new substances before they are put on the market so that suitable precautionary measures can be developed to protect employees, the public and the environment from possible ill effects. NONS support an EU-wide system of notification of new substances.

A new substance is defined as one which is not listed in the European Inventory of Existing Commercial Chemical Substances (EINECS)[11] and may be a substance in its own right or part of a preparation. After notification, new substances are entered in the European List of Notified Chemical Substances (ELINCS)[12]. Both lists give CAS numbers, IOPAC chemical names and some trade names and references.

The Department of Trade and Industry (DTI) is the lead body in the UK for the Chemical Weapons Convention (CWC), which, while primarily an arms control treaty, has implications for both industry and academia since many of the chemicals concerned are dual use goods, i.e. they have legitimate peaceful uses as well as possible military applications. Companies should declare those chemicals which they either produce or use and which are classified under this Convention. A list of those substances can be obtained from the DTI.

4.7.4 Risk assessments

As part of their efforts to ensure a safe working environment, employers are required by the Management of Health and Safety at Work Regulations 1999 (MHSW)[62] to make suitable and sufficient assessments of the health and safety risks arising from their operations as they may affect their employees and others such as contractors, visitors and members of the general public who may be impacted. The assessment should extend to include such aspects as the way in which the work is organised, safety of the product, environmental effects on the local community and ecology.

The object of a risk assessment is to identify all hazards or potential hazards arising from the work so that precautions can be taken to prevent injury or damage to health of employees or anyone else who might be affected.

4.7.4.1 Definitions

The words 'hazard' and 'risk' arise frequently and are often misunderstood. However, they do have precise meanings and should not be confused:

HAZARD – is the inherent property of a substance to cause harm.
RISK – is a combination of the probability of the hazard causing harm and the severity of the resulting harm or damage.
RISK ASSESSMENT – is a comprehensive quantitative or qualitative evaluation of the probability and degree of possible injury or damage to health from identified hazards with a view to implementing preventive measures.

A risk assessment takes account of all the significant factors that can affect the chance and extent of harm. It should conclude on action needed to manage the risk for the benefit of employees, the company, and others who may be affected.

4.7.4.2 Carrying out a risk assessment

Risk assessments need to be carried out by experienced and competent people who are knowledgeable about the hazards of the substances, the equipment and the operations being reviewed. It is a subjective process which cannot normally be validated mathematically, and is essentially practical. Those involved in the assessment need to think logically and laterally, look beyond the obvious and have good interactive skills to tease out crucial information.

Risk assessments for a small unit or department can be carried out by one person but for larger and more complex areas it may be appropriate to involve a small team of people, e.g. safety adviser, area engineer, production team leader, and an operator. To ensure that all facets that could be affected by the hazards are considered a risk assessment should follow a series of logical steps based on a thought-out strategy. A generic risk assessment process will:

1 Define the task or process to be assessed and identify the boundaries.
2 Identify the hazards and eliminate or reduce them as far as possible.
3 Evaluate the risks from the residual hazards by:
   (a) assessing the extent of the hazard
   (b) estimating the probability of harm occurring
   (c) assessing likely extent of the harm or injury.
4 Decide on precautions or control measures to be taken.
5 Train local operators.
6 Implement precautionary measures.
7 Monitor the effectiveness of the measures and adjust as necessary.

By giving empirical values to items 3(a), (b) and (c) and multiplying them together an indicative risk rating can be obtained which can be used to determine priorities for action.

When carrying out a risk assessment all types of hazard should be considered – injury, fire, explosion, pollution, damage to neighbours, damage to equipment etc. A typical risk assessment form is shown in *Figure 4.7.1(a)*.

4.7.4.3 COSHH assessment

An assessment under COSHH is concerned only with hazardous substances and follows the same procedure outlined above. Its aim is to assess the risks to the health of employees from the various substances used or handled. The assessor should be knowledgeable in

Site:		**Area:**
Operations covered by this Assessment:		
Maximum No. of people exposed: **Frequency & durations of exposure:**		
Hazards		
Actions already taken to reduce the risk		
Assessment of residual risk:		
Further actions required:		
Signed: **Position:**	**Date:**	**Review Date**

Figure 4.7.1(a) Risk assessment form. (Courtesy British Sugar plc)

the process and in the likely health effects of the substances. The assessment should include not only the process operators but those working in the vicinity who could be affected.

The technique requires:

- Knowledge of the hazards associated with the substance in order to determine the potential of the hazard to harm health.
- Assessment of the level of exposure to, or contact with, the hazardous substance, i.e. the RISK.
- An estimate of the frequency and duration of exposure.

BRITISH SUGAR PLC

COSHH ASSESSMENT AREA: SITE:

SUBSTANCES/MAIN COMPONENTS		STOCK CODE:	
TASK/EXPOSURE POINTS	HAZARD:	EH 40 EXP LIMIT:	
FREQUENCY/DURATION OF EXPOSURE		NO. OF PEOPLE EXPOSED:	
CONTROL MEASURES	W. I. REFERENCES:		
MONITORING REQUIRED: YES/NO	RISK: LOW MEDIUM HIGH		
ACTION REQUIRED		RESPONSIBILITY	BY

SIGNATURE: DATE FOR NEXT ASSESSMENT:

Figure 4.7.1(b) COSHH assessment form. (Courtesy British Sugar plc)

- A comparison of the level of exposure against the current occupational exposure standard (OES) or maximum exposure limit (MEL) for inhalation and absorption risks.
- Assessment of the numbers of workpeople who may be exposed.
- Agreement for the control measures to be adopted.

By using a numerical rating a measure of the priority ranking for remedial action can be made. In cases where the level of exposure to a material is not clear cut, an occupational hygiene study should be carried out. *Figure 4.7.1(b)* is an example of the documentation used to record the findings of a COSHH assessment.

4.7.4.4 Manual handling risk assessment

In the use of chemicals it may be necessary to handle them manually, often in sacks weighing 25 kg or more. Where this occurs, the risks from the manual handling should be included in the risk assessment. This is also a requirement of the Manual Handling Operations Regulations 1992. This assessment should be carried out by a person trained in handling techniques who should consider:

- whether the task can be done mechanically
- the load – its weight, shape, condition, centre of gravity
- the working environment – condition of the floor, lighting, temperature etc.
- the individual's physique and capacity to carry out the task
- the repetitive nature of the task
- other factors such as clothing and personal protective equipment.

Where the task cannot be done mechanically, the risks identified should be reduced as far as possible, such as by dividing the load or getting assistance, to prevent injuries from lifting, repetitive strains and poor posture.

4.7.5 Minimising the risk

Having identified the hazards and assessed the risks the action needed to prevent injury and harm should be agreed with the manager and operators of the area concerned and implemented. While nothing can be absolutely and unequivocally safe and free from risk, the aim must be to achieve, so far as is reasonably practicable (see *Edwards* v. *National Coal Board*[13]), a standard that reduces risks to a minimum whilst maintaining the viability of the process.

Within the strategy outlined in section 4.7.4.2 above, there are a number of techniques that can be used to ensure the hazards, and consequently the risks, are reduced to a minimum.

4.7.5.1 Prevention of exposure

With hazardous materials, the first and most basic question that must be asked is 'Do we have to use this substance?' Depending on the answer, there are a hierarchy of factors to consider before deciding on the action necessary to ensure safe working:

(a) *Substitution.*
If a substitute can be found it must meet the following criteria:

(i) be safer than the original material
(ii) be suitable for the process to ensure the required quality of finished product – check with chemist and process controller
(iii) be economically viable
(iv) not produce intermediates or daughter products that nullify the benefits of substitution.

(b) *Minimise the quantity*
Ensure that only the minimum quantity of the substance is used in the process. This has the additional benefit of reducing wastage to a minimum. Also the plant should be designed and maintained to minimise the generation of dust, vapours, micro-organisms etc. and, in case of spillage or leakage, be provided with an area for containment.

(c) *Total enclosure*
Total enclosure will ensure that escapes of gases, fumes or liquids are completely contained. Any vents from the enclosure should be through scrubbers or absorbers. The enclosed space may need to be provided with cooling.

(d) *Partial enclosure with local ventilation*
If access is necessary to the plant during parts of the process, local air extraction should be provided and arranged so its exhaust is scrubbed or absorbed then vented to a safe location away from air intake ducts and work areas. For infrequent brief visits, PPE may be an option.

(e) *Local exhaust ventilation* (LEV)
Where regular attention to the process is necessary LEV should be provided. The system must ensure that the hazardous materials do not enter the work area beyond the particular process plant. A laminar flow booth may be a more suitable alternative. Discharge from either system must be through scrubbers or absorbers with final outlet at a safe height and location not to affect neighbours or cause environmental contamination.

(f) *Personal protective equipment* (PPE)
The Personal Protective Equipment at Work Regulations 1992 require employers and the self-employed to provide suitable protective equipment to operators wherever the risks to their health cannot be controlled by other means. PPE should only be used as a last resort and only if engineering control measures are insufficiently effective. PPE should:

- be suitable for the hazardous material involved and the environmental conditions likely to be met
- fit properly, be acceptable to and effective on the wearer allowing for his/her state of health
- not interfere with the work to be carried out by the wearer

- be of an approved type
- not generate further hazards.

A range of PPE, that meets the above criteria, should be tried out by the plant operators before deciding on a particular type. All PPE must be properly looked after and maintained in good working condition. Ideally PPE should be an individual issue, but if this is not possible it must be cleaned between uses and operators instructed accordingly.

Work instructions should detail the type of PPE to be worn when carrying out specific tasks and operators should receive adequate information and training in its use and limitations. With respiratory protective equipment (RPE) each operator should be 'fit tested' during

Figure 4.7.2 Protective clothing to provide protection against contact and dust. (Courtesy Rhone-Poulenc-Rorer)

which they make a number of facial movements in an atmosphere of strong saccharin. If they can taste the saccharin, other more suitable RPE should be tried. Facial hair can also cause failure of the 'fit test'. The testing should be repeated for each type of RPE that has to be worn. *Figure 4.7.2* shows an operator wearing adequate RPE to protect against chemical dust inhalation.

The Personal Protective Equipment at Work Regulations do not apply to work areas covered by the more specific requirements of:

The Control of Lead at Work Regulations 2002[27]
The Ionising Radiations Regulations 1999[28]
The Control of Asbestos at Work Regulations 2002[58]
The Control of Substances Hazardous to Health Regulations 2002[1]
The Noise at Work Regulations 1989[59]
The Construction (Head Protection) Regulations 1989[60]

4.7.5.2 Training and information

An essential feature of any risk management system involving hazardous substances is the training provided to those who handle or use those substances. Such training should include:

(a) Chemical data and information on the substances involved, their hazards and the action to be taken should contamination occur.
(b) The system of work to be followed to ensure that:

- the correct PPE is worn
- the substance is handled properly
- the material is transported safely
- the finished product can be handled safely.

(c) Techniques for dealing with spillages
(d) The operation of special plant or equipment and how to deal with deviations from normal operation or performance.
(e) The purpose and correct use of PPE, its limitation of use, cleaning, care and the reporting of any damage to or faults with any item.
(f) Monitoring techniques and comparison of results with the limits set in EH 40[14].
(g) Checking of the concentrations of substances in finished products and waste.

4.7.5.3 Monitoring

Monitoring of the atmospheric concentrations to which an employee is exposed when undertaking a task should be carried out where:

(a) The COSHH assessment indicates the level of risk warrants it.
(b) The effectiveness of the corrective environmental control measures in the work area need to be evaluated.

(c) There is a need to demonstrate that the MEL has not been exceeded.
(d) Confirmation is needed that exposure levels are below the appropriate OES.
(e) The effectiveness of the LEV needs to be assessed.

Where an occupational hygiene study has been carried out to check on the effectiveness of the measures to prevent operator contact with hazardous substances, the results should be shared with the operators. Where the exposure risk warrants it, COSHH requires that exposed employees be offered medical surveillance. COSHH also requires that LEV systems are regularly monitored for their effectiveness in controlling the emissions into the work area and that the system still operates within the original design parameters.

4.7.5.4 Records

COSHH assessment records and occupational hygiene reports should be kept for at least 40 years. These records should include the name and works number of the employees who have been monitored. Health surveillance records should be kept for 5–40 years, dependent upon the substance involved and the degree of exposure.

4.7.6 Legislative requirements

Following the Flixborough incident[15] which had a major impact on the local community and caused considerable damage to property, the HSE appointed a committee of experts, the Advisory Committee on Major Hazards, to consider the health and safety problems posed by major chemical sites, and to make recommendations. This they did in three reports which identified a need for three basic elements of control:

(i) Identification of the site.
(ii) The location of the site.
(iii) An assessment of the potential hazards on the site.

It had been the intention to implement these three recommendations through a single set of regulations. However, following two major incidents at Seveso and Manfredonia in Italy in 1976 the EU adopted a directive on major hazards. To incorporate this directive into UK law resulted in the preparation of two slightly conflicting sets of regulations, the Notification of Installations Handling Hazardous Substances Regulations 1982 (NIIHHS)[17] and the Control of Industrial Major Accident Hazards Regulations 1984 (CIMAH)[19] with its subsequent amendments. In addition, continuing concern about the effects on workers of exposure to chemicals resulted in, *inter alia*, COSHH and CHIP. The Control of Major Accident Hazards Regulations 1999 (COMAH)[16] incorporate into UK law the contents of an amended EU directive which addresses some current anomalies and expands the type of installation that comes within its scope.

4.7.6.1 The Notification of Installations Handling Hazardous Substances Regulations 1982 (NIIHHS)

The main thrust of these Regulations is to identify those sites which handle or store more than specified quantities of hazardous substances and it has resulted in the compilation of a central register of all major chemical and other potentially dangerous sites. The Regulations require only that the sites be notified and place no further obligations on employers. Guidance on these Regulations is given in an HSE publication[17].

4.7.6.2 The Control of Major Accident Hazards Regulations 1999

The Control of Industrial Major Accident Hazards Regulations 1984 (CIMAH) represented an early attempt to govern major hazard sites in the UK and incorporate requirements contained in EU directive no. 82/501/EEC[18] (often referred to as the Seveso Directive). The Regulations place duties on the owners of hazardous sites to demonstrate safe operation and to notify any major accidents that occur. Where the more dangerous activities, listed in the various schedules, occur the sites have to prepare a 'Safety Case' which describes their activities and their likely impact on the surrounding area. Further, the owners have to prepare an on-site emergency plan and provide information to the local authority who are themselves required to prepare an off-site emergency plan. Finally, information has to be given to the local population who may be affected by the site operations. Guidance on the application of these Regulations is given in two HSE publications[19,20].

These Regulations had a significant impact on the chemical industry in general causing it to be much more overt in its operations. An amendment in 1990 widened the application of the Regulations to include aggregated amounts of substances rather than relating to single substances only. Thus, if a site contains a total of more than 200 tonnes of various substances, which in themselves are not listed but are classified as toxic, the site becomes subject to CIMAH.

The Control of Major Accident Hazards Regulations 1999[16] (COMAH) incorporate the requirements of an amended directive (Seveso II) which, whilst similar to Seveso I and following the same two-tiered format for duties, differs in a number of important ways in that it:

(a) Emphasises the importance of safety management systems.
(b) Allows the directive to keep up to date with technical progress.
(c) Provides more details aimed at ensuring a more uniform implementation by Member States.
(d) Covers a wider range of industries.

The main criteria for determining if the COMAH Regulations apply to a site depends if any substance stored or used is specified in Schedule 1 and is present above a qualifying quantity. Where there are a number of Schedule 1 chemicals present, each one should be calculated as a fraction of the qualifying amount. All the fractions should then be summed and if the result is greater than 1 for the Lower Tier quantities then the site is subject to the requirements for Lower Tier sites. Higher Tier sites are checked in a similar manner. In part 2 of Schedule 1 of the Regulations, 39 substances are listed by name and in part 3 ten generic categories are listed. The categories are based on the Chemical (Hazard Information and Packaging for Supply) Regulations 2002 (CHIP 3).

The main features of the COMAH Regulations include:

(i) Application to the establishment rather than the individual activities as under CIMAH.
(ii) Dependence on the presence on site of threshold quantities of dangerous substances.
(iii) Widened scope to include explosives and chemical hazards at nuclear installations.
(iv) Greater use of generic categories of substances, i.e. toxic.
(v) The requirement for Lower Tier sites to produce a Major Accident Prevention Policy.
(vi) Higher Tier sites to produce a Safety Report which is to be set out more precisely and be made available to the public.
(vii) Land use planning requirements are to be introduced to ensure environmental risks are assessed.
(viii) Carrying out risk assessments on the storage and use of dangerous substances held on the site. Such assessments should also consider the 'domino' effect of incidents involving these dangerous substances and those held by neighbouring companies. It should be noted that such sites need not necessarily be COMAH registered sites to cause a domino effect on a COMAH site.
(ix) Consulting employees and/or safety representatives.
(x) Emergency planning arrangements for on- and off-site incidents.
(xi) Introduction of a new ecotoxic category for materials deemed 'dangerous for the environment'.
(xii) Demolition and clean-up plans.
(xiii) Overlap with Integrated Pollution Control regulations.

4.7.6.3 The Chemicals (Hazard Information and Packaging for Supply) Regulations 2002 (CHIP 3)[3] and the Carriage of Dangerous Goods by Road Regulations 1996 (CDG) and associated regulations

These major pieces of legislation replace the Classification, Packaging and Labelling of Dangerous Substances Regulations 1984 with the object of increasing the protection of people and the environment from the ill effects of chemicals. The Regulations do this by requiring the suppliers to:

(a) Identify the hazards of the chemicals they supply.
(b) Give information about the hazards to the people who are supplied.
(c) Package the chemicals safely.
(d) Label the packaged chemicals to identify the contents.

These are known as the *supply requirements*. A supplier is someone who supplies chemicals as part of a transaction and includes manufacturers, importers and distributors. Similar duties, known as the *carriage requirements*, are imposed on those who consign chemicals for transport by road. The transfer of chemicals between sites, even if under the same ownership, come within the scope of the Regulations which define the term 'chemical' to include pure chemicals, such as ethanol, as well as preparations and mixtures of chemicals, such as paints and pharmaceutical compounds.

The full package of legislation and supporting guidance for CHIP consist of:

(i) the Regulations[3]
(ii) the Approved Supply List[4]
(iii) the Approved Carriage List[21] (Note: The Approved Carriage List is not part of CHIP but of the Carriage of Dangerous Goods (Classification, Packaging and Labelling) and Use of Transportable Pressure Receptacles Regulations 1996 (CDG-CPL) but is included here for completeness.
(iv) approved guide to the classification and labelling of substances dangerous for supply[22] (CHIP 3)
(v) approved code of practice on safety data sheets[5].

Wherever dangerous chemicals are supplied for use at work, CHIP 3 requires that safety data sheets must be provided when the chemical is first ordered and that they must contain sufficient information to enable the recipient to take the precautions necessary to protect his employees' health. This information should also include environmental data. The approved code of practice gives detailed advice on the information to be made available to employers. This could be useful when carrying out a COSHH assessment, but the data sheet should not be considered as a substitute for an assessment.

Dangerous chemicals supplied in a package must carry a label giving information about its hazards and the precautions to be taken in its use or in the event of a spillage. This label information should supplement the information given to employees in their training. For others, such as the emergency services, it allows them to take the correct precautionary action in the event of an incident.

The label on the package must contain details of:

(a) the supplier
(b) the chemical name
(c) the category of danger including the environmental hazards
(d) the appropriate risk and safety phrases.

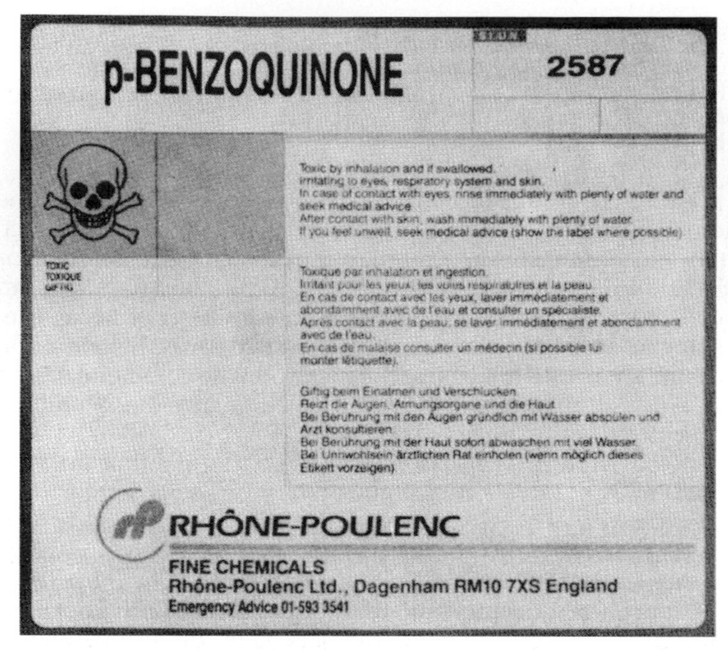

Figure 4.7.3 An example of a label for hazardous substance package showing warning sign. (Courtesy Ciba-Geigy Plastics)

Risk phrases summarise the main danger of the chemical, e.g. 'R45 – may cause cancer' or 'R23 – toxic by inhalation' or 'R50/53 – very toxic to aquatic organisms, may cause long term adverse effects in the aquatic environment. Safety phrases tell the user what to do, or what not to do, e.g. 'S2 – keep out of reach of children' or 'S 29 – do not empty into drains'. A warning symbol is also required on most labels, e.g. a skull and cross bones or a picture of an explosion. Suppliers are responsible for ensuring that the label carries the correct information. These Regulations also require labels to be provided that warn those who handle packages of chemicals during transit on a public road. These transit labels are similar to, but not as comprehensive as, the packaging labels since they concentrate on the immediate information needed should the vehicle carrying them be involved in an accident. Where a package contains only a single substance, the requirements of both the supply and the carriage regulations may be combined into a composite label. An example of a label is shown in *Figure 4.7.3*.

When one or more receptacles are contained in a common outer packaging, the labelling may be in accordance with either of the two regulations (see section 4.7.8). Finally CHIP Regulations require chemicals to be packaged safely and in a manner to withstand the foreseeable conditions of supply and carriage.

In addition, NONS require manufacturers and importers to carry out a certain amount of testing before a quantity of new substance is placed on the market:

- 10 kg to 100 kg per year – very limited testing required
- 100 kg to 1 tonne per year – limited testing
- > 1 tonne per year – full testing required and a technical dossier to be produced.

When > 1 tonne of product is to be put on the market, the appropriate authority (Department of the Environment, Transport and Regions (DETR) and HSE – and in Northern Ireland, the appointed authority) must be notified. The information to be included in the technical dossier is listed in schedule 2 to the Regulations and includes a hazard classification.

4.7.6.4 The Control of Substances Hazardous to Health Regulations 2002 (COSHH)[1]

These Regulations provide for a complete package of duties on health issues relating to the use, handling and storage of hazardous substances at work. In this, they place onerous duties on the employer as well as obligations on the employee.

Substances hazardous to health are defined in reg. 2 as any substance which:

- is listed in the approved supply list[4] as being very toxic, toxic, harmful, corrosive or irritant
- has been assigned an MEL or OES
- is a biological agent
- is dust of any kind in substantial concentrations in air
- any other substance creating like hazards.

Under these Regulations employers are required to:

1 Establish which of the substances they use are hazardous; what those hazards are and, if a new substance is proposed, what hazards it may present.
2 Carry out an assessment of the risks to their employees who may be exposed to hazardous substances.
3 Before a substance is used, assess the potential risk to their employees.
4 Design and install control measures to prevent or control, as far as is reasonably practicable, the exposure of employees to hazardous substances, except for carcinogens and biological agents where it is recognised that some exposure may not be preventable so a strict hierarchy of measures to be taken is laid down.

   Employees must not be exposed to levels above the MEL, but if this occurs the employer must investigate and take corrective action to prevent a repetition.
5 Ensure that the control measurers provided are properly used and looked after. Where these measures include plant or equipment, it must

be properly maintained, regularly examined and tested and records kept of such tests and examinations.

6 Where there is a risk that employees are liable to be exposed to dangerous concentrations of a substance, that exposure must be monitored, recorded and the records kept for between 5 and 40 years, depending on the circumstances of exposure.
7 Where exposure to substances listed in schedule 5 occurs, health surveillance must be provided and medical records kept for at least 40 years.
8 Ensure that employees are:
   (a) Given information about the substances being used;
   (b) Made aware of the hazards those substances present;
   (c) Instructed in the procedure and techniques necessary for the safe use of those substances and any precautions that should be taken;
   (d) Provided with information on the results of monitoring and health surveillance; and
   (e) Given access to the COSHH Risk Assessment and any Occupational Hygiene Reports issued.

A major feature of COSHH is the duty to carry out assessments of the health risks associated with hazardous substances. There is no universally accepted approach to assessing risks and each company should do what is appropriate in the circumstances of their particular operations. An example of an assessment form is shown in *Figure 4.7.1(b)*.

The assessment should be carried out by someone who is familiar with the workplace, the process and the substances being used and who has sufficient knowledge to be able to interpret the implications of what is found. With the more complex processes, it has been found helpful to have a team carrying out these COSHH assessments; the team consisting of safety adviser with representatives from the chemists, production staff, engineers, operators and safety representatives. The assessment should consider the risks from not only those chemicals that are brought in to be stored, used or worked on in the workplace, but also by-products, intermediates and substances that are given off during the process or work activity as well as the finished product, residues, waste, scrap etc.

As part of an assessment, it is necessary to discover where and in what circumstances the substances are stored, used, handled, generated, released, etc., also what people do when handling the substances, as opposed to relying on the assumption that all work is carried out according to instructions. When carrying out a COSHH assessment the whole operation should be reviewed, i.e. from the preparation of the area to start the tasks, through handling the substances, to cleaning the area, equipment or plant. During the assessment, consideration should also be given to non-standard events that are likely to occur. The workforce should be able to provide historical evidence to help with this part of the review. The effectiveness of the measures taken to control and minimise exposures should be monitored since these are part of the overall system of work. The assessors should have a knowledge of the effect that the various substances have on the body and of the routes of entry, such as inhalation, ingestion, skin absorption, etc. Measured exposure levels should be compared with published exposure limits. The number of people exposed and the duration of exposure should also be noted.

The assessment should be recorded to document the conditions found – both good and bad – and the actions that were judged to be necessary to reduce any risk. If conditions change within the plant or process, the COSHH assessment should be reviewed and may need to be repeated. Advice and guidance on carrying out risk assessments are contained in HSE publications[23–25].

In endeavouring to meet the obligations posed by these Regulations, it is helpful to break the work into manageable parts, remembering the less visible people such as cleaners, storekeepers, and maintenance staff who can be exposed to substances often without check or control. A risk assessment should be carried out before any new maintenance activity is undertaken to ensure that correct control measures, training and system of work are agreed and followed.

4.7.7 Storage of substances

The HSE has issued a number of guidance notes on the storage of specific substances such as chlorine[29], LPG[30], highly flammable liquid[31] etc. Further guidance can be obtained from the distributive trade association[32] or from the supplier of the substance who will offer practical advice on standards of storage of his products. Some suppliers provide a 'Duty of Care Service' by visiting sites to assess the bulk storage conditions of the materials they supply.

An overriding principle in the storage of chemicals is that they should not be adversely affected by other adjacent substances or operations. An HSE Guidance Note gives suitable information on this topic[33]. The proposed COMAH Regulations[16] also require consideration to be given to the quantities of hazardous materials that are stored at chemical sites located near to a COMAH site because of the possible 'domino effect' of an incident at one company affecting adjacent companies.

4.7.7.1 Drum compounds and storage tanks

A common problem met in factories using chemicals is where to store flammable solvents, or substances that could give off a hazardous vapour if a leak occurred. Whether in drums or bulk tanks, outside storage is preferable since this:

(a) Allows dispersion of any flammable or hazardous vapour from vents, leaks, spillages etc.
(b) Minimises the potential of ignition.
(c) Provides secure storage to minimise damage to containers.

External storage of chemicals and hazardous substances require particular facilities which, for bulk tanks, drums and intermediate bulk containers (IBCs), include:

(a) a bunded impervious hard-standing (e.g. a concrete plinth) capable of containing any spillage or rainwater,
(b) drainage of the impervious bunded area to a sump, which can be pumped out for authorised disposal, dependent on the hazardous nature of the liquors,
(c) a separation of at least 4 metres from boundaries, building and sources of ignition,
(d) except for bulk tanks, a ramp to allow trolleys and fork-lift trucks easy access for handling the drums or IBCs,
(e) segregation of stocks of materials which if mixed, due to a spillage or incident, could cause a hazardous reaction,
(f) measures to ensure that containers of flammables are not stored under overhead electric power lines.
(g) Water reactive materials should be stored under cover but with plenty of ventilation.

By ensuring that all drums and IBCs are stored in a bunded area, the risk of ground or surface water pollution is reduced. Such a facility also ensures that flammable substances are contained, reducing the risk of the spread of fire should a leak occur.

A responsible person should be put in control of the drum compound with duties that should include:

- regular housekeeping checks
- arranging for water in sumps to be pumped out for appropriate disposal
- arranging for spillages and leaks to be dealt with
- maintaining an inventory of materials stored on the compound.

In the case of bulk storage, the tank should be mounted above ground level and be provided with:

(a) Safe access and egress for delivery or waste disposal tankers.
(b) Adequate separation distances from other tanks and sources of ignition.
(c) Hazardous electrical zone classification when handling flammable materials.
(d) An impervious bund capable of containing 110% of the tank capacity.
(e) A tank contents gauge arranged for local or remote reading.
(f) A pressure/vacuum relief valve or vent to allow the tank to breathe during filling and emptying.
(g) Secure attachment to the ground to prevent movement due to high winds.
(h) Fixed installation for fire fighting to be considered, e.g. sprinklers, drench or foam.
(i) Adequate HAZCHEM labelling.

Figure 4.7.4 shows a bulk storage tank for non-flammable liquids.

Underground storage tanks such as those installed at petrol stations are required to be fitted with a means for detecting leaks[34].

Figure 4.7.4 Storage tank for non-flammable liquids.
(Courtesy Rhône-Poulenc-Rorer)

The tanker off-loading bay – ideally sited off road – should be drained to a sump to prevent water pollution. Safety shower and eye wash facilities should be provided in the area. The location of the emergency equipment should not be within the range of a potential leak or spray from the transfer hose.

4.7.7.2 Tanker off-loading

The tanker off-loading point should be clearly labelled and locked off to prevent unauthorised access. There have been a number of incidents involving tanker loads being off-loaded into the wrong storage tanks. It is important that a representative of the company owning the storage tank is present during the off-loading in order to:

- Check delivery notes and paperwork.
- Unlock the off-loading points access pipe.
- Ensure correct procedures and PPE are adhered to.
- Operators should be aware of how to stop the delivery in the event of an emergency.

Where non-compatible substances are delivered to the same area of a site, the transfer hoses should be arranged to be unique to the particular substance, for example by using different threads on the hose fittings.

Accidents have occurred when sampling road tankers before they are off-loaded. Where this is necessary, the responsibility and the procedure for carrying out such an operation should be agreed in writing with the supplier of the material. The sampling procedure must clearly define the responsibilities of both the customer and the supplier, paying particular attention to:

- Personal protective equipment.
- Sampling equipment and method of taking samples.
- Safe system of work when working on top of the tanker. (Ideally, a walkway adjacent to, or on the top of, the tanker should be provided.)

4.7.7.3 Warehousing

The first step when chemicals and substances are to be stored is to know what the substances are and their characteristics. This will enable storage to be arranged so that incompatible substances are either suitably separated, have an intermediate fire wall or a mutually compatible substance between them to provide such separation. The effects of storing incompatible chemicals in the same area have been described in HSE reports[35,36].

Fire detection and sprinkler systems are beneficial but if used with high bay racking consideration should be given to incorporating inter-rack sprinklers which counteract the chimney effect of a fire. Care should be taken that no water-reactive chemicals are stored below sprinkler installations. The advice of the local fire prevention officers, sprinkler suppliers, insurance surveyors and safety advisers should be sought on storage and rack layouts and on the arrangement of the sprinklers.

Following an incident at Basle, Switzerland, in which contaminated fire-fighting water from a warehouse fire entered the Rhine and caused severe ecological damage, the chemical industry and storage companies have been studying better methods of run-off water control and their recommendations are contained in a booklet[32].

4.7.7.4 Storage of gas cylinders

Gases stored in cylinders under high pressure are used extensively in industry for burning, welding and as process material. The contents of the containing gas cylinders are potential sources of fire and explosion if the cylinders are not stored safely[37]. They should be stored upright and checked regularly, particularly the regulators. Gas cylinders should be stored away from lifts, stairs, gangways, underground rooms, and in an area that is free from fire risks and sources of heat and ignition. In order to prevent the bottom of a stored upright cylinder from corroding, cylinders should be stored under cover on a well-drained surface. Manufacturer's recommendations should also be consulted when storing different types of gases in the one area. This information will include segregation and separation distances for cylinders in storage.

Where gases such as nitrogen and carbon dioxide are used in a small enclosed room (e.g. a laboratory or switch room) any leakage can create an asphyxating atmosphere. Therefore, precautions need to be taken to ensure the atmosphere is checked before anyone enters the area.

4.7.8 Transport

Chemicals are transported either in discrete packages or in bulk, often by tanker. Careful selection must be made of the material and construction of the package or container: for example, a mild steel road tanker will not last long if used to transport 20% sulphuric acid, whereas it is perfectly satisfactory for carrying fuel oil. Where the substance is packaged in small quantities, the packaging and labelling must comply with the Carriage of Dangerous Goods by Road Regulations (1996). These Regulations are supported by Guidance Notes and Approved Codes of Practice[38,39].

> *Note*: The vehicles must carry a suitable warning sign. These Regulations and the Approved Carriage List[21] cover the suitability of the vehicle and container and the information to be made available about the substance(s) being carried. Regulations[40] provide information about the type and standard of instruction and training of drivers of vehicles carrying dangerous goods. Other Regulations[41] lay down the requirements for the safe transport of explosives by road.

Vehicles carrying more than 500 kilograms of dangerous substances should carry a hazard warning board at the front and rear of the units showing the following information:

- HAZCHEM code.
- Substance identification number.
- Hazard warning sign.
- Contact telephone number for technical advice.

The Regulations for transport in packages and in bulk have similar requirements for the proper stowage of the load, adequate instruction and training for drivers, information on the load being carried and proper labelling of the vehicles. A review of the safe handling of chemicals and dangerous substances in road transport operations is contained in a chemical industry publication[42].

The Carriage of Dangerous Goods by Rail Regulations 1996[43], and Amending Regulations[44,45], together with the Packaging, Labelling and Carriage of Radioactive Material by Rail Regulations (1996)[46] cover the suitability of the rail freight/containers and information to be made available when transporting dangerous goods and radioactive material by rail.

The concerns over transporting dangerous materials have led to the Transport of Dangerous Goods (Safety Adviser) Regulations 1999[47] which requires the appointment by companies who transport such materials of a trained and competent person to ensure the safety and integrity of the load during transportation and related loading and unloading. While the duties are placed on the employer, some functions must be assigned to the Dangerous Goods Safety Adviser.

The primary functions are:

1 To provide advice to the employer on health, safety and environmental issues as it relates to the transportation of dangerous goods.
2 To provide accident reports to the employer of any incident that occurs during the transportation of dangerous goods which affect the health and safety of any person or causes damage to the environment. This requirement is a specific function of the appointed Dangerous Goods Safety Adviser.
3 To provide an annual report to the employer of the company's activities with respect to the transport of dangerous goods.
4 To monitor practices and procedures in relation to dangerous goods transportation.

Consideration should be given to the safe transportation of substances on company premises:

1 If drums are carried long distances, they should be transported on pallets on a towed trailer.
2 Drum handling equipment should be regularly maintained.
3 Road surfaces should be free of obstacles that could cause a spillage or a load to become unstable.
4 Winchesters of liquid should be carried in protective containers fitted with a handle.

4.7.9 Plant and process design

The chemical manufacturing process should be operated and maintained in a safe manner. This presumes that the design of the plant is safe. To design a plant and operating process that will be safe requires knowledge of the chemical process and its limiting parameters. The design of the plant should consider not only matters concerned with safety, but also health, the environment, fire, and its effect on neighbours. Once the plant has been installed, regular checks should be carried out to ensure proper compliance with agreed standards. Such checks should include housekeeping, scheduled maintenance and health, safety and environmental audits. Changes to the plant should only be made after discussion and investigation of any likely effects the proposed changes may have on the plant safety.

The results of such assessments should be recorded.

When new processes are proposed, the safety adviser should be involved at the earliest stage and should ensure that adequate chemical data, environmental, fire and safety information are available. Discussions on the plant may involve the enforcing authorities and waste disposal agencies but their early involvement could save delays at a later date.

4.7.9.1 Process safety

This section relates to the safe design of process plant which starts with the chemist developing a chemical reaction or series of reactions that produce the required end product. Initially the proposed process will be tried out in the laboratory in gram quantities. The chemist will be looking not only for product purity and quality but also at the physical and thermal chemistry aspects within the process, such as heats of reaction, temperature rises, rates of pressure rise, temperature of exotherms, on-set temperatures, etc.

When the feasibility of the process is proved a pilot scale plant will be built which will carry out the synthesis in kilogram quantities to establish the manufacturing viability of the process. At this scale of operation, companies need to ensure that an assessment of the chemical reaction hazards of the process has been thoroughly carried out. Such an assessment should include selecting and specifying the basis of safety[26] for each reaction. The basis of safety is identified by using the physical and thermal chemistry information in a HAZOP study on the chemical process and identifying the control measures necessary to ensure a safe reaction. This work can help prevent and control thermal runaway reactions. The basis of safety information should be reviewed whenever a change of process is considered.

The preferred option to make a safe process is to eliminate the hazard completely or reduce its magnitude sufficiently to avoid the need for elaborate safety systems and procedures. In order to design an inherently safer process there is need to:

(a) Substitute – replace the material with a less hazardous substance, or a hazardous reaction with a less hazardous reaction.
(b) Minimise – use smaller quantities of hazardous substances in the process at any one time.

(c) Moderate – use less hazardous conditions, a less hazardous form of a material or facilities that minimise the impact of a release of hazardous material or energy.
(d) Simplify – design facilities that eliminate unnecessary complexity, make operating errors less likely and are forgiving of errors.

At an early stage in designing a safe process the chemist should consider:

(i) Avoiding the use of highly exothermic reactions or thermally unstable reactants and intermediates.
(ii) Replacement of batch reactions with semi-batch or continuous processes. This enables the quantity of reactant present to be reduced and by controlling the addition, may stop the reaction.
(iii) Use water as the solvent instead of a more hazardous organic solvent, providing that the chemical does not react with the water.

The information gathered from the pilot plant study is compiled into a process dossier which provides the data for the chemical engineer to use in the design of the full-scale production plant. It is at this stage that the safety adviser should become involved.

The initial design will be in the form of engineering line diagrams which show the vessels, pumps, interconnecting pipework, etc., but not the detailed instrumentation or control systems. Sufficient information should be available to enable a preliminary safety analysis to be carried out drawing on the expertise of the process chemist, process engineer, control systems engineer, production engineer and the safety adviser. The object of this safety analysis is to identify possible hazards arising from the proposed plant and to define the measures necessary to eliminate or reduce them to an acceptable level. Some of the more obvious hazards to be looked for will be low or high pressures, low or high temperatures and highly exothermic reactions which could cause a runaway reaction with potentially catastrophic results.

In the analysis, each process step is looked at for deviations in temperature, pressure, level etc., and the possible effect of physical problems such as pipe blockage, valve failure, corrosion etc. All the details of this analysis should be recorded. On the findings of this analysis the chemical engineer can refine his design and produce detailed plans including pipework layouts and instrumentation (P & I) diagrams.

Once the detailed plans have been prepared, a more detailed safety analysis can be carried out which could be in the form of a Hazard and Operability Study (HAZOP)[48,49].

4.7.9.2 Hazard and Operability study (HAZOP)

A HAZOP study is a technique used to identify health, safety, environmental and fire hazards and operating problems. The basic study deals with the mechanical, electrical and chemical aspects of the plant operation together with electromechanical control systems. Where the control equipment incorporates computers additional studies are needed.

A HAZOP study requires a multi-disciplinary approach by a team made up of technical specialists, i.e. chemical engineer, chemist, production manager, instrumentation engineer, safety adviser etc. It is co-ordinated by a leader who guides the systematic investigation into the effect of various faults that could occur. The success of this study relies heavily on the quality of the leader and the positive and constructive attitude of the team members. It is essential that the team have all the basic data plus line diagrams, flow charts etc., and understand how a HAZOP study works.

The HAZOP study breaks the flow diagram down into a series of discrete units. Various failure and fault conditions are then considered using a series of 'guide words' to structure the investigation of the various circumstances that could give rise to those faults. Each deviation for each guide word is considered in detail and team members are encouraged to think laterally and to ask questions especially about the potential for causing a fault condition. *Table 4.7.1* shows how each of the guide words can be interpreted to highlight possible deviations from normal operation and *Figure 4.7.5* shows a HAZOP report form that could be used to record the findings of the study.

In the example in *Table 4.7.1*, under the first guide word, 'None', we could ask:

- What could cause no flow?
- How could the situation arise?
- What are the consequences of the no-flow situation?
- Are the consequences identified hazardous or do they prevent efficient operation?
- If so, can we prevent no-flow (or protect against the consequences) by changing the design or method of operation.

Table 4.7.1 Showing typical interpretations of HAZOP guide words

Guide word	*Deviations*
None	No forward flow, no flow, reverse flow.
More of	Higher flow than design, higher temperature, pressure or viscosity etc.
Less of	Lower flow than design, lower temperature, pressure or viscosity etc.
Part of	Change in composition, change in ratio of components, component missing.
More than	More components present in the system, extra phase, impurities present (air, water, solids, corrosion products).
Other than	What else can happen that is not part of the normal reaction, start-up or shutdown problems, maintenance concerns, catalyst change etc.

HAZARD AND OPERABILITY STUDY

Date of Study / / Sheet of sheets

Study title ________ Study team ________

Prepared by ________ Project number ________

Line diagram number ________ Procedure number ________

Step Number/ Guide Word	Deviation	Cause	Consequence	Action

Figure 4.7.5 Report from a HAZOP Study

- If so, does the size of the hazard (i.e. severity of the consequences multiplied by the probability of the occurrence) justify the extra expense?

Similar questions are applied to each the other guide words, and so on. Each time a component is studied the drawing or diagram should be marked. Not until all components have been studied can the HAZOP study be considered complete. Where errors occur on the drawing or more informations needed the drawing should be marked (using a different colour) and the points noted in the report.

To be effective the team needs to think laterally and there should be no criticism of other team members' questions. A strange or oblique question may spark off a train of investigation which could lead to the identification of potentially serious fault conditions.

A well-conducted HAZOP study should eliminate 80–85% of the major hazards, thereby reducing the level of risk in the plant. In safety critical plant another HAZOP study, carried out when the detailed design has been finalised, could increase the probability of safe operations.

When the HAZOP study has been completed the necessary remedial actions should be agreed for implementation by the project or process manager. Records of the changes in the design should be kept and checks made to ensure that the modifications have been carried out during the construction of the plant.

With plant that is controlled by computer, the HAZOP study needs to include consideration of the effects of aberrant computer behaviour and the team carrying out the study may need to be reinforced by the software designer plus an independent software engineer able to question the philosophy of the installed software program. A technique known as CHAZOP has been developed for such plant which also highlights the safety critical control items.

4.7.9.3 Plant control systems

Many small, simple, and relatively low hazard plants are fully manually operated. However, with more complex plant automated controls using electronic control systems are employed This does not necessarily make it safe since faults can, unknowingly, be built into the controlling software. To achieve optimum levels of safe operation, computer software for plant control systems should be devised jointly by the software specialist and the production staff. All operational requirements must be covered to ensure that the software designer does not make assumptions which could result in faulty or even dangerous operation of the plant. The software must be designed to accommodate plant failures and any testing or checking necessary during or following maintenance.

Before installing it, the computer program must be challenged in all possible situations to ensure that it matches operational requirements. Any review of software should include an independent software engineer who can challenge the philosophy behind the software. All software changes must be fully described and recorded and plant operators fully trained in the effects of the changes.

When automated computer control systems are incorporated into a plant, operators tend to rely on them completely to the extent that there is a risk that they forget how to control the plant manually. This can be critical in an emergency and it may be prudent to switch the computer off occasionally, and, under supervised conditions, ensure that the operators are still able to control the plant manually.

Control panels should not be provided with too many instruments since this can confuse the operator and prove counterproductive. However, sufficient instrumentation is needed to enable the operators to know what is going on inside closed vessels, pipes, pumps etc. Critical alarms should be set into separate parts of control panel to highlight their importance. This will reduce the potential for their being confused with others, and possibly overlooked. The tone of audible critical alarms should be different from that of process alarm systems to prevent confusion.

Computer-controlled plant will frequently have three levels of operational and safety control:

Level 1: Will mostly focus on process control of the plant and give indicative warnings of possible safety concerns when, for example, a rapid temperature rise may trigger a warning panel indicator.

Level 2: Control occurs when computer software initiates changes to control reaction kinetics. If a reaction temperature continues to rise, the software would initiate the application of cooling water to the vessel to regain control and continue production.

Level 3: Is entirely a safety system when the process is out of control. It will rely on hard-wired trips that shut the plant down safely and abandon production. The hard-wired trips work independently of the computer system.

There is no universal formula for control systems and a control strategy must be developed for each plant based on the operating parameters. A small batch plant consisting of two chemical reactors having a mixture of manual and automatic controls is shown in *Figure 4.7.6*.

4.7.9.4 Assessment of risk in existing plants

A review of existing chemical facilities should be undertaken to identify possible faults and so avoid acute and/or catastrophic loss. The assessment should focus on 'instantaneous failure prevention' of plant such as:

- bulk oil or chemical storage facilities
- multi-chemical 200 litre drum store (especially if large-scale dispensing is carried out)
- chemical processes or mixing facilities
- solvent recovery plant

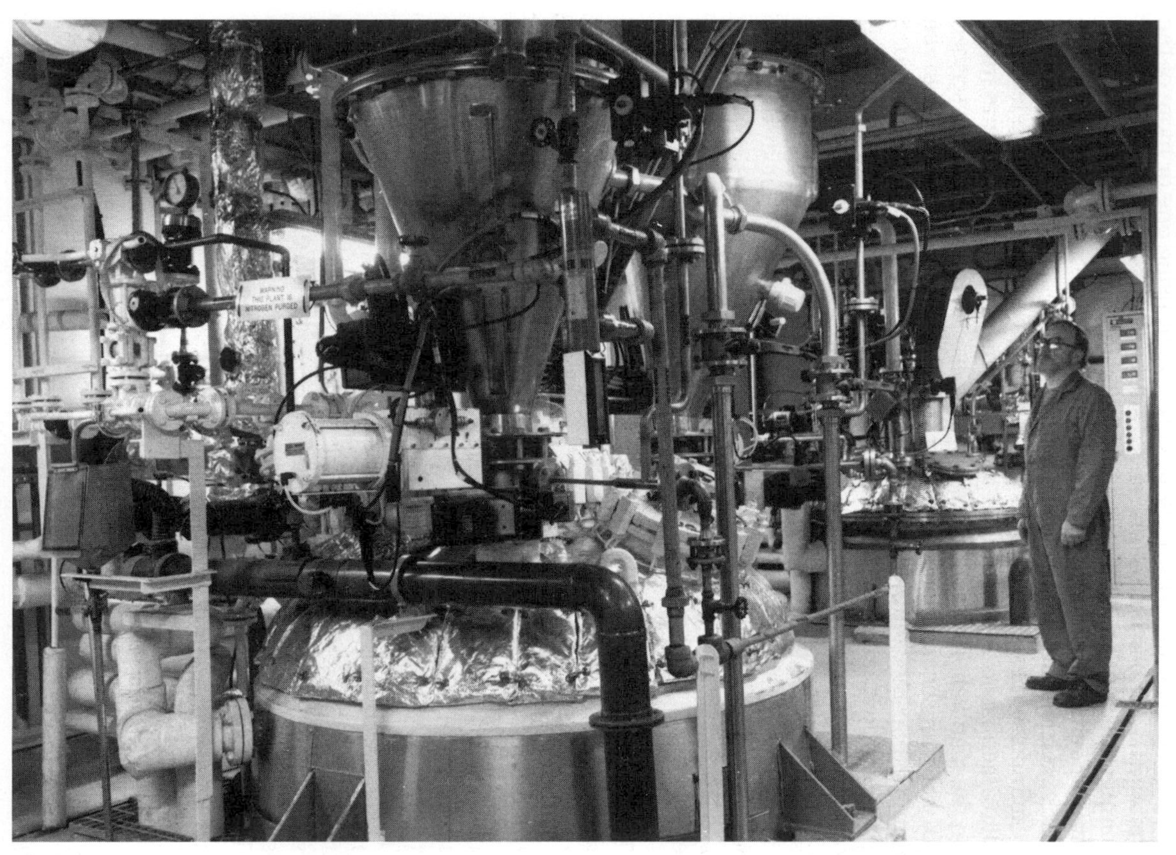

Figure 4.7.6 Two chemical reactors having manual and automatic controls. (Courtesy Rhône-Poulenc-Rorer)

- pipelines and pipework that contain oils or chemicals in quantity and/ or under pressure.

This review will identify those systems or processes that require a further detailed study which can be carried out using one or more of the techniques described below. The end result of the assessment should be a position statement which describes the level of risk from the plant and identifies which facilities require additional measures to ensure they remain both physically and environmentally safe.

A number of techniques have been developed to identify the hazards and to assess the risks from plant and equipment. These techniques range from the relatively simple to the highly complex. A number are described in a BS EN standard[50]. Whichever technique is used it should be appropriate to the complexity of the plant and the materials involved.

4.7.9.4.1 Simpler techniques

The simpler techniques are aimed primarily at determining a ranking order of the risks from the chemical processes carried out in the area. They should clarify which facilities create insignificant risks and require no further action. The position statement for these facilities should record the reasons for this decision. The simpler techniques include:

1. The 'What-if method' is the simplest method to assess chemical process safety risks and is based on questions such as 'What if the mechanical or electrical integrity of the process, the control systems and work procedures all fail, . . . what consequences could arise in the worst case?' While the potential consequences are largely determined by the inherent hazard of the material and the quantity involved, the reviewer is focused on safety concerns, e.g. those arising from fire, explosion, toxic gas release, and environmental protection.
2. The 'Checklist method' is a structured approach whereby the reviewer responds to a predetermined list of questions. This method is less flexible than the 'What-if method' and its effectiveness relies on the strengths and weaknesses of a predetermined checklist. Examples of checklists can be found in chemical process safety literature.
3. The 'Dow-Mond Index' is a more structured approach than the previous two techniques and takes into account quantities and hazards to arrive at a basic risk classification. This method provides a level of quantification of risk and considers the 'off-setting' factors which exist to control intrinsic hazards.

4.7.9.4.2 More complex techniques

Where the ranking process, described above, identifies facilities that warrant an assessment in greater depth, one of the techniques described below should be used:

1. HAZOP study (see section 4.7.9.2.1).
2. Failure modes and effects analysis (FMEA). FMEA is an inductive method for evaluating the frequency and consequence of failures. It

involves examining every component and considering all types of failure for each. It can indicate generic components that may have a propensity to fail.

3 Fault tree analysis (FTA)[51] FTA is a deductive method which starts by considering a particular fault or 'top event' and works backwards to form a tree of all the events and circumstances that could lead to the happening of that top event. By assessing the probability of each individual event, an estimate of the probability of the top event occurring can be obtained. If that probability is unacceptable the major components contributing to it can easily be identified and a cost-effective replacement of them implemented. This method lends itself to assessing the impact of changes in the system and has been useful in determining the causes of accidents.

4.7.9.5 Functional safety life cycle management (FSLCM)[52]

FSLCM is a new technique designed to enable plant safety systems to be managed in a structured way. The technique has been designed to accommodate computer-controlled plants from start-up to shutdown, including emergency shutdowns. It aims to ensure that the safety related systems which protect and control equipment and plant are specified, engineered and operated to standards appropriate to the risks involved. The key concepts of this technique are:

(a) *The safety life cycle* – begins with a clear definition of the equipment and processes for which functional safety is sought and by a series of phases provides a logical path through commissioning, operation to final decommissioning.
(b) *Safety management* – sets a checklist for the things that need to be in place in order to prepare for and manage each phase of the *safety life cycle*. These are incorporated into a formal safety plan.
(c) *Design of safety related systems* – puts the design of safety related control and protective systems into the overall context of the safe operation of equipment or facilities. It requires that such systems are designed to meet specific risk criteria.
(d) *Competencies* – provides guidance on the appropriate skills and knowledge required by those people who will be involved in the technique.

By following a structured life cycle approach the hazards inherent in the operation of equipment or processes can be clearly identified. The standards to which protection is provided can be demonstrated in an objective and constructive way.

4.7.10 Further safety studies

Having carried out a HAZOP study on the plant and incorporated its findings into the design, it is prudent to carry out a further review during the commissioning period to check that the design modifications have produced the desired results. This is necessary since the final details of the physical installation are often left to the installing engineers to decide and these could produce unforeseen hazards. Finally, once the plant is commissioned and operational there should be routine safety checks carried out on a regular basis.

4.7.11 Plant modifications

Plant modifications, even apparently simple ones, can have major consequential effects[15]. It is crucial that the plant is not modified without proper authorisation and, for safety critical parts, the completion of a HAZOP study of the possible effects of the proposed changes. A 'process change form' should be used which should include the reasons for the change. Use of such a form also ensures a degree of control on the modifications made, especially if it has to be sanctioned by a senior technical specialist such as a process engineer, safety adviser, production manager and maintenance manager. There needs to be clear guidance as to when the process change form has to be used so that there can be no misunderstanding. After the plant has been modified it may be necessary to retrain the operators in the changed operation techniques.

4.7.12 Safe systems of work

Since human beings are necessary in the operation of chemical plants there is always the likelihood of errors being made that could result in hazards. It is, therefore, important that operators are trained in the safe way to run the plant. Such training, based on safe systems of work, should include the carrying out of risk assessments. Errors in operation and misunderstandings can be reduced if the system of work is in writing.

4.7.12.1 Instruction documentation

There should be detailed written operating instructions for every chemical plant which can conveniently be considered in three parts:

1 *Operator's instructions* that give specific instructions on how to operate the plant and handle the materials safely. The instructions should contain information on the process, quantities and types of materials used, and any special instructions for dealing with spillages, leaks, emergencies and first aid. The instructions should also contain information on the expected temperatures, pressures and conditions, and provide information on the actions to be taken if they are exceeded, the type of PPE to be worn, a copy of the safety data sheet for each of the materials involved, techniques for taking samples and cleaning instructions.

2 *Manufacturing procedures* aimed at the operators and the supervisor in charge of the plant should explain the process and provide a synopsis of the chemical process undertaken. The procedure should refer to likely problems such as exotherms and give details of actions to take. The sequence of operations, quantities of materials used, temperature and pressure ranges, methods for dealing with spillages and leaks, disposal of waste, etc., should be included.
3 A process dossier should be compiled containing detailed information about the process, the plant and equipment design specifications and the basis of safety for the process. This document should be a major reference source for the process engineer and be consulted and updated whenever a change is made.

4.7.12.2 Training

Both operators and supervision should be trained in the techniques for operating the plant, the process, materials used, their hazards and precautions to be taken, emergency procedures and first aid. The training can be based on the content of the Operator Instructions and the Manufacturing Procedures and should include a study of the safety data sheets. The importance of following the safe methods of work and the reporting of any deviations from the stated operating parameters should be emphasised.

Staff should be made aware of the potential hazards that could be encountered in the process if mistakes were made. For example, what could happen if:

- Equipment was not bonded to earth and a fire started.
- Another chemical was mistakenly added.
- The agitator had been stopped and restarted when it should have been on all the time.
- The reaction was allowed to get too hot and an exothermal reaction took place.
- The reaction got out of control and pressure developed resulting in a two-phase emission.

It is important that the operating staff are regularly re-trained in the operating instructions and that they are briefed on any changes made.

4.7.12.3 Permits-to-work

Permits-to-work are required where the work to be carried out is sufficiently hazardous to demand strict control over both access and the work itself. This can occur when maintenance and non-routine work is being carried out in a chemical plant or for any normal operation where the risks faced make clear and unequivocal instructions necessary for the safety of the operators.

The essential elements of a permit to work are:

(a) The work to be carried out is described in detail and understood by both the operators of the plant and those carrying out the work.
(b) A full explanation is given to those carrying out the work of the hazards involved and the precautions to be taken.
(c) The area in which the work is to be carried out is clearly identified, made as safe as possible and any residual hazards highlighted.
(d) A competent, responsible and authorised person should specify the safety measures, such as electrical isolation, pipes blanked off etc. to be taken on the plant, check that they have been implemented and sign a document confirming this and that it is safe for workmen to enter the area.
(e) The individual workmen or supervisor in charge must sign the permit to say they fully understand the work to be done, restrictions on access, the hazards involved and the precautions to be taken.
(f) The permit must specify any monitoring to be carried out before, during and after the work and require the recording of the results.
(g) When the work is complete, the workmen or supervisor must sign the permit to confirm that the work is complete and it is safe to return the plant to operations.
(h) A competent, responsible and authorised person must sign the permit, cancelling it and releasing the plant back to operations.

The format of a permit to work will be determined by the type of work involved but a typical permit is shown in *Figure 4.7.7*.

Typical work requiring a permit to work includes hot work, entry into confined spaces, excavations, high voltage electrical work, work involving toxic and hazardous chemicals etc. For a permit to work to be effective it is essential that all those involved understand the system, the procedure and the importance of following the laid down procedure. Before the work starts all those concerned should be trained in the system and their individual responsibilities emphasised.

4.7.13 Laboratories

The use of chemicals in laboratories poses totally different problems from those met in a production facility. The scale is much smaller, the equipment generally more fragile and, while the standard of containment for bench work is often less, the skill and knowledge of those performing the reactions are very high.

Work in quality control laboratories is normally repetitive using closely defined analytical methods. Research laboratories are far wider in the scope of the reactions they investigate, sometimes dealing with unknown hazards, and in the equipment they use. The principal hazards met in laboratories are fire, explosion, corrosion, and toxic attacks. A limit should be specified for the total amount of flammables allowed in a laboratory at any one time, which should be enough for the day's work but not exceed 50 litres.

X Y Z Company Limited

PERMIT-TO-WORK

NOTES:
1 Parts 1, 2 and 3 of this Permit to be completed before any work covered by this permit commences and the other parts are to be completed in sequence as the work progresses.
2 Each part must be signed by an Authorized Person who accepts responsibility for ensuring that the work can be carried out safely.
3 None of the work covered by this Permit may be undertaken until written authority that it is safe to do so has been issued.
4 The plant/equipment covered by this Permit may not be returned to production until the Cancellation section (part 5) has been signed authorizing its release.

PART 1 DESCRIPTION

(a) Equipment or plant involved ______________________________

____

(b) Location ______________________________

(c) Details of work required ______________________________

____

Signed ______________________________ Date ______________
person requesting work

PART 2 SAFETY MEASURES

I hereby declare that the following steps have been taken to render the above equipment/plant safe to work on: ______________________________

____

Further, I recommend that as the work is carried out the following precautions are taken: ______

____

Signed ______________________________ Date ______________
being an authorized person

PART 3 RECEIPT

I hereby declare that I accept responsibility for carrying out the work on the equipment/plant described in this Permit-to-Work and will ensure that the operatives under my charge carry out only the work detailed.

Signed ____________________ Time __________ Date __________

Note: After signing it, this Permit-to-Work must be retained by the person in charge of the work until the work is either completed or suspended and the Clearance section (Part 4) signed.

PART 4 CLEARANCE

I hereby declare that the work for which this Permit was issued is now completed/suspended* and that all those under my charge have been withdrawn and warned that it is no longer safe to work on the equipment/plant and that all tools, gear, earthing connections are clear.

Signed ____________________ Time __________ Date __________
* delete word not applicable

PART 5 CANCELLATION

This Permit-to-Work is hereby cancelled

Signed ____________________ Time __________ Date __________
being a person authorized to cancel a Permit-to Work

Figure 4.7.7 Permit-to-work

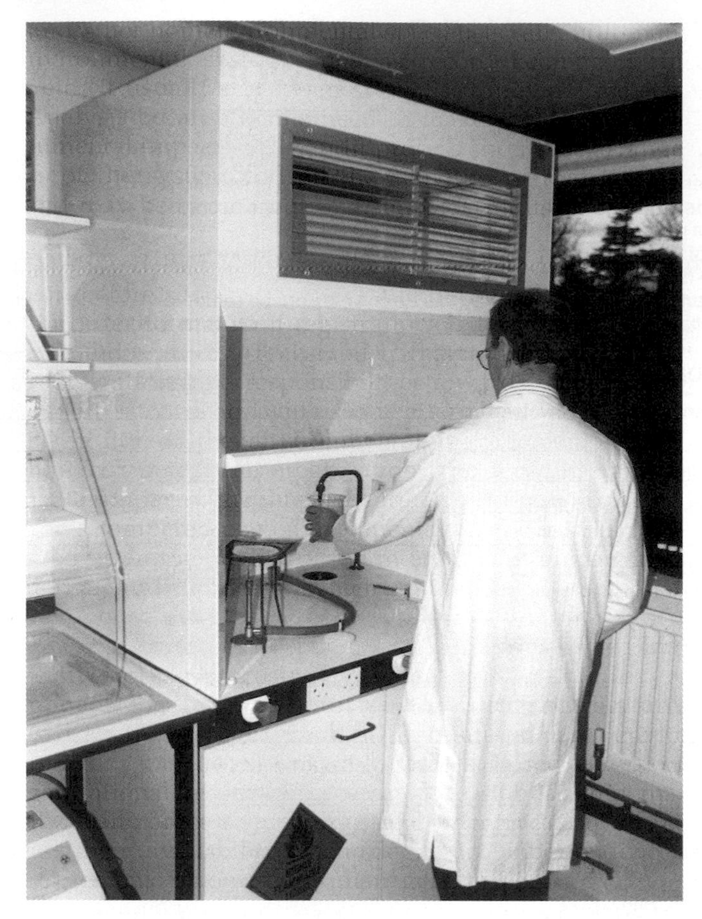

Figure 4.7.8 Well ordered fume cupboard. (Courtesy British Sugar plc)

Hazardous and potentially hazardous reactions should be carried out in a fume cupboard. The effectiveness of the fume cupboard's extraction should be checked regularly in line with COSHH requirements. The fume cupboard should not be used for extra storage space since this can reduce the efficiency of the extraction system. A well-ordered and tidy fume cupboard is shown in *Figure 4.7.8*.

Further measures that can improve laboratory safety include:

(a) Instituting a 'peer review' assessment by asking a competent colleague to review the proposed reaction before allowing experiments to be carried out.
(b) Regular checks of laboratory storage areas to ensure old stocks and out-of-date reactive chemicals (e.g. chemicals which can degrade to form peroxides) are removed for disposal. Only minimum inventories of chemicals should be held.

(c) Producing a laboratory safety manual and regularly training staff in its contents.
(d) Providing spillage cleaning equipment and adequate training in its use.
(e) Establishing safe waste disposal procedures.
(f) Maintaining a high standard of housekeeping.
(g) Not storing liquids at high level over the workbench.

Laboratory safety is a very wide subject and there are a number of publications giving sound guidance[53–55]. Many of the larger chemical manufacuring companies produce their own practical guidance and are pleased to supply copies.

4.7.14 Emergency procedures

The Management of Health and Safety at Work Regulations 1999 (MHSWR)[62] imposes on employers an explicit duty to have in place effective procedures to be followed in the event of serious or imminent danger to people at work. The COMAH Regulations also require affected manufacturers to prepare on-site emergency plans. In addition, COMAH requires employers to co-operate with the local authority in developing off-site emergency plans. (See the publication 'Emergency Planning for Major Accidents'[61].) Irrespective of these statutory requirements it is prudent for every user and storer of hazardous substances to prepare an emergency plan to cover all reasonably foreseeable events such as fire, major spillage or toxic release. The plans can be at two levels, one for the immediate production or storage area and the second for the site as a whole taking account of the likely effects on the local community.

It is very important that employees and the local emergency services know exactly and unambiguously what to do should an incident occur. The Dangerous Substances (Notification and Marking of Sites) Regulations 1990[63] require that the entrances to sites are labelled such that the emergency services have pre-warning that there are hazardous chemicals on site. Additionally, the Planning (Hazardous Substances) Regulations 1992[64] require notification to the local authority of the amounts of hazardous substances held on site. A clear drawing or sketch showing the layout of the site should be available for the emergency services. It should also contain details of the buildings and highlight fire extinguishers, emergency exits, spillage control equipment, etc. All employees should be properly instructed, fully trained and rehearsed in those emergency plans. The local emergency services should be encouraged to familiarise themselves with the site.

Where there is a potential for a major emergency, which would involve the local emergency services and local authority, there must be an agreed plan of action to co-ordinate all the services including managers and employees on the site with their specialised knowledge of the site and its processes. The emergency plans should include a list of emergency contacts including such bodies as the Fire Authority, Local Authority

(Environmental Health Department), the local water utility, the Health and Safety Executive, the Environment Agency, the police, etc.

Advice and guidance on preparing emergency plans are contained in publications by The Society of Industrial Emergency Services Officers (SIESO)[56] and the Chemical Industries Association (CIA)[57]. It must be emphasised that all emergency plans must be regularly practised and reviewed with all personnel who may be actively involved in the process – there is no substitute for actually doing it!

4.7.15 Conclusions

This chapter has summarised some of the health, safety and environmental problems posed by the use of chemicals. A systematic review has been applied in an attempt to clarify the issues and facilitate an understanding of legislative requirements and good practices. Those with responsibilities for handling and using chemicals should study the relevant laws and guidance to ensure that their areas of responsibility meet the highest standards. Management commitment, leadership and setting a good example play important roles in achieving high standards in health, safety and the environment which, in turn, lead to a successful enterprise. To quote the HSC's slogan 'Good health is good business'.

References

1. *The Control of Substances Hazardous to Health Regulations 2002*, The Stationery Office, London (2002)
2. Health and Safety Executive, Legal Series Booklet No. L5, *General COSHH ACOP and Carcinogens ACOP and Biological Agents ACOP*, HSE Books, Sudbury (2002)
3. *The Chemicals (Hazard Information & Packaging for Supply) Regulations 2002*, The Stationery Office, London (2002)
4. Health and Safety Commission, Legal Series Booklet No. L124, *Approved Supply List. Information approved for the classification & labelling of substances and preparations dangerous for supply* (7th edn), HSE Books, Sudbury (2002)
5. Health & Safety Executive, Legal Series Booklet No. L130, Approved Code of Practice: *Safety data sheets for substances and preparations dangerous for supply.* 3rd edn. HSE Books, Sudbury (2002)
6. *The Merck Index*, 10th edn, Merck & Co. Inc. (1983)
7. Sax, N.I., *Dangerous Properties of Industrial Materials* (7th edn), Van Nostrand Reinhold (1989)
8. Bretherick, L., *Handbook of Reactive Chemical Hazards*, Butterworth, Oxford (1979)
9. Health & Safety Executive, Guidance Series Booklet No. HSG 117, *Making Sense of NONS. A Guide to the Notification of New Substances Regulations 1993*, HSE Books, Sudbury (1994)
10. *The Chemical Weapons Act 1996*, The Stationery Office, London (1996)
11. European Union, *European Inventory of Existing Commercial Substances*, EU, Luxembourg
12. European Union, *European List of Notified Chemical Substances*, EU, Luxembourg
13. Edwards *v.* National Coal Board (1949) IKB 704; (1949) 1 All ER 743
14. Health and Safety Executive, Environmental Hygiene Series Guidance Note No. EH 40, *Occupational Exposure Limits*, HSE Books, Sudbury, updated annually
15. Health & Safety Executive, *Investigation Report: Flixborough Disaster*, HSE Books, Sudbury (1975)
16. *The Control of Major Accident Hazard Regulations 1999* (COMAH), The Stationery Office,

London (1999) also The Health and Safety Executive, booklet no: L 111, *A Guide to the Control of Major Accident Hazards Regulations*, HSE Books, Sudbury (1999)
17. *The Notification of Installations Handling Hazardous Substances Regulations 1982*, The Stationery Office, London (1982), also The Health and Safety Executive, booklet no: HSR 16, *Guide to the Notifications of Installations Handling Hazardous Substances Regulations 1982*, HSE Books, Sudbury (1982)
18. European Union, Directive No. 82/501/EEC, *Council Directive on Major Accident Hazards of Certain Industrial Activities*, EU, Luxembourg (1982)
19. Health & Safety Executive, Health and Safety Regulation Booklet No. HSR 21, *Guide to the Control of Industrial Major Accident Hazards Regulations 1984*, HSE Books, Sudbury (1984)
20. Health & Safety Executive, Health and Safety Guidance Booklet No. HSG 25, *Control of Industrial Major Accident Hazards Regulations 1984: Further guidance on emergency plans*, HSE Books, Sudbury (1985)
21. Health and Safety Commission, Legal Series Booklet No. L90, *Approved Carriage List, Information approved for the carriage of dangerous goods by road and rail other than explosives and radioactive material*, HSE Books, Sudbury (1999)
22. Health & Safety Executive, Legal Series Booklet No. L100, *Approved Guide to the Classification & Labelling of Substances and Preparations Dangerous for Supply*, HSE Books, Sudbury (1999)
23. Health & Safety Executive, Health and Safety Guidance Series Booklet No. HSG 97, *A step by step guide to COSHH assessment*, HSE Books, Sudbury (1992)
24. Health and Safety Executive, Legal Series Booklet No. L5, *General COSHH ACOP and Carcinogens ACOP and Biological Agents ACOP* (2002), HSE Books, Sudbury (2002)
25. Health & Safety Executive, Legal Series Booklet No. L86, *Control of Substances Hazardous to Health in Fumigation Operations: Approved Code of Practice: COSHH '94*, HSE Books, Sudbury (1996)
26. Health and Safety Executive, booklet no: HSG 143, *Designing and Operating Safe Chemical Reaction Processes*, HSE Books, Sudbury (2000)
27. *The Control of Lead at Work Regulations 2002*, the Stationery Office, London (2002)
28. *The Ionising Radiations Regulations 1999*, The Stationery Office, London (1999)
29. Health & Safety Executive, Health and Safety Series Guidance Booklet No. HSG 40, *Chlorine from drums and cylinders*, HSE Books, Sudbury (1999)
30. Health & Safety Executive, Chemical Series Guidance Note No. CS4, *Keeping of LPG in cylinders and similar containers*, HSE Books, Sudbury (1986)
31. Health & Safety Executive, Health & Safety Guidance Series Booklets Nos HSG 50, *The Storage of Flammable Liquids in Fixed Tanks (up to 10,000 m^3 total capacity)* (1990); HSG 51, *The Storage of Flammable Liquids in Containers* (1990); HSG 52, *The Storage of Flammable Liquids in Fixed Tanks (exceeding 10,000 m^3 total capacity)* (1991); HSE Books, Sudbury
32. British Distributors' & Traders' Association, *Warehousing of Chemicals Guide*, British Distributors' & Traders' Association, London (1988)
33. Health & Safety Executive, Health and Safety Guidance Series Booklet No. HSG 71, *Chemical warehousing. Storage of Packaged Dangerous Substances*, HSE Books, Sudbury (1998)
34. Health and Safety Commission, Consultative Document No. CD120, *Proposals for new petrol legislation*, HSE Books, Sudbury
35. Health & Safety Executive, Investigation Report (not numbered), *Fire and explosions at B & R Hauliers, Salford, 25 September 1982*, HSE Books, Sudbury (1983) (ISBN 0 11 883702 8)
36. Health & Safety Executive, Investigation Report (not numbered), *Fire and explosions at Cory's Warehouse, Toller Road, Ipswich, 14 October 1982*, HSE Books, Sudbury (1984) (ISBN 0 11 883785 0)
37. British Oxygen Company Ltd, *Safe Under Pressure, Guidelines for all who use BOC Gases in Cylinders*, British Oxygen Company, Guildford, Surrey (1993)
38. Health & Safety Executive, Legal Series Booklets Nos L89, *Approved Vehicle Requirements* (1999); L91, *Suitability of vehicles and containers and limits on quantities for the carriage of explosives: Carriage of Explosives by Road Regulations 1996–Approve Code of Practice* (1996); L92, *Approved requirements for the construction of vehicles for the carriage of explosives by road* (1999); L93, *Approved Tank Requirements: the provisions for bottom loading and vapour recovery systems of mobile containers carrying petrol* (1996); HSE Books, Sudbury.
39. Health & Safety Executive, Health and Safety Regulations Booklet No. HSR 13, *Guide to the Dangerous Substances (Conveyance by Road in Road Tankers and Tank Containers) Regulations 1981*, HSE Books, Sudbury (1981)

40. *The Carriage of Dangerous Goods by Road (Driver Training) Regulations 1996*, The Stationery Office Ltd, London (1996)
41. *The Carriage of Explosives by Road Regulations 1996*, The Stationery Office Ltd, London (1996)
42. Chemical Industries Association, *Hauliers Safety Audit*, Chemical Industries Association, London (1986)
43. *The Carriage of Dangerous Goods by Rail Regulations 1996*, The Stationery Office, London (1996)
44. *The Carriage of Dangerous Goods (Amendment) Regulations 1998*, The Stationery Office, London (1998)
45. *The Carriage of Dangerous Goods (Amendment) Regulations 1999*, The Stationery Office, London (1999)
46. *The Packaging, Labelling and Carriage of Radioactive Material by Rail Regulations 1996*, The Stationery Office, London (1996)
47. *The Transport of Dangerous Goods (Safety Adviser) Regulations 1999*, The Stationery Office, London (1999)
48. Kletz, T., *HAZOP & HAZAN – Identifying and Assessing Process Industry Hazards*, The Institution of Chemical Engineers, Rugby (ISBN 0 85 295285 6)
49. Chemical Industries Association, *A Guide to Hazard and Operability Studies*, Chemical Industries Association, London (1992)
50. British Standards Institution, BS EN 1050, *Safety of Machinery – Principle for Risk Assessment*, BSI, London (1997)
51. British Standards Institution, BS IEC 61025, *Fault Tree Analysis*, BSI, London
52. British Standards Institution, BS IEC 61508, *Safety of machinery – Functional safety of electrical, electronic and programmable electronic safety related systems*, BSI, London
53. Bretherick, L., *Hazards in the Chemical Laboratory*, 4th edn, The Royal Society of Chemistry, London (1986)
54. Weston, R., *Laboratory Safety Audits & Inspections*, Institute of Science & Technology, London (1982)
55. The Royal Society of Chemistry, *Safe Practices in Chemical Laboratories*, The Royal Society of Chemistry, London (1989) (ISBN 0 851 86309 4)
56. The Society of Industrial Emergency Services Officers, *Guide to Emergency Planning*, Paramount Publishing Ltd., Boreham Wood (1986)
57. Chemical Industries Association, *Be prepared for an emergency – Training & Exercises*, Chemical Industries Association, London (1992) (ISBN 0 900623 73 X)
58. *The Control of Asbestos at Work Regulations 2002*, The Stationery Office, London (2002)
59. *The Noise at Work Regulations 1989*, The Stationery Office, London (1989)
60. *The Construction (Head Protection) Regulations* 1989, The Stationery Office, London (1989)
61. Health and Safety Executive, booklet no: HSG 191, *Emergency Planning for Major Accidents*, HSE Books, Sudbury (1999)
62. *The Management of Health and Safety at Work Regulations 1999*, The Stationery Office, London (1999)
63. *The Dangerous Substances (Notification and Marking of Sites) Regulations 1990*, The Stationery Office, London (1990)
64. *The Planning (Hazardous Substances) Regulations 1992*, The Stationery Office, London (1992)

PART V

The environment

Health and safety have for long been recognised as important aspects of working life and there is a long record of legislation and of the part played by caring employers. In the past two decades concern about the environment has become a major issue as scientists have developed ways to measure the damage done to the ecology and the quality of life and to identify the cause of it. Natural disasters have, over the eons, had their adverse effects on the environment but, in the main, nature has been able to accommodate them. What nature cannot accommodate is the gross misuse of the environment by man. This point is being increasingly recognised, both nationally and globally, and there are growing bodies of legislation and standards aimed at checking those abuses.

Within the workplace, responsibility for ensuring compliance with environmental standards and legislation is often delegated to the safety adviser. Suddenly the health and safety professional is to be found in a new front line without training or experience. This part of the book sets out to outline the standards and legislation concerning the environment and to explain how, with goodwill and the right approach at the right level, high environmental standards can, like safety, materially contribute to the well-being and profitability of the enterprise.

Chapter 5.1

The environment: issues, concepts and strategies

J. E. Channing

5.1.1 Introduction

The word 'environment' generates many different responses. To some it is a question of survival of the world as an inhabitable planet. To others it is an over-hyped scare founded on myth rather than fact. The 'environment' does evoke considerable emotion and trying to establish a logical rational position in the midst of scientific uncertainty and strongly held feelings is a significant challenge. Some facts are not in dispute. The huge growth in the human population is the major driver of today's environmental concerns. The number of human beings inhabiting the planet has mushroomed over the last one hundred years. The demands they make on the resources of the planet have grown exponentially. United Nations population data[1] is shown in *Table 5.1.1* demonstrating the actual and predicted growth of the human population.

To the increased birthrate, caused by advances in nutrition and a reduced child death rate, must be added the longevity of the average human being. Both result from the success of the species in developing technologies, medical and social, which have increased life expectancy. More people have added to the strain on resources. The growth in the

Table 5.1.1 Population growth

Year	*Actual or predicted global population*
1800	1 billion
1930	2 billion
1960	3 billion
1974	4 billion
1987	5 billion
1999	6 billion
2050 (low estimate)	7.3 billion
2050 (middle estimate)	8.9 billion
2050 (high estimate)	10.7 billion

human population has not been matched by a growth in the available land. The inhabitable land mass is unchanged and technologies are not yet available to allow large populations to comfortably inhabit the inhospitable deserts of the Sahara or the icy tundra of Canada or Siberia. The consequences are predictable – less space and more competition for available resources leading to local tensions in crowded parts of the world with geo-political tensions and conflicts between nation states. In particular tensions arise between wealthy nations, which consume more resources per capita and want to continue to do so, and less developed nations which seek a fair share. The response from political leaders is to find ways to accommodate the imbalances, so far as their electorates will allow, by introducing regulations to change the behaviour of societies, businesses, and individuals. The basis of the decisions and the directions taken is the known facts of the issue. Herein lies a major problem. The facts of environmental life are constantly changing. The scientific community struggles to make sense of emerging and often conflicting data. Political leaders base their policy decisions on their prognostications. The potential consequence of a wrong decision, given the worldwide scale of the environmental issue, is either to permit an environmental catastrophe or to waste money (another resource) on a huge scale. The challenge of correct decision-making is daunting. Nevertheless it is clearly sensible to take steps to conserve resources. Governments around the world have, to differing degrees, risen to this challenge.

5.1.2 Environmental predictions

The Inter-governmental Panel on Climate Change (IPCC)[2] is a body set up to study the environment which produces data to assist international policy development on global warming. The data are collected from such organisations as the World Meteorological Organisation[3] which has air pollution stations in such remote places as Cape Grim, Tasmania, Australia, Barrow in Alaska, and Ushuaia near Cape Hope, to name but a few. These stations measure temperature, airflow, and the composition of greenhouse gases such as carbon dioxide, methane and nitrous oxide. Their data, from the purest sea air in these remote locations, shows that carbon dioxide levels have risen 10% over the last 20 years. The IPPC has also provided data from air bubbles trapped in samples brought up from undersea bore holes in the Antarctic and Greenland. It shows the amount of carbon dioxide now in the air is the highest it has been for the past 400 000 years. The effect of the increasing carbon dioxide levels is to raise the temperature of the earth. From the end of the last Ice Age, around 14 000 years ago, to the beginning of the Industrial Age, around 1800 AD, the carbon dioxide level remained constant at around 280 parts per million. It now stands at 370 parts per million and is rising. The increase is due to human beings burning fossil fuels and removing forests. (Forests act as 'sinks' which absorb carbon dioxide through photosynthesis.) The possible consequences of unrestrained global warming were outlined in a review in *The Times*[4] that is summarised in *Table 5.1.2*.

Table 5.1.2 Observations and predictions arising from global warming

Year	*Observation or prediction*
900 to 1250 AD	Medieval warm period. Vineyards in Britain.
1550 to 1750 AD	'Little Ice Age'. The River Thames in England regularly freezes with ice so thick that fires could be lit on it to roast animals at fairs.
1800 to 1900 AD	Volcanic activity cooled the earth by throwing into the atmosphere sulphur dioxide that absorbed heat.
1900 to 2000 AD	Increasing use of coal and oil for power generation raises carbon dioxide levels.
By 2050 AD	A 1°C rise in global temperature increases ocean temperatures affecting fish breeding grounds; glaciers shrink and ice caps melt; low lying areas flooded.
By 2100 AD	Increased sea levels (up to 88 cm in worst case scenario) overwhelms many islands in Indonesia and coastal cities such as New York, London and Sydney. Birds who thrive on tundra become extinct; tropical disease and agricultural pests migrate north.
By 3000 AD	Sea levels rise up to 7 metres.

But predictions are just that – predictions. There are alternative views which challenge these predictions. Some scientists believe the current warming is merely the ongoing cyclical change in the Earth's atmosphere. Others believe the model that produced the predictions is flawed. For example, it doesn't take into account cloud formation. When the Earth's surface heats up cloud cover changes in such a way that more energy is released into space.

5.1.3 Sustainable development

Amidst the uncertainty of the predictions, the trend to promote sustainable development policies is soundly based. The broad argument is that human beings must use their intelligence to find ways to improve their well-being, which includes both health and lifestyle, by living in harmony with the planet that sustains them. We should use resources wisely in a manner that safeguards the atmosphere, the oceans, and the land mass we live on. Such an objective may appear too large to be met, but the concept of the Waste Management Hierarchy in *Figure 5.1.1* is a way of achieving this.

The concept is to encourage those activities which ascend the Hierarchy. The top of the Hierarchy is 'Reduce'. This is the only option which does not use up initial resources such as raw materials and energy from fossil fuels to make the product or supply the service in the first place. In sophisticated applications the Waste Management Hierarchy is applied to all stages of a product cycle as illustrated in *Figure 5.1.2*.

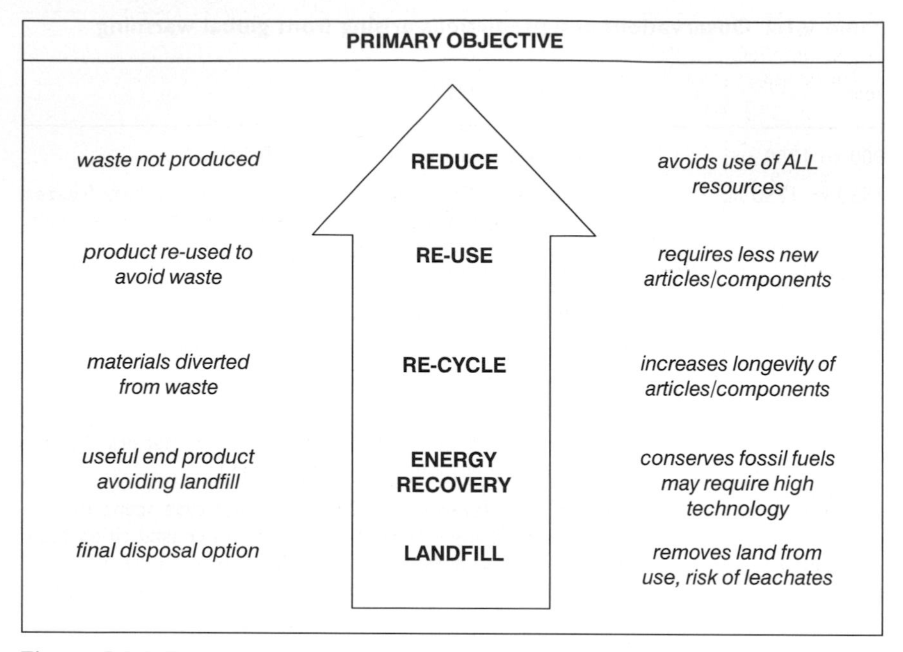

Figure 5.1.1 The Waste Management Hierarchy

The product designers are most influential in reducing the raw materials used in the product. For example, the quantity of metal used in car chassis and panel manufacture has declined as sophisticated engineering has shaped metal parts to provide strength that was once only achieved by using thicker slabs of metal. Production managers can adopt lean manufacturing techniques to cut waste in operations. Waste generated in transportation and distribution (the use of fuel, wear and tear on roads and tyres etc.) can be reduced if less distances were travelled between producer and consumer. This at once raises the economic and political aspects of environmental issues. Relocating a factory to another part of the same country, or even to another country or continent altogether, means for the original location a loss of jobs, less tax revenue and can also mean more costly goods for the consumer if the benefits of large-scale production are lost.

Sachs *et al.*[5] argue that many of today's norms can be successfully altered if there is the social and political will for change. Ideas include:

- focus on total door-to-door time from producer to consumer and not just speed of product distribution from warehouse to supermarket. Limit vehicle speed, acceleration and fuel consumption, and introduce graduated distance charges for vehicles. These factors will promote regional sourcing of products;
- encourage the greater use of rail transportation bearing in mind that a majority of people live within a few miles of a station;
- reconstruct the tax regime so that profits rise as energy consumption declines;

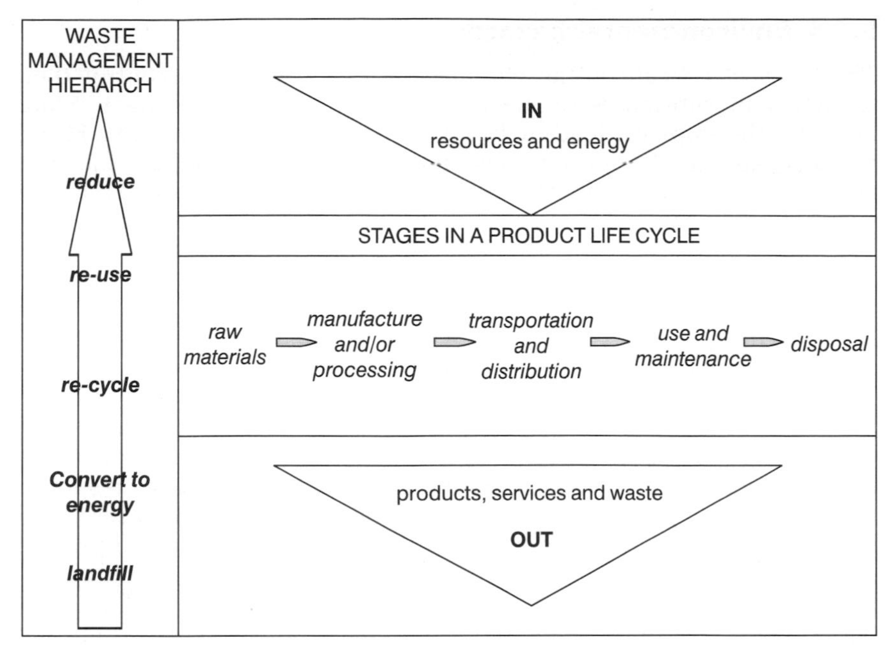

Figure 5.1.2 The product life cycle and sustainable development

- encourage new building on existing or previously used sites rather than 'greenfield' sites;
- promote healthy eating – more fruit and vegetables – which can be produced locally;
- adjusting international trade and loan criteria from satisfying primarily northern (wealthy) countries to satisfying southern (less wealthy) countries;
- shifting towards fairer trade practices with an emphasis on sustainability rather than just structural development.

These ideas are finding support. The Climate Change Levy[6] in the UK taxes high and inefficient energy users. The Contaminated Land Regulations[7] seek to identify areas of contaminated land so that those who cause it can be made to pay for its subsequent clean up. Contaminated Land is defined as '. . . land which appears to the local authority to be in such a condition, by reason of substances in, on or under the land, that significant harm is being caused or that pollution of controlled waters is, or is likely to be, caused'. Clearly there is a considerable scope for interpretation with the major difficulty of proving a 'negative', i.e. at what concentration some trace metal or non-biodegradable chlorinated solvent, for example, will not cause harm. A good review of this topic is available from Butterworth[8]. The intention is however clear – to make so-called 'brownfield' sites available for further use and lessen the need to use up virgin 'greenfield' sites.

5.1.4 Environmental hazards

The focus of concern is the elimination, or, at least, the control then reduction of substances or agents that harm the environment. More narrowly the emphasis tends to be upon harm to the human being. Harm to other fauna and flora are relevant to most people only insofar as it affects, or may affect, human beings now or in the future. Thus the simplest organisms of life are a concern because they are among the first links in the food chain that eventually supplies homo sapiens. Concern over the environment is essentially homo sapiens centred.

5.1.4.1 The appliance of science

The starting point for the identification of environmental hazards is the impact on mankind. In nearly every instance an impact has been observed only where it occurs at much higher doses than normally exist in the environment at large. Scientific evaluation of the dose–response relationship at higher levels is usually extrapolated linearly to zero. The result is to suggest that any dose above zero will cause harm and present a risk. The issue has been considered in the context of radiation by the United Nations Scientific Committee on the Effects of Atomic Radiation (UNSCEAR)[10]. It took epidemiological data on survivors from the Hiroshima and Nagasaki atomic bombs. They were irradiated with high doses and at high dose rates – equivalent to annual doses in the region of 500 to 5000 milli-sieverts (mSv). From this data UNSCEAR tried to judge the impact of nuclear weapons tests generating what are considered 'safe' doses of 0.01 mSv per year. Assuming a linearity of effect from the Hiroshima and Nagasaki data, and no threshold before any effect is incurred, they estimated a risk factor for leukemia of 0.52% per 1000 mSv. This translates into 60 000 leukemia cases worldwide. However, if there is a threshold level of 4000 mSv before leukemia is triggered, then zero cases would arise. In its conclusions UNSCEAR stated:

(1) 'Linearity has been assumed primarily for purposes of simplicity' and
(2) 'There may or may not be a threshold dose. Two possibilities of threshold and no-threshold have been retained because of the very great differences they gender.'

The quest for better scientific data may be a long way off. The difficulties of improving our knowledge is demonstrated by Wienberg's[10] example. He stated that to determine experimentally at a 95% confidence level that a 1.5 mSv dose will increase the mutation rate by 0.5%, as predicted by the linearity assumption, will require tests on 8000 million mice! At this point the predicted effects clearly transcend science. Amidst all this uncertainty, difficult decisions must still be made. Following the Chernobyl incident some 400 000 people were forcibly re-settled elsewhere in Belarus, Ukraine and Russia. They had exceeded an evacuation intervention level set by the International Commission on Radiological Protection (ICRP)[11] of a radiation dose of 70 mSv over a 70 year lifetime. However, a subsequent study by Sohrabi[12] estimated that the Chernobyl fallout in Central Europe in the first year generated an additional dose of 0.3 mSv/year compared with the average national dose estimated at 2.4 mSv per year. This data should also been seen in the context that the average lifetime dose in Norway is 365 mSv (Herrikson and Saxebol)[13], and 2000 mSv in regions of India (Sunta)[14], and the inhabitants of these regions are not relocated. The decision to relocate the Chernobyl victims may now seem to be irrational but it does demonstrate the considerable difficulties which arise when scientific knowledge reaches a frontier and situations arise when social and political decisions must be made. The example of radiation has been dealt with at length because it can be replicated to many other substances or agents which are subjected to environmental control but about which much less detailed data are available.

The health effects of lead at high levels are well known and include anaemia and alimentary symptoms. There is uncertainty about the effects at blood concentrations in the range 35 to 80 g/dl as stated in the Lawther[15] report to the Royal Commission on Environmental Pollution[16]. Nevertheless the UK Health Department recommended that blood lead levels should not exceed 25 g/dl especially in children[17].

These examples demonstrate the problems faced by regulators. The scientific basis of many decisions is uncertain. The impact of low concentrations or doses over extended periods of time on people, flora and fauna are difficult to establish. In such circumstances decisions are often made under pressure from the public or pressure groups, to adopt a *precautionary principle*.

To most practitioners who work in the day to day issues of environmental control these uncertainties are irrelevant. The decision on what constitutes an acceptable level of control for a particular substance or agent has already been made by national or international bodies. The daily task in practice is to manage the consequences. However, for some areas of activity the fact of data uncertainty is of very real concern. In the chemical business, for example, researchers develop new chemicals which have to be tested to demonstrate the point at which toxic effects occur (most chemicals are toxic at some dose rate). Once a toxic effect is observed the precautionary principle can be applied so that environmental concentrations are 10 times to 100 times below the known effect level. This becomes the *predicted no-effect concentration* (PNEC). The more toxic a chemical appears, the more sensitive the species upon which the tests are performed before the precautionary principle is applied. The Notification of New Substances Regulations[18], dealt with in another chapter, enshrines this process in law.

Existing chemicals and processes face similar problems. For example, cadmium is toxic and has been severely controlled. Silver, which is in the same family of elements, is guilty by association even though only ionic silver – which does not occur in nature (silver ions rapidly combine in water to form non-toxic chloride, oxide or sulphate salts) – is toxic.

5.1.4.2 Hazard identification in practice

For most day-to-day practical purposes governments and their agencies have listed the environmental materials and substances which require control. They fall into four broad categories:

(1) *Direct effects* on people in the community whose health and/or safety can be affected following the release of hazardous substances as a result of a significant loss of containment. This may occur through spillage, fire, explosion or a toxic gas cloud release. There is a full interaction with health and safety issues, procedures and practices in these circumstances.
(2) *Indirect effects* on people in the community whose health and safety can be impaired by a persistent low level loss of containment such as may occur if the contents of an underground storage tank leaks into an aquifer from which drinking water is abstracted. Food safety can be compromised if poor production control permits contaminants to arise. One example was the contamination of Perrier Water by traces of benzene. Indirect effects arise from pesticides used in the food chain. In today's international commercial markets produce arrives on supermarket shelves from around the world. The impact of pesticides used on fruit or vegetables in one part of the world may affect the inhabitants of another continent.
(3) *Quality of life*, which may not affect health and safety also constitutes an environmental hazard. These include the use of land, perhaps turning a green and pleasant valley alongside a motorway into a commercial business park with a loss of amenity value to local residents. Odours from factories, noise from late night bars or discos, late night or early morning landings and take-offs at airports all fall within this category. Some will argue that these activities do have a direct impact on health by raising stress levels. Perhaps more complex is the location of waste collection stations, landfill sites and incinerators. Whilst necessary, many people object to such facilities being sited next to their homes because of the increase in the size and number of vehicles they bring to the area with the consequent odours and litter. Such is the outcry on developments that the 'NIMBY' syndrome has become well known – Not In My Back Yard!
(4) The *eco-system* concerning local flora and fauna. Into this category comes the acid rain issue which arises when sulphur dioxide emissions from power stations fall as precipitates to acidify lakes causing fish deaths and affecting afforestation.

5.1.5 Evaluating environmental risks

It must be remembered that *hazard* and *risk* are not the same.

'Hazard' is the intrinsic property of the material to cause harm or damage.

'Risk' is the probability of the hazard actually causing the harm or damage and the severity or consequence of it. Some definitions add a time component. The Second Report of the Advisory Committee on Major Hazards[19] defines risk as 'the probability that a hazard may be realised at any specified level in a given span of time'. The time component adds a dimension to an assessment of an environmental risk compared to an assessment of a safety risk. For example a chemical or combustion process may have a designed life span of 30 to 40 years before being dismantled.

The risk assessment of the design may identify immediate hazards but should be extended to consider the effects of emissions, fuel leakages, etc., over the time span of the plant. An environmental risk from, for example, a leachate escaping from a toxic landfill site is likely to have an impact over many more years. These complex issues of environment risk estimation are generally considered by regulatory bodies when control limits and strategies are set by them. However, as there are often several control options available, suitable techniques must be employed to assist the decision makers in arriving at the most environmentally friendly answer.

5.1.5.1 Cost benefit analysis

The problem of dealing with environmental risks is often approached by identifying the most suitable strategy that controls the risk at least cost. *Cost benefit analysis* can be applied to this task. In fact cost benefit analysis is advocated in UK legislation[20] in the form of control systems which are the *best available techniques not entailing excessive cost* (BATNEEC).

Guidance on cost benefit analysis has been produced by a UK Government/Industry Working Group[21]. The 'costs' are those incurred by industry, government and society whereas the 'benefits' are defined as reductions in risks to health and the environment arising from regulation. The process consists of:

- identifying the dose–response data or other information that quantifies the environmental impact
- estimating the current or baseline exposure level
- calculating the harm or damage caused by the current exposure impact
- estimating the exposure following the application of the proposed regulatory controls
- calculating the remaining or residual harm or damage
- translating the harm or damage prevented into cash terms.

The process is applied to new regulations under consideration and can be applied to the choice of abatement technologies to reduce emissions. In the latter case the benefits are probably easier to identify, mainly to control emissions to below a regulated level. However, judgment and flexibility is still needed to determine the regulatory level. For example, an environmental impact is frequently caused by a total burden of material emitted from a process over a period of time. Regulatory controls, however, tend to be absolute limits which should not be exceeded even momentarily. Consequently reduced emissions arising from less production during weekends or holidays cannot be offset by above-the-limit emissions at other times even if the time weighted average remains below the control limit.

Costs are easy to quantify ... or are they? Costs of abatement technologies are not capital costs alone. Maintenance and servicing costs must be taken into account. Over a span of time they can outweigh the initial capital costs. Furthermore cash spent buying abatement technology amounts to cash that cannot be invested. These costs (opportunity costs) may be calculated and discounted cash flow techniques used to arrive at

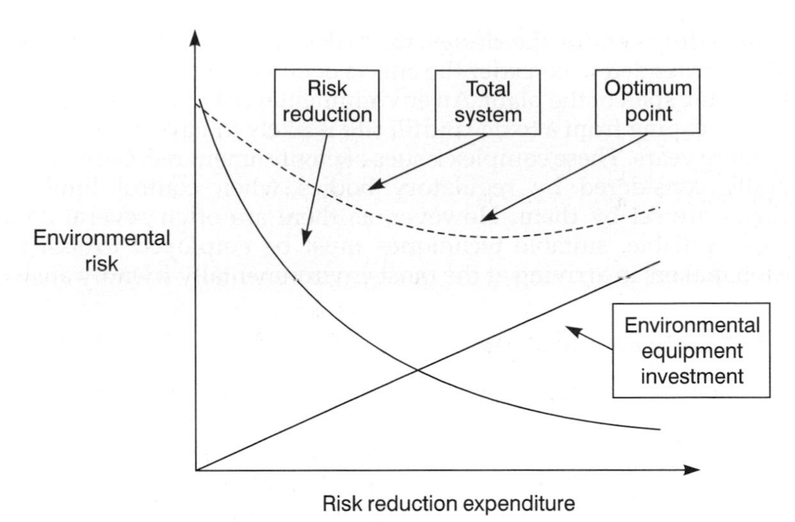

Figure 5.1.3 The cost benefit analysis model

a present day cost estimate[22]. Cost benefit analysis is a beguiling idea and is a useful tool for gaining insights into the variability of costs with benefits so that as sound a decision as possible, covering economic, practical and political issues can be reached by the regulators.

Figure 5.1.3 illustrates the basic concept of cost benefit analysis. The objective is to identify the optimum solution to an environmental risk that balances the gain (or risk reduction) against the cost.

5.1.5.2 Environmental risk perception

In all areas of risk assessment the confounding factor is *perception*. However rigorously a cost benefit analysis is undertaken to reduce the bandwidth of uncertainty, the decision to proceed in a chosen direction may be made or will certainly be influenced by how the risk is perceived. An oft-quoted study by Slovic *et al.*[23] asked four groups of people to rank in order of perceived risk 30 activities and technologies. The results from six of the elements of the study are listed in *Table 5.1.3*.

A review of this partial list demonstrates the wide range of risk perceptions between groups of people. Most striking are the differences of view that exist between experts and other groups. Experts rate 'surgery' and 'x-rays' as more threatening risks than the other groups. Risks from pesticides generate widely divergent points of view. Most fascinating is the data on power generation. Experts rate nuclear power as significantly less risky than all other groups. Nuclear power is one of the most highly regulated activities on the planet yet there is great resistance to the construction of new nuclear power stations even though many do not emit into the environment harmful gases which contribute to global warming. Paradoxically, experts rate non-nuclear electric power generation as more risky – once again differing from all other groups. Slovic identifies the reason for this difference in further work.

Table 5.1.3 Risk ranking of hazards and technologies by four groups (Slovic)

Activity or technology	*Group 1**	*Group 2**	*Group 3**	*Group 4**
Nuclear power	1	1	8	20
Alcoholic beverages	6	7	5	3
Pesticides	9	4	15	8
Surgery	10	11	9	5
Electric power	18	19	19	9
X-rays	22	17	24	7

Group 1 is the 'League of Women Voters'
Group 2 is 'Students'
Group 3 is 'Active Group Members'
Group 4 is 'Experts'

Figure 5.1.4 places the hazards and technologies in a two-by-two matrix. The vertical axis looks at the extent to which the hazard or technology is 'familiar' or 'known'. The horizontal axis considers the extent to which they are seen as local and controllable risks with acceptable consequences or, at the other end of the scale, risks with high potential consequences where there is seen to be a lack of control.

The conclusion is that the newer the risk and the less known about it, the greater is the concern. This becomes greater when the activity is perceived as having high consequences if control of it is lost and catastrophic results ensue.

Perception is shaped by the media. Sandman[24] has studied and published this aspect extensively and his conclusions include:

- The amount of coverage accorded to an environmental risk topic is not related to the seriousness of the risk in health terms. Instead, it relies on traditional journalistic criteria like timelines and human interest.
- Within individual risk stories, most of the coverage is not about the risk. It is about blame, fear, anger, and other non-technical issues such as 'outrage' rather than 'hazard'.

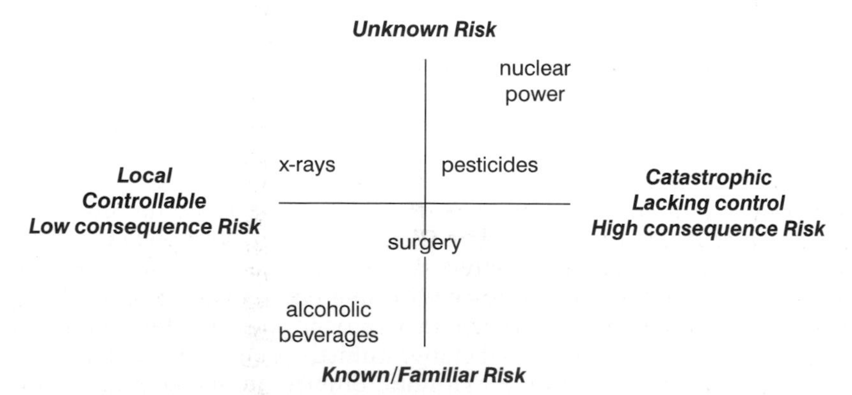

Figure 5.1.4 Risk perception matrix (Slovic)

- When technical information about risk is published in news stories it has little if any impact on the audiences.
- Alarming content about risk is more common than reassuring content or intermediate content except, perhaps, in crisis situations, where an impulse to prevent panic seems to moderate the coverage.
- Exactly what information is alarming or reassuring is very much a matter of opinion. The media audience tends to be alarmed even by information the expert world considers reassuring.

Examples of the above occur weekly if not daily in the media. Occasionally, a more reasoned media view appears. An article in the London Times on the scare that reusable trial contact lenses (after sterilisation) can lead to the spread of Creutzfeldt-Jakob Disease was entitled 'Panic is now the plague'[25]. It sought to provide a greater degree of rationality and science in reporting. Such articles are rare.

While the above looks at perception on a nation-wide level, each local or business decision also has an element of perception. There is a need to make rational, professional environmental judgements and follow up with suitable care being given to the presentation of the findings taking into account the likely viewpoint of the receptor – whether they be local company management, a local regulator, or a community group.

5.1.6 Environmental control strategies

Control strategies apply on global, regional, national and local levels. At a global level, international conferences such as the Kyoto and Rio conferences were convened to address the issue of global warming. At a regional level the European Union has produced directives with which member states must comply covering areas such as:

- Transfrontier shipments of waste
- Packaging regulations
- Regulations which encourage re-use and re-cycling by compelling businesses to take back old equipment and strip down components for re-use
- Elimination of chlorofluorocarbons.

Increasingly, European countries follow the target set by EU directives and adapt their time scales accordingly rather than go-it-alone with national policies. The EU also introduced the *Eco-management and Audit Scheme* (EMAS) to encourage companies to improve their environmental performances.

In spite of the directives and encouragement emerging from the European Commission, some member states are slow to implement directives and to rigorously apply them. This has prompted the EU to consider a directive[26] that would make it a criminal offence in the European Union to cause serious environmental damage. The discharge, emission, or introduction of a substance into air, soil, or water that causes death or serious injury to any person, animals, or plants, or damage to property, would constitute a criminal offence under the new laws. Incitement to break environmental laws would also be a criminal offence and environmental damage would be deemed a criminal offence when committed intentionally or with serious negligence.

5.1.6.1 Environment management systems

The primary environmental management system, however, is not EMAS but ISO 14000[27]. This is an international standard for which companies can seek accreditation from recognised independent auditors. Central to the standard is the requirement to undertake an *Aspects and Impacts Analysis*. This requires companies to evaluate which aspect of their operations have the greatest impact on the environment (air, water, work etc.). Those with highest impacts are expected to be tackled by the development of suitable control strategies.

For example, waste from a manufacturing operation may, at the time of the initial baseline study, be sent to landfill. The 'Aspect' is 'waste' and the 'Impact' is 'landfill' and 'cost'. It is expected that the company will then seek to ascend the Waste Management Hierarchy to reduce the Impact. The process is not a one-off exercise but is ongoing. It is expected that the entire Aspects and Impacts assessment will be done periodically, each time generating new ideas to reduce 'Impacts'. This is a continuous improvement cycle. This approach falls within the scope of the UK Government requirement to consider the Best Practical Environmental Option ('BPEO')[28] which is a concept requiring businesses to consider environmental issues in a wide rather than a narrow perspective. It encourages business to aim for the top of the Waste Management Hierarchy (i.e. not produce waste in the first instance). It encourages lateral thinking, whereas BATNEEC focuses more on end-of-pipe control.

Key to all management systems is the need to develop a company policy and the organisational and administrative arrangements necessary to implement the policy and apply the management system. The organisational aspects will include the delegation of responsibilities and the allocation of tasks. Supporting this will require an administrative process covering routine meetings with notes, action and record systems to monitor performance. The whole structure to be audited by external auditors in the case of ISO 14000 registered companies.

The process is similar to that of ISO 9000 Quality Standards[29]. A safety management system, ISO 18000[30] is being developed to follow the same structure. Because of the similarity of the procedures, these ISO Standards covering quality, environment and safety management can be integrated into one comprehensive management system.

5.1.6.2 Environment and the citizen

The thrust of environmental protection does not lie solely with political representatives and business enterprises. Individual citizens are encouraged to play a part and they are enjoined to separate household waste to permit various recycling or reuse options. Typical recycling uses are shown in *Table 5.1.4*.

Table 5.1.4 Recycling and reuse options for household waste

Waste category	*Recycle/Reuse options*
'Green waste' (grass cuttings, etc.)	Composting
Paper and cardboard waste	Repulp for low grade board and paper
Household metal goods	Re-smelt
Glass bottles	Re-melt and recast
Residual waste	Landfill

There is also encouragement to use motor vehicles less by organising car-share schemes. Some areas are developing cycling lanes to encourage less car use. If such exhortations are the 'carrot' then the 'stick' takes the form of taxation. For example, in the UK annual car tax arrangements have been changed to penalise vehicles which emit greater levels of pollution.

5.1.7 Conclusion

The environment, like the poor, is always with us. In fact the environment in one form or another will exist as long as the Earth does. The thrust of all the environmental initiatives is to permit the planet to be hospitable to human beings for centuries to come. The focus on the environment has grown significantly, decade by decade since the middle of the 20th century. Wilson[31] believes that at least 20% of the current species of animals and plants will be gone or face early extinction by 2030. These extinctions can be compared with that caused by the impact of an asteroid 65 million years ago that ended the reign of the dinosaurs. The failure arises from an inability to recognise the interdependence of life. He says that we do not appreciate, for example, the dependency of the human being on bacteria to breathe and digest. Consequently, in our ignorance we tear into the 'biospheric membrane' which covers the Earth removing, consuming or destroying that which sustains us.

It is likely that the thirst for knowledge and understanding of the environment will grow and impact on our lives and our social structures to an ever increasing amount for the foreseeable future.

References

1. United Nations population Fund, Population Issues Briefing kit 2001, www.unfpa.org/modules/briefkit
2. Inter-governmental Panel on Climate Change, IPCC Secretariat, Geneva, Switzerland *www.ipcc.ch/about*
3. World Meteorological Organisation, Geneva, Switzerland
4. *The Times*, Monday 9 July 2001, London
5. Sachs, W., Loske, R. and Linz, M. (eds) *Greening the North: a Post-Industrial Blueprint of Ecology and Equity*, Zed Books Ltd. www.zedbooks.demon.co.uk/home.htm
6. HM Customs and Excise, Notice CCL1, *A general guide to the climate change levy*, The Stationery Office, London (2002) *www.hmce.gov.uk/forms/notices/ccl1*
7. *The Contaminated Land (England) Regulations 2000*, The Stationery Office, London (2000)

8. Butterworth, J., Contaminated Land, *The Safety and Health Practitioner*, March 2000, vol 18 No.3 pp. 50–53
9. United Nations, *Report of the United Nations Scientific Committee on the Effects of Atomic Radiation*, United Nations, General Assembly Official Records: Thirteenth Session, Supplement No.17 (A/3838) Geneva (1958)
10. Weinberg, A.M., *Science and Trans-Science*, 10, 209–222, Minerva, Pasadena (1972)
11. International Commission on Radiological Protection, Publication No.1 *Recommendation of the International Commission on Radiation Protection*, (1959) and Publication No. 40 *Protection of the Public in the Event of Major Radiation Accidents: Principles for Planning*, (1984), Pergamon Press, Oxford
12. Sohrabi, M., Recent radiological studies of high level natural radiation areas of Ramser. Proc. International Conference on High Levels of Natural Radiation, pp. 39–45, IAEA, Vienna (1990)
13. Henrikson, T. and Saxebol, G., Fallout and radiation doses in Norway after the Chernobyl accident. In Z. Jaworowski *Chernobyl Accident: Regional and Global Impacts*, Environment International, Special Issue, 14(2), 157–163 (1988)
14. Sunta, C.M., A review of the studies of the high background areas of the S-W coast of India. Proc. International Conference on High Levels of Natural Radiation, pp. 71–86, IAEA, Vienna (1990)
15. Lawther, *Lead and Health: the Report of a DHSS working party on levels of lead in the environment*, The Stationery Office, London (1980)
16. Royal Commission on Environmental Pollution, *Lead in the Environment*, Cmnd 8852, The Stationery Office, London (1983)
17. Department of the Environment, Circular no: 22/82, *Lead in the Environment*, The Stationery Office, London (1982)
18. *The Notification of New Substances Regulations 1993*, The Stationery Office, London (1993)
19. Major Hazards Committee, *Second Report of Advisory Committee on Major Hazards*, The Stationery Office, London (1979)
20. *The Control of Pollution Act 1990*, The Stationery Office, London (1990)
21. Gramlich, E.M., *A Guide to Cost-Benefit Analysis*, Waveeland Press, Prospect Heights, USA
22. Channing, J., *The Cost of Control Measures in Safety*, University of Aston in Birmingham, M.Sc. Thesis (1980)
23. Slovic, P., Fischoff, B. and Lichtenstein, S., Perceived risk: psychological factors and social implications, *Proceedings of the Royal Society*, A376, 17–34, London (1981)
24. Sandman, P. M., *Mass Media and Environmental Risk: Seven Principles*, Summary on www.fplc.edu/RISK/vol5/summer/sandman
25. Jenkins, S., Panic is now the Plague, *The Times*, 25 June 1999, London
26. Europe/Middle East, EU mulls over criminal sanctions for Eco-Crimes, *Chemical Week*, Business and Finance News, 17 April 2002
27. British Standards Institution, Generic series of standards on environmental management systems:
    BS ISO 14000: *Environmental management systems;*
    BS ISO 14001: *Environmental management systems. Specification with guidance for use*
    BS ISO 14004: *Environmental management systems. General guidelines, principles and supporting Techniques*, BSI, London
28. Environment Agency, Technical Guidance Note E1: *Best Practical Environmental Option – Assessments for Integrated Pollution Control*, The Stationery Office, London (1997)
29. British Standards Institution, Generic series of standards on quality management systems:
    BS ISO 9000, *Quality Management Systems*
    BS ISO 9001, *Quality Management Systems. Requirements*
    BS ISO 9004, *Quality Management Systems. Guidelines for performance improvement*. BSI, London
30. British Standards Institution:
    BS ISO 18001, *Occupational Health and Safety Management Systems. Specification*
    BS ISO 18002, *Occupational Health and Safety Management Systems. Guidelines for the implementation of ISO 18001*, BSI, London
31. Wilson, E.O., *The Future of Life*, Knopf, London

Chapter 5.2

Environmental management systems

J. E. Channing

5.2.1 Introduction

Controlling environmental risks effectively requires a management system. Two systems have been developed and both emerged during the 1990s. The European Union put forward its Eco-management and Audit System (EMAS) in 1993 and it was subsequently amended in 2001[1]. This system has found favour in some countries such as Germany where more companies are registered by EMAS than by the alternative system. Other countries in the European Union, such as the UK, and countries on other continents have preferred the alternative system, ISO 14001[2] – the system developed by the International Standards Organisation from British Standard BS 7750. ISO 14001 was available in 1996 and BS 7750 was withdrawn in 1997. There is considerable similarity between the two systems. Registration under EMAS can be achieved if ISO14001 is chosen as the environmental management system and certain additional requirements are satisfied.

The benefits of an environmental management system to an organisation are:

- It focuses the organisation on wastes generated by its operations. The term 'wastes' has a wide interpretation and includes the excessive use of water and energy, and solid waste disposal costs. Opportunities to reduce costs by reducing these waste streams can be identified by the application of such a system.
- Registration under ISO 14001 or EMAS is becoming a necessity for doing business. More governments and companies are requiring their suppliers achieve registration.
- Shareholders in publicly quoted companies require to be satisfied that there are no significant environmental liabilities that may affect the share price.
- Communities that live in the vicinity of an organisation's premises gain confidence in an organisation's operations because the registration under an environmental system requires external verification.

An organisation needs to choose which system to adopt. ISO 14001 has the following benefits:

- It is the more commonly used system worldwide.
- It more readily aligns with the international quality standard ISO 9000[3] and the forthcoming international safety standard currently available as OHSAS 18001[4]. This alignment allows a company management to manage quality, safety and environmental issues under one process. Chapter 2.6 has demonstrated how these systems align. Trade associations such as the Chemical Industries Association[5] have also linked them under one management process for chemical businesses.
- EMAS registration can be achieved after ISO registration is gained by undertaking a few additional requirements.

EMAS has an explicit requirement to actively involve employees (whereas the ISO 14001 requirement extends only to train employees) and also to place the organisation's environmental performance within the public arena. Some organisations may wish to have experience of and gains from using ISO 14001 internally before making the data public.

However, it may be advantageous for data to be public if expansion on the premises requires public discussion during the planning application process. Public knowledge of environmental performance may assist to allay fears about the adequacy of the company's ability to control site risks. *Table 5.2.1* shows the alignment between the two systems (giving the relevant section number in brackets).

An essential common feature of registration under both of these schemes is the regular checking of compliance by an independent external environmental management systems auditor. Where non-compliance is found, these auditors have authority to require that changes are made to the environmental management system to bring it into line with the standard's requirements.

5.2.2 Establishing an environmental management system

The requirements of environmental management systems are to establish and then maintain arrangements that cover all the relevant environmental issues, set in place suitable measures to control them, and get them verified by an external competent body. Six aspects to be covered in establishing an environmental management system are considered below.

5.2.2.1 Environmental policy

Top management of the organisation need to define the environmental policy so that it:

- Encompasses the nature, scale and environmental impacts of its activities, products and services. The term 'environmental impact'

Table 5.2.1 Alignment of EMAS and ISO 14001

EMAS: 2001 *Registration requirements*	*ISO 14001: 1996* *Registration requirements*
General requirements (I-A.1) Environmental policy (I-A.2) Planning (I-A.3) – environmental aspects – legal and other requirements – objectives and targets – environmental management programmes	General requirements (4.2.1 part) Environmental policy (4.2) Planning (4.3) – environmental aspects – legal and other requirements – objectives and targets – environmental management programmes
Implementation and operation (I-A.4) – structure and responsibility – training, awareness and competence – communication – environmental management system documentation – document control – operational control – emergency preparedness and response	Implementation and operation (4.4) – structure and responsibility – training, awareness and competence – communication – environmental management system documentation – document control – operational control – emergency preparedness and response
Checking and corrective action (I-A.5) – monitoring and measurement – non-conformance and corrective and preventive action – records – environmental management system audit	Checking and corrective action (4.5) – monitoring and measurement – non-conformance and corrective and preventive action – records – environmental management system audit
Management review (I-A.6)	Management review (4.1.3)
Environmental statement for public and stakeholders (III-A) – annual publication of data – data to be verified by external body – local data for each participating site	No equivalent
External communications and relations (I-B.3)	No equivalent
Employee involvement (I-B.4)	Not a condition for registration
Environmental logo (Article.8)	No equivalent

means the organisation must understand the change to the environment caused by what it does or produces. Changes that are beneficial to the environment should be recognised, but the major thrust of the system is, obviously, to control adverse impacts.

- Commits the organisation to a cycle of continuous improvement and prevention of pollution.

- Commits the organisation to comply with environmental legislation.
- Establishes a process for setting and reviewing environmental objectives and targets.

The policy must not simply be 'window dressing'. It must be properly documented and updated. It should be signed by the Chief Executive, state how it will be implemented and made available to all employees. There is implicit recognition that adverse environmental impacts cannot be corrected immediately. There may be constraints on the cash to fund improvements or the technology may not have developed sufficiently properly to control the environmental impacts. Consequently it is recognised that organisations can align the policy with its business goals. However, great emphasis is placed upon the benefits to be gained from an environmentally conscious organisation, notably lower costs, fewer fines, and greater acceptance of the organisation and its products in the community and the marketplace. Consequently, the Environmental Policy is expected to shape and influence business goals rather than simply recognise and comply with them.

5.2.2.2 Planning

The planning process is a key component of the environmental management system. Four elements are necessary.

1. *Environmental aspects and impacts* affected by the activities, products or services of the organisation need to be identified. The term 'environmental aspect' means understanding what environmental media (or 'aspect') can be affected.

The major direct 'aspects' are:

- emissions to air which range from the flue gases from industrial boilers to aerosol propellants from hair sprays
- releases to water which include regulated discharges to public sewers and storm water run-off which may go into freshwater streams
- waste control and reduction which includes defined hazardous wastes and non-hazardous wastes
- use of natural resources and raw materials including energy
- effects on biodiversity which include, for example, the extraction of mineral deposits from sites where rare creatures inhabit
- community issues which include noise, transportation, and emergency issues.

Indirect environmental aspects can also arise. These refer to those which management may not have total control over and may include the choice of freight routes that are more circuitous to avoid troubled regions of the world. Another example arises from the choice of contractors or suppliers and, in turn, their subcontractors. The environmental practices they follow become more difficult to monitor and influence as the chain extends.

The term 'environmental impacts' essentially refers to *significance*. An organisation's activities may involve the generation of a material that is discharged to water (i.e. the *aspect* of the environment affected is *water*) but its *impact* may not be significant. An example could be the washings from glasses in a bar containing residual amounts of alcohol. Not only are the volumes of alcohol small but also the alcohol is quickly degraded in the treatment works. Whether an impact is of high or low significance is to some extent affected by the opinion of the various stakeholders. A waste material that might be seen as having a low environmental impact to the producer may be seen differently by the householder who lives adjacent to the landfill site.

Essentially an *Environmental Aspects and Impacts Analysis* is a risk assessment taking into account the issue of perception. They can be given a mathematical ranking in the same way as a safety risk assessment. *Table 5.2.2* provides an example in which *impacts* are considered by hazard or severity to the environment (ranked 3 = high, 2 = medium, and 1 = low) and adequacy of existing control systems or technologies (ranked 3 = poor control, 2 = medium control and 1 = good control). The relative environmental impact or significance is the product of these two numbers.

In the example shown, the high relative impact figures draw attention to the consequences of using oil-based paints that produce air pollution,

Table 5.2.2 Example of environmental aspects and impacts analysis

Aspect	*Material, quantity and source*	*Potential consequence*	*Hazard or severity ranking*	*Control ranking*	*Relative impact ranking*
Air emission	1. Low amounts of CO_2 from boiler	Global warming	2	1	2
	2. Large amounts of paint solvents	Photochemical smog	3	3	9
Water release	1. Residual solvents to public sewer	Sewage treatment	1	1	1
	2. Site spillages to local stream	Damage to aquatic life	3	2	6
Waste	1. Rags, wipes and filters from paint booths	Landfill contamination	3	1	3
Natural resources	1. Oil-based paints	Resource use	3	2	6
	2. Electricity	Resource use	2	1	2
Community issues	1. Noise from spray generator	Quality of life	2	2	4

deplete natural resources and can damage the local stream if uncontrolled spillage occurs. This should prompt the organisation to seek alternative materials such as aqueous-based paints or invest in better control technology.

An analysis, similar to that illustrated in *Table 5.2.2* should be undertaken at each site which should produce its own ranking through its own assessment and be internally consistent for other risks on its site. The process can be applied to multi-site organisations providing the organisation as a whole with a comprehensive environmental impact ranking. This can assist in prioritising capital spending over all of its sites. However, since assessments tend to be by individual site, it may mean that one site shows little continuous improvement because capital has been directed to a more needy site. The ranking process is subjective and does not allow for comparison with sites from other organisations. Consequently, one site may rank its own risks highly, while another with similar hazards may rank them much lower. Since the discharges from all sites affect the same environment (which is one way that environment issues are markedly different from safety issues) the sensible course would be to channel capital to where the best overall environmental impact can be achieved. Environmental management systems can only address this as a local company issue. Global impacts are matters for political decisions by governments.

2. *Legal and other requirements* are the second element that an organisation must take into account during the planning stage. It must have in place a process to keep it up to date on the environmental regulations that apply to its activities. If it belongs to a trade association, it should also be aware of any requirements placed upon it by virtue of its membership. In some cases municipal authorities establish local guidelines and practices that extend beyond regulative demands in order to set a good example in the community and to promote environmental awareness. They may, for example, require a contractor to have a published environmental policy prior to considering a tender bid from them.

3. *Objectives and targets* must be set as part of the planning process. In setting these the organisation must take into account shortcomings in its compliance with statutory requirements and begin to remedy them. It should also consider the results of its Aspects and Impacts Analysis and use the results as criteria for setting priorities and targets. As far as possible targets should be:

- Specific – the end point and intermediate stages should be set out in detail.
- Measurable – quantifiable measures should be established where possible.
- Agreed – with the staff responsible for achieving them.
- Realistic – in terms of being achievable. To set impossible goals demotivates staff.
- Programmed – clear time scales should be set for achieving each stage.

Objectives and targets can take the forms of:

- *Improvement Objectives* that are established to make specific defined improvements. They can be responding to high risks identified by the Aspects and Impacts Analysis or to a recognised lack of legal compliance.
- *Monitoring Objectives* relate to those risks that rank less highly but need to be kept under continual observation. An example may be the monitoring of discharges under a 'Consent to Discharge' document that allows wastewater to flow through public sewers to treatment plants. Ongoing monitoring of the effluent stream may be a condition of the consent but it also allows the organisation to understand what it is discharging and the costs incurred. As the Improvement Objectives are achieved an Effluent Monitoring Objective may be 'promoted' and become the focus of a future project to reduce waste and cost.
- *Management Objectives* are of a more overarching nature and focus on educating employees, contractors or suppliers and on the organisation's environmental objectives. To achieve this may require conducting surveys into the use and waste of energy and product and to developing guidelines on environmental matters for distribution among the organisation's own staff or component suppliers.

4. A *programme for the management of the environmental matters* is the final component in the overall planning process. It is the means by which the plan is held together and should include the designation of responsibility for each aspect of the plan, specify the means for achieving it and state the arrangements for releasing the necessary financial resources.

5.2.2.3 Implementation and operation

Once an environmental management system has been agreed, it needs to be implemented and arrangements put in place to ensure its continuing operation. These aspects can conveniently be split into seven components: structure and responsibility; training, awareness and competence; communication; environmental management system documentation; document control; operational control and emergency preparedness and response.

1. The *structure and responsibility* for environmental management must integrate with the normal organisational arrangements with specific environmental roles being allocated and added to existing job specifications. In this way achieving environmental objectives can become a normal part of the annual review of the job holders performance. For effective day-to-day operation, the arrangements should encompass the establishment of a meeting structure to review the progress of the system. This could be incorporated into the normal management meetings or into meetings of the safety committee. Supporting organisatioinal functions such as purchasing strategies, product development and transportation strategies, etc. should be included so that an activity that is beneficial to the environment is embedded into the procedural fabric of the organisation. A director or senior manager should be appointed to co-ordinate

environmental management activities and to report progress to the board. This is required by both the ISO and EMAS standards.

2. *A successful environmental management system will include training, awareness and the development of competence* in those whose work has a significant impact upon the environment. The training should cover:

- an awareness of the importance of conforming with those specific aspects of the environmental system over which the individual has control;
- the effect that their task will have upon the environment;
- how their roles fit into the overall system and how they should respond if an emergency arises in their areas;
- the importance of carrying out their task properly and efficiently, highlighting the consequences for the environment of a failure to do so.

Training should be provided throughout the organisation including the operator who takes samples of site discharges, the scientist who develops new organic molecules and the administrator who purchases components. Some degree of monitoring of the task as it being performed may be with recording of the observations so there is evidence to satisfy the auditor. However, care must be exercised to ensure the system does not become bureaucratic. Getting the balance right just needs a little careful thought.

Inevitably this step requires good *communications* to be effective. Communication extends to regular updates on performance to managers and workgroups to maintain impetus and focus.

3. The *environmental management system documentation* may be kept either in paper or electronic form but it must describe fully the organisation's system and should be such as to enable a documentation trail to be followed by an auditor. To avoid this process becoming bureaucratic it can be integrated into the quality system. Even so, careful thought must be given to ensure the creation of only those documents that are necessary and can subsequently be managed. Both EMAS and ISO 14001 have specific requirements on the *documentation control* in that they must be:

- Reviewed, revised when necessary and approved by an authorised person. In practice, even if no changes are necessary to an existing document, they should be reviewed and re-authorised every two to three years to show that the system is alive and healthy and has been monitored.
- Relevant and up-to-date for existing operations. In multi-site organisations key documents must be available on each site, especially for EMAS registration which is a site-based process. ISO 14001 does allow a single registration over several sites but as auditors still attend individual sites, the central documentation must still be available. This is facilitated if electronic documentation is used. However, there is one side effect; since auditors can challenge individual sites with regard to their application of the requirements of the company's corporate documents. This may occasion changes to the central document. The revised central document with its additional provisions is then applied

to all the company's sites. This can have far reaching implications for international companies where the net effect can be to 'ratchet up' the requirements for all of their sites as a result of a local auditor's opinion of the shortcomings of one site.

- Obsolete documents must be removed or replaced as soon as they are no longer relevant. Removed or replaced documents should be retained on file as evidence that this requirement has been complied with. Some obsolete documents may need to be retained for legal or record purposes to explain why certain courses of action were followed in the past. For example, if a new waste contractor is used, the documentation relating to the previous waste contractor must meet the requirements of the Duty of Care Regulations[6]. It will also assist new supervisors on the site to understand what they should examine when they perform *Duty of Care* audits on the waste contractor.

4. *Operational control* refers to the day-to-day activities that are necessary to achieve the environmental objectives contained within the environmental policy of the organisation. It should ensure that operating procedures adequately control the environmental risks. A test that can be applied is *'if there is no procedure in place, is a deviation from the policy objectives likely to occur?'*. This also provides a check on the *quality* of the operating procedure. Such procedures should extend to maintenance activities, service activities on a customer's premises, and contractors' activities on site and the range of raw materials and finished goods held on site. Training is essential to ensure competence in achieving proper operational control.

5. The final component under this part requires that an organisation has in place arrangements for *emergency preparedness and response*. The arrangements should be tested regularly and the fact recorded. Many organisations carry out practice fire drills for personnel safety and similar procedures and practices are required for environmental incidents. These may include practice drills to test the effectiveness of the response to significant spillages of chemicals to surface waters or to the malfunctioning of air emission control equipment requiring that a process be shut down quickly and safely before further materials are emitted to the environment. Environmental emergencies can extend beyond the site fence to include spillage occurring during the transportation of goods to a client's premises, or the removal of suspect materials from a client by a sales representative. An organisation that has a culture of environmental responsibility can be identified by how rigorously it examines its operations to identify the circumstances that can lead to an environmental emergency.

5.2.2.4 Checking and corrective action

To ensure an environmental management system continues to meet the policy objectives, regular checking of its effectiveness is necessary and, where found wanting, suitable corrective action should be initiated.

There are four aspects that contribute to the maintenance of environmental standards: monitoring and measurement; corrective and preventive action for non-compliance; records and an environmental management system audit.

1. The *monitoring and measuring* of the action and the degree to which they give compliance with the standards demonstrates that the identified significant environmental impacts are controlled. Control can be achieved through 'end-of-pipe' abatement methods, as in the case of efforts to prevent excess airborne emissions escaping from a stack, when records of the stack control parameters (temperature and air flow velocity, etc.) must be kept. Different monitoring and measuring strategies apply to waste management where separate records of waste weights or volumes and source department should be maintained. The objective of the system is to provide management with data so that any necessary remedial or improvement measures can be set in motion to reduce the environmental impacts. Data that is collected should be collated and retained for later reference, either as evidence of compliance or as guidance on future actions. Where monitoring identifies shortcomings in the system itself, there should be a procedure for initiating changes to the system.

2. Where non-compliance or inadequate performance is revealed by monitoring there should be in place arrangements to take corrective and preventive action for non-compliance and to ensure that the issues are addressed properly. For example, if a new operator has not been trained in the organisation's waste segregation procedures six months after starting the job, the arrangements should initiate actions to ensure training is provided as early as possible. The non-conformance observed and the corrective actions taken should be recorded.

3. *Records* are the lifeblood of any organisation aiming to improve its performance. Records relevant to an environmental management system include compliance checks, remedial action taken, training and competency, equipment and system maintenance records, the results of audits, meeting records, responsibilities, etc. While some of the records may originate from other facets of the company's operations, a copy should be kept in the environmental management file. The records do not need to be hard copies and may be kept electronically provided that they are readily available for inspection when required and that they are 'read only' so that they cannot be interfered with.

4. Finally the organisation must establish and maintain a programme for periodic *environmental management systems audits*. These are internal audits to establish that:

- the system conforms to the requirements of the standard;
- procedures are properly implemented and maintained;
- information on performance is fed back to the managers responsible.

These audits are conducted by members from the organisation itself who have been trained and are competent in audit processes. While quality auditors and other non-environmental specialists can audit many aspects of the environmental system, such as record keeping, training programmes etc., it usually requires a specialist to review the Aspects and Impacts Analysis because of the need to make professional judgements on environmental impacts.. The audit process should be arranged so that all aspects of the system and all the activities of the organisation are covered and may be carried out in a series of audits known as an audit cycle.

Both EMAS and ISO 14001 require external auditors to examine an organisation before registration is granted. Thereafter, they must audit annually in order to maintain registration. External auditors will check the internal audit process to establish that the organisation seriously accepts its obligations and responsibilities towards the environment and devotes resources to it. They will use the internal audit reports as a means to test the health of the environmental management system and to indicate how competent the organisation is to recognise and control its environmental impacts. The internal audit system goes to the heart of the organisation's competence in environmental management. Consequently, emphasis is put on the quality of the internal audit arrangements.

The internal audit process, which should be documented, should cover:

- the objectives of the process;
- the scope of the internal audits in terms of subject areas to be addressed, activities, geographical areas, and the environmental criteria;
- the organisational resources necessary to establish and maintain an internal audit process including the identification of competent auditors, their availability, their degree of independence and the support they enjoy from senior managers;
- the planning and preparation for audits including sequence, duration and time scales;
- a specification of the activities covered by the audit itself and the granting of authority to examine all appropriate records necessary to test the scope and robustness of the system;
- a procedure for reporting to senior managers the findings of each audit and a list, including a list of corrective actions needed;
- the means by which internal audits are followed up and corrective actions initiated;
- the frequency of internal audits.

5.2.2.5 Management review

The final element in the system is an obligation on senior managers to review the whole system periodically. The period between reviews is flexible and will depend on the extent of the organisation's environmental impacts. An organisation that deals extensively with chemicals may need to review at least quarterly, if not monthly, whereas a dry goods transportation company could opt for half-yearly reviews. The purpose of the senior manager's review is to ensure that the system is working, that it complies with environmental standards, and that the targets that have been set are being met. Where the review highlights the need for changes in the system itself or the resources that need to be allocated, immediate authorisation can be given. Relevant records from other aspects of the company's operations should be considered and their environmental importance assessed.

5.2.2.6 Links with other management processes

Both EMAS and ISO 14001 recognise the advantages of integrating the environmental management system with other larger management systems. A popular option is to place the environmental management system with an established quality system. However, there is a significant difference in that the existing quality standards do not make *continuous improvement* a requirement whereas both environmental standards do. It is essential that this vital aspect is put in place if integration of environmental with quality standards is considered.

5.2.3 Additional EMAS requirements

Table 5.2.1 shows the additional requirements over the ISO standard that are demanded for registration under EMAS. These include the production of an *environmental statement* which must be made available for inspection by the public and other interested parties, and give information on the environmental impact and performance of the activities undertaken by the organisation. It should also demonstrate ongoing efforts to improve environmental performance. The statement should be seen as a tool to promote a public dialogue and should contain a comprehensive summary that addresses all relevant issues. It must be validated by a named external environmental verifier.

The organisation must also be able to demonstrate that it has effective communications links with members of the public and other interested parties. The organisation must be proactive in disseminating the information in writing and through arranging public meetings with community members, local media and other interested parties. EMAS further requires that arrangements are in place that actively promote employee involvement. These can be in the form of suggestion schemes, employee participation in committees and projects on environment matters and any other form of participation.

An organisation that attains EMAS registration gains the right to use the EMAS logo on its letterheads and in its advertising, but not on its product packaging.

5.2.4 Conclusions

The two environmental management systems, ISO 14001 and EMAS provide comprehensive frameworks that enable organisations to contribute to improving the environment through controls on their site activities and the services and products they supply. It also provides them with means to spur their contractors to become registered under ISO 14001 or EMAS. The standards are designed to integrate with the established quality system and are compatible with safety management systems. The key element that makes them effective in addressing the environmental issues is the need for continuous improvement coupled with an Aspects and Impacts Analysis. This means that no matter how good an organisation is, if it uses energy for its manufacturing process, packaging for its products, if it produces waste or uses transport to provide a service, all of which have an environmental impact, it will face an ongoing challenge to improve for the benefit of the environment.

References

1. European Union, *Regulation No 761/2001 of the European Parliament and of the Council of 19 March 2001 allowing voluntary participation by organisations in a Community eco-management and audit scheme (EMAS)*, European Union, Luxembourg (2001)
2. British Standards Institution, BS EN ISO 14001, *Environmental Management Systems, Specification with guidance for use*, BSI, London (1996)
3. British Standards Institution, BS EN ISO 9000, *Quality Systems, Specification for the design/ development, production, installation and servicing*, BSI, London (1994)
4. British Standards Institution, OHSAS 18002: 2000, *Occupational health and safety management systems*, BSI, London
5. Chemical Industries Association, *Responsible Care Management Systems – Guidance*, Chemical Industries Association, London (1998)
6. *Environmental Protection (Duty of Care) Regulations 1991*, The Stationery Office, London (1991)

Chapter 5.3

Waste management

Samantha Moss

5.3.1 Introduction

In prehistoric times, waste was mainly composed of ash from fires, wood, bones and vegetable or bodily waste. It was disposed of into the ground, where it would act as a compost and help improve the soil. Waste began to be a problem as the transition came from hunter-gatherer to farmer. The industrial revolution led to a massive population shift from rural areas to the city between 1750 and 1850 in the UK. The growing population living in towns led to an increase in the volume of domestic waste arising, which was matched by the production of industrial waste from new large-scale manufacturing processes. As city populations expanded, space for disposal decreased and societies began to develop waste management systems.

Prior to 1972, there were few controls on waste disposal. The *power* to inspect first appears in the Public Health Act 1848 and the *duty* to do so in the Sanitary Act 1866. These powers and duties could not prevent the nuisance arising from wastes but at least they should have ensured there were no unknown major toxic waste deposits. In 1875 the Public Health Act charged local authorities with the duty to arrange the removal and disposal of waste and signified the start of significant local authority power. This Act also ruled that householders had to keep their waste in a 'movable receptacle' (the early dustbin) which local authorities were required to empty every week.

Most waste in the UK had been sent to landfill, a practice that had little impact until concern arose over uncontrolled dumping of toxic waste forced the introduction of the Deposit of Poisonous Waste Act 1972[1]. However, growing concern about the environmental effects of waste in the 1960s led the Government to set up two working groups. The resulting reports paved the way for the Control of Pollution Act 1974[2] which aimed for much wider control on waste disposal and regulation of sites.

Several provisions from the Control of Pollution Act have been transferred to the Environmental Protection Act 1990[3]. Part 2 of this Act introduced a new statutory regime for the control of waste, introducing stricter controls and placing responsibility (the 'Duty of Care') on waste producers and persons who may handle waste. More recently, the Environment Act 1995[4] repealed a number of areas of the 1990 Act and from both these key pieces of legislation, a number of UK regulations have been enacted to address the implementation of good waste management practices.

5.3.2 Waste authorities

The Waste Disposal Authorities continue to exist mostly as County Councils in England, District Councils in Wales and Islands or District Councils in Scotland. These bodies do not undertake disposal themselves but will contract this service to private operators or Local Authority Waste Disposal Companies.

The Environment Agencies of England and Wales (EA) and the Environment Protection Agency of Scotland (SEPA) were established by the Environment Act 1995 and now have responsibility for waste regulation.

Together they manage and monitor the environment through the enforcement of a full range of environmental legislation dealing with site-based emissions, and increasingly seeking to influence the controls, checks and balances that are evolving with emissions that arise from the use of products.

5.3.2.1 Regulatory waste management responsibilities of the environment agencies

The key aspects of legislation from which the agencies with responsibility for environmental protection derive their waste management duties and powers are:

- Control of Pollution (Amendment) Act 1989[5]
- Controlled Waste (Registration of Carriers and Seizure of Vehicles) Regulations 1991 and amendment 1998[6]
- Waste Management Licensing Regulations 1994 and amendments in 1995, 1996, 1997, and 1998[7]
- Transfrontier Shipment of Waste Regulations 1994[8]
- Environment Act 1995[4].

In addition, the Environment Agency has taken on new responsibilities since it was established in 1996 as a result of the following legislative provisions:

- The Special Waste Regulations 1996 and amendments in 1996 and 1997[9]
- Finance Act 1996 and Landfill Tax Regulations 1996 and amendments in 1996, 1998, 1999 and 2002[10]

- Producer Responsibility Obligations (Packaging Waste) Regulations 1997 and amendments in 1999 and 2000[11].

5.3.3 National waste strategies

The waste management industry faces a tremendous challenge in the 21st century as it seeks to develop and implement sustainable waste management practices. Waste managers need to ensure that their activities conform to a range of government policies reflecting both EU and UK legislation and national waste strategies including the Government's vision on sustainability. This vision is based on four broad objectives:

- social progress which recognises the needs of everyone;
- effective protection of the environment;
- prudent use of natural resources; and
- maintenance of high and stable levels of economic growth and employment.

Waste has two main impacts on sustainability: first, the amount of waste produced is a consequence of how efficiently resources to produce goods are used and of the quantity of goods produced and consumed and, second, once waste has been produced, the aim should be to manage the waste to minimise the impact on the environment.

The waste strategies for England, Scotland, Wales and Northern Ireland, seek to ensure that waste management plays a major role in the search for increased sustainability. They focus on measures to achieve waste minimisation and re-use and for the management of wastes to be carried out in an environmentally responsible way. Under the Kyoto Protocol, the Government has agreed to legally binding targets to reduce greenhouse gas emissions to 12.5% below 1990 levels over the period 2008 to 2012 and has a domestic aim to reduce CO_2 emissions by 20% by 2010.

Waste managers need to strive to work within a sustainable waste management framework, with regulatory regimes that prescribe the rules for environmentally responsible operations and practices. To achieve this task it is necessary to look forward and consider the impacts that these products have during their use downstream.

5.3.3.1 Waste strategy for England and Wales 2000

This strategy was published as a requirement of the Environmental Act 1995, and describes the vision for sustainable waste management by preparing a strategy for the recovery and disposal of waste. The strategy clearly identifies landfill as the least desirable option to tackle waste and identified an urgent need to stimulate waste minimisation, recycling and greater resource efficiency to reduce amounts of waste disposed to landfill. One of the key targets in this strategy is that, by 2005, the amount of commercial and industrial waste sent to landfill is reduced to 85% of the 1998 level. This is to be achieved by focusing on recovering value, reducing environmental impacts and addressing growth in wastes. Based on figures from the National Waste Production Survey[12] (commissioned by the Environment Agency in 1998 and covering 20 000 businesses in England and Wales) this translates into a reduction of around 5 million tonnes of waste going to landfill per year. In addition targets have been set that increase the amount of municipal waste that is recycled, aiming to recycle or compost at least 25% of household waste by 2005, 30% by 2010 and 33% by 2015.

5.3.3.2 Waste strategy for Scotland 1999

This strategy, which was published by SEPA in 1999[13], sets out the system for sustainable waste management through the implementation of several key tools including statutory functions, education, economic instruments and focused waste management research and development.

5.3.4 Defining waste

The first key task is to ascertain whether or not the material in question is actually a waste according to the legal definition before any attempt is made to classify it appropriately. This is a complex process that begins with the definition of waste which is given in European Waste Framework Directive[14] and with detailed guidance given by the Department of the Environment in Joint Circular 11/94[15]. The main questions to be considered are outlined in *Figure 5.3.1* and should assist in reaching a view on the status of the material.

It is the responsibility of the waste producer to decide whether the substance or object in his possession is waste. This decision is not always straightforward and final interpretation may be a matter for the courts.

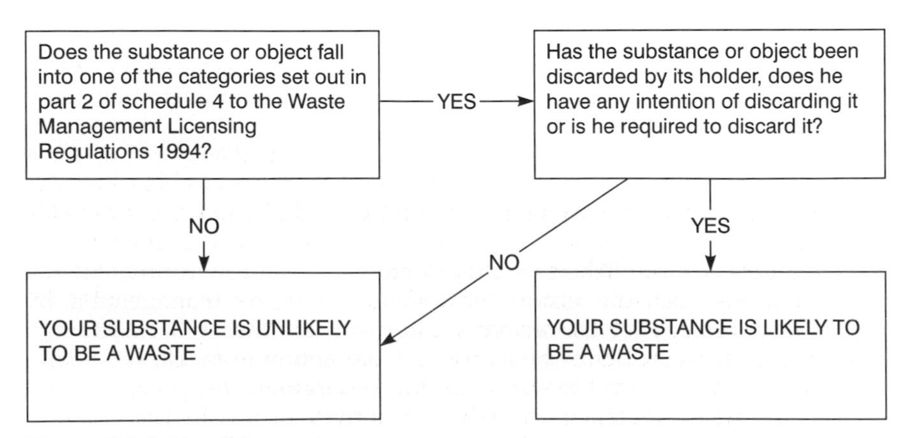

Figure 5.3.1 Identifying waste

However, in a case where the holder is in doubt about the status of a substance or object as waste the regulatory authority should be consulted. In order for material to be classified as a certain type of waste, such as a controlled waste, a substance or object should meet the definition of 'directive waste' as set out in the Waste Framework Directive and its amendment and reflected in section 22 of the Environment Act 1995 and the Waste Management Licensing Regulations 1994[16]. This definition is based on whether the producer or person in possession of the material in question has discarded it, intends to or is required to discard it. Guidance on how this definition should be interpreted is given in Department of the Environment Circular 11/92[17] now replaced by Circular 04/2000. However a number of questions can be posed in order to decide whether or not something is a waste:

- Is the substance or object still part of the commercial cycle (i.e. the producer's main activities) or chain of utility (i.e. if an object has been given away but is still being used for the purpose for which it was made)? If it is part of these, then it is fair to assume that it is not a waste.
- Has the item been consigned to a disposal operation? If so it is almost certainly a waste.
- Has the item been abandoned or dumped illegally? If so, it is a waste.
- Has the item been consigned to a specialised recovery operation? If so it is likely to be a waste. (A specialised recovery operation is one which recovers or recycles materials which otherwise could not be used.)
- Can the item be used in its present form without being subjected to a specialised recovery operation? If so it may well not be a waste (e.g. bottles returned for refilling).
- Does the owner have to pay for the item to be taken away? If so, it is likely to be a waste.
- Will the person receiving the item regard it as something to be disposed of rather than as a useful product? If so it is a waste.
- Has the item been reprocessed so that it can re-enter the commercial cycle (e.g. a recycled solvent which can be sold back to a purchaser)? This is no longer a waste.

There are a number of terms that describe a waste in legislation of which the main terms of reference used in UK law are:

- **Controlled waste** is defined in the Controlled Waste Regulations 1992[18] and section 75 of the Environment Protection Act 1990 as wastes from households, commerce and industry. Wastes excluded from this definition include explosives, wastes from mines and quarries or agricultural wastes. Wastes included are:
  - Household waste: from a domestic property, caravan, residential home, educational establishment, hospital or nursing home.
  - Industrial waste: from a factory or from a premises used for or in connection with, the provision of public transport, the public supply of gas, water, electricity or sewerage services, and the provision of postal or telecommunication services.

  - Commercial waste: from premises used for trade or business or for the purposes of sport, recreation or entertainment.
- **Hazardous (special) waste**: is controlled waste which, because of its hazardous properties, is subject to additional controls under the Special Waste Regulations 1996 (see a later section for further details). There are a number of guidance documents produced by regulatory authorities that further explain aspects of the definitions and the UK interpretations. The Environment Agency Special Waste Explanatory Notes[19] are particularly helpful.
- **Difficult waste**: is a term used to describe wastes that could in certain circumstances be harmful to human health or the environment in the short or long term due to their chemical or biological properties. This term incorporates wastes whose physical properties present handling problems at the point of disposal.
- **Clinical waste**: the definition of clinical waste is given in the Controlled Waste Regulations 1992 in terms of two hazardous properties – infectivity and toxicity. Clinical waste includes:
  - Any waste which consists wholly or partly of human or animal tissue, blood or other bodily fluids, excretions, drugs or pharmaceutical products, swabs, dressings, syringes, needles or other sharp instruments, being a waste which unless rendered safe may prove hazardous to any persons coming into contact with it.
  - Any other waste arising from medical, dental, nursing, veterinary, pharmaceutical or similar practice, investigation, treatment, care or research, or the collection of blood for transfusion, being waste which may cause infection to any person coming into contact with it.

5.3.5 The waste hierarchy

The government's strategy for waste management promotes the 'waste hierarchy' as a guiding principle in the development of a more sustainable waste management system. Following the hierarchy is also a cost-effective and environmentally responsible approach to managing waste. The hierarchy ranks methods of waste management, defining elimination and reduction as the most desirable options followed by re-use, then recovery (through recycling, composting or energy recovery) and finally the least desirable option, disposal.

The terms are covered in Article 3 of the Framework Waste Directive which states that member states of the European Union shall take appropriate measures to encourage the prevention or reduction of waste production and its harmfulness. With reference to industry, actions which will not necessarily lead to waste prevention but reductions in product and/or packaging which lead to reductions in waste generated are to be encouraged.

It should be recognised however, that it is not always feasible (economically, technically or environmentally) to follow this strategy. Other means and principles may be necessary to formulate local waste strategies such as the precautionary principle, the proximity principle, life

Table 5.3.1 The six-stage waste hierarchy

Hierarchy	*Waste option*	*Description and/or examples*
Most desirable	Elimination	Complete elimination of the waste at source.
↓	Reduction	The avoidance, reduction or elimination of waste generally within the confines of the production unit, through changes in industrial processes or procedures. This may involve using technology which requires less material in products and produces less waste in manufacture and by producing longer-lasting products with lower pollution potential.
	Re-use	Examples are: returnable bottles and reusable transit packaging.
	Recovery	Involves finding beneficial uses for waste such as recovering energy by burning it; recycling it to produce a useable product or composting to create products such as soil conditioners and growing media for plants.
	Treatment (Energy recovery)	The destruction, detoxification or neutralisation of wastes into less harmful substances.
Least desirable	Disposal	By incineration or landfill without energy recovery. Secure land disposal may involve volume reduction, encapsulation, leachate containment and monitoring techniques.

cycle analysis and the Best Practicable Environmental Option (BPEO), which all apply to decisions about waste management. The BPEO is the option that provides the most benefits or least damage to the environment as a whole, at an acceptable cost, in the long term as well as the short term.

The Waste Hierarchy (*Table 5.3.1*) captures these concepts. It is usually portrayed in a five- or six-stage list of options and strategies.

5.3.6 Waste management in practice

Most pollution incidents are avoidable and the costs for cleaning up a pollution incident can be very high. Careful planning of facilities and effective operational procedures can reduce the risk of a loss of containment and simple precautions can prevent such a loss becoming a pollution incident. This section covers the good practice and pollution prevention measures necessary to achieve compliance with legal requirements and minimise the likelihood of incidents. Waste management processes are fundamental aspects of good business management and should be integrated into systems set up to run established business

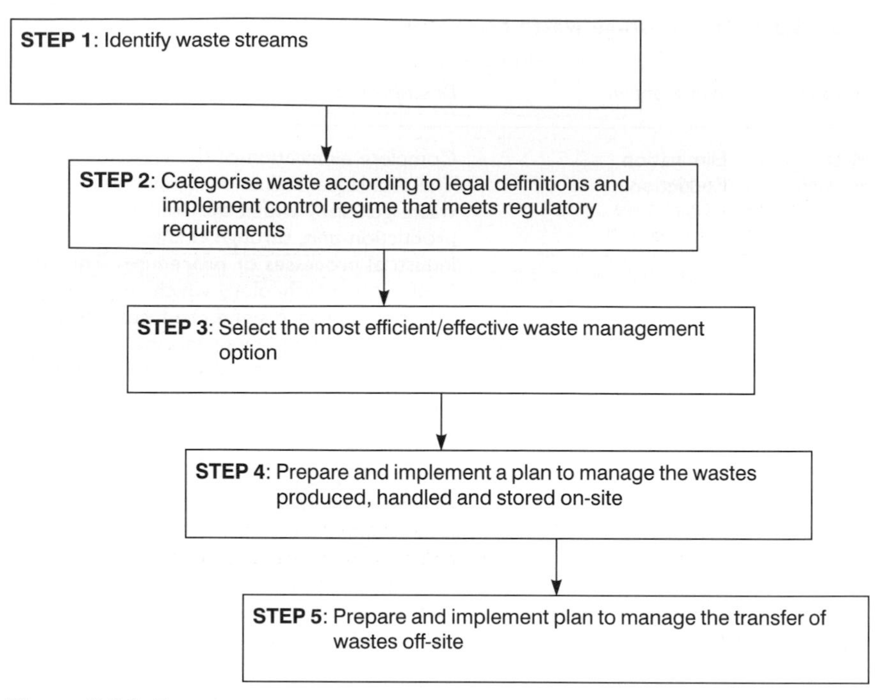

Figure 5.3.2 Five steps in waste management

operations. There are five steps involved in practical waste management as shown in *Figure 5.3.2*. They are discussed more fully in subsequent sections.

5.3.6.1 STEP 1: Identify waste streams

One of the first tasks of a waste minimisation programme is to identify and characterise the facility waste streams. This information can be used to assess inputs of raw materials to the waste stream, how much raw material can be accounted for through fugitive losses, distinguishing large single waste streams from smaller constant flows and tracking wastes that may be subject to seasonal variations. This is important data needed to establish effective and legally compliant waste management procedures as well as providing useful business information that can be used to improve process efficiency and minimise costs. Further details on the practical methods for carrying out this type of assessment are covered in a later section on waste minimisation.

5.3.6.2 STEP 2: Categorise waste according to legal definitions

Once the list of waste streams have been identified and the decision has been made that the materials are in fact wastes according to the legal

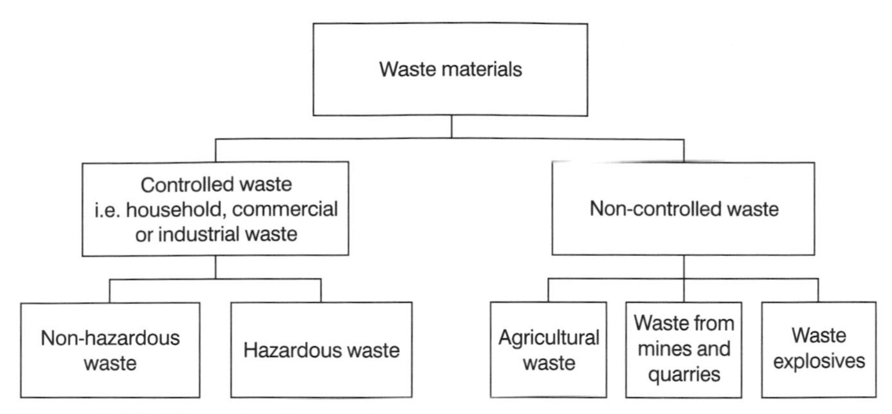

Figure 5.3.3 Waste classification chart

definition, the next stage is to ensure that the wastes are correctly classified. *Figure 5.3.3* is a waste classification chart demonstrating the process.

5.3.6.3 STEP 3: Select the best waste management option

The generation of waste is a common product of physical or chemical processing. However, the quantity and toxicity needs to be minimised in order to protect the environment and the safety of the working population and public. Waste minimisation can be achieved in many ways, through, for example, the design of processes and the selection of suitable raw materials which produce less waste, to the re-use and recovery of materials to avoid sending them for final disposal. This decision-making process is referred to as the waste management hierarchy and has been covered previously. The implementation of a waste minimisation programme to plan for waste elimination or source reduction options is covered in a later section.

The business considerations that form part of the decision to select the best environmental option for the waste material requires that an evaluation of the costs of waste minimisation, treatment and disposal be carried out on an equitable basis. Alternative routes should be costed using as far as possible actual estimates and realistic assumptions about market values of recovered product. The costs of all the elements involved in the process from collection to transfer, treatment and final disposal needs to be considered. An analysis should therefore take account of the following:

- labour costs of collection, delivery and disposal
- administrative costs of collection, delivery and disposal
- site purchase costs
- capital costs of equipment
- operating costs of equipment i.e. fuel, maintenance, insurance
- disposal costs i.e. landfill charges including tax, incinerator fees.

The costs of any recycling schemes that may be associated with the overall strategy will also include:

- labour and administrative costs of collecting and delivering recyclables either to a market or a processing facility
- residual disposal costs
- labour and administrative costs of processing
- capital costs of the recycling equipment such as vehicles, materials handling and processing facilities
- operating costs of recycling equipment, such as fuel, maintenance and insurance
- administrative costs of the recycling scheme including the costs of promotional materials, advisory and marketing services.

The revenues and savings from a recycling scheme include:

- revenues from the sale of materials
- any revenue from commercial sponsorship of the scheme.

The process of financial appraisal should take into account both short- and long-term costs taking note that some of the figures will be estimated as the market value may fluctuate with supply and demand.

5.3.6.4 STEP 4: Prepare and implement a plan to manage wastes on-site

Waste poses a threat to the environment and to human health if it is not managed properly and recovered or disposed of safely. It is critical to good business practice to ensure that environmental legislation is not breached by a lack of understanding of key legislative requirements or poor implementation of waste management procedures. To avoid such risks arising, the responsibility for managing this process should be identified and appropriate resources allocated. There are many aspects to be considered such as waste containment, segregation and storage, waste treatment, handling and the preparations necessary to transfer the wastes off-site to the next stage of the management cycle. All these areas have clearly defined expectations set out in various legal statutes and Codes of Good Practice. The key requirements in this area are set out in this section and should be considered in the development of a strategy to manage the company's wastes while on-site.

5.3.6.4.1 The Duty of Care

The basic guiding principles and controls required for good on-site waste management practice are set out in the Duty of Care concept which places a duty on anyone who in anyway has a responsibility for controlled waste. The Environmental Protection (Duty of Care Regulations) 1991[20], issued under section 34 of the Environmental Protection Act 1990, detail the strict waste management controls to be implemented by the waste holder. The waste holder is defined as *'any person who imports, produces, keeps, treats or disposes of controlled waste or as a broker has control over it.'* A clear understanding of this regime is necessary to implement the different elements that apply. These are clearly set out in the guidance document 'Waste Management, The Duty of Care A Code of Practice'[21] issued by the Department of the Environment, which recommends a series of steps which should normally be sufficient to meet the Duty of Care.

The Duty of Care places the holder of the waste under a duty to take all measures, applicable to him in that capacity, as are reasonable in the circumstances. The steps to be taken are:

1. *To prevent any contravention by any other person from disposing, treating or storing the waste without a licence, breaching the conditions of the licence or in a manner likely to cause environmental pollution or harm to health.*

In practice this means that the waste producer must take all reasonable steps to ensure that all the waste is disposed of to suitably licensed sites and not fly-tipped at an unauthorised site. If the waste was not handled correctly and the producer has failed to make reasonable checks, such as that the recipient's waste management licence permits them to take the type and quantity of the waste involved, then the producer could be liable for breach of the Duty of Care. Full checks do not need to be repeated on each occasion if transfers of the same type of waste follow the same path, however licences should be examined when changes occur. It is reasonable to go beyond licence checking for larger quantities of more hazardous waste which is likely to involve periodic audits at the disposal site to check that disposal takes place according to the conditions of the site licence.

2. *To prevent the escape of waste from his or another person's control.*

The waste must be appropriately packaged in order to prevent foreseeable escape or leakage whilst on site, in transit or in storage. Holders should protect waste against the risks posed by:

- Corrosion or wear of waste containers.
- Accidental spilling or leaking or inadvertent leaching from waste unprotected from rainfall.
- Accident or weather breaking open contained waste and allowing it to escape.
- Waste blowing away or falling while stored or transported.
- Scavenging of waste by vandals, thieves, children, trespassers or animals.

3. *On the transfer of waste, to secure that the transfer is only to an authorised person or to a person for authorised transport purposes.*

4. *On the transfer of waste, to secure that there is also transferred a written description of the waste which is good enough to enable each person receiving it to avoid committing any offences under section 33 and to comply with the duty of care relating to the escape of the waste.*

Points 3 and 4 are discussed in section 5.3.6.5 concerning off-site waste management.

Each holder in the chain has a responsibility to ensure that the duty is discharged while the waste is under his control. Different measures are reasonable for each role. For example, the holder acting as the waste carrier would not normally be expected to provide a description of the waste he collects from the producer unless for some reason he alters the composition of the waste.

5.3.6.4.2 Waste management licensing

This sets out a licensing system designed to regulate the deposit, keeping, treatment or disposal of controlled waste (i.e. industrial, commercial or household wastes) on land. Its objective is to prevent unacceptable environmental emissions by specifying the management system for a site or plant.

This regime was introduced by the Waste Management Licensing Regulations 1994 as required by the EC Framework Directive on Waste and is regulated and enforced by the Environment Agency. All producers of waste need to understand the requirements of waste management licensing in order to assess whether or not they need a licence for the activities carried out on their site. Although there is a broad requirement for licensing, it is equally important to establish whether or not the waste activities are categorised within an extensive list of activities exempt from licensing as set out in Schedule 3 to the Waste Management Licensing Regulations 1994. Exemptions are permitted in circumstances where there are other adequate controls such as consents given under the Water Resources Act 1991[22] and for a number of specified activities carried out at the place of production of the waste (where waste is to be reused or recovered) or where small quantities of waste are being managed. Exemptions must be registered with the Environmental Agency, giving them the business name and address, details of the activity which is exempt and the place where the activity takes place.

The Environment Agency are likely to specify appropriate terms and conditions in a licence and conditions will vary depending on the activity being considered. Clearly conditions for a large and complex landfill site will be much more detailed with higher expectations set than for a simple treatment operation. The conditions are likely to address the types and quantities of waste involved and operational issues such as monitoring, record keeping and controls which are set out in a work plan provided by the site operator. The applicant for the licence must be a 'fit and proper person' to hold a licence and this will be judged according to the following considerations:

- If he or another relevant person has been convicted of a relevant offence.
- The management of the activities would be in the hands of a technically competent person (holding approved waste management qualifications).

- The potential licence holder has made financial provision for the adequate discharge of the obligations arising from the licence.

The regulations further define requirements and procedures for submission of licence applications, surrender or transfer of licences, assessment of 'fit and proper persons' appeals and public registers.

5.3.6.4.3 Good waste management practices

To ensure that waste is properly controlled and managed the following techniques should be practised:

A. Waste storage

Wastes will often have to be stored at some stage for varying periods of time. Underground storage tanks are susceptible to damage and corrosion and therefore above ground storage facilities are preferred. Any waste should be stored in a container that is appropriate for the volume of waste produced and the fabric of the container should be compatible with the nature of the waste, i.e. its chemical or physical properties. This means that wastes must not be kept in corroded, worn or damaged containers. Metal drums are not always suitable for acid wastes and open-top drums are not suitable for liquid wastes.

Where waste is stored it should be placed under cover within a secure designated area that is regularly inspected. Suitable secondary containment (bunding) should be provided for waste chemical substances to prevent any accidental release to the environment in the event of leakage, a spill or leaching. Any bunding provided should be designed to contain 110% of the contents of the stored volume within it, without any drains or valves and should be routinely checked to ensure its integrity is maintained.

B. Waste location

All waste should be placed in a suitable location in order to avoid accidental contamination and to minimise its handling and transportation thus reducing the potential for spillage.

C. Labelling of containers

The labelling of all waste products is essential for two reasons:

- to ensure the correct segregation of wastes in order to avoid incompatible storage;
- to provide the appropriate labelling to indicate any hazardous properties that may be associated with it.

If containers have been re-used, it is important that they have been cleaned prior to refilling in order to remove any residue of previous contents and that old misleading labels are removed and replaced as appropriate.

D. Waste handling and deliveries

The handling of materials involves risks of spillage and accidents. It is important to identify these risks so they can be appropriately minimised. The following aspects should be considered:

- Loading and unloading areas should be designated, marked and isolated from the surface water drainage system.
- The routes for the movements of materials on site should be identified so that any necessary precautions can be taken.
- High-risk areas such as refuelling points should be isolated from the surface water system using ramped bunds or some form of restricted access.
- Deliveries of oil and potentially hazardous materials should be supervised. If there is a spillage it should be contained and reported immediately.
- Tankers should discharge via a lockable fixed coupling within a bunded area.
- Automatic cut-off valves should be fitted to delivery pipes to prevent overfilling.
- Pipe couplings should be unique to the liquid being handled.

E. Spillage and emergency response procedure

Operators are expected to have in place procedures for responding to emergency environmental incidents. *Figure 5.3.4* provides an example of an emergency procedure. Clearly set out and, depending on the waste and its properties, may need additional precautions to be added.

5.3.6.4.4 Managing liquid wastes on-site

There are three methods available for the disposal of trade effluent (liquid waste):

- The transfer off-site for treatment and disposal by a recognised waste contractor.
- Obtain a 'Consent to Discharge Trade Effluent' from the local Water Services Companies so that waste can be discharged to the drains on-site. Authorisation is required in the form of a 'Consent to Discharge Trade Effluent' document issued by the sewerage undertaker (England and Wales) or Water Authorities (Scotland). These bodies are empowered to set conditions and levy charges on those who discharge effluent to the sewerage system.
- Obtain a 'Discharge Consent' from the local Environment Agency if waste is to be discharged to controlled water (fresh and saline natural waters including rivers, streams, lochs, estuaries, coastal waters or groundwater). A consent to discharge to this media, is under the control of the Environment Agency (England and Wales) and the Scottish Environmental Protection Agency (Scotland). These bodies will set conditions and levy charges. (These bodies also regulate the sewerage undertaker since they discharge to controlled waters following treatment that takes place at a sewage works.)

EMERGENCY SPILLAGE PROCEDURE

Names (and telephone numbers) of staff members trained to respond to an emergency spill:

Response Procedure:

1. Notify one of the above persons.
2. Locate the Safety Data Sheets for the spilled material.
3. Review the information given on potential risks to health, safety and the environment.
4. Put on the recommended personal protective equipment.
5. Locate the spill response supplies and equipment.
6. Isolate all near-by drains with absorbent containment materials or other measures.
7. Stop the leak or spill e.g. turn off the valve, stand up a knocked over container etc.
8. Isolate the spill with containment materials.
9. Absorb spilled liquid chemicals with absorbent material or a wet vacuum cleaner.
10. Spills of dry materials must be swept up and not washed to the drain.
11. Place the contaminated absorbent material in a container suitable for storage on-site and transportation off-site. Consideration should be given as to whether it is hazardous waste or not.
12. Wash any equipment or floors so that no traces of the chemical remain.
13. Absorb the wash water and place it in the same container with the contaminated absorbent materials.
14. Label the container with the appropriate workplace and shipping labels.
15. Notify the manager and fill out the incident report form.
16. Send a copy of the report to the manager for evaluation so that measures can be put in place to prevent a recurrence of the incident.

If the spill reached the drainage system which exits the site, call the Water Service Company, or Environment Agency if appropriate (for spills into the surface water drains), immediately to report it. Identify the drains clearly and the drainage route the contaminant followed from the spill to the site perimeter.

Figure 5.3.4 Outline emergency procedure for loss of containment (spillage) of waste materials

Trade effluent, which is fully defined in the Water Industry Act 1991[23], is a liquid that is produced in the course of any trade or industry on trade premises. It does not include domestic sewage or other domestic liquid waste. Waste materials classified as hazardous under the special waste regime may be discharged subject to agreement with the Water Service Provider. It is an offence to discharge trade effluent without the formal consent of the Water Services Companies or when in breach of any condition of such a consent.

The Urban Wastewater Treatment Regulations 1994[24] requires Water Services Companies to exercise their powers under the Water Industry Act 1991 and to write conditions into trade effluent consents and agreements to control trade effluents, either alone or in combination with other wastes.

Controls on trade effluents are necessary to prevent them causing:

- Corrosion of the material of construction of the sewer, especially concrete pipes and mortar joints.
- Blockage or hydraulic overloading of sewers leading to foul flooding of property or pollution of watercourses via network overflows.
- The formation of explosive, flammable or poisonous gases in the sewerage system which may be prejudicial to the health of personnel maintaining the sewerage system or the cause of danger or a nuisance to adjoining properties.

In addition, the treatment works must be protected to ensure that:

- The available plant can treat the sewage arriving at the works effectively and economically.
- Damage does not occur to the structure of the works or to mechanical, electrical and instrumentation equipment.
- Personnel employed in the operation of the treatment works are not harmed.
- Biological treatment processes are not poisoned or inhibited by toxic substances.
- Sewage treatment works effluent does not adversely affect the environment or prevent receiving water from complying with European Union Directives.
- Sewage sludge can be re-cycled or disposed of safely in another environmentally acceptable manner.

The Water Services Companies are themselves subject to controls in the form of consents issued by the Environment Agency to control the release from the treatment works to controlled waters of effluent and sludges produced in the sewage purification process.

Many chemicals will be completely degraded into harmless natural products in the sewage purification process. Their concentrations at discharge need not usually be limited as long as the local sewerage system and associated sewage treatment works provide adequate dilution. Advice will be given by the relevant Water Services Companies on request. However, certain constituents of effluent will be subject to specific controls, for example:

- acidity – to prevent corrosion of concrete pipes
- ammonia – to control toxicity and nuisance in sewers, also toxicity and oxygen demand in receiving waters
- heavy metals – to control toxicity
- sulphates and sulphides – to prevent corrosion of pipes and control nuisance

- excessive oxygen demand – to prevent stagnation effect in sewers and overload of the treatment process.

The trade effluent consents issued by Water Services Companies will usually contain the following requirements:

1 No trade effluent shall be discharged having a pH < 6.0 or > 10.0.
2 Trade effluent shall not contain:

- Sulphides, hydrosulphides, polysulphides and substances producing hydrogen sulphide on acidification in excess of 1.0 mg/l.
- Sulphates as SO_4 in excess of 1000 mg/l.
- Toxic metals in excess of 5 mg/l either individually or in total, e.g. antimony, arsenic, beryllium, chromium, copper, lead, nickel, selenium, tin, vanadium and zinc.
- Cyanides and cyanogen compounds, which produce hydrogen cyanide on acidification in excess of 1.0 mg/l.
- Total suspended solids at pH 7.0 and dried at 110°C in excess of 400 mg/l.
- Settled chemical oxygen demand in excess of 1000 mg/l.
- Ammonia in excess of 250 mg/l.

It is important to understand that limit values may vary from place to place depending upon the size of the sewage treatment works, dilution in the receiving watercourse and other industrial discharges in the catchment area.

The Water Services Companies are empowered to charge for the services they provide. Companies who discharge trade effluent into the sewage system are generally charged by the Water Services Company for the reception, conveyance, treatment and disposal of the trade effluent. Charges may be levied in three ways:

1 A 'Charges scheme' whereby a fixed charge is levied for the services provided.
2 An independent charging agreement between the Water Service Company and the discharger.
3 As a condition of the trade effluent licence where charges are calculated according to the 'Mogden formula'.

Trade effluent charges – Mogden Formula

All Water Services Companies use a Mogden-type formula to charge for trade effluents. The basic Mogden formula is:

$$C = R + V + \frac{Ot}{Os} B + \frac{St}{Ss} S$$

Where:
C = total charge per cubic metre of trade effluent
R = reception and conveyance charge
V = volumetric and primary treatment cost

Ot = the chemical oxygen demand (COD) of the trade effluent
Os = the average COD of settled sewage for the region
B = biological oxidation cost
St = the suspended solids in the trade effluent
Ss = the average suspended solids in crude sewage for the region
S = treatment and disposal costs of primary sludge.

The purpose of the formula is to match as closely as possible the trade effluent charge to the cost of providing the services of reception and conveyance of the trade effluent and its subsequent biological oxidation and sludge treatment and disposal.

The four unit costs (*R*, *V*, *B* and *S*) together with the COD and solids content of sewage are fixed annually in advance by the Water Services Companies. The only remaining variable in the formula is the volume of trade effluent actually discharged that should be determined from the water meter for the incoming supply or the submeter to the process. The use of standard strength will ensure consistency within the area covered by each Water Services Company. There will still be regional differences in charging owing to different tariffs between the Water Services Companies. An extra charge may be made depending on local circumstances.

There are cost-effective on-site treatments to reduce the strength of an effluent leaving the premises to the public sewer. Together with water recycling and chemical management systems, operators can reduce the overall cost of their effluent. Water recycling systems in particular offer a large reduction in water usage (often in excess of 90%), as well as savings on water heating costs and reduction of emergency water storage. From a water and energy conservation viewpoint, their use is to be supported. However, it must be realised that the trade effluent concentration is likely to be radically altered making the waste more concentrated and possibly more hazardous to handle and transport. Consequently it may be necessary to have much of the trade effluent taken off-site for treatment in order to meet discharge consent levels. Anyone considering installation of a water-recycling unit is advised to discuss the matter with their local Water Services Companies.

5.3.6.5 Step 5: Managing wastes off-site

The waste producer's duties and responsibilities in law do not cease once the waste has been transferred off-site as he maintains a shared responsibility through to its final treatment or disposal. There are a number of matters to address as preparations are made to remove the waste from the premises depending on the type of waste being transferred and its final destination. These considerations are covered by:

- Duty of Care requirements relating to contractor selection and the provision of appropriate information.
- An alternative regime for the control of hazardous wastes covered by the Special Waste Regulations 1996.
- The system to follow if wastes are to be exported from the UK.

5.3.6.5.1 The Duty of Care

The Duty of Care applies to all holders of the waste from the producer to the waste manager of the final disposal operation. Duty of Care responsibilities for off-site waste transfer are covered by the following statements in the regulations

- *On the transfer of waste, to secure that the transfer is only to an authorised person or to a person for authorised transport purposes*

The waste producer or holder should ensure that waste is only transferred to an authorised person such as a Waste Collection Authority (WCA) or a Waste Disposal Authority in Scotland, a holder of a waste management licence (or someone who is exempt from holding a licence) or a registered carrier of controlled waste (or someone who is exempt from registration). Details of carrier's requirements are set out in the Controlled Waste (Registration of Carriers and Seizure of Vehicles) Regulations 1991.

The enforcing agencies maintain a register of carriers as a reference list of companies who are authorised to transport waste, bearing in mind that the list is not a recommendation or guarantee of a carrier's suitability to accept a certain type of waste. The registered carrier's authority for transporting waste is a certificate of registration or an official copy, and a check should be made of the expiry date, certificate number and marking to confirm that it has been issued by the enforcing agency and that it covers the particular waste for disposal. Photocopies of a certificate of registration must not be taken as proof of registration.

The phrase 'Authorised transport purposes' refers to the transport of controlled waste within the same premises, the importation of waste to the place of disposal and the exportation by air or sea or by rail from a place in the UK.

- *On the transfer of waste, to secure that there is also transferred a written description of the waste which is good enough to enable each person receiving it to avoid committing any offences under section 33 and to comply with the duty of care relating to the escape of the waste.*

A transfer note (an example of which is given in the Duty of Care Code of Practice) must be completed, signed and kept by the parties involved if waste is to be transferred. A single transfer note may cover multiple consignments of waste transferred at the same time or over a period of time provided that all the details are the same. The transferor and transferee should keep the written description of the waste and a copy of the note for at least two years during which time the Environment Agency may request a copy.

In addition, the description should include the following:

- the source of the waste – reference can be made to the use of the premises or the occupation of the waste producer for wastes that do not require special arrangements for handling or disposal;

- the name of the substance – refers to an explanation of what the waste is, either by its chemical or common name;
- the process producing the waste – an explanation of how the waste was produced including details of the materials used and types of processes operated;
- a chemical and physical analysis – if wastes from different processes are mixed or if the activity or process alters the properties or composition of the material, it may be necessary for a physical and chemical analysis of the waste to be undertaken.

As a general rule, the waste description must provide enough information to enable subsequent holders of the waste to manage it responsibly.

5.3.6.5.2 Hazardous wastes

This group of wastes is referred to as 'special wastes' in the UK, with a proposal to change the classification to 'hazardous wastes'. The Special Waste Regulations 1996 (as amended) sets out an effective 'cradle to grave' system of control which ensures that hazardous wastes (which were initially defined as controlled wastes) are soundly managed from the moment they are first moved as waste until they reach their final destination for recovery or disposal. The regime for managing hazardous waste is however, stricter than the regime for controlled wastes, adding to the requirements a notification procedure to ensure that the regulatory authority (the appropriate Environment Agency) has advanced notification of the impending transfer of waste together with a description of the hazards associated with the waste being transferred.

The starting point of any assessment of special waste is the definition of hazardous waste given in Regulation 2 of the Special Waste Regulations as amended. It states that a controlled waste, other than a household waste, is considered hazardous if:

- it appears on the EC list of hazardous wastes[25] **and** displays one or more of the hazardous properties set out in *Table 5.3.2*;
- it is not listed on the EC hazardous waste list but displays one or more of a sub-set of six hazardous properties in *Table 5.3.2*. These hazards are H3A (first category) and H4 to H8;
- it is a prescription-only medicine.

The Special Waste Regulations (schedule 2, part 4) indicates that there are three methods for assessing a waste for a hazardous property:

- Using the Approved Supply List[27] (made under the Chemicals (Hazard Information and Packaging for Supply) Regulations 2002[26]) which prescribes hazard information and classifications for many common chemicals.
- Using the Approved Guide to the Classification and Labelling of Substances and Preparations Dangerous for Supply[27] which sets out

Table 5.3.2 Classification of hazardous waste properties

- H1 Explosive
- H2 Oxidising
- H3A Highly flammable (5 sub-categories)
- H3B Flammable
- H4 Irritant
- H5 Harmful
- H6 Toxic
- H7 Carcinogenic
- H8 Corrosive
- H9 Infectious
- H10 Teratogenic
- H11 Mutagenic
- H12 Substances releasing toxic/very toxic gas in contact with air/water/acid
- H13 Substance capable of yielding, after disposal, another substance which has abovementioned properties
- H14 Ecotoxic

general principles of classification and informs suppliers how to correctly classify and label chemicals.

- Using the Approved Code of Practice on Test Methods[28] where the latest amendment to Annex 5 to Directive 67/548/EEC has been adopted.

The Approved Supply List provides a risk phrase and classification for substances. The risk phrase(s) can then be linked to the hazards. In cases where a substance is found to exhibit a hazard, there is still a further stage of assessment to determine whether or not it is a hazardous waste. This next step is carried out to ascertain whether the waste *'displays'* the hazardous property and is established by determining the concentration of the hazardous component in the waste. The Special Waste Regulations in Schedule 2, Part 3 set out the concentration thresholds that make a waste hazardous. These are set at different levels depending on the nature of the hazard, i.e. the total concentration of substances classified as 'harmful' in the waste needs to be present at 25% or above to classify it as 'hazardous'.

The hazardous waste transfer must follow the correct consignment procedure for the movement of the waste. This can be established with the help of the waste carrier who can advise on whether a simplified process can be followed for regular consignments of the same type of waste from the same premises. In order to cover the costs of administrating the procedure, the Environment Agency will charge a fee for the consignment of the hazardous waste.

The Environment Agency provides full details on the hazardous waste regime in 'Technical Guidance WM1 Special Wastes: A technical guidance note on their definition and classification'[29]. Further related documents can be obtained from the Environment Agency website.

5.3.6.5.3 International waste transfers

Waste importers and exporters must follow the system of control set out in the Transfrontier Shipment of Waste Regulations 1994 which directly implemented the EC Waste Shipment Regulation 259/93/EEC[30]. It establishes a complex system of control on waste shipments, with detailed provisions varying according to the country of destination and means of transit, the intended purpose of the waste shipment, whether for disposal or recovery, and the type of waste involved. Anyone intending to ship waste across the frontier of a member state for disposal, where this is permitted, must notify the competent authorities in the countries through which the waste may pass by sending a consignment note obtained from the UK Environment Agency. After receipt of the note is acknowledged, the competent authorities of these countries have a limited period of time in which they may request further information, raise objections, impose conditions and authorise shipment. Once shipment has proceeded, the consignee (the person receiving the waste) must send the notifier (the person sending the waste) and the relevant competent authorities a copy of the completed consignment note and, after disposal or recovery, a copy of a certificate of disposal or recovery. Other requirements cover the details to be provided in the consignment notes and obligations which must form part of contracts for shipments.

Wastes destined for recovery are divided into three categories – the Red, Amber and Green Lists – according to the severity of the hazards they present. Red listed wastes are the most hazardous and are subject to the full notification procedure, i.e. written prior consent of the authorities is required for each waste shipment. For Amber listed wastes, shippers and importers can take advantage of a more streamlined procedure whereby the consent of the competent authorities is assumed provided no objections are made. Green listed wastes are normally not subjected to regulation although shipments should be accompanied by a transfer note.

5.3.7 Waste minimisation

It has been estimated that the average product sold to the consumer contains just 5% of the materials that went into its manufacture. The rest has been lost along the way. At each stage of the life cycle of the product, waste is created including mining, refining raw materials, processing and transport. The fate of the 95% inevitably adds to the pollution load on the Earth. Much of this waste can be eliminated by resource efficiency which could be achieved by choosing better materials or more efficient energy sources. The concept of waste minimisation is based on the rationale that by using materials carefully to reduce the generation of waste, pollution is reduced, resources are conserved and charges for waste disposal are minimised. A good waste minimisation programme should identify all significant process emissions to all media (air, water or land) and its aim should be a reduction in the total amount of waste generated.

5.3.7.1 The benefits of waste minimisation

The relative importance of waste minimisation varies in terms of reduced environmental impacts, financial savings, reduced energy and materials usage, achieving regulatory compliance thereby reducing potential liabilities. An improved environment profile of the company can enhance investor confidence and environmentally responsible products may promote customer acceptance of the product. Furthermore waste costs money through a reduction in yield, decrease in productivity and the direct costs of disposal. Therefore economic benefits of waste minimisation can be identified as:

- reduction in waste results in savings in waste disposal costs;
- more economical processes through the reuse and recycling of materials;
- a reduction in environmental costs in the short term through better process control making it less likely for fines to be imposed for pollution incidents;
- a reduction in potential liabilities in the longer term;
- a reassessment of processes which may make them more efficient.

5.3.7.2 Senior management commitment

It is essential that the senior management of a company is engaged in the waste minimisation programme and provides leadership to the process. This ensures interest in:

- a company environmental policy together with a strategy for its implementation
- provision of adequate resources
- a programme for training staff
- strong visible support and leadership with high level accountability.

5.3.7.3 Waste minimisation techniques

Waste can be reduced by the applications of a variety of techniques individually or in tandem including those discussed below.

5.3.7.3.1 Source reduction

Source reduction is a term given to the activity that reduces or eliminates the generation of waste from a process. The techniques used may be outlined as follows:

- Plant management – relates to the alteration of procedures or organisational aspects of the production process. The generation of waste is reduced by implementing improvements to the overall management of the production process in key areas. It is one of the easiest ways to reduce waste and may also result in overall health and safety improvements.

- Employee information and awareness – simple techniques such as training employees to use equipment efficiently and report process malfunctions such as leaking pipes or containers.
- Management accountability – the process of making production managers take ownership of the waste minimisation process and be made accountable for reporting quantities of waste generated and progress towards meeting waste reduction targets.
- Inventory control – focusing on the efficient control of inventories of raw materials or products to prevent wastes arising from poor storage management such as exceeded shelf life or incorrect storage conditions.
- Waste segregation – this technique is very important for chemical wastes since by separating hazardous from non-hazardous wastes the total volume of mixed hazardous waste can be reduced. It may also avoid legal non-compliance related to the mixing of such wastes. By providing separate collection containers for the different waste streams, extra reclamation costs may be avoided by the elimination of a reprocessing step.
- Material handling – improves techniques associated with waste movements to minimise the chance of material wastage due to spillage or other losses.
- Spill and leakage prevention – implemented through regular checks of plant and machinery for leaks in order to avoid material losses as well as reducing employee exposure. In the event of a spillage or loss of containment a spillage procedure should be implemented to ensure a timely response to the unplanned event to minimise loss and prevent other environmental damage occurring.
- Planned preventative maintenance – is the term given to the technique of maintaining equipment before it needs maintenance rather than reacting to equipment failure.
- Process modifications should be considered in order to optimise the efficiency of a particular operation where technically feasible. These can be divided into four categories:

  - Process changes involving the use of alternative low-waste process pathways to obtain the same quality of product. Process changes are typically made during process redesign or the development of a new plant.
  - Equipment modifications which can be introduced to perform existing operations more efficiently in order to eliminate or reduce waste that is generated at start up and shut down or during product changes and maintenance operations.
  - Changes to operational settings involving adjustments to equipment to optimise the process and minimise the production of waste materials.
  - Process automation involving the provision of automated systems that are more efficient at handling or producing a product.

- Raw material substitution – which refers to replacing or substituting hazardous materials used in production processes by less hazardous or non-hazardous alternatives. Changes in these materials may also lead to the reduction or elimination of hazardous wastes. However it should be noted that material changes to a less hazardous material may have an adverse effect on the production process, product quality or waste generation and therefore any changes should be fully evaluated before implementation. Product substitution is achieved by reformulation of the final or intermediate products in order to reduce the quantity of waste arising during its manufacture or use. Since the nature of the finished product may change, the issues around customer acceptability should be addressed.

5.3.7.3.2 Re-use and recycling

Recycling differs from re-use because it involves a processing step. Recycling can be defined as *the collection and separation of materials from waste and subsequent processing to produce marketable products*. Recycling waste materials for future use and reclamation may provide a cost-effective alternative to treatment and disposal. It should be noted however that elimination and minimisation of the waste at source are the preferred options as set out in the waste management hierarchy. Success in the recycling of wastes depends upon the ability to use the materials taking into account the cost-effectiveness of the technology and the ease of implementation of the option. Schemes for recycling may range from those requiring only minor procedural changes that incur no capital investment to more significant projects requiring considerable expenditure.

5.3.7.4 The practical implementation of a waste minimisation programme

The seven stages necessary to implement a successful waste implementation programme are:

- Identify sources of waste using process flow charts to monitor inputs and outputs
- Prioritising significant waste streams
- Assess feasibility of options
- Project implementation
- Set waste reduction targets and time scales
- Involve and educate staff
- Monitor performance against targets and communicate results.

5.3.7.4.1 Identify waste sources

In order to implement the most efficient systems for waste minimisation, it is necessary to understand fully the production processes and business operations and how materials flow through these operations, starting from the acquisition of raw materials, through transport and distribution to final use. Flow diagrams provide the basic means for identifying and organising information that is useful in the assessment. Flow diagrams should be prepared to identify important process steps and sources of waste generation. They are also the foundation upon which material balances, sometimes termed 'mass balances', are developed. Material balances are important for waste minimisation projects since they allow quantifying the losses or emissions that were previously unaccounted for and assist in developing the following information:

- baseline data for tracking progress during the implementation of the project;
- data to estimate the size and cost of additional equipment and other modifications;
- data to evaluate economic performance.

In its simplest form, the material balance is represented by the mass conservation principle:

MASS IN = MASS OUT + MASS ACCUMULATED

Material balances can assist in determining concentrations of waste constituents where analytical test data is limited. These are particularly useful where there are points in a production process where it is difficult or uneconomical to collect analytical data. A material balance can help determine if fugitive losses are occurring, for example, the evaporation of solvents from a tank for cleaning parts can be estimated as the difference between solvent put into the tank and solvent removed from the tank.

To characterise waste streams by material balance can require considerable effort. However, by doing so, a more complete picture of the waste situation can be established. This helps to establish the focus of the waste minimisation activities and provides a baseline for measuring performance.

Mass balance information (materials entering and leaving a process) can be obtained from a variety of sources:

- Samples, analyses and flow measurements of feedstocks, products and waste streams
- Raw material purchase records
- Material inventories
- Emission inventories
- Equipment cleaning procedures
- Product specifications
- Production records
- Operating logs
- Standard operating procedures
- Waste records.

Material balances can be easier, more meaningful and more accurate when they are completed for individual units, operations or processes.

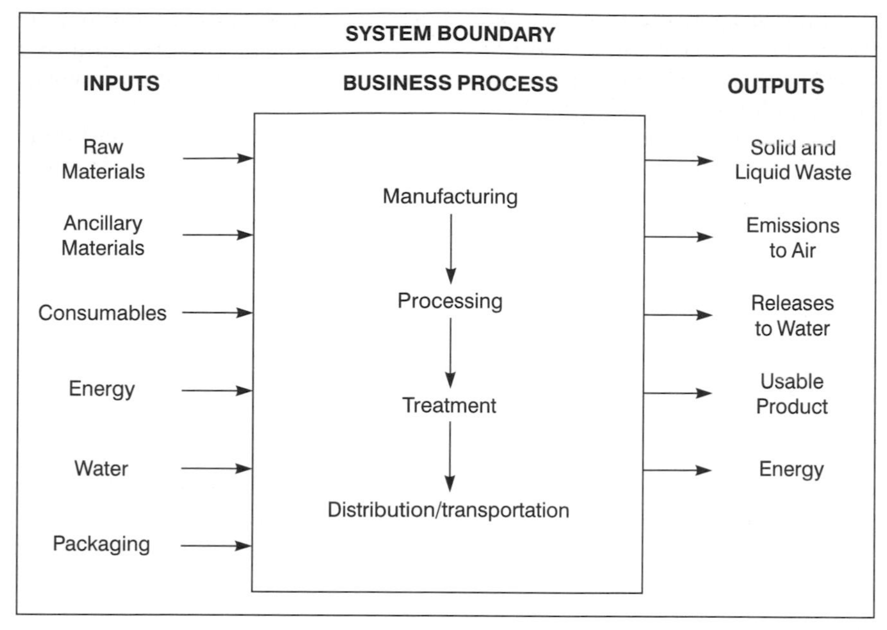

Figure 5.3.5 General material flow diagram

For this reason it is important to define carefully the system boundary for each unit being evaluated. The overall material balance for a facility can then be developed from individual unit material balances. Diagrammatically, any industrial system or process can be represented within a system boundary that encloses all the operations of interest. Several diagrams may be required to map out complex processes, joining each diagram with its separate operations together to form the full system being considered. There are several pieces of software, which can be used to produce flow diagrams. Once established, it provides the basis to understand what ancillary materials, consumables and energy are consumed (the 'inputs') and where wastes are generated (process emissions) at various stages. A general overview of a material flow diagram is provided in *Figure 5.3.5*

Inputs have been defined in terms of:

- Raw materials: essential materials used directly in production.
- Ancillary materials: materials used indirectly for production such as materials for cleaning, maintenance or treatment processes.
- Consumable materials: materials for offices (paper, printer cartridges); sales materials (brochures, envelopes), personal protective equipment.
- Packaging: materials used to deliver goods (plastic wrapping, wooden pallets) and to supply finished products.
- Energy: fuel and electricity for workplace or production heating, company vehicles.
- Water: water from company mains, boreholes, or rivers.

Outputs from the system are essentially the generation of products (a product is a marketable commodity and is the principal outcome of a manufacturing process) and wastes in the form of emissions to air, waste water, solid and liquid waste and energy losses.

Small and medium sized businesses or even large businesses with only a few waste-generating operations should attempt to assess their entire facility. It is also beneficial to look at an entire facility where there are a large number of similar operations. Similarly, the implementation of good operating practices that involve procedural or organisational measures should be implemented on a facility-wide basis, such as employee awareness programmes, inventory or maintenance procedures.

5.3.7.4.2 Prioritising significant waste streams

While all waste streams should be assessed, prioritising is necessary when available funds and or other resources are limited. Prioritising the waste streams should focus on identifying the most important areas first allowing the assessors to move onto lower priority areas as time, resources and funding become available. A variety of criteria can be used to prioritise waste streams including:

- regulatory considerations
- hazardous properties of the waste including toxicity
- quantity
- management costs
- safety and health risks
- potential for success
- concerns about potential environmental liability.

There may be some areas where it will be easy to implement a waste minimisation improvement project without making a large capital investment. Such changes may include:

- segregation of wastes to prevent mixing of hazardous and non-hazardous wastes
- improved material handling and inventory practices to reduce the amount of inventory that has expired
- preventative maintenance
- production scheduling to reduce quantities of batch-generated wastes or unused raw materials
- minor operational changes.

Small adjustments will often not only reduce waste production but can save resources and costs.

After waste streams have been ranked, the next step is to identify which waste minimisation options are technically and economically justifiable.

5.3.7.4.3 Assess feasibility of waste minimisation options

Prioritising key areas for improvement is essential as feasibility exercises can be time consuming and expensive. This part of the programme should consider both the technical feasibility and economic viability of the waste minimisation options considered in section 5.3.7.3.

The technical evaluation determines whether or not a waste management option will work in a particular application.

The following aspects should be considered:

1 Safety considerations related to any new system of work.
2 Compatibility of new equipment with existing plant and operating procedures.
3 Location of the new equipment in terms of available space and any requirements for water and energy supplies.
4 Impacts an option would have on production such as any stoppage that may be required in order to install a new system.
5 Additional technical expertise that may be required to install, operate or maintain a new system.
6 Any other environmental matters related to the operation of a new system such as the generation of modified or new waste streams. The environmental effects of any proposed changes need to be carefully considered to ensure that the overall environmental burden is minimised.

Procedures that can be followed in this evaluation include the following:

(a) reviews of technical literature;
(b) visits to an existing installation to observe its operation and discuss practical aspects of operation;
(c) pilot-scale demonstration of equipment prior to purchase.

When decisions are being made on the feasibility of options, the discussions should involve consultation with key groups of staff that have technical and practical expertise in the areas being reviewed such as purchasing, production, engineering, maintenance to ensure viability and acceptance of an option.

Any solutions that are considered technically feasible need to be assessed for economic viability using the company's preferred method. When performing the economic evaluation, various costs and savings should be considered. A project's profitability is normally estimated from a cost–benefit analysis, where the costs are evaluated by looking at both capital and operating costs. If a waste management option has no significant capital costs, then its profitability can be judged by whether or not an operational cost saving can be made. If such an option does reduce overall operating costs, this becomes a recommendation for it to be implemented.

An economic assessment of a waste solution may include the following:

1 Capital costs of new equipment
2 Additional (or reduced) operating costs (including raw materials, labour, energy and water, inspection and auditing)

3 Training costs for employees covering the new system
4 Disposal costs for any wastes.

Reducing or avoiding current and future operating costs associated with waste treatment, storage and disposal are a major element of the project economic evaluation. The financial benefits from improved operational and environmental performance will include:

1 Capital savings derived from avoiding the installation of emission control and monitoring equipment.
2 Savings in waste disposal costs.
3 Savings due to more efficient use of raw materials.
4 Savings from reductions in product waste, energy and water use.
5 Savings based on the reuse of materials previously sent for disposal or income from sale of recyclable materials.
6 Improved company environmental performance leading to marketing opportunities.

5.3.7.4.4 Project implementation

The assessment and evaluation report will provide the basis for obtaining funding for a waste minimisation project. Waste projects that only involve operational, procedural or material changes that do not involve alterations to plant or equipment should be implemented as soon as the cost savings have been verified. The implementation of projects involving capital expenditure and/or significant process modifications will clearly need to be integrated into a business plan. The plan should take into consideration the time scales for release of the capital and the manpower to manage the project.

The responsibilities and time scales for implementation should be clearly assigned and good communications are essential to maintain momentum and commitment. Appropriate training and supervision should be provided for staff involved in any new scheme.

5.3.7.4.5 Set waste reduction targets and time scales

Goals should be established that are consistent with the policy adopted by the company. Waste minimisation goals can be qualitative, for example, *'a significant reduction of toxic emissions into the environment'*. However, it is better to establish measurable, quantifiable goals since qualitative goals can be subjective and interpreted ambiguously. Quantifiable goals establish a clear guide to the benefits expected from the waste minimisation programme. A major company could adopt a corporate wide goal of (say) 5% waste reduction per year, with each facility in the company setting its own waste minimisation goals. The goals should be challenging enough to motivate staff to be realisable and practical. Attributes of effective goals are listed in *Table* 5.3.3. Goals should be reviewed and refined periodically to reflect any changes such as improved technologies, expanding experiences, new legislation or economic climate.

Table 5.3.3 Attributes of effective goals

Attributes of effective goals
Acceptable to those who will work to achieve them
Flexible and adaptable to changing requirements
Measurable over the allocated project time scales
Motivational to all employees
Suitable for the overall corporate goals and mission
Understandable by all the organisation's stakeholders
Achievable with a practical level of effort and resource

5.3.7.4.6 Staff training

Training and communications programmes for all levels of employee are an essential feature of the plan. Staff need to understand the company's waste management strategy, policies and practices in order to take responsibility for providing support to these activities. A training programme could include the following aspects:

- Company policy and expectations for waste minimisation strategy
- Overview of the programme being implemented
- Key areas of waste identified and associated cost of waste management
- Key environmental impact of wastes generated
- Benefits of efficient waste minimisation and the techniques being used
- Case studies to demonstrate key messages and successes.

5.3.7.4.7 Monitor performance and communicate results

Having set specific reduction goals, these should be tracked and the results reported at regular intervals both to management and employees. Monitoring performance will assist in the overall management of a project such as where material or utility consumption can be compared to a measure of work activity such as production output, i.e. the amount of electricity used per item produced. Good communication will help maintain an interest in the programme being implemented and provide the motivation and commitment to continue, particularly if the projects are reporting successful outcomes. Recognition of the efforts of individuals or groups may also be considered appropriate to reinforce good environmental achievements and help develop and sustain an environmentally responsible culture.

There are many ways of presenting data using bar charts, line graphs or pie charts. Whichever method is chosen the data should be displayed in a clear and simple format, noting any trends and with key messages highlighted.

5.3.8 The cost of failure to manage waste effectively

Where waste is not managed effectively, enforcement action may be taken with the purpose of securing the protection of the environment and compliance with regulatory requirements. The regulators have a range of enforcement options before prosecution is considered. Powers to issue notices requiring actions such as the clean up of land, compliance with discharge consents and compliance by waste operators with their licences. Under the Pollution Prevention Control Regulations[31], prohibition notices can also be issued on premises where processes are carried out. In many cases, the worst consequence of a conviction for an offence relating to waste may be from the adverse publicity for the business as most offences are widely reported in journals, the local press and in serious cases, the national press. These reports can have a damaging effect on the company's reputation and can attract the attention of environmental interest groups.

If an enforcement agency decides to prosecute or caution a defendant, it would consider the following factors:

- the environmental impact of the non-compliance
- foreseeability of the events leading up to an incident
- the offender's previous history, attitude, intentions and personal circumstances.

The most common offences lie in the following areas:

- *Water pollution*

Section 85 of the Water Resources Act 1991 makes it an offence to cause or knowingly permit the pollution of waters. This offence is normally prosecuted under the 'cause' category as this is an offence of strict liability and requires only proof that the offence occurred and the defendant caused it. The penalty can be 6 months imprisonment or a fine of up to £20 000 in the magistrates court or 2 years imprisonment or an unlimited fine in the Crown Court.

- *Water abstraction*

Section 24 of the Water Resources Act 1991 prohibits abstraction of water from streams or the ground except in pursuance of a licence granted by the Environment Agency. This offence is punishable by a fine of up to £5000 or on indictment, an unlimited fine.

- *Waste regulation*

Under Section 33 of the Environmental Protection Act 1990, it is an offence to deposit waste, or knowingly cause or permit waste to be deposited, in or on any land except in accordance with a waste management licence. This places on producers of waste a duty to take reasonable care to know and monitor how others in the waste chain manage their waste if they are to avoid charges of 'knowingly causing' or 'knowingly permitting' unlawful deposit or treatment of waste. A producer has a defence to those causing or permitting offences if he can show he took all reasonable precautions and exercised due diligence.

Penalties for offences under section 33 are, on summary conviction (in a magistrates court), a fine of up to £20 000 or a term of imprisonment of up to 6 months, or both and on indictment (in a Crown Court), a term of imprisonment of up to 2 years, or a unlimited fine or both.

Offences under section 34 of the above Act which places a 'Duty of Care' on any person who handles controlled waste, can on summary conviction, lead to a fine of up to £5000 and on indictment, to an unlimited fine.

- *Waste packaging*

Under regulation 34 of the Producer Responsibility Obligations (Packaging Waste) Regulations 1997 it is an offence to:

- fail to register with the appropriate Agency or join a compliance scheme
- fail to carry out the recycling/recovery obligation
- fail to provide the Agency with a certificate of compliance.

Offences are punishable by a fine of up to £5000 or on indictment, an unlimited fine.

5.3.9 Conclusion

Over the last ten years there have been significant developments in the area of environmental management as Government and business work to drive activities towards sustainable waste management practices.

Success in meeting this vision requires the implementation of different facets such as regulatory obligations, market based benefits (economic, fiscal or voluntary) and Government incentives as set out below.

Legislators have introduced a large number of environmental laws to control the management of wastes and more recently, the concept of producer responsibility has become an important element of the Government's waste strategy, as seen in the packaging laws which place obligations on businesses to recover and recycle specified amounts of packaging and in the proposed 'end of life' laws for vehicles and electronic equipment setting out requirements aimed at controlling final disposal and avoiding landfill.

While regulation plays a key role in environmental protection, other measures are equally important.

Environmental management systems (such as ISO 14001 and the EC's Eco-management and Audit Scheme) have been implemented by organisations to demonstrate their commitment to good environmental management practices in minimising environmental impacts leading in turn to an overall improvement in environmental performance.

The Department of the Environment Food and Rural Affairs (DEFRA) has run a series of initiatives that are intended to encourage business to employ best practice by demonstrating cost-effective actions that can improve both business and environmental performance through programmes such as 'Envirowise' and the Energy Efficiency Best Practice Programme.

Fiscal instruments have more recently begun to emerge as market based forces focus on environmental improvements through taxes and levies such as the landfill tax introduced in 1996, and the Climate Change Levy[32] that came into effect on 1 April 2001.

There is still a fair way to go and the next ten years will see further advancement in environmental management practices as we strive to 'meet the needs of the present without compromising the ability of the future generations to meet their own needs' (a definition of sustainable development by the World Commission on Environment and Development in the Brutland Report)[33].

References

1. *Deposit of Poisonous Waste Act 1972*, repealed by the *Control of Pollution Act 1974*
2. *Control of Pollution Act 1974*, (c.40) replaced by Part II of the *Environmental Protection Act 1990*
3. *Environmental Protection Act 1990*, The Stationery Office, London (1990)
4. *Environment Act 1995*, The Stationery Office, London (1995)
5. *Control of Pollution (Amendment) Act 1989*, The Stationery Office, London (1989)
6. *Controlled Waste (Registration of Carriers and Seizure of Vehicles) Regulations 1991*, The Stationery Office, London (1991)
7. *Waste Management Licensing Regulations 1994* and amendments SI 1995 No 288, SI 1995 No 1950, SI 1996 No 634, SI 1996 No 1279, SI 1997 No 351, SI 1997 No 2203, SI 1998 No 606, The Stationery Office, London.
8. *Transfrontier Shipment of Waste Regulations 1994*, The Stationery Office, London (1994)
9. *The Special Waste Regulations 1996* and amendments SI 1996 No. 2019, SI 1997 No. 251, SI 2001 No. 3148, The Stationery Office, London
10. *Finance Act 1996* and *Landfill Tax Regulations 1996* SI No. 1527 and amendments SI 1996 No. 2100, SI 1998 No. 61, SI 1999 No. 3270 and SI 2002 No. 1, The Stationery Office, London
11. *Producer Responsibility Obligations (Packaging Waste) Regulations 1997* and amendments SI 1999 No. 1361, SI 1999 No. 3447, SI 2000 No. 3375, The Stationery Office, London
12. National Waste Production Survey, www.environmentagency.gov.uk
13. Scottish Environmental Protection Agency (SEPA), *National Waste Strategy*, Scotland (1999), www.sepa.org.uk
14. Council Directive 75/442/EEC of 15 July 1975 on Waste, amended 91/156/EEC and 91/692/EEC
15. Department of the Environment, Transport and Regions, Circular 11/94: *Environmental Protection Act 1990: Part II, Waste Management Licensing, the Framework Directive on Waste*, The Stationery Office, London (1994)
16. *Waste Management Licensing Regulations 1994*, The Stationery Office, London (1994)
17. Department of the Environment, Transport and Regions, Circular 04/2000: *Planning controls for hazardous substances*, replacing DoE Circular 11/92. The Stationery Office, London (2000)
18. *Controlled Waste Regulations 1992*, The Stationery Office, London (1992)
19. Environment Agency, *Special Waste Explanatory Notes*, www.environmentagency.gov.uk
20. *Environmental Protection (Duty of Care) Regulations 1991*, The Stationery Office, London (1991)
21. Department of the Environment, Transport and Regions, *Waste Management, the Duty of Care, A Code of Practice*, The Stationery Office, London
22. *Water Resources Act 1991*, The Stationery Office, London (1991)
23. *Water Industry Act 1991*, The Stationery Office, London (1991)
24. *The Urban Wastewater Treatment Regulations 1994*, The Stationery Office, London (1994)
25. European Union, Decision 94/904/EC establishing a list of hazardous waste pursuant to Article 1(4) of Council Directive 91/689/EEC on hazardous waste (last updated February 2001), EU, Luxembourg

26. *Chemicals (Hazard Information and Packaging for Supply) Regulations 2002*, The Stationery Office, London (2002)
27. Health and Safety Executive, Publication no: L124, *Approved Supply List* (Sixth edition) (2000) gives classification and labelling information for several thousand commonly-supplied chemicals. Publication no: L199 *Approved Classification and Labelling Guide* (Fourth edition) (1999) provides information on how to classify and label chemicals not listed in the Approved Supply List. HSE Books, Sudbury
28. European Union, *Classification, packaging and labelling of dangerous substances in the European Union, Part 2 (1997)*, a publication of the European Commission detailing test methods, EU, Luxembourg (1997)
29. Environment Agency, A guide to the Special Waste Regulations 1996, The Stationery Office, London. www.environment agency.gov.uk/commondata/105385/specwaste
30. European Union Council Regulation 259/93 of 1 February *1993 on the Supervision and control of shipments of waste within, into and out of the European Community*, EU, Luxembourg (1993)
31. *Pollution Prevention and Control [England and Wales] Regulations 2000*, The Stationery Office, London (2000)
32. www.hmce.gov.uk/business/othertaxes/ccl
33. Bruntland, G. (ed.), *Our Common Future: The World Commission on Environment and Development*, Oxford University Press, Oxford (1987).

Further reading

Environmental Compliance Manual: Practical Guidance for Managers – Gee Publishing Limited
Croners Waste Management, Croners Publications Limited
Croners Environmental Management, Croners Publications Limited
Napier, C., *Waste Management: Legal Requirements and Good Practice for Producers of Waste*
Waste Minimisation: An Environmental Good Practice Guide for Industry, Environment Agency (2001)
Waste Minimisation: A Guide for Industry to their Reduction of Wastes and their Pollutant Effects, Loss Prevention Council (1994)
Waste Minimisation, Institution of Wastes Management (1996)
Practical Advice on Best Value and the Waste Management Industry, IWM Business Services Limited
Resource Productivity, Waste Minimisation and the Landfill Tax, Advisory Committee on Business and the Environment (2001)
The US EPA Manual for Waste Minimisation Opportunity Assessments, The US Environment Protection Agency Report number EPA/600/2–88/025 (1988)
Crittenden, B.D. and Kolaczkowski, S.T., *Waste Minimisation Guide*, The Institution of Chemical Engineers, Rugby (1992)

Chapter 5.4

Chemicals and the environment

J. L. Adamson

5.4.1 Introduction

As people have enjoyed the benefits of economic growth, they have become increasingly preoccupied with the quality of their daily lives. This can be seen in aspirations for healthier living, in the desire for cleaner air, water and streets, and in popular enthusiasm for protecting the best of our urban and rural surroundings. We also have the intense hope that we can pass on to our children what we value most about our own heritage.

Under common law a duty of care is placed on people that can be construed as that degree of care necessary to ensure injury is not caused to a neighbour. The term 'neighbour' refers to those persons who are so closely and directly affected by an act that the person causing the act ought reasonably to have foreseen that they would be so affected. With these issues in mind it is important to ensure that industry and commerce aim for the highest possible standards using legal standards as a minimum requirement.

5.4.2 Chemical data

In order to ensure that an environmental effect is not created, the hazards and properties of the chemicals used must first be determined. Such information can be used in controlling the potential pollution or nuisances at source. Controlling pollution at source is always more effective and cheaper than subsequent remedial action at a later date. 'End of pipe' engineering control measures are usually expensive to install and maintain.

The range of information required on the materials to be handled includes:

- toxicity to aquatic organisms
- toxicity to flora
- toxicity to fauna
- toxicity to soil organisms

- toxicity to bees
- the short- and long-term effects on the environment
- the effect on the ozone layer
- the effect on the atmosphere in an enclosed space
- the potential to cause contamination
- the potential to react with other materials to cause contamination
- the potential to cause a change in pH.

Manufacturers and suppliers must provide information on the properties and hazards of chemicals. With these characteristics established strategies for use and suitable methods of control can be developed.

5.4.3 Risk reduction

Where hazardous chemicals are to be used or disposed of, a priority consideration must be to reduce the risks from them to a minimum. To enable this to be achieved it is necessary to know the hazards presented by the substances and the possible environmental effects if they escape. When the use of a hazardous substance is proposed the following questions need to be addressed:

- Does the material have to be used? Can its use be avoided?
- Can the material be substituted with a less hazardous material?
- Can the quantity of material to be used be reduced? In a chemical reaction one of the constituents is usually present in excess in order to make the reaction proceed to completion. If the chemical that is in excess is the polluting material, development work may be needed to see if the concentrations of the reactants can be reversed or a non-polluting or less hazardous material used instead.

Where hazardous chemical(s) are used they should be assessed to determine the impact on the environment. Such an assessment should also consider the effect of intermediates and breakdown products that may be produced. The assessment should include consideration of the effect of the materials on the atmosphere, the sewage treatment system, controlled waters, land and groundwater, and waste disposal routes. The risk assessment should include an estimation of the risk from the hazardous materials, the magnitude of the effects, and the probability of an occurrence. Included should be an evaluation of the significance of the hazards and the consequences should the environment be affected. It is a structured approach to setting priorities for controlling the hazards. The risk assessment will provide information of where control measures will need to be applied to prevent an impact on the environment.

5.4.4 The Environmental Protection Act 1990 (EPA)[1]

This piece of legislation is the most profound development in British environmental law this century and contains many fundamental changes to pollution control regimes that are having far-reaching effects on industry. The Act is split into nine parts with Parts 1, 2 and 3 having the greatest implications for industry. It is an enabling Act that specifies broad environmental objectives to be achieved and relies on subsidiary regulations to provide the details of the requirements.

5.4.4.1 Part 1: Integrated pollution control and air pollution control

This section deals largely with the administration of the Act outlining the authorisation of the bodies for enforcing it and the powers vested in them. It considers enforcement in two parts:

(i) Integrated pollution control (IPC) by the Environmental Agency (EA).
(ii) Air pollution control (APC) by local authorities (LA).

The Environmental Protection (Prescribed Processes and Substances) Regulations 1991[2] allocate the enforcement of standards between the Environment Agency and the local authorities.

The Environment Agency was formed under the Environment Act 1995 by combining Her Majesty's Inspectorate of Pollution, the National Rivers Authority, and the Waste Disposal Authority. The Environment Agency enforces the larger, more complex and polluting industries such as power stations, the chemical industry and heavy metal manufacture. Such processes are known as the Part A Processes and come under the IPC legislation. IPC deals with the environmental aspects of all stages of a process, aiming to achieve the least environmentally damaging combination of options and thus minimise pollution. It also requires that the end disposal option must not be damaging to the environment. For example, an air pollutant may be removed by scrubbing the gas stream with water. However, what effect does the substance in water have on the sewage works or the end disposal source, namely, the river? If the effect is detrimental then another disposal route must be chosen, e.g. thermal oxidation of the air stream. *Figure 5.4.1* shows a modern scrubbing plant for cleaning vapour discharges.

The conditions in an IPC authorisation cover emissions to water, land and air and extend to the prevention of persistent offensive odours at or beyond the boundary of the premises. The standards demanded by the authorisation include the application of the 'best available techniques not entailing excessive costs' (BATNEEC). This is a new enforcement concept seen as equivalent to 'so far as is reasonably practicable' of the Health and Safety at Work Act, 1974 (HSW). The Environment Agency has produced numerous process- and sector-specific guidance notes on Part A Processes[3].

By contrast, APC deals only with emissions into the atmosphere. The local authorities who enforce it are required to seek and abide by the decisions of the Health and Safety Executive on matters affecting human

Figure 5.4.1 Modern vapour-scrubbing plants to ensure compliance with atmospheric emissions standards (Courtesy Rhone-Poulenc-Rorer)

health and, where there is a possibility of discharges to water courses, of the Environment Agency. APC is a more limited approach than IPC and relates to processes having a supposedly lower pollution potential. Such processes are known as Part B Processes.

Employers who are required to seek authorisation for processes involving discharges, whether to river, land or atmosphere, should not discharge until an authorisation has been granted. However, where delays occur in granting authorisation, existing processes may continue to be run but must meet current environmental standards. For new processes it is an offence under Section 6 of the Environmental Protection Act 1990 to operate a process without an authorisation. When granted, the authorisation will state the discharge standards to be achieved by the process. Failure to meet those standards may attract formal caution, an enforcement notice, a prohibition notice, or prosecution with, on conviction, penalties similar to those under the HSW.

The EPA empowers inspectors to issue Enforcement and Prohibition Notices, which must specify the remedial steps to be taken. There is an appeal procedure against such notices, which is to the Secretary of State who will appoint an arbiter. Included in this part of the Act is a power to recover the cost of making the authorisations and of enforcement – on the principle of 'the polluter pays'. Fees will be charged for initial applications and for the annual renewal of the authorisations.

When Environment Agency Inspectors monitor discharges, they compare their findings with the appropriate authorised discharge limits and take action accordingly. They may employ specialist consultants to monitor the concentrations of air emissions from Part A Processes.

The Integrated Pollution Control Regulations are being replaced by the Pollution Prevention and Control Regulations 2000 (PPC)[4]. These regulations cover a larger range of industries and a wider number of environmental impacts. For example:

- New industries: intensive farming, the food and drink industry, and many waste management facilities.
- Environmental impacts: energy efficiency, noise, waste reduction, condition of the site when the facility closes, consumption of raw materials, accident prevention.

The PPC Regulations apply to the complete installation or the whole site whereas the IPC Regulations apply only to a process. The PPC require the application of the 'best available techniques' (BAT) which is more onerous than the BATNEEC principle and does not give exemption to installations that produce trivial emissions. Because PPC requires that the condition of the site should be left in an unpolluted state when the occupier finally vacates it, the occupier must demonstrate the environmental condition of the site when the PPC Authorisation was first applied for. This is to show that the company as occupier had not polluted the ground during its period of occupancy.

The Environment Agency will provide information on the requirements of PPC and a timetable for its implementation.

5.4.4.2 Part 2: Waste disposal[5,6,7]

Waste is any substance which is surplus to the process or company's requirements. It can be a scrap material, an effluent, other unwanted substances, or any substance or article to be disposed of because it is broken, worn out, contaminated or otherwise spoiled. Under the Special Wastes Regulations 1996 certain substances, listed in Schedule 1 of the Regulations, have been classified as special wastes and their disposal must follow a strict laid down procedure.

A novel feature of the Environmental Protection Act is the introduction of the concept of a 'duty of care'. It places on all involved in the production and handling of waste a number of duties which cover five aspects:

1 To prevent the keeping, treatment or disposal of waste without a licence or in breach of a licence.
2 To prevent the escape of waste.
3 To transfer waste only to an authorised person.
4 To ensure that there is clear information on, and labelling of, the waste.
5 To retain documentary evidence.

The major responsibility for ensuring that these duties are fulfilled rests with the creator of the waste. Thus, the producers of waste have an obligation to prevent the escape of waste along the whole length of the disposal chain through the use of reputable disposal companies and registered hauliers. Transporters of waste have to be registered with the county council for the area in which they are based. Producers must ensure that waste is carried by authorised carriers and disposed of at facilities which have planning permission, are licensed and that the licence allows that type of waste to be deposited. Producers must also satisfy themselves that the waste can be stored securely without loss of containment and that the method of disposal is appropriate. There must be tight control on the documentation for the movement of the waste to demonstrate fulfilment of the duty of care. Producers must periodically audit each stage of the waste disposal operation and there are penalties for non-compliance.

The standards for the keeping and disposal of waste are being significantly tightened with landfill licence conditions becoming more stringent, usually by requiring better leachate control, methane monitoring, aquifer monitoring, etc. Inspectors of the Environment Agency, which has taken over the waste regulation function previously carried out by the county councils, will carry out more frequent inspections. As with IPC, recovery of the cost of enforcement and administration of waste control is introduced. Regulations may be made to extend the definition of 'special wastes' to encompass many of the known hazardous substances in current use. Other regulations may provide for a list of exemptions from licensing for substances even though they are still classified as special waste.

5.4.4.3 Part 3: Statutory nuisances

A statutory nuisance is the emission of smoke, fumes, gases, dust, steam, smells, other effluvia and noise at a level which is judged to be prejudicial to health of, or a nuisance to, the community or anyone living in it. The legal interpretation is somewhat flexible.

Where a local authority is satisfied that either a statutory nuisance exists or is likely to occur or recur, it can serve an Abatement Notice requiring the abatement, prohibition or restriction of the occurrence or recurrence. The notice will specify the date by which the notice is to be complied with and may also specify the remedial action to be taken. Appeals against the notice can be made within 21 days. Failure to comply with the notice can attract a fine on conviction of up to £20 000. It is a defence to a prosecution to demonstrate that the best practicable means had been used to prevent or counteract the effects of the nuisance. A private individual can start a proceeding in a magistrates' court and, if the court is satisfied that a nuisance exists, it can require the accused to abate or terminate it. The court may also specify the remedy to be followed.

5.4.4.4 Summary of EPA

Governmental and public pressure is demanding a significant tightening of environmental standards. Much of the onus for achieving this will fall on industry which may be involved in considerable expense in carrying out the work necessary to meet the new standards.

5.4.5 Minimising environmental harm

The fundamental concept to controlling environmental pollution is that it is better to prevent environmental damage rather than have to cure the harm after an event has occurred. Apart from the environmental improvement there are other considerations and possible advantages to be considered. These include the legal and regulatory implications and the potential of commercially based incentives.

By definition there is an interaction between a pollutant and the environment. It may be obvious or not, for example, because a pollutant is emitted to the atmosphere does not mean that there will not be effects on the ground or on water such as occurs when oxides of sulphur (SO_x) produce acidic deposition on soil or in rivers. This is the basis for IPC, in the UK. IPC is the consideration of the overall environmental effects of discharges on air, water and land. There may be several options for disposing of waste material such as deposition on to land, discharge to water, or incineration with exhaust emissions being discharged to the atmosphere. Under IPC the environment must be considered as a whole with each disposal route being examined to ensure that the ultimate destination of the substance being discharged does not cause environmental pollution. This approach forms the basis of the concept of 'Best

Practicable Environmental Option' (BPEO). The Environment Agency has published guidance on assessing BPEO for IPC processes[8]. The Chemical Industries Association has also published a booklet on the subject[9] which recommends that, in order to determine the BPEO the following data is needed:

(a) The synthesis route options available.
(b) For each synthesis route options:
   - Its technical viability
   - Environment, health and safety issues
   - Capital and revenue costs
   - Likely development times
   - Site policy
   - Implications of related legislation.
(c) Select the preferred process synthesis route and identify:
   - The theoretical amount of waste produced per year
   - Amount of special waste handled per year
   - Chemical oxygen demand (COD) of the effluent stream
   - Toxicity of the effluent stream on sewage treatment
   - Concentration of harmful non-biodegraded material that would pass through the sewage works treatment and into the river
   - Heavy metals in effluent streams
   - Total dissolved solids in effluent steams
   - Inventory and storage of hazardous substances
   - Energy consumption
   - The identifiable circumstances where an unplanned release would have severe effects
   - Quantities of the different air emissions produced by the process
   - Fugitive emissions.
(d) Select the preferred process route based on the data collected in (c) above.
(e) Generate waste management options for the preferred choice.
   This assessment should identify all the waste streams from all stages of the process. Such work should include all minor releases from storage tank vents, effluent pit emissions, etc. For each waste stream consideration should be given to the waste hierarchy of reduction, reuse, recycle as well as end of pipe treatment and disposal.
(f) Review waste management options.
   The list of possible options for the management of each waste stream should be reviewed using an approach similar to step (b) above. The waste management hierarchy of '*Avoid – Minimise – Reuse – Recycle – Recover energy – Dispose*' should be followed.
(g) Assess waste management options.
   This action involves assessing in detail each of the options identified in (f) above where releases to the environment have been identified and quantified. It may be necessary to carry out dispersion calculations for releases to air and water. From this data comparisons are made with local environmental quality limits. The option creating the least overall environmental impact can then be identified.

(h) Select the preferred waste management option.
Section (g) identifies the BPEO for each waste stream. Other aspects of the waste management options such as energy use, technical feasibility, cost, etc., are likely to influence the final selection of the best practical waste management option for each stream. The review should also include consideration of other waste treatment facilities that are already in existence on the site so that the most cost-effective combination of waste treatment methods can be identified.
(i) Check that the environmental impact is acceptable.

The preferred process route and associated waste management options identified by the above procedure is the Best Practical Environmental Option (BPEO).

The results of the estimated discharges should be compared with the levels permitted by the enforcing authority and then collated and documented as part of the preparation of the application for an authorisation under the IPC requirements of Part 1 of the Environment Act.

5.4.5.1 Controlling emissions to the environment

One of the key elements of the Environmental Protection Act 1990 is the concept of Best Available Techniques Not Entailing Excessive Costs (BATNEEC). 'Techniques' includes technology as well as hardware and operational issues such as housekeeping. BATNEEC applies to processes that are subject to Environment Agency and Local Authority control. It is up to industry to ensure that they maintain the BATNEEC standards. However, it is good for other companies who do not come under IPC Regulations, to consider the concepts as a means to demonstrate their commitment to a green image.

The definition of 'Best Available Techniques' is given in 'Integration Pollution Control: A Practical Guide'[3]. A brief description is:

(a) **Best** is the most effective method for preventing, minimising or rendering harmless, polluting emissions.
(b) **Available** means that the technique of control is generally accessible to companies. The source of the control technique is not restricted to the UK but it may be in use elsewhere in the world.
(c) **Techniques** embraces both the plant used in the process and how the process is operated. It includes matters such as staff numbers, their qualifications and experience, working methods, training, supervision and the manner of operating the process. It also includes the design, construction, lay-out and maintenance of the plant and buildings.
(d) **Not entailing excessive cost** presumes that BAT will be used. However, it also means that BAT can be modified by economic considerations where it can be shown that the relative costs of applying BAT would be excessive compared with the degree of environmental protection achieved. For example: if one technology

reduces pollution by 90% and another reduces the pollution by 95% but at four times the cost, then it may be reasonable to assume that because of the small additional benefit derived from the enormous extra cost, the second technology would not be justified and that BATNEEC would be satisfied by the original technology. However, if the pollutant involved was particularly dangerous then the extra cost may not be considered to be excessive but necessary on health grounds.

It should be noted that the Environment Agency or Local Authority Inspector must decide in each individual case what is the optimum BATNEEC option and translate it into requirements in the conditions of the IPC authorisation.

5.4.5.2 Legal and regulatory requirements

In many instances there are fixed limits for discharging materials to air, to the sewer or to controlled waters. These limits need to be agreed with the enforcing authorities prior to starting the process. These criteria need to be given to the design team prior to starting the design of the process and its control system. For chemical reactions it is important that the design team are made aware of the permitted rates of emission against time of the reaction process since not all reactions are linear in their emission rates.

Limits for discharges to the atmosphere can be obtained from the Environment Agency or the Local Environmental Health Department. Advice on the criteria to be applied for aqueous discharges to controlled waters can be obtained from the Environment Agency. Limits for discharges to the sewer can be obtained from the local Water Company.

If limits need to be applied to the discharges then it is likely that the particular enforcing authority will require regular monitoring by the company to demonstrate compliance with the requirements. The enforcing authority may also occasionally monitor the discharge themselves to check compliance with the standards. The permitted limits will be written into the IPC authorisation, or Trade Effluent Consent document issued by the respective enforcing authority.

5.4.6 Air pollution: control measures and abatement techniques

The main aims of environmental protection are to prevent the pollution in the first instance or to reduce the pollution at source before an environmental effect occurs. If, having taken all steps to reduce pollution at source and still not achieving the target levels, it may be necessary to employ 'end of pipe' techniques as well. Such technologies do not result in an intrinsically cleaner process, but rather reduce to an acceptable level the emissions to atmosphere from the process. There are a number of different types of abatement equipment available on the market, and their application to a specific process is a matter for advice from an expert consultant or contracting company. Such abatement equipment includes:

(a) *Mechanical separators*. These are designed for large particles of greater than 300 microns. The removal mechanism employed is deposition. They are of low initial cost and of low efficiency. The residues collected can be removed using a screw conveyor for disposal. If fine particles are present in the air stream then the equipment must be used in conjunction with other types of abatement equipment.
(b) *Cyclones*. This is a more advanced type of mechanical separator. The equipment is designed to remove coarse dusts of 20 to 75 microns. Below 20 microns the efficiency drops off rapidly. The removal mechanism employed is centrifugal force. The residues collected can be removed from the base of the unit using a collecting bag or a screw conveyor for disposal. Cyclones are not very efficient when used with fabric filters.
(c) *Fabric filters and bag filters*. This type of filter can be 99% efficient in the removal of particles greater than 10 microns, or 90% efficient for the removal of particles greater than 1 micron. The removal mechanism is by impingement of the dust on the fabric of the bags. By mechanically shaking the filter the material falls off and is collected at the base of the unit for removal and disposal. These units are efficient but have a high running cost. Fabric filters and bag filters are affected by moisture so it is important that the temperature of the unit is maintained above the dew point.
(d) *Wet scrubbers*. These types of filter can be used for particles, soluble gases and some organic compounds. They are good for high temperature moisture laden gas streams and are 99% efficient for removing particles of greater than 5 microns. The process gas stream is bubbled through a liquid, usually water but it can be passed through an acid or alkaline liquid to remove corrosive gases. Units that work this way produce a large pressure drop across the filter. A common type of wet scrubber system passes the gas stream through a fine spray of liquid which offers little resistance to the air flow. The collected water is likely to be contaminated and may need separation and chemical treatment before disposal.
(e) *Electrostatic precipitators*. These units can be used to remove particles down to 2 microns with a removal efficiency of 95%. They are very efficient for removing dry dust particles but are very expensive and often require a pre-cleaner in the gas stream. The removal mechanism is by electrostatic attraction. They operate at a flow rate of less than 1.5 m/s over a wide range of temperatures but can be sensitive to moisture. The residues separated out fall to the base of the unit from where they can be removed.
(f) *Gas absorbers*. Gas absorbers are used to remove highly odourous or toxic gases. They work by passing the gases through a packed column through which is passed an absorbing liquid. The absorption liquid will require renewal to maintain its effectiveness. At the end of its useful life it may need treatment prior to its disposal.

(g) *Incineration*. These types of unit are commonly used for destroying organic compounds and odours. They can be very efficient but are expensive to install and operate. Incinerators work by oxidising the pollutant to simple combustion products (carbon dioxide, water vapour, etc.). It should be remembered that when using natural gas and air as the fuel oxides of nitrogen would also be produced. Catalytic combustion can be used to reduce oxides of nitrogen. The heat generated can be used to pre-heat the incoming gases or to generate steam in a boiler for use elsewhere on the plant.
(h) *Absorption columns*. Absorption materials such as carbon can be used to remove organic vapours. However, they have the disadvantage that eventually the media becomes saturated and needs to be replaced or regenerated.
(i) *Bio-technology*. These units are used for removing organic vapours and odours. The gas stream is passed through a media containing micro-organisms that metabolise the organic vapours to produce harmless carbon dioxide and water. They operate within a temperature range of 15 and 40°C and require a high humidity with a constant supply of organic compounds and nutrients. The micro-organisms can become damaged if the flow of their feedstock is not constant. They can also be poisoned by certain chemicals, such as sulphur dioxide.

5.4.7 Monitoring atmospheric pollution

In order to assess the level of pollution discharged to the atmosphere it is necessary to monitor the emissions at the discharge point and compare the results with the allowed environmental limits. Information on such limits can be obtained from the Environment Agency, or the Local Environmental Health Department. Where limits do not exist it will be necessary to assess the maximum ground level concentration, which occurs when the emitted plume inverts and returns to ground level. Computer modelling can be used to calculate the anticipated concentration levels, which should be less than one-fortieth of the 8-hour time weighted average (TWA) Occupational Exposure Standard[10].

The choice of measuring equipment depends on a number of factors including:

- The pollutant to be monitored
- The source of the pollutant
- Accuracy required
- Whether the pollutants cause interference with the method of test
- Size of stack or exhaust duct
- Velocity and temperature of discharge
- Whether continuous or periodic sampling is required
- Ease of operation
- Cost.

Typical measures taken of the emissions include temperature, flow rates, particulates and the concentrations of pollutants. Various of the techniques are used are described below.

5.4.7.1 Temperature

Normally a mercury in glass thermometer is insufficiently robust and cannot be read remotely so are not generally used.

(a) A *thermocouple* is the simplest device for measuring temperature. When two dissimilar metals are joined they generate an electromotive force (emf) that is proportional to the temperature. Two thermocouples are used in the measuring circuit, one located at a reference temperature and the other in the gas being measured. As the temperature of the system alters so the electromotive force generated changes. Thermocouples have the advantage that they are robust, react quickly to temperature changes, can be used remotely and, because they generate an electrical current, can be used in control circuits.
(b) *Resistance thermometers* work on the principle that the resistance of an electrical conductor varies with the temperature. While they can be useful for remote reading of temperature, their response to temperature changes is relatively slow.
(c) *Acoustic temperature measurement* is based on the principle that the speed of sound is proportional to the temperature of the gas through which the sound pressure wave passes.

5.4.7.2 Flue gas flow rates

(a) The pitot tube is the standard method for measuring air velocity in ductwork and can accommodate the presence of solvent vapours and high temperatures. However, dust laden air may cause false readings. In use the head of the pitot tube must face into the gas flow. The standard pitot tube consists of a pair of concentric tubes. The inner tube has an open end that faces into the gas stream and measures the 'total' gas pressure. The outer tube, which is sealed to the inner tube at its leading end has a series of holes at right angles to the gas flow that measure the 'static' pressure of the gas. The tubes are connected to the opposite ends of a manometer that measures the differential pressure which is proportional to the gas velocity.
(b) A *rotating vane anemometer* gives a direct reading of the velocity of the flue gas. It comprises a rotating vane that is driven by the gas flow. The rate of gas flow is indicated by the speed of rotation of the vanes but a correction factor may need to be applied to allow for different gas densities.
(c) *Hot wire anemometer* comprises an electrically heated wire that is inserted into the gas stream. The flow of gas cools the wire and the change in electrical characteristics indicates the rate of gas flow. Allowances may need to be made for the temperature of the gas stream.

5.4.7.3 Particulate concentration

(a) *Transmissiometers* are the most commonly used devices for monitoring the particulate concentration in a duct or stack. They work by passing a light across the duct and reflecting the beam at a mirror. The light beam is attenuated by the presence of particulate matter in the duct. The measured reduction in light intensity is converted to an electrical signal, which can be used to monitor the concentration of particulates in the gas stream. Transmissiometers are used to monitor particulate concentrations in the range of 50 to 200 microns and are most efficient when measuring particles of a small diameter.
(b) *Beta attenuation monitors* are used to measure the dust concentration that has been collected on a filter paper. A sample of air from the duct is drawn out of the duct isokinetically and through a filter paper. The layer of particulates that are collected on the filter attenuates the intensity of the beam of beta rays that is passed through the paper. The attenuation is approximately proportional to the surface concentration of dust on the filter paper. The monitors can be used for particulate measurement in the range of 2 to 2000 microns.

5.4.7.4 Gas and vapour concentrations

(a) *Infra-red absorption spectrometry* is used to monitor a wide range of organic vapours. A twin beam instrument works by passing the gas stream through a sample cell. The radiation from the source in the cell is split by means of mirrors into two beams. One beam is passed through a reference cell and the other beam through the sample cell. The beams are brought together onto a half silvered mirror or 'chopper' that rotates at a constant speed allowing each beam alternately to reach the detector. With the unit sited in a clean atmosphere the instrument can be zeroed since the intensities of the two beams reaching the two halves of the detector will be equal and the heating effect on the gas within the two halves of the cell is the same. The capacitance of the detector is therefore at its baseline level. The higher the concentration of pollutant in the measuring cell, the less will be the amount of radiation reaching the sample half of the detector. This will cause the capacitance of the cell to change. The quantity of pollutant is proportional to the amount of infra-red energy and so the concentration of pollutant can be measured.
(b) *Ultra-violet spectrometry* works in a similar way to infra-red absorption spectrometry except that the detector is replaced with a UV detector. Since glass cannot be used because it is opaque to UV, quartz is used instead. The instrument can be used to monitor mercury vapour.
(c) *Flame ionisation* is used to monitor organic vapours with the exception of formic acid. The instrument is most sensitive to hydrocarbons and is less sensitive when oxygen, sulphur and chlorine are present. An electric field applied across an ionised gas cloud causes a current to flow between electrodes depending on the charged species present and the structure of the molecule. The current generated is approx-

imately proportional to the number of carbon atoms being measured. The flame ionisation meter consists of a cell containing the hydrogen and air burner. The vapour sample is introduced into the flame together with the fuel gas. Orifices which have been calibrated to give the correct flow at an indicated gas pressure control the flow rates of hydrogen and combustion air. When no hydrocarbons are present in the gas, the hydrogen flame produces a small ion current. When a constant flow of sample gas contains hydrocarbons, a current is generated which is proportional to the concentration of methane or propane equivalents, dependent on which gas is used to calibrate the equipment.

(d) *Colorimetry paper tape* equipment works on the principle of a colour being produced when a gas reacts with a reagent. When a gas stream is passed through a paper tape that has been impregnated with a reagent, a stain is produced. The intensity of the stain is measured using reflectometry and hence the concentration of pollutant can be determined. Hydrogen sulphide and hydrogen chloride can be measured using this equipment.

(e) *Gas chromatography* is used to monitor organic vapours. Components in a gas stream are separated when they are passed through a column containing an absorption medium. The separation of the mixture is dependent on the selective absorption of the various components onto the column medium. The components are carried through to a flame ionisation detector. An electron capture detector is used if chlorinated gases are present.

(f) *Conductimetry* is a technique that relies on the change in conductivity of a solution resulting from increasing the concentration of the gases it contains. The conductivity of the solution is measured at the start of the sampling period and again after a known time has elapsed. The technique is temperature dependent so maintenance of a constant temperature is important. The method is used for monitoring sulphur dioxide.

5.4.8 Control of water pollution

Industrial activities can produce a large amount of water pollution. The Water Industries Act, 1991[11], sets out the provisions regarding the duties of water and sewage undertakings to set consent limits for the discharge of trade effluent. Discharges of industrial wastewater to a sewer must comply with consent limits set by the sewage undertaking. These limits are set to protect the sewage works, the drains and its environment, and also the river into which the final treated effluent is discharged. Section 5 of the Environmental Protection (Prescribed Processes and Substances) Regulations 1991[12], describes a list of substances that are strictly controlled. This list is also known as the RED LIST and includes mercury, cadmium, and pesticides. The EEC Directive 76/464/EEC[13], describes the BLACK LIST of substances that are highly toxic, persistent, carcinogenic or liable to bioaccumulate. A GREY LIST of substances is also described which includes metals, cyanides, biocides and ammonia.

The Water Resources Act 1991[14], is designed to prevent the pollution of inland waters by providing the Environment Agency with powers to initiate remedial works to prevent pollution by companies and by water and sewage undertakings who discharge directly to a controlled water. Pollution includes materials that are poisonous and noxious to the controlled water and also any solid matter. Solid matter can stifle life in the aquatic media. Controlled waters include:

(1) *Relevant territorial waters*. These are waters that extend three nautical miles seaward from base lines from which the breadth of the territorial sea is measured.
(2) *Coastal waters*. These waters extend from landward of the base lines to the high water limit and to the freshwater limits of any river.
(3) *Inland fresh waters* include rivers and water courses above the freshwater limit and any lake or pond.
(4) *Groundwaters* are waters that are contained in underground strata.

Companies need to consider many aspects when discharging trade effluent to the sewer including:

(a) The effect of the discharge on the material of construction of the drain. For example a high sulphate content can cause the deterioration of concrete drains.
(b) The potential hazards of vapours or gases that may be evolved and collect in the headspace above the liquors in the drain. Such a situation can be dangerous for personnel who have to enter the drain. For example solvent vapours can create an explosive atmosphere, and hydrogen sulphide can create a toxic atmosphere in the drain and can cause nuisance odours for neighbours.
(c) High concentrations of suspended solids, toxic materials and oils/greases in the discharge can adversely affect the sewage treatment works process.
(d) Toxic non-biodegradable materials may pass through the sewage treatment process unchanged and create harm to the controlled waters into which the treated liquors are discharged.

Companies need to consider what is in their effluent streams to ascertain the hazardous condition of the discharge. There may be a need to pass the liquors through a settlement tank that contains a system of weirs to remove solid materials. Water-immiscible solvents will also be removed using this technique. The liquors may be corrosive and require treatment to keep the pH level within consent limits. Effluent discharges containing a high chemical oxygen demand (COD) may cause an overload of the sewage treatment process particularly if the discharges are very irregular in COD strength. Solutions to the problem should be discussed with the local sewage treatment works. To reduce the level of pollution discharged the liquors can be collected in a storage tank for a controlled release to the sewer at an acceptable rate. Such a system could also be used to control the release of materials to the sewer that are toxic to the sewage works. Alternatively, companies could treat their own trade effluent or make a contribution to the sewage treatment works to help them meet the extra cost of improving the sewage treatment needed. It is important that companies train their employees in environmental awareness so that they are made to realise the harm that can be created when pollutants are tipped down the drain without any thought of the consequences.

5.4.9 Groundwater pollution[15]

Surface water is water that flows across the land and drains into land drains, ditches, streams, and rivers. The concentration of pollution in such waters can change many times dependent on environmental conditions. Groundwater is water that soaks into the aquifers. An aquifer is a layer of porous rock, which is able to store water. It is from the aquifers that fresh water can be abstracted for use. Water movement in the aquifers is slow and therefore any pollution that reaches such a source will remain there for decades. It is vital that the aquifers are protected from pollution, which can occur at a specific point or from a diffuse source. Examples of point sources are landfill sites, or a leaking chemical storage tank, or historical pollution from beneath a chemical plant. A diffuse source is pollution from over a large area such as the application of a pesticide to agricultural land or it can be from a number of point sources over a wide area of land.

It is good policy for any company handling chemicals to know the condition of the groundwater beneath its site. This information will be needed for PPC applications. Such an assessment looks at the historical condition of the groundwater. It is also prudent policy for the company to review the current and future site conditions to enable them to take action to prevent groundwater pollution. Such a review should include a physical survey of the site to look for potential groundwater and surface water contamination areas. The review should include:

- Chemical storage areas, which should be bunded to 110% of the gross contents or maximum sized vessel.
- Chemical storage tank off- or on-loading points should be sited within the bunded area or over a collection bund.
- Tankers off- or on-loading chemicals should be sited in a bunded area that is located off the main road system on the site.
- All chemical containers should be labelled and stored in approved bunded areas.
- Materials stored in bunded areas should be stored away from the edge, or in such a position that if they fell they would remain in the bund.
- Fire sprinkler water, if activated in an emergency may become polluted with site chemicals and provision should be made to collect it in a lagoon or collection pit.
- Chemical production areas should be drained to a sump to prevent the escape of polluted liquid.

When considering the erection of a new facility on the site planning permission must be obtained from the local authority. The authority may request an environmental impact assessment of the proposal. The assessment should address the potential questions that are likely to be asked by the local authority political representatives when they consider the planning application. Providing this detail will save time in getting approval. The Environmental Health Authority may also request information on the state of the groundwater in the area and include a list of chemical pollutants that they wish to have monitored. Another area of concern is the potential for gas (such as methane or hydrogen sulphide), to seep into basement areas of the facility. If this is a risk, an impervious membrane may need to be incorporated into the foundations for a new building.

5.4.10 Waste disposal and duty of care

The general definition of waste is that it is material that nobody wants. Waste not only covers material that is disposed on a landfill, by chemical treatment, or by incineration, but also disposal to a sewer or to the air. This section deals with the former.

What is waste to one company could be a raw material to another. Companies who need to apply for Pollution Prevention and Control (PPC) Authorisation will need to demonstrate what action has been taken to minimise waste on their sites. They need to manage their waste without endangering human health or the environment and should follow the 'Waste Hierarchy':

- waste minimisation (the most desirable option);
- reuse of materials;
- recycle materials;
- recovery of materials (including energy from waste, and composting);
- disposal by landfill, or chemical treatment, or incineration without energy recovery (the least desirable option).

In the UK, waste is classified into two types – special waste and controlled waste.

Special waste or hazardous waste is waste which meets the hazard criteria defined in the 'Special Waste Regulations 1996'[16]. The material can be described as dangerous, hazardous, or toxic waste. Because of the hazards of special waste the Regulations require that movement of such waste is tracked from source to final disposal by means of a consignment note system. Examples of Special Waste are asbestos, waste from a laboratory or waste from a hospital.

Controlled waste is household, Industrial and Commercial Waste that is produced on factory, trade or business premises, or as a result of sport, recreation or entertainment activities. Controlled waste can be collected by the local authorities for disposal to landfill sites.

5.4.10.1 Developing a waste management strategy

The first step in developing a waste management strategy is to assess the type, quantity and origin of all wastes and emissions on site. This task requires a mass balance of all materials that are brought onto and sent off the site and should include lists of:

- All the incoming, intermediates and final products on the site including packaging.
- All utilities used: water, gas, electricity, oil etc.
- A list of all the waste that is produced on the site, including discharges to the sewer, emissions to air, and waste sent off site for disposal.
- Materials sent off site for recovery; including materials used to generate energy.

Each item on the list needs to be quantified and a hazard assigned. The waste generated then needs to be reviewed and the following questions asked:

- Does the hazardous material need to be used?
- Can a less hazardous material be used?
- Can the quantity of material be reduced by changes to the process? (Waste minimisation.)
- Can the waste materials be recovered, re-used or recycled?
- Can the waste be detoxified or otherwise rendered less hazardous through a chemical, thermal, physical or microbial process?

If wastes cannot be eliminated or reduced by the above techniques the material may need treatment, either in-company or by specialist contractors, as part of the disposal process. The methods of disposal include:

- *Physical treatment* which requires the segregation of the waste into its components. Such processes include solid–liquid separation using techniques including coagulation, centrifugation, sedimentation and filtration.
- *Chemical treatment* which involves processes that alter the chemical structure of the waste's constituents. Neutralisation is an obvious technique but other methods will depend on the materials involved.
- *Biological treatment* for organic wastes, including treatment with bacterial products, anaerobic digestion, and by composting. These techniques operate on the principle of microbial breakdown of the waste and use specific techniques for particular waste streams.
- *Incineration* is an expensive disposal option. The technique involves the use of a purpose built plant that operates at high temperatures. Incinerators can be used to destroy highly toxic waste and flammable solvents. The process needs a support fuel (which can be the waste solvent) to destroy solid and non-combustible waste. Some companies send their solvent waste to specialist disposal companies who produce a solvent blend, which is suitable for use as a fuel in a cement works.

In this case, the composition of the blended solvent must be strictly controlled to ensure that the cement works does not exceed its emission limits.

- *Landfill* is the cheapest and the most commonly used disposal route in the UK. However there are increasing restrictions on the type of material that can be sent for disposal by landfill. Well-managed landfill sites are suitable for less hazardous waste. The water from the site's digestion process is sent to the sewage works for disposal and the methane, which is produced as the materials break down, is often collected and used to generate electricity.

5.4.10.2 Duty of care

Legislation places a duty of care on all those involved in the handling of waste, from the creator to the final disposer. The extent of the duty depending on degree of hazard of the waste.

(a) Controlled waste

The Environmental Protection (Duty of Care) Regulations 1991[17] require that the movement of waste is performed under a system of transfer notes and records. The transferor (the giver), and transferee (the receiver), of controlled waste must complete and sign a transfer note. This note must contain the following information:

1 Identification of waste including:

   - the quantity of waste
   - whether the waste is loose or in a container when transferred
   - if it is in a container, the type of container
   - the place and time of transfer.

2 Name and address of the transferor and transferee.
3 Whether the transferor is the producer or importer of the waste.
4 If the transferee is authorised for transfer purposes which of those purposes apply.
5 The category of authorised person of the transferee (and transferor, where applicable), or if exempt, which exemption applies and why.

The code of practice contains a suggested form to cover this procedure, but alternatives can be adopted provided all the relevant information is covered. The 1991 Regulations do not require each transfer of waste to be individually documented but allow a single transfer note to cover multiple consignments of waste transferred at the same time or over a period of time provided that the details of the waste remain unchanged.

(b) Special Waste

The transfer of special waste is regulated by a system of consignment notes to ensure that waste can be tracked at every stage of its movement, from the time it is originally transferred to the point where it is finally disposed of at a suitably licensed site. The system requires the use of a pre-printed five-copy self-carboning pad of consignment notes with each of the five sheets coloured differently. The use of the sheets is as follows:

1st sheet (white)	Pre-notification copy – consignor to send to consignee's Agency office
2nd sheet (yellow)	Copy for consignee to send to own Agency office.
3rd sheet (pink)	Consignee's copy – keep for site lifetime
4th sheet (orange)	Carrier's copy – keep for 3 years
5th sheet (green)	Consignor's copy – keep for 3 years

A consignment note can be used for a single notification; for repeated shipments of similar material, and for a carrier's round.

When the waste is first identified for disposal the waste producer must notify the local Environment Agency office responsible for the waste's final destination, by sending them the top white Pre-notification sheet of the consignment note. The waste cannot be moved until at least three clear days have elapsed after notification. The remaining four copies of the consignment note remain with the waste and the different copies are completed, as appropriate, by the waste producer, the carrier and the waste manager. It is important that the consignment note contains accurate information on the description and quantity of the waste. The waste must be labelled with its name and the hazard it presents, and be transported safely. This means that waste containers must be of an approved standard, sealed and in good condition.

5.4.10.3 Selection of a waste contractor

When selecting a competent carrier and waste contractor it is important that the producer identifies companies who will handle and eventually dispose of the waste legally. Discovery of illegally disposed waste could result in prosecution of the waste producer, even if the waste was transferred to a contractor in good faith. The waste producer must be able to prove that he took steps to check the credentials of the waste carrier and waste contractor. Such steps are especially important when dealing with special waste and could include trailing the consignment of waste to the disposal site to ensure the correct procedures are followed.

When disposing of waste, the following steps describe a simple process to ensure selection of competent contractors:

- Provide a description of the waste, hazard, quantity, and the type of container in which it is stored.
- Select a registered waste carrier and obtain a copy of his registration certificate.
- Approach a number of disposal contractors to tender for the disposal business.
- Select an appropriate disposal system and licensed contractor. Obtain a copy of the waste contractors licence.

5.4.10.4 Duty of Care audit

The Environment (Duty of Care) Regulations 1991[17] require that the waste generator take responsibility for the waste from 'cradle to grave'. The duty does not end when the waste leaves the premises but extends to ensure that it is treated in a proper manner throughout its transportation and final disposal. It is important that a potential waste contractor is subjected to a Duty of Care audit prior to final selection.

An example of a suitable set of questions to ask when carrying out a Duty of Care audit is as follows:

DUTY OF CARE AUDIT REPORT

Disposal Company's Address	
Date of Visit	
Type of producers waste handled	
Description of producers waste handled	
Type of producers containers and labelling	
Waste management licence number	
Waste carriers licence number	
IPC Authorisation number	
Registered waste broker	
Company contact name	
Other regulations applicable to the site	
Auditor	
Circulation of Report:	

Typical audit questions

1 *Site ownership*
- 1.1 Who owns the site?
- 1.2 Name of the site manager.
- 1.3 How long has the waste company been on the site?
- 1.4 What operations were on the site before the waste company started their disposal activities?
- 1.5 Are there any outstanding or potential future problems associated with the previous operations on the site?
- 1.6 Are any other companies currently working on the site?
- 1.7 Do these companies generate waste; and if so what is it?
- 1.8 Do these other companies' operations impact on the waste company's site and operations?
- 1.9 What arrangements are in place to segregate the different operations?
- 1.10 What environmental insurance cover is there and what is its scope?

2 *Security*
- 2.1 Is the whole site fenced and locked when not in use?
- 2.2 Is the waste company's operations segregated and locked off from the rest of the site?
- 2.3 Is the site manned 24 per day and 7 days per week?
- 2.4 Can vandals, scavengers or children get onto the site?
- 2.5 Have there been any such events, what happened and what was the follow-up action? Were the police involved?

3 *Type of waste handled on the site*
- 3.1 Controlled waste.
- 3.2 Special waste: type of hazards: flammable/explosive/toxic/corrosive.
- 3.3 Does the site waste disposal licence, and/or IPC authorisation, cover the waste handled?
- 3.4 Does the site analyse the waste received to ensure that it complies with the producers specification and the site disposal licence, etc?
- 3.5 Is the waste site exempt a licence? Check why and ratify.
- 3.6 What are the special waste hazards and associated problems?

4 *Storage conditions*
- 4.1 Is the waste stored under cover as protection against the weather?
- 4.2 Are there any sunlight concerns?
- 4.3 Could the waste generate peroxides and if so what checks/safeguards are carried out as protection against possible incidents?
- 4.4 Is the waste stored on hard standing?
- 4.5 Are there any cracks in the concrete?
- 4.6 Is the waste storage area bunded?

4.7 Can stored material fall outside the bund?
4.8 What is the allowed maximum drum stack height? Is this a safe figure?
4.9 How is rainwater dealt with on the site?
4.10 Is there a spillage procedure and how are spills dealt with?
4.11 What is the standard of labelling of the waste containers?
4.12 What is the condition of the stored waste containers?
4.13 Are there any legal requirements for the storage of waste and materials on the site?

5 *Other activities undertaken by the waste contractor on the site*
5.1 Are there any other work activities on the site that are not connected with the waste disposal operations?
5.2 What waste does this work generate and how is it dealt with?

6 *Waste disposal arrangements on the site*
6.1 What waste is generated by the disposal operations on the site?
6.2 How is this waste disposed of?
6.3 Is the producer's own waste involved in this disposal?
6.4 Have any complaints been raised in connection with this disposal?
6.5 Have the enforcing authorities raised any issues or legal issues with these disposal arrangements?
6.6 What follow-up actions have been taken to resolve the issues raised?
6.7 What recycling or reuse of waste is undertaken on the site?

7 *Chemical effluent*
7.1 What chemical effluent is generated on the site?
7.2 Is the effluent treated before it is discharged?
7.3 Is the producer's waste involved in this disposal route?
7.4 Are any chemical effluent discharge limits applied to the site's effluent stream?
7.5 Is chemical effluent monitored and are the results in compliance with the limits?
7.6 Have any concerns or legal issues been raised by the enforcing authorities about effluent discharges from the site?
7.7 What remedial plans have been introduced as a result of the concerns raised by the enforcing authorities?

8 *Atmospheric emissions*
8.1 What atmospheric discharges take place on the site?
8.2 Are there any limits on emissions to the atmosphere from the site?
8.3 What treatment system is in place to control the discharges to atmosphere?
8.4 Are the emissions monitored and how is this done?
8.5 Is the monitoring continuous?
8.6 Are the discharge results in compliance with the emission limits?

8.7 Are there any steam emissions or odours from the plant?
8.8 Have any concerns or legal issues been raised by the enforcing authorities about emissions from the site?
8.9 What remedial plans have been introduced as a result of the concerns?
8.10 Is the producer's waste involved in the processes that produce discharges to the atmosphere?

9 *Surface waters and groundwater*
9.1 Are there any discharges from site operations to surface waters, controlled waters or groundwater on the site?
9.2 Are these discharges polluted?
9.3 Have any concerns or legal issues been raised by the enforcing authorities about discharges from the site?
9.4 Are there any legal limits on discharges from the site?
9.5 Is there a collection tank for all surface waters on the site so that the liquors can be tested before releasing to the river?
9.6 If a fire occurred what would happen to the fire water run-off? Is there a potential for surface water or groundwater pollution?
9.7 What steps are taken to prevent pollution of surface water and groundwater systems on the site?
9.8 Is there any evidence on the site of pollution of the surface water or groundwater?

10 *Documentation and records*
10.1 Does an authorised person on behalf of the receiver of the waste check the transfer note for compliance with the Regulations, and that it meets the requirements of what has been agreed with the sender of the waste?
10.2 Do the records identify the sender of the waste?
10.3 Have the sender, the carrier and the receiver of the waste signed the transfer note?
10.4 Is the date and time of arrival of the waste recorded?
10.5 Is the name and address of sender of the waste recorded?
10.6 Is the waste analysed and are the results recorded?
10.7 Does the company have a good record system for maintaining waste information from previous years?
10.8 What happens to waste that fails the waste specification?
10.9 What records are kept of waste that is shipped off site for onward disposal?
10.10 For blended loads, is the waste analysed and the results compared with the required specification?

11 *Checking of waste disposal arrangements of other disposals down the chain*
11.1 What checks are carried out by the waste disposal company to ensure that the waste sent off site meets legal requirements?
11.2 Does the waste disposal company carry out Duty of Care visits to the disposal companies that they deal with? Examine reports produced.

11.3 Have any issues been raised and are there any outstanding actions?

12 Issues raised by the waste disposal chain members

12.1 Issues raised by the waste producer about the carrier.
12.2 Issues raised by the waste producer about the waste disposer.
12.3 Issues raised by the carrier about the waste producer.
12.4 Issues raised by the carrier about the waste disposer.
12.5 Issues raised by the waste disposer about the waste producer.
12.6 Issues raised by the waste disposer about the carrier
12.7 Issues raised by the waste disposer about the disposer down the chain.

13 Complaints

13.1 Has the site received any complaints from or been in legal action with:

- Enforcing Authorities
- Police
- Local Authorities
- Neighbours
- Public
- Contractors.

13.2 What action has been taken to resolve the issues raised?
13.3 Are there any outstanding complaints?

14 Health and safety issues

14.1 Are there issues with regards to unsafe situations on site in respect of:

- Chemical exposure
- Potential fire or explosion situations
- Dangerous handling activities
- Tanker off-loading concerns
- Internal and external noise.

15 Enforcing authorities

15.1 Who are the enforcing authorities for air, water, groundwater and waste disposal?
15.2 Is there a procedure for informing them if something goes wrong?

16 Other matters

16.1 Are there any concerns with the site, the operations carried out, or the documentation systems, that would cause the waste producer not to deal with the waste company?

5.4.10.5 Responsibilities of the waste producer

Once the waste contractor has been selected, the waste producer should regularly inspect the disposal site. Some companies carry out a Duty of Care visit annually; others are less frequent at three years. It is important for the waste producer to ensure they have a current copy of the carriers registration and waste contractor's licence. The most important licence condition that the waste producer should check is that the disposal site is licensed to dispose of the type of waste generated by the producer, and that the carrier can handle that waste. The Environment Agency has the responsibility for enforcing the site licence conditions. Any queries, concerns or unanswered questions should be directed to the Agency.

5.4.11 Reuse or recycling of industrial waste

Consideration of the cost of investigating or developing methods for the re-use or recycling of industrial waste may appear prohibitive but many companies that have carried out investigations have discovered that not only do they achieve an environmental improvement but the exercise proves economically viable.

Examples of the recovery and re-use of waste materials include:

(a) Solvents distilled for re-use, for resale, or for use as a fuel in a boiler or a regenerative thermal oxidiser for the destruction of volatile organic compounds (VOCs).
(b) Waste solvents blended with other waste solvents to meet an agreed specification for use as a fuel in a cement kiln.
(c) Recovery of used oil for sale as recovered oil, or blended with new oil to produce premium grades, or as a fuel.
(d) The collection of cardboard and waste paper for recycling.
(e) The collection of metal waste for resmelting.
(f) Metal drums being refurbished and sold as second hand drums.
(g) Waste plastic can be shredded, cleaned and mixed as a filler in the production of new products. Such products may not be used in the food industry.
(h) Scrap wood recycled chipboard.

5.4.12 Environmental management systems

There are two internationally recognised environmental management systems: ISO 14001 – Environmental Management Systems[18]; and EMAS – Eco-Management and Audit Scheme[19]. Both standards have the basic aim of minimising the environmental impacts and the continual improvement of overall environmental performance. There is no legal requirement to conform with an Environmental Management System (EMS) but they do provide a basis for understanding the impact of the company's operations on the environment and demonstrate a systematic approach to recording environmental information. By following the principles of an EMS it is possible to identify shortcomings in environmental compliance. Identifying these shortcomings will assist in reducing the environmental impact of operations and will demonstrate a commitment to continual improvement in environmental standards. The Environmental Agency encourage companies to attain an EMS registration.

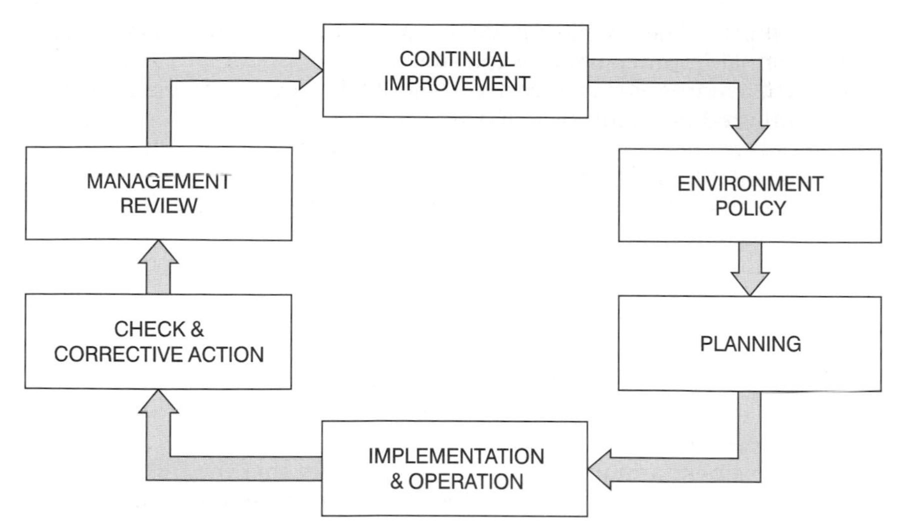

Figure 5.4.2 ISO 14001 Environmental Management System

5.4.12.1 EMS requirements

The ISO 14001 EMS was designed to be compatible with ISO 9000 Quality System. Both systems share many common management principles, techniques and procedures. However, the main distinctions between the systems are described in the policies of both management systems.

Figure 5.4.2 describes the ISO 14001 EMS.

The suggested steps in the development of an EMS are:

1 Environmental policy
 (i) The policy should specify the company's commitment to comply with the relevant legislation and to maintain a continual environmental improvement. The policy should be in writing and a copy available for inspection. It should also be communicated to the workforce and reviewed annually.

2 Planning
 (i) The relevant legislation and regulatory requirements that apply should be listed.
 (ii) The planning of an environmental system should involve the identification of those aspects of the company's operations that impact on the environment as potential pollution sources. The effect of the aspects should be assessed to determine the extent of the impact on the environment. Impacts can be ranked, either qualitatively as one, two or three or qualitatively as high, medium and low, based on the quantity of pollutant involved and the seriousness of its effect on the environment.
 (iii) Based on the ranking of the environmental impacts, a set of manageable environmental objectives should be identified and

adopted. These will provide the motivation for continual environmental improvement on the site.

(iv) The environmental objectives should be reviewed regularly and adjusted as conditions change.

3 Implementation and operation

(i) A management structure organised to implement the EMS should be established.

(ii) The workforce should be trained in the requirements of the EMS and effective methods of communication with employees established.

(iii) Information that describes the organisational structure, responsibilities, practices, and procedures of the EMS should be produced and distributed to encourage the implementation of the policy and processes.

(iv) Implement a monitoring system to ensure that the EMS operates as required. This should include a plan to demonstrate that the workforce are trained and that the system is regularly tested.

(v) Prepare an emergency scheme to cover unexpected environmental issues that may arise.

4 Checking and corrective action

(i) Provide a documented system for monitoring operational activities and recording the results so they can be checked against the set objectives.

(ii) Establish an audit system for the EMS.

5 Management review

(i) The EMS should be reviewed at least annually to determine the adequacy, suitability, and effectiveness of the system in meeting the objectives and achieving environmental improvements. The review should include the policy and procedures to see if there is a need for change to faciliate achieving the EMS objectives.

Quality Systems and Environmental Management Systems require the same attitude to achieving and maintaining standards and it is common practice to integrate the two. This can beneficially be extended to Health and Safety Management Systems[20] (OHSAS 18001). This approach has a number of advantages including a more committed management and workforce, reduced bureaucracy, improved health, safety and environmental standards and a better relationship with the enforcing authorities.

5.4.13 Conclusion

The implementation of Environmental Management Systems allows a reasoned and logical approach to improving environmental standards, makes managers aware of the company's environmental position and creates environmental gains. An effective auditing system helps maintain the EMS and identifies opportunities for environmental improvements. All those involved, from senior manager to operator, must understand the operation of the EMS so that the full potential for environmental improvements and the economic benefits derived from it, can be realised. EMS presents opportunities to improve the environment for our children and, by reducing environmental pollution, benefits the community at large.

References

1. *Environmental Protection Act 1990*, The Stationery Office, London (1990)
2. *Environmental Protection (Prescribed Processes and Substances) Regulations 1991*, The Stationery Office, London (1991)
3. *Integration Pollution Control: A Practical Guide*, The Stationary Office, London (ISBN 1 85112 021 1)
4. *Pollution Prevention and Control [England and Wales] Regulations 2000*, The Stationery Office, London (2000)
5. *The Control of Pollution Act 1974* (as amended by *the Environmental Protection Act 1990*), The Stationery Office, London (1990)
6. *The Special Waste Regulations 1996*, The Stationery Office, London (1996)
7. *The Control of Pollution (Supply and Use of Injurious Substances) Regulations 1986*, The Stationery Office, London (1986).
8. *Best Practicable Environmental Option Assessments for Integrated Pollution Control*, Volume 1 – *Principles and Methodology*, and Volume 2 – *Technical Data*, The Stationery Office, London (ISBN 0 11 310126 0)
9. Chemical Industries Association, *Assessment procedure for determining BPEO*, Chemical Industries Association, Rugby
10. Health & Safety Executive, *Guidance Note EH40/latest edition, Occupational Exposure Limits*, HSE Books, Sudbury
11. *The Water Industries Act 1991*, The Stationery Office, London (1991)
12. *Environmental Protection (Prescribed Processes and Substances) Regulations 1991* (SI 1991 No 472), The Stationery Office, London (1991)
13. European Union, Directive 76/464/EEC on *Pollution Caused by Certain Dangerous Substances Discharged into the Aquatic Environment of the Community* (OJ L129, 18 May 1976) (the Framework Directive), European Union, Luxembourg (1976)
14. *The Water Resources Act 1991*, The Stationery Office, London (1991)
15. *The Groundwater Regulations 1998* (SI 1998 No 2746), The Stationery Office, London (1998)
16. *Special Waste Regulations 1996* (SI 1996 No 972). The Stationery Office, London (1996)
17. *Environmental Protection (Duty of Care) Regulations 1991* (SI 1991 No 2839), The Stationery Office, London (1991)
18. British Standards Institution, BS ISO 14001: *Environmental Management Systems*, British Standards Institution, London
19. European Union, Regulation 1836/93: *Eco-Management and Audit System (EMAS)*, Published in the Official Journal L 168, 10 July 1993, EU, Luxembourg (1993)
20. British Standards Institution, BS OHSAS 18001: *Occupational Health and Safety Management Systems*, BSI, London

Chapter 5.5

The environment at large

G. N. Batts

5.5.1 Introduction

Traditional health and safety issues are increasingly being associated with other aspects of business such as environmental issues. Fairly large companies have combined Health, Safety and Environment Departments rather than having environmental concerns as 'add-on' functions. Organisations in the UK with more than five employees are required by law to have a health and safety policy and many appoint a safety adviser, but there is no such requirement for having an environmental policy.

There is no legal obligation for any organisation to publish health and safety or environmental reports. However, many large organisations incorporate these reports in their Annual Reports to shareholders.

The protection of the workforce has not always been extended to protecting the environment. Some companies with excellent health and safety performances have poor environmental records. Perhaps this is because the structure and scale of legal penalties for damaging the environment have in the past been fairly small by comparison with serious breaches of health and safety laws and the civil actions that frequently follow.

In recognising that environmental fines have been much less than the cost of complying, national governments are tending to increase environmental fines substantially. The societal response on environmental issues is coming into line with health and safety.

This chapter deals with introducing and highlighting environmental issues on both a global and a local scale as they impinge on health and safety. It shows how environmental concerns and controls are increasingly merging with health and safety issues and that this will increase as current proposals become reality. Sustainable development and corporate social responsibility are issues that are of concern to larger companies now but probably are of less concern to small and medium enterprises. In the longer term these issues are likely to impact on businesses of all sizes.

5.5.2 Environmental issues

Comments on environmental issues appear regularly in the media. Often these quote arguments in favour of or against some theory concerning 'the environment'. First, what is 'the environment'? A popular definition is *everything around me plus me*. Apart from natural catastrophes such as earthquakes and asteroids hitting Earth, the major long-term problems with our environment can be traced to human activities.

5.5.2.1 What are the problems with our environment?

It is perhaps contentious to suggest that the animal world would be a model of sustainability if there were no modern human beings on the planet. Our ancestors lived in harmony with their surroundings and loss of biodiversity could simply be attributed to survival of the fittest or evolution. Left alone Nature is wonderful for establishing a balance of raw materials, food chains and food webs with producers and consumers and predators playing their part. Any imbalance observed in Nature can usually be attributed to the activities of human beings, many of whom are greedy and take and consume more than they really need. Throughout most of the Earth's existence there has been a tolerable balance between production and consumption or supply and demand. As a result there were normally sufficient resources to satisfy the needs of economic activity. There have been some temporary problems but early economic activity including farming, agriculture, mining, transport, trading, textiles, tanning, metalworking, alchemy, even perhaps warfare, appeared to have continued without causing too much damage to the resources of our planet.

Following the industrial revolution and towards the middle of the 20th century after the Second World War, it became apparent that the lifestyle of the human race could not continue in the same way indefinitely. The planet could not take the prevailing rate of resource consumption and pollution. Politicians worldwide had to take some action. Either that or the human race would need to find another planet to inhabit and pollute.

The problem has often been, perhaps unfairly, attributed solely to the 'Western World'. It has been accused of consuming the Earth's resources at an alarming rate while simultaneously increasing pollution. The USA has been singled out on many occasions with claims that if everyone in the world lived like the average USA citizen the human race would need three of our planets to provide the natural resources and space for waste disposal. It is true that the USA produces about 25% of the world's carbon dioxide emissions but it also has large areas of forests (known as 'sinks') for absorbing these emissions through photosynthesis. The issue is not so simple that it can be solved by blaming one group alone. The problems and their solutions rest with many parties and interrelated functions.

Meadows[1] published an article in the early 1970s showing how resource limits would affect the global economy if nothing was done to

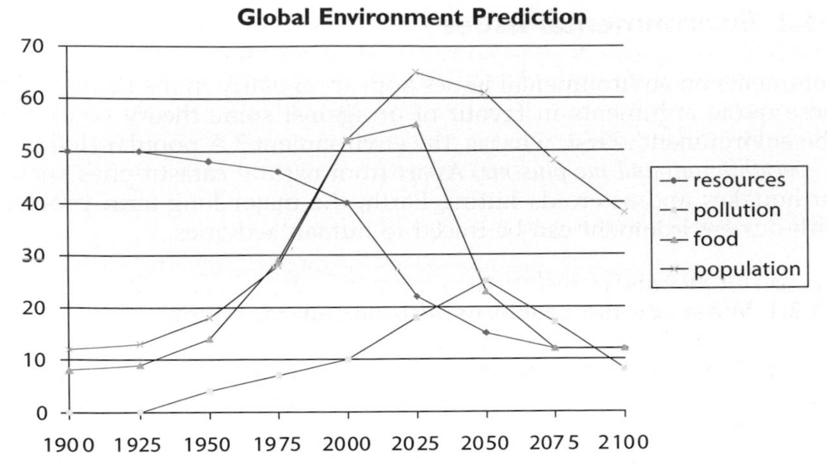

Figure 5.5.1 Schematic of Meadow's prediction

halt the decline. This was discussed at the 1972 Stockholm Conference. *Figure 5.5.1* is a schematic of this 'Doomsday Scenario'.

It has been suggested that many of the international agreements around this time and during the subsequent decade were due to this analysis. The question remains, however, whether these problems can be prevented from getting any worse or, better still, can they be reversed? All countries have a role to play. The challenge is to help the less developed nations so they do not make the same mistakes that have been made by the wealthy countries. To do this, the wealthy nations must help the developing countries more than at present and this creates stumbling blocks and national self-interests. Why should Brazil stop cutting down its rainforests to provide timber it can export and improve its economy if other countries don't help Brazil by developing alternative exports?

The environment is effectively a balance between the three media of atmosphere, soil and water. They co-exist and sustain our lives. Improvements designed to protect one can actually make matters worse for another. The underlying principle of the Design for the Environment with its environmental impact indices is to achieve a proper balance between sustainable development and economic growth.

5.5.2.2 Acid rain and acid precipitation

During the burning of fossil fuels, such as coal, oil derivatives and gas, traces of sulphur and nitrogen react with oxygen and are oxidised to a mixture of acidic oxides known as SOx and NOx respectively. These oxides, which are gaseous emissions and are soluble in water, rise into the atmosphere and then dissolve in rain, sleet or snow to fall as acid rain (acidic precipitation). This type of pollution clearly does not recognise national boundaries. For example SOx and NOx emissions from the UK can be carried across the North Sea to fall as acid rain in Scandinavia. Acid rain has destroyed forests and vegetation and reduced the pH of some lakes in Sweden to as low as 3.5 (i.e. close to dilute acetic acid or vinegar) destroying much aquatic life since very few species can live in such extreme conditions. On the other hand traffic fumes from northern France and Belgium can be carried in the opposite direction into the south-east of England.

5.5.2.3 Global warming or the greenhouse effect

This phenomenon is attributed to the build-up of carbon dioxide in the atmosphere leading to an increase in the ambient temperature of the Earth. The build-up of carbon dioxide is caused by the combustion of fuels and other organic matter in which carbon is oxidised. This would not be a problem if there was enough vegetation to absorb the extra carbon dioxide through the process of photosynthesis. Unfortunately Man is also destroying large areas of vegetation, in effect, destroying the 'lungs' of the Mother Earth. The consequence of the earth warming up is that the polar ice caps melt causing a general rise in sea level.

An alternative name to Global Warming in popular terms is the 'Greenhouse Effect'. The layer of carbon dioxide around the Earth acts like a blanket keeping the radiated heat inside in much the same way as glass does in a garden greenhouse. Without this carbon dioxide layer the Earth would be on average 20 to 30 degrees centigrade colder. Electromagnetic radiation coming from the Sun passes through the outer atmosphere, including the carbon dioxide layer, and strikes the Earth where it is absorbed. The water and land warm up emitting thermal radiation that is of a different wavelength from the electromagnetic energy. The increase in the carbon dioxide layer prevents the escape of this thermal energy and the Earth warms up.

However, there is considerable debate among some academics and politicians as to whether Global Warming is a problem. Some argue that the current rise in global temperature is just the result of the earth following an apparently normal cycle of warming up and cooling down that has occurred over centuries. Furthermore, carbon dioxide is not the only cause of Global Warming. Methane and some halogenated hydrocarbons are more efficient at trapping heat in the Earth and are not broken down as easily. There is a scale known as the GWP (Global Warming Potential) that shows methane to be about ten times worse than carbon dioxide in this respect. Other materials are far worse, as shown in *Table 5.5.1*.

5.5.2.4 Ozone depletion

In the outer atmosphere, known as the stratosphere, there is a gaseous layer of ozone (a triatomic rather than the usual diatomic molecule of oxygen) that used to be about 50 km thick. In this layer, ozone exists in equilibrium with normal diatomic oxygen. However, over the second half of the 20th century significant releases of various substances into the atmosphere have disturbed this equilibrium causing ozone to be destroyed. Details of the chemical reactions involved in this process are to

Table 5.5.1 Global Warming potential of greenhouse gases

Gas	*Global Warming Potential*	*Source*
Carbon dioxide	1	Fossil fuels, deforestation, road vehicles, forest fires
Methane	10	Agriculture, natural decay of vegetation, landfill
Nitrous oxide	100	Fossil fuels, agriculture, road vehicles
Chlorofluorocarbons (CFCs)	3000–7000	Refrigerants, aerosol sprays, solvents

be found in *The Chemistry of Pollution* by Ellenberg[2]. The ozone depleting potential of various gases is shown in *Table 5.5.2*.

The consequences of releasing these substances into the atmosphere were not recognised for some time. Their effect, which is now recognised, is to create partial 'holes' in the ozone layer and in some extreme cases it has disappeared completely. The most famous ozone holes are over Antarctica but there are signs that northern Europe also has an ozone hole over it.

Ozone protects the Earth by absorbing harmful electromagnetic rays emanating from the Sun. Without the ozone these rays can reach the earth and damage vegetation and can cause skin cancer in people. This is particularly a problem when fair-skinned non-indigenous people move into a country that receives high levels of solar radiation, such as Australia.

5.5.2.5 Photochemical smog and particulates

The production of smog is a complex reaction between NOx, volatile organic compounds (VOCs) and sunlight at ground level. The majority of the components come from traffic fumes and industrial activity. Some

Table 5.5.2 Ozone depleting potential of various gases

Gas	*Ozone depleting potential*	*Source*
Chlorofluorocarbons	1	Refrigeration, aerosol sprays, solvents
Carbon tetrachloride	1.1	Manufacturing intermediates
Halons	3–10	Fire extinguishants
1,1,1 trichloroethane	0.1	Degreasing agent

reactions also produce ozone at ground level (where it is not beneficial) together with a darkening and discolouring of the low atmosphere. One cause of the darkening is polymerisation of hydrocarbon VOCs in the atmosphere. On occasions, it is possible to see from an aircraft a pollution haze hanging over a major city. The air may be dull yellow where the sulphur content in the pollutants is high.

Particulates are also generated. They affect our breathing, with the very young and old and those suffering lung problems such as asthma being especially susceptible. The most significant problems arise from particulates with a size of less than 10 microns (known as PM10s). These can affect the alveoli of the lungs and reduce oxygen transfer to the blood. A major source of particulates is the combustion of diesel fuel in vehicles not fitted with efficient filters or traps. Some governments have reacted to this risk. For example in the mid-1990s, the Swedish government decided that all public diesel vehicles must be fitted with a suitable trap. They were fitted retrospectively to every vehicle.

Other sources of particulates arise from the combustion of solid fuels in power stations and generated as dust particles in a number of industrial operations such as slitting and cutting, coating operations, plastic extrusion processes, etc.

5.5.2.6 Resource depletion

The Earth only has a finite supply of many resources. We are all guilty of taking them for granted. For example known coal resources may last for about 250 years, oil perhaps only 50 years and natural gas as little as 30 years. These are known as finite or non-renewable resources. They were produced by chemical reactions under certain natural conditions over extended periods of time and their production cannot be repeated. Once used they will be gone forever. Another example is fish stocks where, in some parts of the world, over-fishing has interfered with the natural reproduction cycle causing serious depletion of stocks. This problem has been recognised globally and moves are being made to limit catches.

5.5.2.7 Eutrophication

Fertilisers have without doubt changed the face of agriculture dramatically since the Haber Bosch process first ‘fixed’ nitrogen into a form that could be used for crop growth. But while farmers can produce more crops than ever before there is a price to pay. Fertilisers have to be soluble so that they can be absorbed into the soils and pass into the root system of plants to support growth. However, if a farmer applies too much fertiliser or it rains at the wrong time, they are washed into rivers and lakes. This causes the aquatic plants, especially algae, to grow faster using up the dissolved oxygen in the water. The result is that there is insufficient oxygen for fish to breathe and many die.

5.5.2.8 Waste disposal

Human activity generates an enormous amount of waste that has for years simply been discharged to the air, water or soil. Unfortunately nature cannot always cope and improper treatment can store up serious problems for future generations. Landfill site capacity is becoming scarce in many countries with some countries banning landfill completely. Waste disposal has polluted seas and waterways, and caused heavy metals, such as lead and cadmium, to enter the food chain. Uncontrolled or ill-considered waste disposal has killed fish and other aquatic life, and destroyed coral reefs. Until quite recently the disposal of wastes at sea without any treatment was a common occurrence. There are now an increasing number of international agreements controlling such activities. The North Sea is one area of water that has been the subject of controls and this has been achieved because this sea is largely within the control of one political group – the European Union.

Apart from regular waste disposal activities there is always the potential for accidents at sea, such as oil spills that can cause extensive immediate damage to marine life and whose effects may last for decades.

5.5.2.9 Renewable versus non-renewable energy

Most of the energy used comes from the destruction of natural resources (coal, oil, natural gas, etc.) but there is a growing range of renewable energy sources being developed in various parts of the world. An obvious example is the use of wood as a fuel in that it can be replaced simply by growing more trees. Hydroelectric power, solar energy, wind power, and tidal power are other examples of renewable energy that are being developed to decrease our reliance on fossil fuels. Some governments are giving financial incentives to encourage the development of these technologies.

Table 5.5.3 Renewable energy targets

Country	*Output in 1997 (%)*	*Target in 2010 (%)*
Sweden	49.1	60.0
Finland	24.7	31.5
Spain	19.9	29.4
Italy	16.0	25.0
France	15.0	21.0
Denmark	8.7	29.0
Germany	4.5	12.5
UK	1.7	10.0
USA*	2.0	10.0

*Data for 2002 and 2020 correspondingly

Table 5.5.3 shows the renewable energy targets for several countries for the year 2010 compared with the percentage of energy from renewable sources in 1997.

The renewable energy issue raises the controversial matter of nuclear energy as the source of power generation. However, the process produces long-lasting radioactive wastes that need to be disposed of. While much research and development work has gone into finding safe means of disposal, there remains the risk of long-term escape and contamination of other resources. Even so, nuclear energy appears to be a suitable alternative source of power. Nuclear fusion, where nuclei are merged, releasing energy, does not produce the same waste problems but the process is still under development and has not yet become commercially available.

5.5.3 The environment and the media

Most company's success is based on their good reputation for products and/or the effectiveness of their services. But this does not necessarily mean that their environmental performance is of a high order. Naming and shaming of companies for bad environmental performances is taking place and government ministers have made threats to publicise examples. This could affect a company's public standing and should act as a strong stimulus for them to look to their environmental laurels. In a recent incident in the UK the senior managers of a large chemical company were summoned to meet the Environment Minister to discuss their poor environmental record. This made national news, affected their share price and was very embarrassing for the company.

Data on releases and emissions are being placed on the Internet by enforcing authorities under such schemes as 'Your Right to Know' in the UK. These data are provided from public records under the conditions of permits and consents and enable individuals to find out what is happening environmentally in their local neighbourhood.

The danger for industry is misrepresentation of these data by the media to make a 'newsworthy' story. Although discharges may be well below legal limits, the fact that certain substances, such as carcinogens or heavy metals are being released into their local environment can cause a public outcry. An example is mercury that can be discharged legally to the environment in very small amounts (milligram quantities) from some sites. Although the amounts are probably less than in one old dental filling, the mention of mercury does have an emotive effect.

It is important for those who find themselves responsible for environmental issues to be able to handle media enquiries. It is essential to work with the Public Relations Department (if there is one) and if possible to be trained to answer media questions. Even those companies that have kept within legal environmental limits are not necessarily free from adverse publicity.

There is increasing pressure to increase fines for anyone breaking environmental laws, and also a trend to seek out large organisations and make examples of them. In some cases the bad publicity is deserved, such as the fine of over £1million for the Milford Haven Port Authority disaster. This was a turning point in the UK that brought environmental matters to the public attention, even though the fine was reduced on appeal. It only takes one lapse on an environmental issue to destroy the carefully nurtured reputation of even the largest multinational company. For smaller companies the risk is not so much of public image as their business supplying larger companies who are increasingly demanding high levels of both environmental performance and health and safety performance from their suppliers.

5.5.4 The global impact of environmental issues

Excluding terrorist activities, there have been a number of large-scale industrial and maritime accidents in recent decades that have become household names. Examples include Chernobyl, Bhopal, Flixborough, Sveso, Piper-Alpha, Exxon Valdez, Torrey Canyon, Milford Haven, etc. These involved significant impacts on the health and safety of people and the environment and, in some cases, led to significant changes in regulations and operating procedures.

Smaller, more localised incidents may not reach the national media because in some way they are not of interest to the whole country. Even so, the effects of small incidents can be felt far away from the original source. Incidents resulting in acid rain, oil spills, toxic releases and ozone depletion can have an impact many miles from the release point. On a more local level, contamination of a local stream or river can impact for many miles downstream.

Members of the public are becoming increasingly aware of and concerned about their local environment as well as about the global issues. They no longer accept emissions from a factory as the price for providing employment. Neighbours of a factory become very alarmed if there is news about possible environmental issues such as chemical emissions, contamination of water or soil, release of chemical fumes, etc. Such incidents can be made very newsworthy by local newspapers, radio and TV stations but become 'one day wonders' if the ramifications are essentially local.

Even the best-laid plans can backfire on companies when environmental issues are being covered. This was demonstrated by the events surrounding Shell and its attempt to dispose of the Brent-Spar oilrig in the late 1990s. Shell had, with third parties, carried out an environmental impact assessment for the disposal of the structure that recommended deep-sea burial. However, the pressure group, Greenpeace, decided against this idea and mounted a very aggressive demonstration programme. Live broadcasts from the Greenpeace ship alongside the Brent-Spar were handed to national and international news media and the issue escalated.

There were widespread protests against Shell in many countries and particularly in Germany where garage forecourts were subjected to blockades and calls for customer boycotts that hurt Shell economically. This resulted in a climbdown by Shell who sought alternative disposal methods even though the least damaging environmental choice was their original proposal. One lesson of this event is that companies need to consider the whole life cycle of their products and processes, including their end-of-life treatment or disposal, rather than leaving this as an afterthought.

5.5.5 Ethical investing and green procurement

Concepts such as ethical investments are likely to be low on the agenda of senior management but in recent years there has been a flurry of proposals, strategies, suggestions etc. that their investments should be linked with responsible or ethical funds. The aims of these are, in theory, to loan money or invest only in activities that are ethical. Non-ethical activities typically include the use of child labour in emerging countries, poor health and safety records or a history of environmental problems. The Co-operative Bank is perhaps the best known large UK institution that claims to embrace ethical investing. According to their policy[3] they only put money into projects and companies that meet certain socially responsible criteria. It is difficult to predict whether this is going to be the usual approach in the future, or whether it is just appeasing Greenpeace and Friends of the Earth. Even if ethical investing does increase in the future, it may continue only if the world economy remains healthy. In periods of recession when companies and even governments are faced with economic survival, it will be tempting to simply maximise profits and only meet legislative requirements.

In 2001 the FTSE4Good[4] index was launched internationally with the aim of showing potential investors the most ethical companies based on a range of criteria. There have been some criticisms of the basis of the index and the presence or absence of certain companies. The criteria used to compare companies on the index include:

- working towards environmental sustainability;
- developing positive relationships with stakeholders;
- upholding and supporting universal human rights.

However limited the uptake at present, this is clearly a political initiative that is unlikely to disappear. Their Internet site is being updated regularly and extended to cover different parts of the world.

Many claims have been made that government purchasing departments should buy 'green' products or services under the concept 'Green Procurement'. This attempts to encourage industry and service providers to consider the overall environmental impact of their products, processes, activities etc. over their complete life cycle, i.e. the 'cradle to grave' concept or Life Cycle Assessment. In theory the benefit to organisations following this approach is that they will have preferential treatment when tendering for government contracts. However, government departments have to operate within financial constraints that can predispose orders being placed with those who undercut prices rather than those with 'Green Products'.

Life Cycle Assessment and Design for Environment principles are still comparatively new concepts and can have widely variable and unpredictable outcomes. In the competitive real world, few companies can produce sufficient 'environmentally-friendly' products at a cost that would sell in sufficient numbers to make them profitable or sustainable.

Even so, there are proposals in the European Union (EU) to bring the principles of Life Cycle Assessment and Design for Environment in for a range of products. These proposals are currently being resisted by industry as unworkable, although political trade-offs are likely to result in some controlling legislation before 2010. Responsible companies are beginning to incorporate these principles in their research into new products. However, the incorporation of green issues into product design is only practised if there is a potential market advantage.

In manufacturing operations protecting the environment is often a win/win situation bringing reduced costs through reduced waste, using less raw materials and reduced energy demand. Many attempts have been made to persuade industry to adopt the recognised environmental management systems of ISO 14000 and the Eco Management Audit Scheme (EMAS) as a means of enhancing environmental performance. This topic is dealt with in section 5.5.10.

5.5.6 Increasing environmental legislation, controls and public reporting

Compared with health and safety legislation, effective environmental controls are a relatively recent innovation. Most environmental legislation has appeared over the past 30 years and although it seems to increase each year, the focus appears to be changing. The trend is shifting from monitoring and controlling at point sources of emissions and releases towards reducing impacts at source by making changes to processes. This doesn't mean that controlling emissions will be abandoned. Recently the EU adopted a Directive on Environmental Liability[5] that deals with identifying those responsible for damaging the environment and holding them accountable. With increasing accountability comes the need to report. There are already mandatory reporting requirements in many countries for certain regulated processes. For example, in the UK, a site that operates a Part A Process under the 1990 Environmental Protection Act has certain environmental data reporting obligations. In Ireland since the mid-1990s certain sites have Integrated Pollution Prevention Control Licences with extensive reporting obligations and this approach has been widely adopted across the European Union.

Many large responsible organisations produce voluntary annual environmental reports as part of their public relations policy. Some are stand-alone environmental reports while others are integrated with a health and safety report and presented as part of the company's Annual Report. There is pressure on government to put a legal obligation on companies above a certain size to publish an annual environmental report[6].

5.5.7 End-of-pipe control

'End-of-pipe' controls are those schemes aimed at ameliorating, at the point of emission, the worst effects of atmospheric pollution and were the basis of most of the early environmental legislation which concentrated on setting absolute emission limits in the pious hope that it would maintain the status quo. With hindsight this approach was flawed since, to reduce environmental damage emissions should be either eliminated or reduced to as low as is technically possible. An example is the setting of SOx emission limits on power stations. In the electricity industry, a generating station could only meet these limits by either reducing output or by putting scrubbers on the top of the emission stacks to absorb most of the SOx. Problems then arose when there was pressure to increase output resulting in exceeding the capacity of the scrubbers. In these circumstances end-of-pipe controls were no longer sufficient. Even if these moves were successful, there was little pressure to reduce the consumption of oil or coal and thus the environmental impact associated with their extraction. Another example is the reduction by incineration of organic solvents to carbon dioxide and water with the aim of preventing VOCs entering the atmosphere and creating smog. While smog may be eliminated at ground level more carbon dioxide would be released to the atmosphere and add to the Global Warming problem. This represents a partial environmental gain at best.

The problem relates to setting limits that are absolute and are not linked to production levels. Hence, a site could reduce its emissions, discharges or waste per unit of production yet still exceed its legal limits because its overall production had to be increased in an effort to remain more profitable in a highly competitive world. Controls that set fixed absolute limits do not lead to a culture embracing continuous improvement.

If new technologies became available, the BATNEEC principle (best available technology not entailing excessive cost) needs to be applied when setting revised emission limits. It is important that economic factors are considered when setting environmental limits. Unfortunately, with the setting of absolute emission levels there is no real incentive for a company to reduce impacts as far as possible because there is no reward based on a sliding scale. On the contrary, the pressure is to maintain operations at a level that just meets legal requirements. Ed Gallagher (former Chief Executive of the UK Environmental Agency) lamented on many occasions that some companies are only motivated to operate the 'CATNAP' principle (cheapest available technology neatly avoiding prosecution)!

This attitude of industry is not surprising when there is no obvious economic incentive to continuously reduce environmental impacts. The issue of a statutory notice will only halt operations until emissions can be brought down to agreed levels but will not result in any ongoing improvements. If bringing emissions down to set levels results in the operation no longer being viable it poses the dilemma of closing down local businesses and putting local people out of work. Emission limits also require recording to ensure that they are not being breached.

Responsible site owners will monitor and record their levels of emission to ensure they are operating within the agreed limits with enforcers visiting premises only to check or validate the data. The costs of monitoring are an on-cost on the product.

5.5.8 Polluter pays

Legislative arrangements vary from setting absolute limits to introducing charges linked to the amount of environmental impact caused by emissions. This latter arrangement is cost-based and gives a financial incentive to companies to reduce their impacts on the environment. Examples are the Water Industries Act 1991[7] and the Water Resources Act 1991[8]. The former relates to discharges to sewers of trade effluent where the concern is to control the quality of the discharge to ensure it can be accommodated by the sewage treatment works. It is enforced by the water utility companies who make a charge depending on the type, content and quantity of the discharge. This gives an incentive to companies to improve the quality of their discharges. The water utility companies decide which sites require consents and also have some control on deciding what parameters to apply. Criteria in these consents can be influenced by the limits imposed by the Environment Agency on the water utility company itself. Monitoring data have to be obtained, available and checked periodically at locations chosen by the water utility company.

The Water Resources Act 1991 relates to water abstraction and to discharges to controlled waters. It is administered by the Environment Agency who apply stricter standards than those required under the Water Services Act since the discharges are to natural water courses that have no means of treatment. Usually the water utility companies have consents from the Environmental Agency to discharge their treated effluent into controlled waters.

Charges paid by the polluter are based on the 'Mogden Formula'[9] which allows for suspended solids, concentration of certain components such as ammonia, BOD (biological oxygen demand) and/or COD (chemical oxygen demand), and volume discharged. There are also 'standing charges' which vary among the Water Service Companies. The actual charge finally levied increases with the amount of polluting material discharged and requiring treatment. The output from the treatment works is a water stream that is usually discharged directly to controlled waters and a solid residue that may be spread on land, treated as solid waste for landfill disposal or burnt in an incinerator.

A Landfill Tax was introduced in the UK in the mid-1990s in the attempt to encourage industry to reduce the amount of waste disposed of by landfill burial. Reports suggested that landfill costs in the UK were far too low and did not reflect the true cost of this treatment method. Other countries such as The Netherlands had already banned landfill disposals. By imposing a tax based on the weight and related to the hazardous properties of the wastes, the intent was to encourage producers of waste to reduce the amounts. This tax is a direct charge on a company's profits but the revenue from this tax can be ploughed back into environmental improvement projects by providing grants to worthy causes. Critics of this tax fall into two camps which claim, either that it is too low to make any difference on industry's practices of sending wastes to landfill or that it is too high and so leads to an increase in 'fly-tipping' that actually has a far more harmful impact on the environment. Whatever the issues of landfill taxes, the problem is that many countries simply have less and less landfill space for burying their wastes. The logical solution seems to be to generate less waste in the first place.

5.5.9 Producer or shared responsibility

During the 1990s environmental laws were introduced into the European Union that made those responsible for generating the adverse environmental impacts also responsible for their treatment and control. Thus there is or soon will be legislation on packaging waste, tyres, batteries, end-of-life vehicles and electronic products, placed on the supplier of the product. However, there is a major problem with this approach because the key question becomes who is or who are the producer(s)? Is it just one person or company that makes a product or provides a service? Do distributors have a responsibility? Furthermore, do the general public or the consumers and customers have an element of responsibility? It can be argued that they contribute to any environmental damage by wanting the products and/or services thereby creating the demand for the producer to meet. In Europe much of the 'Producer Responsibility' legislation is focused on the role of industry. An alternative approach could be to consider shared responsibility where all those involved in the manufacture, supply and use of products have some role to play which should relate to their part in producing the waste.

There is an example of a shared approach in the UK. The Producer Responsibility Obligations (Packaging Waste) Regulations 1997[10] requires each player to carry a certain percentage of the cost of collecting and recycling the wastes. Companies have to obtain Packaging Recovery Notes (PRNs) to cover their obligated packaging wastes. However, companies having a turnover less than £2 million or handling less than 50 tonnes of packaging per year are exempt. It is interesting that of all the European Union Member States only the UK has this 'shared' approach. A possible explanation is that it has proved quite complex to implement even though the format was agreed by a cross-section of UK industry. The complexity arises because of the attempt to include all parties in sharing the costs and prevent 'free riders'. The problem faced in Europe is to achieve an equitable balance between all the contributors within a complex legal system.

The lawmakers, to justify the introduction of legislation, use the term 'Producer Responsibility'. Without really involving all types of producers, the responsibility normally focuses on industry. This approach, however, tends to go beyond controlling traditional environmental impacts and is starting to impinge on product design via Design for Environment principles. There are some attempts in Europe to bring in legislation that would control the actual design of products but these go beyond current legislative requirements and, arguably, beyond what the customer wants[11].

5.5.10 Environmental management system (EMS) and sustainable development

Environmental management became topical in the mid-1990s as an addition to the suite of management quality systems. International Standards such as ISO 14001[12] and the predominantly European Eco Management Audit Scheme[13] became very popular and the number of accredited sites grew rapidly. Some companies such as IBM announced in 1997 a single registration for their worldwide operations[14]. Other companies like Eastman Kodak introduced a site-by-site programme over several years until all their manufacturing sites were registered[15].

There was some resistance at first to an imposed EMS as a result of experience with the ISO 9000 Quality Standards[16] which had, in some instances, become just a bureaucratic process. The benefit to a company of following another standard were not widely appreciated, however, the central concept of an EMS is that a company or site should identify the environmental impact of its operations by carrying out an 'aspects and impacts analysis'. Subsequently it can rank these and thereby focus its management programmes on those areas having the highest adverse impact. Essential to achieving recognition under these schemes is that the whole assessment process is repeated periodically. A continuous improvement culture is thus generated. This emphasis on continuous improvement was missing from the early ISO 9000 Quality Standards.

As companies began to introduce their EMS it became apparent that protecting the environment need not be seen as an add-on cost but rather as an opportunity to save money. For example reducing energy, using less fuel, water, raw materials, producing less waste and introducing vehicle route optimisation all lead to environmental protection and improved business performance. This is a win-win situation. Hence in some organisations it became relatively easy to convince senior managers of the benefits of having an EMS. Because of the similarity in the structure of quality and environmental management systems there was also a great opportunity to merge the two. Batts in his book[17] explains in practical terms how to implement an environment management system to meet the requirements of ISO 14001.

Environmental management has moved on to embrace the concept of Sustainable Development. The 1987 World Commission on Environment/ Development under the guidance of Gro Harlem Brundtland (Prime Minister of Norway) made Sustainable Development the theme of its entire report called 'Our Common Future'[18]. It defined Sustainable Development as 'development that meets the needs of the present generation without compromising the ability of future generations to meet their own needs'. In the UK in the mid-1990s the Department of Trade and Industry (DTI) modified the definition to 'development which ensures a better quality of life for everyone, now and for generations to come'.

The basic concept of Sustainable Development is that the Earth does not have infinite resources of all the raw materials that are consumed by people around the world during their lives. However, it also recognises that economic success is also important and that in theory there is no reason why the human race cannot co-exist with its environment without damaging it irrevocably. Sustainable Development says there is nothing wrong with making money by economic activity as long as the well-being of the environment is a factor in that economic success. An obvious example is sustainable forests supplying wood and pulp where trees are continuously planted to replace those cut down. Finland for example plants three trees for every one cut.

In fact it is not surprising that most of the 'wealthy' or developed countries around the world have good environmental records because they can afford to treat the environment with respect. There are of course some exceptions. The key is to move from situations where the Earth's resources are plundered for a quick profit to one where the long-term effect of human activity on the planet is considered.

Sustainable Development is at the centre of the 6th European Union Environmental Impact Strategy Action Plan as outlined in a keynote speech by Peter Johnston of the Information Society DG[19]. It notes that Sustainable Development does not really focus enough on end-products observing that the rapid development of global communications 'shortens the timetable for progress on sustainable development and offers a potential 'win-win' alternative to the traditional trade-off between growth and environmental sustainability'. The Action Plan demonstrates that real improvements can be made. It cites IBM as having achieved a 4% per year improvement in its use of energy corporately in its manufacturing sites. The report states that many of the improvements needed to reach Sustainable Development have come from the service sector. International conferences such as the 'Rio+10' conference in South Africa[20] seek solutions to these difficult issues. It is predominantly a challenge to the 'rich nations' that have not always been as responsible as they might have been. It has recently become apparent that certain European countries plus the USA tried to block environmental agreements at the 1972 Stockholm Conference for economic reasons[21]. Even today problems of balancing sustainable development issues remain and some countries will always place economic factors above environmental issues[22].

But it is not all bad news. Good examples can be found in global industries and small companies. The 2001 Queen's Award for Sustainable Development recognises large companies such as Kodak Limited and smaller ones for their environmental responsibility[23].

5.5.11 Corporate social responsibility

Sustainable Development links environmental protection with economic prosperity and this link is being extended into Corporate Social Responsibility to include moral and ethical issues. This is sometimes called 'triple bottom line reporting' and means that more and more is expected of large organisations to conduct their business according to certain principles. This direction brings more accountability to report different performances against certain criteria. The European Commission published a Green Paper 'Promoting a European Framework for Corporate Social Responsibility'[24]. Some forward thinking companies and organisations already consider Corporate Social Responsibility to be an investment for the future, rather than a cost, in the same way that protecting the environment can also be good business. The underlying principle of Corporate Social Responsibility outlined in the Green Paper is the 'triple bottom line' concept. The term 'triple' is used to merge economic, social and environmental issues into management strategies to create a better business. The European Commission believes that by adopting this strategy, businesses can contribute to the strategic European Union goal decided at the Lisbon Summit[25]. This goal is 'to become the most competitive and dynamic knowledge-based economy in the world, capable of sustainable economic growth with more and better jobs and greater social cohesion'.

It is hoped that the concept of Corporate Social Responsibility will lead business beyond conventional legal requirements to a greater responsibility that matches the expectations and interests of their stakeholders.

These stakeholders include employees, shareholders, suppliers, customers, local authorities, non governmental organisations, and local communities. Together they are providing the pressures that are shaping the political climate. The first steps along a tortuous road have been taken. The pace will increase with implications to all professionals who have a role to play in protecting the environment.

References

1. Meadows, D., *The Limits to Growth*, Report to the Club of Rome and the United Nations Conference on the Human Environment, Stockholm (1972)
2. Fellenberg, G., *The Chemistry of Pollution*, J Wiley & Sons, New York (2000), ISBN 0471613916
3. The Co-operative Bank plc, *Ethical Investment Policy*, The Co-operative Bank plc, Skelmersdale (2001)
4. The Financial Times, Internet Site *www.ftse4good.com*, The Financial Times, London (2001)
5. European Union, White Paper COM92000/66, *On Environmental Liability*, EU, Luxembourg (2000). It can be downloaded from europa.eu.int/eur-lex/en
6. Meacher, M., UK Environment Minister reported in *Industrial Environmental Management* December (1998)
7. *The Water Industries Act 1991*, The Stationery Office, London (1991)
8. *The Water Resources Act 1991*, The Stationery Office, London (1991)
9. The British Photographic Association, *The Mogden Formula: A Code of Practice for the Photoprocessing Industry for the Care of the Environment*. The British Photographic Association now the Photo Imaging Council (PIC), Caterham (1997)
10. *The Producer Responsibility Obligations (Packaging Waste) Regulations 1997*, The Stationery Office, London (1997)
11. European Union, *Proposal for a Draft Directive on the impact on the environment of electrical and electronic equipment (EEE)*, EU, Luxembourg (2000)
12. British Standards Institution, BS ISO 14001, *Environmental Management Systems. Specifications for guidance for use*, BSI, London (1996)

13. European Union, Regulation (EC) No: 761/2001, *Allowing voluntary participation by organisations in a Community eco-management audit scheme* (EMAS), published in the Official Journal L114 dated 24 April 2001, EU, Luxembourg (2001)
14. IBM, *IBM and the Environment, Progress Report*, IBM, New York (1997). See also www.ibm.com/ibm/environment
15. Eastman Kodak, Press Release: *Kodak Reaches Goal: All 28 facilities worldwide now certified to ISO 14001*, Eastman Kodak, Rochester, US (January 2002)
16. British Standards Institution, BS ISO 9000, *Quality management systems. Fundamentals and Vocabulary*, BSI, London (2000)
17. Batts, G., *An Essential Guide to ISO 14001*, Chandos Publishing, Oxford (1999), ISBN 1 902375 30 0
18. Brundtland, Gro Harlem, *Our Common Future*, World Commission on Environment and Development, Oxford University Press, Oxford (1987)
19. Johnston, P., h:/C1/Johnston/Speeches/Amsterdam IT for EH&S June 2001 (7–8 June 2001), *The Knowledge Economy Can We Have Growth, More and Better Jobs and Sustainable Development?*
20. Rio-declaration signatory countries committed themselves, at the 19th Special Session of the United Nations' General Assembly (1997), to draw up strategies for sustainable development for the 2002 World Summit on sustainable development
21. Hammer, M., Filthy rich, *New Scientist*, No: 2324, p. 7, 5 January 2002
22. EC Conference, Brussels 27&28/11/2001 accessible from http://www.socialresponsibility.be
23. UK Government, *Queen's Award for Enterprise: Sustainable Development*, The Stationery Office, London accessible from www.queensawards.org.uk or www.dti.gov.uk
24. European Union, COM(2001) 366; *Promoting a European framework for Corporate Social Responsibility*, EU, Luxembourg (2001). This document is downloadable from http://europa.eu.int/eur-ex/en/com/availability/cpi_avail_year2001_351_400_en.html
25. European Union Member State Summit in Lisbon, Portugal (2000)

Appendices

Appendices

Appendix 1

The Institution of Occupational Safety and Health

The Institution of Occupational Safety and Health (IOSH) is the leading professional body in the United Kingdom concerned with matters of workplace safety and health. Its growth in recent years reflects the increasing importance attached by employers to safety and health for all at work and for those affected by work activities. The Institution provides a focal point for practitioners in the setting of professional standards, their career development and for the exchange of technical experiences, opinions and views.

Increasingly employers are demanding a high level of professional competence in their safety and health advisers, calling for them to hold recognised qualifications and have a wide range of technical expertise. These are evidenced by Corporate Membership of the Institution for which proof of a satisfactory level of academic knowledge of the subject reinforced by a number of years of practical experience in the field is required.

Recognised academic qualifications are an accredited degree in occupational safety and health or the Diploma Part 2 in Occupational Safety and Health issued by the National Examination Board in Occupational Safety and Health (NEBOSH). For those assisting highly qualified OSH professionals, or dealing with routine matters in low risk sectors, a Technician Safety Practitioner (SP) qualification may be appropriate. For this, the NEBOSH Diploma Part 1 would be an appropriate qualification.

Further details of membership may be obtained from the Institution.

Appendix 2

Reading for Part I of the NEBOSH Diploma examination

The following is suggested as reading matter relevant to Part 1 of the NEBOSH Diploma examination. It should be complemented by other study.

Module 1A:	The management of risk	Chapters	9–all 10–paras. 8–11 11–all 12–paras. 1–3 22–paras. 1–6 29–para. 11
Module 1B:	Legal and organisational factors	Chapters	1–all 2–all 3–paras. 1–6 7–para. 2 8–all 10–paras. 13 and 14 14–paras. 1–4
Module 1C:	The workplace	Chapters	7–para. 2 20–all 21–all 24–all 26–paras. 1–8 28–paras. 2 and 4 29–paras. 1, 2, 7 and 11
Module 1D:	Work equipment	Chapters	25–all 26–all 27–all
Module 1E:	Agents	Chapters	15–all 16–all 17–all 19–paras. 1–6 20–all 22–paras. 4–7 29–paras. 1–4
Module 1CS:	Common skills	Chapter	12–para. 7

Additional information in summary form is available in *Health and Safety ... in brief* by John Ridley published by Butterworth-Heinemann, Oxford (1998).

Appendix 3

List of abbreviations

ABI	Association of British Insurers
AC	Appeal Court
ac	Alternating current
ACAS	Advisory, Conciliation and Arbitration Service
ACGIH	American Conference of Governmental Industrial Hygienists
ACoP	Approved Code of Practice
ACTS	Advisory Committee on Toxic Substances
ADS	Approved dosimetry service
AFFF	Aqueous film forming foam
AIDS	Acquired immune deficiency syndrome
ALA	Amino laevulinic acid
All ER	All England Law Reports
APAU	Accident Prevention Advisory Unit
APC	Air pollution control
BATNEEC	Best available technique not entailing excessive costs
BLEVE	Boiling liquid expanding vapour explosion
BOD	Biological oxygen demand
BPEO	Best practicable environmental option
Bq	Becquerel
BS	British standard
BSE	Bovine spongiform encephalopathy
BSI	British Standards Institution
CAW	Control of Asbestos at Work Regulations 2002
CBI	Confederation of British Industries
cd	Candela
CD	Consultative document
CDG	The Carriage of Dangerous Goods by Road Regulations 1996
CDG-CPL	The Carriage of Dangerous Goods by Road (Classification, Packaging and Labelling) and Use of Transportable Pressure Receptacle Regulations 1996
CDM	The Construction (Design and Management) Regulations 1994
CEC	Commission of the European Communities

CEN	European Committee for Standardisation of mechanical items
CENELEC	European Committee for Standardisation of electrical items
CET	Corrected effective temperature
CFC	Chlorofluorocarbons
CHASE	Complete Health and Safety Evaluation
CHAZOP	Computerised hazard and operability study
CHIP 2	The Chemical (Hazard Information and Packaging for Supply) Regulations 1994
Ci	Curie
CIA	Chemical Industries Association
CISBE	The Chartered Institution of Building Services Engineers
CLAW	Control of Lead at Work Regulations 2002
CIMAH	The Control of Industrial Major Accident Hazards Regulations 1984
CJD	Creutzfeldt–Jacob disease
COD	Chemical oxygen demand
COMAH	The Control of Major Accident Hazards Regulations 1999
COREPER	Committee of Permanent Representatives (to the EU)
COSHH	The Control of Substances Hazardous to Health Regulations 2002
CPA	Consumer Protection Act 1987
CTD	Cumulative trauma disorder
CTE	Centre tapped to earth (of 110 V electrical supply)
CWC	Chemical Weapons Convention
dB	Decibel
dBA	'A' weighted decibel
dc	Direct current
DEFRA	Department of the Environment, Food and Rural Affairs
DG	Directorate General
DNA	Deoxyribonucleic acid
DO	Dangerous occurrence
DSEAR	Dangerous Substances and Explosive Atmospheres Regulations 2002
DSE(R)	The Health and Safety (Display Screen Equipment) Regulations 1992
DSS	Department of Social Services
DTI	Department of Trade and Industry
EA	Environment Agency
EAT	Employment Appeals Tribunal
ECJ	European Courts of Justice
EC	European Community
EEA	European Economic Association
EEC	European Economic Community
EcoSoC	Economic and Social Committee
EHRR	European Human Rights Report

EINECS	European inventory of existing commercial chemical substances
ELF	Extremely low frequency
ELINCS	European list of notified chemical substances
EMAS	Employment Medical Advisory Service, also (European) Eco-Management and Audit System
EN	European normalised standard
EP	European Parliament
EPA	Environmental Protection Act 1990
ERA	Employment Rights Act 1996
ESR	Essential safety requirement
EU	European Union
eV	Electronvolt
EWA	The Electricity at Work Regulations 1989
FA	Factories Act 1961
FAFR	Fatal accident frequency rate
FMEA	Failure modes and effects analysis
FPA	Fire Precautions Act 1971
FP(W)	Fire Precautions (Workplace) Regulations 1997
FSLCM	Functional safety life cycle management
FTA	Fault tree analysis
GEMS	Generic error modelling system
Gy	Gray
HAVS	Hand-arm vibration syndrome
HAZAN	Hazard analysis study
HAZCHEM	Hazardous chemical warning signs
HAZOP	Hazard and operability study
HELA	HSE/Local Authority Enforcement Liaison Committee
hfl	Highly flammable liquid
HIV+ve	Human immune deficiency virus positive
HL	House of Lords
HSC	The Health and Safety Commission
HSE	The Health and Safety Executive
HSI	Heat stress index
HSW	The Health and Safety at Work, etc. Act 1974
Hz	Hertz
IAC	Industry Advisory Committee
IB	Incapacity Benefit
IBC	Intermediate bulk container
ICC	International Criminal Court
ICRP	International Commission on Radiological Protection
IEC	International Electrotechnical Committee (International electrical standards)
IEE	Institution of Electrical Engineers
IOSH	Institution of Occupational Safety and Health
IPC	Integrated pollution control

IQ	Intelligence quotient
IRLR	Industrial relations law report
ISO	International Standards Organisation (mechanical standards)
ISRS	International Safety Rating System
JHA	Job hazard analysis
JP	Justice of the Peace
JSA	Job Safety Analysis
KB	King's Bench
KISS	Keep it short and simple
LA	Local Authority
LEL	Lower explosive limit
$L_{EP,d}$	Daily personal noise exposure
LEV	Local exhaust ventilation
LJ	Lord Justice
LOLER	Lifting Operations and Lifting Equipment Regulations 1998
LPC	Loss Prevention Council
LPG	Liquefied petroleum gas
LR	Lifts Regulations 1997
lv/hv	Low volume high velocity (extract system)
mcb	Miniature circuit breaker
MEL	Maximum exposure limit
MHOR	The Manual Handling Operations Regulations 1992
MHSW	The Management of Health and Safety at Work Regulations 1992
MOSAR	Method organised for systematic analysis of risk
MPL	Maximum potential loss
M.R.	Master of the Rolls
NC	Noise criteria (curves)
NDT	Non-destructive testing
NEBOSH	National Examination Board in Occupational Safety and Health
NI	Northern Ireland Law Report
NIHH	The Notification of Installations Handling Hazardous Substances Regulations 1982
NIJB	Northern Ireland Judgements Bulletin (Bluebook)
NLJ	Northern Ireland Legal Journal
NONS	The Notification of New Substances Regulations 1993
npf	Nominal protection factor
NR	Noise rating (curves)
NRPB	National Radiological Protection Board
NZLR	New Zealand Law Report

OJ	Official journal of the European Community
OECD	Organisation for Economic Development and Co-operation
OES	Occupational exposure standard
OFT	Office of Fair Trading
OR	Operational research
P4SR	Predicted 4 hour sweat rate
Pa	Pascal
PAT	Portable appliance test
PC	Personal computer
PCB	Polychlorinated biphenyl
PER	Pressure Equipment Regulations 1999
PHA	Preliminary hazard analysis
PMNL	Polymorphonuclear leukocyte
PPE	Personal protective equipment
ppm	Parts per million
PSSR	Pressure Systems Safety Regulations 2000
ptfe	Polytetrafluoroethylene
PTW	Permit to work
PUWER	The Provision and Use of Work Equipment Regulations 1992
PVC	Polyvinyl chloride
QA	Quality assurance
QB	Queen's Bench
QMV	Qualifies majority voting
QUENSH	Quality, environment, safety and health management systems
r.	A clause or regulations of a Regulation
RAD	Reactive airways dysfunction
RCD	Residual current device
RF	Radio frequency
RGN	Registered general nurse
RIDDOR	The Reporting of Injuries, Diseases and Dangerous Occurrences Regulations 1995
RM	Resident magistrate
RoSPA	Royal Society for the Prevention of Accidents
RPA	Radiation protection adviser
RPE	Respiratory protective equipment
RPS	Radiation protection supervisor
RR	Risk rating
RRP	Recommended retail price
RSI	Repetitive strain injury
s.	Clause or section of an Act
SAFed	Safety Assessment Federation
SC	Sessions case (in Scotland)
Sen	Sensitiser

SEN	State enrolled nurse
SIESO	Society of Industrial Emergency Services Officers
Sk	Skin (absorption of hazardous substances)
SLT	Scottish Law Times
SMSR	The Supply of Machinery (Safety) Regulations 1992
SPL	Sound pressure level
SRI	Sound reduction index
SRN	State registered nurse
SRSC	The Safety Representatives and Safety Committee Regulations 1977
SSP	Statutory sick pay
Sv	Sievert
SWL	Safe working load
SWORD	Surveillance of work related respiratory diseases
TEU	Treaty of European Union 1991
TLV	Threshold Limit Value
TSS	Toxic shock syndrome
TUC	Trades Union Congress
TWA	Time Weighted Average
UEL	Upper explosive limit
UGR	Unified glare rating system
UK	United Kingdom
UKAEA	United Kingdom Atomic Energy Authority
UKAS	United Kingdom Accreditation Service
v.	versus
VAT	Value added tax
VCM	Vinyl chloride monomer
vdt	Visual display terminal
VWF	Vibration white finger
WATCH	Working Group on the Assessment of Toxic Chemicals
WBGT	Wet bulb globe temperature
WDA	Waste Disposal Authority
WHSWR	The Workplace (Health, Safety and Welfare) Regulations 1992
WLL	Working load limit
WLR	Weekly Law Report
WRULD	Work related upper limb disorder
ZPP	Zinc protoporphyrin

Appendix 4

Organisations providing safety information

Institution of Occupational Safety and Health, The Grange, Highfield Drive, Wigston, Leicester LE18 1NN 0116 257 3100

National Examination Board in Occupation Safety and Health, NEBOSH, Dominus Way, Meridian Business Park, Leicester LE3 2QW 0116 263 4700

Royal Society for the Prevention of Accidents, Edgbaston Park, 353 Bristol Road, Birmingham B5 7ST 0121 248 2222

Employment National Training Organisation, Kimberley House, 47 Vaughan Way, Leicester LE1 4SG 011 6251 7979

British Standards Institution, 389 Chiswick High Road, London W4 4AL 020 8996 9000

Health and Safety Commission, Rose Court, 2 Southwark Bridge, London SE1 9HF 020 7717 6000

Health and Safety Executive, Enquiry Point, Magnum House, Stanley Precinct, Trinity Road, Bootle, Liverpool L20 3QY 0151 951 4000 or any local offices of the HSE

HSE Books, PO Box 1999, Sudbury, Suffolk CO10 6FS 01787 881165

Employment Medical Advisory Service, at local HSE Area office

Institution of Fire Engineers, 148 New Walk, Leicester LE1 7QB 0116 255 3654

Medical Commission on Accident Prevention, 35–43 Lincolns Inn Fields, London WC2A 3PN 0171 242 3176

The Asbestos Information Centre Ltd, PO Box 69, Widnes, Cheshire WA8 9GW 0151 420 5866

Chemical Industry Association, King's Building, Smith Square, London SW1P 3JJ 020 7834 3399

School of Industrial and Manufacturing Sciences, Cranfield University, Cranfield, Bedford MK43 0AL 01234 750111

National Institute for Occupational Safety and Health, 5600 Fishers Lane, Rockville, Maryland, 20852, USA

Noise Abatement Society, Chalkhurst, Upper Austin Lodge Road, Eynsford, Dartford, Kent DA4 0HT

Home Office, 50 Queen Anne's Gate, London SW1A 9AT 0171 273 4000
Fire Services Inspectorate, Horseferry House, Dean Ryle Street, London SW1P 2AW 0171 217 8728

Department of Trade and Industry: all Departments on 020 7215 5000
Consumer Safety Unit, General Product Safety: 1 Victoria Street, London SW1H 0ET
Gas and Electrical Appliances: 151 Buckingham Palace Road, London SW1W 9SS
Manufacturing Technology Division, 151 Buckingham Palace Road, London SW1W 9SS

Department of Transport
Road and Vehicle Safety Directorate, Great Minster House, 76 Marsham Street, London SW1P 4DR 0171 271 5000

Advisory, Conciliation and Arbitration Service (ACAS), Brandon House, 180 Borough High Street, London SE1 1LW 020 7210 3613

National Radiological Protection Board (NRPB), Chilton, Didcot, Oxfordshire OX11 0RQ 01235 831600

Northern Ireland Office
Health and Safety Inspectorate, 83 Ladas Drive, Belfast BT6 9FJ 028 9024 3249
Agriculture Health and Safety, 14 Fairhill Road, Cookstown 028 8676 7874
Employment Medical Advisory Service, Royston House, 34 Upper Queen Street, Belfast BT1 6FX 01232 233045 ext: 58

Commission of the European Communities, 2 Queen Anne's Gate, London SW1H 9AA 020 7227 4300

British Safety Council, National Safety Centre, 70 Chancellor's Road, London W6 9RS 020 7741 1231

Confederation of British Industry, Centre Point, 103 New Oxford Street, London WC1A 1DU 020 7379 7400

Safety Assessment Federation (SAFed), Nutmeg House, 60 Gainsford Street, Butler's Wharf, London SE1 2NY 020 7403 0987

Railway Inspectorate, Rose Court, 2 Southwark Bridge, London SE1 9HS 020 7717 6000

Pollution: local office of the Environmental Agency

Appendix 5

List of Statutes, Regulations and Orders

Appendix 6

List of Cases

A & Others v. National Blood Authority & Others, (26 March 2001 unreported), *133*
Abouzaid v. Mothercare (UK) Ltd, (21 December 2000 unreported), *133*
Ashdown v. Samuel Williams & Sons (1957) 1 All ER 35, *92*
Ashington Piggeries Ltd v. Christopher Hill Ltd (1971) 1 All ER 847, *91*
Austin v. British Aircraft Corporation Ltd (1978) IRLR 332, *112*

Baker v. T.E. Hopkins & Sons Ltd (1959) 1 WLR 966, *167*
Balfour v. Balfour (1919) 2 KB 571, *85*
Ball v. Insurance Officer (1985) 1 All ER 833, *142*
Beale v. Taylor (1967) 3 All ER 253, *90*
Beckett v. Cohen (1973) 1 All ER 120, *122*
Berry v. Stone Manganese Marine Ltd (1972) 1 Lloyd's Reports 182, *148*
Bett v. Dalmey Oil Co. (1905) 7F (Ct of Sess.) 787, *305*
Bigg v. Boyd Gibbins Ltd (1971) 2 All ER 183, *84*
Bowater v. Rowley Regis Corporation (1944) 1 All ER 465, *147*
British Airways Board v. Taylor (1976) 1 All ER 65, *122*
British Coal v. Armstrong and Others (*The Times*, 6 December 1996, CA), *149*
British Gas Board v. Lubbock (1974) 1 WLR 37, *120*
British Home Stores v. Burchell (1978) IRLR 379, *112*
British Railways Board v. Herrington (1971) 1 All ER 897, *150*
Buck v. English Electric Co. Ltd (1978) 1 All ER 271, *148*
Bulmer v. Bollinger (1974) 4 All ER 1226, *28*
Bunker v. Charles Brand & Son Ltd (1969) 2 All ER 59, *160*

Cadbury Ltd v. Halliday (1975) 2 All ER 226, *120*
Carlill v. Carbolic Smoke Ball Co. (1893) 1 QB 256, *84*
Close v. Steel Company of Wales (1962) AC 367, *41*
Cunningham v. Reading Football Club (1991) *The Independent*, 20 March 1991, *44*

Queensway Discount Warehouses v. Burke (1985) BTLC 43, *120*
Quintas v. National Smelting Co. Ltd (1961) 1 All ER 630, *146*

R. v. Bevelectric (1992) 157 JP 323, *122*
R. v. British Steel plc (1995) ICR 587, *40*
R. v. Bull, *The Times*, 4 December 1993, *123*
R. v. Francois Pierre Marcellin Thoron, CA (Criminal Division) 30 July 2001, *29*
R. v. George Maxwell Ltd (1980) 2 All ER 99, *8*
R. v. Kent County Council (6 May 1993, unreported), *123*
R. v. Liverpool City Council ex parte Baby Products Association, The times, 1 December 1999, *131*
R. v. Secretary of State for Transport v. Factortame Ltd C 221/89; (1991) 1 AC 603; (1992) QB 680, *27*
R. v. Sunair Holidays Ltd (1973) 2 All ER 1233, *122*
R. v. Swan Hunter Shipbuilders Ltd and Telemeter Installations Ltd (1981) IRLR 403, *821*
Rafdiq Mughal v. Reuters (1993), *606*
Readmans Ltd v. Leeds City Council (1992) COD 419, *26*
Ready-mixed Concrete (South East) Ltd v. Minister of Pensions and National Insurance (1968) 1 All ER 433, *88*
Richardson v. LRC Products Ltd, (2 February 2000 unreported), *132*
Roberts v. Leonard (1995) 159 JP 711, *122*
Rowland v. Divall (1923) 2 KB 500, *90*
Rylands v. Fletcher (1861) 73 All ER Reprints N. 1, *150*

Scammell v. Ouston (1941) All ER 14, *84*
SCM (UK) Ltd v. W. J. Whittle and Son Ltd (1970) 2 All ER 417, *154*
Scott v. London Dock Company (1865) 3 H and C 596, *147*
Shepherd v. Firth Brown (1985) unreported, *150*
Sillifant v. Powell Duffryn Timber Ltd (1983) IRLR 91, *107*
Smith (or Duncan) v Hazards Ltd, *22*
Smith v. Baker (1891) AC 325, *167*
Smith v. Crossley Bros. Ltd (1951) 95 Sol. Jo. 655, *162*
Smith v. Leach Brain & Co. Ltd (1962) 2 WLR 148, *160*
Smith v. Stages (1989) 1 All ER 833, *145*
Spartan Steel and Alloys Ltd v. Martin and Co. (Contractors) Ltd (1972) 3 All ER 557, *154*
Spencer v. Paragon Wallpapers Ltd (1976) IRLR 373, *109*
Stark v. Post Office, (2000) ICR 1013, *40*
Stevenson, Jordan and Harrison v. Macdonald & Evans (1951) 68 R.P.C. 190, *88*
Systems Floors (UK) Ltd v. Daniel (1982) ICR 54; (1981) IRLR 475, *88*

Tesco Supermarkets v. Nattrass (1972) AC 153, *121*
Thompson v. Smiths Ship Repairers (North Shields) Ltd (1984) 1 All ER 881, *149*
Toys R Us v. Gloucestershire County Council (1994) 158 JP 338, *123*
Trotman v. North Yorkshire County Council, *147*

Index